W9-BVY-895

Technology and Society
Issues for the 21st Century and Beyond

Third Edition

Linda S. Hjorth
Barbara A. Eichler
Ahmed S. Khan
John A. Morello
DeVry University–DuPage Campus

Upper Saddle River, New Jersey
Columbus, Ohio

Library of Congress Cataloging in Publication Data
Technology and society : issues for the 21st century and beyond / edited by Linda S. Hjorth ... [et al.]. — 3rd ed.
p. cm.
Includes index.
ISBN 0-13-119443-7
1. Technology—Social aspects. I. Hjorth, Linda S.
T14.5 .T44182008
303.48'3—dc22

2006036247

Editor in Chief: Vernon Anthony
Editor: Jeff Riley
Editorial Assistant: Lara Dimmick
Production Editor: Stephen C. Robb
Design Coordinator: Diane Y. Ernsberger
Cover Designer: Diane Y. Ernsberger
Cover art: Corbis
Production Manager: Matt Ottenweller
Production Coordination: Julie Hotchkiss, Custom Editorial Productions, Inc.
Director of Marketing: David Gesell
Marketing Manager: Ben Leonard
Marketing Assistant: Les Roberts

This book was set in Times Roman by Laserwords. It was printed and bound by R. R. Donnelley & Sons Company. The cover was printed by R. R. Donnelley & Sons Company.

Copyright © 2008 by Pearson Education, Inc., Upper Saddle River, New Jersey 07458.
Pearson Prentice Hall. All rights reserved. Printed in the United States of America. This publication is protected by Copyright and permission should be obtained from the publisher prior to any prohibited reproduction, storage in a retrieval system, or transmission in any form or by any means, electronic, mechanical, photocopying, recording, or likewise. For information regarding permission(s), write to: Rights and Permissions Department.

Pearson Prentice Hall™ is a trademark of Pearson Education, Inc.
Pearson® is a registered trademark of Pearson plc
Prentice Hall® is a registered trademark of Pearson Education, Inc.

Pearson Education Ltd.
Pearson Education Singapore Pte. Ltd.
Pearson Education Canada, Ltd.
Pearson Education—Japan
Pearson Education Australia Pty. Limited
Pearson Education North Asia Ltd.
Pearson Educación de Mexico, S.A. de C.V.
Pearson Education Malaysia Pte. Ltd.

10 9 8 7 6 5 4 3 2 1

ISBN-13: 978-0-13-119443-4
ISBN-10: 0-13-119443-7

If you are thinking a year ahead, sow seed. If you are thinking 10 years ahead, plant a tree. If you are thinking 100 years ahead, make people aware. By sowing seed once, you harvest once. By planting a tree, you will harvest tenfold. By opening the minds of people, you will harvest a hundredfold.

CHINESE PROVERB

To the students who, in their quest for knowledge, consistently reinforce our desire to present varied ideas within the field of technology and society.

LINDA S. HJORTH

To my family, my students, and colleagues whose very capable adoption of technology, along with their arguments for the value of humankind, always inspire me.

BARBARA A. EICHLER

To Tasneem, my parents, my students, and all seekers of truth and wisdom.

AHMED S. KHAN

To our students and our families for their patience, support, and prayers.

JOHN MORELLO

Preface

ISSUES EXPLORED IN THE TEXT

One of the underlying issues explored in *Technology and Society: Issues for the 21st Century and Beyond* is whether we are in charge of technology or whether technology controls us. At what point does technological dependency cause social problems? And to what extent are we, as caring social beings, concerned about the impact of technology? This text encourages readers to analyze and reflect on technology's effects on the global village's economy, politics and environment. The new century will usher in an urgent challenge to resolve the conflicts among our technological, environmental, and social worlds. The ability to understand the impact of technology on our lives and on succeeding generations will be essential to reaching the goals of survival, peaceful coexistence, ethical living, safety, and prosperity.

The chapters in this text are designed to stimulate, inspire, and provoke awareness of technology's impact on society. They are supported by a variety of features intended to supplement and complement learning, critical analysis, and social awareness.

ORGANIZATION OF THE TEXT

This work has been divided into nine parts, covering nine separate topics, but is united by a single idea: that technological change has been a constant companion to changes in society, ethics, energy, the environment, population, conflict, the third world, health, and even the future.

Part I, "History of Technology," takes a look at how technology has changed the way humanity has developed, from the first time a rock was fastened to a stick, sharpened to a point, and used for security or to hunt game for a meal, through modern methods that can measure the effects of human presence on Earth on the air we breathe and the water we drink. In all the intervening centuries, technology has been there. It has even been a factor in determining whether human actions are right or wrong.

Part II, "Ethics and Technology," examines some of the ethical problems that technological development has presented for humanity. For example, advances in nuclear power and medicine seem, on the surface, to be achievements in which the world might rejoice. However, whether and how society chooses to use these tools as well as the disparity in their availability around the world, makes for some hard decision making.

Part III, "Energy," explores the role energy technology has played over time. The readings in this section provide an informative examination of various energy choices and also shed some light on why our society has not taken full advantage of some alternative sources of energy.

Part IV, "Ecology," is dedicated to the struggle to find balance between technological growth and ecological safety. The articles in this section suggest that there is a price to be paid for technological growth, and that price might be clean air, clean water, and good health. But Part IV also provides views that suggest that the balance between technology and ecology can be struck, and both can flourish.

Part V, "Population," looks at a topic in which everyone should be interested, because it is about us. In less than 100 years, the world's population jumped from a little over one billion to well over six billion people. Can technology feed and house all of us? Or do we need to use technology to help control population growth before it gets out of control?

Part VI, "War and Technology," deals with the impact of technology on conflict. In this section, the reader is asked to consider if there is any way to restraint weapons technology when the technology seems to offer few, if any, restraints of its own. The articles in this section also suggest that future technologies may change the face of war but perhaps not its consequences.

Part VII, "Health and Technology," seeks to strengthen our understanding of the links between human health and technology. The last few decades have seen many technological advances that have improved the human condition; doctors can perform surgeries today that the previous generation could only imagine. But some health issues, basic nutrition, and of course, AIDS are daunting issues that have heretofore escaped a technological solution. Is it possible that technology does not hold all the answers when it comes to health? The readings in this section pose that and other questions.

Part VIII, "Technology and the Third World," presents readings about another critical area of concern to our increasingly interconnected planet. The nations that make up the Third World continue to fall behind their First World neighbors in virtually every imaginable category. How did this happen? The readings in this part explore how this disparity came about, focus on some specific cases, and offer some interesting glimpses of how the Third World is taking matters into its own hands when dealing with the gap between its level of technology and that of the developed world.

Finally, Part IX, "Technology and the Future," allows the reader a glimpse of what we can expect from technology in the years to come, and considers how far advances in technology can take us. It offers an ambitious look at how new ideas will affect us in the 21st century and beyond, from health to ethics, and yes, even war.

FEATURES

- *Flowcharts:* One of the unique features of the text is the use of flowcharts as logical, interactive maps that emphasize the problems, possible solutions, and points of direction and significance of the chapters and case studies. Many of the flowcharts have been class-tested, and we have found that students like them because they appeal to their kinesthetic and problem-solving learning style. The visual process of flowcharting the information presented in the readings seems to increase insight and critical thinking skills.
- *Internet Exercises:* The Internet exercises enable students to (a) become familiar with the knowledge dissemination mode of the ever-expanding Internet; and (b) incorporate the Internet's multimedia resources to enhance learning.

- *Useful Web Sites:* Considering the vast scope and rate of change of technological growth, it is difficult to cover all facets of technology and related issues of its impact on society. Therefore, each part concludes with a list of Web sites for the reader to use to supplement and enhance the content.
- *Questions:* Each article concludes with questions to integrate knowledge and synthesize understanding of social issues impacted by technology. The questions also create excitement and wonderment about the tenacity of technology and its impact on the quality of life.
- *Statistics:* Boxes containing statistics, percentages, and bar graphs enhance many readings. They allow the reader to comprehend the magnitude of the social issue presented from a numerical format.

INSTRUCTOR RESOURCES

An *Online Instructor's Resource Manual* (0-13-119444-5) features PowerPoint® presentations including figures from the text, along with Lecture Notes.

To access supplementary materials online, instructors need to request an instructor access code. Go to **www.prenhall.com,** click the **Instructor Resource Center** link, and then click **Register Today** for an instructor access code. Within 48 hours after registering, you will recieve a confirming e-mail including an instructor access code. When you have recieved your code, go to the site and log on for full instructions on downloading the materials you wish to use.

ACKNOWLEDGMENTS

We gratefully acknowledge the valuable input of the following reviewers for this edition: Gertrude Abramson, Nova Southeastern University; Harold E. Laubach, Ph.D., Nova Southeastern University; Chris Merrill, Illinois State University; and Kurt Rosentrater, Northern Illinois University and South Dakota State University. We also thank the reviewers of previous editions: Julian Thomas Euell, Ithaca College; Raymond A. Eve, University of Texas at Arlington; Samuel A. Guccione, Eastern Illinois University; and Gerald Harris, DeVry University, Chicago. In addition, thanks go to Vernon Anthony, Jeff Riley, and Steve Robb at Pearson Education, Inc., and Andrea Edwards at Triple SSS Press Media Development, Inc.

Contents

Technology and Society

History of Technology

The Transmission of Science from Antiquity to the Middle Ages

Nearly half a century ago, science historian George Sarton used a diagram similar to this one to show how the Arabic scientific effort revived not only Greek science, but also many scientific concepts of Iranian and Hindu origin. The transmission of scientific traditions never takes a single direct route. Alternate paths and crosscurrents were involved in all of the great eras of early scientific progress.

Adapted from: Turner, Howard R. Science in Medieval Islam: An Illustrated Introduction. *New Delhi: Oxford University Press, 1997.*

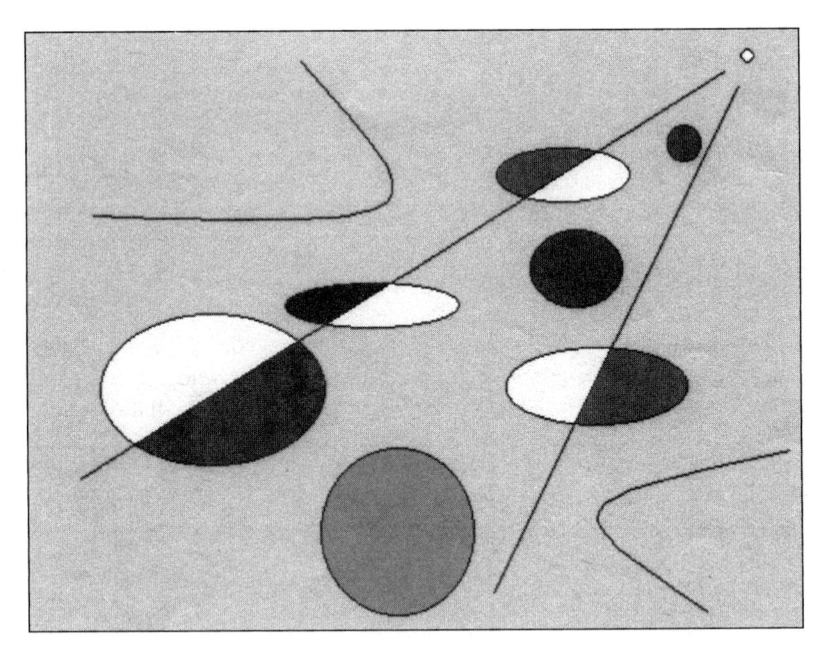

OBJECTIVES

After reading Part I, History of Technology, *you will be able to*

1. Distinguish the importance of technological development and human development over time.
2. Assess the impact of technology on society.
3. Formulate opinions about the influence of technology on social development.
4. Assess the state of technology growth in developed and developing societies.
5. Identify the conditions necessary to foster combined social and technological growth.

INTRODUCTION

This section on the history of technology begins with an apology and an explanation. Anyone attempting to write anything about the history of technology is bound to be guilty of grievous errors. Claiming to offer a comprehensive view of technology's development over time is guaranteed to arouse suspicions that something has been missed. On the other hand, to cover everything—and by that I mean everything—remotely connected with the history of technological development is equally sure to inspire charges of superficiality. Issues concerning the *what* about technology are just part of the problem. What about the *who*? Why do the Wright Brothers get all the credit when it comes to the development of air travel? Does Tesla receive too many raves for his work in developing alternating current? And why does it always seem that only white European males get all the publicity, while everyone who is not white, male, or European could be languishing in obscurity? Therefore, I want to apologize here and now for any errors of commission or omission found within Part I, for most of them are unintentional. Those omissions that have slipped by were committed out of ignorance or because of a lack of space.

My apologies having been said, let me explain what I hope this section will accomplish. The history of technology and the history of human development seem to have traveled the same path. Consequently, when we talk about technology's evolution over time, it seems as if we are talking about human development as well. Just as tracing the history of human development in its entirety is intimidating, so too is any attempt to completely cover the course of technology development. Therefore, the selection of readings in this section will try to identify the high points in technology history and connect them to high points in human history. And, like most other histories, this section will also fall victim to the subjective whims of the author, who will pick what he thinks are the most obvious and important events, only to realize they're after all just his opinions, but offer them anyway.

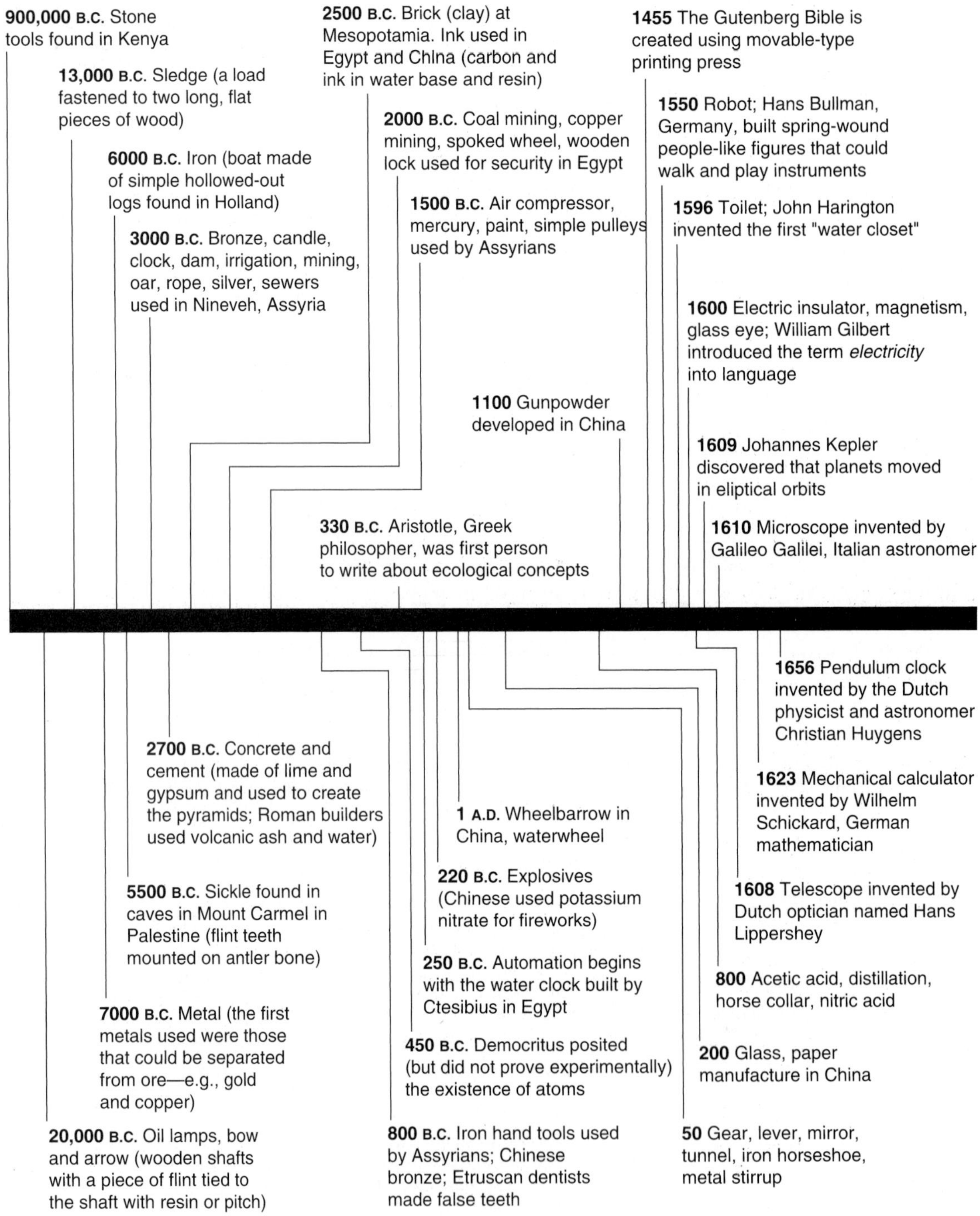

900,000 B.C. Stone tools found in Kenya

20,000 B.C. Oil lamps, bow and arrow (wooden shafts with a piece of flint tied to the shaft with resin or pitch)

13,000 B.C. Sledge (a load fastened to two long, flat pieces of wood)

7000 B.C. Metal (the first metals used were those that could be separated from ore—e.g., gold and copper)

6000 B.C. Iron (boat made of simple hollowed-out logs found in Holland)

5500 B.C. Sickle found in caves in Mount Carmel in Palestine (flint teeth mounted on antler bone)

3000 B.C. Bronze, candle, clock, dam, irrigation, mining, oar, rope, silver, sewers used in Nineveh, Assyria

2700 B.C. Concrete and cement (made of lime and gypsum and used to create the pyramids; Roman builders used volcanic ash and water)

2500 B.C. Brick (clay) at Mesopotamia. Ink used in Egypt and China (carbon and ink in water base and resin)

2000 B.C. Coal mining, copper mining, spoked wheel, wooden lock used for security in Egypt

1500 B.C. Air compressor, mercury, paint, simple pulleys used by Assyrians

800 B.C. Iron hand tools used by Assyrians; Chinese bronze; Etruscan dentists made false teeth

450 B.C. Democritus posited (but did not prove experimentally) the existence of atoms

330 B.C. Aristotle, Greek philosopher, was first person to write about ecological concepts

250 B.C. Automation begins with the water clock built by Ctesibius in Egypt

220 B.C. Explosives (Chinese used potassium nitrate for fireworks)

1 A.D. Wheelbarrow in China, waterwheel

50 Gear, lever, mirror, tunnel, iron horseshoe, metal stirrup

200 Glass, paper manufacture in China

800 Acetic acid, distillation, horse collar, nitric acid

1100 Gunpowder developed in China

1455 The Gutenberg Bible is created using movable-type printing press

1550 Robot; Hans Bullman, Germany, built spring-wound people-like figures that could walk and play instruments

1596 Toilet; John Harington invented the first "water closet"

1600 Electric insulator, magnetism, glass eye; William Gilbert introduced the term *electricity* into language

1608 Telescope invented by Dutch optician named Hans Lippershey

1609 Johannes Kepler discovered that planets moved in eliptical orbits

1610 Microscope invented by Galileo Galilei, Italian astronomer

1623 Mechanical calculator invented by Wilhelm Schickard, German mathematician

1656 Pendulum clock invented by the Dutch physicist and astronomer Christian Huygens

1661 Pollution control (John Evelyn of England) proposed methods for preventing smoke from polluting the air

1666 Isaac Newton discovered gravity

1676 Andre Felibien invented the screwdriver

1681 Barograph, a device for recording atmospheric pressure, invented by English physicist Robert Hooke

1698 Steam pump invented by Thomas Savery, an English military engineer (pump used to drain water from mines)

1714 Thermometer invented by Gabriel Danile Fahrenheit in Poland

1718 Machine gun invented by James Puckle in England; fired 63 rounds in seven minutes

1725 Punched card (Basile Bouchon of France used punched paper to control a loom)

1743 First passenger elevator invented for use by Louis XV of France at Versailles palace

1744 "Franklin stove" designed by Benjamin Franklin in United States; widely adopted in Europe decades later

1745 Electric capacitor (The first storage device for an electric charge was invented by Edwald von Kliest of Germany); first working electric capacitor (Leyden jar invented by Peter van Musschenbroek at University of Leyden)

1748 Electroscope, an instrument that detects electric charges, invented by Jean Antoine Nollet of France; steel-nibbed pen invented by Johann Janssen in france

1749 Electricity (Benjamin Franklin discovered that electricity has two states: positive and negative)

1752 Diving suit (Englishman John Smeaton invented an air pump that fit into a diving bell)

1756 Dentures (Prussian dentist Philip Pfaff described steps to make wax impressions and cast models for missing teeth)

1757 Achromatic lens (John Dolland of England invented lenses made of two kinds of glass for the telescope)

1758 First railroad passenger car made of wood (Great Britain)

1760 First bifocals were designed by Benjamin Franklin (U.S.)

1764 Spinning jenny (Invented by English weaver Hames Hargreaves, the machine could do the work of 30 spinning wheels)

1769 First automobiles had three wheels, was powered by a steam engine, and had a top speed of 3 mph; they were bulky and inefficient

1776 David Bushnell (U.S.) built a hand-cranked submarine

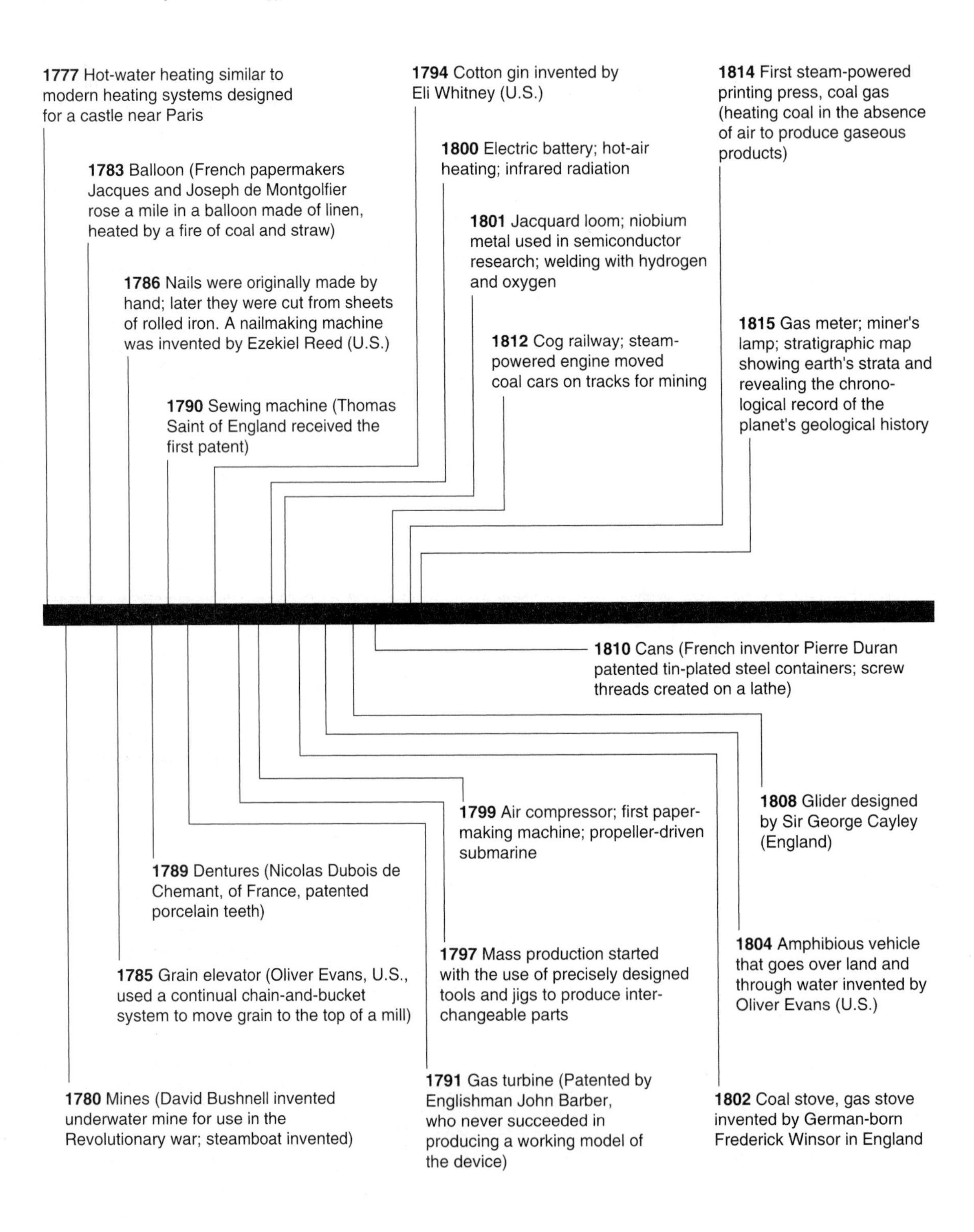
1777 Hot-water heating similar to modern heating systems designed for a castle near Paris
1783 Balloon (French papermakers Jacques and Joseph de Montgolfier rose a mile in a balloon made of linen, heated by a fire of coal and straw)
1786 Nails were originally made by hand; later they were cut from sheets of rolled iron. A nailmaking machine was invented by Ezekiel Reed (U.S.)
1790 Sewing machine (Thomas Saint of England received the first patent)
1794 Cotton gin invented by Eli Whitney (U.S.)
1800 Electric battery; hot-air heating; infrared radiation
1801 Jacquard loom; niobium metal used in semiconductor research; welding with hydrogen and oxygen
1812 Cog railway; steam-powered engine moved coal cars on tracks for mining
1814 First steam-powered printing press, coal gas (heating coal in the absence of air to produce gaseous products)
1815 Gas meter; miner's lamp; stratigraphic map showing earth's strata and revealing the chronological record of the planet's geological history
1810 Cans (French inventor Pierre Duran patented tin-plated steel containers; screw threads created on a lathe)
1808 Glider designed by Sir George Cayley (England)
1799 Air compressor; first paper-making machine; propeller-driven submarine
1789 Dentures (Nicolas Dubois de Chemant, of France, patented porcelain teeth)
1804 Amphibious vehicle that goes over land and through water invented by Oliver Evans (U.S.)
1797 Mass production started with the use of precisely designed tools and jigs to produce interchangeable parts
1785 Grain elevator (Oliver Evans, U.S., used a continual chain-and-bucket system to move grain to the top of a mill)
1791 Gas turbine (Patented by Englishman John Barber, who never succeeded in producing a working model of the device)
1780 Mines (David Bushnell invented underwater mine for use in the Revolutionary war; steamboat invented)
1802 Coal stove, gas stove invented by German-born Frederick Winsor in England

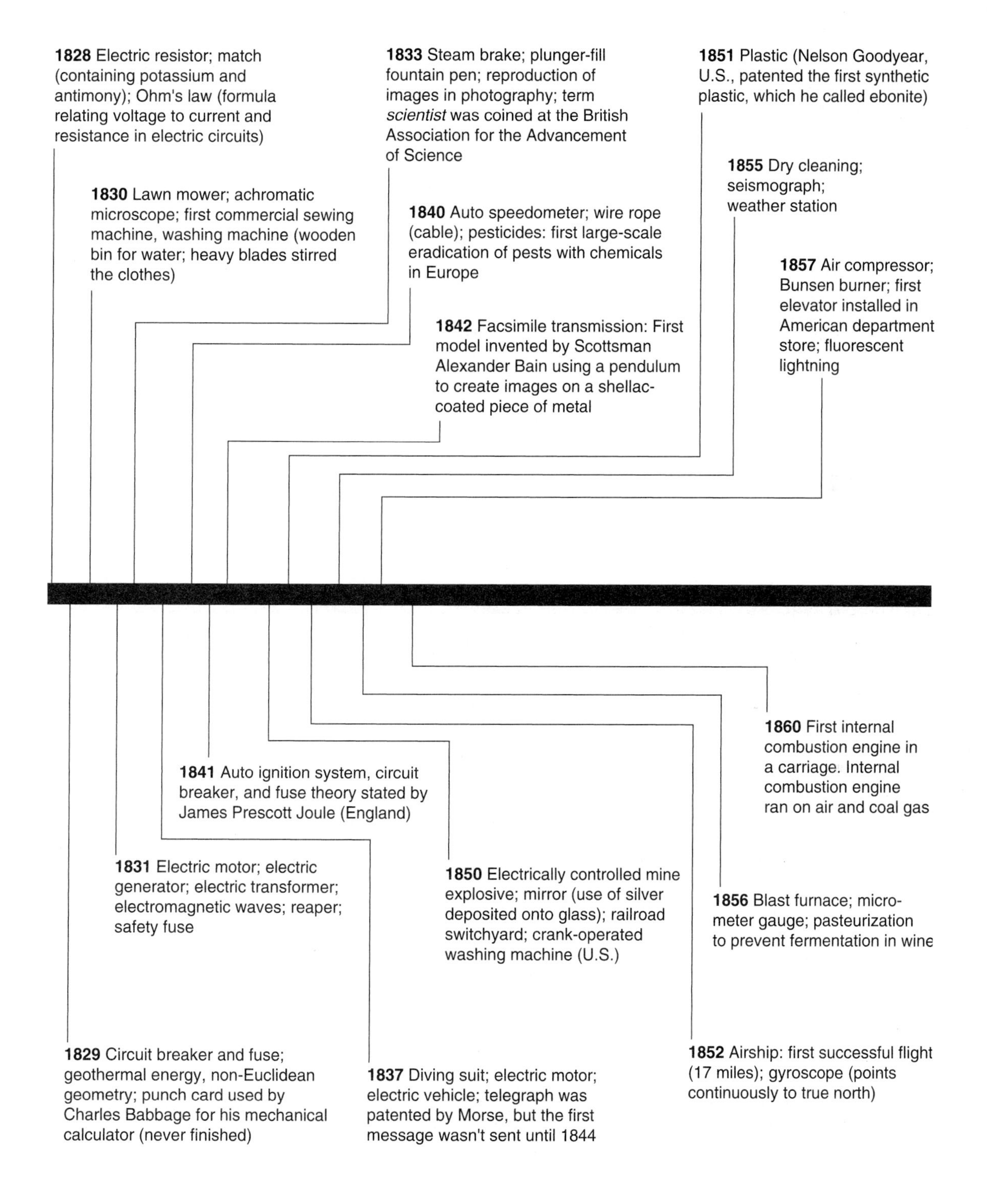
1828 Electric resistor; match (containing potassium and antimony); Ohm's law (formula relating voltage to current and resistance in electric circuits)
1829 Circuit breaker and fuse; geothermal energy, non-Euclidean geometry; punch card used by Charles Babbage for his mechanical calculator (never finished)
1830 Lawn mower; achromatic microscope; first commercial sewing machine, washing machine (wooden bin for water; heavy blades stirred the clothes)
1831 Electric motor; electric generator; electric transformer; electromagnetic waves; reaper; safety fuse
1833 Steam brake; plunger-fill fountain pen; reproduction of images in photography; term *scientist* was coined at the British Association for the Advancement of Science
1837 Diving suit; electric motor; electric vehicle; telegraph was patented by Morse, but the first message wasn't sent until 1844
1840 Auto speedometer; wire rope (cable); pesticides: first large-scale eradication of pests with chemicals in Europe
1841 Auto ignition system, circuit breaker, and fuse theory stated by James Prescott Joule (England)
1842 Facsimile transmission: First model invented by Scottsman Alexander Bain using a pendulum to create images on a shellac-coated piece of metal
1850 Electrically controlled mine explosive; mirror (use of silver deposited onto glass); railroad switchyard; crank-operated washing machine (U.S.)
1851 Plastic (Nelson Goodyear, U.S., patented the first synthetic plastic, which he called ebonite)
1852 Airship: first successful flight (17 miles); gyroscope (points continuously to true north)
1855 Dry cleaning; seismograph; weather station
1856 Blast furnace; micro-meter gauge; pasteurization to prevent fermentation in wine
1857 Air compressor; Bunsen burner; first elevator installed in American department store; fluorescent lightning
1860 First internal combustion engine in a carriage. Internal combustion engine ran on air and coal gas

1861 Color photography (Thomas Sutton and James Clerk Maxwell took pictures, made positive transparencies, and projected them on a screen); tanker named *Elizabeth Watts* was the first ship built in the United States to carry oil from Pennsylvania to London

1862 Rapid-firing machine gun; acoustics of music

1864 Nitroglycerine used as a detonator by Alfred Nobel. James Clerk Maxwell created a set of four equations that described the relation between electricity and magnetism

1865 Genetics (Gregor Mendel of Austria conducted experiments crossbreeding various strains of garden peas.) Electrolytic refining of copper

1866 Ernst Heinrich Haeckle was the first to use the term *ecology*

1867 Barbed wire patented. Dynamite patented. First successful torpedo invented. First commercial typewriter invented

1868 First tilting dental chair invented. First electric dental drill invented. (The drill wasn't put on the market until 1872 and wasn't used until the 1890s due to the lack of electricity)

1869 Sir William Herschel (England) devised fingerprint identification system; steam-powered monorail system invented in Syria; periodic table of elements created; stock ticker (printer of stock prices) invented by Thomas Alva Edison; DNA discovered by Johann Friedrick Miescher; automatic air brake invented for railroad cars

1870 Refrigerator invented and manufactured using ammonia

1876 Telephone patent was granted to Alexander Graham Bell by the U.S. Supreme Court. Elisha Gray filed for, and failed to receive, the patent two hours after Bell

1876 Carpet sweeper invented by Melville Bissell (U.S.); J. Willar Gibbs, American physicist, wrote about the general principles of physical chemistry

1877 Duplicating was invented by Thomas Alva Edison. Carbon microphone created by Emile Berliner and Thomas Alva Edison. Moving-coil microphone invented by Charles Cuttris and Werner Siemens; phonograph was patented by Thomas Alva Edison. Emile Berliner invented a wax phonograph record that produced a fuzzy sound

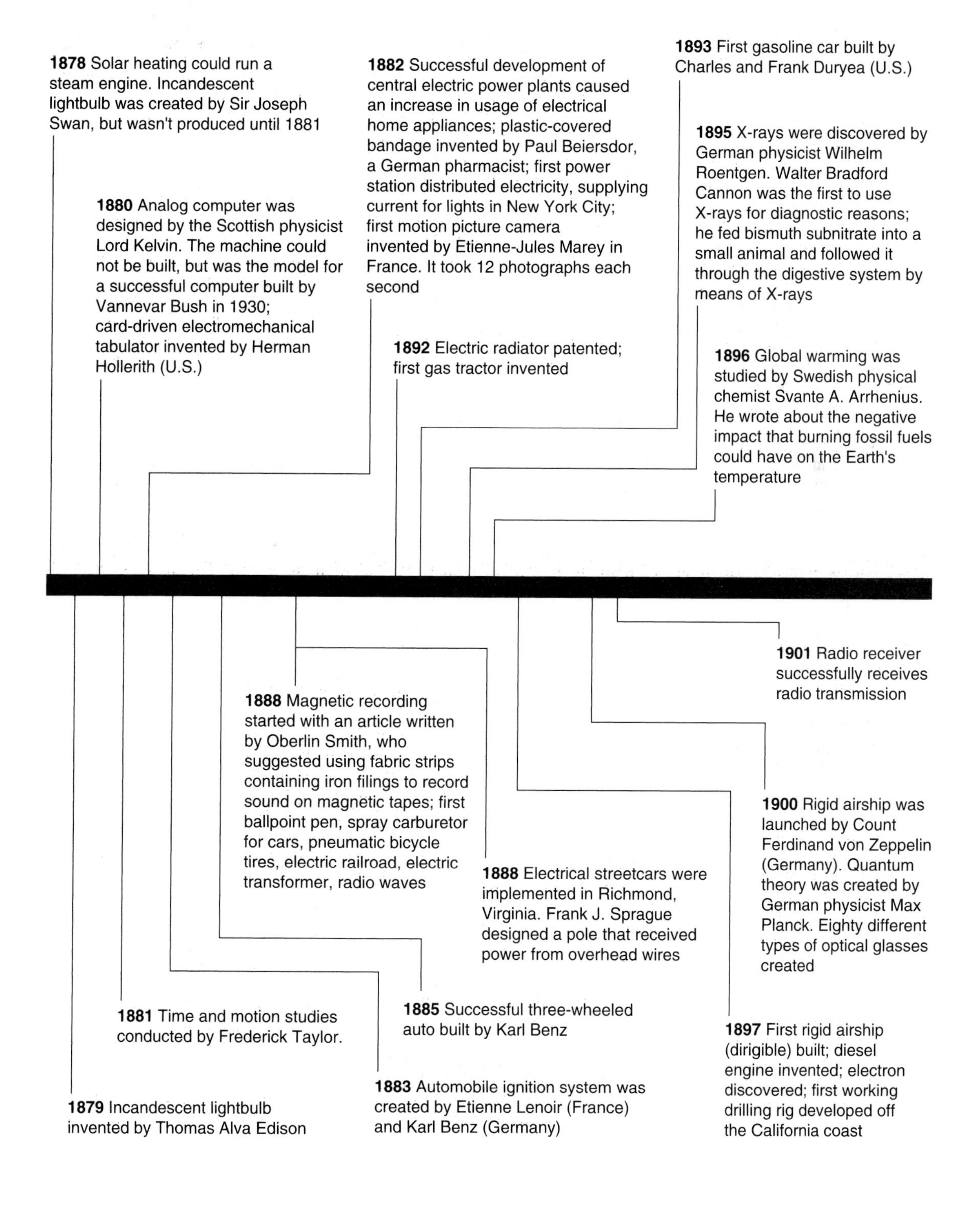

1878 Solar heating could run a steam engine. Incandescent lightbulb was created by Sir Joseph Swan, but wasn't produced until 1881
1880 Analog computer was designed by the Scottish physicist Lord Kelvin. The machine could not be built, but was the model for a successful computer built by Vannevar Bush in 1930; card-driven electromechanical tabulator invented by Herman Hollerith (U.S.)
1882 Successful development of central electric power plants caused an increase in usage of electrical home appliances; plastic-covered bandage invented by Paul Beiersdor, a German pharmacist; first power station distributed electricity, supplying current for lights in New York City; first motion picture camera invented by Etienne-Jules Marey in France. It took 12 photographs each second
1892 Electric radiator patented; first gas tractor invented
1893 First gasoline car built by Charles and Frank Duryea (U.S.)
1895 X-rays were discovered by German physicist Wilhelm Roentgen. Walter Bradford Cannon was the first to use X-rays for diagnostic reasons; he fed bismuth subnitrate into a small animal and followed it through the digestive system by means of X-rays
1896 Global warming was studied by Swedish physical chemist Svante A. Arrhenius. He wrote about the negative impact that burning fossil fuels could have on the Earth's temperature
1901 Radio receiver successfully receives radio transmission
1888 Magnetic recording started with an article written by Oberlin Smith, who suggested using fabric strips containing iron filings to record sound on magnetic tapes; first ballpoint pen, spray carburetor for cars, pneumatic bicycle tires, electric railroad, electric transformer, radio waves
1888 Electrical streetcars were implemented in Richmond, Virginia. Frank J. Sprague designed a pole that received power from overhead wires
1900 Rigid airship was launched by Count Ferdinand von Zeppelin (Germany). Quantum theory was created by German physicist Max Planck. Eighty different types of optical glasses created
1881 Time and motion studies conducted by Frederick Taylor.
1885 Successful three-wheeled auto built by Karl Benz
1897 First rigid airship (dirigible) built; diesel engine invented; electron discovered; first working drilling rig developed off the California coast
1879 Incandescent lightbulb invented by Thomas Alva Edison
1883 Automobile ignition system was created by Etienne Lenoir (France) and Karl Benz (Germany)

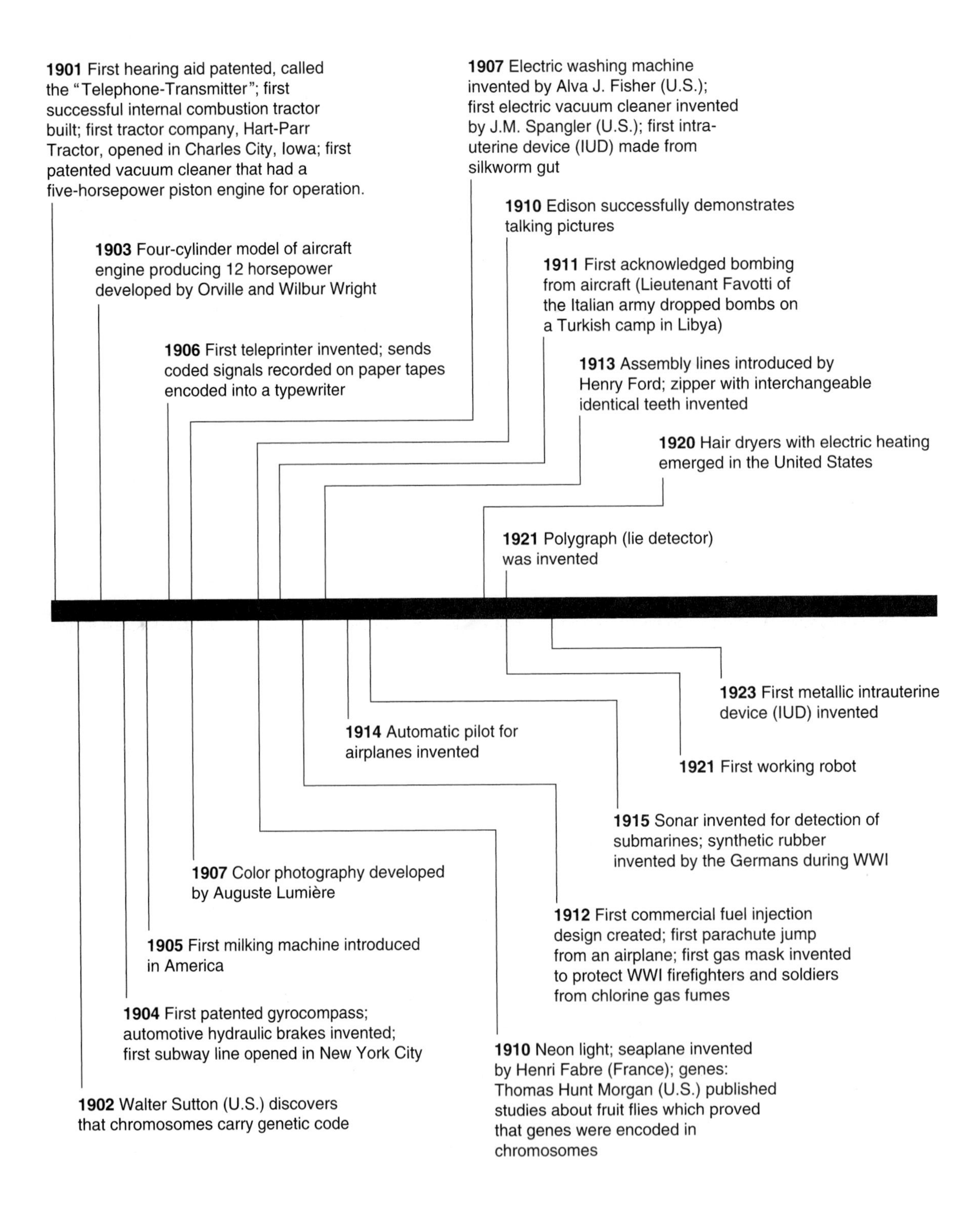
1901 First hearing aid patented, called the "Telephone-Transmitter"; first successful internal combustion tractor built; first tractor company, Hart-Parr Tractor, opened in Charles City, Iowa; first patented vacuum cleaner that had a five-horsepower piston engine for operation.
1902 Walter Sutton (U.S.) discovers that chromosomes carry genetic code
1903 Four-cylinder model of aircraft engine producing 12 horsepower developed by Orville and Wilbur Wright
1904 First patented gyrocompass; automotive hydraulic brakes invented; first subway line opened in New York City
1905 First milking machine introduced in America
1906 First teleprinter invented; sends coded signals recorded on paper tapes encoded into a typewriter
1907 Color photography developed by Auguste Lumière
1907 Electric washing machine invented by Alva J. Fisher (U.S.); first electric vacuum cleaner invented by J.M. Spangler (U.S.); first intra-uterine device (IUD) made from silkworm gut
1910 Neon light; seaplane invented by Henri Fabre (France); genes: Thomas Hunt Morgan (U.S.) published studies about fruit flies which proved that genes were encoded in chromosomes
1910 Edison successfully demonstrates talking pictures
1911 First acknowledged bombing from aircraft (Lieutenant Favotti of the Italian army dropped bombs on a Turkish camp in Libya)
1912 First commercial fuel injection design created; first parachute jump from an airplane; first gas mask invented to protect WWI firefighters and soldiers from chlorine gas fumes
1913 Assembly lines introduced by Henry Ford; zipper with interchangeable identical teeth invented
1914 Automatic pilot for airplanes invented
1915 Sonar invented for detection of submarines; synthetic rubber invented by the Germans during WWI
1920 Hair dryers with electric heating emerged in the United States
1921 Polygraph (lie detector) was invented
1921 First working robot
1923 First metallic intrauterine device (IUD) invented

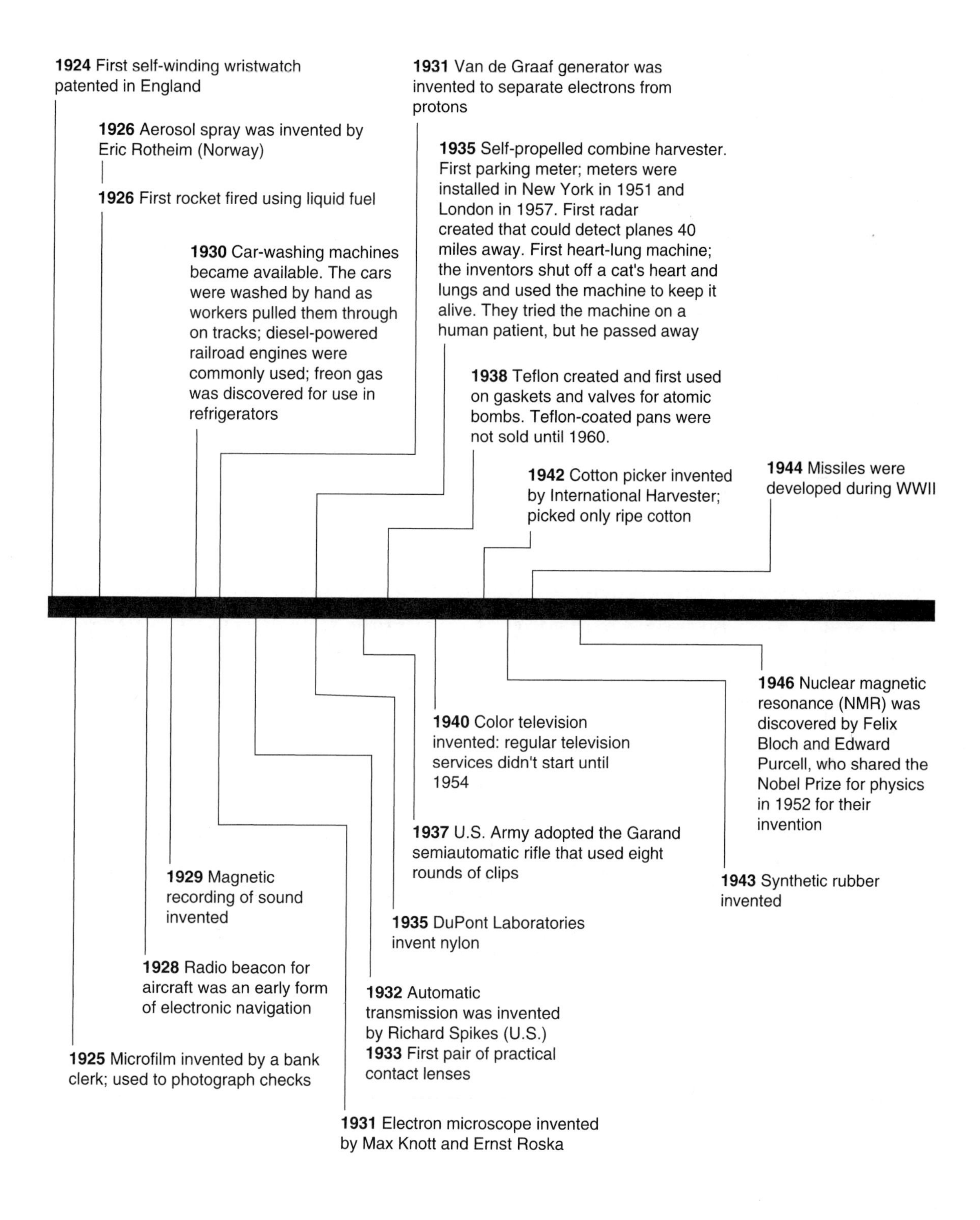

1924 First self-winding wristwatch patented in England
1926 Aerosol spray was invented by Eric Rotheim (Norway)
1926 First rocket fired using liquid fuel
1930 Car-washing machines became available. The cars were washed by hand as workers pulled them through on tracks; diesel-powered railroad engines were commonly used; freon gas was discovered for use in refrigerators
1931 Van de Graaf generator was invented to separate electrons from protons
1935 Self-propelled combine harvester. First parking meter; meters were installed in New York in 1951 and London in 1957. First radar created that could detect planes 40 miles away. First heart-lung machine; the inventors shut off a cat's heart and lungs and used the machine to keep it alive. They tried the machine on a human patient, but he passed away
1938 Teflon created and first used on gaskets and valves for atomic bombs. Teflon-coated pans were not sold until 1960.
1942 Cotton picker invented by International Harvester; picked only ripe cotton
1944 Missiles were developed during WWII
1925 Microfilm invented by a bank clerk; used to photograph checks
1928 Radio beacon for aircraft was an early form of electronic navigation
1929 Magnetic recording of sound invented
1931 Electron microscope invented by Max Knott and Ernst Roska
1932 Automatic transmission was invented by Richard Spikes (U.S.)
1933 First pair of practical contact lenses
1935 DuPont Laboratories invent nylon
1937 U.S. Army adopted the Garand semiautomatic rifle that used eight rounds of clips
1940 Color television invented: regular television services didn't start until 1954
1943 Synthetic rubber invented
1946 Nuclear magnetic resonance (NMR) was discovered by Felix Bloch and Edward Purcell, who shared the Nobel Prize for physics in 1952 for their invention

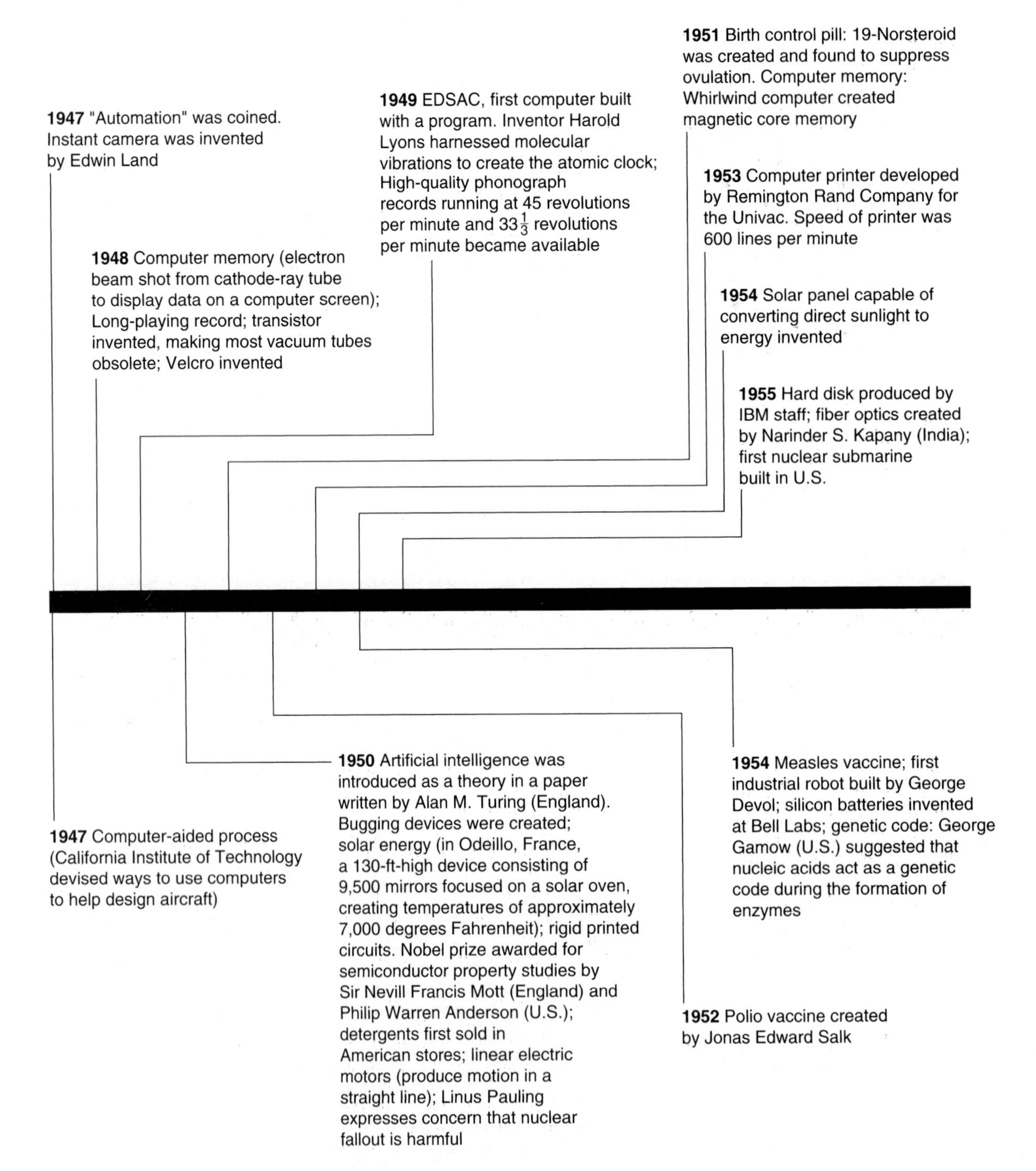
1947 "Automation" was coined. Instant camera was invented by Edwin Land
1948 Computer memory (electron beam shot from cathode-ray tube to display data on a computer screen); Long-playing record; transistor invented, making most vacuum tubes obsolete; Velcro invented
1949 EDSAC, first computer built with a program. Inventor Harold Lyons harnessed molecular vibrations to create the atomic clock; High-quality phonograph records running at 45 revolutions per minute and $33\frac{1}{3}$ revolutions per minute became available
1951 Birth control pill: 19-Norsteroid was created and found to suppress ovulation. Computer memory: Whirlwind computer created magnetic core memory
1953 Computer printer developed by Remington Rand Company for the Univac. Speed of printer was 600 lines per minute
1954 Solar panel capable of converting direct sunlight to energy invented
1955 Hard disk produced by IBM staff; fiber optics created by Narinder S. Kapany (India); first nuclear submarine built in U.S.
1947 Computer-aided process (California Institute of Technology devised ways to use computers to help design aircraft)
1950 Artificial intelligence was introduced as a theory in a paper written by Alan M. Turing (England). Bugging devices were created; solar energy (in Odeillo, France, a 130-ft-high device consisting of 9,500 mirrors focused on a solar oven, creating temperatures of approximately 7,000 degrees Fahrenheit); rigid printed circuits. Nobel prize awarded for semiconductor property studies by Sir Nevill Francis Mott (England) and Philip Warren Anderson (U.S.); detergents first sold in American stores; linear electric motors (produce motion in a straight line); Linus Pauling expresses concern that nuclear fallout is harmful
1952 Polio vaccine created by Jonas Edward Salk
1954 Measles vaccine; first industrial robot built by George Devol; silicon batteries invented at Bell Labs; genetic code: George Gamow (U.S.) suggested that nucleic acids act as a genetic code during the formation of enzymes

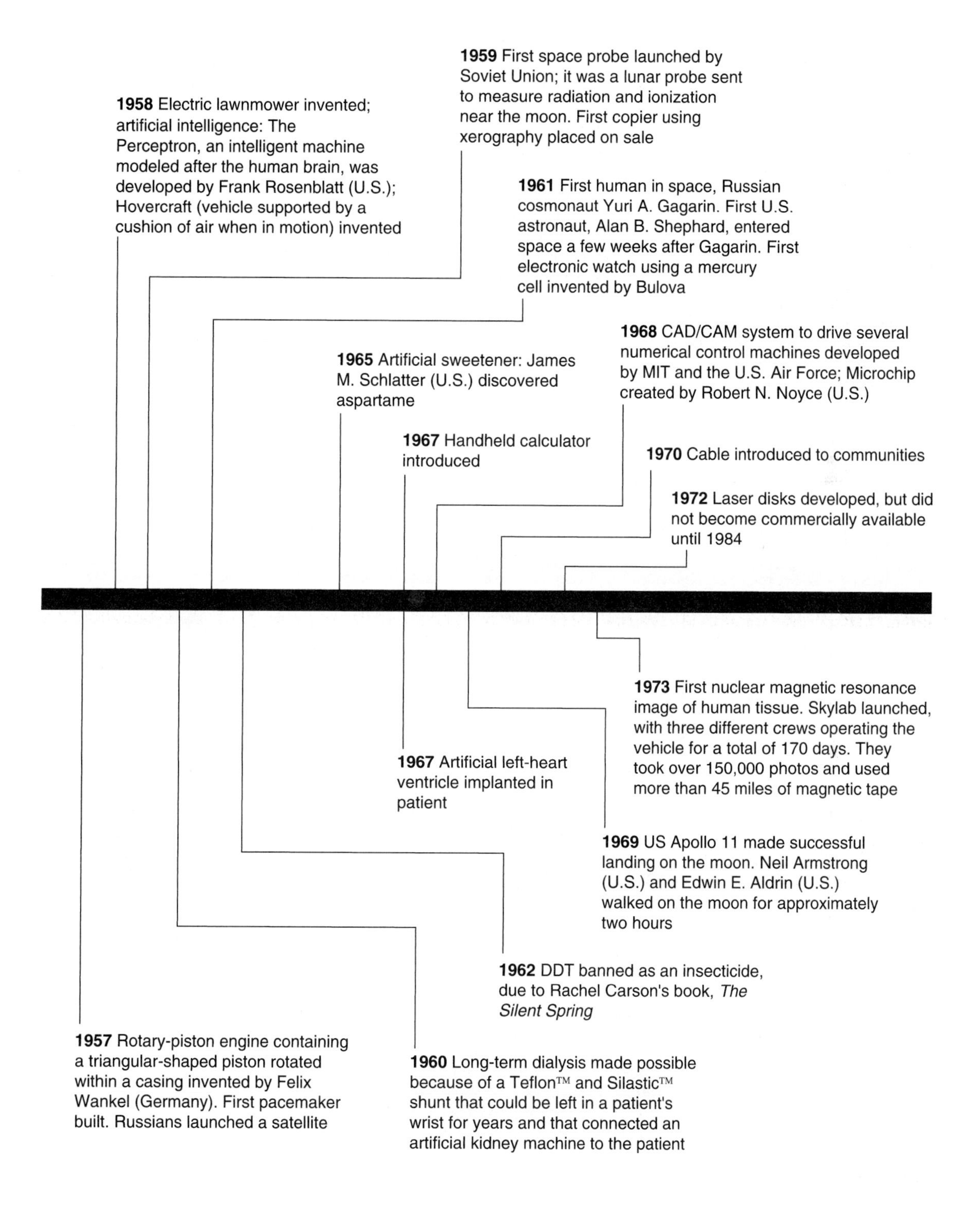

1959 First space probe launched by Soviet Union; it was a lunar probe sent to measure radiation and ionization near the moon. First copier using xerography placed on sale
1958 Electric lawnmower invented; artificial intelligence: The Perceptron, an intelligent machine modeled after the human brain, was developed by Frank Rosenblatt (U.S.); Hovercraft (vehicle supported by a cushion of air when in motion) invented
1961 First human in space, Russian cosmonaut Yuri A. Gagarin. First U.S. astronaut, Alan B. Shephard, entered space a few weeks after Gagarin. First electronic watch using a mercury cell invented by Bulova
1968 CAD/CAM system to drive several numerical control machines developed by MIT and the U.S. Air Force; Microchip created by Robert N. Noyce (U.S.)
1965 Artificial sweetener: James M. Schlatter (U.S.) discovered aspartame
1967 Handheld calculator introduced
1970 Cable introduced to communities
1972 Laser disks developed, but did not become commercially available until 1984
1973 First nuclear magnetic resonance image of human tissue. Skylab launched, with three different crews operating the vehicle for a total of 170 days. They took over 150,000 photos and used more than 45 miles of magnetic tape
1967 Artificial left-heart ventricle implanted in patient
1969 US Apollo 11 made successful landing on the moon. Neil Armstrong (U.S.) and Edwin E. Aldrin (U.S.) walked on the moon for approximately two hours
1962 DDT banned as an insecticide, due to Rachel Carson's book, *The Silent Spring*
1957 Rotary-piston engine containing a triangular-shaped piston rotated within a casing invented by Felix Wankel (Germany). First pacemaker built. Russians launched a satellite
1960 Long-term dialysis made possible because of a Teflon™ and Silastic™ shunt that could be left in a patient's wrist for years and that connected an artificial kidney machine to the patient

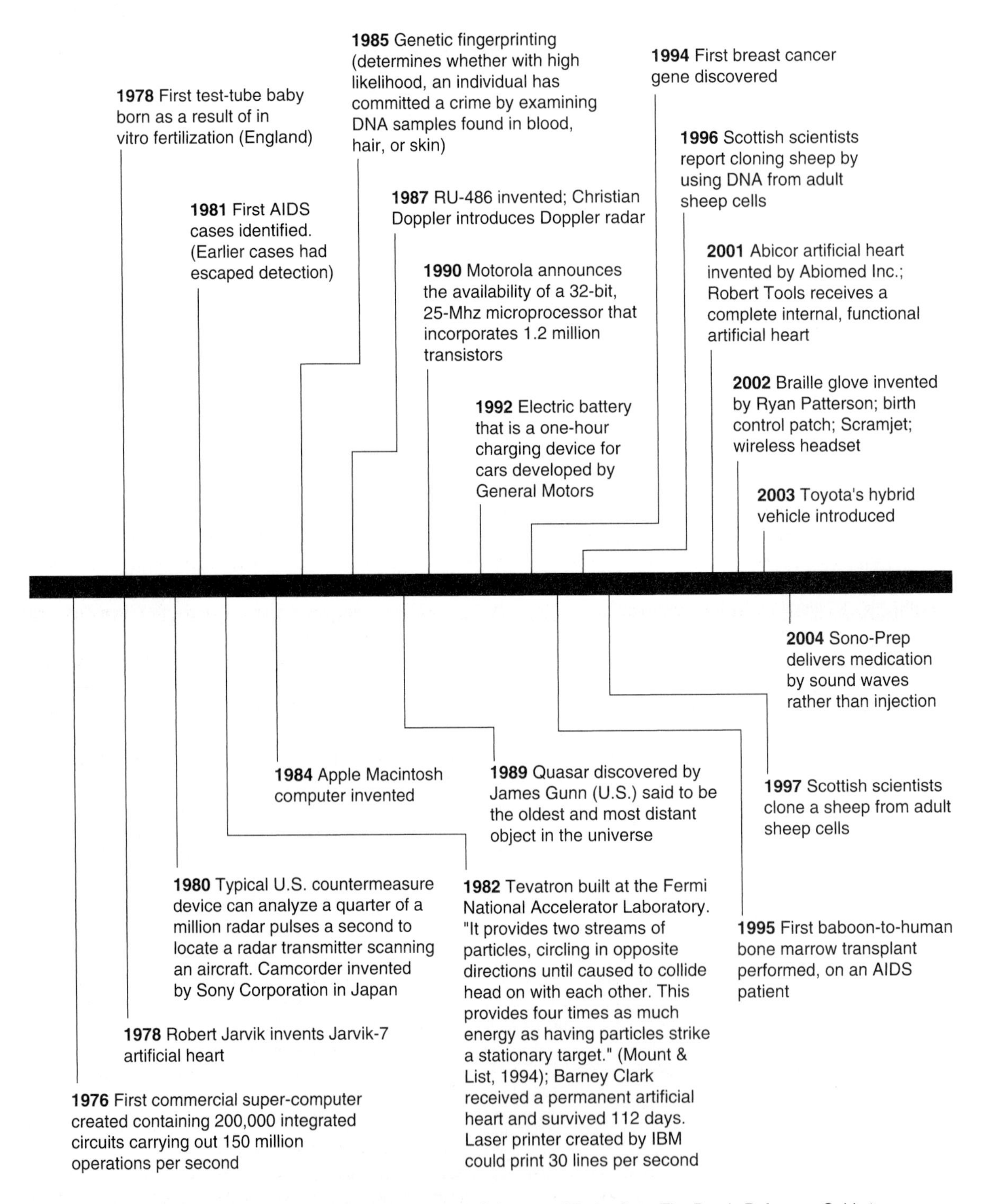

1985 Genetic fingerprinting (determines whether with high likelihood, an individual has committed a crime by examining DNA samples found in blood, hair, or skin)
1994 First breast cancer gene discovered
1978 First test-tube baby born as a result of in vitro fertilization (England)
1996 Scottish scientists report cloning sheep by using DNA from adult sheep cells
1987 RU-486 invented; Christian Doppler introduces Doppler radar
1981 First AIDS cases identified. (Earlier cases had escaped detection)
2001 Abicor artificial heart invented by Abiomed Inc.; Robert Tools receives a complete internal, functional artificial heart
1990 Motorola announces the availability of a 32-bit, 25-Mhz microprocessor that incorporates 1.2 million transistors
2002 Braille glove invented by Ryan Patterson; birth control patch; Scramjet; wireless headset
1992 Electric battery that is a one-hour charging device for cars developed by General Motors
2003 Toyota's hybrid vehicle introduced
2004 Sono-Prep delivers medication by sound waves rather than injection
1984 Apple Macintosh computer invented
1989 Quasar discovered by James Gunn (U.S.) said to be the oldest and most distant object in the universe
1997 Scottish scientists clone a sheep from adult sheep cells
1980 Typical U.S. countermeasure device can analyze a quarter of a million radar pulses a second to locate a radar transmitter scanning an aircraft. Camcorder invented by Sony Corporation in Japan
1982 Tevatron built at the Fermi National Accelerator Laboratory. "It provides two streams of particles, circling in opposite directions until caused to collide head on with each other. This provides four times as much energy as having particles strike a stationary target." (Mount & List, 1994); Barney Clark received a permanent artificial heart and survived 112 days. Laser printer created by IBM could print 30 lines per second
1995 First baboon-to-human bone marrow transplant performed, on an AIDS patient
1978 Robert Jarvik invents Jarvik-7 artificial heart
1976 First commercial super-computer created containing 200,000 integrated circuits carrying out 150 million operations per second

This timeline represents information adapted from Milestones in Science and Technology: The Ready Reference Guide to Discoveries, Inventions, and Facts, *Second Edition, by Ellis Mount and Barbara A. List. Copyright © 1993. Reproduced with permission of Greenwood Publishing Group, Inc., Westport, CT.*

The Great Leap Forward

JARED DIAMOND

World population figures around 1 A.D. have been estimated at about 200 million people. One million years prior to 1 A.D., population figures of early humankind of *Homo sapiens* or *Homo erectus* have been estimated at about 125,000. In order for us to understand and appreciate the very long path of this development and growth of humankind, it is helpful to discuss early hominoid history and its development, expansion, and explosion along with a discussion of the relationship of tools, concepts, and technology to that expansion. "The Great Leap Forward" gives some perspectives on the early struggles of humankind and exponential growth toward modern creativity, expansion, and technology.

One can hardly blame nineteenth-century creationists for insisting that humans were separately created by God. After all, between us and other animal species lies the seemingly unbridgeable gulf of language, art, religion, writing, and complex machines. Small wonder, then, that to many people Darwin's theory of our evolution from apes appeared absurd.

Since Darwin's time, of course, fossilized bones of hundreds of creatures intermediate between apes and modern humans have been discovered. It is no longer possible for a reasonable person to deny that what once seemed absurd actually happened—somehow. Yet the discoveries of many missing links have only made the problem more fascinating without fully solving it. When and how did we acquire our uniquely human characteristics?

We know that our lineage arose in Africa, diverging from that of chimpanzees and gorillas, sometime between 6 and 10 million years ago. For most of the time since then we have been little more than glorified baboons. As recently as 35,000 years ago western Europe was still occupied by Neanderthals, primitive beings for whom art and progress scarcely existed. Then there was an abrupt change. Anatomically modern people appeared in Europe and, suddenly, so did sculpture, musical instruments, lamps, trade, and innovation. Within a few thousand years the Neanderthals were gone.

Insofar as there was any single moment when we could be said to have become human, it was at the time of this Great Leap Forward 35,000 years ago. Only a few more dozen millennia—a trivial fraction of our 6- to 10-million year history—were needed for us to domesticate animals, develop agriculture and metallurgy, and invent writing. It was then but a short further step to those monuments of civilization that distinguish us from all other animals—monuments such as the *Mona Lisa* and the Ninth Symphony, the Eiffel Tower and Sputnik, Dachau's ovens and the bombing of Dresden.

What happened at that magic moment in evolution? What made it possible, and why was it so sudden? What held back the Neanderthals, and what was their fate? Did Neanderthals and modern peoples ever meet, and if so, how did they behave toward each other? We still share 98 percent of our genes with chimps; which genes among the other 2 percent had such enormous consequences?

Understanding the Great Leap Forward isn't easy; neither is writing about it. The immediate evidence comes from technical details of preserved bones and stone tools. Archeologists' reports are full of such terms as "transverse occipital torus," "receding zygomatic arches," and "Chatelperronian

Jared Diamond © 1989. Reprinted with permission of *Discover Magazine*.

backed knives." What we really want to understand—the way of life and the humanity of our various ancestors—isn't directly preserved but only inferred from those technical details. Much of the evidence is missing, and archeologists often disagree over the meaning of the evidence that has survived.

I'll emphasize those inferences rather than the technical details, and I'll speculate about the answers to those questions I just listed above. But you can form your own opinions, and they may differ from mine. This is a puzzle whose solution is still unknown.

To set the stage quickly, recall that life originated on Earth several billion years ago, the dinosaurs became extinct around 65 million years ago, and, as I mentioned, our ancestors diverged from the ancestors of chimps and gorillas between 6 and 10 million years ago. They then remained confined to Africa for millions of years.

Initially, our ancestors would have been classified as merely another species of ape, but a sequence of three changes launched them in the direction of modern humans. The first of these changes occurred by around 4 million years ago: the structure of fossilized limb bones shows that our ancestors, in contrast to gorillas and chimps, were habitually walking upright. The upright posture freed our forelimbs to do other things, among which toolmaking would eventually prove to be the most important.

The second change occurred around 3 million years ago, when our lineage split in two. As background, remember that members of two animal species living in the same area must fill different ecological roles and do not normally interbreed. For example, coyotes and wolves are obviously closely related and, until wolves were exterminated in most of the United States, lived in many of the same areas. However, wolves are larger, they usually hunt big mammals like deer and moose, and they often live in sizable packs, whereas coyotes are smaller, mainly hunt small mammals like rabbits and mice, and normally live in pairs or small groups.

Now, all modern humans unquestionably belong to the same species. Ecological differences among us are entirely a product of childhood education: it is not the case that some of us are born big and habitually hunt deer while others are born small, gather berries, and don't marry the deer hunters. And every human population living today has interbred with every other human population with which it has had extensive contact.

Three million years ago, however, there were hominid species as distinct as wolves and coyotes. On one branch of the family tree was a man-ape with a heavily built skull and very big cheek teeth, who probably ate coarse plant food; he has come to be known as *Australopithecus robustus* (the "robust southern ape"). On the other branch was a man-ape with a more lightly built skull and smaller teeth, who most likely had an omnivorous diet; he is known as *Australopithecus africanus* (the "southern ape of Africa"). Our lineage may have experienced such a radical division at least once more, at the time of the Great Leap Forward. But the description of that event will have to wait.

There is considerable disagreement over just what occurred in the next million years, but the argument I find most persuasive is that *A. africanus* evolved into the larger-brained form we call *Homo habilis* ("man the handyman"). Complicating the issue is that fossil bones often attributed to *H. habilis* differ so much in skull size and tooth size that they may actually imply another fork in our lineage yielding two distinct *habilis*-like species: *H. habilis* himself and a mysterious "Third Man." Thus, by 2 million years ago there were at least two and possibly three protohuman species.

The third and last of the big changes that began to make our ancestors more human and less apelike was the regular use of stone tools. By around 2.5 million years ago, very crude stone tools appear in large numbers in areas of East Africa occupied by the protohumans. Since there were two or three protohuman species, who made the tools? Probably the light-skulled species, since both it and the tools persisted and evolved. (There is, however, the intriguing possibility that at least some of our robust relatives also made tools, as recent anatomical analyses of hand bones from the Swartkrans cave in South Africa suggest. See "The Gripping Story of Paranthropus," by Pat Shipman, in the April 1989 issue of *Discover Magazine*.)

With only one human species surviving today but two or three a few million years ago, it's clear that one or two species must have become extinct. Who was our ancestor, which species ended up instead as a discard in the trash heap of evolution, and when did this shakedown occur?

The winner was the light-skulled *H. habilis*, who went on to increase in brain size and body size. By around 1.7 million years ago, the differences were sufficient that anthropologists give our lineage the new name *Homo erectus* ("the man who walks upright"—*H. erectus* fossils were discovered before all the earlier ones, so anthropologists didn't realize that *H. erectus* wasn't the first protohuman to walk upright). The robust man-ape disappeared somewhat after 1.2 million years ago, and the Third Man (if he ever existed) must have disappeared by then also.

As for why *H. erectus* survived and *A. robustus* didn't, we can only speculate. A plausible guess is that the robust man-ape could no longer compete: *H. erectus* ate both meat and plant food, and his larger brain may have made him more efficient at getting the food on which *A. robustus* depended. It's also possible that *H. erectus* gave his robust brother a direct push into oblivion by killing him for meat.

The shakedown left *H. erectus* as the sole protohuman player on the African stage, a stage to which our closest living relatives (the chimp and gorilla) are still confined. But around 1 million years ago, *H. erectus* began to expand his horizons. His stone tools and bones show that he reached the Near East, then the Far East (where he is represented by the famous fossils known as Peking man and Java man) and Europe. He continued to evolve in our direction by an increase in brain size and skull roundness. Approximately 500,000 years ago, some of our ancestors looked sufficiently enough like us—and sufficiently different from earlier *H. erectus*—to be classified as our own species, *Homo sapiens* (the "wise man"). However, they still had thicker skulls and brow ridges than we do today.

Was our meteoric ascent to sapiens status half a million years ago the brilliant climax of Earth's history, when art and sophisticated technology finally burst upon our previously dull planet? Not at all. The appearance of *H. sapiens* was a non-event. The Great Leap Forward, as proclaimed by cave paintings, houses, and bows and arrows, still lay hundreds of thousands of years in the future. Stone tools continued to be the crude ones that *H. erectus* had been making for nearly a million years. The extra brain size of those early *H. sapiens* had no dramatic effect on their way of life. That whole long tenure of *H. erectus* and early *H. sapiens* outside Africa was a period of infinitesimally slow cultural change.

So what was life like during the 1.5 million years that spanned the emergence of *H. erectus* and *H. sapiens*? The only surviving tools from this period are stone implements that can, charitably, be described as very crude. Early stone tools do vary in size and shape, and archeologists have used those differences to give the tools different names, such as hand-ax, chopper, and cleaver. But these names conceal the fact that none of these early tools had a sufficiently consistent or distinctive shape to suggest any specific function. Wear marks on the tools show that they were variously used to cut meat, bone, hides, wood, and nonwoody parts of plants. But any size or shape tool seems to have been used to cut any of these things, and the categories imposed by archeologists may be little more than arbitrary divisions of a continuum of stone forms.

Negative evidence is also significant. All the early stone tools may have been held directly in the hand; they show no signs of being mounted on other materials for increased leverage, as we mount steel ax blades on wooden handles. There were no bone tools, no ropes to make nets, and no fishhooks.

What food did our early ancestors get with those crude tools, and how did they get it? To address this question, anthropology textbooks usually insert a long chapter entitled something like "Man the Hunter." The point they make is that baboons, chimps, and some other primates prey on small vertebrates only occasionally, but recently surviving Stone Age people (like Bushmen) did a lot of big-game hunting. There's no doubt that our early ancestors also ate some meat. The question is, how much meat? Did big-game hunting skills improve gradually over the past 1.5 million years, or was it only since the Great Leap Forward—a mere 35,000 years ago—that they made a large contribution to our diet?

Anthropologists routinely reply that we've long been successful big-game hunters, but in fact there is no good evidence of hunting skills until around 100,000 years ago, and it's clear that even then humans were still very ineffective hunters. So it's reasonable to assume that earlier hunters were even more ineffective.

Yet the mystique of Man the Hunter is now so rooted in us that it's hard to abandon our belief in its long-standing importance. Supposedly, big-game hunting was what induced protohuman males to cooperate with one another, develop language and big brains, join into bands, and share food. Even women were supposedly molded by big-game hunting: they suppressed the external signs of monthly ovulation that are so conspicuous in chimps so as not to drive men into a frenzy of sexual competition and thereby spoil men's cooperation at hunting.

But studies of modern hunter gatherers, with far more effective weapons than those of early *H. sapiens*, show that most of a family's calories come from plant food gathered by women. Men catch rats and other small game never mentioned in their heroic campfire stories. Occasionally they get a large animal, which does indeed contribute significantly to protein intake. But it's only in the Arctic, where little plant food is available, that big-game hunting becomes the dominant food source. And humans didn't reach the Arctic until around 30,000 years ago.

So I would guess that big-game hunting contributed little to our food intake until after we had evolved fully modern anatomy and behavior. I doubt the usual view that hunting was the driving force behind our uniquely human brain and societies. For most of our history, we were not mighty hunters, but rather sophisticated baboons.

To return to our history: *H. sapiens*, you'll recall, took center stage around half a million years ago in Africa, the Near East, the Far East, and Europe. By 100,000 years ago, humans had settled into at least three distinct populations occupying different parts of the Old World. These were the last truly primitive people. Let's consider among them those whose anatomy is best known, those who have become a metaphor for brutishness: the Neanderthals.

Where and when did they live? Their name comes from Germany's Neander Valley, where one of the first skeletons was discovered (in German, *thal*—nowadays spelled *tal*—means "valley"). Their geographic range extended from western Europe, through southern European Russia and the Near East, to Uzbekistan in Central Asia, near the border of Afghanistan. As to the time of their origin, that's a matter of definition, since some old skulls have characteristics anticipating later "full-blown" Neanderthals. The earliest full-blown examples date from around 130,000 years ago, and most specimens postdate 74,000 years ago. While their start is thus arbitrary, their end is abrupt: the last Neanderthals died around 32,000 years ago.

During the time that Neanderthals flourished, Europe and Asia were in the grip of the last ice age. Hence Neanderthals must have been a cold-adapted people—but only within limits. They got no farther north than southern Britain, northern Germany, Kiev, and the Caspian Sea.

Neanderthals' head anatomy was so distinctive that, even if a Neanderthal dressed in a business suit or a designer dress were to walk down the street today, all you *H. sapiens* would be staring in shock. Imagine converting a modern face to soft clay, gripping the middle of the face from the bridge of the nose to the jaws, pulling the whole mid-face forward, and letting it harden again. You'll then have some idea of a Neanderthal's appearance. Their eyebrows rested on prominently bulging bony ridges, and their nose and jaws and teeth protruded far forward. Their eyes lay in deep sockets, sunk behind the protruding nose and brow ridges. Their foreheads were low and sloping, unlike our high vertical modern foreheads, and their lower jaws sloped back without a chin. Yet despite these startlingly primitive features, Neanderthals' brain size was nearly 10 percent greater than ours! (This does not mean they were smarter than us; they obviously weren't. Perhaps their larger brains simply weren't "wired" as well.) A dentist who examined a Neanderthal's teeth would have been in for a further shock. In adult Neanderthals front teeth were worn down on the outer surface, in a way found in no modern people. Evidently this peculiar wear pattern resulted from their using their teeth as tools, but what exactly did they do? As one possibility, they may have routinely used their teeth like a vise, as my baby sons do when they grip a milk bottle in their teeth and run around with their hands free. Alternatively, Neanderthals may have bitten hides to make leather or wood to make tools.

While a Neanderthal in a business suit or a dress would attract your attention, one in shorts or a bikini would be even more startling. Neanderthals were more heavily muscled, especially in their shoulders and neck, than all but the most avid bodybuilders. Their limb bones, which took the force of those big muscles contracting, had to be considerably thicker than ours to withstand the stress. Their arms and legs would have looked stubby to us because the lower leg and forearm were relatively shorter than ours. Even their hands were much more powerful than ours; a Neanderthal's handshake would have been bone crushing. While their average height was only around 5 feet 4 inches, their weight was at least 20 pounds more than that of a modern person of that height, and this excess was mostly in the form of lean muscle.

One other possible anatomical difference is intriguing, although its reality as well as its interpretation are quite uncertain—the fossil evidence so far simply doesn't allow a definitive answer. But a Neanderthal woman's birth canal may have been wider than a modern woman's, permitting her baby to grow inside her to a bigger size before birth. If so, a Neanderthal pregnancy might have lasted one year, instead of nine months.

Besides their bones, our other main source of information about Neanderthals is their stone tools. Like earlier human tools, Neanderthal tools may have been simple hand-held stones not mounted on separate parts such as handles. The tools don't fall into distinct types with unique functions. There were no standardized bone tools, no bows and arrows. Some of the stone tools were undoubtedly used to make wooden tools, which rarely survive. One notable exception is a wooden thrusting spear eight feet long, found in the ribs of a long-extinct species of elephant at an archeological site in Germany. Despite that (lucky?) success, Neanderthals were probably not very good at big-game hunting; even anatomically more modern people living in Africa at the same time as the Neanderthals were undistinguished as hunters.

If you say "Neanderthal" to friends and ask for their first association, you'll probably get back the answer "caveman." While most excavated Neanderthal remains do come from caves, that's surely an artifact of preservation, since open-air sites would be eroded much more quickly. Neanderthals must have constructed some type of shelter against the cold climate in which they lived, but those shelters must have been crude. All that remain are postholes and a few piles of stones.

The list of quintessentially modern human things that Neanderthals lacked is a long one. They left no unequivocal art objects. They must have worn some clothing in their cold environment, but that clothing had to be crude since they lacked needles and other evidence of sewing. They evidently had no boats, as no Neanderthal remains are known from Mediterranean islands nor even from North Africa, just eight miles across the Strait of Gibraltar from Neanderthal-populated Spain. There was no long-distance overland trade: Neanderthal tools are made of stones available within a few miles of the site.

Today we take cultural differences among people inhabiting different areas for granted. Every modern human population has its characteristic house style, implements, and art. If you were shown chopsticks, a Schlitz beer bottle, and a blowgun and asked to associate one object each with China, Milwaukee, and Borneo, you'd have no trouble giving the right answers. No such cultural variation is apparent for Neanderthals, whose tools look much the same no matter where they come from.

We also take cultural progress with time for granted. It is obvious to us that the wares from a Roman villa, a medieval castle, and a Manhattan apartment circa 1988 should differ. In the 1990s, my sons will look with astonishment at the slide rule I used throughout the 1950s. But Neanderthal tools from 100,000 and 40,000 years ago look essentially the same. In short, Neanderthal tools had no variation in time or space to suggest that most human of charactcristics, *innovation*.

What we consider old age must also have been rare among Neanderthals. Their skeletons make clear that adults might live to their thirties or early forties but not beyond 45. If we lacked writing and if none of us lived past 45, just think how the ability of our society to accumulate and transmit information would suffer.

But despite all these subhuman qualities, there are three respects in which we can relate to Neanderthals' humanity. They were the first people to leave conclusive evidence of fire's regular, everyday use: nearly all well-preserved Neanderthal caves have small areas of ash and charcoal indicating a simple fireplace. Neanderthals were also the first people who regularly buried their dead, though whether this implies religion is a matter of pure speculation. Finally, they regularly took care of their sick and aged. Most skeletons of older Neanderthals show signs of severe impairment such as withered arms, healed but incapacitating broken bones, tooth loss, and severe osteoarthritis. Only care by young Neanderthals could have enabled such older folks to stay alive to the point of such incapacitation. After my litany of what Neanderthals lacked, we've finally found something that lets us feel a spark of kindred spirit in these strange creatures of the Ice Age—human, and yet not really human.

Did Neanderthals belong to the same species as we do? That depends on whether we would have mated and reared a child with a Neanderthal man or woman, given the opportunity. Science fiction novels love to imagine the scenario. You remember the blurb on a pulpy back cover: "A team of explorers stumbles on a steep-walled valley in the center of deepest Africa, a valley that time forgot. In this valley they find a tribe of incredibly primitive people, living in ways that our Stone Age ancestors discarded thousands of years ago. Are they the same species as us?" Naturally, there's only one way to find out, but who among the intrepid explorers— male explorers, of course—can bring himself to make the test? At this point one of the bone-chewing cavewomen is described as beautiful and sexy in a primitively erotic way, so that readers will find the brave explorer's dilemma believable: Does he or doesn't he have sex with her?

Believe it or not, something like that experiment actually took place. It happened repeatedly around 35,000 years ago, around the time of the Great Leap Forward. But you'll have to be patient just a little while longer.

Remember, the Neanderthals of Europe and western Asia were just one of at least three human populations occupying different parts of the Old World around 100,000 years ago. A few fossils from eastern Asia suffice to show that people there differed from Neanderthals as well as from us moderns, but too few have been found to describe these Asians in more detail. The best characterized contemporaries of the Neanderthals are those from Africa, some of whom were almost modern in their skull anatomy. Does this mean that, 100,000 years ago in Africa, we have at last arrived at the Great Leap Forward?

Surprisingly, the answer is still no. The stone tools of these modern-looking Africans were very similar to those of the non–modern-looking Neanderthals, so we refer to them as Middle Stone Age Africans. They still lacked standardized bone tools, bows and arrows, art, and cultural variation. Despite their mostly modern bodies, these Africans were still missing something needed to endow them with modern behavior.

Some South African caves occupied around 100,000 years ago provide us with the first point in human evolution for which we have detailed information about what people were eating. Among the bones found in the caves are many of seals and penguins, as well as shellfish such as limpets; Middle Stone Age Africans are the first people for whom there is even a hint that they exploited the seashore. However, the caves contain very few remains of fish or flying birds, undoubtedly because people still lacked fishhooks and nets.

The mammal bones from the caves include those of quite a few medium-size species, predominant among which are those of the eland, an antelope species. Eland bones in the caves represent animals of all ages, as if people had somehow managed to capture a whole herd and kill every individual. The secret to the hunters' success is most likely that eland are rather tame and easy to drive in herds. Probably the hunters occasionally managed to drive a whole herd over a cliff: that would explain why the distribution of eland ages among the cave kills is like that in a living herd. In contrast, more dangerous prey such as Cape buffalo, pigs, elephants, and rhinos yield a very different picture. Buffalo bones in the caves are mostly of very young or very old individuals, while pigs, elephants, and rhinos are virtually unrepresented.

So Middle Stone Age Africans can be considered big-game hunters, but just barely. They either avoided dangerous species entirely or confined themselves to weak old animals or babies. Those choices reflect prudence: their weapons were still spears for thrusting rather than bows and arrows, and—along with drinking a strychnine cocktail—poking an adult rhino or Cape buffalo with a spear ranks as one of the most effective means of suicide that I know. As with earlier peoples and modern Stone Age hunters, I suspect that plants and small game made up most of the diet of these not-so-great hunters. They were definitely more effective than baboons, but not up to the skill of modern Bushmen and Pygmies.

Thus, the scene that the human world presented from around 130,000 years ago to sometime before 50,000 years ago was this: Northern Europe, Siberia, Australia, and the whole New World were still empty of people. In the rest of Europe and western Asia lived the Neanderthals; in Africa, people increasingly like us in anatomy; and in eastern Asia, people unlike either the Neanderthals or Africans but known from only a few bones. All three populations were still primitive in their tools, behavior, and limited innovativeness. The stage was set for the Great Leap Forward. Which among these three contemporary populations would take that leap?

The evidence for an abrupt change—at last!—is clearest in France and Spain, in the late Ice Age around 35,000 years ago. Where there had previously been Neanderthals, anatomically fully modern people (often known as Cro-Magnons, from the French site where their bones were first identified) now appear. Were one of those gentlemen or ladies to stroll down the Champs Elysées in modern attire, he or she would not stand out from the Parisian crowds in any way. Cro-Magnons' tools are as dramatic as their skeletons; they are far more diverse in form and obvious in function than any in the earlier archeological record. They suggest that modern anatomy had at last been joined by modern innovative behavior.

Many of the tools continue to be of stone, but they are now made from thin blades struck off a larger stone, thereby yielding roughly ten times more cutting edge from a given quantity of raw stone. Standardized bone and antler tools appear for the first time. So do unequivocal compound tools of several parts tied or glued together, such as spear points set in shafts or ax heads hafted to handles. Tools fall into many distinct categories whose function is often obvious, such as needles, awls, and mortars and pestles. Rope, used in nets or snares, accounts for the frequent bones of foxes, weasels, and rabbits at Cro-Magnon sites. Rope, fishhooks, and net sinkers explain the bones of fish and flying birds at contemporary South African sites.

Sophisticated weapons for killing dangerous animals at a distance now appear also—weapons such as barbed harpoons, darts, spear-throwers, and bows and arrows. South African caves now yield bones of such vicious prey as adult Cape buffalo and pigs, while European caves are full of bones of bison, elk, reindeer, horse, and ibex. Several types of evidence testify to the effectiveness of late Ice Age people as big-game hunters. Bagging some of these animals must have required communal hunting methods based on detailed knowledge of each species' behavior. And Cro-Magnon sites are much more numerous than those of earlier Neanderthals or Middle Stone Age Africans, implying more success at obtaining food. Moreover, numerous species of big animals that had survived many previous ice ages became extinct toward the end of the last ice age, suggesting that they were exterminated by human hunters' new skills. Likely victims include Europe's woolly rhino and giant deer, southern Africa's giant buffalo and giant Cape horse, and—once improved technology allowed humans to occupy new environments—the mammoths of North America and Australia's giant kangaroos.

Australia was first reached by humans around 50,000 years ago, which implies the existence of watercraft capable of crossing the 60 miles from eastern Indonesia. The occupation of northern Russia and Siberia by at least 20,000 years ago depended on many advances: tailored clothing, as evidenced by eyed needles, cave paintings of parkas, and grave ornaments marking outlines of shirts and trousers; warm furs, indicated by fox and wolf skeletons minus the paws (removed in skinning and found in a separate pile); elaborate houses (marked by postholes, pavements, and walls of mammoth bones) with elaborate fireplaces; and stone lamps to hold animal fat and light the long Arctic nights. The occupation of Siberia in turn led to the occupation of North America and South America approximately 11,000 years ago.

Whereas Neanderthals obtained their raw materials within a few miles of their home, Cro-Magnons and their contemporaries throughout Europe practiced long-distance trade—not only for the raw materials for tools, but also for "useless" ornaments. Tools of obsidian, jasper, and flint are found hundreds of miles from where those stones were quarried. Baltic amber reached southeast Europe, while Mediterranean shells were carried to inland parts of France, Spain, and the Ukraine.

The evident aesthetic sense reflected in late Ice Age trade relates to the achievements for which we most admire the Cro-Magnons: their art. Best known are the rock paintings from caves like Lascaux, with stunning polychrome depictions of now-extinct animals. But equally impressive are the bas-reliefs, necklaces and pendants, fired-clay sculptures, Venus figurines of women with enormous breasts and buttocks, and musical instruments ranging from flutes to rattles.

Unlike Neanderthals, few of whom lived past the age of 40, some Cro-Magnons survived to 60. Those additional 20 years probably played a big role in Cro-Magnon success. Accustomed as we are to getting our information from the printed page or television, we find it hard to appreciate how important even just one or two old people are in preliterate society. When I visited Rennell Island in the Solomons in 1976, for example, many islanders told me what wild fruits were good to eat, but only one old man could tell me what other wild fruits could be eaten in an emergency to avoid starvation. He remembered that information from a cyclone that had hit Rennell around 1905, destroying gardens and reducing his people to a state of desperation. One such person can spell the difference between death and survival for the whole society.

I've described the Great Leap Forward as if all those advances in tools and art appeared simultaneously 35,000 years ago. In fact, different innovations appeared at different times: spear-throwers appeared before harpoons, beads and pendants appeared before cave paintings. I've also described the Great Leap Forward as if it were the same everywhere, but it wasn't. Among late Ice Age Africans, Ukrainians, and French, only the Africans made beads out of ostrich eggs, only the Ukrainians built houses out of mammoth bones, and only the French painted woolly rhinos on cave walls.

These variations of culture in time and space are totally unlike the unchanging monolithic Neanderthal culture. They constitute the most important innovation that came with the Great Leap Forward: namely, the capacity for innovation itself. To us, innovation is utterly natural. To Neanderthals, it was evidently unthinkable.

Despite our instant sympathy with Cro-Magnon art, their tools and hunter-gatherer life make it hard for us to view them as other than primitive. Stone tools evoke cartoons of club-waving cavemen uttering grunts as they drag women off to their cave. But we can form a more accurate impression of Cro-Magnons if we imagine what future archeologists will conclude after excavating a New Guinea village site from as recently as the 1950s. The archeologists will find a few simple types of stone axes. Nearly all other material possessions were made of wood and will have perished. Nothing will remain of the multistory houses, drums and flutes, outrigger canoes, and world-quality painted sculpture. There will be no trace of the village's complex language, songs, social relationships, and knowledge of the natural world.

New Guinea material culture was until recently "primitive" (Stone Age) for historical reasons, but New Guineans are fully modern humans. New Guineans whose fathers lived in the Stone Age now pilot airplanes, operate computers, and govern a modern state. If we could carry ourselves back 35,000 years in a time machine, I expect that we would find Cro-Magnons to be equally modern people, capable of learning to fly a plane. They made stone and bone tools only because that's all they had the opportunity to learn how to make.

It used to be argued that Neanderthals evolved into Cro-Magnons within Europe. That possibility now seems increasingly unlikely. The last Neanderthal skeletons from 35,000 to 32,000 years ago were still full-blown Neanderthals, while the first Cro-Magnons appearing in Europe at the same time were already anatomically fully modern. Since anatomically modern people were already present in Africa and the Near East tens of thousands of years earlier, it seems much more likely that such people invaded Europe rather than evolved there.

What happened when invading Cro-Magnons met the resident Neanderthals? We can be certain only of the result: within a few thousand years no more Neanderthals. The conclusion seems to me inescapable that Cro-Magnon arrival somehow caused Neanderthal extinction. Yet many anthropologists recoil at this suggestion of genocide and invoke environmental changes instead—most notably, the severe Ice Age climate. In fact, Neanderthals thrived during the Ice Age and suddenly disappeared 42,000 years after its start and 20,000 years before its end.

My guess is that events in Europe at the time of the Great Leap Forward were similar to events that have occurred repeatedly in the modern world, whenever a numerous people with more advanced technology invades the lands of a much less numerous people with less advanced technology. For instance, when European colonists invaded North America, most North American Indians proceeded to die of introduced epidemics; most of the survivors were killed outright or driven off their land; some adopted European technology (horses and guns) and resisted for some time; and many of those remaining were pushed onto lands the invaders did not want, or else intermarried with them. The displacement of aboriginal Australians by European colonists, and of southern African San populations (Bushmen) by invading Iron Age Bantu-speakers, followed a similar course.

By analogy, I suspect that Cro-Magnon diseases, murders, and displacements did in the Neanderthals. It may at first seem paradoxical that Cro-Magnons prevailed over the far more muscular Neanderthals, but weaponry rather than strength would have been decisive. Similarly, humans are now threatening to exterminate gorillas in central Africa rather than vice versa. People with huge muscles require lots of food, and they thereby gain no advantage if less muscular people can use tools to do the same work.

Some Neanderthals may have learned Cro-Magnon ways and resisted for a while. This is the only sense I can make of a puzzling culture called the Chatelperronian, which coexisted in western Europe along with a typical Cro-Magnon culture (the so-called Aurignacian culture) for a short time after Cro-Magnons arrived. Chatelperronian stone tools are a mixture of typical Neanderthal and Cro-Magnon tools, but the bone tools and art typical of Cro-Magnons are usually lacking. The identity of the people who produced Chatelperronian culture was debated by archeologists until a skeleton unearthed with Chatelperronian artifacts at Saint-Césaire in France proved to be Neanderthal. Perhaps, then, some Neanderthals managed to master some Cro-Magnon tools and hold out longer than their fellows.

What remains unclear is the outcome of the interbreeding experiment posed in science fiction novels. Did some invading Cro-Magnon men mate with some Neanderthal women? No skeletons that could reasonably be considered Neanderthal–Cro-Magnon hybrids are known. If Neanderthal behavior was as relatively rudimentary and Neanderthal anatomy as distinctive as I suspect, few Cro-Magnons may have wanted to mate with Neanderthals. And if Neanderthal women were geared for a 12-month pregnancy, a hybrid fetus might not have survived. My inclination is to take the negative evidence at face value, to accept that hybridization occurred rarely if ever, and to doubt that any living people carry any Neanderthal genes.

So much for the Great Leap Forward in western Europe. The replacement of Neanderthals by modern people occurred somewhat earlier in eastern Europe and still earlier in the Near East, where possession of the same area apparently shifted back and forth between Neanderthals and modern people from 90,000 to 60,000 years ago. The slowness of the transition in the Near East, compared with its speed in western Europe, suggests that the anatomically modern people living around the Near East before 60,000 years ago had not yet developed the modern behavior that ultimately let them drive out the Neanderthals.

Thus, we have a tentative picture of anatomically modern people arising in Africa over 100,000 years ago, but initially making the same tools as Neanderthals and having no advantage over them. By perhaps 60,000 years ago, some magic twist of behavior had been added to the modern anatomy. That twist (of which more in a moment) produced innovative, fully modern people who proceeded to spread westward into Europe, quickly supplanting the Neanderthals. Presumably, they also spread east into Asia and Indonesia, supplanting the earlier people there of whom we know little. Some anthropologists think that skull remains of those earlier Asians and Indonesians show traits recognizable in modern Asians and aboriginal Australians. If so, the invading moderns may not have exterminated the original Asians without issue, as they did the Neanderthals, but instead interbred with them.

Two million years ago, several protohuman lineages existed side-by-side until a shakedown left only one. It now appears that a similar shakedown occurred within the last 60,000 years and that all of us today are descended from the winner of that shakedown. What was the Magic Twist that helped our ancestor to win?

The question poses an archeological puzzle without an accepted answer. You can speculate about the answer as well as I can. To help you, let me review the pieces of the puzzle: Some groups of humans who lived in Africa and the Near East over 60,000 years ago were quite modern in their anatomy, as far as can be judged from their skeletons. But they were not modern in their behavior. They continued to make Neanderthal-like tools and to lack innovation. The Magic Twist that produced the Great Leap Forward doesn't show up in fossil skeletons.

There's another way to restate that puzzle. Remember that we share 98 percent of our genes with chimpanzees. The Africans making Neanderthal-like tools just before the Great Leap Forward had covered almost all of the remaining genetic distance from chimps to us, to judge from their skeletons. Perhaps they shared 99.9 percent of their genes with us. Their brains were as large as ours, and Neanderthals' brains were even slightly larger. The Magic Twist may have been a change in only 0.1 percent of our genes. What tiny change in genes could have had such enormous consequences?

Like some others who have pondered this question, I can think of only one plausible answer: the anatomical basis for spoken complex language. Chimpanzees, gorillas, and even monkeys are capable of symbolic communication not dependent on spoken words. Both chimpanzees and gorillas have been taught to communicate by means of sign language, and chimpanzees have learned to communicate via the keys of a large computer-controlled console. Individual apes have thus mastered "vocabularies" of hundreds of symbols. While scientists argue over the extent to which such communication resembles human language, there is little doubt that it constitutes a form of symbolic communication. That is, a particular sign or computer key symbolizes a particular something else.

Primates can use as symbols not just signs and computer keys but also sounds. Wild vervet monkeys, for example, have a natural form of symbolic communication based on grunts, with slightly different grunts to mean *leopard*, *eagle*, and *snake*. A month-old chimpanzee named Viki, adopted by a psychologist and his wife and reared virtually as their daughter, learned to "say" approximations of four words: *papa*, *mama*, *cup*, and *up*. (The chimp breathed rather than spoke the words.) Given this capability, why have apes not gone on to develop more complex natural languages of their own?

The answer seems to involve the structure of the larynx, tongue, and associated muscles that give us fine control over spoken sounds. Like a Swiss watch, our vocal tract depends on the precise functioning of many parts. Chimps are thought to be physically incapable of producing several of the commonest vowels. If we too were limited to just a few vowels and consonants, our own vocabulary would be greatly reduced. Thus, the Magic Twist may have been some modifications of the protohuman vocal tract to give us finer control and permit formation of a much greater variety of sounds. Such fine modifications of muscles need not be detectable in fossil skulls.

It's easy to appreciate how a tiny change in anatomy resulting in capacity for speech would produce a huge change in behavior. With language, it takes only a few seconds to communicate the message, "Turn sharp right at the fourth tree and drive the male antelope toward the reddish boulder, where I'll hide to spear it." Without language, that message could not be communicated at all. Without language, two protohumans could not brainstorm together about how to devise a better tool or about what a cave painting might mean. Without language, even one protohuman would have had difficulty thinking out for himself or herself how to devise a better tool.

I don't suggest that the Great Leap Forward began as soon as the mutations for altered tongue and larynx anatomy arose. Given the right anatomy, it must have taken humans thousands of years to perfect the structure of language as we know it—to hit on the concepts of word order and case endings and tenses, and to develop vocabulary. But if the Magic Twist did consist of changes in our vocal tract that permitted fine control of sounds, then the capacity for innovation that constitutes the Great Leap Forward would follow eventually. It was the spoken word that made us free.

This interpretation seems to me to account for the lack of evidence for Neanderthal–Cro-Magnon hybrids. Speech is of overwhelming importance in the relations between men and women and their children. That's not to deny that mute or deaf people learn to function well in our culture, but they do so by learning to find alternatives for an existing spoken language. If Neanderthal language was much simpler than ours or nonexistent, it's not surprising that Cro-Magnons didn't choose to associate with Neanderthals.

I've argued that we were fully modern in anatomy and behavior and language by 35,000 years ago and that a Cro-Magnon could have been taught to fly an airplane. If so, why did it take so long after the Great Leap Forward for us to invent writing and build the Parthenon? The answer may be similar to the explanation why the Romans, great engineers that they were, didn't build atomic bombs. To reach the point of building an A-bomb required 2,000 years of technological advances beyond Roman levels, such as the invention of gunpowder and calculus, the development of atomic theory, and the isolation of uranium. Similarly, writing and the Parthenon depended on tens of thousands of years of cumulative developments after the Great Leap Forward—developments that included, among many others, the domestication of plants and animals.

Until the Great Leap Forward, human culture developed at a snail's pace for millions of years. That pace was dictated by the slowness of genetic change. After the Great Leap Forward, cultural development no longer depended on genetic change. Despite negligible changes in our anatomy, there has been far more cultural evolution in the past 35,000 years than in the millions of years before. Had a visitor from outer space come to Earth before the Great Leap Forward, humans would not have stood out as unique among the world's species. At most, we might have been mentioned along with beavers, bowerbirds, and army ants as examples of species with curious behavior. Who could have foreseen the Magic Twist that would soon make us the first species, in the history of life on Earth, capable of destroying all life.

BOX 1 Perspectives

After reading "The Great Leap Forward," one can reflect on these human phenomena. The human body has not changed biologically in the last 50,000 years. The *Homo sapiens* brain has the potential of 10 (to the 11th power) neurons and 10 (to the 14th power) connectors, which are capable of storing the equivalent of 20.5 million volumes of information (Pytlik 1985, 286)! With this brain power, humans have the capability to think abstractly, intuitively, and creatively and to project into the unknown. "Science and technology have fostered this potential dramatically. Ten thousand years ago humans learned to write and so to store information outside their bodies. This landmark moved the human drama from pure biological evolution to a cultural revolution." (Pytlik 1985, 286).

Excerpted from Pytlik, E., Lauda, D., and Johnson, D. (1985). *Technology, Change and Society.* Worcester, Massachusetts: Davis.

QUESTIONS

1. What three changes took place in human development to distinguish humans from apes?
2. What physical characteristics are evident in the development of *Homo erectus*?
3. Approximately when did the Great Leap Forward occur?
4. What misconceptions does the author mention regarding popular beliefs about Neanderthals?
5. What conclusions might be drawn from the fact that tools used 40,000 years ago by Neanderthals differed little from those used by them 100,000 years ago?
6. What human behaviors have been found in the investigation of Neanderthal societies?
7. What are some basic differences between the Neanderthal and the Cro-Magnon of the Great Leap Forward?
8. Speculate why today's society would find a more comfortable link with Cro-Magnon society than with Neanderthal society.
9. Discuss why a possible encounter between Cro-Magnon and Neanderthal societies might have meant an end to Neanderthal life.

DISCUSSION

1. When you read the above paragraph, what are your reflections about the unique development of humankind?
2. After reading "The Great Leap Forward," what are your reflections on the role of technology, language, and writing to humankind's development? Discuss the contributions of language, writing, and information processing to the continued development of humankind.

2

"George"

PAUL ALCORN

He was a small creature, no more than inches in height, and he was, as usual, hungry. His brain was about half the size of modern *Homo sapiens*, about 650 cubic centimeters, and he lived on a wide, flat plain in Africa, where the tall grass and clutches of low-branched trees made a hunter's paradise for him and his fellow creatures. His kind would one day be known as *Homo habilis*, but that was nearly 2 million years in the future. For our purposes, we will call him George.

As I have said, he was hungry. It was a characteristic of his species. Warm blooded and a hunter, George and his companions spent much of their time ranging out across the plain near their most recent campsite in search of food. They were omnivorous, as likely to devour the tough nutty fruit of a nearby berry bush as the raw flesh of some small reptile or insect that failed to escape their notice in time. From day to day, George and his fellows satisfied their internal furnace with the fuel of whatever they could find, always searching for the great kill that would allow them to gorge themselves and replenish the dwindling supply of protein gathered from the last great kill some days or even weeks earlier.

Today they were near the high rock carapace to the east, though they had no concept of direction in those terms. It was merely the "high place over there, where the sun rises." George was scouting ahead of the pack, a chore he seemed to relish. A loner, he would often run ahead, somehow enjoying the prospect of being the first to sight a potential prey, hoping to be the first to wrestle it to the ground, to pound it to death with his clenched hands, or to tear its throat with his teeth.

The band was in the narrow passage that led into the center of the mound of rocks, close to where a fellow hunter had perished only days before at the hands of another predator, a huge cat creature with claws to tear at the throat and jaws to sink deeply into the flesh and break the victim's neck. George had seen it happen. He remembered it all too well. He was cautious, listening and sniffing the air, remembering what had happened to . . . who was it? His simple consciousness forgot those things easily, but the memory of the danger remained solidly in his mind.

The others were far behind him and out of sight as he turned into the natural bowl formed by the circle of high, flat rocks near the center of the carapace. He could feel the eyes on him, almost smell death in the air. Instinctively, he knew he was not alone. Turning quickly, he scanned the rocks above, seeking any telltale clue of whatever was lurking there. He spun so quickly and jerked his head about so violently in his panic that he nearly missed the low, flat, black-furred head, the huge yellowish eyes that stared back at him.

Above George and a little to his left was the same sleek creature that had made a meal of his fellow hunter only days before. George panicked. He turned and leaped to the side of a sheer rock, clinging with his toes and fingertips as the cat made a lunging pass at him. The panther missed the small ape-man by inches. George scrambled toward the summit of the rock, churning his legs wildly in search of some foothold, reaching out blindly with his hands for any

Excerpted from *Social Issues in Technology: A Format for Investigation,* 2nd ed., by Paul A. Alcorn, © 1997. Reprinted by permission of Prentice Hall, Inc., Upper Saddle River, NJ.

purchase further up the rock face. Lacerations appeared on his knees and thighs as he slid against the sharp black obsidian. His fingers numbed as they bit again and again into the narrow, knifelike crevices above. But he was making progress. Below him, the cat yowled and paced, panting heavily and leaping toward the fleeing figure.

Springing with all its might, the huge panther nearly reached George, who pulled forward with a last great effort and reached a wide ledge nearly halfway up the rock face. As he slid himself onto the strip of rock and flattened himself against the wall, a single stone slipped over the edge and fell, striking the huge cat squarely on the nose. With a howl, the panther retreated. George kicked another rock toward the beast, then another, missing both times. In panic, he grabbed several more and hurled them toward the beast, striking him again, this time dead center at the skull. The panther slumped to the ground, stunned by the blow. George grabbed for another loose rock and another, improving his aim with each throw, grasping larger and larger rocks until at last he found himself holding heavy slabs of obsidian over his head with both hands and hurling them down on the lifeless victim. He struck the creature again and again and again.

In the night, belly taut, legs splayed out before him, George lay with the other hunters in the natural bowl of the rocks stuffed with the meat of the dead panther. He was smiling, staring up into the night sky at the bright starlit veil, a swath of white that spread across the sky like a river in the firmament. Absently, he licked his hand and passed the thick saliva over the crusted scratches on his belly and legs. Around him, the sound of night creatures echoed off the surrounding walls as predator and prey continued the struggle for survival.

In his right hand, George clutched at a single round stone, about three inches long across its short axis and weighing nearly half a pound. He felt safe now. He knew that he could fend off any attack. Tomorrow he would try his luck again with his newfound weapon. Tomorrow he would try it against one of the doglike scavengers of the plain or use it to bring down a bird near the river. Perhaps he would never need to be hungry again. Had he not slain the mighty panther single-handedly? Had he not proven himself the greatest hunter of them all? Who knew what he might be able to do the next time? Who could really know?

QUESTIONS

1. Who was George? Describe him in detail.
2. Reflect on George's life. How was his life different from that of humans today? Write a minimum of three paragraphs on the differences.
3. How does the last paragraph of this article correlate to the importance of technology in the life of *Homo habilis*?

The "New" George

JOHN A. MORELLO

In a previous article, you met "George," who lived millions of years ago. Paul Alcorn captured George's life in vivid detail. He depended on plants and animals for his survival, picking what he could from the trees around him and hunting animals that sometimes were in the process of hunting him. He seemed to be living on the edge of extinction and in the constant grip of fear. His less-than-modest mental capabilities, braced by his even more breathtaking lack of tools and the skills to use them, made George's odds of survival long. But one day, while fighting for his life, he discovered the power of a rock and the skills and technique to use it. The blending of technology and human need changed George's life. Yet, while technology seemingly made George's life a little easier, what lurked below the surface of that discovery was something most of us already know: technology isn't always a blessing. Consider for a moment what would happen if we provided George with a few of the technological marvels he was forced to do without so many years ago. Many of the technologies George will use can be found in the timeline in the introduction to this part. As you'll see, each one comes with implications and obligations that George, or the rest of us for that matter, probably didn't count on when they first entered our lives.

For starters, let's go back to when we first met George. Alcorn says George and his companions spent much of their time looking for food. It was necessary in order to survive. Without proper storage, Monday's dinner was Tuesday's carrion, so every day was a struggle. But suppose George had refrigeration? The first refrigerator was invented around 1870. It used ammonia to help keep food fresh. Later, ammonia would be replaced with freon. Now George could kill his prey, drag it back to his lair, eat what he wanted that day, and store the rest. George's life would now be less tenuous. However, like the rest of us, George would learn that this particular technology is not without its headaches. Refrigeration depends on chemicals, specifically ammonia and freon, to keep things cold. These can be nasty if they come into contact with human flesh or the respiratory system. Maintaining such a device can be tricky. There are laws and regulations governing the recharging and maintenance of refrigerators, and George could find himself in trouble with the law regarding the proper disposal of hazardous materials and, if the electricity goes off, he's got a brand new problem to deal with. Equally problematic could be the way refrigeration affects food. Some foods become dry and lose their flavor if not packaged properly. George will need Tupperware or disposable containers to keep foods fresh. And, if he opts for disposable containers, where is he going to put them when he wants to dispose of them? George's companions might not appreciate the mountain of disposable containers in the back of the cave. Last, but not least, how about cleanliness? Like everybody else, George is going to have to clean out the refrigerator. And, when he reaches all the way to the back and finds something that was left there a couple of millennia ago, he may decide that the daily grind of looking for food wasn't so bad after all. Of course, he could wait until the 1950s when the first McDonald's opened.

George also liked plants. Alcorn said he was omnivorous. But the berries George feasted on may have been spoiled by insects that ate their way into the fruit, and who either left or decided to make that blueberry a weekend retreat. Pesticides could have helped. The first pesticides began to appear in large quantity in Europe around 1840. They would have helped George handle his bug problems and made sure other plants, namely weeds, didn't grow up around his berries, blocking the sun and choking off their water supply. But George would have to remember which pesticides he was using. DDT would be pulled off the market by 1962, and others containing dioxin would be linked with all sorts of diseases. He would have to wash his fruits and vegetables carefully in order to avoid ingesting any pesticides directly. He would also have to make sure he knew how much pesticide to use. Too much and he might harm the soil. Add way too much and the first heavy rain would wash the excess into his drinking water. These would prove to be fairly strict limitations for someone like George, who could not read; they are tough even for those of us who can. Perhaps George is better off taking his chances without pesticides.

Alcorn says that George was also a hunter. He enjoyed the prospect of being the first to sight a potential prey, wrestling it to the ground, and pounding it to death with his fists or using his teeth to tear its throat. If George made a practice of using his teeth as a weapon—and it sounds like he did—those teeth would soon be worn to the gums, making him an excellent candidate for dentures. The Etruscans produced the first set around 800 B.C.; more fashionable and reliable models came out in the eighteenth century. Still, dentures require some getting used to. There would be no heat-of-the-moment throat ripping for someone who keeps his teeth in a jar at night or, in George's case, on the nearest rock. Chemicals are needed to keep them clean, adhesives to keep them firmly in place, insurance companies to decide whether they are covered, and regular visits to the dentist to ensure that everything is opening and shutting the way it should. George would soon learn what many of us already know: technological substitutions are nice, but nothing beats the real thing.

During his encounter with the panther, George discovered that sometimes the hunter can become the hunted. In his desperation, he reached for a rock, and then another, and then another after that. Giving his hand–eye coordination some on-the-job training, George discovered that a well placed rock was better than a fist. In time, George would come to depend on more sophisticated weapons. Gunpowder will come on the scene about 1100 A.D. By the 1700s, machine guns will be introduced and will be perfected in the nineteenth century. But in George's first battle for dominance among species, the rock allowed him to be the dominator. That night, as he and his companions gorged themselves on panther meat, George held a rock in his hands. It was now a symbol of power in man's confrontation with the dangerous world around him. At first, George might only use his rock to kill for food. But would he stop there? What would stop him from using his new technology to establish and defend exclusive hunting rights in areas he shared with other humans? Would he resist the notion of using his new technology to settle disputes? Or would he use it to acquire property belonging to another? How long would it be before others learned to use a rock, and then a stick, and so on?

Yes, the night of George's first panther kill presented unlimited possibilities. Who knew what he or someone else might do with this new tool the next time? Who could really know?

QUESTIONS

1. Describe how George might use refrigeration and how the use of refrigeration might affect his life.
2. Make a list of technologies on which you depend. How many of them have made your life significantly easier? How many of them have made your life more complicated?
3. Suppose for a moment that you were George. What other problems might you encounter in living his life with some of the technologies that suddenly confront him?

4

Sticks and Stones May Break Your Bones . . . But Bronze and Iron Can Create a Civilization

JOHN A. MORELLO

Technologies are frequently assumed to occur autonomously from the culture or society surrounding it. The truth of the matter is that technologies are deeply intertwined with both. They may even have a role in defining societies, as they did millions of years ago in periods such as the Stone, Bronze, and Iron Ages. While it may be true that "no technology has an internal logic that fully explains why it ends up looking and behaving the way it does" (Smith and Clancey 1997), it is just as likely that the way a technology is made can define the period in which it is used.

In the age before metals, tools and weapons were made of stone, thus giving rise to the period in human history called the Stone Age. According to the best estimates, the Stone Age began in Europe, Africa, and Asia over 2 million years ago. Stone was obviously a big part of the lives of the humanoid creatures that lived at that time. Humanoids were hunters and gatherers and depended on tools made of stone, such as chipped pebbles or flaked stone implements, which they used for a variety of purposes. Those tools became part of the signature of the Paleolithic, or Old Stone Age. Then the Stone Age version of the Swiss army knife gave way to specialization. Humanoids developed a variety of tools for very specific purposes. By the end of the Paleolithic period, *Homo sapiens*, the descendants of the humanoids, were making tools for specific purposes, including needles and harpoons. The sophistication of the tools helped to establish the semblance of a pecking order in society. Caves found in Europe that date back to the Cro-Magnon period of human history contained wall paintings featuring tools in use at the time and may possibly have revealed a social hierarchy built, in part, by tools.

Historians studying the Mesolithic, or Middle Stone Age, have found tools modeled after those found in the Paleolithic Age, but modified to the new conditions. The Mesolithic period seems to have been more tropical and temperate in terms of climate. Therefore, tools—which were still chipped from stone—were modified to function in more heavily forested regions. Perhaps the stone, chipped from a larger one, was shaped into a blade to cut down trees. No matter. Time marched on. Before you could say Mesolithic three times fast, the Neolithic period, or New Stone Age, was in full swing. Stone was still the preferred material when it came to making tools, but tools increased in variety and looked much better than their predecessors. As the Neolithic period opened the door to increased farming techniques, tools began reflecting the needs of farming societies. By 6000 B.C., the Paleolithic and Mesolithic Ages were ancient history, so to speak. The Neolithic Age, with its dependence on sharpened stone tools for farming and chopping wood, seemed to be firmly in command.

If it is true that time waits for no man, it is also true that time does not allow technology much down time either. By 4500 B.C., the residents of what today is known as Thailand were reportedly making use of bronze in their daily routine, kicking off what was to become the Bronze Age. At first, bronze was only used for decorative purposes. Yet by the time bronze became available in the Middle East, especially during the days of the Sumerian civilization, it was the stuff of legends and treasure. It has been said that the wealth of Troy had been built on bronze and that Babylon's importance rose significantly during the Bronze Age. However, the Bronze Age lost its luster just as quickly as it had originated and was replaced by a more durable alloy—iron.

With the introduction of iron came the role that technology would play in working iron into different shapes for different purposes. Iron soon replaced bronze as the premier material in the manufacturing of tools and weapons. Thanks in part to the Greek civilization, the world was soon introduced to the widespread use of iron. By approximately 500 B.C., furnaces were built that could generate enough heat to melt iron and shape it into tools for peace and war. But the Greeks were not through just yet. They discovered that by adding small amounts of carbon to iron as it was hammered over a charcoal fire, a new substance could be produced—steel. Steel could be used to build catapults, improved swords, and even body armor, as well as pots and dishes.

Thus, the Paleolithic, Mesolithic, and Neolithic Ages were intertwined with the Stone, Bronze, and Iron Ages. Each one assisted and advanced the other. In subsequent years, historians would say the same thing about steam during the Industrial Revolution and the automobile during the twentieth century. But everything first started with stone.

REFERENCE

Smith, Merritt Roe, and Clancey, Gregory. (1997). *Major Problems in the History of American Technology*. Boston: Houghton Mifflin.

QUESTIONS

1. Place the following in chronological order:
   a. Mesolithic
   b. Paleolithic
   c. Neolithic
2. In terms of technological but also social development, what do you think was the significance of the Bronze Age?
3. Suppose that we all lived in the Stone Age and everything we needed could be made from stone. Would there be any point in going on?

The History of Energy

John A. Morello

Some people like to say that things were so much easier in the old days. If you were cold, there was the sun to warm you. When that did not work, you looked for wood to burn. If you were in the mood for a trip, horsepower was usually at your disposal, but it was the kind that ran on oats, not gasoline. For the more adventurous, there was always the wind—provided that it could fill your sails long enough to get you to your destination. Backbreaking labor could be alleviated somewhat by getting animals to do the heaviest work.

Things were simpler—or at least they seemed to be—for generations. Yet time and ingenuity are unstoppable forces. Sun and wood, while seemingly plentiful energy sources, can sometimes be hard to find, and wind power can be fickle. Just ask any sailor who has found his voyage delayed by becalmed seas. Something more reliable and easily controlled by man seemed to be necessary. Something like steam.

James Watt and Thomas Newcomen, each in their own way, contributed to the rise of steam power. When Watt introduced an effective steam engine in 1769 (effective in the sense that he raised efficiency from 5 to 25 percent), a world of possibility was opened (dKospedia.com). The "steam economy," as it was called, introduced civilization to steam engines, steamboats, steamships, and steam locomotives. Yet what was really driving the steam economy was coal. Coal-powered trains and boats dotted landscapes and seascapes and also provided heat for smelting iron into steel and generating heat for houses (ucsusa.org). It seemed an inevitable successor to all previous energy sources and was generally considered to also be an inexhaustible source. Reserves were estimated to last more than 300 years (dKospedia.com). Coal seemed to be everywhere, and so was its residue. The soot from coal-fired plants and coal-heated homes were belched into the air from chimneys around the industrial world in the mid- to late-nineteenth century, turning blue skies gray and sometimes even black. What did not remain in the sky came back to earth in the form of black rain, coating things and sometimes people with dingy dust. It made housecleaning a never-ending task and made breathing, especially for the young, old, and sick, downright difficult. An alternative needed to be found to reduce this dependence on coal.

This alternative turned out to be petroleum. What was seen by many to be a useless thing (farmers complained that it seeped into their wells and contaminated their water) and later a dangerous thing (as hucksters and con artists tried to pass it off as medicine) became a useful thing with which to light houses during the wake of the dimming whale oil industry (ucsusa.org). The first commercial well was built in Petrolia, Canada, in 1858 and was a leading oil producer in the world until 1901, when a large deposit was found in Texas (dKosopedia.com). At that point, fate seemed to intervene. At approximately the same time that oil was being discovered, the internal combustion engine—better known as the automobile—was being introduced in large numbers, and oil producers were discovering ways to refine petroleum into gasoline. As they became more efficient, the price of gasoline dropped; as car makers became more efficient, the price of automobiles also dropped. The combination of cheap cars and cheap gas put the world on wheels. Cars became bigger, but still affordable. And if cars were not terribly fuel efficient, who cared? A gallon of gas often cost less than a quarter (ucsusa.org).

By the end of World War II, the planet was dotted with coal-fired power plants and hydroelectric dams. Utility lines stretched for thousands of miles, bringing light to both cities and rural areas alike. Gasoline made it possible for people to escape the city for the country and vice versa. At this time, a new source of energy was introduced. Nuclear power had proven itself to be a tool of war. Could the "peaceful atom" be of use to a world in search of energy? Hundreds of nuclear plants were built around the world to replace the coal-fired ones that had produced electricity. Over 200 such plants were planned for the United States alone (ucsusa.org). Because it seemed so safe and so cheap, thousands of American homes during the post-war years forsook oil and gas in favor of electricity for heating and cooking. No one cared about the spinning meter in the back of the house.

That all changed during the 1970s. First, Arab states effectively restricted the flow of oil to industrial nations in an effort to influence Middle East foreign policy. Oil prices skyrocketed. There were lines in front of filling stations, and suddenly gasoline was no longer cheap. When the Shah of Iran was forced to leave office in 1979, another oil embargo resulted in longer lines and even higher prices. If that wasn't enough, nations around the world began to have second thoughts about nuclear power. The disasters at Three Mile Island in the United States in 1977 and at Chernobyl in the Soviet Union in 1986 sent clear signals to consumers that nuclear power may be cheap, but not always safe. The United States responded by pulling the plug on all plant construction. There has not been a new nuclear plant built in the United States since 1978.

As the twentieth century came to a close, nations large and small looked for ways to reduce their dependence on oil, use coal more cleanly, and reduce fears about nuclear power. Alternatives considered to date have included ethanol, a corn-based additive that can be added to gasoline to improve mileage and reduce emissions, and two others that may have some people wondering whether the more things change, the more things stay the same: solar power and wind power.

REFERENCES

"History of Energy." Accessed July 30, 2005. Available at www.dKosopedia.com/index.php/History_of_Energy.

"Renewable Energy Sources." Clean Energy. Accessed July 30, 2005. Available at www.ucsusa.org/clean_energy/*renewable_energy_basics*.

Yergin, Daniel. (1992). *The Prize: The Epic Quest for Oil, Money and Powe*r. New York: Touchstone Press.

QUESTIONS

1. It is widely known that James Watt helped produce steam power, but he was not alone in this endeavor. Who else made a contribution?
2. Coal was and is considered an important energy source. Why is it still so popular, and why is it also considered by some to be a hazard?
3. Before it became an energy source, what was petroleum's main use?

The History of Ecology

JOHN A. MORELLO

Quick . . . who was the first ecologist? Stumped? Don't feel badly because it is probably a trick question. Actually, there may have been more than one. Aristotle and his student, Theophrastus, both demonstrated an interest in animal species, and Theophrastus even went so far as to record the interrelationship between animals and their environment (Ramaley 1940). Nice work, given the fact ecology had yet to be classified as a science!

What exactly *is* ecology, and why has it become so important? Ecology is the scientific study of the distribution and abundance of living organisms and how these properties are affected by interactions between those organisms and their environment (en.wikipedia.org). Ecology's importance as a field of inquiry grew as the impact of human activity on the plant and animal world became more pronounced.

Advances in the study of ecology occurred at about the same time that Britain, Spain, and Portugal were conducting their global explorations. Ecological work was done during the eighteenth century with the discovery of new lands as well as new species of plants and animals. Scientists accompanied the explorers on later missions, including Alexander von Humboldt, who worked on behalf of the British. He was the first to study organisms and their environment, as well as the relationship between plant species and climate. His "Idea for a Plant Geography," published in 1805, raised global awareness about ecology and earned him the unofficial title as the father of ecology (en.wikipedia.org).

The importance of the field was boosted further with Charles Darwin's *The Origin of Species*, published in 1859. Suddenly, ecology was seen as a repetitive, mechanical inquiry and more of an organic evolutionary exercise. By the late nineteenth century, the field had been enlarged to incorporate studies of the conditions promoting life on earth. It was recognized that the quality of life depended on more than just the quality of the air, water, and soil, but on plants and minerals as well. Consequently, terms such as atmosphere, hydrosphere, and lithosphere seemed inadequate—or at least they did to Austrian geologist Eduard Suess, who proposed the term biosphere in 1875. The idea went a step further during the 1920s with the publication of Russian geologist Vladimir Vernadsky's work, *The Biosphere.* In it, he redefined the biosphere as the sum of all ecosystems and linked the biosphere's viability to that of the combined ecosystems. Anything that negatively affected the state of a single ecosystem had an equally negative effect on the state of the biosphere. It was a timely concept since assaults on various ecosystems had been going on throughout the eighteenth and nineteenth centuries as the multiplication of colonies caused deforestation and the Industrial Revolution led to serious concerns about the human impact on the environment. Thus, ecology became concerned not simply with the vitality of life, but with the effects of humanity to undermine it.

Since 1945, ecology has been an important part of many nations' social and philosophical movements. It is a major political force in many nations and, after 1971, took on global status when the United Nations' Economic, Social, and Cultural Organization (UNESCO) inaugurated a research program titled "Man and Biosphere." The program sought to increase knowledge about the mutual relationship between humans and nature. Since then, international conferences on ecology have been held in Stockholm (1972) and Rio de Janeiro (1992). It was at the Rio conference that the biosphere concept was officially recognized by major international organizations, as were the risks associated with reductions in biodiversity. In 1997, the pronouncement of the Kyoto Protocol warned of the dangers of greenhouse gases and their effect on climate around the world. In fact, among its participants, the Kyoto Protocol established the importance of looking at ecology from not only a national perspective, but from a global perspective as well. It also heightened awareness of the impact of human activity on the earth's environment.

REFERENCES

"History of Ecology." Accessed July 31, 2005. Available at http://en.wikipedia.org/wiki/History_of_ecology.

Ramaley, Francis. (1940). "The Growth of a Science." *University of Colorado Studies, 26,* 3-14.

QUESTIONS

1. What is ecology?
2. How was the development of ecology as a science affected by global exploration?
3. Discuss the post–World War II importance placed on ecology and explain why it has become so important. What have various nations done, acting together or separately, to promote it?

The History of Climatology

JOHN A. MORELLO

If you think interest in the weather was piqued by the arrival of The Weather Channel, think twice. The scientific as well as non-scientific community has been interested in hot, cold, rainy, and dry weather for a long time. In fact, serious students of the weather have called their work the study of climatology, which is the science that studies climates and investigates their phenomena and causes. The study of climatology involves weather information depicting up-to-the-moment events, seasonal changes, long-term effects of weather, and how climates change over time ("Climatology," en.wikipedia.org). The pioneers of climatology go back in time further than the old joke that everyone talks about the weather, but nobody does anything about it. Climatologists may be just as powerless to do anything about it as the rest of us, but they understand it better than most people and can explain why things happened the way they did.

One of the first inductees to the climatology Hall of Fame was probably Edmund Halley of Britain, the same person for whom Halley's Comet has been named ("Edmund Halley," en.wikipedia.org). After leaving Oxford in 1676, where he did research on sunspots and solar systems, Halley sailed the south Atlantic, compiling research for a number of papers he would publish upon his return. One of them dealt with trade winds and monsoons. In that same paper, he claimed that atmospheric motions were caused by solar heating and that there was a relationship between barometric pressure and sea level. He returned to Oxford to teach in 1703 and published his theory that comets sighted in 1456, 1531, 1607, and 1682 were actually the same comet and that it would return in 1758. The comet returned just as Halley predicted it would, and when it left earth's view, it left with Halley's name on it.

Following Halley's discoveries was the work of Wladimir Köppen, born in Russia to German parents in 1846. He established the Köppen Climate System, which is still used today to group climates into similar types. The system was prompted by Köppen's curiosity about the relationship between plants and the climate in which they grew. He had the opportunity to do some research while working for the Russian Meteorological Service in the early 1870s. He moved to Germany in 1875 and established a weather forecasting system service in Hamburg. Building on the data he developed, he devoted the rest of his professional career to private research. In 1884, he released a map of climatic zones, in which seasonal temperature ranges were plotted ("Wladimir Köppen," en.wikipedia.org). The map was the foundation for Köppen's Climate System, first published in 1900 and modified until its final version was published in 1936.

The system far outlived its creator, who died in 1940, but continued on thanks to the efforts of two other climatologists, fellow German Rudolf Geiger and Hungarian Milutin Milankovic, for whom the Milankovitch Cycles are named. The cycles stemmed from Milankovic's earlier work on solar climates and temperatures prevailing on the planets ("Milutin Milankovic," en.wikipedia.org). Wladimir Köppen concurred with Milankovic's idea of a "curve of insolation" at the earth's surface and used it in his research dealing with climatic change over time. His own work grew to include a theory of the secular motion of the earth's poles as well as his theory of glacial periods, or ice ages, which were ultimately known as part of the Milankovitch Cycles.

Milankovic's Ice Age theory was controversial and hotly debated even after his death in 1958. Meteorologists argued that multiple ice ages could not exist on earth because changes in the Earth's orbit were too small to significantly affect the climatological system. Yet by the 1970s, samples of deep-sea sediment helped to confirm Milankovic's Ice Age theory. Public opinion also played a role, as people around the world began to observe Earth Day and governments invested more resources into better understanding their respective climates and how to best preserve them.

REFERENCES

"Climatology." Accessed August 28, 2006. Available at http://en.wikipedia.org/wiki/Climatology.

"Edmund Halley." Accessed August 1, 2005. Available at http://en.wikipedia.org/wiki/Edmund_Halley.

"Milutin Milankovic." Accessed August 28, 2006. Available at http://en.wikipedia.org/wiki/Milutin_Milankovic.

"Wladimir Köppen." Accessed August 1, 2005. Available at http://en.wikipedia.org/wiki/Wladimir_K%C3%B6ppen.

QUESTIONS

1. What was Edmund Halley's contribution to the study of climatology?
2. What is the Köppen Climate System?
3. What was the role of the Russian government in helping establish Köppen as a climatologist?

The History of Ethics

John A. Morello

What is ethics? Generally speaking, any discussion of ethics will inevitably turn to a discussion of human conduct. Perhaps, when this text turns to the discussion of ethics and technology, human conduct regarding the use of technology will emerge. But that's for later. What is now at hand is to determine what ethics is and how it has become such a factor in our lives.

Ethics is a general term used to describe the study and evaluation of human conduct in the light of moral principles (bartleby.com). Things become tricky when trying to figure out whose moral principles human conduct should be measured against. There are those who think that every human being is guided by his or her own built-in set of principles—a moral compass, so to speak. On the other hand, there are those who think that we are driven to do not what we think is right, but what society requires of its members. Consequently, problems arise when someone exhibits what he considers to be acceptable human conduct, only to discover that he is out of step with society as a whole. Furthermore, this situation is not merely a contemporary dilemma, but rather is one that has been debated for centuries.

The Greeks—Socrates, Aristotle, and Plato, to name a few—were among the first to begin a serious inquiry into the study of ethical behavior. Since that time, ethics has become the process by which people evaluate the intentions and consequences of their actions. The question of motivation has been a key to understanding human conduct. Who or what motivates us to behave the way we do? Philosophers, such as Rousseau, argued that ethical behavior was the result of conscience, which people equate with the little voice that they heard in their head as kids when they did something wrong, felt badly about it, and decided to confess. In more elegant terms, Rousseau and others defined it as an innate moral sense that serves as the basis of ethical decision.

Rousseau may have been flattering us by choosing to believe that people are intrinsically good and know the difference between right and wrong. Others have not been so kind. John Stuart Mill and John Locke, for example, argued that humans are guided by no such innate principle, but rather by experience. By this they meant that we as humans know what is right only because we saw someone else do it, or that we did it wrong the last time and do not want to make the same mistake twice. Thus, on the one hand, ethics is believed to be the result of conscience. On the other hand, ethics is defined as what results when people remember what happened the last time a similar situation arose. Confused? It gets better, or worse, depending on your point of view.

Ethics has also endeavored to deal with the notion of goodness. In effect, can or should people aspire to the principle of absolute goodness or just to relative goodness? Some philosophers have argued that ethics should be tied to a system of absolute beliefs, such as religion, or upheld through the strictures of law. Other philosophers have held that whatever the individual believes to be good behavior will satisfy the needs of the greater society. In summary, commandments and constitutions dictate good behavior according to some, while others cling to the idea that what is good for one is good for all.

How do ethics and technology use coexist? That is a difficult question to answer, and there does not seem to be a clear consensus. Most technologies do not come with guidelines regarding their ethical use, so humans are left to their own devices. They can rely on their faith to tell them what is right or wrong. They can let their conscience be their guide, or they can contemplate the consequences of their actions before a court of law. What happens when an individual employs technology while acting under the impression that he is doing the right thing, only to discover that he has erred in the eyes of society? Homeowners who use a handgun to defend loved ones and property, only to discover that they have violated some civil statute, find themselves in a type of ethical Catch-22. One might also consider the decision to deploy nuclear weapons at the end of World War II, knowing that it would likely end the fighting, but only at the cost of thousands of lives. Stem cell research presents another ethical dilemma. Embryonic stem cells may hold the key to solving both illness and injury, but the price paid for such achievements is considered by some to be the death of life in its most delicate and defenseless state.

Whatever the situation, it is apparent that humanity's quest to create technologies that are capable of making life easier has made their ethical use a good deal harder.

REFERENCES

"Ethics." (2001) Columbia Encyclopedia, 6th Edition. New York: Columbia University Press, 2001–04. Accessed August 8, 2005. Available at www.bartleby.com/65/et/ethics/html.

"Ethics." (2005) Encyclopedia Britannica. Accessed August 28, 2006. Available at http://www.britannica.com/eb/article-9106054.

QUESTIONS

1. Define ethics.
2. If ethics relies on the importance of understanding the consequences of one's actions, how do you think technology use may affect that process?
3. Discuss Rousseau's approach to ethics. How was his approach challenged by John Stuart Mill and John Locke?

Conclusion

This section has attempted to involve readers in considering the connection that humanity has established with technology. It would be hard to imagine a world without technology and perhaps even harder to live in that world, if it ever existed. Can any one of us imagine life without the car, the phone, or even the computer? When stories are published about societies who have managed to avoid these items, many readers can only shake their heads and wonder at how hard and primitive such a life would be. Yet people in these societies might rightly turn to us and wonder how we have become so dependent on these things for which they have no use. We jump in a car without a moment's hesitation to run an errand that could have been done on foot or delayed. The phone offers instant communication and feeds the irrepressible urge to call someone simply to say hello, but not much else. Meanwhile, the stationery and finc pcns in the writing desk go unused, and everyone misses the opportunity for a thoughtful exchange of ideas.

And so it goes. This is not to say that human existence has not benefited from technology. We often live longer as a result of technology. But do we live better? That's the real question.

INTERNET EXERCISE

Use any of the Internet search engines (e.g., Google, Alta Vista, Yahoo, Infoseek) to research and compare the following countries and ethnic/cultural groups in terms of technological achievement according to the criteria in the table that follows.

Countries

- United States
- Myanmar (formerly Burma)
- Amazonian Indians
- England
- Pakistan

- Australian Aborigines
- Japan
- Brazil
- Hottentots
- Singapore
- Vietnam
- Amish

Country	*Population*	*Education*	*Politics*	*Religion*	*Communications*	*Technological Level*	*State of Environment*	*Language*
United States								
Myanmar (Burma)								
Amazonian Indians								
Pakistan								
Japan								
Australian Aborigines								
Brazil								
Hottentots								
Singapore								
Vietnam								
Amish								
England								

USEFUL WEB SITES

www.ac.wwu.edu/~gmyers/eheassign1.html	Leading thinkers in the field of ethics
www.oilonline.com	Current oil production technologies

Ethics and Technology

Balancing Technology, Ethics, Morals, and Values.

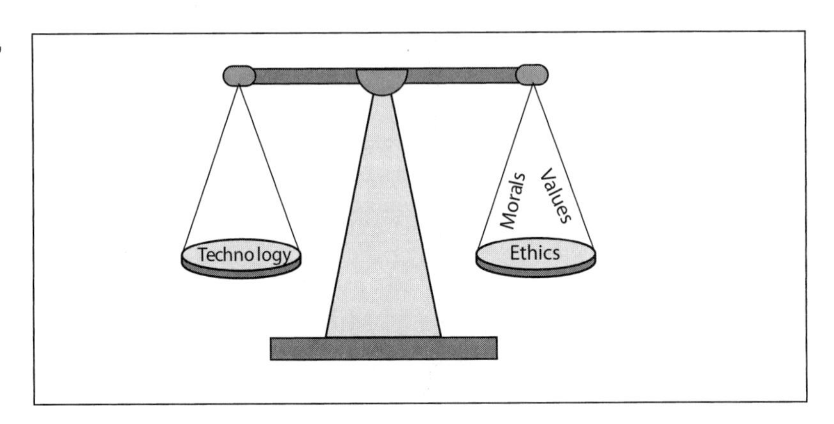

OBJECTIVES

After reading Part II, Ethics and Technology, *you will be able to*

1. Examine the relationship between ethics and technology.
2. Distinguish between consequentialist ethical theories, utilitarianism, and deontological ethical theories.
3. Describe science- or technology-based ethical issues and conflicts.
4. Define technology.
5. Explain the significance of science- or technology-based rights.
6. Evaluate the challenge of contemporary science and technology to traditional ethical theories.
7. Explain why human beings might embrace and be resistant to technological change.
8. Discuss different ways that technology could create positive impacts on society.

INTRODUCTION

Every technology is both a burden and a blessing; not either-or, but this-and-that.
NEIL POSTMAN, Author, Technopoly: The Surrender of Culture to Society

As technologies are embraced as "blessings," Part II resonates with concerns that techno-users need to undertake with immense responsibility so that newly developed technologies will not mature into societal burdens. Technology is more than a mechanism or tool; it is also a catalyst for societal changes. Once technology is introduced, life, thinking, behavior, and social norms change. When cars, microwaves, computers, birth control pills, penicillin, and other technologies were invented, life as we knew it drastically changed. Our point of reference changed, and our behaviors were altered. With the advent of technology, we arrive at places faster, communicate globally, cook quickly, and even look to medical technologies to get us to sleep swiftly. Technologies have been invented, for the most part, to make life "better," but do they also carry a societal or psychological cost?

For hundreds of years, the Kaiapo Indians of the lower Amazon Basin were considered to be "skilled farmers and hunters and the fiercest warriors of central Brazil" (Henslin 1995; Simons 1995, 463). They eventually sold gold and mahogany and became a wealthy village. With the newfound wealth, Chief Kanhonk bought a small satellite, which the Indians called the "big ghost," so that they could watch television. Prior to the satellite purchase, the Kaiapos would meet at night to tell stories and share ancestral customs. After the purchase of the satellite, however, children were found straying from their ancestral storytelling traditions to watch Western cartoons instead (Henslin 1995, 464). When studying various cultures, we often find that each has its own story to tell about the impact of technology on the lives, culture, and values of its people.

As technology permeates through all cultures and time spans, sociological and behavioral adaptations develop. For example, in the 1930s, many American children surrounded the radio in their living room to let their imaginations go while listening to "Orphan Annie," "Terry and the Pirates," or "Jack Armstrong, the All American Boy." The words emitted from the radio allowed children to create the characters and backgrounds within the world of their imaginations. No visual cues limited what their minds created. This all changed for children when television dominated the living rooms of the 1950s. Children no longer relied on their imaginations, for the television presented every visual and auditory detail. Additionally, parents worried that their kids would end up with damaged eyes or receive large amounts of radiation from sitting too close to the television.

Part II of this book, "Ethics and Technology," discusses the link between technology and its implications for society. Many of the issues presented are more serious than those faced by the Kaiapos or American children in the 1930s and 1950s. But the issues remain the same: With each technology introduced at any time in any culture, adaptations seem to follow. Many of the readings presented in this section encourage users of technology to evaluate the technologies with respect and with their own analyses of personal and social responsibility.

Most technologies work to provide progress to those who use them. However, there are times when technologies are misused and consequently harm or death befalls their users. Therefore, it is befitting to start this part of the book with ethics and its place in the development of technology.

As we have bridged our way into a new century and prepare to hand our world over to a younger generation, it will be important to consistently evaluate the technologies that we use and their impact on society and the world in which we live. This part of the book supports that endeavor.

REFERENCES

Henslin, J. (1995). *Sociology: A Down-to-Earth Approach.* Needham Heights, MA: Allyn and Bacon.

Simons, Marlise. (1995). The Amazon's Savvy Indians. *Down to Earth Sociology: Introductory Readings*, 8th ed. New York: Free Press.

Ethics

ROBERT MCGINN

INTRODUCTION

For at least the last two decades, many of the most divisive ethical issues debated in Western societies have been precipitated by developments in science and technology, including advances in reproductive, genetic, weapons, and life-prolonging technologies. The adoption and alteration of public policy for regulating science- or technology-intensive practices, such as the provision of access to exotic medical treatment, the disposal of toxic waste, and the invasion of individual privacy, have also raised perplexing ethical issues. This chapter is devoted to the analysis of such conflicts.

There is no universally shared criterion for deciding when a conflict of values falls within the province of ethics rather than, say, law. However, the issues and conflicts discussed in this chapter involve values widely regarded as integral to the enterprise of ethics in contemporary Western societies; such values include freedom, justice, and human rights, such as privacy. Disputes over whether an agent's freedom should be limited prospectively or its prior exercise punished, whether justice has been denied or done to some party, or whether someone's human rights have been protected or violated are widely regarded in Western societies as specifically *ethical* disagreements, thus marking them as human value issues or conflicts of special importance in these societies.

As a prelude to analysis of science- or technology-based ethical issues, we will describe a quartet of basic considerations centrally involved in judgments about and the playing out of such conflicts. We will then characterize and analyze a number of important *kinds* of ethical issues and conflicts associated with contemporary science and technology. Where appropriate, we will indicate noteworthy sociocultural factors that, in concert with the technical developments in question, help generate the issue or conflict. All this will pave the way for a key conclusion reached in this chapter: that developments in contemporary science and technology are calling into question the adequacy of traditional ethical thinking. A more comprehensive and sensitive kind of ethical analysis is now required, one more adequate to the complexity and consequences of contemporary scientific and technological processes and products.

CLARIFICATION OF ETHICAL ISSUES AND CONFLICTS

Ethical issues and conflicts, whether or not they are associated with developments in science and technology, can often be usefully clarified if four kinds of considerations pertinent to ethical decision- and judgment-making about controversial actions, practices, and policies are kept in mind.[1]

The Facts of the Matter

One consideration is that of determining, as scrupulously as possible, the facts of the situation underlying or surrounding the conflict in question. Doing so may require ferreting out and scrutinizing purportedly factual assumptions and allegedly empirical claims made by disputants

Science, Technology and Society by McGinn, Robert © 1991. Reprinted by permission of Pearson Education, Inc., Upper Saddle River, NJ.

about the conflict situation in question. It may also require ascertaining whether any persuasive accounts of the facts of the matter have been developed by neutral parties. In such efforts, important concerns include unmasking pseudo-facts and factoids posing as bona fide facts, ensuring that the credibility attributed to an account of the facts reflects the reputation and interests of its source, and setting the strength of the evidence required to warrant acceptance of an account of the facts at a level proportional to the gravity of the issue or conflict in question.

Affected Patients and Their Interests

A second kind of clarificatory consideration in thinking about an ethical issue or conflict is that of identifying all pertinent "patients"—that is, all affected parties with a legitimate stake in the outcome of the dispute. Further, all protectable interests of each stakeholder should be delineated and their relative weights carefully and impartially assigned.

Key Concepts, Criteria, and Principles

A third kind of consideration is that of identifying the key concepts, criteria, and principles in terms of which the ethical issue or conflict in question is formulated or debated. For example, the abortion issue hinges critically on the protagonists' respective concepts of what it is to be a "human being" and a "person" as well as what is meant by a "viable" fetus. The ethical (and legal) debate over the withdrawal of technological life-support systems turns sharply on what is meant by "killing" someone as well as on which criteria implicitly or explicitly govern protagonists' use of the key terms "voluntary consent" (the withdrawal or withholding of treatment) and "death."

Ethical Theories and Arguments

A fourth kind of basic consideration to be kept clearly in mind is that ethical disputes often involve two quite distinct kinds of decision-making theories and arguments. Consequentialist ethical theories and arguments make determination of the rightness or wrongness of actions and policies hinge exclusively on their estimated *consequences*. The most familiar consequentialist ethical theory is utilitarianism—the view that an action or policy is right if and only if it is likely to produce at least as great a surplus of good over bad, or evil, consequences as any available alternative. There are different versions of utilitarianism, depending on, among other things, what a given thinker understands by "good" and "bad," or "evil," consequences. For example, so-called hedonic utilitarians, of whom nineteenth-century British reformer Jeremy Bentham is perhaps the best known, took pleasure to be the only good, and pain the only bad, or evil. "Ideal utilitarians," such as the early twentieth-century British philosopher G. E. Moore, construed "good" and "bad" quite differently, including things like friendship and beauty among goods and their absence or opposites, such as alienation and ugliness, among bads, or evils.[2]

The second kind of ethical theory and argument that often enters into ethical disputes is called deontological. *Deontological* ethical theories and judgments hold that certain actions or practices are inherently or intrinsically right or wrong—that is, right or wrong in themselves, independent of any consideration of their consequences. Different deontological theorists and thinkers point to different things about actions and policies, in light of which they are judged to be right or wrong. Some point to supposedly intrinsic moral properties of actions and policies falling into one or another category. For example, actions such as telling a lie or breaking a promise may be regarded as intrinsically wrong. Others emphasize that a certain course of action is obligatory or impermissible because it is approved or disapproved, or unconditionally mandated or prohibited by some authority, perhaps a deity. As we will see in this chapter, several of the most important kinds of ethical issues and conflicts engendered by developments in science and technology arise from or are exacerbated by the fact that partisans of one position on an issue are consequentialists while their opponents are deontological thinkers (for convenience, deontologists).*

KINDS OF SCIENCE- OR TECHNOLOGY-BASED ETHICAL ISSUES AND CONFLICTS

We now turn to examination of science- and technology-based ethical issues and conflicts. Given the purpose of this book, our objective here will not be to provide detailed discussions of—much less solutions to—even a select number of the long list of such problems. Rather, we will *identify and critically analyze the limited number of qualitatively distinct*

*In reality, the ethical thinking of denizens of contemporary industrial societies is rarely so black and white. It often incorporates consequentialist, deontological, and perhaps other considerations in uneasy or unstable combinations.

kinds of such disputes. (Eight are considered here.) The aforementioned quartet of basic concerns—facts; patients and interests; concepts, criteria, principles; and ethical theories and arguments—will be used to shed light on the sociotechnical roots and intractability of many of these problems.

Violations of Established World Orders

Some ethical conflicts arise from the fact that scientific or technological breakthroughs make possible actions or practices that, in spite of what some see as their benefits, others believe violate some established order of things whose preservation matters greatly to them. The order of things in question may be regarded as "natural" or as "sacred."

For example, much opposition to recent achievements in biomedicine and genetic engineering flows from beliefs that employing such techniques is *unnatural*. Thus some oppose the technique of *in vitro* fertilization as involving the unnatural separation of human reproduction from sexual intercourse. In a similar vein, the production of farm animals with genes from at least two different animal species ("transgenic animals") is viewed by some critics as a transgression of natural animal-species boundaries, while the use of genetically mass-produced bovine growth hormones to substantially increase the volume of milk produced by cows is opposed by some as "chang[ing] the natural behavior of animals" or as "interrupt[ing] the naturalness of [farmers'] environment."[3]

Opposition to technological violations of natural orders is, however, sometimes based on concern about the long-term consequences of intervention for human or other animal well-being or for ecosystem integrity. For example, some oppose the production of transgenic animals because they believe that the resultant animals will suffer physically as a result of being maladapted. Similarly, some critics of the use of bovine growth hormone to raise milk production levels are primarily concerned with the safety of such milk for young children. The plausibility of such consequentialist ethical thinking hinges on the details of the particular case under consideration, including the estimated magnitudes, likelihoods, and reversibility of the projected consequences of intervention.

Deontological ethical arguments against such intervention as intrinsically wrong take two forms. First, the intervention-free order of nature is regarded as natural and intrinsically "good," while technology is not viewed as part of the natural order but rather as artificial. Therefore, it is concluded, attempts to use technology to intervene in the natural order are improper. A second argument notes that something has existed or has been done in the so-called natural way from time immemorial and concludes that therefore it should be done or continue to be done in that same way—without technological intervention. Is either the "unnaturalness" or the "longevity" argument persuasive?

It is unclear why the development and transformative use of technology on nature should be seen as "unnatural." The claim that because God created the natural order, it should not be "tampered with" by humans is suspect for two reasons. Those holding this idea presumably also believe that the human being was created by the deity, in which case the human is no less "natural" a creation than the "natural order" and indeed is properly regarded as part of that order. Moreover, they also presumably believe that God endowed humans with creative powers, including the ability to devise technics, thereby enabling them to intervene in the natural order. If so, why is it unnatural for natural creatures to use their God-given powers to intervene in the natural order? It seems implausible that the deity would endow its natural creatures with an unnatural power or with a power whose use was unnatural. If it is not the very use of technology to transform nature that is unnatural but only certain uses of it, then these opponents of technological violations of natural orders must clarify what it is that makes some technological interventions "violations" of those orders and others simply harmonious interventions in them.

As for the argument that the existing way of doing something is the proper way because of its longevity, it too fails to convince. That a practice is long-standing may make it familiar or seem natural. But long-standingness cannot by itself justify the view that the practice is proper. That would be drawing an *evaluative* conclusion from a purely *factual* premise. Conversely, a particular technological intervention in a long-established natural order might initially be resisted because it is unfamiliar or deemed unnatural. However, the fact that something runs counter to a long-standing practice does not suffice to show that it is improper. If that were so, then the abolition of slavery would have been improper. In fact, opposition to a practice initially regarded as unnatural because of its novelty or strangeness often diminishes over time as the new way becomes increasingly familiar. Some innovative technical practices eventually come to seem natural and quite proper, as has been the case with the use of antibiotics.

Is there, then, nothing to the concerns about technological violations of established orders as unnatural? Even if the deontological arguments examined here fail to hold water, the concerns they express still warrant serious consideration, for deontological thinking and argument are sometimes disguised or compressed versions of what are at bottom consequentialist thinking and argument. Reference to an innovative practice as unnatural and therefore as intrinsically wrong may be a powerful if deeply misleading way of expressing concern over its possible elusive long-term consequences.

Other scientific and technological developments have made possible practices that are seen by some groups as violations of world orders viewed not so much as natural but as *sacred*. Thus, in the Hasidic community centered in Brooklyn, New York, birth control is forbidden, supposedly on the basis of the Torah.[4] For the Wahabi, a fundamentalist Arabian Muslim sect, television, with its image-reproducing capacity, violates the sacred order related in the Koran. Opposition to certain technologies or technological ways of doing things as violations of sacred orders is less likely to ebb in the minds of such opponents, for the sacred way is apt to be regarded as God's way and, as such, as immutable and thus as something that ought not adapt itself to human technological change.

Violations of Supposedly Exceptionless Moral Principles

Other ethical issues arise from the fact that the use, failure to use, or withdrawal of particular scientific procedures or items of technology is seen by some as violating one or another important moral principle that its adherents believe to be exceptionless. For example, some people are categorically opposed to nuclear weapons on the grounds that their use will inevitably violate the supposedly exceptionless principle that any course of action sure to result in the destruction of innocent civilian lives in time of war is ethically impermissible.

Similarly, the supposedly exceptionless principle that "life must never, under any circumstances, be taken"—put differently, that "life must always be preserved"—is clearly an important ground of the categorical judgment that withdrawal of life-sustaining technologies, whether mechanical respirators or feeding and hydration tubes, is ethically wrong. A third example is the opposition by some to the "harvesting" of fetal tissue for use in treating Parkinson's or Alzheimer's disease, even in a relative. This opposition is often rooted in the supposedly exceptionless principle that "a human being must never be treated merely as a means to some other end, however worthwhile in itself."

Sociologically, opposition to certain scientific and technological developments on the grounds that they involve or may involve violations of some special order of things or some supposedly exceptionless moral principle reflects a fundamental fact about modern Western industrial societies. While much has been written about their secularization, there remain in such societies significant numbers of people whose ethical thinking is deontological in character, whether or not religiously grounded.

The worldviews of such individuals contain categories of actions that are strictly forbidden or commanded. For them, the last word on a particular science- or technology-based ethical issue or conflict sometimes hinges solely on whether the action or practice in question falls into one or another prescribed or proscribed category. While deontological thinkers may resort strategically to consequentialist arguments in attempting to change the views of consequentialist opponents, the latter's arguments against their adversaries' deontologically grounded positions usually fall on deaf ears, however well documented the empirical claims brought forward as evidence. Deontological fundamentalists and consequentialist seculars are, in their ethical judgment and decision making, mutual cultural strangers.

An interesting consequence of deontological appeals to supposedly exceptionless moral principles in the context of potent contemporary technologies is the appearance of moral paradoxes. For example, Gregory Kavka has shown that the situation of nuclear deterrence undermines the venerable, supposedly exceptionless "wrongful intention principle"—namely, the principle that "to form the intention to do what one knows to be wrong is itself wrong."[5] Launching a nuclear strike might well be ethically wrong (because of the foreseeable loss of innocent civilian lives). But what about forming the intention to do so if attacked? According to the wrongful intention principle, forming the intention to carry out that wrong action would itself be *wrong*. However, since forming that intention might well be necessary to deter an attack and thus to avoid launching an impermissible retaliatory strike, it might well be ethically *right*. One and the same action—that of forming the intention to retaliate—is therefore both right and wrong, a moral paradox. Thus can technological developments compel reassessment of supposedly exceptionless ethical principles.

Distributions of Science- or Technology-Related Benefits

Some contemporary ethical issues and conflicts arise from the fact that the benefits of developments in science and technology are allocated in ways that do not seem equitable to one or another social group. This is particularly so with respect to medical benefits, whether they be diagnostic tests, surgical procedures, or therapeutic drugs, devices, or services.

Concerns over whether an allocation of such benefits is "distributively just" often emerge when demand for the benefit exceeds its supply and decisions must be made about who will receive the benefit and who will not—sometimes tantamount to deciding "who shall live and who shall die." For example, in the early 1960s, the supply of dialysis units available to the Artificial Kidney Center in Seattle, Washington, was insufficient to meet the needs of those with failed kidneys.[6] Criteria were selected to use in deciding who would be granted access to this beneficial scarce technical resource. Today, the demand for various kinds of human organs often exceeds available supplies. The criterion of "need" is thus by itself insufficient to make allocation decisions. Criteria such as "likelihood of realizing a physiologically successful outcome" seem promising, but are quite problematic, for, as Ronald Munson has argued,

> [T]he characteristics required to make someone a "successful" dialysis patient are to some extent "middle-class virtues." A patient must not only be motivated to save his life, but he must also understand the need for the dialysis, be capable of adhering to a strict diet, show up for scheduled dialysis sessions, and so on. As a consequence, where decisions about whether to admit a patient to dialysis are based on the estimates of the likelihood of a patient's doing what is required, members of the white middle class have a definite edge over others. Selection criteria that are apparently objective may actually involve hidden class or racial bias.[7]

Other criteria, such as "probable post-treatment quality of life" and "past or likely future contribution of the treatment candidate to the community" are no less problematic. Hence, some believe that for such allocations to be distributively just, once need and physiological compatability have been established, a lottery should determine access to the scarce benefit.

On other occasions, it is not the shortage but the high cost of a medical treatment and the inability of all needy patients who want the treatment to afford it that engenders ethical conflict. Science and technology are often central factors in these high costs, for such costs may reflect the high purchase price of a machine or drug paid by a care unit, something which may in turn reflect the device's or substance's high research and development costs. Such situations help pose the contentious ethical issue of whether access to some needed expensive drug or procedure should be permitted to hinge on whether a prospective patient can afford to pay the going market price.

Deontological ethical thinkers who have come to think of medical care as a basic human right may find it morally unthinkable that a person be denied access to such treatment simply because of not being able to afford it (or because, for example, of being "too old"). In contrast, consequentialists, some of whom find the concept of an "absolute right" potentially dangerous, may believe that a particular technically exotic treatment is so expensive that granting everyone in life-or-death need unlimited access to it will effectively preclude many more individuals from getting less expensive, more beneficial, non–life-or-death treatments. Diverging accounts of "the facts of the matter" and different criteria for what makes a treatment "exotic" often bulk large in such judgments.

Consequentialists are apt to believe that individuals do *not* have a moral right to draw without limit on public or insurance company funds to have their or their family members' lives extended regardless of the quality of the sustained life and the prognosis for its improvement. They may even hold that the financial and social consequences of doing so create a moral obligation to *terminate* such treatment, or at least to cease drawing on public funds to pay for continued treatment. In such ways have advances in science and technology as well as people's varying concepts (e.g., of a life worth living) and divergent ethical theories become intertwined in complex ethical disputes over distributive justice, rights, and obligations.

Infliction of Harm or Exposure to Significant Risk of Harm Without Prior Consent

A fourth category of ethical issues and conflicts engendered by developments in science and technology arises from activities that, while undertaken to benefit one group, inflict harm or impose significant risk of harm on another without the latter's prior consent. Examples of this sort of phenomenon abound and include some research on animals; production of cross-border and multigenerational pollution; the maintenance of carcinogen-containing workplaces; and the operation of "hair-trigger" military defense systems. As with earlier categories, the ethical issues and conflicts here have both technical and social roots.

Most parties to such disputes would agree that, other things being equal, it is always unethical to subject a morally pertinent party to undeserved harm or serious risk of same without the party's freely given prior consent. Let us examine how the consent issue plays out in the four abovementioned kinds of cases.

In laboratory experimentation on sentient animals, the issue of consent bulks large in the persistent ethical conflict. Consequentialist proponents hold, on cost-benefit grounds, that activities that promise future benefits (including avoidance of suffering) for humans but that (supposedly unavoidably) inflict suffering on existing animals are ethically permissible, perhaps even obligatory. Those carrying out such activities may proceed because, since animals cannot consent to anything, they are different in a morally relevant respect from human beings.[8] Hence the consent condition, precluding similar treatment of humans, is, in the case of animals, legitimately waived. Some opponents of such research, often deontologists, also subscribe to the prior-consent principle, but they see animals, such as rabbits and monkeys, as no less morally relevant patients than are humans. They draw a diametrically different conclusion: Since the consent of laboratory animals cannot be obtained, research activity that produces suffering for animals is ethically wrong or impermissible, even if the suffering is "unavoidable"—computer models that make animal tests unnecessary may not be available—and the benefits of the research could plausibly be shown to exceed its costs. It is not difficult to see why resolution of this disagreement is unlikely to be forthcoming.

Explanation of the rise of ethical conflict over cases of cross-border and multigenerational pollution (e.g., acid rain, dumping toxic chemical or metal waste into multinational bodies of water, and the possibly insecure disposal of nuclear waste) must heed technical factors as well as the problematic issue of consent. But for the capacity of contemporary scientific and technological activities to produce potent geographically and temporally remote effects, these ethical disagreements would simply not arise. Moreover, this "action-at-a-distance" capacity contributes to the tendency either to neglect or to assign modest weights to the legitimate interests of affected patients at considerable geographical or temporal remove. This in turn facilitates proceeding with the activities in the absence of consent of such endangered parties, or, in the case of not-yet-born members of future generations, impartial reflection on whether they would consent if they were informed and in a position to give or withhold it. The facts that the human capacity for empathy tends to diminish rapidly the more remote the injured or endangered party and that the world is organized into a weak international system of sovereign states both contribute to the genesis of such ethical conflicts.

In ethical disputes over workplaces made dangerous because of scientific or technological activities or products, the issue of consent rears its head in a different form. Historically, employers or their representatives argued that maintenance of a dangerous workplace was not unethical because a worker who accepted a job in one thereby voluntarily consented to exposure to all its attendant risks. To the extent that workplace hazards in the early industrial era were primarily threats to worker safety and that a worker had other less dangerous employment alternatives, such a viewpoint might seem at least minimally plausible. However, as twentieth-century industrial workplaces became pervaded with thousands of industrial chemicals of uncertain bearing on worker health, the notion that workers could meaningfully consent to the imposition of any and all risks that their workplaces posed to their health began to ring hollow. Workers had to make decisions to take or keep jobs in ignorance of what, if any, risks they would be or were being exposed to. Put differently, management could proceed with its risk-imposing activities without their workers' informed consent.

This situation came to be viewed by some as a violation of the prior-consent principle, and hence as unethical. Others saw it as ethically permissible because the benefits (to both company and workers) of proceeding in this way supposedly outweighed the (typically undervalued) costs of doing so. The main attempt to mitigate this situation has taken the form of right-to-know legislation: Workers have a right to a safe workplace, but not to a risk-free (in particular, carcinogen-free) one. They are, however, entitled to that which is deemed necessary for their giving informed consent to imposition of workplace risks; specifically, they are entitled to "be informed about" all carcinogenic and other toxic substances used in their workplaces. Whether the extensive technical information provided and the way in which it is communicated to workers suffice to ensure their "informed consent" remains an open factual and criteriological question at the core of a persistent ethical issue.

Ethical conflict over the operation of "hair-trigger" military defense systems is also driven by both technical and social factors. Such systems are called "hair-trigger" because they can be set to "fire" on being subjected to slight pressures. Their risk arises from the enormity of destruction that can be unleashed by slight pressure on the sensitive firing mechanism; such pressure can be brought by mistaken "sightings" or misinterpretations of data. The

1988 downing of an Iranian commercial aircraft by the high-tech-equipped U.S.S. *Vincennes* because of misinterpretation of radar and electronic data is a tragic case in point, albeit on a relatively small scale. There have been numerous occasions on which American retaliatory nuclear forces have been activated and on the verge of being unleashed because of what turned out to be mistaken technological indications that a Soviet attack had been launched.

While technological "progress" is partly behind such ethical conflicts, consent is also a factor. In the case of hair-trigger military defense systems (e.g., ones operating on a "launch-on-warning" basis), the public has not been afforded an opportunity to explicitly give or withhold its consent, informed or otherwise, to the imposition of such grave risks. For opponents of such systems, this alone makes them ethically objectionable. For proponents, the astonishing speed of current or emerging offensive military technologies makes the risk of *not* employing hair-trigger defense systems exceed the risk of relying upon them. Moreover, such systems are morally justified, proponents argue, since the people have indirectly consented to such risks by voting democratically for the government that imposes them. The fact that civilian aircraft are permitted to fly over populous areas without their residents having first voted on whether to accept the associated risks does not suffice to show that the people do not consent to the risks imposed on them by this practice. Hence, it would be argued, the consent condition has not been violated and ethical impropriety has not been demonstrated. The same would be true in the case of hair-trigger defense systems. However, the greater the magnitude of the danger involved—enormous in the case of the nuclear war—the lower the risk of its accidental occurrence must be if the explicit securing of consent is to be reasonably set aside as having been implicitly given. The upshot is that the lack of shared criteria for deciding whether citizen consent has been effectively obtained in such cases is central to this acute ethical dispute.

Two morals of this kind of ethical conflict deserve attention. First, the problematics of consent are, to a significant degree, science- and technology-driven. Second, the potency of much contemporary scientific and technological activity is pressuring sensitive ethical analysts to enlarge the domain of morally pertinent patients whose interests are to be taken into account in assessing the ethical propriety of current or proposed actions or policies. This situation is reflected in ongoing struggles over whether to include various kinds of previously excluded stakeholders, such as those far afield who are nevertheless deleteriously affected by potent "spill-over" effects and future citizens whose legitimate interests may be jeopardized by activities designed first and foremost to benefit the presently living. Here, too, the contours of the evaluative enterprise of ethics are being subjected to severe stress by developments in contemporary science and technology.

Science- or Technology-Precipitated Value Conflicts

A fifth kind of science- or technology-based issue or conflict arises when a scientific or technological advance allows something new to be done that precipitates a value conflict, not necessarily between the values of opposed parties, but *between two or more cherished values of one and the same party*. For example, life-extending technologies have engendered situations in which family members are compelled to choose between two values, to both of which they owe allegiance: human life preservation and death with dignity. The crucial point about this increasingly frequent kind of value conflict is that the parties plagued by such conflicts would not be so but for scientific or technological advances.

Most recently, genetic tests allowing those with access to their results to know something of a sensitive nature about the health-related state or genetic predispositions of the person tested have proliferated. This has given rise to value conflicts between testers' or policymakers' concern for the protection of human *privacy* regarding disclosure of test results and their concern for *fairness* to one or another interested party other than the testee.

For example, in 1986 an adoption agency was trying to place a 2-month-old girl whose mother had Huntington's disease, a progressive, irreversible neurological disorder. The prospective adoptive parents indicated that they did not want the girl if she was going to develop the disease. The agency asked a geneticist to determine whether the child had the gene that would sooner or later manifest itself in the disease. The geneticist, while presumably sympathetic to the would-be adoptive parents' seemingly reasonable request, declined to do the testing. He reasoned that since many victims of the disease have claimed that they would have preferred to have lived their lives without knowing they had the "time-bomb" gene for the disease, it would be unethical to test someone so young, that is, at a point before she could decide whether to exercise her right to privacy in the form of *entitlement not to know* that she had the fatal gene.[9]

Tests for various genetic disorders, such as Down syndrome, sickle cell anemia, and Tay-Sachs disease, have been available for some time. In the foreseeable future, however, it is expected that tests will become available for identifying genetic traits that predispose people to more common health problems, such as diabetes, heart disease, and major forms of mental illness. Thus, according to Dr. Kenneth Paigen, "We are going to be able to say that somebody has a much greater or much less than average chance of having a heart attack before age 50 or after age 50."[10]

The potential implications of such advances for matters such as employment eligibility, life insurance qualification, and mate selection are formidable. Will employers with openings for positions with public safety responsibilities (e.g., commercial airline pilots) be permitted to require that applicants take genetic tests that will disclose whether they are predisposed to heart disease or to a genetically based mental disorder such as manic depression? Will insurance companies be permitted to require applicants for life or health insurance to take genetic tests predictive of life expectancy or diabetes? Will prospective spouses come to expect each other to be tested to determine their respective genetic predispositions and whether they are carriers of certain traits of genetic diseases and to disclose the test results?

In the case of companies recruiting for jobs with public safety responsibilities, prohibition of such tests to protect applicant privacy could impose significant safety costs on society. In the insurance case, preventing mandatory disclosure of test results in the name of individual privacy would spread the cost of defending this cherished value over society at large in the form of increased insurance premiums for those *without* life- or health-threatening genetic traits or predispositions. In the case of mate selection, declining to pursue and disclose the results of reliable genetic tests could set up partners for severe strains on their relationship should presently identifiable genetic disorders manifest themselves in the partners or their offspring at a later date.

It remains to be seen how society will resolve the public policy questions raised in such cases by the ethical value conflict between privacy and fairness. However, it is already clear that advances in genetic science are going to pose powerful challenges to society's commitment to the right of individual privacy. The knowledge about the individual afforded by such tests is likely to be of such pertinence to legitimate interests of other parties that the protection afforded individual privacy may be weakened out of concern for fairness to those parties, perhaps to the point of recognizing that under certain conditions they have a right to that knowledge. Put differently, technology is here bringing micro, or personal, justice and macro, or societal, justice into conflict.

Science- or Technology-Engendered "Positive" Rights

Besides conflicts over the values of freedom and justice, issues of "rights," especially "human rights," bulk large in contemporary Western ethics. In recent decades, advances in science and technology have engendered a new ethical issue: that of how best to respond to newly recognized, so-called *positive rights*.

In the modern era, some claims have come to be widely recognized as "human rights"—that is, as irrevocable entitlements that people supposedly have simply because they are human beings. Human rights are thus contrasted with civil or institutional rights—rights that people have because they are delineated in specific revocable legal or institutional documents. Among the most widely recognized human rights are "life" and "liberty."

These rights, and some derived from them—privacy, for example, is widely thought to be a kind of special case of liberty—have traditionally been viewed as what philosophers call "negative" or "noninterference" rights—that is, as entitlements *not to be done to* in certain ways. Thus, the right to life is construed as entitlement not to be deprived of one's life or physical integrity. The right to liberty is construed as entitlement not to have one's freedom of action physically constrained or interfered with, unless its exercise has unjustifiably harmed another's protectable interests (e.g., those in life, limb, property, reputation) or poses an unreasonable risk of doing so.

As scientific and technological progress have gathered momentum in recent decades, several rights traditionally viewed as negative have given birth to a number of correlative "positive" rights—in other words, entitlements *to be done to* in certain ways. Consider three examples. Many believe that the right to life, traditionally construed as a negative human right, must, *in the context of new life-preserving scientific and technological resources,* be regarded as having taken on a positive component: entitlement to be provided with whatever medical treatment may be necessary to sustain one's life (independently of ability to pay for it). According to this way of thinking, affirmation of the right to life in the contemporary scientific and technological era requires affirmation of a positive right of access to whatever means are necessary to sustain life. Thus, for example, denial of costly life-sustaining drugs to a patient in need of them on any grounds save scarcity, including concern over the aggregate high cost to society of providing them, is seen by many as a violation of the patient's right to life. Hence, depending on whether the ethical analyst is a deontological or consequentialist thinker, failure to provide these drugs would be deemed categorically or prima facie ethically impermissible.

The right to privacy has traditionally been viewed as a noninterference right, entailing, among other things, entitlement not to have one's home broken into by government authorities without a search warrant issuable by a court only upon proof of "probable cause." However, the computer revolution has put individual privacy interests at risk. To compensate for this, legislation in the United States and other countries entitles citizens to be provided with certain categories of information being kept on them in computerized files. For example, the U.S. Fair Credit Reporting Act of 1970 entitles each citizen to review and correct credit reports that have been done on them and to be notified of credit investigations undertaken for purposes of insurance, mortgage loans, and employment. Given the exceptional mobility of this information and the potential for severely damaging individual privacy that this creates, protection of the right of individual privacy in the computer era is held to require acknowledgement of countervailing positive rights: entitlement to know what exactly about oneself is contained in computerized government files and to have one's record rectified if it is shown to be erroneous.

A third, somewhat more speculative example involves a special case of the general right to liberty—namely, the traditional negative right of reproductive freedom: entitlement not to be interfered with in one's procreative undertakings, be they attempts to have or to avoid having offspring, including via "artificial" contraceptive means. It remains to be seen whether, in the context of the recent and continuing revolution in reproductive science and technology, the negative right of individual reproductive freedom will engender a positive counterpart: entitlement of those with infertility problems to be provided with (at least some) technical reproductive services enabling them to attempt to have offspring, where access does not hinge on a client's ability to pay or even perhaps on marital status.

As such scientifically and technologically generated positive rights expand, ethical tension will mount. Some deontologists, believing that human rights are "absolute," may conclude that their corresponding positive rights are likewise, hence inviolable. Others, including most consequentialists, while treating rights as claims that always deserve society's sympathetic consideration, may conclude that they cannot always be honored without regard to the social consequences of doing so. The most important consequence of this ethical tension may be that the day is drawing closer when society will have to come decisively to grips with the consequences of philosophical commitment to a concept of rights as "absolute" and "immutable."

Public Harms of Aggregation

Suppose that each of a large number of people carry out essentially the same action. Suppose further that, considered individually, each of these actions has at worst a negligible negative effect on a social or natural environment. Finally, suppose that the aggregated effect of the large number of people doing the same thing is that substantial harm is done to the environment in question. Let us call such outcomes public *harms of aggregation*. Many, if not most, such harms are possible only because, to an unprecedented degree, modern production, communication, and transportation methods have made many scientific and technological processes and products available on a mass basis. A curious moral aspect of such situations is that, as the individual acts were assumed to be of negligible negative impact, they are typically regarded as being ethically unobjectionable. Hence, the aggregate effect of a mass of ethically permissible actions may nevertheless turn out to be quite ethically problematic. In ethics, sometimes numbers *do* count.[11]

Consider, for example, the pollutants emitted by each of the approximately 400 million automobiles in the world. The aggregate negative environmental effect of individually innocuous, hence seemingly ethically unproblematic, effects is known by now to be substantial. To the extent that this aggregate effect can be shown to harm people's health, particularly groups at special risk of being affected (e.g., the elderly, young children, and those with respiratory problems), the aggregate effect would begin to be judged as ethically unacceptable and unjust, and the individual pollution-emitting activities might begin to appear as something other than ethically neutral.

An analogy may help clarify this novel ethical situation. Suppose that a populous nation experiences a devastating depression in which many of its people suffer. Suppose that after the fact, it is plausibly shown that an important contributing cause of the depression's occurrence was the fact that each family in the country had accumulated a substantial but individually manageable level of consumer debt. If the country was fortunate enough to recover its economic health, would not the new accumulation of a substantial but still manageable amount of consumer debt by an individual family be likely to be regarded as an ethically irresponsible thing to do? If so, the same could be said of an individual car owner's emission of pollution or an individual consumer's failure to recycle.

A somewhat futuristic example of the same pattern from the biomedical realm is that of predetermination of the sex of one's offspring. Given the fact that reproductive freedom is widely viewed as a human right, it is safe to assume that attempts of individual couples—at least married ones—to avail themselves of the latest scientific or technological means to determine the sex of their offspring will be regarded as ethically permissible. However, suppose that in a populous society with a culture biased in favor of male offspring, a significant number of couples opt for predetermination and that a significant sexual imbalance of male over female offspring results. Suppose further that at least some of the ethically problematic consequences envisioned as resulting from this state of affairs come to pass (e.g., increased crime committed by men or heightened male aggressiveness in competition for scarcer female mates).

The upshot of this situation is that twentieth-century science and technology may be pushing society toward reevaluation of "permissive" ethical judgments traditionally made about individual actions that are at worst "negligibly harmful." Consequentially speaking, the threshold of harm necessary to activate negative ethical judgments may be in the process of being reduced by the aggregative potential of modern science and technology in populous societies.

Practitioner Problems

The kinds of science- and technology-based ethical problems considered thus far have something in common: while spawned by developments within the spheres of science and technology, the resultant issues and conflicts unfold, not primarily within those spheres, but in society at large. In contrast, problems in the final category considered here, while related to concerns of society at large, arise primarily *inside* the communities of practitioners of science and technology. They are ethical problems associated with the concrete processes and practices of scientific and technological activity, both those in which these activities unfold and those in which their results are communicated. Such problems are sometimes viewed as falling within the province of "professional ethics," meaning that they are ethical problems that arise in the course of professional practice.

Problems of Execution Edward Wenk has identified three kinds of ethical issues faced by practicing engineers in their work.[12]

(1) Distributive Justice. The first is essentially an issue of distributive justice, involving as it does an allocation of costs, benefits, and risks. This kind of problem arises when an engineer must decide whether to embark upon or proceed with a feasible project that he or she recognizes is likely to expose people to a non-trivial degree of risk to their safety, health, or property without their consent. Beyond answering the question, "Can it be done?" about the contemplated project, the would-be ethical engineer must confront the quite different question, "Ought it be done?" For example, from an ethical point of view, should an engineering company accept a lucrative contract to erect a potentially hazardous structure, such as a hydroelectric dam, in a geologically unstable area near a rural village in the absence of the informed consent of its inhabitants?

Other things being equal, if the degree of risk—understood as a function of the estimated magnitude of the harm that could occur and the estimated likelihood of its occurrence—is substantial, then it would be ethically wrong to proceed. If it is negligible, then it would be ethically permissible, perhaps even obligatory, to do so. One problem with this kind of situation is that determination of what constitutes an "acceptable risk" is not a strictly technical question but a social and psychological one. Among other things, the answer to it depends on *what* members of the population at risk believe to be at stake, on *how highly* they value it at the time in question, and on *how seriously* they would regard its loss.

Meridith Thring has extended this analysis in the case of engineers who are independent operatives doing work in research and development. For years, Thring, a university professor of mechanical engineering, had been doing research on industrial robots. However, he eventually came to believe that "the primary aim [of such work] is to displace human labour." For this reason, he abandoned work on industrial robots and decided to work only on

> applications where the aim is to help someone to do the job he does now without actually exposing his body to danger or discomfort; or where we need to amplify or diminish his skill and strength. A good example is "telechirics," . . . artifacts that allow people to work artificial hands and arms and operate machines in hazardous or unpleasant environments as if they were there, while they are in fact in comfortable and safe conditions.[13]

Thring implies that it is also ethically incumbent on engineers to consider whether the work they contemplate—here, a research and development project—poses an unacceptable risk to any important nonsafety interest of patients likely to be affected by it—for example, that of not being rendered redundant. At bottom, this too is an issue in distributive justice. For Thring, the ethical engineer must first carefully estimate the costs, benefits, and risks likely to be associated with a possible technological endeavor and then ask, "Are those benefits, costs, and risks likely to be allocated to the affected parties in a way that is distributively just?" The engineer may then proceed with the work only if he or she can answer that question in the affirmative.

Similar ethical constraints apply to the initiation or continuance of scientific experiments that pose significant undisclosed risks to the safety, health, or property interests of people participating in or likely to be affected by them. Three of the most ethically repugnant scientific experiments carried out in or on behalf of the United States during this century are of this character and warrant brief description.

Beginning in 1932, U.S. Public Health Service researchers administered placebos to 431 black men in Tuskegee, Alabama. Each experimental subject, induced to come in for blood tests supposedly as part of an area-wide campaign to fight syphilis, was tested for and found to have syphilis. However, *the subjects were neither told that they had the disease nor treated for it.* The purpose of the experiment was to obtain scientific knowledge about the long-term effects of syphilis on mental and physical health. Nontreatment continued for 40 years, long after it became known that penicillin was a cure for syphilis and was widely available. Following press exposure in 1972, the experiment was terminated.[14] A Public Health Service investigation disclosed that of 92 syphilitic patients examined at autopsy, 28 men (30.4 percent) died from untreated syphilis—specifically, from syphilitic damage to the cardiovascular or central nervous systems. Hence, the total number of men who died as a result of nontreatment may have exceeded 100.[15]

In the 1950s, the U.S. Central Intelligence Agency solicited and funded a series of "mind-control experiments." Among the techniques used on experimental subjects were sensory deprivation, electroshock treatment, prolonged "psychological driving," and the administering of LSD and other potent drugs. By one estimate, at least 100 patients went through one series of brain-washing procedures.[16] Many participants in the experiments suffered long-term physical and mental health problems. In 1953, one subject was given a glass of liquor laced with LSD. He developed a psychotic reaction and committed suicide a week later.[17]

As part of its Biological Warfare program, the U.S. Army secretly sprayed bacteria and chemicals over populated areas of the United States (and Panama) during a 20-year period beginning in 1949. At least 239 such tests were carried out. The objective was to determine the country's vulnerability to germ warfare by simulating what would happen if an adversary dropped certain toxic substances on the United States. One frequently used microorganism was *Serratia marcescens.* Its safety was questioned prior to 1950, and strong evidence that it could cause infection or death existed in the late 1950s. Nevertheless, it continued to be used in tests over populous areas for the next decade. Four days after a 1950 spraying over the San Francisco Bay Area, a patient was treated at the Stanford University Medical Center for infection caused by *Serratia*, the first case ever recorded at the hospital. Within the next five months, ten more patients were treated at Stanford for the same infection. One of them died.[18]

Their ethically reprehensible character aside, such cases serve the useful purpose of showing that "freedom of scientific inquiry" is not an absolute, unconditional, or inviolable right. While clearly an important human value, "freedom of scientific inquiry" may, under certain conditions, have to take a backseat to other important values, such as protection of the dignity and welfare of each and every individual human being.

(2) Whistle-Blowing. Wenk's second kind of ethical issue in engineering is that of "whistle-blowing." Engineers—or scientists—may become aware of deliberate actions or negligence on the part of their colleagues or employers that seem to them to pose a threat to some component of the public interest (e.g., public safety, the effective expenditure of public monies, and so on). If the worried practitioner's "in-house" attempts to have such concerns addressed are rebuffed, then he or she must decide whether to "go public" ("blow the whistle") and disclose the facts underlying the concerns.

Problematic phenomena of the sort that impel some practitioners to consider such a course of action are often driven by the huge profits and professional reputations at stake in modern research and development activity. These phenomena can be associated with any of a number of phases of engineering or science projects. Consider, for example, misleading promotional efforts to secure public funding; cheap, unreliable designs; testing shortcuts;

misrepresented results of tests or experiments; shoddy manufacturing procedures; intermittently defective products; botched installation; careless or inadequate operational procedures; or negligent waste disposal. A significant number of such cases have come to light in recent years, of which three follow.

At Morton Thiokol, Inc. (MTI), several engineers working on the Challenger Space Shuttle booster project tried to convince management that the fateful January 1986 launch should not be authorized since the already suspect O-ring seal on the booster rocket had not been tested at the unusually cold temperatures prevailing on the day of the tragedy. MTI senior engineer Roger Boisjoly testified before the presidential committee investigating the disaster about what led up to the decision by the company and the National Aeronautics and Space Administration (NASA) to authorize the launch, a process at whose turning point MTI's general manager told his vice president of engineering to "take off his engineering hat and put on his management hat."[19] For his candid testimony, Boisjoly was allegedly subjected to various forms of mistreatment within the company and was placed on extended sick leave.[20] [See Case Study 1 at the end of this part for more information about Roger Boisjoly and Morton Thiokol, Inc.]

In 1972, three engineers employed by the San Francisco Bay Area Rapid Transit (BART) system, after receiving no response to their in-house memos of concern, went public about subsequently confirmed safety-related deficiencies that they had detected in the design of BART's Automated Train Control System. They were summarily dismissed for their trouble.[21]

A senior engineer at the Bechtel Corporation, part of a task force assigned to plan the removal of the head of the failed nuclear reactor vessel at Three Mile Island after the famous 1979 accident, became concerned about shortcuts allegedly being taken by his company in testing the reliability of the crane to be used to remove the vessel's 170-ton lid. When he protested the alleged shortcuts, he was relieved of many of his responsibilities. He then went public, was suspended, and later was fired.[22]

Sociologically speaking, several things are noteworthy about such cases. Technical employees who find themselves in situations in which they are asked or required to do things that violate their sense of right and wrong are not an endangered species. Results of a survey of 800 randomly selected members of the National Society of Professional Engineers published in 1972 disclosed that over 10 percent felt so constrained. A "large fraction" were sufficiently fearful of employer retaliation that they acknowledged they would rather "swallow the whistle" than become whistle-blowers.[23] A major 1983 study of technical employees found that 12 percent of respondents "reported that, in the past two years, they have been in situations in which they voiced objection to, or refused to participate in, some work or practice because it went against their legal/ethical obligations as engineers or their personal senses of right and wrong."[24]

For a number of reasons, engineers have traditionally been loath to criticize their employers publicly. Most obviously, those who feel compelled to "go public" enjoy precious little legal or professional-association protection against employer retaliation, often in the form of firing. However, as Wenk argues, some reasons that discourage whistle-blowing are sociocultural in nature:

> For engineers, a problem arises because most work in large organizations and many find excitement of participation in mammoth undertakings. They learn to value formal social structures, an attitude akin to "law and order." They adopt the internal culture and values of their employer and are likely to be allied with and adopt the perspectives of the people who wield power rather than with the general population over whom power is held. As a consequence, there is less tradition and inclination to challenge authority even when it is known to be wrong in its decisions which imperil the public.[25]

Not without reason, engineers—and, increasingly, scientists in large industrial organizations—tend to see themselves as employees with primary obligations to their employers, not the public. Moreover, the notion that employees retain certain citizen rights—for example, freedom of expression—when they enter the workplace is a relatively new notion in American industrial history. In a classic 1892 opinion, Oliver Wendell Holmes, then Massachusetts Supreme Court justice, wrote:

> There are few employments for hire in which the servant does not agree to suspend his constitutional rights of free speech as well as idleness by the implied terms of the contract. The servant cannot complain, as he takes the employment on the terms which are offered him.[26]

Only in the late twentieth century has this traditional attitude begun to be reversed, partly because the costs of such enforced silence are now viewed as unacceptable to society.

Ethically speaking, the obligation of technical professionals to blow the whistle when it is appropriate to do so arises from several factors. First, much contemporary research and development is supported by public money, as is the graduate education of many scientists and engineers (through government fellowships and loans). Second, the scale of the possible harm to the public interest at stake in many contemporary technical activities is large. The third factor is the following ethical principle of harm prevention: "[W]hen one is in a position to contribute to preventing unwarranted harm to others, then, other things being equal, one is morally obliged to attempt to do so."[27] An engineer or scientist sometimes possesses personal, specialized, "insider" knowledge about a troubling facet of a technical activity or project. Coupled with the credibility attached to testimony provided by authoritative technical professionals (as opposed to claims made by nonspecialist activists), that knowledge puts the scientist or engineer in a special position to make a possibly decisive contribution to preventing unwarranted harm to others or at least to preventing its repetition. This gives rise to a moral obligation to blow the whistle—once all other reasonable steps to rectify the situation "in house" have been taken and failed.

Some have urged that the obligation to responsibly blow the whistle be emphasized during the formal education of scientists and engineers—by the use of actual case studies, for example.[28] Others have stressed the importance of effecting structural and policy changes in the organizations in which technical professionals work and in their professional associations so that whistle-blowers are not required to choose between remaining silent and becoming self-sacrificing "moral heroes."[29] A third approach is that of legislation. A measure of protection for whistle-blowers has been built into some federal environmental and nuclear laws, and roughly half the states prohibit the firing of employees who have blown the whistle on their employers for practices violating existing public policies. However, some advocates for whistle-blowers see the need for comprehensive federal legislation allowing whistle-blowers who suffer reprisals to initiate legal action against their employers up to two years after such occurrences.[30]

(3) Consideration of Long-Term Effects. Wenk's third and final category of ethical issues confronting engineers in daily practice involves "managing the future." He argues that engineers have a tendency to focus on designing, producing, or installing "hardware" without adequately "anticipat[ing] . . . longer term effects." This is an abdication of the engineer's "professional responsibility." In terms of the "quartet of basic concerns" that we have utilized in this chapter, given the scale and scope of the effects of many contemporary engineering products and projects, engineers who fail to scrutinize their projects with comprehensive critical vision, both with respect to its likely consequences (including possible longer-term ones) and its likely patients or "stakeholders" (including, where appropriate, members of future generations) are guilty of unprofessional and ethically irresponsible conduct. Uncritical allegiance to the deontological dictum "if it can be done, it should be done" no longer confers immunity from charges of ethical impropriety on technical practitioners.

Problems of Communication Other ethical issues faced by technical practitioners have to do not with possible effects of scientific and technological projects on the safety, health, or property of those who may be affected by them, but with problematic aspects of *practitioners' communication* of and about their work.[31] Issues in this category, most notably ones involving *fraud* and *misrepresentation*, pertain to publication or presentation of claimed findings and to work-related interactions with nontechnical funding or policy-making officials. Cases of fraud have come to light in recent years in which experiments reported on in published papers were in fact never carried out, crucial data were fabricated, and conclusions were drawn from data allegedly known not to support them.[32]

(1) Fraud. Falsification of scientific data may not be as infrequent as normally supposed. June Price Tangney surveyed researchers in the physical, biological, behavioral, and social sciences at a large American university. Of 1,100 questionnaires distributed, 245 were completed and returned. Half of the respondents were senior researchers with the rank of full professor. Not surprisingly, the survey revealed that 88 percent of the respondents believe that scientific fraud is uncommon. However, 32 percent reported that they had a colleague in their field whom they had at some time suspected of falsifying data.[33] This figure, while suggestive, must be interpreted cautiously. It is not proof that one third of all scientists engage in such misconduct, for not only may suspicions be mistaken, but multiple respondents could have had the same individual in mind.

Whatever the extent of fraud in science, scientists see a number of factors contributing to the phenomenon. Tangney's respondents identified the following as major motivations for fraud: desire for fame and recognition (56 percent), job security and promotion (31 percent), firm belief in or wish to promote a theory (31 percent), and "laziness" (15 percent).[34] Underlying many such factors, she contends, is *the highly competitive nature of contemporary science*: the pressure to publish, the shortage of desirable jobs, and the fierce competition for funds. Beyond contributing to fraud, Tangney argues that such pressures can negate "what might otherwise be a fairly adequate self-policing mechanism in the scientific community." Indeed, the results of her survey call into question the common wisdom about the self-correcting nature of science, via processes like refereeing and publication; of the aforementioned 32 percent who had suspected a colleague of falsifying data, over half (54 percent) reported that they had taken no action to confirm or disconfirm their suspicions.[35] The competitive nature of contemporary science may have biased the reward system in the profession *against* undertakings aimed at uncovering fraud.

In a highly competitive academic environment, many researchers may feel that, if they raise questions about serious misconduct, their own reputations will be tarnished and their own chances for resources and advancement will be diminished. A scientist may be rewarded for uncovering "legitimate" flaws or shortcomings in a rival's work. However, there generally is little to be gained and much to be lost by attempting to expose a fraud.[36]

(2) Misrepresentation. Misrepresentation takes a number of forms in the communication of research findings, including both failure to credit or fully credit deserving contributors and crediting or overcrediting undeserving contributors. It might seem that such acknowledged species of misconduct, however regrettable, do not deserve to be called unethical, except perhaps by deontologists, for whom they fall into forbidden categories of actions regardless of the gravity of their consequences for science or society. However, consequentialists may also selectively regard such practices as unethical since they can in fact result in serious public harm. In May 1987, a scientist was accused by an investigative panel of scientists appointed by the National Institutes of Health of "knowingly, willfully, and repeatedly engag[ing] in misleading and deceptive practices in reporting results of research."[37] The pertinence of these practices to consequentialist ethical judgment making becomes clear from the panel's finding that the scientist's publications had influenced drug treatment practices for severely retarded patients in facilities around the United States.

Presentations of research findings to groups of peers also offer opportunities for ethically problematic behavior. Such presentations are sometimes used to establish claims of priority in the conduct of certain kinds of research. However, given the intensely competitive nature of the contemporary scientific research enterprise, if the research work is still in progress, it is understandable that scientists may opt to disclose just enough of their findings to serve their priority interests but not enough to reveal their overall strategies or the next steps in their "battle plans." However, quests for priority and resultant recognition may go beyond being unprofessional and become unethical if deliberately misleading or outright false information is disseminated in hopes of sending rivals "off on wild goose chases," diverting them from paths believed potentially fruitful. While making such an ethical judgment may appear open only to a deontological thinker, doing so can also be defended on consequentialist grounds by, for example, appealing to the harm that such deception can inflict on knowledge-sharing institutions, like the peer seminar, which have usefully served scientific progress and thus, indirectly, human welfare.

The interaction of scientists and engineers with public funding agencies or policy-making officials can also be ethically problematic. Institutional or organizational pressures to obtain funding for research and development ventures or units with significant prestige or employment stakes can induce applicants to resort to various forms of misconduct in hopes of increasing the chances of favorable action by a funding agency. Among these are use of false data, misrepresentation of what has been accomplished to date on a project in progress, and gross exaggeration of what can be expected in the grant period or of the scientific or social significance of the proposed work.

Dealings with makers of public policy often lend themselves to such hyperbole, for policymakers are typically individuals with nontechnical backgrounds who are unable to assess critically the plausibility of the claims made about current or proposed research or development projects. If a prestigious researcher deliberately misrepresents the potential or state of development of a pet project to an influential policymaker in order to enhance the project's funding prospects, then insofar as approval is secured through this deception and at the cost of funding for other worthwhile projects, consequentialist thinkers may join deontologists in judging the practitioner guilty of unethical conduct. This they may do not least on grounds of the long-term consequences for the welfare of society of undermining the integrity of the research funding process.

THE CHALLENGE OF CONTEMPORARY SCIENCE AND TECHNOLOGY TO TRADITIONAL ETHICAL THEORIES

The foregoing discussion of categories of ethical issues and conflicts engendered by developments in science and technology strongly suggests that these forces are putting growing pressure on traditional absolutistic ethical thinking. There are several reasons that the validity and utility of such thinking are being called into question in an era of rapid scientific and technological development.

As we saw, many such theories condemn or praise particular kinds of actions if they but fall into one or another category of supposedly intrinsically good or bad deeds. However, as noted, an action or practice condemned as "unnatural" can come to seem less so over time, especially if the original ethical judgment was predicated on the act's or practice's being unusual or unfamiliar when it first came to attention. Similarly, traditional absolutistic ethical outlooks are sometimes based on static worldviews born of their subscribers' limited experience. As a culture or subculture dominated by such a worldview overcomes its geographical or intellectual isolation and interacts more with the rest of the world, supposedly immutable categories or rules pertaining to "sacred" things or ways tend to loosen up. Adherents of such worldviews may come to recognize that respectable members of different cultures or subcultures think and act differently than they do about the same matters. Further, new products, processes, and practices can, as noted, have long-term hidden effects. Their eventual eruption and empirical confirmation sometimes call for revision of ethical judgments made before recognition that such subterranean effects were at work. However, such reevaluation is not congenial to absolutistic thinking, which purports to base its unwavering ethical judgments and decisions on something other than consequences. Considerations such as these make it increasingly difficult to sustain absolutistic ethical theories and outlooks in contemporary scientific and technological society.

Besides challenges to its intellectual tenability, contemporary science and technology are calling into question the utility of traditional absolutistic ethical theories—that is, their ability to serve as intelligent guides to action in a world of rapid and profound technical and social change. Such categorical theories and outlooks are helpless when confronted with ethical issues and conflicts of the sorts discussed in the third section of this chapter. For example, such theories shed no light on cases involving the distribution of benefits, costs, and risks associated with scientific and technological developments; intrapersonal conflicts between two venerable ethical values; public harms of aggregation; or the situations of technical professionals torn between loyalty to employers, concern for their families' well-being, and devotion to promoting the public interest. Finally, traditional deontological approaches to ethical thinking offer no incentives to agents to consider whether, in the face of possible unforeseen effects of a technical innovation, expansion of the domain of pertinent patients or the list of their protectable interests might be in order.

This is not to imply that consequentialist theories and thinking are immune from difficulties when confronted by contemporary scientific and technological developments. For example, uncertainty about possible elusive or projected long-term consequences of scientific and technological innovations and developments makes ethical judgments based on such assessments provisional and open to doubt. However, that is a price that must be paid if ethical judgments and decisions are to be made on an empirical rather than an a priori basis and are to be focused on the bearing of scientific and technological developments on human harm and well-being.

One conclusion of this chapter, then, is that developments in contemporary science and technology call for revisions in traditional ethical thinking and decision making. One kind of ethical theory deserving serious consideration we will call *qualified neo-consequentialism*. Under this theory, ethical judgments about actions, practices, and policies hinge first and foremost on assessments of their likely consequences. In particular, these assessments must have the following *neo-consequentialist* qualities. They should be

1. *Focused on harm and well-being*—directed to identifying and weighing the importance of consequences likely to influence the harm or well-being of affected patients;*
2. *Refined*—designed to detect or at least be on the lookout for subtle effects that, although perhaps hidden or manifested only indirectly, may nonetheless significantly influence stakeholder harm or well-being;

*The reader will note that no account has been presented here of exactly what is meant by human "harm" and "well-being." That substantial task must be left for another occasion. Suffice it to say here that for this writer, "harm" is not reducible to considerations of physiological deprivation, physical injury, and property damage or loss; nor is "well-being" reducible to considerations of material abundance, financial success, and high social status.

3. *Comprehensive*—designed to attend to *all* harm- and well-being-related effects—social and cultural as well as economic and physical in nature—of the candidate action, policy, or practice on *all* pertinent patients, remote as well as present, "invisible" as well as influential;
4. *Discriminating*—designed to enable scientific and technological options to be examined critically on a case-by-case basis, in a manner neither facilely optimistic nor resolutely pessimistic, and such that any single proposal can emerge as consequentially praiseworthy and be adopted or as consequentially ill-advised and be rejected in its present form if not outright; and
5. *Prudent*—embodying an attitude toward safety that, as long as a credible jury is still out or if it has returned hopelessly deadlocked, is as conservative as the magnitude of the possible disaster is large.

Further, the assessments sanctioned by our proposed ethical theory must also meet certain conditions. If an action, policy, or practice is to earn our theory's seal of approval, its projected outcome must not only be likely to yield at least as large a surplus of beneficial over harmful consequences as that of any available alternative, but it must also meet certain additional *qualifications*, two of which will now be briefly discussed.

It is scarcely news that contemporary scientific and technological activities unfold in societies in which those who stand to benefit greatly from their fruits are rarely the same as those likely to bear their often weighty costs and risks. We therefore stipulate that to be ethically permissible or obligatory, the allocation of the scrupulously projected benefits, costs, and risks of a technical undertaking among the various affected stakeholders must also be *distributively just*.

Various criteria have been put forth for evaluating the justice of such distributions.[38] One that deserves serious consideration is John Rawls's famous "difference principle."[39] Imagine, says Rawls, a group of people in "the original position"—that is, convened to formulate from scratch the rules that will govern the first human society, one shortly to come into being. Suppose that these deliberations take place behind "a veil of ignorance"—that individual group members have no knowledge whatsoever of whom or what they will turn out to be (e.g., male or female, black or white, Asian or North American, physically handicapped or not) or of their eventual economic well-being or social status. Then, Rawls contends, the group would eventually reach agreement that it was in each member's best interest that the following rule of justice be adopted: an unequal distribution of any social or economic "good" will be permitted in the society-to-be only if there is good reason to believe that it will *make everyone, including the less fortunate, better off*. Indeed, Rawls eventually offered a stronger version of his principle according to which an unequal distribution of such a good is just only if there is good reason to believe that it will make everyone better off *and* that it will *yield the greatest benefit to those currently worst off*.[40] If either version of this rule is found compelling, it would have to be applied to each predominantly beneficial but unequal distribution of projected science- or technology-based benefits, costs, and risks before the conclusion could be reached that it was ethically permissible or obligatory to proceed with the action, project, or practice in question.

Our neo-consequentialist ethical theory has a second qualification. A science- or technology-related course of action may sometimes be denied ethical approval even if all of the foregoing conditions are met. Even then, it may be proper to withhold ethical approval if the projected harmful consequences (1) *exceed some substantial quantitative threshold*—either in a single case or when aggregated over multiple cases of a similar sort—and (2) are not *greatly* outweighed by their positive counterparts. In such situations, the decision-making party may decide that it would be prudent to decline the admittedly greater projected benefits offered by the option under consideration.

Ethical decision making that takes no account of the absolute magnitude of an option's projected negative consequences even if they are outweighed by their positive counterparts, or of how the outweighed negative consequences of an individual course of action may aggregate over multiple instances, is deeply flawed. Indeed, allowing "yielding a positive balance of benefit over harm" by itself to compel ethical approval of individual courses of action may even be unjustified on consequentialist grounds, for following that criterion consistently in multiple instances may over time lead to unacceptable public harms of aggregation. For example, assessing the impact on traffic of individual proposed downtown high-rise office buildings solely in terms of the modest number of additional cars each structure may attract into the city may allow the aggregate effect on traffic of approval of a large number of such projects to go unreflected in individual decision-making processes.

CONCLUSION

Hopefully, the reader will find the qualified neo-consequentialist approach to thinking about ethical issues and conflicts just sketched more adequate to the realities of contemporary scientific and technological practice. In any event, in this chapter we have characterized a number of different kinds of science- and technology-based ethical issues and conflicts, indicated some noteworthy technical and social roots of such problems, and argued that important traditional ethical concepts and modes of thinking are being subjected to increasing pressure by scientific and technological changes in contemporary society. While this stress is being strenuously resisted in some quarters, it is likely in the longer run to lead to major transformations of ethical ideas and thinking.

ENDNOTES

1. I owe my initial awareness of this framework to a 1972 lecture at Stanford University by ethicist Dr. Karen Lebacqz.
2. See, e.g., William Frankena, *Ethics*, Foundations of Philosophy series, 2nd ed. (Englewood Cliffs, N.J.: Prentice-Hall, 1973), chaps. 2 and 3. See also Mary Warnock, *Ethics Since 1900* (London: Oxford University Press, 1960), chap. 2, pp. 48–52.
3. *Wall Street Journal,* May 4, 1989, p. B4.
4. Stephen Isaacs, "Hasidim of Brooklyn: Fundamentalist Jews Amid a Slumscape," *Washington Post,* "Outlook" section, February 17, 1974, Section B, p. 1.
5. Gregory Kavka, *Moral Paradoxes of Nuclear Deterrence* (Cambridge: Cambridge University Press, 1987), pp. 19–21.
6. Ronald Munson, *Intervention and Reflection* (Belmont, Calif.: Wadsworth, 1979), p. 398.
7. *Ibid.*, pp. 399–400.
8. For example, Carl Cohen argues that animals are not members of any "community of moral agents." Incapable of, among other things, giving or withholding consent, animals, unlike humans, cannot have rights, thereby precluding involuntary experimentation on them. See Cohen's "The Case for the Use of Animals in Biomedical Research," *New England Journal of Medicine,* 315, no. 14, 1986, p. 867.
9. Gina Kolata, "Genetic Screening Raises Questions for Employers and Insurers," *Science,* 232, no. 4748, April 18, 1986, p. 317.
10. *New York Times,* August 19, 1986, p. 21.
11. John M. Taurek, "Should the Numbers Count?" *Philosophy and Public Affairs,* 6, 1977, pp. 293–316.
12. Edward Wenk, Jr., "Roots of Ethics: New Principles for Engineering Practice," American Society of Mechanical Engineers Winter Annual Meeting, Boston, Massachusetts, December 1987, 87-WA/TS-1, pp. 1–7.
13. Meredith Thring, "The Engineer's Dilemma," *The New Scientist,* 92 no. 1280, November 19, 1981, p. 501.
14. *New York Times,* July 26, 1972, p. 1.
15. *New York Times,* September 12, 1972, p. 23. For a detailed account of this episode, see James H. Jones, *Bad Blood: The Tuskegee Syphilis Experiment* (New York: Free Press, 1981).
16. Harvey Weinstein, *A Father, a Son, and the CIA* (Toronto: James Lorimer and Co. Ltd., 1988).
17. Leonard A. Cole, *Politics and the Restraint of Science* (Totowa, N.J.: Rowman and Allanheld, 1983), p. 111.
18. *Ibid.*, pp. 112–114.
19. Roger Boisjoly, "Ethical Decisions: Morton Thiokol and the Space Shuttle *Challenger* Disaster," American Society of Mechanical Engineers Winter Annual Meeting, Boston, Massachusetts, December 1987, 87-WA/TS-4, p. 7.
20. *Ibid.*, p. 11.
21. Stephen H. Ungar, *Controlling Technology: Ethics and the Responsible Engineer* (New York: Holt, Rinehart and Winston, 1982), pp. 12–17.
22. Rosemary Chalk, "Making the World Safe for Whistle-Blowers," *Technology Review,* January 1988, p. 52.
23. Rosemary Chalk and Frank von Hippel, "Due Process for Dissenting 'Whistle-Blowers.'" *Technology Review,* June/July 1979, p. 53.
24. Chalk, "Making the World Safe," pp. 56–57.
25. Wenk, "Roots of Ethics," p. 3.
26. Chalk and von Hippel, "Due Process," p. 53.
27. Compare this principle with Kenneth Alpern's "principle of due care" and "corollary of proportionate care" in his "Moral Responsibility for Engineers," *Business and Professional Ethics Journal,* 2, Winter 1983, 40–41.
28. Boisjoly, "Ethical Decisions," p. 12.
29. Richard DeGeorge, "Ethical Responsibilities of Engineers in Large Organizations: The Pinto Case," *Business and Professional Ethics Journal,* 1, no. 1, 1981, p. 12.
30. Chalk, "Making the World Safe," pp. 55–56.
31. For a useful bibliography on this aspect of the problem, see Marcel Chotkowski LaFollette, "Ethical Misconduct in Research Communication: An Annotated Bibliography," published under NSF Grant No. RII-8409904 ("The Ethical Problems Raised by Fraud in Science and Engineering Publishing"), August 1988.
32. See, e.g., William Broad and Nicholas Wade, *Betrayers of the Truth* (New York: Simon & Schuster, 1982), pp. 13–15; and Nicholas Wade, "The Unhealthy Infallibility of Science," *New York Times,* June 13, 1988, p. A18.
33. June Price Tangney, "Fraud Will Out—Or Will It?" *New Scientist,* 115, no. 1572, August 6, 1987, p. 62.
34. *Ibid.*
35. *Ibid.*
36. *Ibid.*, p. 63.
37. *New York Times,* April 16, 1988, p. 6.

38. For cogent discussion of various criteria of distributive justice, see Joel Feinberg, *Social Philosophy*, Foundations of Philosophy Series (Englewood Cliffs, N.J.: Prentice Hall, 1973), chap. 7.
39. John Rawls, *A Theory of Justice* (Cambridge, Mass.: Harvard University Press, 1971), chaps. 1–3, especially pp. 75–78.
40. For discussion of alternate versions of Rawls's difference principle, see Robert Paul Wolff, *Understanding Rawls* (Princeton, N.J.: Princeton University Press, 1977), pp. 40–41.

QUESTIONS

1. Examine the relationship between ethics and technology by comparing consequentialist, deontological, and utilitarian (hedonic and ideal) theory to three specific correlated technologies.
2. Compare and contrast four kinds of considerations that are pertinent to ethical decisions and judgment making.
3. Evaluate the significance that the "violation of world order" dispute has to scientists developing stem cell research.
4. Use the chart below to provide specific technological examples for the eight technology-based ethical conflicts listed in this reading.

Dispute	*Technological Example*

The Relationship Between Ethics and Technology

PAUL ALCORN

In an evolving universe, who stands still moves backward.
R. ANTON WILSON

INTRODUCTION

In this chapter, we will develop a precise definition of technology to avoid misunderstanding when the term is used in this text. We will discuss the way in which technology is created and functions in a culture and then develop an understanding of how this concept of technology relates to ethical behavior. Finally, we will discuss how to use technology ethnically, that is, do technology in a way that works.

DEFINITION OF TECHNOLOGY

Essentially, technology is that whole collection of methodology and artificial constructs created by human beings to increase their probability of survival by increasing their control over the environment in which they operate. Technology includes and is essentially a means of manipulating natural laws to our benefit by constructing objects and methodology that increase our efficiency and reduce waste in our lives. The objects we create are artifacts, literally artificial constructs, that have been manufactured for specific uses and purposes. Everything that we use that is not as it comes to us in nature falls under the heading of technology. This is a very broad definition. All of the physical objects of our lives that were in any way altered from the way they appeared in nature represent technology. A sharpened stick is technology, as is a dollar bill or a Caterpillar™ tractor; they merely have different functions and have been produced through a different series of steps, usually through the use of other technology.

It may be noted that human beings are not the only animals that create artifacts, and for that reason, the mere creation of artifacts does not in and of itself constitute technology. Birds build nests, chimpanzees use sticks as tools to gather food, and bees build elaborate hives. What is missing in these artifacts that separates them from what we mean by technology is the matter of choice. A bee contributes to the development of a hive because of genetic encoding. It is a process that is "hardwired," as an electrical engineer would say. It has no choice about what it is doing. The same is true of a bird building a nest or an otter using a rock to open a clam by resting the clam on its stomach as it floats and hammering it with a stone. Such behavior is instinctual. But not all methodology used by living creatures other than humans is instinctual. Some higher primates—chimpanzees, for example—are capable of reasoning through problems and using objects to create methodology for solving those problems. They have been observed experimentally under controlled conditions learning to attach telescoping rods together to gather food that is otherwise out of reach. Yet they have very limited capacity in this regard and do not pass this information on to others in a cultural way. What truly separates humans from the other members of the animal kingdom in this regard is our incredible power of choice.

TECHNOLOGY AND CHOICE

With humans, the technology we choose to build and the manner in which we use it is totally a matter of choice. We have an infinite capacity to produce technological goodies,

PRACTICAL ETHICS FOR A TECHNOLOGICAL WORLD by Alcorn, Paul A., © Reprinted by permission of Pearson Education, Inc., Upper Saddle River, NJ.

within the boundaries of natural law, and we can accept or reject an idea as we choose. Thus, at one point in time, we may choose to develop the use of fire for cooking and at another decide to develop the art or science of architecture for the purpose of providing ourselves with shelter. Additionally, at one point we may decide to use dome-shaped hovels as shelter and at another time and place opt for alabaster palaces or multistory office buildings. The choice is all ours. It is in that choice of what artifacts to produce and the range of artifacts that we are capable of producing that we find the true nature of technology. And, as nearly as we can tell, that choice seems to be the sole province of human activity.

TECHNOLOGY AND EVOLUTION

In *Social Issues in Technology: A Format for Investigation,* I offered a detailed explanation of technology and the technological process. In this book I offer a general understanding of technology and why it exists in our lives. Technology is a vital part of what it is to be human; in order to understand our world, it is necessary to understand the purpose, the source, and the processes of our technological world.

For a human being, doing technology is a natural process. It represents one of the chief capacities with which nature has provided us for our survival. As with any other creature, *Homo sapiens* has certain characteristics that allow the species to perpetuate itself and successfully compete with other species for a niche in the natural world. Ecologically, we are an integral part of a much larger system that is designed to grow, develop, and maintain itself as an extensive living structure.

Every element in that system has the capacity to survive based on certain characteristics. For human beings, those *survival traits,* as these characteristics are called, include our capacity to create and use technology. There are specific and overwhelming advantages to this ability. Because we use artificial structures for our survival rather than develop the necessary characteristics through genetic alteration to our being, we are able to develop and adapt at a much higher rate than other animals or plants. We have effectively externalized the process of evolutionary development.

As an example, consider the characteristics of other animals versus those of a human being. Other animals have the advantage of speed, or claws, or special poisons that they can inject into their prey. Herbivores have specially designed digestive systems that allow them to consume large amounts of cellulose, a very difficult substance to break down, and turn it into useful energy. Some animals fly, others are very fleet of foot, others have incredible capacities to blend into the environment, and still others design complex living environments (e.g., hanging basket nests or colonized networks of tunnels). Each species has specific characteristics that offer it an advantage.

Now compare this with a human being. We do not have armored bodies covered with scales or shells. We cannot run particularly fast (although genetically we do have incredible stamina compared to most animals, a characteristic that allowed our hunter ancestors to follow game for days until the game was exhausted). Nor can we take to the air, with wings on our backs, or glide on membranes built into our bodies as bats or flying squirrels do. Yet we are capable of moving at a rate of speed far beyond that of a cheetah or other fleet-footed animal. We are able to fly across the face of the planet and into the outer reaches of our world and beyond. We can live underwater in craft that outperform the largest fish and exist in environments in which the extremes of temperature or altitude would kill most other creatures. We do it all in spite of the fact that we have at our disposal not a single physical trait that allows us to do so.

That is because the nature of our evolution has been external to our bodies. Instead of developing the eyes of a hawk, we develop binoculars and telescopes. Instead of becoming fleet of foot, we build automobiles and locomotives and airplanes. Instead of wings on our back, we have the wings of air transports and helicopters and the lifting power of balloons and dirigibles. Our characteristics are external to our physical being. It is in this ability to artificially create what we need for survival that we find our chief advantage. Like other animals, we use the laws of nature to aid us in our survival, but whereas other species do this through genetic alteration, a process that takes thousands if not millions of years, we manufacture the alterations quickly and efficiently. We find ourselves at last at a point at which we do not adapt to nature, we adapt nature to us! Such capacity is unparalleled in nature.

But with this capacity comes a problem. Nature is an experimenter. Nature will try numerous variations on a theme to find the combination of characteristics that allow a given organism to survive in a competitive world. If one alteration does not work, such as growing extra wings or limiting the number of eyes of a species to one, then that version fails and does not survive long enough to create progeny or pass on the undesirable trait. If a variation offers superior opportunities for survival, many more of that version survive to pass on the characteristics to offspring, and eventually that version predominates. Thus, through evolutionary mutation and survival of the fittest, we arrive at a creature that is perfectly adapted to its environment.

This is also true of humans, but with one exception. Since we are producing change through the creation of technology rather than trial-and-error mutation, we can very quickly generalize a new "trait" over the entire population in a relatively short period of time. In a matter of generations rather than millennia, a new technological device, such as the bow and arrow or the chariot, can come into general use by everyone who sees it. If it offers a very great advantage to those who have it, everyone either perishes or soon learns to use the new technology. There is little time for experimentation and testing here.

This has been seen often in the past with sometimes devastating results. The practice of agriculture is an excellent example if we look at the relationship between climatic change and the extensive use of agriculture in a region. Some of the most arid regions of the globe were once great forests or grasslands that were cleared for agriculture. Unfortunately, with the deforestation came a host of environmental changes that led to everything from soil erosion to changes in weather patterns. This is just a single example of the problems that can arise from moving too quickly to embrace a technology. Other examples include the virtual lack of forests in Lebanon today where once stood vast woodlands of cedar, a prized wood traded all over the Mediterranean from North Africa to Egypt to ancient Israel, and the cliff dwellers of the southwestern United States, who flourished toward the end of the first millennium and then abandoned their cities when they could not adjust to climactic changes in growing cycles.

What if the governments of the world in the last half of the twentieth century had decided that, since nuclear weapons were the ultimate in destructive power, they would embrace that technology as is and abandon other means of war? We would have been left with no alternative but to create a nuclear holocaust in case of threat or attack. We are perhaps now in a similar predicament with biological and chemical weapons of mass destruction; they are cheap, effective, and easily produced and delivered. A single strain of a deadly bacterium or virus could cause a reduction of population around the world that would bring civilization as we know it to an end. And the tragic event would be the result of industrial and technological processes at work.

TECHNOLOGY AND RESISTANCE TO CHANGE

Because of this danger to our well-being—these seeds of destruction within our success—nature has also equipped us with another trait. That other trait is a resistance to changes in our culture. *Homeostasis,* as it is known, represents a fear of the unknown that extends to any technological device that may come along. Any new idea or new technology is initially suspect to most of the population because it is untested, unfamiliar, and therefore considered a potential threat. This is as much a survival mechanism as the capacity to create that technology in the first place. Because of homeostasis, time is a necessary ingredient for a given advance in technology to be generalized over the whole society. It is first embraced by a small section of the population eager to try new things and ideas, but the rest of society either initially ignores it or cautiously watches to see where it will lead. Should the new idea not be a particularly good one, that is, should it not increase the probability of individual and group survival, it tends to go by the wayside without much further ado. On the other hand, if it is actually a valuable idea, the new technology will continue to exist long enough for people to get used to it or to lose their initial fear of it, and then they are more willing to try this new gizmo. This is particularly true if those who first accept it have illustrated its value. Eventually, the acceptance and use of the new technology spreads throughout the culture.

This process can be easily seen in the case of the computer. Less than a century old, this device, once a curiosity used for certain esoteric operations by scientists and government, has become one of the primary tools of a modern technological society. It has been viewed as an oddity, feared, mystically couched in arcane terminology and given unrealistic assumptions of power by the uninitiated, seen as the subject of hobbyists and gadgeteers, embraced by big business and then small business, and finally accepted as an unavoidable way of life. The process took time while the population figured out how to use the new technology and how to configure it so that it was useful for their needs. It took time to gain acceptance and overcome the natural tendency of human beings to do things in the "same old way." It grew in popularity and use as a solution to a range of problems over the life of its development. All of that time was a gestation period for society to absorb and gain benefit from the new technology. Every invention goes through the same process, affected by a number of factors such as complexity, range of application, expense, and the degree of societal resistance.

The point to remember is that that resistance is necessary and natural, a safety net built into us by nature that allows us to take time to differentiate between new ideas that are truly beneficial and those that are potentially or truly dangerous to our survival. It is all part of the same natural process of creation and use of technology.

Human beings cannot help being creative. It is an element of our makeup that cannot be changed. Creativity and technological expertise require nurturance, but the tendency to learn the laws of nature and apply them to creating artificial constructs to enhance our lives comes as natural to us as breathing.

TECHNOLOGY AND ETHICS

Given that creating technology is natural and that within the limits of our understanding as to the nature of the universe we can choose what technology to use and how to use it, where do the ethics of the process arise? If you remember back to our working definition, ethics is the process of doing what works. Apparently, from the history of the human race, using technology tends to work. This is evidenced, if in no other way, by the predominance and domination of our species over the face of the earth. We are incredibly successful as a species, reflecting incredibly successful natural traits, and that includes technology and its use. Apparently, technology works for us, or we would not include the capacity to create it in our repertoire of survival traits in the first place. By definition, then, in and of itself, it must be ethical.

That's a nice idea, and it would certainly be a blessing for all of us if that were true. Unfortunately, it is not as simple as that. Technology, as it turns out, is neither ethical nor unethical; it is merely a tool to be used or misused as we choose. Thus, we are back to the choice of action again, the one control we have in our lives.

Each technology and each application of technology raises ethical issues with which we must deal. Each new device or application of what we know requires some consideration of whether the use of that device will work for us or not. To further muddy the issue, we often cannot even say with certainty whether a technology will benefit us or not. In fact, in most cases, technology turns out to be a double-edged sword, with both costs and benefits in its use, and this in turn requires us to determine whether or not the benefits are worth the costs. And that's assuming we can even actually determine the costs accurately in the first place.

Also, we need to consider the idea that the use of technology may benefit some while costing others. This is not an uncommon occurrence, particularly where one technology replaces another, as in the case of the automobile replacing the horse-drawn buggy or the word processor replacing the "steno pool."

As you can see, this cost-benefit situation creates quite a dilemma. Just knowing that the ethical thing to do is to do what works is not very useful as a guide to behavior if we do not know what works in the first place. This is not a new idea. It is a problem that we as a species have been wrestling with off and on for ten thousand years or more, particularly when new technologies and new ways of manipulating the world present themselves. A few examples will clarify this point nicely.

When the automobile was first introduced, it was hailed not only as a solution to transportation problems within cities but also as a defense against growing pollution. That may seem quite confusing from our perspective as citizens of the world at the beginning of the twenty-first century, but a century ago, the pollution problems faced by industrial urban dwellers was decidedly different. At that time, at the birth of the automobile age, the chief means of transportation was the horse. Anyone who wasn't walking or traveling by train within an urban environment was traveling on foot or by horse. Carriages, drays (freight wagons), and specialized coaches were all horse drawn. With the horses came horse dung, and it was everywhere. The streets were pocked with piles of dung to be cleaned up, dung that ran into the sewers and that produced a prodigious number of flies. And with the flies came disease. We do not think of horses and horse dung as being a major health hazard in our lives today, but a hundred years ago, it was a major problem. Thus, the "horseless carriage" was hailed as the eliminator of the "hay burner" technology of equestrian transportation.

Yet today, we view the automobile as a chief air pollution source; it dumps tons of carbon monoxide and other pollutants into the atmosphere, promoting global warming and creating smog in any city of size. Hence the solution becomes the issue. At the present time, there is a movement toward nonpolluting electric cars. California has gone so far as to mandate a 10 percent non-combustion engine vehicle quota for the state. Electric cars are the obvious non-combustion engine choice, and as the number of electric vehicles rises, replacing gasoline engine automobiles, it is believed substantial improvement in the environment will result. And so another solution has been found.

This being the case, should we not expect these electric vehicles to create other dilemmas? At the present time, nearly all electric automobiles are powered by heavy lead-acid batteries, deep charged and able to deliver power at sufficient rates for a reasonable amount of time. And much research is being done to develop better and more powerful batteries that will charge more quickly and deliver more power for even longer periods of time. Thus it appears, at least for the foreseeable future, that a dependence on lead-acid batteries will be dominant. But a new problem arises: what will we do with the spent batteries? Batteries are already seen as a pollution problem, with only one per car. What will happen when the number of batteries per vehicle rises to twelve or twenty? Could we be exchanging one form of pollution for another? It is not just lead-acid batteries that present this type of dilemma as we progress and change technology.

With any technological change and any acceptance of a new technology as standard, there is always a cost. There is never a free lunch, although payment can be deferred for some length of time. Yet in the end, someone has to pay, and I'm sure it comes as no surprise that delaying payment until our children or grandchildren are making the rules is not a very efficient way to operate. Intuitively, it is unethical to use this approach, although economically or politically it may be expedient.

To what extent should we consider the future payment for our exploitation of technology and technological possibility? Although we do not always know (indeed, seldom do we know) the true cost of a technological development, there are certainly some issues that we do know will need to be handled. History offers numerous examples of what to expect from technological change. How far does our responsibility go? One school of thought says not to worry about the future consequences because we have always been able to deal with what comes along. Still newer technology will solve the problem. New ideas and alternative ways of handling the issues will arise naturally out of necessity. We need only utilize what is available to us now and let the future generations worry about how to handle the problems that arise. These are the attitudes that led to the destruction of environments in the ancient world. As agriculture and population exploded beginning some ten thousand years ago, whole civilizations were destroyed by resultant drought and crop failure. Whole ecosystems were altered, turning fertile plains into deserts and lush forests into arid wasteland. Solutions were found, but what was the cost? The people of these transitional periods endured starvation and being uprooted as their productivity collapsed.

On the other hand, consider the approach of the Five Nations of the Iroquois Confederation. These Native Americans of the northeastern United States banded together in a peaceful structure that allied independent nations, building a greater confederation. The Cuyahoga, Seneca, Onondaga, Mohican, and Oneida nations agreed to work together for the betterment of all and for their mutual defense against their unfriendly neighbors, chiefly the Algonquin. This amazing group of people elected fifty men from among their number to collectively make the decisions for the whole group. (Interestingly, it was the women of the tribe who actually chose the fifty men to head the joint council.) They always considered the future consequences of those decisions, *for seven generations hence*! No decision that was merely expedient was acceptable. Compare this approach with the political process present in most industrialized countries today. How many decisions are made on the basis of how the people will be affected a century and a half in the future? It appears we could learn a great deal from these Native American tribes. (Incidentally, it would not be a wise idea to embrace the wisdom of the Native Americans without exception. The Iroquois, for example, are noted for their horrific treatment of prisoners of war, whom they first honored and then tortured for as long as possible without killing them, then ritually ate them, not for the food value, but to absorb some of their bravery and strength. It was considered a pity if the prisoner could not be kept alive in a state of agony for at least twenty-four hours before he or she died.)

Numerous other examples can be cited describing the failure of humans to include negative future circumstances in their deliberations. Again and again we see in the industrialized world the adoption of a technology that results in future problems. This is not to paint a dark portrait of technology or to suggest that we should abandon our technological ways. Our whole history as a nation has been one of progress and growth. It is merely a reminder that every new opportunity brings with it an obligation to consider the consequences of our actions, and this we seem rather reluctant to do.

COUNTERPOINT AND APPLICATION

If technologizing, that is, creating and using technology, is so natural to being human, then it would appear that it is always an ethical process, as it always works. It is not very useful for any species to go against its nature in the quest for survival, except as a part of the evolutionary process, and natural selection would seem to be quite adequate to this end. Why all the fuss about the ethical nature of technology? It is neutral. It is what it is. Talking about the ethical nature of technology is like talking about the ethical nature of a stick. Isn't that true?

Of course, that is not true at all. It must be remembered that the drive to create technology and thus evolve externally to our bodies is indeed a natural process, yet it still entails free will, or choice, on our part. An almost infinite array of technological possibilities is available to us, depending on how we choose to apply the basic principles that constitute our understanding of the physical universe (physical laws). It is because of that choice that we must consider ethical content.

Surely, the homeostatic tendencies of the species go a long way toward allowing us to adjust if we make mistakes in our choices in technological design and creation. But considering the speed at which the world changes and the far-reaching effects of even the seemingly most insignificant changes in methodology, it becomes critical to consider the usefulness of technological change in the broadest of terms, and that is a matter of what is the ethical thing to do. Technology has ethical content by virtue of the free will with which we create it. What do we choose to do and not do? We make those choices in a desire to improve our position in life, either individually, collectively, or both. Do we know that our choices are sound ones, and do they truly work to achieve the goals that they are designed to achieve? Therein lies the ethical issue.

When looking at technology and creating technological change through the modification, production, or application of technology, it is wise to think in a broader context. It is best to consider why exactly we are doing whatever it is we are doing, what our goals are, and whether the process undertaken actually achieves those goals. Additionally, we must consider what other goals or conditions are affected by the new creation or application and how that affects our overall goals in life. In other words, what would happen if we all behaved like the Native American confederacy mentioned earlier and considered the consequences of our actions for the next seven generations, or 150 years? How would we behave differently?

QUESTIONS

1. Define technology by integrating your own definitions from the reading.
2. Assess Alcorn's statement that "technology . . . increases our efficiency and reduces waste in our lives." How do you interpret his statement? What is an example of a technology that reflects your analysis of his words? Explain your answer by integrating details or additional examples.
3. Describe two technologies that illustrate homeostasis.
4. It can be argued that every technology has both positive and negative potential consequences. However, the technology itself is neutral. The truth may be that it is in the application of technology that ethical issues emerge. To explore this issue, list the benefits and costs of five technologies.

BOX 1 Technology: Ethical Questions

Designer Children

Should parents have the choice to design their own children? If parents are having fertility problems and decide to use in vitro fertilization, should they be able to use technology to pick the sex of their child? Is it ethical for a doctor to spend more time deciding which embryo is healthy and less time about changing the natural laws of trait and gender selection? In January 2001, the American Society for Reproductive Medicine, a group that establishes ethical guidelines for fertility clinics, stated, "The potential for inherent gender discrimination, inappropriate unnecessary medical burdens and costs for parents, and inappropriate and potentially unfair use of limited medical resources are serious ethical issues." Do you agree or disagree? Why?

Gender Selection: Ethical Considerations and New Technologies. Accessed July 16, 2005. Available at http://www.about.cohttp://atheism.about.com/library/weekly/aa100301a.htm.

Snuppy

Should animals be cloned? "After arduous labor involving thousands of eggs, 1,095 embryos and 123 canine surrogates, South Korean scientists reported that they succeeded in creating the world's first dog . . . the name is Snuppy." The hope is that scientists can create reliable treatments for diseases such as diabetes, heart disease, or cancer through stem cell research and cloning of a variety of animals. Do you agree or disagree? Why?

Gorner, Peter. August 4, 2005. "In Huge Advance, 1st Dog Cloned." *Chicago Tribune*.

Part-Human Animals

Should mice have brains that consist of mostly human brain cells? "At the University of Nevada-Reno, researchers created pigs with human blood, fused rabbit eggs with human DNA and injected human stem cells to make paralyzed mice walk." The goal is to use animals and stem cell research to improve life for humans. At Stanford University, a proposal has been made that mice brains be created from mostly human brain cells. Do you agree or disagree? Why?

"Scientists Create Animals that Are Part-Human." April 29, 2005. *Associated Press*. Accessed June 8, 2005. Available at http://msnbc.msn.com/id/7681252.

Bosses Monitor Your Internet Use

Should employers monitor Internet use by employees at work? Companies are concerned that the time employees spend on the Internet is time that should be spent working. They are also concerned that intellectual property or private company information might get into the wrong hands. Do you agree or disagree? Why?

Frauenheim, Ed. "Companies Ramping Up E-Mail Monitoring." June 8, 2005. *CNET Networks, Inc*. Accessed July 20, 2005. Available at
http://news.com.com/Companies+ramping+up+e-mail+monitoring/2100-1022_3-5738134.html

11

The Unanticipated Consequences of Technology

TIM HEALY

1. INTRODUCTION

When the successful cloning of a lamb called Dolly was announced in February of this year [1997] by Scottish researchers, it set off a spate of anxious questions. Many of them concerned the ethics of cloning, but another set asked about the unanticipated consequences. If we go down the cloning road, where will it lead? The answer is that we don't know. All of our technological roads twist and turn, and we can never see around the bend or through the fog.

The purpose of this paper is to investigate the ubiquitous phenomenon of unanticipated consequences. We begin with a look at some definitions which shed light on the matter and then consider the nature of change. This leads to a broadening of the definition of the word "technology" and a look at what was one of our earliest examples of unanticipated consequences. We then address the crucial question of why we have such consequences. Some additional examples follow, and we then look at what society does in the face of unanticipated consequences. The paper concludes with a discussion of some of the ethical implications of acting when we know that there can be unanticipated consequences to our actions.

2. SOME DEFINITIONS

It is important here to distinguish between unanticipated and undesired consequences. The former are consequences which are not foreseen and dealt with in advance of their appearance. Undesired consequences are those which are harmful, but which we are willing to accept or accept the risk of occurring. Consequences may be:

Anticipated

- intended and desired
- not desired, but common or probable
- not desired and improbable

Unanticipated

- desirable
- undesirable

As an example, consider the development of a nuclear power plant at an ocean site. The anticipated and intended goal or consequence is the production of electric power. The undesired but common and expected consequence is the heating of the ocean water near the plant. An undesired and improbable consequence would be a major explosion, and we would associate the term "risk" with this outcome, but not with the heating of the water.

An unanticipated and desirable consequence might be the discovery of new operating procedures which would make nuclear power safer. An unforeseen and undesirable consequence might be the evolution of a new species of predator fish in the warmed ocean water, which destroy existing desired species.

This paper concentrates on unanticipated consequences of our technologies. Anticipated negative consequences have been dealt with extensively in the literature on risk; see, for

Tim Healy. Reprinted with permission of the Markkula Center for Applied Ethics at Santa Clara University (www.scu.edu/ethics).

example, Margolis (1996) and Bernstein (1996). The latter emphasizes the role of mathematics in risk assessment.

Two brief points should be made before we proceed. The first is that change is always with us. Even without the intervention of human beings, nature changes constantly. Continents move, weather changes, species evolve, new worlds are born, and old ones die. The second point is that all change seems to involve unanticipated consequences. Hence, the unanticipated is a part of life. There is no absolute security. Unanticipated consequences can be mitigated, largely through the gaining of additional information or knowledge, but not eliminated. That's the nature of our life, natural and human.

3. A BROADER DEFINITION OF TECHNOLOGY

Although we focus here on the term "technology" as it is usually taken, it is worth pointing out that human beings do much that has unanticipated consequences in all areas of life, certainly including, for example: medicine, business, law, politics, religion, education, and many more. Because of the parallels among these fields, it is useful to think of a broader definition of technology, such as "...that which can be done, excluding only those capabilities that occur naturally in living systems." (Benziger) This matter is dealt with in some detail in *Technopoly: The Surrender of Culture to Technology* (Postman).

Seen in the light of this broader definition, writing is one of our first technologies. Postman recalls the story of Thamus and the god Theuth, from Plato's *Phaedrus*, as an example of unanticipated consequences. Theuth had invented many things, including: numbers, calculation, geometry, astronomy, and writing. Theuth claimed that writing would improve both the memory and the wisdom of humans. Thamus thought otherwise.

> Theuth, my paragon of inventors, the discoverer of an art is not the best judge of the good or harm which will accrue to those who practice it. So it is in this: you, who are the father of writing, have out of fondness for your off-spring attributed to it quite the opposite of its real function. Those who acquire it will cease to exercise their memory and become forgetful; they will rely on writing to bring things to their remembrance by external signs instead of by their internal resources. What you have discovered is a receipt for recollection, not for memory. And as for wisdom, your pupils will have the reputation for it without the reality: they will receive a quantity of information without proper instruction, and in consequence be thought very knowledgeable when they are for the most part ignorant. And because they are filled with the conceit of wisdom instead of real wisdom they will be a burden to society.

It is true in all of our technologies that the discoverer of an art or the designer of a new system is not usually the best judge of the good or harm which will accrue to those who practice it. And yet, paradoxically, it is often the designer to whom we must go to ask for the likely outcomes of her work. This is a problem with which society must wrestle, and we shall discuss later how it does so.

4. WHY DO WE HAVE UNINTENDED CONSEQUENCES?

Dietrich Dorner has recently analyzed systems in a way that can help us see why they can be so difficult to understand and hence why consequences are unanticipated. Dorner has identified four features of systems which make a full understanding of any real system impossible. These are:

- complexity
- dynamics
- intransparence
- ignorance and mistaken hypotheses

Complexity reflects the many different components which a real system has and the interconnections or interrelations among these components. Our system models necessarily neglect many of these components or features, and even more so their interrelations, but there is always a danger in doing so because it is from such interrelations that the unanticipated may arise. Our economic system is an example of a highly complex system. Not only are there many players, but the players are also interrelated in many ways which are difficult to identify and define. If Player A sets this price, how will Player B respond, and what will Player C think and do when she observes the actions of A and B?

Many devices and systems exhibit dynamics, that is, the property of changing their state spontaneously, independent of control by a central agent in charge of the system. One of the most fascinating examples of our time is the Internet, an extraordinarily dynamic system with no one in charge. There is no way to model the Internet system in a way which will predict its future and the future of the people and things which will be impacted by the Internet. Many of our complex technological systems have this property. Examples might include: a new freeway system, nuclear power, high-definition television, genetic engineering. For example, a freeway system is dynamic because a large number of players initiate actions beyond any central control. Driver A slows down to observe an accident; Driver B responds in an unpredictable way depending on his skills, state of mind, sobriety perhaps, and other factors. The system, although structured to some degree, is in many ways on its own.

Intransparence means that some of the elements of a system cannot be seen, but can nevertheless affect the operation of the system. More complex systems can have many contributors to intransparence. In the Internet, for example, the list would include almost all of the users at a particular time, equipment failures at user sites, and local phenomena, such as weather, which affect use of the Internet at other locations. We need to understand that what you can't see might hurt you.

Finally, ignorance and mistaken hypotheses are always a possibility. Perhaps our model is simply wrong, faulty, misleading. This last problem is particularly interesting and important because it is the one we can do something about. We can take steps to reduce our ignorance and to increase our understanding, as we shall discuss in Section 6. And in Section 7 we argue that we are obliged to do so.

Let's look next at some other perspectives on this problem. Peter Bernstein has addressed the matter from the viewpoint of probabilities and economics. He points out that economists have sometimes believed that deterministic forces drive our societies and their enterprises. More contemporary economists have seen less order. Bernstein puts it this way.

> The optimism of the Victorians was snuffed out by the senseless destruction of human life on the battlefields (of the First World War), the uneasy peace that followed, and goblins let loose by the Russian Revolution. Never again would people accept Robert Browning's assurance that "God's in his heaven: /All's right with the world." Never again would economists insist that fluctuations in the economy were a theoretical impossibility. Never again would science appear so unreservedly benign, nor would religion and family institutions be so unthinkingly accepted in the western world....
>
> Up to this point, the classical economists had defined economics as a riskless system that always produced optimal results....
>
> Such convictions died hard, even in the face of the economic problems that emerged in the wake of the war. But a few voices were raised proclaiming that the world was no longer what once it had seemed. Writing in 1921, the University of Chicago economist Frank Knight uttered strange words for a man in his profession: "There is much question as to how far the world is intelligible at all...It is only in the very special and crucial cases that anything like a mathematical study can be made."

Edward Tenner takes still another perspective on the phenomenon of unanticipated and unintended consequences. He sees in some of our technologies a "revenge effect" in which our perverse technologies turn against us with consequences which exceed the good which had been planned.

> Security is another window on revenge effects. Power door locks, now standard on most cars, increase the sense of safety. But they have helped triple or quadruple the number of drivers locked out over the last two decades—costing $400 million a year and exposing drivers to the very criminals the locks were supposed to defeat.

We shall return to this issue of perversity in Section 6 when we see how society attempts to deal with unintended consequences.

For Dorner on engineering, for Bernstein on economics, for Tenner's perverse technologies, the message is the same. The world is not knowable and predictable. Its complexities are too great, its uncertainties beyond our understanding. Some unanticipated consequences are a necessary feature of all of our enterprises. But this does not mean that we should give up the effort to reduce uncertainty. We shall return to this in Section 7 on ethical implications.

In the next section we turn to some examples of such consequences. Then in Section 6, we consider how society responds to the problem of unanticipated consequences.

5. SOME EXAMPLES

In this section we consider some anecdotes, some cases of consequences which were not anticipated. We follow a historical path in this effort. We have already reached back to a time before the dawn of human history for a story of the invention of writing. Now we jump forward to the last two hundred years, touching on some of the effects of the Industrial Revolution and moving on to questions which are being asked today about newly proposed technologies.

James Beniger's, *The Control Revolution: Technological and Economic Origins of the Information Society*, traces in some detail the evolution of technological development over the past two centuries, particularly in the United States. While Beniger stresses the role and need for control in technology, he does not pay a great deal of explicit attention to consequences. But, the implicit implications of the changes in speed brought on by the Industrial Revolution are clear.

> Speeding up the entire societal processing system...put unprecedented strain on...all of the technological and economic means by which a society controls throughputs to its material economy. Never before in history had it been necessary to control processes and movements at speeds faster than those of wind, water, and animal power—rarely more than a few miles an hour. Almost overnight, with the application of steam, economies confronted growing crises of control throughout the society. The continuing resolution of these crises, which began in America in the 1840s and reached a climax in the 1870s and 1880s, constituted nothing less than a revolution in control technology. Today the Control Revolution continues, engine of the emerging Information Society.

The twentieth century was to bring still another quantum leap in speed with the development of aviation. Perhaps no American is a better metaphor for the growth of technology in this century than Charles A. Lindbergh. Lindbergh's fascination with emerging technologies in the first decades of this century mirrored that of the nation as a whole, although in Lindbergh's case it was tempered by a love of nature.

> I loved the farm, with its wooded river and creek banks, its tillage and crops, and its cattle and horses. I was fascinated by the laboratory's magic: the intangible power found in electrified wires, through which one could see the unseeable. Instinctively I was drawn to the farm, intellectually to the laboratory. Here began a conflict between values of instinct and intellect that was carried through my entire life, and that I eventually recognized as inherent in my civilization.

In 1927, Lindbergh symbolized the triumph of technology when he flew alone across the Atlantic Ocean and electrified the world. But the euphoria did not last, as it never does. Two years later the great depression hit, and the decade to follow saw the rise of Hitler and the terrible destruction brought on by the Second World War, with its technologies so dependent on aviation. Lindbergh began to question the idea of progress.

> Sometimes the world above seems too beautiful, too wonderful, too distant for human eyes to see, like a vision at the end of life forming a bridge to death. Can that be why so many pilots lose their live? Is man encroaching on a forbidden realm?...Will men fly through the sky in the future without seeing what I have seen, without feeling what I have felt? Is that true of all things we call human progress—do the gods retire as commerce and science advance?
>
> As a college youth, I thought civilization could never be destroyed again, that in this respect our civilization was different from all others of the past. It had spread completely around the world; it was too powerful, too universal. A quarter-century later, after I had seen the destruction of high-explosive bombs and flown over the atomic-bombed cities of Hiroshima and Nagasaki, I realized how vulnerable my profession—aviation—had made all peoples. The centers of civilization were the centers of targets.

In the end, Lindbergh found reconciliation between the world of nature and spirit and the world of technology, a balance between what Eliade has called the Sacred and the Profane. He came to see that a balance was essential and that technology is good when it helps to preserve that balance.

> Decades spent in contact with science and its vehicles have directed my mind and senses to areas beyond their reach. I now see scientific accomplishments as a path, not an end; a path leading to and disappearing in mystery...Rather than nullifying religion and proving that "God is dead," science enhances spiritual values by revealing the magnitudes and the minitudes—from cosmos to atom—through which man extends and of which he is composed.

As the undesired consequences of much of twentieth-century technology became evident to Lindbergh, his response was not a rejection of technology, but rather a turning to the fundamental questions of why we are here. After his death, Susan Gray put it this way:

> Of all the man's accomplishments—and they were impressive—the most significant is that he spent most of his life considering and weighing the values by which he should live.

By the second half of the twentieth century, we had become painfully aware that our technologies are not unmixed blessings, that they can have fundamental effects on the way we live. Let's look at some more prosaic examples from the past couple of decades. We'll start with one from Charles Handy's *The Age of Unreason*.

> Microwave ovens were a clever idea, but their inventor could hardly have realized that their effect would ultimately be to take the preparation of food out of the home and into the, increasingly automated, factory; to make cooking as it used to be into a matter of choice, not of necessity; to alter the habits of our homes, making the dining table outmoded for many, as each member of the family individually heats up his or her own meal as and when they require it.

Tenner raises an example which is particularly interesting for two reasons. First, it is not clear which of a number of technologies is causing the unanticipated effects. Second, the issue is intensely political and interpersonal, partially because of the first reason. This is the matter of the effect which various erosion control technologies have on the condition of coastal beaches. In Tenner's words:

> People concerned about the coasts are likely to dispute when and where environmental revenge effects are happening. Whatever more rigorous research may show, it is clear that the shoreline is a zone of chronic technological difficulty. Just as logging and fire suppression alter the forest's composition and fire ecology, compelling more and more vigilance, so beach protection feeds on itself by establishing a new order that needs constant and ever more costly maintenance.

Now let's turn very briefly to two examples of emerging technologies whose major unanticipated consequences we have yet to experience. We shall look at the Internet and at cloning.

Actually, the Internet has already had a very significant impact on human life, involving the ways in which we meet each other, the ways we transact business, the ways we share information, and many more. Still, all of this is surely only the small tip of a huge iceberg, which seems very likely to change our lives in ways which we cannot today imagine. We cannot begin to anticipate the consequences of this technology.

The other technology which has stirred the public imagination in the waning years of this century is the cloning of animals and the possibility that we may eventually be able to clone human beings. There has been no dearth of questions about the future raised by this subject. Here are just a few.

1. Will cloning of human beings change what it means to be human?
2. What good might come from the development of cloning?
3. Should we halt research on cloning animals because it might lead to human cloning?
4. If the government takes no action to control cloning, could that decision be worse than a decision to take some specific action?
5. Is it possible to control cloning effectively?

Each of these questions speaks to the uncertainty inherent in the actions which we might take in this field.

We have surveyed here just a few examples which make concrete the concerns which we may have about the unintended and unanticipated consequences of our actions and our enterprises. In the next section, we ask what society as a whole does in the face of uncertainty.

But before we get to society, we really must say something about the individual. It is clear, but nonetheless worth stating, that each of us often has the opportunity and the right to reject the unanticipated consequences of a technology by refusing to use the technology in ways which have undesirable consequences. Whether a technology is for good or for ill must be our choice. We can use it to enrich our lives or to let our lives lose all meaning. Sometimes a technology is so pervasive that we cannot escape it, but often we have the freedom to choose.

The microwave oven is a good example of a technology whose unanticipated consequences can be rejected if we so choose. We don't always have to accept the fast food approach if we choose not to. The violence of television and the pornography of the Internet are not forced on us. The contribution which the automobile makes to a sedentary life can often be rejected. If we become a slave to our telephone or other like media, it is not the telephone which should accept the blame. Discipline is still a virtue, for ourselves and for our children.

6. SOCIETY'S RESPONSE

In this section, we turn to the question of what individuals and societies do in light of the fact that their actions will have unanticipated consequences. We begin with an expansion of the discussion begun in Section 4 on why we fail to anticipate consequences fully. Then we ask what specific steps can be taken to reduce uncertainty. Finally, we ask how people respond to proposals for new actions and how this helps set the course of our actions.

The first part of this section is based largely on an outstanding study by Robert K. Merton titled "The Unanticipated Consequences of Purposive Social Action." Merton chose his title carefully. His use of the word "unanticipated" rather than "unintended" helped motivate the brief discussion on terminology in Section 2 above. The word "purposive" is meant to stress that the action under study involves the motives of a human actor and consequently a choice among various alternatives.

Merton begins by stating that there had been up to the time of the article (1932) no systematic, scientific analysis of the subject. He surmises that this may be because, for most of human history, we attributed the unexpected to "the gods," or "fate," or divine interference. With the dawning of the Age of Reason, we began to believe that life could be understood. We didn't have to leave it to "the gods." It is curious in this century that optimism, may we say "faith," in human understanding of the complexities of life began to fade. Later we shall see this view espoused by the economist Frank H. Knight in the first half of this century and by another economist, Kenneth Arrow, writing in the second half of the century. In a sense we have come full circle, although today our uncertainty is not generally attributed to "the gods" as such.

Merton cautions us that there are two pitfalls to be aware of in considering actions and consequences. The first is the problem of causal imputation, that is, the matter of determining to what extent particular consequences ought to be attributed to particular actions. The problem is exacerbated by the fact that consequences can have a number of causes. Let's consider an example.

Periodically, the Federal Reserve Board changes the short-term interest rate in an attempt to maintain a balance between the health of the economy and the inflation rate. What makes the problem difficult is that inflation rates over a given period of time are dependent on many factors, including, for example, the short-term interest rate, consumer confidence, employment rates, technological productivity, and even the weather. So, if the inflation rate follows a certain path over a given one-year period, to what extent should we attribute the path to the actions taken by the Federal Reserve?

The second pitfall is that of determining the actual purposes of a given action. Suppose that in a given year the unemployment rate drops from 8% to 6%, and the President claims that this drop was due to a series of social programs pushed by the Administration. How do we know if this was in fact the cause, or at least a major cause, of the result? This is of course an important question because it helps us decide whether this same action in the future may be desirable. It is of course a very difficult question to answer. Merton suggests a test: "Does the juxtaposition of the overt action, our general knowledge of the actor(s), and the specific situation and the inferred or avowed purpose 'make sense'?"

It is clear that the major limitation to the correct anticipation of consequences is our state of knowledge. Our lack of adequate knowledge can be expressed in a number of classes of factors. The first class derives from the type of knowledge which is obtained in the sciences of human behavior. The problem is that such knowledge tends to be stochastic or conjectural in the sense that the consequences of a repeated act are not constant, but rather there is a range of consequences, any one of which may arise. Consider again the case of the Federal Reserve increasing the short-term interest rate by say 0.25%. This action may be repeated a number of different times over the years, with a number of different consequences to the inflation rate. In this sense the consequence is stochastic or random, although probably within a fairly small range. The reason that we do not get a dependable result is that many factors influence the inflation rate, as we saw earlier. We do not know exactly how these other factors will influence the rate, nor do we know how the various factors will interrelate with each other, with secondary effects on inflation.

Another class of factors is error. We may, for example, err in our appraisal of the situation. We may err by applying an action, which has succeeded in the past, to a new situation. This is a particular common mistake. It has been said that we are creatures of habit, and with good cause. Much of our lives are lived repeating the same or very similar actions (eating, driving, walking, etc.), and it is absolutely necessary that we have in place habitual ways of carrying out these actions to accomplish a desired end. And it is natural that we extend our habit of doing things habitually to areas where situations have changed. This is a problem that we must be aware of constantly. We may also err in the selection of a course of action. We may not choose the correct thing to do. We also may not do what we do well. And finally, we may err by paying attention to only one factor affecting the consequences.

Another of Merton's factors which limits our ability to anticipate consequences is what he calls the "imperious immediacy of interest," which refers to situations where the actor's concern with the immediate foreseen consequences excludes consideration of longer-term consequences. One's actions may be rational in terms of immediate results, but irrational in terms of one's long-term interests or goals.

A related phenomenon has to do with one's basic values and a strange twist of fate that sometimes arises. Suppose that one's immediate values call for frugality and hard work, a "Protestant ethic." Such an individual may well end up accumulating a significant amount of wealth and possessions. On the other side of the coin, one whose values call for spending and conspicuous consumption may well end up with little wealth in the long run. This phenomenon is explored in some detail by Stanley and Danko.

The final point which Merton makes is that the very prediction of a consequence becomes a new factor in determining what will ensue as a result of some action. Prediction is a new variable in the complex of factors which lead to consequences. Consider again, for example, the case of the actions of the Federal Reserve Board. Suppose that the short-term interest rate is increased by 1/2 percent and that a major financial leader predicts that this will cause the stock market to drop by 10 percent. This prediction will almost certainly affect, in one way or another, the stock market.

In a beautiful brief biographical essay, the economist Kenneth Arrow adds a further caution for the anticipator of consequences. Arrow believes that "most individuals underestimate the uncertainty of the world." The result is that we believe too easily in the clarity of our own interpretations. Arrow calls for greater humility in the fact of uncertainty and finds in the matter a moral obligation as well.

> The sense of uncertainty is active; it actively recognizes the possibility of alternative views and seeks them out. I consider it essential to honesty to look for the best arguments against a position that one is holding. Commitments should always have a tentative quality.

A related idea has been expressed recently by Stephen Carter, who believes that "integrity" is more than acting out one's convictions. For Carter integrity has three parts:

- discerning what is right and wrong
- acting on what you have discerned
- saying openly that you are acting on what you have discerned

The process of discerning requires an active search for the truth. We are not free just to act on our beliefs. We are obliged as well to actively challenge our beliefs, search for more appropriate beliefs, and adopt them (tentatively, of course) as we find them.

Next we turn to an analysis of ways in which we can reduce the uncertainties which are a part of our complex lives. This discussion is based largely on the pioneering work of Frank Knight in the first decades of this century. Knight argues that we can decrease uncertainty in four ways.

- Increase knowledge.
- Combine uncertainties through large-scale organization.
- Increase control of the situation.
- Slow the march of progress.

Clearly an increase in knowledge can help us reduce uncertainty. We can carry out additional studies, analyses, and experiments. The major problem with this approach is the cost in money and time. It also requires, of course, as we have just seen, the recognition that we do not already have complete knowledge of the situation.

The typical way in which we combine uncertainties through large-scale organizations is with one form or another of insurance. A group of people come together to protect each other against serious loss in case of a catastrophic consequence. The cost is in money and perhaps freedom.

It is often possible to reduce uncertainty through control. For example, the government might attempt to reduce inflation as a consequence of Federal Reserve actions through price controls. The use of controls generally has a monetary cost and a reduction of freedom.

Finally, we can reduce the level of uncertainty by slowing the march of progress. This might seem draconian at first, but in fact we do this all the time when we take time to study a problem, do some more tests, write an environmental report, and take the matter to the planning commission. Sometimes it is even more dramatic. Two days after the news was released about the successful cloning of a sheep, the President of the United States announced a moratorium on federal funding of any cloning work until the matter could be studied further.

There is a final point to be made, which has been noted by a number of people. As much as we would like to reduce much of the uncertainty in life, we would not choose to eliminate all uncertainty. A life in which everything was predictable—was known before the fact—would be a boring life indeed. We have been given a world to live in which is inherently unpredictable. That's the bad news and the good news, all at once.

In the final part of this section, we consider the question of how we respond or react to proposals for change made by others in the light of the uncertainties or the unanticipated consequences of all of our actions. Such proposals might be projects for massive social change—"let's eliminate welfare programs—"or they might be very personal individual decisions—"let's buy that house by the lake." Whether it be in the halls of Congress or the dining room table, when one person makes a proposal, another reacts.

Albert Hirschman has written a fascinating analysis of negative reactions to proposals titled *The Rhetoric of Reaction: Perversity, Futility, Jeopardy*. The latter three terms are types of reaction. Let's look at each in turn.

An argument from perversity says that the opposite will happen from that which you claim. Suppose you suggest that welfare be eliminated in order to save taxpayers' money. An argument from perversity might be that such an action will in fact have just the opposite consequence. If you eliminate welfare, crime will increase, more prisons will have to be built, and the cost to taxpayers will increase, not decrease. The argument which Thamus makes to Theuth in Section 3 is another example of an argument from perversity.

An argument from futility suggests that your action will have no effect on the situation which you are trying to change. You propose a law to regulate pornography on the Internet. An argument from futility would say that your law will not change anything because it will be ignored.

A jeopardy argument claims that your proposal will place in jeopardy some valuable resource. You propose to put in a new freeway along the creek heading up toward Central City. Your opponent claims that such a freeway will ruin beloved old Riverside Park because of the noise and pollution.

7. ETHICAL IMPLICATIONS

Ethics is about what we ought to do. But how do you decide what you ought to do when the outcomes of your actions are uncertain? In this section we consider this problem. There are no simple answers to this very important question, but there are some general principles which we can set forth and which may or may not be applicable in a particular situation. It is almost certainly not the case that all of them would apply in a given situation since there will in general be conflicts between them. The purpose of this section is to state these principles and say a little about their application.

One should take advantage of the opportunities to reduce uncertainty, which are discussed near the end of Section 6, to the extent that the costs permit. We are not obliged to exhaust all of our resources of money or freedom or other to reduce uncertainty because this could easily outweigh the good to be gained from the action. But to the extent that reductions can take place with reasonable cost, it seems morally prudent to do so. The judgment of what is an acceptable cost to bear may be very difficult to establish. In some cases it may be practical to use formal mathematical techniques, such as decision theory, to help arrive at an appropriate cost. In other cases, it may be so difficult to quantify goods and costs as to render such an approach useless. Of course, the effort to reduce uncertainty is not always an individual task, but is often a community effort, with the individual recognizing the richer and more diverse views brought by other individuals, as well as by the community in a collective sense.

All persons should share equally in the benefits of an action or a project, and they should also share equally in the risks due to unanticipated consequences. This is, of course, an ideal since we cannot in general ensure that such equal distribution of benefits and risks is possible. In such cases other principles must be applied.

Persons who do not share in benefits of an action should not as a rule be subject to costs and risks. Justice suggests that burdens should not be borne by those who cannot benefit. While this is a good ideal, it is often difficult to apply in a particular case. If taken to the extreme, it could make it difficult or impossible to take many of our actions. For example, it would forbid the building of a coal-burning power plant on the grounds that the emissions from the plant could affect the environment of the entire globe, including that of some individuals who could not expect to benefit from the electric power. There is a corollary to this principle.

Persons who gain some benefit from an action should be able to choose their level of cost and risk. There are many situations in which some stand to gain a great amount from a project and others to gain relatively little. To the extent possible, each person should be able to choose their level of cost. It is in this way that one's rights may be preserved. Of course, joint projects do not always allow this. Communities almost always form themselves in ways such that the majority is able to be a tyrant to the minority. For example, a community may decide to initiate a flood control project. A given individual may be concerned that there are possible undesired consequences to him from the project. A rights approach should support his concerns, but common good might not.

Projects affecting more than one individual provide the greatest balance of benefits over harms for all involved. This is the utilitarian principle. This of course provides us with a way to approach the flood control project. On the basis of anticipated consequences, we might well decide to go ahead with the project on utilitarian grounds. But it seems particularly important that we try as much as possible to anticipate as many consequences as possible because of the potential threat to the rights of some.

In the assessment of value to others, we ought to recognize that a resource has greater value to one who is poor than to one who is rich. If you give ten dollars to a poor man, you improve his life much more than if you give ten dollars to a rich man. The nonlinearity of wealth should be considered in decisions on actions. And in fact the principle of justice suggests that our first thoughts should be for those who have the least in our society.

We ought to recognize that consequences of an action may extend over the long term and that the effects of such consequences on the actor or on others must not be ignored. It is not right to assume that our actions and their consequences are necessarily limited to "here and now." In fact their effects may extend over great distances, perhaps over the entire earth, and may extend in time for years, decades or perhaps even centuries. We are obliged to take these factors into consideration to the extent that we can.

Persons should face the truism that life is extremely complex and that all positions should be tentative. It is inherently dishonest to assert that one knows what is best when this is not the case, and it almost never is the case, again because of the complexity of life. Implicit in this principle is the requirement on all of us to seek to refine and improve our positions, to speak with humility about the consequences of our actions, and to act, when necessary, with a clear understanding that we do not "own the truth." Each of us has an inherent right to his or her worldview or mental model of life. I do not have the right to assume that my view of the world is correct and your view is incorrect, to assume that "I am right and you are wrong."

8. CONCLUSIONS

We have attempted in this paper to outline some of the salient features of the problem of unanticipated consequences. Because the problem is so common to all of what we do in life, it should not come as a surprise that the study has led to some very general results or positions which we might apply to all of our lives. We close here with a brief list of some of these results.

- Life is very complex, more so than we admit.
- All of our actions have unanticipated consequences.
- We bear a moral obligation to take our positions tentatively, with humility in the light of our ignorance.
- Short-term and long-term values are often different and contradictory.
- Uncertainty can be reduced, but there is always a cost.
- It is desirable to reduce uncertainty, but not to eliminate it.

In the end we are left with a dilemma, which is hardly surprising. We act with uncertainty about the consequences of our acts, and yet we have to act, for even to do nothing is to act, and there will be consequences. Change is an inherent part of life. Part of that change is natural, part is within our control. We have a right to act, but we also have an obligation to accept some level of responsibility for the unanticipated consequences of our actions. That level of responsibility is as hard to define as the unanticipated consequences themselves, but it is there nonetheless.

REFERENCES

Kenneth Arrow, "I Know a Hawk from a Handsaw," in *Eminent Economists: Their Life Philosophies,* Cambridge University Press, Cambridge, 1992.

James Beniger, *The Control Revolution: Technological and Economic Origins of the Information Society,* Harvard University Press, Cambridge, 1986.

Peter Bernstein, *Against the Gods: The Remarkable Story of Risk,* John Wiley and Sons, New York, 1996.

Stephen Carter, *Integrity,* Basic Books, New York, 1996.

Dietrich Dorner, *The Logic of Failure: Why Things Go Wrong and What We Can Do To Make Them Right,* Metropolitan Books, New York, 1989 (English Translation, 1996).

Susan Gray, *Charles A. Lindbergh and the American Dilemma: The Conflict of Technology and Human Values,* Bowling Green State University Popular Press, 1988.

Charles Handy, *The Age of Unreason,* Harvard Business School Press, Boston, 1990.

Albert Hirschman, *The Rhetoric of Reaction: Perversity, Futility, Jeopardy,* Harvard University Press, Cambridge, 1991.

Frank Knight, *Risk, Uncertainty, and Profit,* Houghton Mifflin Company, Boston, 1921.

Howard Margolis, *Dealing with Risk: Why the Public and the Experts Disagree on Environmental Issues,* University of Chicago Press, Chicago, 1996.

Robert Merton, "The Unanticipated Consequences of Purposive Social Action," *American Sociological Review,* Vol. 1, Dec., 1936, pp. 894-904.

Neil Postman, *Technopoly: The Surrender of Culture to Technology,* Vintage Books, New York, 1993.

Thomas Stanley and William Danko, The *Millionaire Next Door: The Surprising Secrets of America's Wealthy,* Longstreet Press, Atlanta, GA, 1996.

Edward Tenner, *Why Things Bite Back: Technology and the Revenge of Unintended Consequences,* Knopf, New York, 1996.

QUESTIONS

1. Provide an example of an anticipated and unanticipated consequence to technology that is not listed in this reading.

2. Create your own detailed definition of technology. How does it differ from the ones presented in your readings?

3. Edward Tenner discusses the "revenge effect" in this reading. Do you agree or disagree with his theory? Formulate an explanation to support the reasoning behind your point of view.

4. Rank the three most important concepts that you learned from Healy's reading. Explain why you believe the ones that you chose are the most important.

The Future of Food: An Introduction to the Ethical Issues in Genetically Modified Foods

MARGARET R. MCLEAN

Let's begin with a pop quiz. True or False?

1. All plants contain genes.
2. Only genetically modified plants contain genes.
3. Plants can be modified to contain animal genes.
4. A tomato containing a jellyfish gene would taste like squid.
5. Genetically modified foods are available at Safeway.
6. I have never eaten a genetically modified food.

The answers are true, false, true, false, true, and . . . . most likely, false. The truth is that we have been eating genetically modified (GM) foods for a decade. About 75 percent of processed food that is produced in the United States contains some GM ingredients. This includes crackers, breakfast cereals, and cooking oils. Almost everything that contains soy or corn—including the nearly ubiquitous high-fructose corn syrup—has been genetically modified.

Humans were modifying crops long before the advent of genetics and "modern" biotechnology. Once humans began to practice settled agriculture some 8,000 years ago, they selected which plants to plant, grow, and harvest—first choosing from the wild and then from cultivated crops. These first farmers chose plants that grew well *and* demonstrated resistance to disease, pests, and shifting weather patterns. Ever since, farmers have bred, crossed, and selected plant varieties that were productive and useful. These age-old techniques can now be complemented, supplemented, and perhaps supplanted by an assortment of molecular "tools" that allow for the deletion or insertion of a particular gene or genes to produce plants (animals and microorganisms) with novel traits, such as resistance to briny conditions, longer "shelf life," or enhanced nutrient content. A change in a plant's genetic sequence changes the characteristics of the plant. Such manipulation of genes—genetic engineering—results in a genetically modified organism or GMO.

Both "traditional" biotechnology and "modern" biotechnology result in crops with combinations of genes that would not have existed absent human intervention. A drought-resistant crop can be developed through "traditional" methods involving crosses with resistant varieties, selection, and backcrossing. Modern biotechnology can speed up this process by identifying the particular genes associated with drought resistance and inserting them directly. Whether developed through traditional or modern means, the resultant plants will resist drought conditions, but only the second, genetically engineered one is a GMO or, if meant for human consumption, a GMF.

Genetic engineering has both sped up the process of developing crops with "enhanced" or new characteristics and allowed for the transfer of genes from one organism to another, even from great evolutionary distances, such as the insertion of a gene from an African frog into rhododendrons to confer enhanced resistance to root rot. Moving genes between species creates transgenic plants and crops.

Importantly, genetic engineering is not the whole of agricultural biotechnology, which also includes techniques such as tissue and cell culture. This conference primarily

This talk was delivered at the conference held at Santa Clara University on April 15, 2005. Margaret R. McLean. Reprinted with permission of the Markkula Center for Applied Ethics at Santa Clara University (www.scu.edu/ethics).

concerns itself with a small piece of agricultural biotechnology, the genetic engineering of food crops.

The most commonly grown GM food crops are those that have been engineered to withstand herbicide spraying (e.g., Roundup Ready soybeans and corn) or to produce substances toxic to insects (e.g., *Bt* corn). Crops that can tolerate herbicides have been an economic success story—approximately 80 percent of the U.S. market in soybeans and cotton is in plants that can withstand the popular herbicide Roundup.

To date, most of the development of GM crops—dubbed "first-generation crops"—has been aimed at benefiting the farmers' bottom line—increasing yields, resisting pests and disease, and decreasing the use of herbicides. Over 80 percent of the soybeans and 40 percent of the corn grown in the U.S. is genetically modified. Worldwide, close to a billion acres are planted in GM crops, mostly corn and soy for animal consumption.

The first GM food produced was the Flavr Savr tomato in 1994, touted for its flavor and long shelf life. Interestingly, the Flavr Savr tomato did not contain an alien gene; rather, a gene normally present in the tomato was blocked so that a normal protein involved in ripening was not produced giving the tomato a longer shelf life and, theoretically, better flavor. It failed to attract consumers.

Despite the tomato's flop, so-called "second-generation" crops will one day line supermarket shelves. These include products such as Monsanto's Roundup Ready soybeans with reduced trans fats and increased heart-healthy mono-unsaturated fats; Syngenta's StayRipe banana, which ripens slowly and has a prolonged shelf life; potatoes and peanuts less liable to trigger life-threatening allergic reactions; and tomatoes that help prevent cancer and osteoporosis (Stokstad, Eric: "Monsanto Pulls the Plug on Genetically Modified Wheat," *Science* 304:1088, 2004; Associated Press: "Americans Clueless about Gene-altered Foods," MSNBC.com, March 26, 2005).

Also in the pipeline are GM crops designed to produce pharmaceuticals, so-called "pharma crops." Last year [2004], the California Rice Commission advised the state Food and Agriculture Department to allow Ventria Bioscience of Sacramento to grow 50 hectares of GM rice near San Diego. Ventria planned to grow two types of rice modified with synthetic human genes—one to make human lactoferrin to treat anemia and the second to produce lysozyme to treat diarrhea (Dalton, Rex: "California Edges towards Farming Drug-producing Rice," *Nature* 428: 591, 2004). Anemia and diarrhea plague children under 5 in developing countries. But the California Food and Agriculture Department denied Ventria's request after rice growers expressed concern that international customers would refuse their rice out of fear of contamination. Earlier this week (4/12/05), brewer Anheuser-Busch threatened to boycott rice from Missouri if Ventria is allowed to set up its "biopharming" practices there. Again, the concern is the potential that the GM rice could cross-pollinate other crops and introduce foreign genes and proteins into the human food chain.

INB Biotechnologies (Philadelphia) is developing a nontoxic anthrax vaccine through the transgenic modification of petunias, causing the plant to manufacture new proteins, which when eaten prompt the development of anti-anthrax antibodies. So, instead of "eat your peas," the imperative will be to "eat your petunias!"

The advent of GM crops provides new opportunities for increasing agricultural production and productivity, enhancing nutritional value, developing and delivering pharmaceuticals and vaccines, and feeding the world. But, it is far from easy sailing for GM foods in light of the public concern for associated risks—risks to human and animal health; risks to biodiversity and the environment—and intermittent consumer outrage at not knowing if "the breakfast of champions" has had a genetic boost or not. GM foods are not labeled as such and the industry game of "I've Got a Secret" has bred distrust among consumers and fuels an inherent skepticism about the safety of GM foods.

A common approach to thinking about the ethics of the genetic engineering of food crops and the appropriate regulatory environment is by evaluating safety and weighing potential risks and benefits.

The risk side of the ledger includes (Food and Agriculture Organization of the United Nations):

1. First are potential risks to the environment and wildlife.
    a. Genes may "escape" and find their way into other members of the species or other species. Imagine the trouble if herbicide-resistant genes found their way into weeds.
    b. GM crops could compete or breed with wild species threatening biodiversity.
    c. Monogenetic crops may not react sufficiently to environmental stresses, posing the danger of a reenactment of Ireland's potato famine.
    d. What are the risks to birds, insects, and other non-target species that come into contact with or consume GM plants?

2. Second are potential risks to human health.
   a. There is the potential that allergy-producing genes will be inserted into unrelated foodstuffs. Since GM foods are not labeled, a person could suffer a potentially fatal allergic reaction, e.g., an allergenic Brazil nut gene was transferred to a soybean variety, but the resultant modified crop was never released to the public.
   b. GM products may inadvertently enter the human food supply as evidenced by the settlement earlier this month [2004] between Syngenta and the U.S. government over the accidental sale of unapproved GM (*Bt*10) corn seed to farmers.
3. Third are potential socio-economic effects.
   a. Small-scale farmers could be negatively impacted by the market dominance of a few powerful seed companies. Some worry about the potential loss of traditional farming practices such as collecting, storing, and replanting seed.
   b. The proprietary nature of biotechnology may slow basic research, and patent protection may hinder the entry of GM foods into developing countries as has been the case with pharmaceuticals.
4. Fourth is the potential risk to public trust generated in part by industry refusal to label GM foods as such.

The benefit side of the ledger stresses:

1. First, there are potential benefits to agricultural productivity through the development of crops more resistant to pests, disease, and severe weather, decreasing the risk of devastating crop failure.
2. Second are potential benefits to the environment including:
   a. Improved productivity could result in more food from less land and a decreasing reliance on the cultivation of marginal land.
   b. Genetically engineered pest and disease resistance could reduce the need for pesticides and other chemicals, thereby decreasing the environmental load and farmer exposure to toxins.
   c. The potential longer shelf life of fruits and vegetables could decrease the gross wastage associated with transportation and storage.
3. Third are potential benefits to human health and well-being.
   a. Genetic engineering could be used to remove genes associated with allergies, e.g., the blocking of the gene that produces the allergenic protein in peanuts.
   b. The insertion of genes into crops, such as rice and wheat, can enhance their nutritional value, e.g., Golden Rice.
   c. Genetic modification could be used to produce healthier foods, e.g., by eliminating trans fats or caffeine for example.
   d. Genetic engineering could be used to develop pharmaceuticals and vaccines in plants, decreasing the risk of adverse reactions and enabling faster vaccination of large populations.

Although weighing risks and benefits is necessary, it is neither easy nor the sole concern in considering the ethics of agricultural biotechnology. Certainly, both human well-being and environmental safety are of primary concern; but our ethical obligations are not discharged solely by a guarantee of some degree of protection from harm, as important as that is. We also must be concerned with justice and the common good—raising concerns about human and environmental sustainability and the just distribution of nutritious food *and* acknowledging the need for thoughtful regulation that addresses necessary human and environmental protections while pursuing benefit. Such a task might well begin with a good dose of humility.

And so, we approach the "future of food" and the questions we have before us today:

- Should we have genetically modified foods?
- And, since we do, how ought they be regulated?
- How do we weigh values and risk in biotechnology?
- And, finally, is the genetic modification of food necessary to relieve world hunger?

QUESTIONS

1. What did you learn about genetically modified organisms from the quiz that started this reading? What was your reaction to the statistic that 75 percent of processed food produced in the United States contains some genetically modified ingredients? How does this information help you create and support an opinion about genetically modified organisms?

2. Do you believe that there should be global acceptance of genetically modified organisms? What cultures and countries oppose GMOs?
3. List and explain the risk factors that genetically modified organisms have on the environment, wildlife, human health, and the economy.
4. List and explain the benefits that genetically modified organisms have on the environment, shelf life for fruits and vegetables, and human health.
5. Do you believe that genetically modified foods should be labeled? Explain the reasons for labeling GMOs.
6. Should genetically modified foods be regulated by the Food and Drug Administration? Explain why you think that humans might experience less risk with regulations.

BOX 1 Genetically Modified Foods Exercise

Locate research that compares and contrasts the use of genetically modified foods. Answer the following questions based on your research.

1. Are genetically engineered foods safe or unsafe? Support your answer with three examples from your research.
2. What are the benefits of genetically engineered foods? What are the problems?
3. List and explain the positive and negative effects of genetic engineering on humans and the environment.
4. What is the process that is used to make a genetically modified plant?
5. Based on your research, do you believe that labeling genetically modified foods will help consumers make more informed decisions about the foods they eat?

13 Whistle-Blowing

LINDA STEVENS HJORTH

The economic and technological triumphs of the past few years have not solved as many problems as we thought they would, and, in fact, have brought us new problems we did not foresee.

HENRY FORD II

Whistle-blowing occurs when people believe that something is wrong with a system, process, or product and feel personally responsible or morally obligated to let officials know about the wrongdoing. In an article entitled "Blowing the Whistle: The Organizational and Legal Implications for Companies and Employees," published in *Psychology Today* (1996), it was pointed out that traditionally, whistle-blowers have long histories of successful employment and a firm belief in their organizations. When they perceive that something within their organization is potentially harmful, they become morally convicted to correcting, changing, or reporting the misdeed. Personal responsibility and concern about others is at the heart of their conviction. Whistle-blowers believe that by reporting the misconduct or hazard, their concerns will not be ignored and the problem will be solved. Whistle-blowers tend to have high ethical standards, strong religious beliefs, and a faith that superiors share the same desire as themselves to correct wrongs.

Whistle-blowing often occurs when an individual believes that decision making by a company or the government may be breaking the law, financially profitable but morally wrong, potentially dangerous, or a case of exploitation of management authority (VAOIG 1998, 1). Some factors upon which people blow the whistle include the following: "misleading promotional efforts to secure public funding; cheap, unreliable designs; testing shortcuts; misrepresented test results or experiments; shoddy manufacturing procedures; intermittently defective products; botched installations; careless or inadequate operational procedures; negligent waste disposal" (McGinn 1991, 158). The reality is that whistle-blowers have strong ethical and moral beliefs that carry them through the rough times they are often subjected to after they report the wrongdoing. It is only through family and peer support systems or strong religious beliefs that they can come to terms with the isolation, demotion, intimidation, and threat of loss of work that they may experience after blowing the whistle (Rechtschaffen 1996, 38).

In fact, "[W]histle-blowers are part and parcel of the corporate culture on which they blow the whistle. They are often rather senior because it is those issuing orders who usually have the most control over and the most knowledge about what is occurring within the corporation." (Koehn 1998, 3). Even though whistle-blowing can be personally costly, it remains an important part of social and personal decisions in the face of precarious conflicts. Individuals have to go through soul-searching and heart-wrenching decisions when making the decision to blow the whistle and when dealing with the ramifications of this decision. The two examples of whistle-blowing described in the next part of the text is (1) the classic case of Roger Boisjoly, the chief engineer for the *Challenger* Space Shuttle, and (2) the case of the "nuclear warriors."

REFERENCES

Koehn, D. (1998). "Whistleblowing and Trust: Some Lessons from the ADM Scandal." *Business Ethics Magazine* Articles. Available at http://condor.depaul.edu/eumes/adm.nun.

McGinn, R. (1991). Presidential Commission on the Space Shuttle *Challenger* Accident, February 25, 1986. *Science, Technology and Society*. Upper Saddle River, NJ: Prentice Hall.

Rechtschaffen, G., and Yardley, J. (1996). Whistleblowing and the Law: Are You Legally Protected if You Blow the Whistle? *Management Accounting* (March):38–41.

VAOIG Hotline. (1998). "Whistleblowing Information." Available at http://www/va/gov/ VAOIG/hotline/wmsue.nun.

Case Study 1

Roger Boisjoly, Chief Engineer at Morton Thiokol

LINDA STEVENS HJORTH

"Whistle-blowing arises from an unsuccessful attempt to achieve change through the chain of command. When issues are critical and management's response is unsatisfactory, employees have little recourse but to circumvent the chain of command. All too often they are punished in a modern version of blaming the messenger."
ROGER BOISJOLY, Chief Engineer At Morton Thiokol

One of the saddest days in the history of America's space program was January 28, 1986, the day that the *Challenger* Space Shuttle exploded. Six astronauts and a school teacher, Christa McAuliffe, died when the shuttle exploded 73 seconds after liftoff. The nation, transfixed and confused, watched the disaster on their television sets, wondering how something that seemed so good could have pivoted so quickly into terror . . . How could it have happened?

On January 24, 1985, Roger Boisjoly, chief engineer at Morton Thiokol, watched the launch of Flight 51-C and noted that the temperature that day was much cooler than it had been during other recorded launches. When he inspected the solid rocket boosters, he found that "both the primary and secondary-ring seals on a field joint had been blackened" from what he thought might be severe hot gas blowby related to low temperatures (Boisjoly and Curtis, p. 9). Boisjoly presented his findings to NASA's Marshall Space Flight Center and received tough questioning from the Flight Readiness Review committee. He decided that he needed further evidence before he could positively link low temperatures with hot gas blowby. He further studied the effects of temperature on the seals of Flight 51-B and found that, in fact, there was a direct correlation between low temperatures and the chance for a catastrophe to result (Boisjoly and Curtis 1987, 9–10).

During the following months, Boisjoly felt frustration that management did not seem to be listening to him about the problem he had discovered. On January 27 (the day before the *Challenger* disaster), the predicted temperature for the launch was 18 degrees Fahrenheit. This low temperature prompted Morton Thiokol, the Marshall Space Flight Center, and the Kennedy Space Center to discuss their concerns about the safety of the launch since no shuttle had ever been launched at a temperature lower than 53 degrees Fahrenheit. Boisjoly again stated his concern that the O-rings on the shuttle's solid rocket boosters would stiffen in the cold and lose their ability to act as a seal, potentially causing a fatal disaster (Schlager 1994, 611).

After the meeting, Boisjoly returned to his office and created the following journal entry:

> I sincerely hope that this launch does not result in a catastrophe. I personally do not agree with some of the statements made . . . stating that SRM-25 is okay to fly tomorrow (Boisjoly and Curtis 1987, 13).

On January 28, 1986, as Boisjoly and his engineering team had predicted, the *Challenger* exploded, and all of those on board died. As the nation mourned, the Rogers Commission was formed to investigate the reason for the explosion. Research after the fact showed that there was "a failure in the joint between the two lower segments of the right Solid Rocket Motor . . . (i.e.) the destruction of the seals that are intended to prevent hot gases from leaking through the joint during the propellant burn. . ." (Presidential Commission 1986). It was a tragic technological disaster that could have been prevented. The seal had leaked because the booster was used at a temperature below its range of safe operation, something that Roger Boisjoly had tried to warn management about (Westrum 1991, 132).

The experience of Roger Boisjoly is representative of what whistle-blowers often face when they make a moral decision to tell the truth. Their truths are based on their own perceptions, values, and morals; the question is why others do not choose to listen or act upon their recommendations.

REFERENCES

Boisjoly R., and Curtis, E. (1987). *Roger Boisjoly and the* Challenger *Disaster: A Case Study in Engineering Management, Corporate Loyalty and Ethics.* Lowell, MA: University of Lowell. (Reprinted from the Proceedings of the ASEM Eighth Annual Meeting, October 1987.).

Presidential Commission on the Space Shuttle *Challenger* Accident. February 25, 1986, Vol.4.

Schlager, N. (1994). *When Technology Fails: Significant Technological Disasters, Accidents and Failures of the Twentieth Century.* Washington, DC: Gale Research.

Westrum, R. (1991). *Technologies and Society: The Shaping of People and Things.* Belmont, CA: Wadsworth Publishing Company.

QUESTIONS

1. What was the main cause of the *Challenger* explosion? What do you think might be different about the respective answers of NASA, Morton Thiokol, and Roger Boisjoly to this question? Use the Internet or sources from the library to provide additional information to support your answer.
2. Pretend for a moment that you are Roger Boisjoly. Your recommendation that the launch be postponed has been overridden. What should you do? Provide a detailed explanation as to why you would or would not pick each of the following options.
   a. Resign your position in protest.
   b. Blow the whistle by calling a local radio or television station.
   c. Tell the astronauts and their families of your concerns.
   d. Do nothing.
3. What is your opinion of Roger Boisjoly's role as a whistle-blower? How would you have handled this situation differently?

Case Study 2

Nuclear Warriors*

ERIC POOLEY

George Betancourt looked up from his desk as George Galatis burst into the office, a bundle of papers under his arm. On that morning in March 1992, the two men—both senior engineers at Northeast Utilities, which operates five nuclear plants in New England—were colleagues but not yet friends. Apart from their jobs and first names, they seemed to have little in common. Betancourt, 45, was extravagantly rebellious—beard, biker boots, ponytail sneaking out the back of his baseball cap—while Galatis, 42, was square-jawed and devout: Mr. Smith Goes Nuclear. But Galatis respected Betancourt's expertise and knew he could count on him for straight answers.

On this day, Galatis wanted to know about a routine refueling operation at the Millstone Unit 1 nuclear plant in Waterford, Connecticut. Every 18 months the reactor is shut down so the fuel rods that make up its core can be replaced; the old rods, radioactive and 250°F hot, are moved into a 40-ft.-deep body of water called the spent-fuel pool, where they are placed in racks alongside thousands of other, older rods. Because the federal government has never created a storage site for high-level radioactive waste, fuel pools in nuclear plants across the country have become de facto nuclear dumps—with many filled nearly to capacity. The pools weren't designed for this purpose, and risk is involved: the rods must be submerged at all times. A cooling system must dissipate the intense heat they give off. If the system failed, the pool could boil, turning the plant into a lethal sauna filled with clouds of radioactive steam. And if earthquake, human error, or mechanical failure drained the pool, the result could be catastrophic: a meltdown of multiple cores taking place outside the reactor containment, releasing massive amounts of radiation and rendering hundreds of square miles uninhabitable.

To minimize the risk, federal guidelines require that some older plants like Millstone, without state-of-the-art cooling systems, move only one-third of the rods into the pool under normal conditions. But Galatis realized that Millstone was routinely performing "full-core off-loads," dumping all the hot fuel into the pool. His question for Betancourt was, "How long has this been going on?"

Betancourt thought for a minute. "We've been moving full cores since before I got here," he said, "since the early '70s."

"But it's an emergency procedure."

"I know," Betancourt said. "And we do it all the time." What's more, Millstone 1 was ignoring the mandated 250-hr. cool-down period before a full off-load, sometimes

© *1996 Time Inc., Reprinted by permission.*

*In spite of the many changes that have been made in the nuclear industry since 1996, the whistle-blowing and ethical issues presented in this article remain as relevant today as they were in 1996.

moving the fuel just 65 hrs. after shutdown, a violation that had melted the boots of a worker on the job. By sidestepping the safety requirements, Millstone saved about two weeks of downtime for each refueling—during which Northeast Utilities has to pay $500,000 a day for replacement power.

Galatis then flipped through a safety report in which Northeast was required to demonstrate to the Nuclear Regulatory Commission that the plant's network of cooling systems would function even if the most important one failed. Instead, the company had analyzed the loss of a far less critical system. The report was worthless, the NRC hadn't noticed, and the consequences could be dire. If Millstone lost its primary cooling system while the full core was in the pool, Galatis told Betancourt, the backup systems might not handle the heat. "The pool could boil," he said. "We'd better report this to the NRC *now*."

Betancourt saw that Galatis was right. "But you do that," he said, "and you're dog meat."

Galatis knew what he meant. Once a leading nuclear utility, Northeast had earned a reputation as a rogue—cutting corners and, according to critics, harassing and firing employees who raised safety concerns. But if Galatis wanted to take on the issue, Betancourt told him, "I'll back you."

So began a three-year battle in which Galatis tried to fix what he considered an obvious safety problem at Millstone 1. For 18 months, his supervisors denied the problem existed and refused to report it to the NRC, the federal agency charged with ensuring the safety of America's 110 commercial reactors. Northeast brought in outside consultants to prove Galatis wrong, but they ended up agreeing with him. Finally, he took the case to the NRC himself,

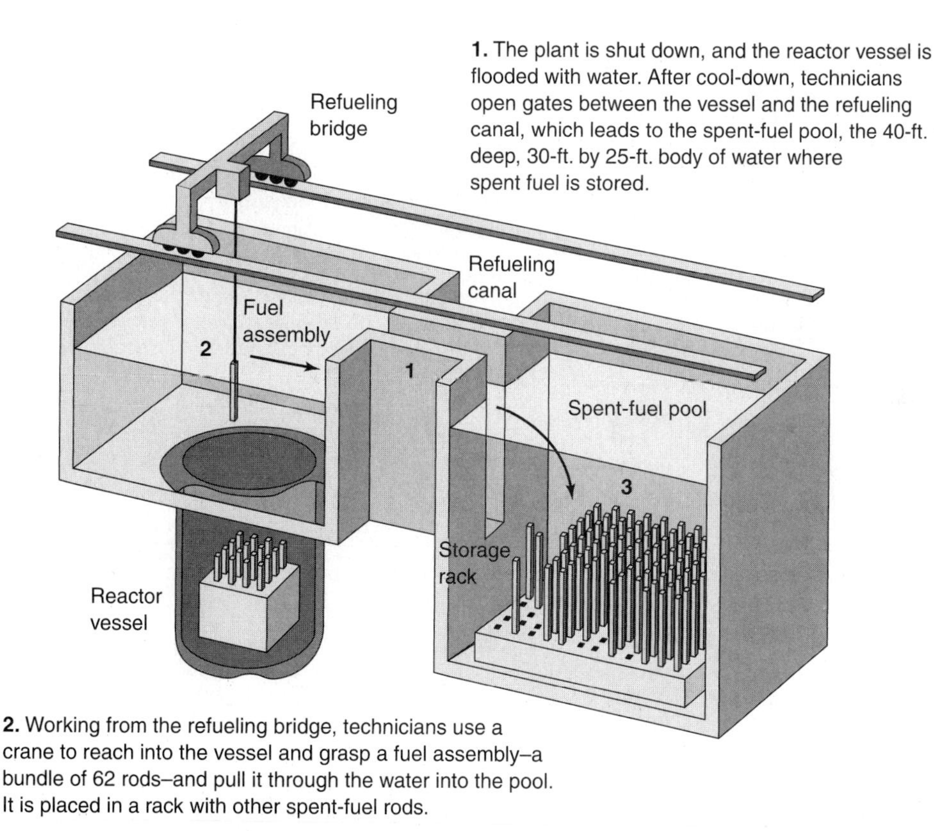

1. The plant is shut down, and the reactor vessel is flooded with water. After cool-down, technicians open gates between the vessel and the refueling canal, which leads to the spent-fuel pool, the 40-ft. deep, 30-ft. by 25-ft. body of water where spent fuel is stored.
Refueling bridge
Refueling canal
Fuel assembly
2
1
Spent-fuel pool
3
Storage rack
Reactor vessel
2. Working from the refueling bridge, technicians use a crane to reach into the vessel and grasp a fuel assembly–a bundle of 62 rods–and pull it through the water into the pool. It is placed in a rack with other spent-fuel rods.

3. The process is repeated for each of the 580 fuel assemblies. When the plant is ready to resume operation, two-thirds of the rods are returned to the core along with 190 fresh assemblies. The rest are stored in the pool for future disposal.

only to discover that officials there had known about the procedure for a decade without moving to stop it. The NRC says the practice is common, and safe—if a plant's cooling system is designed to handle the heat load. But Millstone's wasn't. And when Galatis learned that plants in Delaware, Nebraska, and New Jersey had similar fuel-pool troubles, he realized the NRC was sitting on a nationwide problem.

Ten years after the disastrous uncontained meltdown at Chernobyl, 17 years after the partial meltdown at Three Mile Island, most Americans probably give only passing thought to the issue of nuclear safety. But the story of George Galatis and Millstone suggests that the NRC itself may be giving only passing thought to the issue—that it may be more concerned with propping up an embattled, economically straitened industry than with ensuring public safety. When a nuclear plant violates safety standards and the federal watchdog turns a blind eye, the question arises, How safe are America's nuclear plants?

Though the NRC's mission statement promises full accountability—"nuclear regulation is the public's business," it says—the agency's top officials at first refused to be interviewed by TIME. After repeated requests, Chairwoman Shirley Ann Jackson, a physics professor who was appointed by President Clinton last summer, finally agreed to talk. But the veteran official in charge of the agency's day-to-day operations, executive director James M. Taylor, would provide only written answers to TIME's faxed questions.

"The responsibility for safety rests with the industry," Jackson told TIME. "Like any other regulatory body, NRC is essentially an auditing agency." Jackson argued that her agency is tough—"When we catch problems, it never makes the papers"—but added that with 3,000 employees and just four inspectors for every three plants, "we have to focus on the issues with the greatest safety significance. We can miss things."

In fact, Millstone is merely the latest in a long string of cases in which the NRC bungled its mandate and overlooked serious safety problems until whistle blowers came forward. The NRC's relationship with the industry has been suspect since 1974, when the agency rose from the ashes of the old Atomic Energy Commission, whose mandate was to promote nuclear power. The industry vetoes commission nominees it deems too hostile (two of five NRC seats are vacant), and agency officials enjoy a revolving door to good jobs at nuclear companies such as Northeast. "The fox is guarding the henhouse," says Delaware Senator Joseph Biden, who is pushing legislation to create an independent nuclear safety board outside the NRC. The Democrat, who is also calling for a federal investigation of NRC effectiveness, believes the agency "has failed the public."

It all comes back to money. "When a safety issue is too expensive for the industry, the NRC pencils it away," says Stephen Comley, executive director of a whistle-blower support group called We the People, which has brought many agency failures to light. "If the NRC enforced all its rules, some of the plants we've studied couldn't compete economically."

In a rare point of agreement with activists, the nuclear industry also says regulations threaten to drive some plants out of business, but it argues that many NRC rules boost costs without enhancing safety. "The regulatory system hasn't kept pace with advances in technology," says Steve Unglesbee, a spokesman for the Nuclear Energy Institute, the industry's p.r. unit. "Industry-wide, our safety record is improving. But NRC creates so many layers of regulation that every plant is virtually assured of being in noncompliance with something."

The NRC suggested as much in a 1985 agency directive on "enforcement discretion," which allowed the agency to set aside hundreds of its own safety regulations. Since 1990, Millstone has received 15 such waivers—more than any other nuclear station. In November, Jackson scaled back the policy, but she says this never endangered public safety. Others disagree.

"Discretionary enforcement was out of hand," says NRC acting Inspector General Leo Norton, who investigates agency wrongdoing but has no power to punish. "We shouldn't have regulations on the books and then ignore or wink at them."

Yet the tensions between cost and safety can only increase as deregulation of the nation's utilities ushers in a new era of rate-slashing competition. In some states, consumers will soon choose their electric company the way they now choose a long-distance telephone carrier. Companies with nuclear plants are at a disadvantage because nuclear-generated electricity can cost twice as much as fossil-generated power. No new plants have been ordered in 18 years, and a dozen have been mothballed in the past decade.

For now, however, nuclear power provides 20% of the electricity consumed in the U.S.; New England depends on nuclear plants for more than half its supply. Long-term, says Northeast senior vice president Donald Miller, Millstone and her sisters will survive only "if we start running them like a business [and] stop throwing money at issues." New England's largest power company, with $6.5 billion in assets and $3.7 billion in revenues last year, Northeast is slashing its nuclear work force of 3,000 employees by one-third over the next five years. Company CEO Bernard Fox says the move will not undermine safety.

George Galatis went to work at Northeast Utilities in June 1982 with a degree from Rensselaer Polytechnic Institute and experience with a top manufacturer of nuclear components. At Northeast, he started in the division that oversees the utility's 15 fossil-fuel plants, then moved to the nuclear group, specializing in performance and reliability. Eric DeBarba, Northeast's vice president of technical services, describes him as a solid engineer. "Nobody here ever questioned his honesty or motives," DeBarba says.

Galatis tells it differently. In March 1992 he began working on Millstone 1, one of three nuclear plants perched on a neck of land that juts into Long Island Sound from the shore of southeastern Connecticut. He was checking specifications for a replacement part for a heat exchanger in the spent-fuel cooling system. To order the proper part, he needed to know the heat load. So he pulled a safety report that should contain the relevant data.

But they weren't there.

"The report didn't contain the safety analysis for what we were doing," says Galatis. "No heat-load calculations." It was then he realized the plant had been routinely operating "beyond design basis," putting 23 million BTUs into a pool analyzed for 8 million, which is, he says, "a bit like running your car at 5,000 r.p.m."

Galatis raised the issue with members of Northeast's division of nuclear licensing. "They tried to convince me they had it analyzed," he says. He asked them to produce the documents, and they could not. Galatis sensed trouble when, in later talks, "they began denying that the first discussions had taken place." In June 1992, he spelled out the problem in a memo, calling the fuel pool a license violation and an "unreviewed safety question"—NRC lingo for a major regulatory headache—and adding other concerns he had found, such as the fact that some of the pool's cooling pipes weren't designed to withstand an earthquake, as they were required to do. Northeast sat on the memo for three months until Galatis filed an internal notice-of-violation form, and Betancourt, a leader in the spent-fuel field for years, wrote a memo backing him up.

"When I started in the industry, 20 years ago," Betancourt says, "spent fuel was considered the ass end of the fuel cycle. No one wanted to touch it. Everyone wanted to be on the sexy side, inside the reactor vessel, where the action and danger were. No one noticed fuel pools until we started running out of room in them."

In 1982 Congress mandated that the Department of Energy begin to accept nuclear waste from commercial reactors in 1998. Consumers started paying into a federal fund meant to finance a storage site. Though the Energy Department has collected $8.3 billion, no facility has been completed; in a case of NIMBY writ large, no state wants such a site in its backyard. As the nation's stockpile of spent fuel reached 30,000 tons, activists seized the issue as a way to hobble the industry, and the Energy Department announced that a permanent facility planned for Yucca Mountain, Nevada, wouldn't be ready until 2010; Energy Secretary Hazel O'Leary now puts its chances of opening at no better than fifty-fifty. Bills to create temporary sites are stalled in both houses of Congress.

"Slowly, we woke up to this problem," says Betancourt. The NRC relaxed standards and granted license amendments that allowed plants to "rerack" their rods in ever more tightly packed pools. Sandwiched between the rods is a neutron-absorbing material called Boraflex that helps keep them from "going critical." After fuel pools across the country were filled in this way, the industry discovered that radiation causes Boraflex to shrink and crack. The NRC is studying the problem, but at times its officials haven't bothered to analyze a pool's cooling capacity before granting a reracking amendment. "It didn't receive the attention that more obvious safety concerns got," says Inspector General Norton.

Then, in late 1992, David Lochbaum and Don Prevatte, consultants working at Pennsylvania Power & Light's Susquehanna plant, began to analyze deficiencies in spent-fuel cooling systems. They realized that a problem had been sneaking up on the industry: half a dozen serious accidents at different plants had caused some water to drain from the pools. In the worst of them, at Northeast's Haddam Neck plant in 1984, a seal failure caused 200,000 gallons to drain in just 20 minutes from a water channel next to the fuel pool. If the gate between the channel and the pool had been open, the pool could have drained, exposing the rods and causing a meltdown. Says Lochbaum: "It was a near miss."

The NRC insists that the chance of such an accident is infinitesimal. But the agency's risk-assessment methods have been called overly optimistic by activists, engineers and at least one NRC commissioner. The agency's analysis for a fuel-pool drainage accident assumes that at most one-third of a core is in the pool, even though plants across the country routinely move full cores into pools crowded with older cores. If the NRC based its calculations on that scenario, says Lochbaum, "it would exceed the radiation-dose limits set by Congress and scare people to death. But the NRC won't do it." The NRC's Taylor told TIME that the agency analyzes dose rates at the time a plant opens—when its pool is empty. The law, he said, "does not contain a provision for rereview."

Lochbaum and Prevatte reported Susquehanna to the NRC and suggested improvements to its cooling system. The NRC, Lochbaum says, didn't read the full report. He and Prevatte called Congress members, pushed for a public hearing and presented their concerns to NRC staff. Conceding that Lochbaum and Prevatte "had some valid points," the agency launched a task force and in 1993 issued an informational notice to the 35 U.S. reactors that share Susquehanna's design, alerting them to the problem but requiring no action. One of the plants was Millstone 1.

In 1992, Galatis didn't know about Lochbaum's struggle to get fuel-pool problems taken seriously. He did know he would face resistance from Northeast, where the bonus system is set up to reward employees who don't raise safety issues that incur costs and those who compromise productivity see their bonuses reduced. (Northeast says it has a second set of bonuses to reward those who raise safety issues. Galatis never got one.)

"Management tells you to come forward with problems," says Millstone engineer Al Cizek, "but actions speak louder than words." A Northeast official has been quoted in an NRC report saying the company didn't have to resolve a safety problem because he could "blow it by" the regulators. An NRC study says the number of safety and harassment allegations filed by workers at Northeast is three times the industry average. A disturbing internal Millstone report, presented to CEO FOX in 1991 and obtained by TIME, warns of a "cultural problem" typified by chronic failure to follow procedures, hardware problems that were not resolved or were forgotten, and a management tolerant of "willful [regulatory] noncompliance without justification." The report, written by director of engineering Mario Bonaca, changed nothing. "We've been working at this," says Fox, "but making fundamental change in a complex, technical environment is really hard."

A 1996 Northeast internal document reports that 38% of employees "do not trust their management enough to willingly raise concerns [because of] a 'shoot the messenger' attitude" at the company. In recent years, two dozen Millstone employees have claimed they were fired or demoted for raising safety concerns; in two cases, the NRC fined Northeast. In one, Paul Blanch, who had only recently been named engineer of the year by a leading industry journal, was subjected to company-wide harassment after he discovered that some of Millstone Unit 3's safety instrumentation didn't work properly.

Galatis had watched that case unfold. "George knew what he was getting into," says Blanch. "He knew Northeast would come after him. He knew the NRC wouldn't protect him. And he did it anyway."

In January 1993, Galatis pushed for a meeting with Richard Kacich, Northeast's director of nuclear licensing. Galatis outlined the pool's problems and asked for a consultant, Holtec International, to be brought in. Holtec agreed with Galatis that the pool was an unanalyzed safety question; later the consultant warned that a loss of primary cooling could result in the pool's heating up to 216°F—a nice slow boil.

Galatis sent a memo to DeBarba, then vice president of nuclear engineering, in May 1993. Galatis was threatening to go to the NRC, so DeBarba created a task force to address "George's issues," as they were becoming known. The aim seems to have been to appease Galatis and keep him from going public. DeBarba says the calculations that Holtec and Galatis used were overly conservative and that experience told him there was no problem. The pool hadn't boiled, so it wouldn't boil. If a problem ever developed, there would be plenty of time to correct it before it reached the crisis stage. "We live and work here. Why would we want an unsafe plant? We had internal debate on this topic," DeBarba told TIME. "Legitimate professional differences of opinion." In 1977, he says, the NRC stated, "We could make the choice [of a full-core off-load] if it's 'necessary or desirable for operational considerations.' But that does not mean that what George raised was not an issue. We have rules on this, and we want to get it right."

By October 1993, Galatis was writing to the chief of Northeast's nuclear group, John Opeka, and to Fox, who was then company president. Galatis mentioned the criminal penalties for "intentional misconduct" in dealings with the NRC. Opeka objected to Galatis' abrasive tone but hired another consulting firm, which also agreed with Galatis. Northeast moved on to yet another consultant, a retired NRC official named Jim Partlow.

In December, during a four-hour interview that Galatis calls his "rape case"—because the prosecutor, he says, put the victim on trial—Partlow grilled Galatis about his "agenda" and "motives." After Galatis showed him the technical reports, Partlow changed his mind about Galatis and began questioning Kacich about the apparent violations. In two March 1994 memos to Kacich, Partlow backed Galatis, scolded the utility for taking so long to respond to him and suggested that they should reward Galatis "for his willingness to work within the NU system . . . Let him know that his concern for safety . . . is appreciated."

DeBarba and Kacich created another task force but did not modify the cooling system. Kacich began having conversations with Jim Andersen, the NRC's project manager for Millstone 1, about Galatis' concerns and how to get through the spring 1994 off-load. Andersen, who works at NRC headquarters in Washington, has told the inspector general that he knew all along Millstone was off-loading its full core, but didn't know until June 1993 that it was a problem. Even then he did not inform his superiors. In a bow to Galatis, Millstone modified its off-load procedure, moving all the rods but doing so in stages. Before the off-load, Northeast formally reported to Andersen what he'd known for months: that Millstone might have been operating outside its design basis, a condition that must be reported within 30 days.

During the spring outage, a valve was accidentally left open, spilling 12,000 gallons of reactor-coolant water—a blunder that further shook Galatis' faith. He began to see problems almost everywhere he looked and proposed the creation of a global-issues task force to find out whether Millstone was safe enough to go back online. His bosses agreed. But when the head of the task force left for a golf vacation a few weeks before the plant was scheduled to start up, Galatis says, he knew it wasn't a serious effort. So he made a call to Ernest Hadley, the lawyer who had defended whistle-blower Blanch against Northeast two years before.

An employment and wrongful-termination lawyer, Hadley has made a career of representing whistle blowers, many of them from Millstone. For 10 years he has also worked with Stephen Comley and We the People. Comley, a Massachusetts nursing home operator, is a classic New England character, solid and brusque. He founded We the People in 1986 when he realized the evacuation plans for Seabrook Station, a plant 12 miles from his nursing home, included doses of iodine for those too old and frail to evacuate.

"Some of us were expendable," says Comley. "That got me going." For years he was known for publicity stunts—hiring planes to trail banners above the U.S. Capitol—and emotional outbursts at the press conferences of politicians. The NRC barred him from its public meetings until a judge ordered the ban lifted. But Comley's game evolved: instead of demanding that plants be shut down, he began insisting they be run safely. He teamed up with the sharp-witted Hadley to aid and abet whistle-blowers and sank his life savings into We the People before taking a dime in donations. Comley, says the NRC's Norton, "has been useful in bringing important issues to our attention. Steve can be a very intense guy. I don't think it's good for his health. But people who seem—not fanatical, but overly intense—help democracy work."

In April, 1994, two years after he discovered the problems with Millstone's cooling system, Galatis reported the matter to the NRC. He spoke to a "senior allegations coordinator," waited months, then refiled his charges in a letter describing 16 problems, including the cooling system, the pipes that couldn't withstand seismic shock, the corporate culture. "At Northeast, people are the biggest safety problem," Galatis says. "Not the guys in the engine room. The guys who drive the boat."

Galatis told DeBarba and Kacich that he was going to the NRC. He continued to experience what he calls "subtle forms of harassment, retaliation and intimidation." His performance evaluation was down-graded, his personnel file forwarded to Northeast's lawyers. DeBarba "offered" to move him out of the nuclear group. He would walk into a meeting, and the room would go suddenly silent. DeBarba says he is unaware of any such harassment.

With missionary zeal, Galatis continued to forward allegations to the NRC. Yet four months passed before Galatis finally heard from Donald Driskill, an agent with the NRC's Office of Investigations (the second watchdog unit inside the NRC, this one tracks wrongdoing by utilities). Galatis felt that Driskill was too relaxed about the case. Driskill talked to Northeast about Galatis' charges—a breach of confidentiality that the NRC calls "inadvertent." When Hadley complained to him about Northeast's alleged harassment of Galatis, Driskill suggested he talk to Northeast's lawyer: "He's a really nice guy."

While playing detective—sniffing through file drawers and computer directories—Galitis found items that he felt suggested collusion between the utility and its regulator. Safety reports made it clear that both on-site inspectors and officials from the NRC's Office of Nuclear Reactor Regulation had known about the full-core off-loads since at least 1987 but had never done anything about them. Now, to clear the way for the fall 1995 off-load, NRC officials were apparently offering Northeast what Galatis calls "quiet coaching." One sign of this was a draft version of an NRC inspection report about the spent-fuel pool that had been E-mailed from the NRC to Kacich's licensing department. "What was that doing in Northeast's files?" asks Inspector General Norton.

On June 10, 1995, Jim Andersen visited the site to discuss Galatis' concerns with Kacich's staff. Andersen wouldn't meet with Galatis but huddled with Kacich's team, trying to decide how to bring Millstone's habits into compliance with NRC regulations either by requesting a license amendment—a cumbersome process that requires NRC review and public comment—or by filing an internal form updating the plant's safety reports. This was the easier path, but it could be used only if the issue didn't constitute an unreviewed safety question. Andersen told DeBarba and Kacich that the license amendment "is the cleaner way to go," but they weren't sure there was enough time to get an amendment approved before the next off-load, scheduled for October 1995.

On July 10, Betancourt met with Ken Jenison, an inspector from the NRC's Region 1 office, and gave testimony in support of Galatis' safety allegations. Less than a week later, Betancourt was called to the office of a good-natured human-resources officer named Janice Roncaioli. She complained that he wasn't a "team player," Betancourt says, and ran through the company's termination policies. Rancaioli called Betancourt's account of the meeting "slanted" but would not comment further, citing employee-confidentiality rules.

BOX 1 Near Misses

The Nuclear Regulatory Commission's Office of the Inspector General—a watchdog that can investigate but not punish—has looked into an array of cases in which safety problems were ignored by NRC staff. Some highlights:

- After a 1975 fire knocked out equipment at the Browns Ferry plant in Alabama, the NRC approved a material called Thermo-Lag as a "fire barrier" to protect electrical systems. Between 1982 and 1991, however, the NRC ignored seven complaints about Thermo-Lag; when an engineer testified that fire caused it to melt and give off lethal gases, the NRC closed the case without action. After more complaints and an inspector general's investigation, the NRC "reassessed." Now, it says, "corrective action is ongoing."
- In 1980 workers at Watts Bar 1, a plant then under construction by the Tennessee Valley Authority, flooded the NRC with some 6,000 allegations of shoddy workmanship and safety lapses—enough to halt construction for five years. The NRC breached confidentiality and identified whistle-blowers such as electrical supervisor Ann Harris to the TVA; several were fired. After 23 years and $7 billion, Watts Bar 1 was completed last fall. Although workers say the TVA has abandoned thorough safety inspections in favor of a "random sampling" program, the NRC in February granted an operating license to Watts Bar, the last U.S. nuclear plant scheduled for start-up.
- In the early 1980s, when Northeast Utilities Seabrook Station in New Hampshire was under construction, Joseph Wampler warned the NRC that many welds were faulty. His complaints went unanswered, and he was eventually fired. Blacklisted, he says, Wampler moved to California and revived his career. But in 1991 the NRC sent a letter summarizing Wampler's allegations—and providing his full name and new address—to several dozen nuclear companies. His career was destroyed a second time; he now works as a carpenter. The NRC fined Northeast $100,000 for problems with the welds.
- In 1990 Northeast engineer Paul Blanch discovered that the instruments that measure the coolant level inside the reactor at Millstone 3 were failing. Blanch was forced out, and the problem went uncorrected. In 1993 the NRC's William Russell told the inspector general that the agency had exercised "enforcement discretion," a policy that allows it to waive regulations. Later Russell said the remark had been taken out of context.
- Last December a worker at the Maine Yankee plant in Bath charged that management had deliberately falsified computer calculations to avoid disclosing that the plant's cooling systems were inadequate. The NRC didn't discover this, the Union of Concerned Scientists told reporters, because it didn't notice that Maine Yankee had failed to submit the calculations for review—although they were due in January 1990.
- In 1988 a technician at the Nine Mile Point plant near Oswego, New York, called the NRC with allegations of drug use and safety violations at the plant. The NRC executive director at the time, Victor Stello Jr., took a personal interest in the

matter, but his chief aim seemed to be building a case against Roger Fortuna, the deputy director of the NRC's Office of Investigation, for leaking secrets to the watchdog group We the People. The NRC demanded that We the People head Steve Comley turn over tapes he had allegedly made of conversations with Fortuna. When Comley refused, he was ruled in contempt and fined $350,000 (he still has not paid). The charges against Fortuna were found to be without merit, and when the case came to light—during hearings to confirm Stello as Assistant Secretary of Energy—Stello withdrew his name. "The tension between enforcement and appeasement," a ranking NRC official says, "tugs at this agency every day."

In a July 14 meeting, Jenison, one official who wasn't going to stand for any regulatory sleight-of-hand, told DeBarba and Kacich that if Northeast tried to resolve its licensing problems through internal paperwork alone, he would oppose it. Northeast had to get a license amendment approved before it could off-load another full core, and time was running out. DeBarba and Kacich called on Galatis and Betancourt to help them write the amendment request. The plan included, for the first time, the cooling-system improvements Galatis had been demanding for three years. It was a kind of victory, but he felt disgusted. "The organizational ethics were appalling," he says. "There's no reason I should have had to hire a lawyer and spend years taking care of something this simple."

So Galatis helped Kacich with the amendment request, which was filed July 28. Then he and Hadley drew up another document: a petition that asked the NRC to deny Northeast's amendment request and suspend Millstone's license for 60 days. The petition, filed on behalf of Galatis and We the People, charged that Northeast had "knowingly, willingly, and flagrantly" violated Millstone 1's license for 20 years, that it had made "material false statements" to the NRC, and that it would, if not punished, continue to operate unsafely.

On Aug. 1, Betancourt was called into DeBarba's office; Roncaioli was present, and DeBarba told Betancourt he was being reassigned. "We want to help you, George," Betancourt recalls DeBarba saying, "but you've got to start thinking 'company.' " It was all very vague and, Betancourt thought, very intimidating. On Aug. 3—the day Betancourt was scheduled to meet with the Office of Investigations—Roncaioli called him to her office again. According to Betancourt, she said she wanted to "reaffirm the meaning" of the DeBarba meeting. Betancourt's wife and children began to be worried that he would be fired. "Why don't you just do what they want you to?" his eldest girl asked. Betancourt didn't know quite how to answer. "Your own daughter telling you to roll over," he says.

After Galatis filed his petition, on Aug. 21, he found himself in many of New England's newspapers. As citizen's groups called meetings, Northeast and the NRC assured everyone that the full-core off-load was a common practice that enhanced safety for maintenance workers inside the empty reactor vessel. "We've been aware of how they off-loaded the full core," NRC spokeswoman Diane Screnci told one paper. "We could have stopped them earlier."

At a citizens' group meeting, Galatis met a mechanic named Pete Reynolds, who had left Millstone in a labor dispute two years before. Reynolds shared some hair-raising stories about his days off-loading fuel. He told Galatis—and has since repeated the account to TIME—that he saw work crews racing to see who could move fuel rods the fastest. The competition, he said, tripped radiation alarms and overheated the fuel pool. Reynolds's job was to remove the big bolts that hold the reactor head in place. Sometimes, he said, he was told to remove them so soon after shutdown that the heat melted his protective plastic booties.

Galatis knew that if such things had happened, they would be reflected in operator's logs filed in Northeast's document room. So, on Oct. 6, he appeared in the room and asked for the appropriate rolls of microfiche. The logs backed up what Reynolds had said: Millstone had moved fuel as soon as 65 hrs. after shutdown—a quarter of the required time. The logs noted the sounding of alarms. Galatis wondered where the resident inspector had been.

The deadline came for Millstone's off-load, but the amendment still had not been granted. Connecticut's Senator Chris Dodd, Representative Sam Gejdenson, and a host of local officials were asking about the plant's safety, and Millstone scheduled a public meeting for late October. Senior vice president Don Miller sent a memo to his employees warning them that "experienced antinuclear activists" had "the intention of shutting the station down and eliminating 2,500 jobs." The memo stirred up some of Galatis' colleagues. "You're taking food out of my girl's mouth," one of them told him.

DeBarba assembled a task force to assess what had to be done to get the pool ready for the overdue off-load, but he kept Galatis and Betancourt off the team. The task force came up with six serious problems, most already raised by Galatis. Scrambling to fix the pool in a few weeks, DeBarba hired extra people. The plant shut down, anticipating permission to move fuel.

Galatis and Hadley had been waiting two months for a reply to their petition to deny Northeast's amendment. Finally, on Oct. 26, a letter from William Russell, director of the NRC's Office of Nuclear Reactor Regulation, informed them that their petition was "outside of the scope" of the applicable regulatory subchapter. Two weeks later, the NRC granted Northeast's amendment. Millstone started moving fuel the next morning.

Because of Galatis, the plant is still shut down. "What's especially galling," says Hadley, "is that the NRC ignored my client and denied his motion, then validated his concerns after the fact." In late December, Inspector General Norton released his preliminary report. He found that Northeast had conducted improper full-core off-loads for 20 years. Both the NRC's on-site inspectors and headquarters staff, the report said, "were aware" of the practice but somehow "did not realize" that this was a violation. In other words, the NRC's double-barreled oversight system shot blanks from both barrels. Norton blamed bad training and found no evidence of a conspiracy between Northeast and the NRC to violate the license. He is still investigating possible collusion by the NRC after Galatis came forward. What troubled him most, Norton told TIME, is that agency officials all the way up to Russell knew about the off-loads and saw nothing wrong with them. "The agency completely failed," says Norton. "We did shoddy work. And we're concerned that similar lapses might be occurring at other plants around the country."

In a second investigation, the Office of Investigations is looking into Northeast's license violations and the alleged harassment of Galatis and Betancourt. The intense public scrutiny their case has received will, Galatis says, "make it harder for them to sweep this one under the rug."

On Dec. 12, Russell sent a letter informing Northeast that because "certain of your activities may have been conducted in violation of license requirements," the NRC was considering penalties. In an extraordinary move, Russell demanded a complete review of every system at Millstone 1, with the results "submitted under oath," to prove that every part of the plant is safe—the global examination Galatis asked for two years ago. The results, Russell wrote, "will be used to decide whether or not the license of Millstone Unit 1 should be suspended, modified or revoked."

Now the pressure is on NRC Chairwoman Jackson to prove her commitment to nuclear safety—and her ability to reform an inert bureaucracy. "I will not make a sweeping indictment of NRC staff," Jackson, a straight-talking physicist who in July 1995 became both the first female and the first African American to run the NRC, told TIME. "Does that mean everybody does things perfectly? Obviously not. We haven't always been on top of things. The ball got dropped. Here's what I'm saying now: The ball will not get dropped again."

In response to the problems Galatis exposed, Jackson launched a series of policies designed to improve training, accountability, and vigilance among inspectors and NRC staff. She ordered the agency's second whistle-blower study in two years and a nationwide review of all 110 nuclear plants to find out how many have been moving fuel in violation of NRC standards. The results will be in by April, along with a menu of fuel-pool safety recommendations. (By using a technique called dry-cask storage, utilities could empty their pools and warehouse rods in airtight concrete containers, reducing risk. In the past, the NRC has ruled that the process isn't cost effective.)

Jackson still refuses to meet with Galatis or even take his phone call. "Mr. Galatis is part of an adjudicatory process," she explains. But in a letter turning down Stephen Comley's request that she meet with him and Galatis, Jackson wrote, "The avenues you have been using to raise issues are the most effective and efficient ways. I see no additional benefit to the meeting."

Asked by TIME if she considered three years and two wrecked careers "the most efficient" way to raise the fuel-pool issue, Jackson offered a thin smile. "I'm changing the process," she said. "When all is said and done, then Mr. Galatis and I can sit and talk."

For Galatis, the endgame should have been sweet. On Dec. 20, a Millstone technical manager fired off a frank piece of E-mail warning his colleagues that "the acceptance criteria are changing. Being outside the proper regulatory framework, even if technically justifiable, will be met with resistance by the NRC. Expect no regulatory relief." DeBarba put 100 engineers on a global evaluation of the plant, and they turned up more than 5,000 "items" to be addressed before the plant could go back online. The company announced a reorganization of its nuclear division in which DeBarba and Miller were both promoted. Miller, who told TIME that "complacency" was to blame for the utility's troubles, was put in charge of safety at Northeast's five nuclear plants. On Jan. 29, the NRC, citing chronic safety concerns, employee harassment, "and historic emphasis on cost savings vs. performance," enshrined all three Millstone plants in the agency's hall of shame: the high-scrutiny "watch list" of troublesome reactors. Northeast announced that Millstone would stay down at least through June, at a cost of $75 million. And Standard & Poor's downgraded Northeast's debt rating from stable to negative.

"A hell of an impact," says Betancourt, but "I'm going to lose my job."

"If I had it to do over again," says Galatis, "I wouldn't." He believes his nuclear career is over. (Though still employed by Northeast, he knows that whistle blowers are routinely shut out by the industry.) He's thinking about entering divinity school.

In January, Northeast laid off 100 employees. To qualify for their severance money, the workers had to sign elaborate release forms pledging not to sue the utility for harassment. Four engineers say they were fired in retaliation for their testimony to the NRC four years ago on behalf of whistle-blower Blanch. The company denies any connection between the layoffs and Blanch's case. That makes Blanch chuckle. "The two Georges had better watch their backs," he says. "Up at Northeast, they've got long memories."

Issues for the 21st Century and Beyond

Complete and discuss the following flowchart.

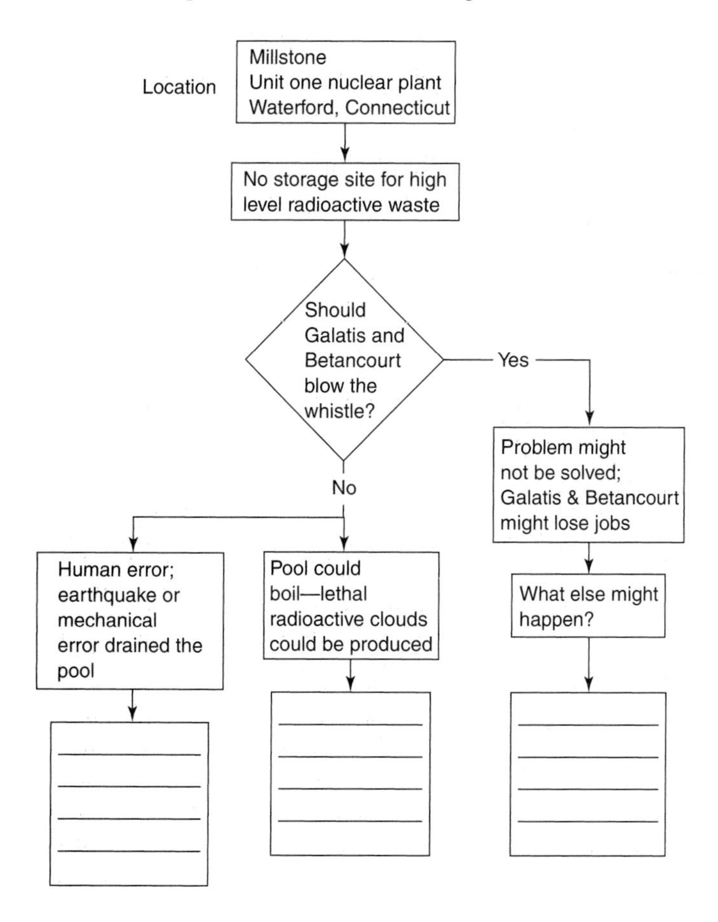

In the end, Galatis believes, the NRC's recent flurry of activity is little more than window dressing. "If they wanted to enforce the law," he says, "they could have acted when it counted—before granting the license amendment. Whatever wrist slap they serve up now is beside the point."

"I believe in nuclear power," he says, "but after seeing the NRC in action, I'm convinced a serious accident is not just likely but inevitable. This is a dangerous road. They're asleep at the wheel. And I'm road-kill."

QUESTIONS

1. What is the solution to the problem of storing nuclear reactive waste?
2. Why was Millstone sidestepping the safety requirements? What would have happened if Millstone lost its primary cooling system while the full core was in the pool?
3. What would you have done if you had been in Galatis's and Betancourt's position? Defend the reasoning for your actions.
4. What part does the NRC play in monitoring the lethality of this situation?

SCENARIO 1: Academic Technological Choices: Ethical? Unethical?

Stevens College: July 8, 20XX

The Integrity Committee at Stevens College is meeting with John, Maria, and Arthur to decide if their behaviors in the classroom warrant disciplinary measures.

John's Case

John's instructor, Dr. Smythe, told the committee: "I was observing John and noticed that he looked perplexed as he struggled to answer test questions. The next thing I knew, he had quickly pulled out his cell phone (trying to hide it under the desk) and started text messaging someone; within seconds, another student in the class pulled out his phone and started sending text messages too." Dr. Smythe talked to both of them and found out that John had text-messaged his friend for the answers to the test questions, and his friend gave him the answers. Once John received answers, he recorded them on his test.

Maria's Case

Maria took Psychology 101 and became frustrated during a test when she could not remember the difference between bipolar disorder and schizophrenia. She picked up her PDA and accessed the Internet, looked up the definitions on Google, and recorded the correct answers. Another student in the class noticed that she was doing this and reported it to Professor DeBoni. Professor DeBoni confronted Maria, and she denied the accusation. The instructor submitted this case to the disciplinary committee because he was sure that Maria had used her PDA to answer the questions, but could not prove it.

Arthur's Case

The English instructor, Mr. Hayes, was grading an essay and thought it might be plagiarized. He compared the essay to an original source on the Internet and found that Arthur had cut and copied the paper word for word without using work cited or in-text source citations. Arthur stated that the paper was his original writing. Mr. Hayes submitted the essay and the original paper from the Internet to the committee.

Response/Action

You are serving on the Integrity Committee. Use Robert McGinn's reading on Ethics to decide which behaviors are ethical or unethical. Apply one ethical theory to each of your answers. Based on your findings, what do you believe the punishment should be for John, Maria, and Arthur?

SCENARIO 2: Intergenerational Justice

Today, 20XX

The life that you lead now will create a legacy for future generations. Is it ethical for future generations to pay emotionally, sociologically, or economically for the decisions that are made in this generation? What actions and choices do you make that could impact the lives of your children and grandchildren? Do we as human beings living in this new century have the right to burden future generations with the choices that we make today? Is it ethically right to expect future generations to clean up the problems that we create today?

Pick two of the following actions and describe how they will impact future societies in positive and negative ways by creating your own scenario. Describe the ethical issues in each scenario.

Actions

- Moving into a new neighborhood without researching the toxicity of the water or air.
- Refusing to recycle or reuse products, believing that the sustainability of the earth is not your concern.
- Failing to check the paint in your house for lead, the basement for radon, or the floors and attic for asbestos.
- Driving large gas-guzzling cars that increase the emission of greenhouse gas and add to global warming.
- Refusing to analyze the part this generation can play in saving energy by buying hybrid cars or assessing potential energy sources such as wind, fuel cell, or hydropower.
- Volunteering time to find ways to preserve land for future generations.

Conclusion

Technology allows us to transport people more quickly, communicate in a variety of formats (e.g., telephone, e-mail, chat room), warm up our food rapidly in the microwave, and access and store data efficiently. Technologies envelop our very existence. As technologies have become a part of everyday life, it is important to ask about their ethical ramifications. For example, what part does our own moral development play in our adult ethical decision making?

Part II has provided theory and practical issues that correlate to moral interactions with technology and within business environments. It has defined technology and explained theories of ethical decision making. It has explained the role that managers, technicians, and individuals play in creating ethical environments in science, technology, and business. The goal of Part II is to ask you to evaluate your role in utilizing technology, always keeping in mind the ethical choices that you make on a daily basis.

The next part of the book will discuss alternative energy forms that may decrease our dependence on oil and open our choices to wind, solar, hydro, and other energy sources.

INTERNET EXERCISES

1. Use any of the Internet search engines, such as Google, to research the following topics:
    a. the definition of whistle-blowing
    b. the definition of cybercrime
    c. the definition of bioethics
2. Using Internet resources, prepare a summary of ethical dilemmas/roles regarding the following people/events:
    a. Roger Boisjoly
    b. Jack Grillum
    c. Daniel Duncan
    d. William LeMessurier
    e. Bhopal
    f. *Challenger* Space Shuttle
    g. DC-10

h. Love Canal

i. Wobuen

j. Times Beach

k. Weapons of mass destruction

l. Enron

m. Erin Brokovich

n. Asbestos in Libby, Montana

"The conscientious effective engineer is a virtuous engineer."
SAMUEL FLORMAN

USEFUL WEB SITES

http://www.cdt.org	Center for Democracy and Technology
http://www.epic.org	Ethical dimensions of current issues
http://www.bhopal.com/facts.htm	Facts about the gas leak in Bhopal, India
http://www.gspjournal.com	Journal of ethical, social, and legal aspects of genomics
http://www.globalethics.org	Web site sponsored by the Institute for Global Ethics. It includes cases, publications, and guidelines with the purpose of finding a global common ground in ethical values.
http://www.ethics.org	Ethics Resource Center. Basic information on writing a code of conduct, plus access to the newsletter *Ethics Today*.
http://www.onlineethics.org	Online Ethics Center for Engineering and Science. Helpline that responds to individuals who face ethical conflicts.

PART
III

Energy

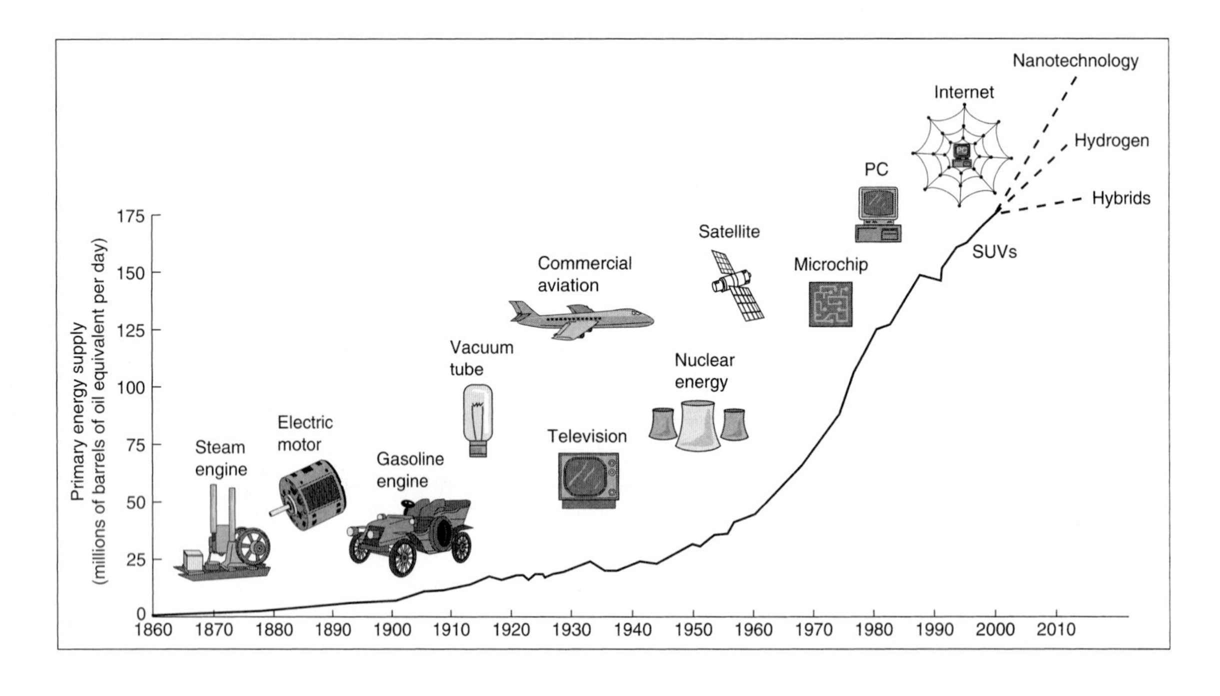
Nanotechnology
Internet
Hydrogen
PC
Hybrids
175
Satellite
SUVs
Microchip
Commercial
aviation
150
125
Vacuum
tube
Nuclear
energy
100
Primary energy supply
(millions of barrels of oil equivalent per day)
Electric
motor
Television
75
Steam
engine
Gasoline
engine
50
25
0
1860
1870
1880
1890
1900
1910
1920
1930
1940
1950
1960
1970
1980
1990
2000
2010

OBJECTIVES

After reading Part III, Energy, *you will be able to*

1. Understand the historical background and major and minor implications of energy sources.
2. Discover various problems, urgencies, and implications of increased use of fossil fuel.
3. Distinguish between developed and developing world's energy consumption issues and potential problems.
4. Analyze and use tables and graphs to promote adoption of appropriate energy policies at the individual, national, and international level for achieving a balance between economic growth and sustainable environment.
5. Understand the impact of renewable and nonrenewable energy sources for a sustainable environment.
6. Examine ethical considerations of increased fossil fuel usage.
7. Realize the importance of the development of new energy sources for a sustainable environment.
8. Appreciate the importance and potential contributions of individual awareness and ethical action to the planning of the future.

INTRODUCTION

The Energy is the go of things.
James Clark Maxwell

Energy is the most basic element of progress for all economies—the pivotal force that sustains life and ensures a standard of living and, ultimately, the standard of life. Energy technologies are society's most basic infrastructure that enables economic growth.

In the pre-Industrial Revolution era, the population of planet Earth was small, and energy needs were limited to cooking and heating. Energy could be exploited without serious damage to the atmosphere, hydrosphere, or geosphere. But the dawn of the Industrial Revolution created an energy-hungry genie whose appetite for hydrocarbons has grown to such a degree that it is jeopardizing future prospects for a sustainable environment (see Figures P3.1, P3.2).

Since the Industrial Revolution, atmospheric carbon dioxide levels have risen from an estimated 280 parts per million to 362 parts per million, the highest in 150 years (see Figure P3.3). The mainstream scientific community, including the Intergovernmental Panel on Climatic Change with 2,500 of the world's leading atmospheric scientists, now finds evidence that human activity is indeed altering the Earth's climate (see Figure P3.4).

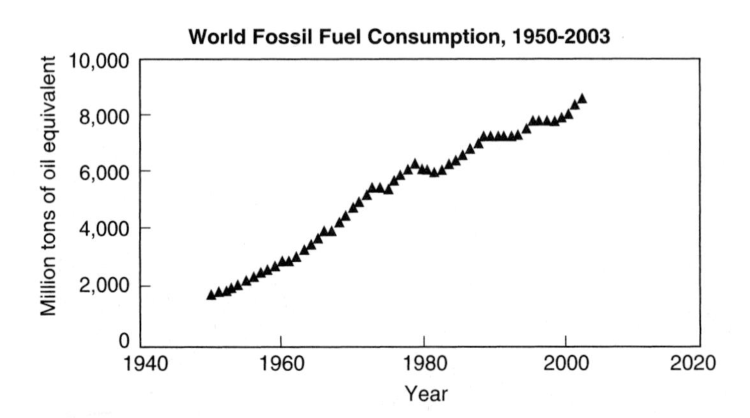

Figure P3.1 World fossil fuel consumption, 1950–2003.

Source: *Signposts 2004*, Worldwatch Institute, Washington, D.C.

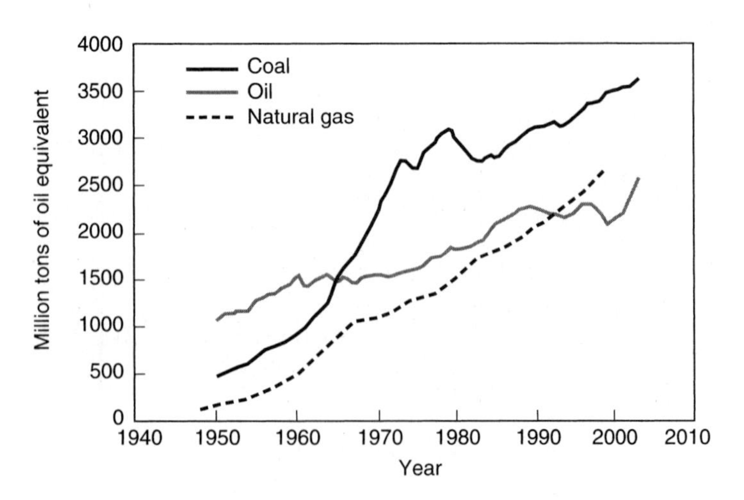

Figure P3.2 World fossil fuel consumption by source, 1950–2003.

Source: *Vital Signs 2003,* Worldwatch Institute, Washington, D.C.

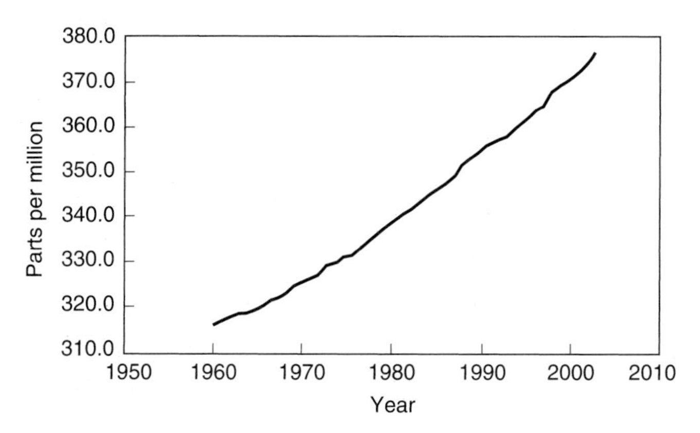

Figure P3.3
Atmospheric concentrations of carbon dioxide, 1960–2003.
Source: *Vital Signs 2003*, Worldwatch Institute, Washington, D.C.

In 1996, worldwide carbon emissions from the burning of fossil fuels climbed to 6.25 billion tons, reaching a new high for the second year in a row (see Figures P3.5, P3.6).

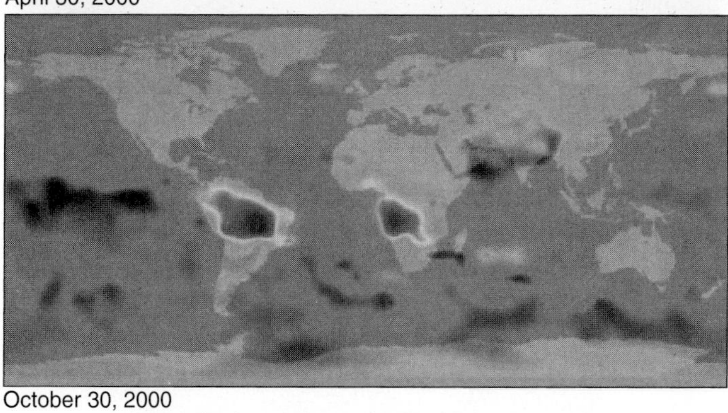

Figure P3.4
Global carbon monoxide concentrations.
Source: National Aeronautics and Space Administration (NASA).

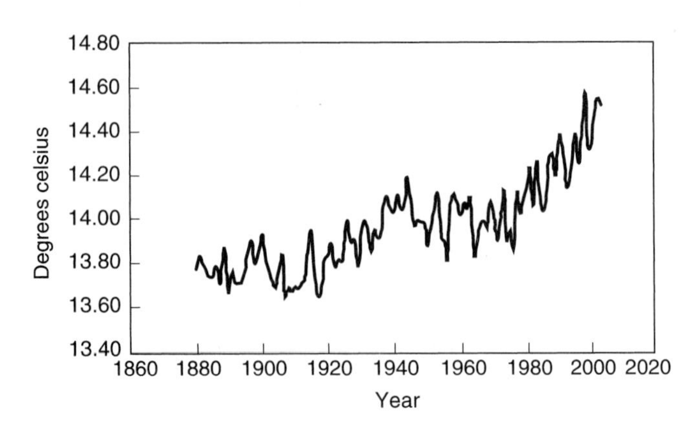

Figure P3.5
Global average temperature at earth's surface (land-ocean index), 1880–2003.
Source: *Vital Signs 2003*, Worldwatch Institute, Washington, D.C.

The world's output of goods and services is growing fast. In 1996, it expanded by 3.8 percent, up modestly from 3.5-percent growth in 1995. The gross world product (GWP) climbed from $26.9 trillion to $28.0 trillion (1995 dollars) over the same period, and GWP per person increased from $4,733 to $4,846. If the world economy keeps a similar growth trend in the 21st century, the increased use of fossil fuels will be catastrophic for the environment of the planet. According to some estimates, the entire biosphere (i.e., the Earth) provides at least $33 trillion worth of free materials every year compared with the GWP of $28 trillion. The economies of the earth would grind to a halt without the services of ecological life-support systems.

Since the Industrial Revolution, the world has become increasingly dependent on fossil fuels. Modern civilization is actually based on nonrenewable resources, which puts a finite

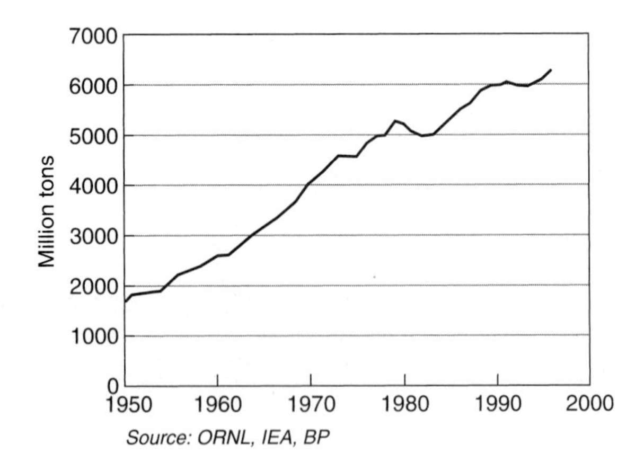

Figure P3.6
World carbon emissions from burning of fossil fuel, 1950–1996.
Source: *Vital Signs 1997*, Worldwatch Institute, Washington, D.C.

limit on the length of time our civilization can exist. The following facts illustrate the magnitude of the energy problems that the world faces today:

- The United States is responsible for almost 25 percent of the world's total energy consumption. Americans use nearly 1 million gallons of oil every two minutes.
- The United States, the largest single source of carbon emissions, is responsible for 24 percent of the emissions of this climate-changing gas (see Figure P3.7).
- The energy currently wasted by U.S. cars, homes, and appliances equals more than twice the known energy reserves in Alaska and the U.S. outer continental shelf.
- Oil production has reached a plateau or declined for 33 of the 48 largest oil producers, including six of the 11 members of the Organization of Petroleum Exporting Countries (OPEC).
- In 2003, world oil consumption increased to 78.1 million barrels a day; by the second half of 2004, it jumped to 82.6 million barrels a day. This increase in oil consumption is a reflection of global economic activity, especially of China and India. China's oil consumption experienced a 26% increase, from 4.95 million barrels per day in 2002 to an estimated 6.24 millions barrels a day in mid 2004.

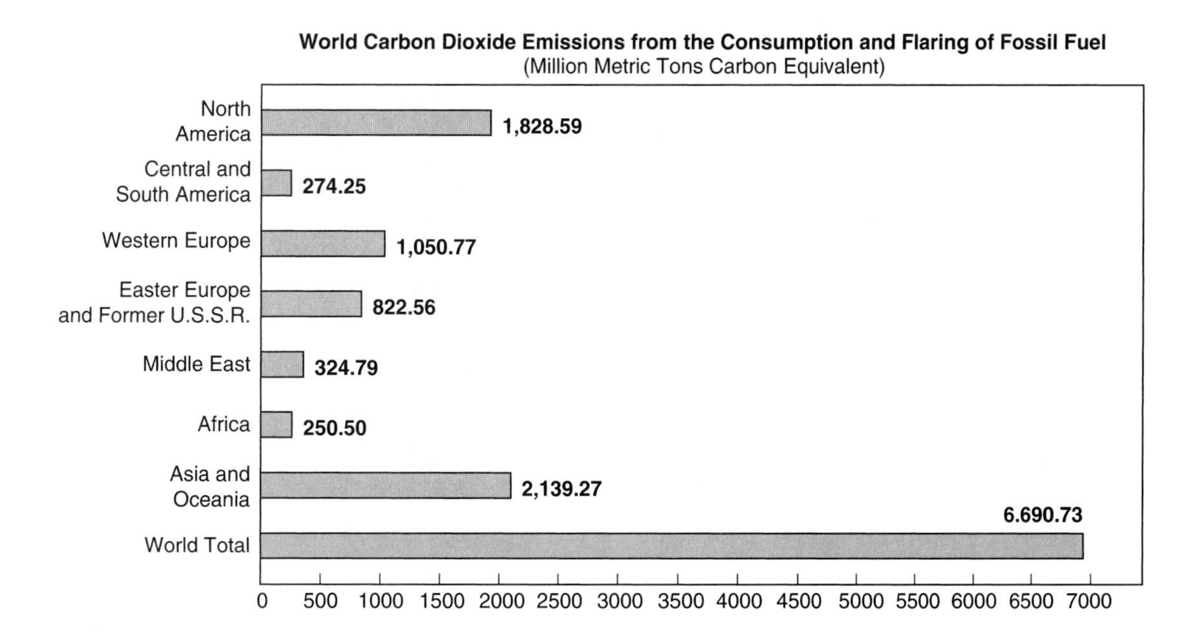

Figure P3.7
Word carbon dioxide emissions from the consumption and flaring of fossil fuels (million metric tons carbon equivalent).
Source: U.S. Department of Energy, http://www.eia.doe.gov/pub/interational/iealf/tableh1.xls

- Americans could cut energy consumption in half by the year 2030 simply by using energy more efficiently and by using more renewable energy sources.
- A human being living at survival level needs about 2,000 kilocalories of energy per day. Americans consume 230,000 kilocalories of energy a day, 115 times the level needed for survival.
- If just 1 percent of America's 243 million vehicle owners (2004) were to tune up their cars, more than a billion pounds of carbon dioxide could be eliminated.
- One-fifth of the world's population now accounts for 70 percent of the globe's energy use.
- China, the world's fastest-growing economy, with 12.7% of world's total, is the second largest emitter of energy-related carbon-dioxide emissions. China's share of world carbon emissions is expected to reach 17.8% by 2025.
- One person in Western Europe uses as much energy as 80 people in sub-Sahara Africa.
- A U.S. citizen uses as much energy as 330 citizens of Bangladesh.
- By the end of 2003, oil consumption in India reached 2.2 million barrels a day. In India, SUV sales represent 10 percent of automobile purchases.
- On July 13, 2006, the global oil price reached $78.40 a barrel, an all-time high. In the United States, experts believe that the gasoline price could reach $5 a gallon within the next few years.
- With only 2 percent of global reserves and 4.5 percent of total production, the United States remains the world's largest oil consumer.
- Cars and gas continue to remain cheaper in the United States because they are not heavily taxed.
- The United States represents 25 percent of current global emissions as well as 36.4 percent of 1990 industrial-country emissions. The March 2001 decision by the United States to withdraw from negotiations on the Kyoto Protocol dealt a blow to

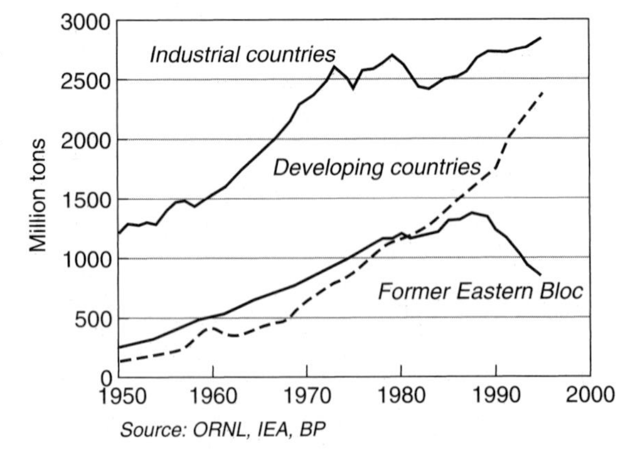

Figure P3.8
Carbon emissions from burning of fossil fuel by economic region, 1950–1996.
Source: *Vital Signs 1997*, Worldwatch Institute, Washington, D.C.

BOX 1 Energy *(noun)*

1. The capacity for vigorous activity.
2. The ability of matter or radiation to do work because of its motion (kinetic energy), its mass (such as released in nuclear fission), or its electric charge.
3. Fuel and other resources used for the operation of machinery.

"The Energy is the go of things."
JAMES CLARK MAXWELL

$E = mc^2$
ALBERT EINSTEIN

The capacity of doing work.
Work is done when a body is moved by a force. The rate of doing work is known as power. Power is energy spent per time unit. The units of energy are as follows:

- Ergs
- Joules
- Calories
- Watts
- Kilowatt hours (kWh)
- British thermal units (Btus)
- Horsepower
- Foot-pounds
- Barrels of oil
- Metric tons of coal

Energy is interchangeable with matter and takes many forms (see Figure P3.9):

Kinetic	Chemical
Potential	Nuclear
Electrical	Sound
Magnetic	Light
Heat	Mass

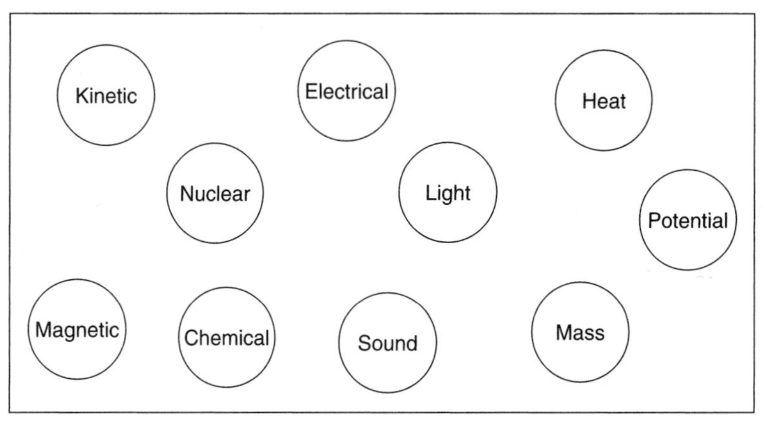

Figure P3.9
Universal set of energy.

The approximate amount of total energy as mass and radiation in the universe is equal to 4.5 X 10^{58} kWh, which is a small fraction of the amount of energy as gravitation in the universe.

The sun's daily output energy is 8.33 X 10^{25} kWh. The earth receives 4.14 X 10^{15} kWh of energy daily.

international efforts to battle climate change, but it also pushed the rest of the world to move forward and reach final agreement on the treaty in July 2001.

- The power of the 16.6 million cars and light trucks sold in the United States in 2003 added up to two and a half times the total U.S. electrical generating capacity.

- Coal has been mined for 800 years, but over half of it has been extracted in the past 37 years.
- Petroleum has been pumped out of the ground for about 100 years, but half of it has been consumed during the past 25 years.
- If the rest of the world used energy at the same rate as the United States, global energy use would be four times as large as it is, and its impact on the environment would be increased manyfold.

The world as a whole could not sustain the rate of energy consumption now enjoyed by the industrial nations (see Figure P3.10). The Third World is trying to industrialize, but the high energy costs are adding to staggering debt loads. A reduction in the use of oil by the industrial nations is essential for world peace and global justice, as well as for environmental protection and long-term sustainability of the planet.

Figure P3.10
World energy by source, 2000.
Source: *Signposts 2004*, Worldwatch Institute, Washington, D.C.

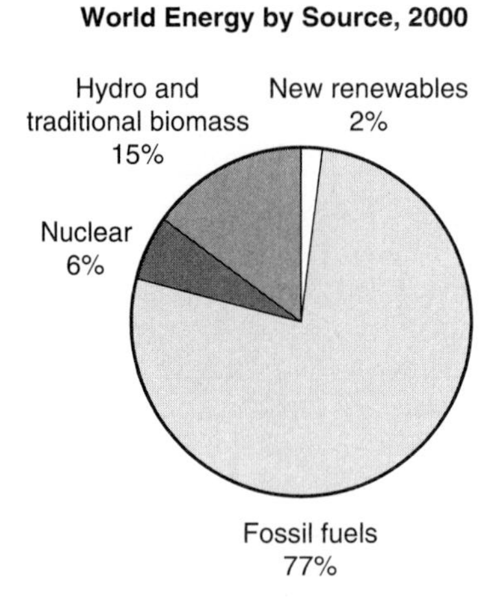

According to some estimates, most of the increases in energy demand will probably occur in the developing world, where population growth rates are high and industrialization and urbanization are under way. In contrast, demand is expected to remain stable or drop in the industrialized countries, where population growth rates are low. Demand could stabilize or decline in eastern Europe and former Soviet republics depending on the success of economic reforms. Much hinges on consensus and whether sustainable policies are enacted.

A report by the World Resources Institute concludes that if efficiency is strenuously pursued in all countries, global energy demand in 2020 will be only 10 percent higher than in 1980, even with the expected growth in population. Energy use in the Northern Hemisphere would have to be cut in half, while that in the Southern Hemisphere would grow, and living standards in the Third World in 2020 could then be comparable to those of Western Europe in the mid-seventies. The use of renewable energy sources (hydroelectric, biomass, wind power, solar thermal, and photovoltaic sources) will contribute to sustainability and justice and, in most cases, the environmental impact will be lower than that of fossil fuels.

As we discuss the energy issue in relation to our future, our challenge will be to support world economic growth and unprecedented energy needs in a sustainable approach, a formula that demands both economic feasibility and preservation of the environment. Part III investigates present energy technologies and concerns plus future strategies that will lead the reader to become aware of both future challenges and their potential solutions.

Man strides over the earth, and deserts follow in his footsteps.

ANCIENT PROVERB

Table P3.1
Units of Conversion Between Energy and Power

1 watt (w) = 1 joule (j) for 1 second
1 kilowatt (kW) = 1,000 watts
1 megawatt (MW) = 1,000 kilowatts (10^6 watts)
1 gigawatt (gw) = 1,000 megawatts (10^9 watts)
1 terawatt (tw) = 1,000 gigawatts (10^{12} watts)
1 twh = 1 billion kWh
1 kW (capacity) = 8,760 kilowatt-hours (kWh; maximum annual production)
1 kWh = 1.34 horsepower-hours (hph)
1 kWh = 3,415 Btu
1 hph = 2,547 Btu
1 hph = 18 ft^3 CH_4 (methane) using internal combustion
1 metric ton coal equivalent = 8,000 kWh (heat equivalent)
1 ton = 2,000 pounds
1 metric ton = 2,205 pounds
1 short ton = .9072 metric tons
1 million metric tons of crude oil of petroleum products = 4×10^{13} Btu
1 million metric tons of coal = 2.8×10^{13} Btu
1000 million m^3 of natural gas = 3.6×10^{13} Btu
1 barrel of crude oil = 5.8×10^6 Btu
1 watt-hour = 3.6×10^3 joules
1 Btu = 1.055107×10^3 joules
1 calorie = 4.184 joules
1 foot-pound = 1.356 joules
1 metric ton coal equivalent (MTCE) = 8,000 kWh (1,000 kWh = .125 MTCE)
1,000 ft hydrogen 79 kWh
1 langley = .21622 kWh per square foot

Fossil Fuel Fundamentals

AHMED S. KHAN
BARBARA A. EICHLER

Energy is the lifeblood of the global economy. Despite tremendous advances in the various technological fields, fossil fuels still remain the number one source of energy, as they were a century ago. The exponential growth of population during the past 50 years, coupled with inadequate research efforts in developing efficient and environment-friendly energy technologies, have led to an increased demand for fossil fuels. The world consumes more than 83 (2004) million barrels of petroleum each day. By 2015, consumption will increase to 99 million barrels per day. Table 1 presents a summary of world crude oil, coal, and natural gas reserves.

Coal, crude oil, and natural gas are fossil fuels. These fuels are considered to be precious resources because they are nonrenewable. Fossil fuels were formed millions of years ago by the decomposition of plants and animal matter. Fossil fuels are burned with oxygen from air to produce heat, which in turn is used in a heat engine to produce mechanical power. The increased use of fossil fuels has led to increased carbon emissions, which are not only endangering the flora and fauna, but are also posing serious health, social, and economic problems.

Due to a lack of technological advances in renewable energy sources, the use of fossil fuels will continue to dominate the world's energy supply in the foreseeable future. Political leaders of the developed world are still more concerned about economic growth without giving serious thought to addressing the issues of acid rain, smog, and global warming. The 2001 United Nations report on global climate change warns that the world's poorest countries would "bear the brunt of devastating changes" from global warming.

ENERGY PRODUCTION AND CONSUMPTION

Most commercial energy is produced from fossil fuels. Developed countries consume high amounts of oil, coal, and natural gas. The United States leads the world in total energy consumption, but in the developed world, Iceland (502.2 million Btu), Luxemburg (435.5 million Btu), Norway (424.2 million Btu), and Canada (418.4 million Btu) are the highest per capita users. With only 5 percent of the world's population, the United States consumes about one-quarter of the world's energy. Developing countries also use one-quarter of the world's energy, but they represent 77 percent of the world's population. Most of the least-developed countries (LDCs) receive a meager energy ration, well below the moderate threshold levels necessary for economic development. They lack resources to buy energy rations to promote development. Poor rural people in developing countries are forced to use wood for their energy needs and thus contribute to the severe problems of deforestation, serious environmental degradation, and pollution.

FOSSIL FUEL ADVANTAGES AND DISADVANTAGES

Fossil fuels have literally fueled highly accelerated growth since the early nineteenth century. Coal, charcoal, oil, and

Table 1
World Crude Oil, Coal, and Natural Gas Reserves, January 1, 2003 (World Oil)

Region/ Country	*Crude Oil (billion barrels)*	*Recoverable Coal (million short tons)*	*Natural Gas (trillion cubic feet)*
North America	45.359	280,464	262.057
Central & South America	75.854	23,977	244.360
Western Europe	17.033	101,343	175.690
Eastern Europe & former U.S.S.R.	81.921	290,183	2,046.953
Middle East	669.757	4,885	2,516.9666
Africa	96.271	61,032	438.875
Asia & Oceania	48.478	322,394	441.731
World total	1034.673	1,081,279	6,126.634

Source: Energy Information Administration, http://www.eia.doe.gov/emeu/iea/.

gas have provided efficient and powerful energy sources that have been fairly easily accessible from the Earth, and they have rapidly changed the technologies and growth patterns of global societies. With these two advantages of accessibility and power efficiency, the world has developed an increasing reliance on fossil fuels despite looming environmental, resource, and political concerns.

In 1850, wood accounted for nearly 90 percent of world energy use. Coal then became more dominant in the 1890s, and wood and coal each accounted for approximately half of the total. In 1910, coal use expanded to about 60 percent and then shrank as oil and natural gas use increased and assumed a more significant role in world energy production and use (Dunn, p. 88). According to Energy Information Administration's (EIA) 2004 statistics, crude oil remains the most heavily used energy source, accounting for 38 percent of global oil energy production. Coal production accounts for 26 percent and natural gas for 23 percent, with hydroelectric power, nuclear power, and renewables accounting for the remaining percentages (www.eia.doe.gov).

Despite the enormous economic and technological growth of the past 200 years—made possible by the use of fossil fuels—most energy experts caution that this fossil fuel mentality must be rethought and changed in the early twenty-first century. There exist urgent concerns regarding supply, rapid world economic growth and energy demands (especially in the LDCs), and heavy air pollution effects with possible global warming.

Although world estimates of fossil fuel supplies differ greatly, Barbour states that "global reserves of oil will last for 44 years, natural gas for 60 years and coal for three centuries at current depletion rates" (Barbour, p. 117). Not only are supplies a problem, but their distribution is very unequal: the Middle East holds approximately two-thirds of the world's oil supply; the former Soviet Union and the United States possesses most of the coal; and the United States, the former Soviet Union, and the Middle East possesses most of the natural gas. Such unequal distribution of resources leads to pricing monopolies, fewer opportunities for the LDCs, and, most especially, political tensions and power struggles when supplies become costly or threatened. Therefore, the central problem is that these fossil fuels are not equally distributed, are not renewable, and cannot support twenty-first-century energy growth projections at their present rates of use. According to *International Energy Outlook 2006* projections, the total world consumption of energy will expand from 421 quadrillion British thermal units (Btu) in 2003 to 563 quadrillion Btu in 2005 and then to 722 quadrillion Btu in 2030, or a 71 percent increase over the 2003 to 2030 period (www.eia.doe.gov).

Another key problem is environmental degradation. Fossil fuels are dirty and contain large quantities of polluting emissions. These cause acidification of rain, lakes, rivers, and soils plus significant air pollution and air particulants. Coal is the worst polluter, followed by oil and then gas (see Table 2).

Barbour states that carbon dioxide (CO_2) emission changes in the atmosphere have been very significant.

Table 2
Emissions from a 1000 Megawatt–Generating Plant (in thousands of tons per year)

Fuel	*Sulfur Oxide*	*Nitrogen Oxide*	*Carbon Dioxide*
Coal	70	25	1600
Oil	30	12	1400
Gas	—	15	1100

Source: United States Environmental Protection Agency (EPA).

The burning of all fossil fuels results in the formation of CO_2, and "the enormous quantities released into the air by fossil fuels have increased the CO_2 content of the air by 25 percent in the past hundred years. At the present emission rate, it will have doubled by 2030. Coal is the worst offender, releasing 24 kilograms of CO_2 per billion joules of heat produced, oil is the next (20 kg.) and natural gas produces least (14 kg.). In the United States, coal-burning utilities account for a third of the CO_2 emitted, oil-burning vehicles another third, and fuels burned in homes and industry, the final third." (Barbour, p. 120) The United States alone is responsible for emitting more than one-fourth of the world's annual emissions of CO_2 (see Table 3).

The world adds about 6 billion tons of CO_2 to the atmosphere from fossil fuel combustion each year, whereas sustainable, stable amounts would be about 1 billion. Concern is that such continual CO_2 concentrations added to the environment will cause severe atmospheric changes and raise global warming temperatures, resulting in problems with climate change, ice melts, and other drastic ecosystem patterns.

As can be seen, these are some of the major disadvantages for continuing the global policy of fossil fuel reliance. There are additional arguments to support this position, but these are some of the fundamental and urgent issues surrounding the continued reliance on a fossil fuel energy and economy framework. In 2001, a group of 250 scientists that included the members of the National Academy of Sciences, in an open letter to the American public, stressed on the role of energy conservation for a sustainable future.

Table 3
World CO_2 Emissions from Consumption and Flaring of Fossil Fuels, 2002

Country/Rank	*Millions of Metric Tons**
1. United States	1,568.59
2. China	906.11
3. Russia	415.16
4. Japan	321.67
5. India	279.87
6. Germany	228.62
7. United Kingdom	150.77
8. Canada	161.37
9. Italy	122.36
10. France	111.08
World total	6,690.73

*2,204.62 pounds equal a metric ton.
Source: *International Energy Annual 2002*, U.S. Department of Energy, Energy Information Administration.

> Conservation must be front and center in our energy future. Unfortunately, energy conservation is painted as a return to the Stone Age, conjuring images of people huddling in the cold of their living rooms in front of lifeless TVs. But in reality, just the opposite is the case. In the last twenty years some of the country's best scientists and engineers have produced great innovations in the efficient use of energy. Cars that get 70 or more miles per gallon, appliances that use half the energy they did ten years ago, lighting fixtures that last for years at a fraction of the energy cost, and new homes that heat and cool with modest amounts of energy are proven winners in energy and economic terms. Just a 3 mile-per-gallon increase in the fuel efficiency of SUVs alone would reduce U.S. oil consumption more than the entire Arctic National Wildlife Refuge could supply. A

Storm brewing over energy resources and environmental effects.
Photo courtesy of Ahmed S. Khan.

study by five national laboratories concluded that a government-led efficiency program emphasizing research and incentives to adopt new technologies could reduce the growth in electricity demand by as much as 47 percent. This would drastically reduce our need to build new power plants.
(http://www.oilcrisis.com/Cleveland/openletter.htm)

THE UNITED STATES DEPENDENCE ON FOSSIL FUEL

In 1993, close to 90 percent of the U.S. energy supply came from coal, oil, and natural gas. As the developing countries continue to industrialize, competition for the world's finite oil supply will increase, resulting in higher prices. U.S. dependence on fossil fuel is evident by the following facts (see Table 4).

- Since 1983, the United States has experienced a steady increase in petroleum consumption. In 2004, the petroleum consumption level reached an all-time high (22.7 million barrels per day).
- The average horsepower of new vehicles has increased steadily by 63 percent (from 102 horsepower [1975] to 166 horsepower [1996]).
- U.S. dependence on imported oil has increased to record levels during the last 25 years. In 2005, net imports provided 48.7 percent of U.S. oil consumption, and the Persian Gulf still supplied one-fifth of U.S. imports.
- In 1972, the American Petroleum Institute (API) estimates of crude oil reserves were 36.3 billion barrels.
- In 1996, the Energy Information Administration (EIA) estimates of crude oil reserves were 22 billion barrels.
- Renewable energy was approximately 8.4 percent of U.S total consumption in 1972. It declined slightly to 7.6 percent in 1997.

Table 4
Energy Consumption in the United States by source 2003 (Quadrillion Btu)

Energy Type	*Consumption Percentage*
Wood (waste, ethanol)	2.93%
Hydroelectric (conventional)	2.83%
Nuclear	8.1%
Coal	23.13%
Natural gas	22.93%
Petroleum	39.8%
Fossil fuel total	85.92%
Renewables total	6.26%

Source: Energy Information Administration, http://www.eia.doe.gov.

An increase in fossil fuel use changed the landscape of the twentieth century.
Photo courtesy of Ahmed S. Khan.

- A new barrel of oil or gas reserves that cost about $15 to find in 1977 cost less than $5 to find today.
- Carbon dioxide emissions grew in all sectors during the past decade. Carbon emissions range between five and six metric tons per person per year.
- In 2001, for the first time in U.S. car sales history, trucks—pickups, sport-utility vehicles, and minivans—outsold cars, taking 50.9 percent of the market.
- Automakers sold 17.2 million cars and trucks in 2001 compared with 17.4 million in 2000 (owners of SUVs receive much higher tax credit than owners of hybrid-electric cars).
- In 2006, global oil prices rose above $78 a barrel, an all-time high. In the United States, experts believe that gasoline prices could rose above $5 a gallon within the next few years.

COAL

Coal is an organic rock present in the Earth's crust. Coal has a very complex molecular structure containing carbon, hydrogen, sulfur, and nitrogen. Based on the presence of these elements, coal is classified into two types: low-rank coal (lignite and subbituminous have low heating value) and high-rank coal (bituminous and anthracite have high heating value). Low-rank coals are roughly 50–100 million years old, and high-rank coals were formed around 350 million years ago. The typical elemental composition of bituminous coal is as follows: 82 percent carbon, 9 percent oxygen, 5 percent hydrogen, 3 percent sulfur, and 1 percent nitrogen. Since bituminous coal has a higher heating value (13,000–15, 000 Btu per pound), most electric power plants prefer it to the low-rank coals. Yet, bituminous coal contains a large amount of sulfur (up to 4 percent), which is converted to sulfur dioxide on combustion and therefore contributes to the problem of acid rain.

U.S. coal reserves are estimated at more than 508 billion short tons of identified resources known as demonstrated reserve base (DRB). Half the coal in the DRB may not be available or accessible for mining. Only about 280 billion short tons of the DRB can be recovered. Table 5 lists global coal reserves and consumption.

Recent advances in ultra-clean fuel technology (coal gasification/liquefaction) can lower the overall amount of greenhouse gases introduced into the atmosphere and reduce dependence on foreign oil. Coal gasification/liquefaction processes utilize carbonaceous matter (coal, coal waste, biomass, refinery waste, and other materials) to produce liquid products that are environmentally friendly, known as ultra-clean fuels.

Table 5
World Coal Reserves and Consumption (2003)

Region /Country	*Total Recoverable Coal (million short tons)*	*Consumption (million short tons)*
North America	279,506	1,182.43
Central & South America	21,928	37.57
Western Europe	36,489	1052.40 (Europe)
Eastern Europe & former U.S.S.R.	279,778	413.30 (Eurasia)
Middle East	462	16.23
Africa	55,486	200.33
Asia & Oceania	327,264	2,795.88
World total	1,000,912	5,698.15

Source: Energy Information Administration, http://www.eia.doe.gov/emeu/iea/.

Advantages

- Coal is one of the most abundant energy resources.
- Coal is versatile. It can be burned directly or transformed into a liquid, gas, or feedstock.
- Coal is inexpensive compared with other energy sources.

Drawbacks

- Coal is a source of pollution. Coal burning emits sulfur dioxide, nitrogen oxide, and particles to the atmosphere and leaves a residue of solid waste.
- Coal mining, especially strip mining, can be very unsightly, and abandoned mines have marred the landscape.
- Coal liquefaction and gasification require large amounts of water.
- Coal is bulky and is therefore more difficult to transport and burn than liquid or gases.
- Coal is porous. Water is trapped in the pores (as much as 30 percent by weight), which reduces the heating value (for low-rank coals, the heating value is less than 8,000 Btu per pound).
- Processes for making liquids and gas from coal are not fully developed.

PETROLEUM (CRUDE OIL)

Petroleum is a complex liquid mixture that contains hundreds of compounds. The majority of these compounds contain carbon and hydrogen and thus have high heating value. A typical elemental analysis of crude oils is as follows: 83–87 percent carbon, 11–16 percent hydrogen, 0–7 percent nitrogen, and 0–4 percent sulfur. The yield of an oil well depends upon the type and age of the oil field. For example, the typical output of an oil well in Saudi Arabia is about 10,000 barrels per day, whereas an oil well in the United States produces an average of 15 barrels per day.

After its recovery from reservoirs in the earth, crude oil is refined by fractional distillation into various fuel products: gasoline (fuel for spark-ignition engines), diesel fuel (fuel for compression-ignition engines), kerosene (fuel for jet engines), and fuel oils (fuel for industrial and residential furnaces). A very small remaining fraction is used to produce petrochemicals (used in such fields as pharmaceuticals, cosmetics, plastics, detergents, textiles). Table 6 presents a summary of world oil production and consumption.

Advantages

- Oil is one of the most abundant energy resources.
- Its liquid form makes it easy to transport and use.
- Oil has high heating value.

Drawbacks

- Oil burning leads to carbon emissions.
- Oil has finite sources (some experts disagree).
- Oil recovery processes need to be developed to provide better yields.

Table 6
World Crude Oil Production and Petroleum Consumption, 2002 (thousand barrels per day)

Region /Country	*Crude Oil Production*	*Petroleum Consumption*
North America	14,497.3	23,842.8
Central & South America	6,791.0	5,238.1
Western Europe	6,595.8	14,698.3
Eastern Europe & former U.S.S.R.	9,592.9	5,256.7
Middle East	21,431.1	5,043.5
Africa	8,082.1	2,674.7
Asia & Oceania	7,940.2	21,452.1
World total	74,930.5	78,206.2

Source: Energy Information Administration, http://www.eia.doe.gov/.

- Oil drilling endangers the environment and ecosystems.
- Oil transportation (by ship) can lead to spills, causing environmental and ecological damage.

NATURAL GAS

Natural gas, also known as methane, is a colorless, odorless fuel that is one of the most commonly used sources of energy today. Natural gas is produced by drilling into the earth's crust where pockets of gas were trapped hundreds of thousands of years ago. One of the biggest advantages of natural gas is that it burns cleaner than many other forms of fossil fuel, such as coal and oil. It can help improve the quality of air and water. Natural gas combustion results in virtually no atmospheric emissions of sulfur dioxide or small particulate matter and produces far lower emissions of carbon monoxide, reactive hydrocarbons, nitrogen oxides, and carbon dioxide than combustion of other fossil fuels (coal and oil). When natural gas (or methane, a molecule made of one carbon atom and four hydrogen atoms) is burned, the principal products of combustion are carbon dioxide and water vapor. The burning of other fossil fuels (oil and coal) produce ash particles, which do not burn and can be carried into the atmosphere. Since the burning of natural gas does not produce reactive hydrocarbons, it can help improve air quality by not producing acid rain and damaging ozone levels. Table 7 presents a summary of world natural gas production and consumption.

The burning of fossil fuels accounts for 75–80 percent of carbon dioxide emissions and 20–30 percent of methane emissions into the environment. Burning of natural gas emits up to 45 percent less carbon dioxide than other fuels. If newer high-efficiency burning techniques are employed, the reduction in carbon dioxide emissions can be nearly 70 percent.

Natural gas can make a significant contribution toward improving air quality if it is employed in the transportation sector. Advanced natural gas–fueled vehicles have the potential to reduce carbon monoxide by as much as 90 percent and reactive hydrocarbon emissions by as much as 85 percent compared with gasoline vehicles. About 130,000 natural gas vehicles are in operation in the United States, with more than 2.5 million in use around the world.

According to the Natural Gas Vehicle Coalition (NGV), compared with most gasoline vehicles, the dedicated natural gas vehicle (NGV) typically reduce emissions CO by 70 percent, non-methane organic gas (NMOG) by 89 percent, and NOx by 87 percent (www.ngvc.org).

Fuel cell technology is one of the most environmentally friendly advances in natural gas technology. NASA first used these cells in the 1960s to generate power in the space capsules. Fuel cells rely on the chemical interaction of natural gas and certain other metals, such as platinum, gold, and other electrolytes, to produce electricity. Their only byproduct is water. The high cost of fuel cell technology has impeded the growth of its implementation. Fuel cells are being used in hospitals to generate power and are also being considered for use in vehicles.

Table 7
World Dry Natural Gas Production and Consumption, 2002 (billion cubic feet)

Region/Country	*Dry Gas Production*	*Consumption*
North America	33,757	27,410
Central & South America	5,910	3,820
Western Europe	12,179	16,427
Eastern Europe & former U.S.S.R.	26,480	24,970
Middle East	11,881	7,862
Africa	9,278	2,554
Asia & Oceania	10,999	12,462
World total	110,485	95,504

Source: Energy Information Administration, http://www.eia.doe.gov/

Advantages

- Natural gas is inexpensive compared with oil.
- Natural gas is clean to burn and less polluting than other fossil fuels.
- Natural gas burning does not produce any ash particles.
- Natural gas has a high heating value (about 24,000 Btu per pound).

Drawbacks

- Natural gas is not a renewable resource.
- Natural gas is a finite resource trapped in Earth (some experts disagree).
- There exists an inability to recover all of the natural gas from a producible deposit because of unfavorable economics and lack of technology.

FUTURE STRATEGIES FOR A SUSTAINABLE ENVIRONMENT

As we approach the end of the cheap fossil fuel age, we must transform our lifestyles from a growth-oriented (infinite resources and energy, local and national outlook) to a balance-oriented lifestyle (finite resources and infinite energy by developing efficient renewable energy technologies, global outlook) by exploring the following strategies.

Conservation Energy conservation should be the central focus of all energy strategies for meeting future energy needs.

Alternate fuels The allocation of funds should be distributed to improve the efficiency of renewable energy technologies, such as solar and wind power.

Development of new technologies Development of high-temperature superconductors could reduce a large amount of losses in the transmission grid and thus provide economical electricity.

Education Teaching the importance of responsible use of energy at personal, national, and international levels should be promoted.

Environment-friendly policies Public policies, which promote conservation and encourage use of alternate fuels, need to be developed and implemented at the national and international level.

An increased consumption of oil in the developed world, coupled with the uneven global distribution of reservoirs, has led to various wars during the last five decades. Unless serious strategies and plans are implemented to develop efficient alternate energy technologies, the deterioration of the environment will continue and the peril of future wars over oil will loom.

In numerous ways, energy has improved our quality of life, but we have paid a heavy price in the form of irreversible damage to the environmental and ecological systems of our biosphere. Earth's atmosphere is a precious environment. In the infinite cosmic ocean, the Earth remains the only planet that sustains life—let's help the Earth to retain its uniqueness.

BOX 1 Oil's Central Place in the Economy

1. Two of the top 10 U.S. corporations by sales (ExxonMobil and Chevron Texaco) and three of the top 20 (ExxonMobil, Chevron Texaco, and ConocoPhillips) are oil companies.
   - In 2002, the top 10 oil companies had revenues of nearly $430 billion.
   - The world automobile fleet grew from 53 million in 1950 to 539 million in 2003.
   - Automobile production surged from 8 million in 1950 to 41 million in 2003.
   - In 2003, more than 2 million cars were sold; by 2010, this number is expected to reach 28 million.

Source: Thomas Prugh, Christopher Flavin, and Janet L. Sawin. "Changing the Oil Economy," *State of the World 2005*. Worldwatch Institute. W.W. Norton., p. 103.

BOX 2 Oil Pipeline in the Amazon Rainforest

Oil pipeline in Ecuador from the Amazon Basin to the Port of Esmeraldas.

Photo courtesy of Mary Jane Parmentier (January 2005).

This pipeline in Ecuador carries oil extracted from the Amazon rain forest to the Pacific coastal port and refineries of Esmeraldas. While Ecuadorian oil is a small percentage of global oil output, it is a major source of revenue for Ecuador. Opinions within the country vary on the economic reliance on oil, ranging from those that maintain it is a key factor for funding economic development to those who decry the environmental damage from spills, impact on indigenous cultures, lack of local benefit from oil revenues, and lack of more sustainable programs for development. It has been estimated that Ecuador's known oil reserves could be depleted in less than 30 years, with some estimates set far sooner, as oil extraction has expanded significantly in recent years.

Sources: Energy Information Administration, http://www.eia.doe.gov/emeu/cabs/ecuador.html, accessed June 26, 2005.
"Ecuador—Oil or Human Rights?", Amnesty International, http://www.amnestyusa.org/business/ecuador.html, accessed June 26, 2005;
"Joint Letter from International Organizations Regarding the OCP Heavy Crude Pipeline in Ecuador, Defending the Amazon, http://www.amazonwatch.org/newsroom/view_news.php?id=646, accessed June 26, 2005.

BOX 3 A 40-gallon barrel of crude oil produces

- 19 gallons of gasoline
- 4 gallons of distillate fuels
- 4 gallons of jet fuel
- 3 gallons of residual (diesel and heating) fuels
- 5 gallons of other byproducts including oil, grease, wax, and asphalt.

BOX 4 Offshore Drilling in Arabian Sea

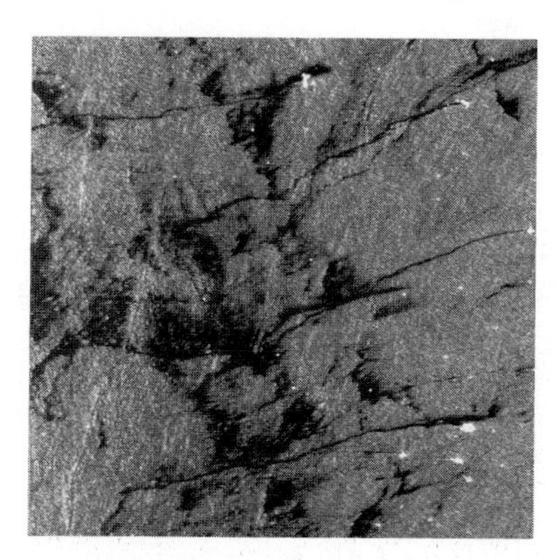

Space radar image of oil slicks in an offshore drilling field about 150 km (93 miles) west of Bombay, India, in the Arabian Sea. The dark streaks are extensive oil slicks surrounding many of the drilling platforms, which appear as bright white spots.
Source: NASA, http://visibleearth.nasa.gov/view_rec.php?id=466.

REFERENCES

"An Open Letter to the American People." (2001). *Scientists for a Sustainable Energy*. Accessed on September 5, 2006: http://www. oilcrisis.com/Cleveland/openletter.htm

Barbour, Ian. (1993). *Ethics in an Age of Technology*. New York: Harper Collins Publishers.

Blair, Cornelia, Alison Landes, and Nancy Jacobs (eds). (1997). *Energy: An Issue of the 90's*. Wylie, TX: Information Plus.

"COAL: The Fuel of America's Industrialization, The Fuel of America's Future." Accessed February 9, 2002. Available at http://www.ultracleanfuels.com/html/about.html

de Souza, Anthony F. (1990). Resources. *Geography of World Economy*. Columbus, OH: Merrill Publishing.

Dunn, Seth. (2001) "Decarbonizing the Energy Economy." In Lester R. Brown, Christopher Flavin, and Hilary French, et.al. *State of the World, 2001*. New York: W. W. Norton & Company.

The Environmental Impact of Natural Gas." Accessed September 5, 2006. Available at http://www.naturalgas.org/environment/naturalgas.asp.

Eldridge, Earle. (2002) "2001 Car Sales Rank Second Best Ever." *USA TODAY*, 01/04/2002. Accessed September 5, 2006. Available at http://www.usatoday.com/money/autos/2002/01/04/autos.htm

"Fossil Fuels." Accessed February 9, 2002. Available at http://www.ems.psu.edu/~radovic/fossil_fuels.html.

"International Energy Outlook 2006." Energy Information Administration. Accessed September 5, 2006. Available at http://www.eia.doe.gov/oiaf/world.html

"Natural Gas Vehicles: The Environmental Solution Now." (2003). The Natural Gas Vehicle Coalition. Accessed September 6, 2006. Available at http://www.ngvc.org/ngv/ngvc.nsf/bytitle/environmentalbenefits.html

"Overview of Natural Gas." Accessed September 5, 2006. Available at http://www.naturalgas.org/overview/overview.asp

"UN Issues Global Warning: Rich Countries Must Cut Fossil Fuels." [February 19, 2001]. Accessed February 9, 2002. Available at http://www.foe.org.au/pr/190201.htm

World Resources 1998–89, A Report by Resources Institute. New York: Basic Books.

QUESTIONS

Use any of the Internet search engines (e.g., Google, Alta Vista, Yahoo, Infoseek) in Questions 1 and 2 to research the following information.

1. Define the following terms.
   a. Bcf
   b. Brine
   c. Btu
   d. CFCs
   e. Fuel cell
   f. Hydrocarbon
   g. Lithography
   h. Mcf
   i. Methane
   j. NGV
   k. NES
   l. Lignite
   m. Subbituminous
   n. Bituminous
   o. Anthracite
2. Research statistics for carbon dioxide emissions from the consumption and flaring of fossil fuels for the following countries.
   a. Canada
   b. United States
   c. Mexico
   d. Chile
   e. Cuba
   f. France
   g. Germany
   h. Ireland
   i. United Kingdom
   j. Turkey
   k. Former U.S.S.R
   l. Oman
   m. Gabon
   n. Zimbabwe
   o. Pakistan
   p. India
   q. Bangladesh
   r. Vietnam
3. What effect does burning fossil fuels have on the environment? What are various strategies to reduce global warming?
4. Refer to Table 8. Estimate how much you could realistically reduce your personal use of electricity and, hence, fossil fuel emissions on a yearly basis. Understand the urgency to help to reduce emissions and try to apply that estimate.
5. Discuss the pros and cons of owning a SUV.

Table 8

Ten Steps You Can Take To Reduce Global Warming

The United States, with 5 percent of the world's population, generates about 25 percent of the "greenhouse gases" that are polluting the atmosphere. This is about 40,000 pounds of CO_2 per person each year. These are some measures you can take to reduce your consumption of fossil fuels, along with the amount of CO_2 you will avoid releasing into the atmosphere.

What You Can Do	*Estimated Pounds of CO_2 Saved per Year*
1. Run your dishwasher with a full load and use the air option rather than heat to dry the dishes.	200
2. Wash clothes in warm or cold water, not hot.	Up to 500
3. Ask your utility company for a home energy audit to find out where your home is poorly insulated or energy inefficient.	Could cut thousands of pounds per year
4. Plant trees next to your home; paint your home a light color if you live in a warm climate and a dark color if you live in a cold climate.	Up to 10,000 pounds per year
5. As you replace home appliances, select the most energy efficient.	Could cut thousands of pounds per year
6. Buy and replace your incandescent light bulbs with compact fluorescent bulbs.	500 pounds per year for each bulb
7. Choose a car with good gas mileage.	2,500 for each 10-mpg improvement
8. Buy minimally packaged goods; choose reusable products over disposable ones; recycle.	1000 if waste is reduced by 25 percent
9. Leave your car at home twice a week (walk, bike, or use mass transit instead).	Up to 1590 pounds per year
10. Turn down your water heater thermostat; 120 degrees is usually hot enough.	500 pounds per year for each 10-degree adjustment

Source: Environmental Defense Fund.

6. List the advantages (+) and disadvantages (–) of coal.

(+)	*(–)*

7. List the advantages (+) and disadvantages (–) of natural gas.

(+)	(–)

8. List the advantages (+) and disadvantages (–) of petroleum (crude oil).

(+)	(–)

9. How can we avoid a future oil crisis and oil wars? List and explain three ways.

BOX 5 Transportation in the Philippines: Necessity Is the Mother of Invention (Plato)

Overloaded passengers in a tricycle, Iloilo, Philippines (Fare for the ride: 1 peso per mile, 55 pesos = 1$)
Photo courtesy of Antonio Abayon.

Tricycles are a common mode of transportation in the Philippines.
Photo courtesy of Antonio Abayon.

An overloaded tricycle on the outskirts of Manila, Philippines.
Photo courtesy of Antonio Abayon.

BOX 6 Best Sellers in the Auto Industry: Enhanced Comfort at the Cost of Increased Carbon Emissions?

To err and not to reform, this indeed is an error. (Confucius)

Since 2001, the sale of trucks, pickups, minivans, and sport utility vehicles (SUVs) account for more than 50 percent of total automobile industry sales.
Photo courtesy of Ahmed S. Khan.

Energy for a New Century

CHRISTOPHER FLAVIN

The stone age did not end because the world ran out of stones, and the oil age will not end because we run out of oil.
DON HUBERTS, Shell Hydrogen (Division Of Royal Dutch Shell)

The age of oil has so dominated social and economic trends for the last 100 years that most of us have a hard time imagining a world without it. Oil is cheap, abundant, and convenient—easy to carry halfway around the world in a supertanker or across town in the tank of a family sport utility vehicle. From Joe Sixpack to the PhD energy economists employed by governments and corporations, we tend to assume that we will burn fossil fuels until they're gone and that the eventual transition will be painful and expensive.

But if you turn the problem around, our current energy situation looks rather different: from an ecological perspective, continuing to depend on fossil fuels for even another 50 years—let alone the century or two it might take to use them up—is preposterous. As the new century begins, the world's 6 billion people already live with the dark legacy of the heavily polluting energy system that powered the last century. It is a legacy that includes impoverished lakes and estuaries, degraded forests, and millions of damaged human lungs.

Fossil fuel combustion is at the same time adding billions of tons of carbon dioxide to the atmosphere each year, an inexorable escalation that must end soon if we are not to disrupt virtually every ecosystem and economy on the planet.

An energy transition in the new century is therefore ecologically necessary, but it is also economically logical. The same technological revolution that has created the Internet and so many other 21st-century wonders can be used to efficiently harness and store the world's vast supplies of wind, biomass, and other forms of solar energy—which is 6,000 times as abundant on an annual basis as the fuels we now use. A series of revolutionary technologies, including solar cells, wind turbines, and fuel cells, can turn the enormously abundant but diffuse flows of renewable energy into concentrated electricity and hydrogen that can be used to power factories, homes, automobiles, and aircraft.

These new energy conversion devices occupy about the same position in the economy today that the internal combustion engine and electromagnetic generator held in the 1890s. The key enabling technologies have already been developed and commercialized, but they only occupy small niche markets—and their potential future importance is not yet widely appreciated. As with the automobile and incandescent light bulb before them, the solar cell and hydrogen-electric car are steadily gaining market share—and may soon be ready to contribute to a third energy revolution. They could foster a new generation of mass-produced machines that efficiently and cleanly provide energy needed to take a hot shower, sip a cold beer, or surf the Internet.

Thanks to a potent combination of advancing technology and government incentives, motivated in large measure by environmental concerns, the once glacial energy markets are now shifting. During the 1990s, wind power has grown at a rate of 26 percent per year, while solar energy has grown at 17 percent per year. During the same

---

Reprinted with permission from *World Watch Magazine*, March 2000, Worldwatch Institute (www. worldwatch.org), Washington, D.C.

period, the world's dominant energy source—oil—has grown at just 1.4 percent per year.

Wind and solar energy currently produce less than 1 percent of the world's energy, but as the computer industry long ago discovered, double-digit growth rates can rapidly turn a tiny sector into a giant. In the past two years, perhaps a dozen major companies have joined Royal Dutch Shell in announcing major new investments in giant wind farms, solar manufacturing plants, and fuel cell development. The "alternative" energy industry is beginning to take on the same kind of buzz that surrounded John D. Rockefeller's feverish expansion of the oil industry in the 1880s—or Bill Gates's early moves in the software business in the 1980s. This January, stocks of solar and fuel cell companies suddenly jumped several-fold in a month, following the pattern of Internet stocks.

The 21st century may be as profoundly reshaped by the move away from fossil fuels as the 20th century was shaped by them. Energy markets, for example, could shift abruptly, drying up sales of conventional power plants and cars in a matter of years and influencing the share prices of scores of companies. The economic health—and political power—of whole nations could be boosted, or in the case of the Middle East, sharply diminished. And our economies and lifestyles are likely to become more decentralized with the advent of new energy sources that provide their own transportation network—for example, the sunshine that already falls on our rooftops.

How quickly the world's energy economy is transformed will depend in part on whether fossil fuel prices remain low and whether the opposition of many oil and electric power companies to a new system can be overcome. The pace of change will be heavily influenced by the pace of international negotiations on climate change and of the national implementation plans that follow. In the 1980s, California provided tax incentives and access to the power grid for new energy sources, which enabled the state to dominate renewable-energy markets worldwide. Similar incentives and access have spurred rapid market growth in several European countries in the 1990s. Such measures have begun to overcome the momentum of a century's investment in fossil fuels.

Earth Day 2000—with its central theme, "Clean Energy Now!"—provides a timely opportunity for citizens to express their desire for a new energy system and to insist that their elected officials implement the needed policy changes. If they do so, smokestacks and cars may soon look as antiquated as manual typewriters and horse-drawn carriages do.

QUESTIONS

1. Comment on the author's proposed new energy system. What key energy technologies does the author recommend for a sustainable environment?
2. What percent of the world's total energy is produced by solar and wind energy technologies?
3. What needs to be done at the personal, community, national, and international levels to create a new energy system?

BOX 1 World Energy Projections: 2003–2030

According to *The International Energy Outlook 2006*:

- Total world consumption of marketed energy will expand from 421 quadrillion British thermal units (Btu) in 2003 to 563 quadrillion Btu in 2015, and then to 722 quadrillion Btu in 2030, or a 71 percent increase over 2003 to 2030 period.
- Net electricity consumption will more than double between 2003 and 2030, from 14,781 billion kilowatt-hours to 30,116 billion kilowatt-hours.
- Worldwide consumption of electricity generated from nuclear power will increase from 2,523 billion kilowatt-hours in 2003 to 2,940 billion kilowatt-hours in 2015 and 3,299 billion kilowatt-hours in 2030.
- Natural gas share of world electricity markets will increase from 19 percent in 2003 to 22 percent in 2030.
- World carbon dioxide emissions will increase from 25,028 million metric tons in 2003 to 33,663 million metric tons in 2030.

Source: International Energy Outlook 2006 (http://www.eia.doe.gov/oiaf/ieo/index.html).

The Immortal Waste

AHMED S. KHAN
BARBARA A. EICHLER

If there ever was an element that deserved a name associated with hell, it is plutonium. This is not only because of its use in atomic bombs—which certainly would amply qualify it—but also because of its fiendishly toxic properties, even in small amounts.
ROBERT E. WILSON

Since December 1942—when Enrico Fermi successfully initiated the first nuclear fission reaction (see Table 1)—millions of tons of nuclear waste have been accumulated as a result of a global race for power generation and weapon production. Today, in 32 nations, more than 440 nuclear power reactors generate one-sixth of the world's electricity production. Twenty-six power reactors are under construction. A typical 1000-MW power reactor generates a large amount of nuclear waste. However, the significant part of the nuclear waste generated worldwide is contributed by nuclear weapons production. The International Atomic Energy Agency (IAEA) has classified nuclear waste into three categories: high-, intermediate-, and low-level nuclear waste (see Table 2). The classification is based on waste's source, temperature and half-life.

For the last 50 years, scientists have proposed various methods and techniques (see Tables 3 and 4) for the permanent storage of nuclear waste, but various governments around the world have not given serious consideration to this problem. As a result of this neglect, 80,000 tons of irradiated fuel and hundreds of thousands of tons of radiated waste are sitting in temporary storage sites.

Unlike chemical waste, whose toxicity can be neutralized or reduced by various techniques to minimize adverse effects on the environment and humans, nuclear waste is radioactive (see Table 5), and the potency of radioactivity can only be eliminated by natural decay. Depending on the type of element, it can take hundreds and thousands or even millions of years for material decay to occur (see Tables 6, 7, and 8).

Accidental or intentionally released radiation can rapidly spread through air and water to contaminate the environment. Radioactive waste from Soviet and U.S. weapons facilities has spread thousands of kilometers from the source, thereby contaminating the environment and people.

According to a 1991 study commissioned by International Physicians of Nuclear War, "The fallout from the atmospheric atomic bomb testing has spread around the globe and will eventually cause an estimated 2.4 million cancer deaths." The accident at Chernobyl released an estimated 50 to 250 million curies into the environment, whereas the bombing of Hiroshima and Nagasaki released an estimated 1 million curies. The radiation released from Chernobyl will be responsible for an estimated 14,000 to 475,000 cancer deaths.

Nuclear energy provides over 16 percent of the world's electricity and displaces approximately 6 million barrels of oil per day. Today, more than 440 nuclear power plants in 32 countries generate 366,913 MW. In the United States, 104 nuclear plants provide 19 percent of the electricity. At first look, nuclear energy appears to be clean (i.e., it does not contribute to atmospheric emissions, but it does generate the immortal nuclear waste that possesses the greatest threat to the environment and people).

Table 1
Nuclear Energy: Chronology of Research and Development

1905	Albert Einstein developed the theory of the relationship between mass and energy. $E = mc^2$ i.e., energy is equal to the mass times the square of the speed of light.
1932	Preliminary work done by Frederic and Joliot-Curie led to the discovery of the neutron by James Chadwick of England.
1934	Enrico Fermi carried out a series of experiments in Rome that showed neutrons could cause the fission of many kinds of elements, including uranium atoms.
1938	German scientists Otto Hahn and Fritz Strassman bombarded uranium with neutrons from a radium–beryllium source and discovered a radioactive barium isotope among residual material. This indicated that a new type of reaction—*fission*—had taken place. Some mass of uranium was converted to energy, thereby verifying Einstein's theory.
Dec. 2, 1941	A group of scientists led by Fermi initiated the first self-sustaining nuclear chain reaction in a laboratory at the University of Chicago.
July 16, 1945	The first atomic bomb was tested at Alamogordo, New Mexico, by the U.S. army under the code name "Manhattan Project."
Aug. 1, 1946	The Atomic Energy Act of 1946 established the AEC to control nuclear energy development and explore the peaceful uses of nuclear energy.
Dec. 20, 1951	The experimental Breeder Reactor I at Arco, Idaho, produced the first electric power from nuclear energy.
Dec. 8, 1953	President Eisenhower delivered his "Atoms for Peace" speech before the United Nations.
Jan. 21, 1954	The U.S. Navy launched the first nuclear-powered submarine, the U.S.S. *Nautilus*, which was capable of cruising 115,000 km (62,500 nautical miles) without refueling.
Aug. 30, 1954	President Eisenhower signed the Atomic Energy Act, which permitted and encouraged the participation of private industry in the development and use of nuclear energy and permitted greater cooperation with U.S. allies.
Jan. 10, 1955	The Atomic Energy Commission (AEC) announced the Power Demonstration Reactor Program. The AEC would cooperate with industry to construct and operate experimental nuclear power reactors.
Aug. 8–20, 1955	The first United Nations International Conference on peaceful uses of atomic energy was held in Geneva, Switzerland.
Sept. 2, 1957	The Price–Anderson Act granted financial protection to the public and to AEC licensees and contractors in the occurrence of a major accident involving a nuclear power plant.
Oct. 1, 1957	The International Atomic Energy Agency (IAEA) is established in Vienna, Austria, by the United Nations to promote the peaceful use of nuclear agency.

Table 1 (*Continued*)

Dec. 2, 1957	The world's first large-scale nuclear power plant began operations in Shippingport, Pennsylvania.
July 21, 1959	The world's first nuclear-powered merchant ship, the N.S. *Savannah*, was launched in Camden, New Jersey.
Nov. 25, 1961	The U.S. Navy commissioned the world's largest ship, the U.S.S. *Enterprise*—a nuclear-powered aircraft carrier capable of cruising up to 20 knots for distances up to 740,800 km (4,000,000 nautical miles) without refueling.
April 3, 1965	The first nuclear reactor in space (SNAP-10A) was launched.
March 5, 1970	The Treaty for Non-proliferation of Nuclear Weapons was ratified by the United States, the United Kingdom, the Soviet Union, and 45 other nations.
1971	A total of 22 commercial nuclear power plants were in full operation in the United States.
August 1974	The federal government released the results of a safety study by Dr. Norman Rasmussen of MIT, which concluded that a meltdown in a power reactor would be extremely unlikely.
Oct. 11, 1974	The Energy Reorganization Act of 1974 divided AEC functions between two newly formed agencies—the Energy Research and Development Administration (ERDA) and the Nuclear Regulatory Commission (NRC).
1976	A total of 61 nuclear power plants with an aggregate capacity of 42,699 megawatts were producing 8.3 percent of electricity generated in the United States.
March 28, 1979	The worst accident in a commercial reactor occurred at the Three Mile Island (TMI) nuclear power station near Harrisburg, Pennsylvania. The cause was the loss of coolant from the reactor core due to mechanical and human errors. Without the cooling water surrounding the fuel, its temperature exceeded 5,000 degrees Fahrenheit, causing melting and damage to the reactor core. Due to the accident, the radioactive material normally confined to the fuel was released into the reactor's cooling water system.
1979	After the Three Mile Island accident, the Nuclear Regulatory Commission (NRC) imposed stricter safety regulations and more rigid inspection procedures to improve the safety of nuclear reactors. The 12 percent of electricity produced commercially in the United States was generated by 72 licensed nuclear reactors.
1981	The Shippingport power station was shut down after 25 years of service.
1984	A total of 83 nuclear power reactors generated 14 percent of the electricity produced in the United States.
1986	The worst accident in nuclear history—the Chernobyl disaster—took place in the former Soviet Union on April 26.

Table 2
Nuclear Waste Classification

Type	*Characteristics*
High Level	Produced in two ways: (a) By reprocessing spent fuel to recover isotopes that can be used again as fuel. (b) Reactor fuel rods that contain long-lived isotopes (with a half-life of 30 years or more) contain transuranic (heavier than uranium) elements. Most are harmful, highly radioactive, and must be handled and transported with shielding. Must be isolated from humans and the environment for thousands of years. By 1990, 26,400 cubic meters of high-level waste was generated. Twenty metric tons of used nuclear fuel is generated by a typical plant in the U.S.; 2,000 metric tons is generated by the nuclear industry per year; over the past four decades about 54,000 metric tons of used nuclear fuel has been produced by the nuclear industry in the U.S.
Intermediate Level	Produced as reactor byproducts and other materials, such as equipment and tools, have become radioactive. Less harmful, but cannot be handled and transported without shielding. By 1990, 3,400 cubic meters of intermediate-level waste was generated. IAEA estimates that, by 1995, the rate of generation will be 3,800 cubic meters per year.
Low Level	Produced due to contamination of metal, paper, rags, and so forth by radioactive material at power reactors, medical equipment, and other non-military sources. Contains no transuranic elements. Least harmful, it can be handled and transported without shielding. By 1990, 370,000 cubic meters of low-level waste was generated. In U.S. approximately 1,419 thousand cubic feet of low-level waste was disposed of in 1998.

Source: "Nuclear Waste: The Challenge Is Global," *IEEE Spectrum*, July 1990, U.S. Nuclear Regulatory Commission (http://www.nrc.gov), and Nuclear Energy Institute (http://www.nei.org).

Table 3
Technical Options for Dealing with Irradiated Fuel

Method	*Process*	*Problems*	*Status*
Antarctica ice burial	Bury waste in ice cap	Prohibited by international law; low recovery potential; concern over catastrophic failure	Abandoned
Geologic burial	Bury waste in mined repository hundreds of meters deep	Difficulty predicting geology, groundwater flows, and human intrusions over long time periods	Under active study by all nuclear countries as favored strategy
Long-term storage	Store waste indefinitely in specially constructed buildings	Dependent on human institutions to monitor and control access to waste for long time periods	Not actively being studied by governments, although proposed by nongovernmental groups

Table 3 (*Continued*)

Method	*Process*	*Problems*	*Status*
Reprocessing	Chemically separate uranium and plutonium from fission products in irradiated fuel; decreases radioactivity by 3 percent	Increases volume of waste 160-fold; poor economics; increases risk of nuclear weapons proliferation	Commercially under way in four countries; total of 16 countries have reprocessed or plan to reprocess irradiated fuel
Seabed burial	Bury waste in deep ocean sediments	Possibly prohibited by international law; transport concerns; nonretrievable	Under active study by consortium of 10 countries
Space disposal	Send waste into solar orbit beyond Earth's gravity	Potential launch failure could contaminate whole planet; very expensive	Abandoned
Transmutation	Convert waste to shorter-lived isotopes through neutron bombardment	Technically uncertain whether waste stream would be reduced; very expensive	Under active study by United States, Japan, Soviet Union, and France

Source: *State of the World, 1992,* Worldwatch Institute, W. W. Norton & Company.

Table 4
Selected Country Programs on High-Level Waste Burial

Country	*Earliest Planned Year*	*Status of Program*
Argentina	2040	Granite site at Gastre, Chubut, selected.
Belgium	2020	Underground laboratory in clay at Mol.
Canada	2020	Independent commission conducting four-year study of government plan to bury irradiated fuel in granite at yet-to-be-identified site.
China	none announced	Irradiated fuel to be reprocessed; Gobi Desert sites under investigation.
Finland	2020	Field studies being conducted; final site selection due in 2000.
France	2010	Three sites to be selected and studied; final site not to be selected until 2006.
Germany	2008	Gorleben salt dome sole site to be studied.

Table 4 *(Continued)*

India	2010	Irradiated fuel to be reprocessed; waste stored for 20 years, then buried in yet-to-be-identified site.
Italy	2040	Irradiated fuel to be reprocessed and waste stored for 50–60 years before burial in clay or granite.
Japan	2020	Limited site studies; cooperative program with China to build underground research facility.
Netherlands	2040	Interim storage of reprocessing waste for 50–100 years before eventual burial, possibly in the seabed or in another country.
Soviet Union	none announced	Eight sites being studied for deep geologic disposal.
Spain	2020	Burial in unidentified clay, granite, or salt formation.
Sweden	2020	Granite site to be selected in 1997; evaluation studies under way at Aspo site near Oskarshamn nuclear complex.
Switzerland	2020	Burial in granite or sedimentary formation at yet-to-be-identified site.
United States	2017	Yucca Mountain, Nevada, site geology being studied and, if approved, will receive 70,000 tons of waste.
United Kingdom	2030	Fifty-year storage approved in 1982; explore options including seabed burial.

Source: *State of the World, 1992*, Worldwatch Institute, W. W. Norton & Company.

Table 5
Types of Radiation

Mode of Radiation	*Sources*	*Penetrating Power*	*Approx. Distance Traveled in Air*	*Shielding Material*
Alpha α	Fission and fission products	Very small	5 cm	Paper
Beta β	Fission, fission products, activation products	Small	300 cm at 1 MeV*	Water, plastic, wood
Gamma γ	Fission, fission products, activation products	Very large		Lead, plastic, paraffin
Neutron n	Fission	Very large		Water, plastic, paraffin

*MeV = Mega electronvolt

Table 6
Radioactive Decay

Type of Radiation	*Nuclide*	*Half-Life*
α	uranium-238	4.47 billion years
β	thorium-234	24.1 days
β	protactinium-234	1.17 minutes
α	uranium-234	245,000 years
α	thorium-230	8,000 years
α	radium-226	1,600 years
α	radon-222	3.823 days
α	polonium-218	3.05 minutes
β	lead-214	26.8 minutes
β	bismuth-214	19.7 minutes
α	polonium-214	0.000164 second
β	lead-210	22.3 years
β	bismuth-210	5.01 days
α	polonium-210	138.4 days
	lead-206	stable

Source: *Radiation Doses, Effects and Risks,* United Nations Environment Programs, Nairobi, Kenya.

Table 7
Half-Lives of Radioactive Elements

Element (symbol-mass no.)	*Half-Life (years)*	*Decay Mode*
Uranium		
U-232	72	α, β
U-233	1.59×10^5	α, β
U-235	7.03×10^8	α, β
U- Uranium		
U-232	72	α, β
U-233	1.59×10^5	α, β
U-235	7.03×10^8	α, β
U-236	2.34×10^7	α, β
U-238	4.46×10^9	α, β
U-239	23.5 minutes	α, β

Table 7 (*Continued*)

Plutonium		
Pu-239	2.41×10^4	α, β
Tellurium		
Te-130	2×10^{21}	β
Indium		
In-115	5.1×10^{14}	β

Table 8
Radioactivity and Thermal Output per Metric Ton of Irradiated Fuel from a Light-Water Reactor

Age (years)	*Radioactivity (curies)*	*Thermal Output (watts)*
At discharge	177,242,000	1,595,375
1	693,000	12,509
10	405,600	1,268
100	41,960	299
1,000	1,752	55
10,000	470	14
100,000	56	1

Sources: Ronnie B. Lipschultz, *Radioactive Waste: Politics, Technology and Risk*, Cambridge, MA: Ballinger Publishing Company, 1980; J.O. Biomeke et al., "Projections of Radioactive Wastes to Be Generated by the U.S. Nuclear Power Industry," Oak Ridge National Laboratory, National Technical Information Service, Springfield, VA, February 1974.

At present, there is no permanent depository for high-level nuclear wastes in the world. Yet there are more than 440 nuclear power plants operating worldwide. Each nuclear power plant yields about 30 tons of high-level waste. Therefore, in very simplistic terms and not accounting for various types of reactors, the world's high-level nuclear waste approaches 15,000 tons annually.

In 1988, the United States proposed its only licensed, permanent, high-level waste dumping site 1,200 feet under Yucca Mountain, Nevada, for its 110 operating nuclear plants. Since the proposal of Yucca Mountain, however, there remains little progress in this site becoming a reality since the plan is fraught with problems and technical difficulties. The U.S. Department of Energy said that it cannot offer a permanent storage site until 2017 at the earliest, and wastes continue to pile up. Some of the problems associated with the Yucca Mountain plan are as follows.

LOCATION PROBLEMS

The Department of Energy did not follow Nevada procedures and did not receive proper state authorization for the site.

- Nevada does not want the repository—NIMBY (Not in My Back Yard)—and passed a resolution in 1990 against storing radioactive wastes anywhere within Nevada's borders.
- The dump site was based entirely on political considerations, and thus science's role in this decision was minimal.
- There is a young active volcano within seven miles of the site and 32 active faults on the site itself. This renders the site unstable and also suggests that, through these faults, contaminated water could escape through

a network of geologic cracks. (Federal requirements prohibit the construction of a nuclear waste repository where water can travel 5 km from the burial site in less than 100 years.)

- The site is probably among the most highly mineralized areas on the continent. Two of North America's largest gold mines are 15–20 miles away, and gold and silver have been found at Yucca Mountain, making the site vulnerable to prospectors.
- Two years of planning have progressed for this site. But many key technical issues remain unresolved.

TECHNICAL PROBLEM

The site will accommodate 70,000 tons of high-level waste. It would take 28 years of working every day to fill the site. By the end of 28 years, there will be tens of thousands more tons of waste that are ready for disposal, and there will be no room for the new waste. Additionally, trucks and other means of moving the waste would be arriving from all parts of the country at 90-minute or more frequent intervals, posing serious safety and transportation questions.

During the past half-century, a number of major nuclear accidents have taken place at various nuclear reactors in the Soviet Union and the United States (see Table 9), resulting in contamination of the environment. Yet it was the accident at Chernobyl that finally destroyed the myth of nuclear energy being a clean energy source. At a little past 1:24 a.m. on April 26, 1986, two mammoth explosions blew apart Unit Four of the Chernobyl nuclear power plant. The plant is located about 70 miles north of Kiev, the capital of Ukraine, a republic of the former Soviet Union. The roof of the plant was blown off and radioactive gases and materials were released into the atmosphere, reaching up to 1,100 meters (Newton 1994). According to recent estimates, more than 50 million curies were released as a result. The cause of the accident was the flawed design of the RBMK reactor (large power-boiling reactor). After the accident, Soviet scientists were reluctant to modify the design of the RBMK reactor, but eventually modified it to make it safer. Forty RBMK-type reactors are still operating in Eastern Europe and former Soviet states. In 32 countries, 440 atomic power plants continue to operate and generate enormous amounts of nuclear waste (see Table 10).

Table 9
Major Nuclear Accidents

United States	*Soviet Union*
1951, Detroit Accident in a research reactor. Overheating of fissionable material because permissible temperatures had been exceeded. Air contaminated with radioactive gases.	**September 1957** Accident at reactor near Chelyabinsk. A spontaneous nuclear reaction occurred in spent fuel, causing a substantial release of radioactivity. Radiation spread over a wide area. The contaminated zone was enclosed within a barbed wire fence and ringed by a drainage channel. The population was evacuated and the topsoil removed; livestock was destroyed and buried in pits.
24 June 1959 Meltdown of part of fuel rods due to failure of cooling system at experimental power reactor in Santa Susanna, California.	**7 May 1966** Prompt neutron power surge at a nuclear power station with a boiling-water nuclear reactor in the town of Melekess. Two people were exposed to severe doses of radiation.
3 January 1961 Steam explosion at an experimental reactor near Idaho Falls, Idaho. Three people died.	**7 January 1974** Explosion of reinforced concrete gasholder for the retention of radioactive gases in No. 1 reactor of Leningrad nuclear power station.
5 October 1966 Partial core melt due to failure of cooling system at the Enrico Fermi reactor, near Detroit.	**6 February 1974** Rupture of intermediate loop in No. 1 reactor at the Leningrad nuclear power station due to boiling of water. Three people were killed. Highly radioactive water with pulp from filter powder discharged into the environment.
19 November 1971 Almost 53,000 gallons (200,000 liters) of water contaminated with radioactive substances from an overflowing waste storage tank at Monticello, Minnesota, flowed into the Mississippi River.	**October 1975** Partial destruction of the core at No. 1 reactor of the Leningrad nuclear power station. About one
28 March 1979 Core melt due to loss of cooling at the Three Mile Island nuclear power station. Radioactive gases released into the atmosphere	

Table 9 (*Continued*)

and 172,000 cubic feet of liquid radioactive waste was discharged into the Susquehanna River. Population evacuated from vicinity of disaster.

7 August 1979 About one hundred people received a radiation dose six times higher than the normal permissible level due to the discharge of highly enriched uranium from a plant producing nuclear fuel near the town of Irving, Texas.

25 January 1982 A broken tube in a steam generator in the R.E. Ginna nuclear power plant, near Rochester, New York. A breakdown in the cooling system caused a leak of radioactive substances into the atmosphere.

30 January 1982 Near the town of Ontario, New York, a breakdown in the cooling system caused a leak of radioactive substances into the atmosphere.

28 February 1985 At the Virgil C. Summer nuclear power station in Jenkinsville, South Carolina, the reactor became critical too soon, leading to an uncontrolled nuclear power surge.

19 May 1985 At the Indian Point 2 nuclear power station near New York City, there was a leakage of several hundred gallons of radioactive water, some of which entered the environment outside the facility.

1986 Webbers Falls, Oklahoma, explosion of a tank containing radioactive gas at a uranium-enrichment plant. One person was killed, eight others injured.

and a half million curies of highly radioactive gases were discharged into the environment.

31 December 1978 No. 2 unit at the Byeloyarsk nuclear power station was heavily damaged by a fire that started when a roof panel in the turbine fell onto a fuel tank. The reactor was out of control. In the effort to supply emergency cooling water to the reactor, eight persons were exposed to severe doses of radiation.

October 1982 Explosion of generator in No. 1 reactor of the Armyanskaya nuclear power station. The turbine hall burned down.

September 1982 Destruction of the central fuel assembly of No. 1 reactor at the Chernobyl nuclear power station due to errors by the operational staff. Radioactivity was released into the immediate vicinity of the plant and into the town of Pripyat, and staff doing repair work were exposed to severe doses of radiation.

27 June 1985 Accident in No. 1 reactor of the Balakovo nuclear power station. During start-up activities, a relief valve burst. Fourteen people were killed. This accident was due to errors made in haste and nervousness by inexperienced operational staff.

April 26, 1986, Chernobyl, eighty miles north of Kiev, Ukraine. Two mammoth explosions blew apart unit four of the nuclear power plant, releasing 50–250 million curies into the atmosphere (the Hiroshima and Nagasaki bombings released an estimated 1 million curies). The accident killed 31 people immediately, and a few days later, Soviets admitted the deaths of 224 others. Total casualties remain unknown. It is the worst nuclear disaster to date.

Accidents in other countries

Dec. 12, 1952 Chalk River, Ottawa, Canada Apartial meltdown of the reactor's uranium fuel core occurred due to accidental removal of four control rods. Millions of gallons of radioactive water accumulated inside the reactor. There were no casualties or injuries.

Oct. 7, 1957 Windscale Pile No. 1, north of Liverpool, England A fire in a graphite-cooled reactor discharged radiation, contaminating a 200-square-mile area.

1976, near Greifswald, East Germany The radioactive core of a reactor in the Lubmin nuclear power plant nearly melted down due to the malfunction of safety systems during a fire.

Sept. 30, 1999 Tokaimura, Japan An uncontrolled chain reaction in a uranium-processing nuclear fuel plant discharged high levels of radioactive gas into the air. Two workers were killed and one seriously injured.

Aug. 9, 2004 Mihama, Japan Leakage of non-radioactive steam in a nuclear power plant occurred. Four workers were killed and many received burn injuries.

For a complete list of nuclear accidents, visit http://archive.greenpeace.org.

Table 10
List of Nuclear Reactors

	Reactors in Operation	*Total MW (e)*	*Reactors under Construction*
North & Central America	2,005		2,005
Canada	17	12,113	0
Mexico	2	1,310	0
United States	104	99,210	0
South America			
Argentina	2	935	1
Brazil	2	1,901	0
Europe			
Belgium	7	5,801	0
Bulgaria	4	2,722	0
Czech Republic	6	3,548	0
Finland	4	2,656	0
France	59	63,363	0
Germany	17	20,339	0
Hungary	4	1,755	0
Lithuania	1	1,185	0
Netherlands	1	449	0
Romania	1	655	1
Slovakia	6	2,442	0
Slovenia	1	656	0
Spain	9	7,585	0
Sweden	11	9,451	0
Switzerland	5	3,220	0
Ukraine	15	13,107	2
United Kingdom	23	11,852	0
Asia			
Armenia	1	376	0
China	9	6,602	2
India	14	2,550	9
Iran	0	0	1
Japan	54	45,468	3
South Korea	20	16,810	0
North Korea	0		1
Pakistan	2	425	0
Russia	31	21,743	4
Taiwan	6	4,884	2
Africa			
South Africa	2	1,800	0
Total	440	366,913	26

MW(e) = Megawatt electric
Source: International Atomic Energy Agency (IAEA), http://www.iaea.org/programmes/a2/index.html.

What appears to be a clean source of energy are a few megawatts of low-cost power for today's consumption. But in the long run, these will cause an immortal radioactive contamination of the environment. With the growing amount of nuclear waste and limited technical options for its storage and disposal, it is just a matter of time until major accidents happen that will result in contamination of the environment.

REFERENCES

Grossman, D., and Shulman, S. A Nuclear Dump: The Experiment Begins. (1989). *Discover* (March):49–56.

Lenssen, N. (1992). Confronting the Nuclear Waste. *State of the World*. New York: W. W. Norton & Company.

Raloff, J. Fallout over Nevada's Nuclear Destiny. (1990). *Science News* 13 (6 Jan):11–12.

Newton, D. (1994). Chernobyl Accident. *When Technology Fails*. Detroit, MI: Gale Research Inc.

QUESTIONS

1. Define the following terms.
   a. Fission
   b. Low-level nuclear waste
   c. Intermediate-level nuclear waste
   d. High-level nuclear waste
   e. Alpha (α) radiation
   f. Beta (β) radiation
   g. Gamma (γ) radiation
   h. Isotope
2. What methods are being employed for the interim storage of commercial spent fuel and high-level nuclear waste?
3. Do you think that Yucca Mountain is a safe place for the permanent storage of nuclear waste? Explain your answer by citing facts.
4. Do you think that nuclear energy is a "clean" source of energy?
5. With more than 440 operational nuclear plants worldwide and 26 plants under construction, is the chance for a nuclear accident growing larger or smaller? What might be the consequences of a major nuclear accident?
6. What alternatives are there to nuclear energy?
7. "Considering the potential risks involved with nuclear technology and the past experiences at Three Mile Island and Chernobyl, the worldwide development of new nuclear plants should be stopped." Do you agree with this statement? Explain your answer.
8. A number of developing countries are constructing nuclear power plants for their growing energy needs. For the majority of these developing countries, nuclear energy appears to be the most viable solution for their growing demand for energy due to scarce resources. The developed countries are trying to prevent the transfer of nuclear technology to third-world countries. Should third-world nations be denied access to nuclear technology? Support your answer. What alternative sources would you recommend?

True Reality
Of this there is no academic proof in the world;
For it is hidden, and hidden, and hidden.

RUMI, *Idries Shah*, The Way of Sufi, *New York: E.P. Dutton, 1970, p. 106.*

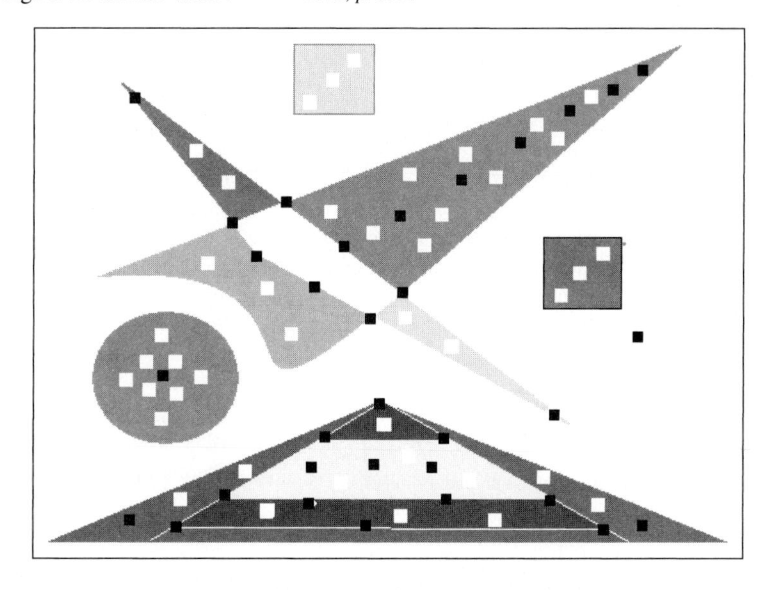

Figure 1
The vanishing environment.
Source: An abstract illustration by Ahmed S. Khan.

Oil & Blood: The Way to Take Over the World

MICHAEL RENNER

In its drive toward war against Iraq, the Bush administration insisted throughout the fall of 2002 that its purpose was to eliminate weapons of mass destruction and establish democracy. No doubt, Saddam Hussein's regime was dictatorial and dangerous, and Iraq's civilian population had suffered grievously. But there was no clear evidence that Iraq posed the immediate and growing threat that the administration depicted.

So, why the renewed focus of U.S. policy on Iraq? Was it a desire to fortify U.S. political domination of the oil-rich Middle East? Not at all, said the White House. "The only interest the United States has in the region is furthering the cause of peace and stability, not [Iraq's] ability to generate oil," contended the president's spokesman, Ari Fleischer. Given U.S. addiction to oil and Washington's long history of intervention in the region, this is a disingenuous, if not downright deceptive, statement.

The Middle East—and specifically the Persian Gulf region—accounts for about 30 percent of global oil production. But it has about 65 percent of the planet's known reserves and is therefore the only region able to satisfy any substantial rise in world oil demand—an increase that the administration's energy policy documents say is inevitable. Saudi Arabia, with 262 billion barrels, has a quarter of the world's total reserves and is the single largest producer. But Iraq, despite its pariah status for the past 12 years, remains a key prize. At 112 billion barrels, its known reserves are second only to Saudi Arabia's. And, given that substantial portions of Iraqi territory have never been fully explored, there is a good chance that actual reserves are far larger.

For half a century, the United States has made steadily increasing investments in keeping the Gulf region in its geopolitical orbit. The investments have included the overthrow of "hostile" governments and support of client regimes, massive arms transfers to allies, acquisition of military bases, and direct and indirect forms of intervention—many of these activities involving shifting alliances and repeated large-scale violence. In Washington's calculus, securing oil supplies has consistently trumped the pursuit of human rights and democracy. This is still the case today, as the Bush administration prepares for a more openly imperial role in the region.

Saudi Arabia has had a close relationship with the United States since the 1940s. But it has long been vulnerable to pressures from the far more populous Iraq and Iran. Iran was brought firmly into the Western orbit by a 1953 CIA-engineered coup against the Mossadegh government, which had nationalized Iran's oil. The coup re-installed the Shah on the Persian throne. Armed with modern weaponry by the United States and its allies, the Shah became the West's regional policeman once the military forces of Britain—the former colonial power—were withdrawn from the Gulf area in 1971.

Iraq, on the other hand, was a pro-Western country until 1958, when its British-installed monarchy was overthrown. Fearing that Iraq might turn communist under the new military regime, the United States dabbled in a temporary alliance of convenience with the Ba'ath (Renaissance)

Michael Renne© 2003. Reprinted with permission of Worldwatch Institute. *World Watch Magazine*, January/February 2003.

Party in its efforts to grab power. CIA agents provided critical logistical information to the coup plotters and supplied lists with the names of hundreds of suspected communists to be eliminated.

Even so, in 1972, the Ba'ath regime signed a treaty of friendship and cooperation with the Soviet Union. Baghdad turned to Moscow both for weapons and for help in deterring any U.S. reprisals against Iraq for nationalizing the Iraq Petroleum Company, which had been owned by Royal Dutch-Shell, BP, Exxon, Mobil, and the French firm CFP. Iraq was the first Gulf country to successfully nationalize its oil industry.

Saddam Hussein, a strongman of the Ba'ath regime who formally took over as President in 1979, was instrumental in orchestrating the pro-Moscow policy. But as it became apparent that the Soviet Union could not deliver the technologies and goods (both civilian and military) needed to modernize Iraq, he shifted to a more pro-Western policy. Western governments and companies were eager to soak up the rising volume of petrodollars and to lure Iraq out of the Soviet orbit. During the 1980s, this eagerness extended to supplying Baghdad with the ingredients needed to make biological, chemical, and nuclear weapons.

The year 1979 turned out to be a watershed, as the Shah of Iran was swept aside by an Islamic revolution that brought Ayatollah Khomeini to power. One of Washington's main geopolitical pillars had crumbled, and the new regime in Teheran was seen as a mortal threat by the conservative Persian Gulf states. The Carter administration responded by pumping rising quantities of weapons into Saudi Arabia and began a quest for new military bases in the region (Bahrain eventually became the permanent home base for the U.S. Fifth Fleet). But there was no escaping the fact that neither Saudi Arabia nor any of the smaller Gulf states were strong enough to replace Iran as a proxy.

Instead, Iraq became a surrogate of sorts. Iran and Iraq had long been at loggerheads. Seeing a rival in revolutionary disarray and sensing an opportunity for an easy victory that would propel him to leadership of the Arab world, Saddam Hussein invaded Iran in September 1980. Eager to see Teheran's revolutionary regime reined in, the United States turned a blind eye to the aggression, opposing UN Security Council action on the matter.

But instead of speeding the Iranian regime toward collapse, the attack consolidated Khomeini's power. And marshalling revolutionary fervor, Iran was soon turning the tide of battle. With the specter of an Iraqi defeat looming, the United States went much further in its support of Saddam.

To facilitate closer cooperation, the Reagan administration removed Iraq from a list of nations that it regarded as supporters of terrorism. Donald Rumsfeld, now secretary of defense, met with Saddam in Baghdad in December 1983. His visit paved the way to the restoration of formal diplomatic relations the following year after a 17-year hiatus.

- The United States made available several billion dollars' worth of commodity credits to Iraq to buy U.S. food, alleviating severe financial pressures that had threatened Baghdad with bankruptcy. The food purchases were a critical element in the regime's attempts to shield the population as much as possible from the war's repercussions—and hence limiting the likelihood of any challenges to its rule. The U.S. government also provided loan guarantees for an oil export pipeline through Jordan (replacing other export routes that had been blocked because of the war).
- Though not selling weapons directly to Iraq, Reagan administration officials allowed private U.S. arms dealers to sell Soviet-made weapons purchased in Eastern Europe to Iraq. U.S. leaders permitted Saudi Arabia, Kuwait, and Jordan to transfer U.S.-made weapons to Baghdad. And they abandoned earlier opposition to the delivery of French fighter jets and Exocet missiles (which were subsequently used against tankers transporting Iranian oil).
- From the spring of 1982 on, the Reagan administration secretly transmitted highly classified military intelligence—battlefront satellite images, intercepts of Iranian military communications, information on Iranian troop deployments—to Saddam Hussein's regime, staving off its defeat.
- As the war went on, the United States took an increasingly active military role. It tilted toward Iraq in the "war on tankers" by protecting oil tankers in the southern Gulf against Iranian attacks, but did not provide security from Iraqi attacks for ships docking at Iran's Kharg Island oil terminal. Later, the United States even launched attacks on Iran's navy and Iranian offshore oil rigs.

Washington's immediate objective was to prevent an Iranian victory. In a larger sense, though, U.S. policymakers were intent on keeping both Iraq and Iran bogged down in war, no matter how horrendous the human cost on both sides—hundreds of thousands were killed. (The Reagan administration secretly allowed Israel to ship several billion dollars' worth of U.S. arms and spare parts to Iran.) Preoccupied with fighting one another, Baghdad and

Teheran would be unable to challenge U.S. domination of the Gulf region. Reflecting administration sentiments, Henry Kissinger said in 1984 that "the ultimate American interest in the war [is] that both should lose."

Oil and geopolitical interests translated into U.S. support for Saddam Hussein when he was at his most dangerous and murderous—not only committing an act of international aggression by invading Iran, but also by using chemical weapons against both Iranian soldiers and Iraqi Kurds. U.S. assistance to Baghdad was provided although top officials in Washington knew at the time that Iraq was using poison gas.

Undoubtedly, U.S. support emboldened Saddam Hussein to invade Kuwait in 1990. But the United States would never consent to a single, potentially hostile, power gaining sway over the Gulf region's massive oil resources. When its regional strongman crossed that line, U.S. policy shifted to direct military intervention.

Following the Gulf War, the United States supplied Saudi Arabia and other allies among the Gulf states with massive amounts of highly sophisticated armaments. Washington and other suppliers delivered more than $100 billion worth of arms from 1990 to 2001. In the late 1980s, Saudi Arabia had imported 17 percent, by dollar value, of worldwide weapons sales to developing countries. In the 1990s, the Saudi share rose to 38 percent.

But rather than becoming independent military powers, Riyadh and the other Gulf states are at best beefed-up staging grounds for the U.S. military: Washington has for many years been "pre-positioning" military equipment and supplies and expanding logistics capabilities to facilitate any future intervention. And although political sensitivities rule out a visible, large-scale U.S. troop presence, more than 5,000 U.S. troops have been continuously deployed in Saudi Arabia and more than 20,000 in the Gulf region as a whole.

Despite insinuations by the Bush administration, there is no evidence that Saddam Hussein's regime is in any way linked to the events of September 11, 2001. However, the terrorist attacks facilitated a far more belligerent, unilateralist mood in Washington and set the stage for the Bush administration doctrine of preemptive war.

Installing a U.S. client regime in Baghdad would give American and British companies (ExxonMobil, Chevron-Texaco, Shell, and BP) a good shot at direct access to Iraqi oil for the first time in 30 years—a windfall worth hundreds of billions of dollars. And if a new regime rolls out the red carpet for the oil multinationals to return, it is possible that a broader wave of de-nationalization could sweep through the world's oil industry, reversing the historic changes of the early 1970s.

Rival oil interests were a crucial behind-the-scenes factor as the permanent members of the UN Security Council jockeyed over the wording of a new resolution intended to set the parameters for any action against Iraq. The French oil company TotalFinaElf has cultivated a special relationship with Iraq since the early 1970s. And along with Lukoil of Russia and China's National Petroleum Corp., it has for years positioned itself to develop additional oil fields once UN sanctions are lifted. But there have been thinly veiled threats that these firms will be excluded from any future oil concessions unless Paris, Moscow, and Beijing support the Bush policy of regime change. Intent on constraining U.S. belligerence, France, Russia, and China nonetheless are eager to keep their options open in the event that a pro-U.S. regime is installed in Baghdad—and accordingly voted in favor of the U.S.-drafted resolution in November.*

But the stakes in all this maneuvering involve much more than just the future of Iraq. The Bush energy policy is predicated on growing consumption of oil, preferably cheap oil. Given rising depletion of U.S. oil fields, most of that oil will have to come from abroad, and indeed primarily from the Gulf region. Controlling Iraqi oil would allow the United States to reduce Saudi influence over oil policy and give Washington enormous leverage over the world oil market.

Both in the Middle East and in other regions, securing access to oil goes hand in hand with a fast-expanding U.S. military presence. From Pakistan to Central Asia to the Caucasus and from the eastern Mediterranean to the Horn of Africa, a dense network of U.S. military facilities has emerged—with many bases established in the name of the "war on terror."

Only in the most direct sense was the Bush administration's Iraq policy directed against Saddam Hussein. In a broader sense, it aimed to reinforce the world economy's reliance on oil—undermining efforts to develop renewable energy sources, boost energy efficiency, and control greenhouse gas emissions. The same administration that decided to slash annual spending for energy efficiency and renewables R&D has no problem with preparing for a war that could cost as much as $200 billion.

By rejecting U.S. participation in the Kyoto Protocol early in his tenure, George W. Bush sought to throw a wrench into the international machinery set up to address

*UN resolution 1441 passed November 8, 2002

the threat of climate change. By securing the massive flow of cheap oil, he may hope to kill Kyoto. In a perverse sense, a war on Iraq reinforces the assault against the Earth's climate.

QUESTIONS

1. Explain the reasons for the renewed focus of U.S. policy on Iraq.
2. Discuss the moral and ethical implications of endangering thousands of human lives in wars fought during 1980–2003 just to secure an oil supply at cheaper prices.
3. How can the United States avoid future oil crises and potential oil wars? List and explain three strategies.
4. What are the ethical, economic, and environmental implications of the U.S. refusal to sign the Kyoto protocol?

Fuel Cells and the Hydrogen Economy

Ahmed S. Khan

A fuel cell is an electrochemical device that converts chemical energy into electrical form. It consists of two porous electrodes—anode and cathode—separated by an electrolyte in a solid or liquid form.

A fuel cell is like a battery, but it is an active battery because it does not require recharging as long as fuel (hydrogen) is supplied. In a battery, the electrodes are chemically dissimilar, which develops an electric potential difference between them, whereas in a fuel cell, the electrodes are chemically similar. One electrode is supplied with a fuel and the other with an oxidant; thus, an electric potential difference is developed between them. By attaching a load to the fuel cell, an electric current can be drawn to generate electrical power. The electrical energy delivered to the external circuit is less than the energy changes due to a chemical reaction in the cell. The electrical power generated can be used in vehicle propulsion and the operation of electrically driven components. Table 1 compares the fuel efficiency of fuel cells with other power-generation technologies. Table 2 presents the history of fuel cell development.

Figure 1 illustrates a basic hydrogen-oxygen fuel cell. Hydrogen- or hydrogen-rich fuel is supplied to the anode, whereby a catalyst separates hydrogen's negatively charged electrons from the positively charged ions (protons). The electrons and protons follow different paths to the electrode. The negatively charged electrons follow the external electric circuit and thus provide the direct current (DC) for external use. If alternating current (AC) is desired at the output of a fuel cell, then the DC output must be connected to a conversion device, called an inverter, which converts the DC signal to an AC form (see Figure 2). The protons move through the electrolyte to reach the cathode, where they recombine with oxygen and electrons to produce water and heat.

Hydrogen-powered fuel cells and systems are becoming increasingly popular to power cars, trucks, and buses and also to provide electricity and heat for homes and industrial units. Table 3 lists the characteristics and applications of various types of fuel cells.

Advantages

- Fuel cells increase thermal efficiency (see Table 1).
- Fuel cells produce electricity for longer periods.
- Fuel cells generate power with little pollution. (Fuel cells release harmless byproducts in the form of exhaust gases, water, and waste heat.)
- Fuel cells operate with a low noise level.
- Fuel cells can be quickly recharged by refueling compared to batteries that require time-consuming recharge.

Disadvantages

- The cost is higher than conventional systems.
- The durability of a fuel cell system is not yet established.
- Platinum-group metals used for catalyzing reactions are expensive.

Table 1
Comparison of Fuel (Thermal) Efficiencies of Various Power Technologies

Technology	*Efficiency*
Fuel cell electric power	45%
Steam electric power plant	
- Steam at 62 bar, 480° C	30%
- Steam at 310 bar, 560° C	42%
Nuclear electric power	
- Steam at 70 bar, 286° C	33%
Gas turbine electric power plant	30%
Automotive gasoline engine	25%
Automotive diesel engine	35%

Source: James Fay and Dan Golomb. (2002). *Energy and the Environment.* New York: Oxford University Press, p. 62.

Table 2
History of Fuel Cell Development

1800	British scientists William Nicholson and Anthony Carlisle described the process of using electric current to decompose water into hydrogen and oxygen.
1839	William Grove (1811–1896), a Welsh scientist, developed a wet-cell battery (Grove cell) that used a platinum electrode immersed in zinc sulfate and a zinc electrode in zinc sulfate to generate about 12 amperes of current at 1.8 volts.
1893	Friedrich Oswald, one of the founders of the field of physical chemistry, experimentally determined the roles of the various components of the fuel cell (electrodes, electrolytes, oxidizing and reducing agents, anions, and cations).
1930s	Experimental studies were conducted to determine the effectiveness of solid oxide electrolytes.
1940s–1950s	Experimental studies were conducted to increase the efficiency of the fuel cell by using new electrolytes, electrodes/catalysts.
1960s	- Major advances in electrode materials - Development of proton exchange membrane (PEM) fuel cell technology - Development of 15-kW fuel cell power plants (1969)
1970s	Major advances were made in the electrode materials.
1980s	Fuel cell power plants with the capacity of generating 5 MW were developed.
1990s	Fuel cells were used to power buses and buildings.
2000s	- Major advances in fuel cell technology led to reduction in the size of the fuel cell and an increase in the output power levels. - Incorporation of fuel cell technology is being used to power cars and buses and to produce electricity and heat for homes and industrial applications. - Breakthroughs in advanced micro fuel cell technology allow fabrication of miniaturized fuel cells for handheld electronic devices such as handheld entertainment system and PDA/smartphone.

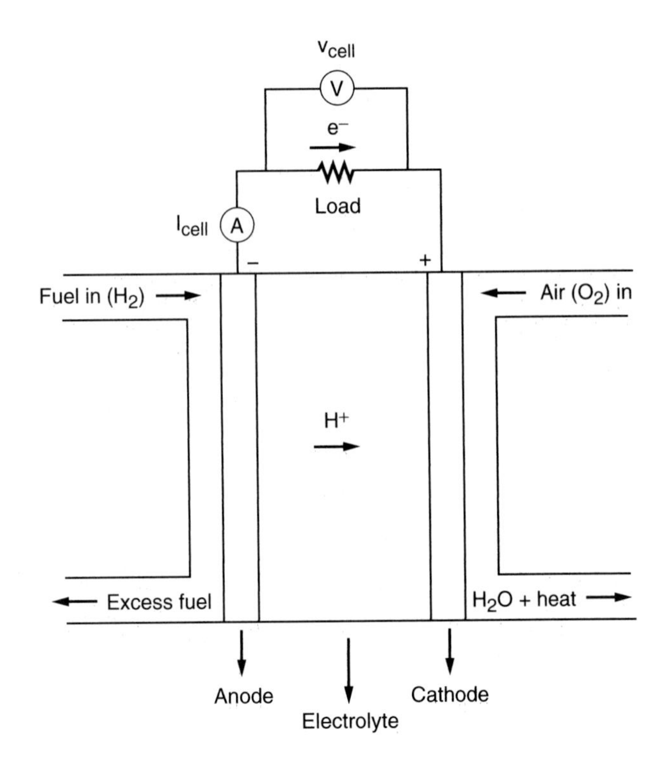

Figure 1
Basic fuel cell construction.

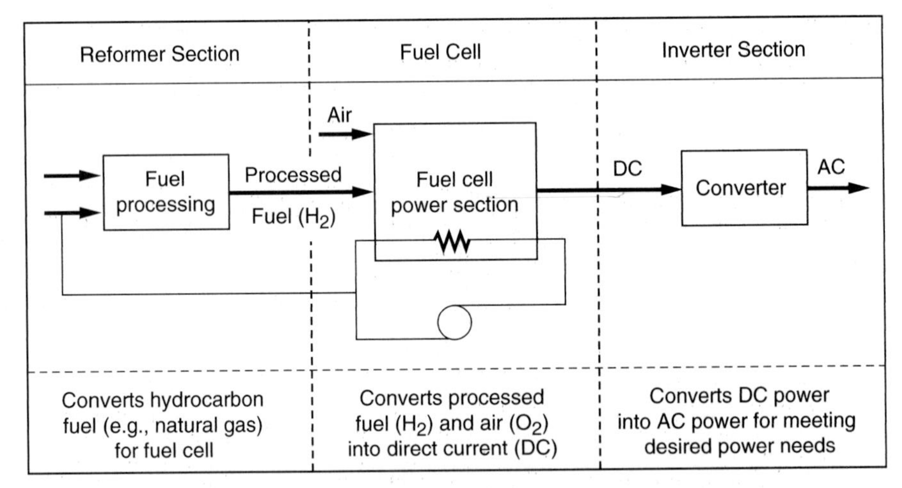

Figure 2
Block diagram of a fuel cell power system.

- Production of hydrogen for fuel cells causes emissions of CO_2 into the atmosphere.
- Power density is low compared to combustion engines and batteries (power density refers to power produced per unit volume or per unit mass).

Fuel cells use hydrogen, which is considered an environmentally friendly fuel because it helps generate electrical power at high efficiency without generating any greenhouse gases, such as CO_2. Hydrogen from fuel cells can be obtained from two sources: methane and the electrolysis of water.

Table 3
Types of Fuel Cells

Cell Type	*Electrolyte/Catalyst*	*Typical Output (power/operating temperature)*	*Efficiency/Application*
Alkali	Alkaline compounds /platinum	0.3–5 kW/150°–200° C	Efficiency is approximately 70 percent Apollo space missions used alkali cells to produce both electricity and water. Hydrogen-powered vehicles
Molten carbonate	Carbonate compounds/ nickel catalyst (inexpensive)	1–100 MW can be designed and constructed/650° C	Efficiency is in the 60–80 percent range High operating temperature makes it unsuitable for homes, but it can be used for industrial applications
Phosphoric acid	Phosphoric acid/ platinum catalyst (expensive)	200+ kW/150–200° C	Efficiency is in the 40–80 per cent range. Can be used to power hotels, hospitals, and office buildings. Can also be used in buses
Proton exchange membrane (PEM)	Thin sheet of permeable polymer/platinum catalyst (expensive) Requires purified hydrogen	50–250 kW/80° C	Efficiency is approximately 40–50 percent Low operating temperature makes it suitable for use in cars and homes
Solid oxide	Metal oxide compounds, such as zirconium oxide, as a solid ceramic electrolyte/ includes a lanthanum manganate cathode and a nickel -zirconia anode	100 kW/1000° C	Efficiency is approximately 60 percent operating temperature. Suitable for high-power industrial applications

kW = kilowatt; MW = megawatt.

$$CH_4 + 2H_2O \text{ -----> } CO_2 + 4H_2$$

$$2H_2O \text{ -----> } 2H_2 + O_2$$

Both of these reactions require additional energy to bring them to completion. To re-form methane into hydrogen, the first reaction requires combustion of additional methane to supply heat energy to re-form methane into hydrogen. The second reaction (electrolysis) requires passing of an electrical current through water. If this electric current is generated by burning a fossil fuel, it makes the reaction very energy inefficient because the heating value of hydrogen will be less than one-third of the heating value of the fossil fuel burned in the electric power plant to generate the electricity. Both of these two methods for obtaining hydrogen generate CO_2; thus, they contribute to carbon emission into the atmosphere.

One way to prevent CO_2 emissions into the atmosphere during the re-forming process of methane into hydrogen is to recover CO_2 and sequester it underground or in an ocean. This would allow 60–80 percent of the heating value of fossil fuel to be utilized while controlling CO_2 emissions. Production, storage, and transportation of hydrogen in an economical and environmentally friendly manner are the main issues for progressing toward a viable hydrogen economy.

"If we really decided that we wanted a clean hydrogen economy, we could have it by 2010."

U.S. National Renewable Energy Laboratory Researcher, April 2001

REFERENCES

Dunn, Seth. (2001). Hydrogen Futures: Towards a Sustainable Energy System. *World Watch Paper*, August.

Fay, James, and Golomb, Dan. (2002). *Energy and the Environment*. New York: Oxford University Press, pp 58–65.

"Fuel Cell Basics." Available at http://www.americanhistory.si.edu/fuelcells/basics.htm. Accessed September 5, 2006.

"Fuel Cell Fundamentals." Available at http://www.batterygeek.net/fuelcell.asp. Accessed September 5, 2006.

"Fuel Cell Technology." U.S. Department of Energy. Available at http://www.eere.energy.gov/hydrogenandfuelcells. Accessed September 5, 2006.

Hunt, V. (1982). *Handbook of Energy Technology: Trends and Perspectives*. New York: Van Nostrand Reinhold Company, pp. 25, 437.

O'Hayre, Ryan et al. (2006). *Fuel Cell Fundamentals*. New York: Wiley, p. 8.

"True Breakthrough in Advanced Micro Fuel Cell Technology for Handheld Devices." Available at http://www.physorg.com/news221.html. Accessed September 5, 2006.

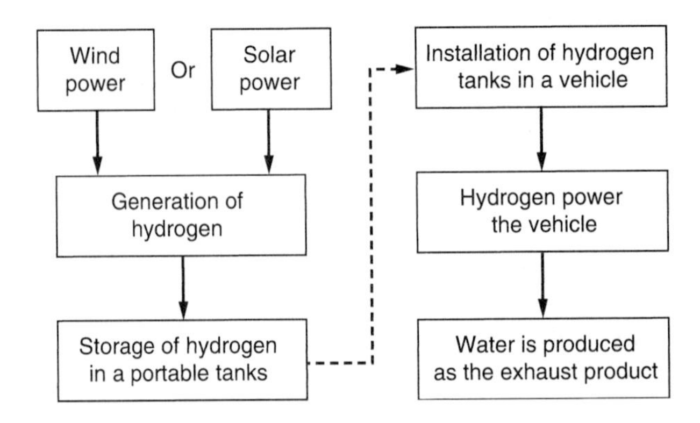

Figure 3
A potential hydrogen fuel system.

QUESTIONS

1. Describe the operation of a basic fuel cell.
2. What is the main difference between a fuel cell and a battery?
3. Can fuel cell technology compete with fossil fuels?
4. What types of fuel cells are the most appropriate for the following applications?
   a. Automobiles
   b. Buses
   c. Spacecraft
   d. Airplanes
   e. Homes
   f. Hotels
   g. Office buildings
   h. Small industrial units
   i. Large industrial units
5. List the advantages (+) and disadvantages (–) of fuel cells.

(+)	(–)

6. Hydrogen has been called "tomorrow's oil by some experts." Do you agree with their assessment? Explain your answer.
7. Use Internet resources to complete the following table.
   a. Which is the most efficient car (make/model) in terms of MPG?
   b. Which is the most environment-friendly car (make/model) in terms of GHG emissions?

Hybrid Car Manufacturer Features/Strengths	*Model Limitations*	*Price*	*City MPG*	*Highway MPG*	*GHG Emissions*
Toyota					
Honda					
Ford					
BMW					

GHG = greenhouse gas; MPG = miles per gallon.

The Hydrogen Experiment

SETH DUNN

In Reykjavík, Iceland, scientists, politicians, and business leaders have conspired to put into motion a grand experiment that may end the country's—and the world's—reliance on fossil fuels forever. The island has committed to becoming the world's first hydrogen economy over the next 30 years.

Riding from the airport to Iceland's capital, Reykjavík, gives one the sensation of having landed on the moon. Black lava rocks cover the mostly barren landscape, which is articulated by craters, hills, and mountains. Other parts of the island are covered by a thin layer of green moss. American astronauts traveled here in the 1960s to practice walking the lunar surface, defining rock types, and taking specimens.

I, too, have traveled here on a journey of sorts to a new world—a world that is powered not by oil, coal, and other polluting fossil fuels, but one that relies primarily on renewable resources for energy and on hydrogen as an energy carrier, producing electricity with only water and heat as byproducts. My quest has brought me to the cluttered office of Bragi Árnason, a chemistry professor at the University of Iceland whose 30-year-old plan to run his country on hydrogen energy has recently become an official objective of his government, to be achieved over the next 30 years. "I think we could be a pilot country, giving a vision of the world to come," he says to me with a quiet conviction and a deep, blue-eyed stare that reminds me of this country's hardy Viking past.

Reprinted with permission from *World Watch Magazine*, November/December 2000. Worldwatch Institute, (www. worldwatch.org) Washington, D.C.

When he first proposed this hydrogen economy decades ago, many thought he was crazy. But today, "Professor Hydrogen," as he has been nicknamed, is something of a national hero. And Iceland is now his 39,000-square-mile lab space for at long last conducting his ambitious experiment. Already, his scientific research has led to a multimillion-dollar hydrogen venture between his university, his government, other Iceland institutions, and a number of major multinational corporations.

I am not alone in my expedition to ground zero of the hydrogen economy: hundreds of scientists, politicians, investors, and journalists have visited over the past year to learn more about Iceland's plans. My journey is also an echo of what happened in the 18th century, when merchants and officials flocked to another North Atlantic island—Great Britain—to witness the harnessing of coal.

Today, many experts are watching Iceland closely as a "planetary laboratory" for the anticipated global energy transition from an economy based predominantly on finite fossil fuels to one fueled by virtually unlimited renewable resources and hydrogen, the most abundant element in the universe. The way this energy transition unfolds over the coming decades will be greatly influenced by choices made today. How will the hydrogen be produced? How will it be transported? How will it be stored and used? Iceland is facing these choices right now, and in plotting its course has reached a fork in the road. It must choose between developing an interim system that produces and delivers methanol, from which hydrogen can be later extracted, or developing a full infrastructure for directly transporting and using hydrogen. Whether the country tests incremental improvements or more ambitious steps will have important economic and environmental implications, not only for Iceland, but for other countries hoping to draw conclusions from its experiment.

Iceland is not undertaking this experiment in isolation. Its hydrogen strategy is tied to three major global trends. The first of these is growing concern over the future supply and price of oil—already a heavy burden on the Icelandic economy. The second is the recent revolution in bringing hydrogen-powered fuel cells—used for decades in space travel—down to earth, making Árnason's vision far more economically feasible than it was just ten years ago. The third is the accelerating worldwide movement to combat climate change by reducing carbon emissions from fossil fuel burning, which in its current configuration places constraints on Iceland that make a hydrogen transition particularly palatable. How the island's plans proceed will both help to shape and be shaped by these broader international developments.

A HEAD START

Straddling the Mid-Atlantic Ridge, Iceland is a geologist's dream. Providing inspiration for Jules Verne's *Journey to the Center of the Earth,* the island's volcanoes have accounted for an estimated one-third of Earth's lava output since A.D. 1500. Eruptions have featured prominently in Icelandic religion and history, at times wiping out large parts of the population. Reykjavík is the only city I know that has a museum devoted solely to volcanoes. There, one can find out the latest about the 150 volcanoes that remain active today.

Iceland's volcanic activity is accompanied by other geological processes. Earthquakes are frequent, though usually mild, which has made natives rather blasé about them. Also common are volcanically heated regions of hot water and steam, most visible in the hot springs and geysers scattered across the island. In fact, the word "geyser" originated here, derived from *geysir*, and Reykjavík translates to "smoky bay." During my visit, the well-known Geysir, which erupts higher than the United States' Old Faithful, was reemerging from years of dormancy, to the delight of Icelanders everywhere.

The country first began to tap its geothermal energy for heating homes and other buildings (also called district heating) in the 1940s. Today, 90 percent of the country's buildings—and all of the capital's—are heated with geothermal water. Several towns in the countryside use geothermal heat to run greenhouses for horticulture, and geothermal steam is also widely harnessed for power generation. One tourist hotspot, the Blue Lagoon bathing resort, is supplied by the warm, silicate-rich excess water from the nearby Reykjanes geothermal power station. Yet it is estimated that only 1 percent of the country's geothermal energy potential has been utilized.

Falling water is another abundant energy source here. Although it was floating ice floes that inspired an early (but departing) settler to christen the island Iceland, the country's high latitude has exposed it to a series of ice ages. This icy legacy lingers today in the form of sizable glaciers, including Europe's largest, which have carved deep valleys with breathtaking waterfalls and powerful rivers. The first stream was harnessed for hydroelectricity in the 1900s. The country aggressively expanded its hydro capacity after declaring independence from Denmark in the 1940s, beginning an era of economic growth that elevated it from Third World status to one of the world's most wealthy nations today. Hydroelectricity currently provides 19 percent of Iceland's energy—and that share could be significantly increased, as the country has harnessed only 15 percent of potential resources (though many regions are unlikely to be tapped due to their natural beauty, ecological fragility, and historical significance).

Iceland is unique among modern nations in having an electricity system that is already 99.9 percent reliant on indigenous renewable energy—geothermal and hydroelectric. The overall energy system, including transportation, is roughly 58 percent dependent on renewable sources. This, some experts believe, prepares the country well to make the transition from internal combustion engines to fuel cells and from hydrocarbon to hydrogen energy. With its extensive renewable energy grid, Iceland has a head start on the rest of the world and is positioned to blaze the path to an economy free of fossil fuels.

PEAT AND PETROLEUM

When Vikings first permanently settled Iceland in the 9th century A.D., they used bushy birchwood and peat reservoirs to make fires for cooking and heating and to fuel iron forges to craft weapons. But deforestation soon led to the end of wood supplies, and the cold climate would freeze the peat bogs, limiting their use as fuel.

Beyond its peat supplies, Iceland has virtually no indigenous fossil fuel resources. As the Industrial Revolution gathered momentum, the nation began to import coal and coke for heating purposes; coal would remain the primary heating source until the development of geothermal energy. In the late 1800s, as petroleum emerged as a fuel, Iceland turned to importing oil. Today, imported oil—about 850,000 tons per year—accounts for 38 percent of national energy use; 57 percent of this is used to run its motor vehicles and the boats of its relatively large fishing industry, the nation's leading source of exports. Dependence on oil imports costs the nation $150 million annually and explains why transport and fishing each account for one-third of its carbon emissions.

The final third of Iceland's greenhouse emissions is found in other industries—primarily the production, or smelting, of metals like aluminum. The availability of low-cost electricity—at $.02 per kilowatt, it is the world's cheapest—has made Iceland a welcome haven for these energy-intensive industries. Metals production, along with transport and fishing, makes the island one of the world's top per-capita emitters of carbon dioxide and offsets much of the greenhouse gas savings Iceland has achieved in space heating and electricity.

These features of Iceland's energy economy—a carbon-free power sector, costly dependence on oil for fishing and transportation, rising emissions from the metals industry—have placed the nation in a difficult situation with regard to complying with international climate change commitments. The 1997 Kyoto Protocol's guidelines for reducing greenhouse gas emissions in industrial nations are based on emission levels from the year 1990, which prevents Iceland from taking credit for its previously completed transition to greenhouse gas-free space heating and electricity generation. Although the government, arguing its special situation, negotiated a 10 percent reprieve between 1990 and 2010 under the Protocol, officials estimate that plans to build new aluminum smelters will cause it to exceed this target. Because of this so-called "Kyoto dilemma," Iceland is among only a few remaining industrial nations that have not signed the agreement.

In 1997, as the Protocol talks gathered momentum and the nation's dilemma was becoming apparent, a recently elected Parliamentarian named Hjalmar Árnason submitted a resolution to the Parliament, or Althing, demanding that the government begin to explore its energy alternatives. Árnason, a former elementary school teacher who says he was "raised by an environmental extremist" father (he is not related to the scientist Bragi Árnason), soon found himself chairing a government committee on alternative energy, which was commissioned to submit a report. One of the first people he tapped for the committee was Professor Hydrogen.

SCIENCE MEETS POLITICS

Bragi Árnason began studying Iceland's geothermal resource "as a hobby," he tells me, while a graduate student pursuing doctoral research in the 1970s. His deep knowledge of the island's circulatory system of hot water flows enables him to explain, for example, why the water you shower with in Reykjavík probably last fell as rain back in A.D. 1000. As he came to grasp the size of the resource, he began to consider ways in which this untapped potential might be used. At the time, the climbing cost of oil imports was beginning to hit the fishing fleet, prompting discussion of alternative fuels—including hydrogen.

Iceland has been producing hydrogen since 1958, when it opened a state fertilizer plant on the outskirts of Reykjavík under the post-war Marshall Plan. The production process uses hydro-generated electricity to split water into hydrogen and oxygen molecules—a process called electrolysis (see Figure 1). The fertilizer plant uses about 13 megawatts of power annually to produce about 2,000 tons of liquid hydrogen, which is then used to make ammonia for the fertilizer industry. In 1980, Bragi Árnason and colleagues completed a lengthy study on the cost of electrolyzing much larger amounts of hydrogen, using not only hydroelectricity but geothermal steam as well—which can speed up what is a very high-temperature process. Their paper found that this approach would be cheaper than importing hydrogen or making it by conventional electrolysis, but it did not find a receptive audience as oil prices plummeted during the 1980s.

The early 1990s saw a reemergence of Icelandic interest in producing hydrogen, both for powering the fishing fleet and for export as a fuel to the European market. In a 1993 paper, Dr. Árnason argued that a transition in fuels from oil to hydrogen may be "a feasible future option for Iceland and a testing ground for changing fuel technology." He also contended that the country could benefit from using hydrogen sooner than other countries. Some of his reasons included Iceland's small population and high levels of technology, its abundance of hydropower and geothermal energy, and its absence of fossil fuel supplies. Another was the relatively simple infrastructural change involved in converting the fishing fleet from oil to hydrogen by locating small production plants in major harbor areas and adapting the boats for liquid hydrogen.

Early on, the plan was to use liquid hydrogen to fuel the boats' existing internal combustion engines. But "then

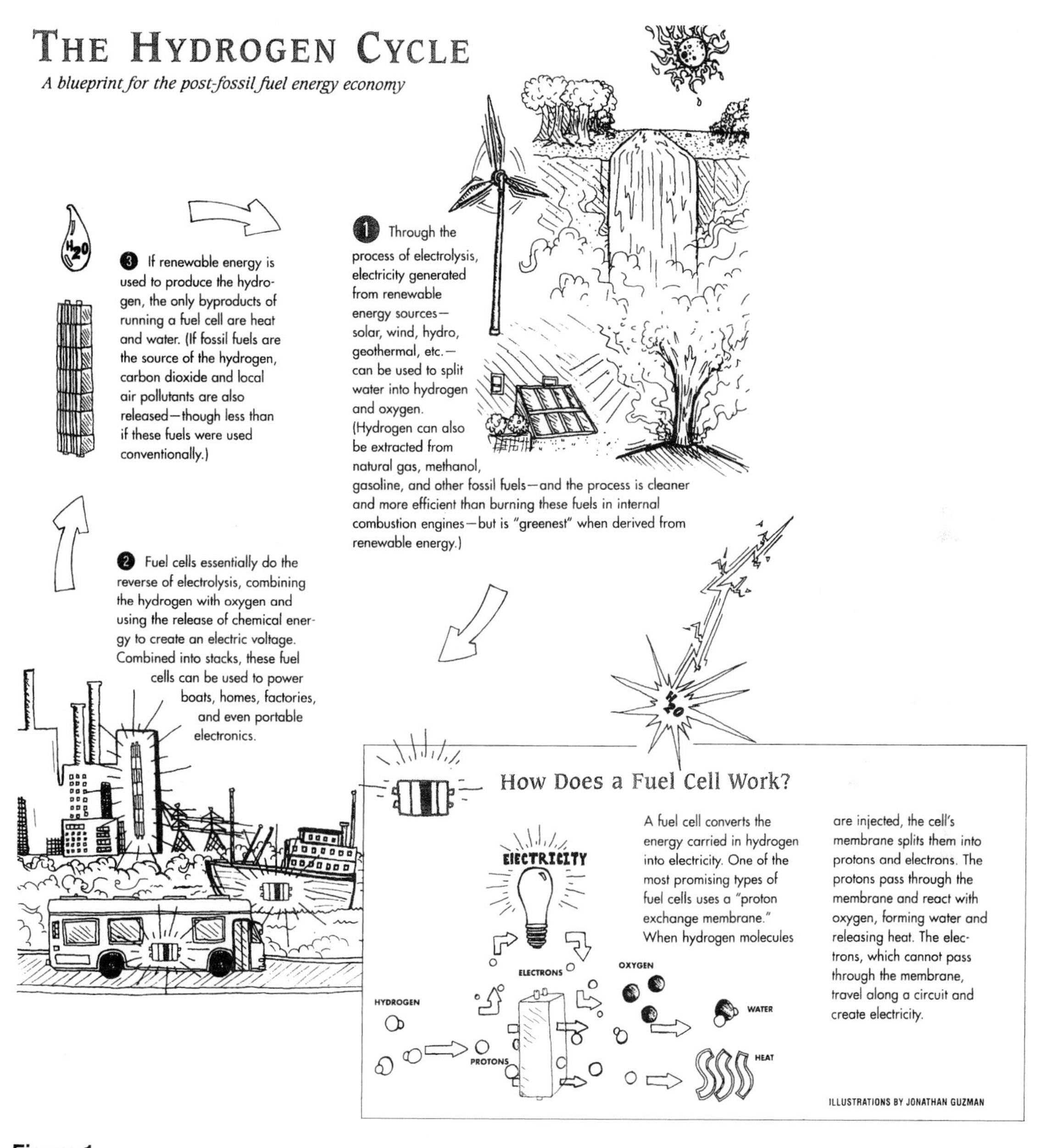

THE HYDROGEN CYCLE
A blueprint for the post-fossil fuel energy economy
1 Through the process of electrolysis, electricity generated from renewable energy sources—solar, wind, hydro, geothermal, etc.—can be used to split water into hydrogen and oxygen. (Hydrogen can also be extracted from natural gas, methanol, gasoline, and other fossil fuels—and the process is cleaner and more efficient than burning these fuels in internal combustion engines—but is "greenest" when derived from renewable energy.)
2 Fuel cells essentially do the reverse of electrolysis, combining the hydrogen with oxygen and using the release of chemical energy to create an electric voltage. Combined into stacks, these fuel cells can be used to power boats, homes, factories, and even portable electronics.
3 If renewable energy is used to produce the hydrogen, the only byproducts of running a fuel cell are heat and water. (If fossil fuels are the source of the hydrogen, carbon dioxide and local air pollutants are also released—though less than if these fuels were used conventionally.)
How Does a Fuel Cell Work?
A fuel cell converts the energy carried in hydrogen into electricity. One of the most promising types of fuel cells uses a "proton exchange membrane." When hydrogen molecules are injected, the cell's membrane splits them into protons and electrons. The protons pass through the membrane and react with oxygen, forming water and releasing heat. The electrons, which cannot pass through the membrane, travel along a circuit and create electricity.
ELECTRICITY
ELECTRONS
OXYGEN
HYDROGEN
WATER
PROTONS
HEAT
ILLUSTRATIONS BY JONATHAN GUZMAN

Figure 1
The hydrogen cycle.

came the fuel cell revolution," as Dr. Árnason puts it. By the late 1990s, the fuel cell, an electrochemical device that combines hydrogen and oxygen to produce electricity and water, had achieved dramatic cost reductions over the previous two decades. The technology had become the focus of engineers aiming to make fuel cells a viable replacement not only for the internal combustion engine, but for batteries in portable electronics and for power plants as well. Demonstrations of fuel cell-powered buses in Vancouver and Chicago, and their growing use in hundreds of locations in the United States, Europe, and Japan, caught the attention of governments and major automobile manufacturers. The fuel cell was increasingly viewed as the "enabling technology" for a hydrogen economy.

One Icelander particularly taken with these developments was a young man named Jón Björn Skúlason, who, while attending the University of British Columbia in Vancouver, became familiar with Ballard Power Systems, a leading fuel cell manufacturer headquartered just outside the city. Upon returning home, Skúlason encouraged the politician, Hjalmar Árnason, in his promotion of energy alternatives and hydrogen; his enthusiasm earned him a position on the expert committee. In 1998, the panel formally recommended that the nation consider converting fully to a hydrogen economy within 30 years.

By then, Hjalmar Árnason had already given the process a push. During a phone interview with a reporter from the *Economist*, he floated the year 2030 as a target date for the government's evolving hydrogen plans. The resulting article, published in August 1997, created a buzz abroad, and the parliamentarian received hundreds of phone calls from around the world. That fall, Iceland's prime minister released a statement announcing that the government was officially moving the country toward a hydrogen economy. The ministers of energy and industry, commerce, and environment signed on, as well as both sides of the two-party Althing. And Árnason obtained permission to start negotiating with interested members of industry.

A PIECE OF THE ACTION

Iceland has a tradition of "stock companies," or business cooperatives that evolved in the eighteenth century to help domestic farmers and fishers compete with the formidable Danish trading companies that at the time controlled fishing and goods manufacturing. The first of these, granted royal support in 1752, brought in weavers from Germany, farmers from Norway, and other overseas experts to teach the Icelanders the best methods of agriculture, boat-building, and the manufacture of woolen goods. Over the years, these long-lasting business associations helped the nation's enterprises survive and sometimes thrive.

The formation of the Icelandic Hydrogen and Fuel Cell Company (now Icelandic New Energy) can be seen as the latest example of this stock company tradition—but with a contemporary twist: German carmakers instead of weavers, Norwegian power companies rather than farmers. The first to contact Hjalmar Árnason after publication of the *Economist* article was DaimlerChrysler. Its roots traceable back to Otto Benz, designer of the first internal-combustion engine car, DaimlerChrysler now aspires to be the first maker of fuel cell-powered cars. The firm has entered into a $800 million partnership with Ballard Power Systems and Ford to produce fuel cell cars and plans to have the first buses and cars on European roads in 2002 and 2004, respectively—making Iceland a potentially valuable training ground, especially for testing fuel cell vehicles in a cold climate.

The second company to touch base with the Iceland government was Royal Dutch Shell, the Netherlands-based energy company that, among those now in the oil business, has perhaps the most advanced post-petroleum plans. Birthplace of the "scenario planning" technique that prepared it for the oil shocks of the 1970s better than most businesses, Shell has posited an Iceland-like future for the rest of the world, with 50 percent of energy coming from renewable sources by 2050. The firm surprised its colleagues in mid-1998 by creating a formal Shell Hydrogen division and then sending its representatives to the World Hydrogen Energy Conference in Buenos Aires.

The third group to establish communications with island officials was Norsk Hydro, a Norwegian energy and industry conglomerate. The company is involved in a trial run of a hydrogen fuel cell bus in Oslo and has considerable experience in hydrogen production: it has its own fertilizer business, and Norsk Hydro electrolyzers run Iceland's hydrogen-producing fertilizer plant. Norsk Hydro is also involved in the politically sensitive issue of Iceland's planned aluminum smelters, having signed commitments with the national power company and the ministries of energy and industry and commerce to construct a new smelter on the island's east coast.

Negotiations among these companies and the Icelandic government culminated in February 1999 with the creation of the Icelandic Hydrogen and Fuel Cell Company. Shell, DaimlerChrysler, and Norsk Hydro each hold shares of the company. The majority partner, *Vistorka* (which means "eco-energy"), is a holding company owned by a diverse array of Icelandic institutions and enterprises: the New Business Venture Fund, the University of Iceland, the National Fertilizer Plant, the Reykjanes Geothermal Power Plant, the Icelandic Technological Institute, and the Reykjavík Municipal Power Company. Also indirectly involved with the holding company is the Reykjavík City-Bus Company.

The stated purpose of the new joint venture is to "investigate the potential for replacing the use of fossil fuels in Iceland with hydrogen and creating the world's first hydrogen economy." On the day of its announcement, Iceland's environment minister stated: "The Government of Iceland welcomes the establishment of this company by these parties and considers that the choice of location for this project is an acknowledgement of Iceland's distinctive status and long-term potential." Like the *Economist* article, the announcement attracted industry attention. But for some companies, it was too late to climb on the bandwagon. Toyota officials reportedly attempted, to no avail, to take over the project by offering to foot its entire bill and supply all the needed engineers.

BUSES, CARS, AND BOATS

Bragi Árnason and a colleague, Thorsteinn Sigfússon, have outlined a gradual, five-phase scenario for the hydrogen transformation. In phase one (an estimated $8 million project that has received $1 million from the government), hydrogen fuel cells are to be demonstrated in Reykjavík's 100 municipal public transit buses. The current plan is to have three buses on the streets by 2002. The fertilizer plant will serve as the filling station for the buses, its hydrogen pressurized as a gas and stored on the roofs of the vehicles. Because enough hydrogen can be stored on board to run a bus for 250 kilometers, the average daily distance traveled by a Reykjavík bus, there is no need for a complicated infrastructure for distributing the fuel.

In phase two, the entire city bus fleet—and possibly those in other parts of the island—will be replaced by fuel cell buses. The Reykjavík bus fleet program has a price tag estimated at $50 million, and this spring [2000] received $3.5 million from the European Community. Phase three involves the introduction of private fuel cell passenger cars—which requires a more complicated infrastructure. At present, storing pressurized hydrogen gas on board a large number of smaller vehicles, with more geographically dispersed refueling requirements, is too expensive to be considered a realistic option. The first fuel cell cars are therefore expected to run not on hydrogen directly, but rather on liquid methanol—which contains bound hydrogen but must be reformed, or heated, on board the vehicle to produce the hydrogen to power the fuel cell.

Methanol is also, at the moment, the preferred fuel for the final two phases: the testing of a fuel cellpowered fishing vessel, followed by the replacement of the entire boating fleet. These trawlers use electric motors that are in the range of one to two megawatts—larger than those for cars and buses, but close to the size of the fuel cells that are now starting to be commercialized for stationary use in homes and buildings. Several European vessel manufacturers have already expressed their interest in becoming involved in this phase, and Dr. Árnason would like to see a fuel cell boat demonstrated no later than 2006.

But using methanol as an intermediate step to hydrogen is not without its problems. Skúlason, who is now president of Icelandic New Energy, notes that Shell is concerned about the use of methanol, particularly its toxicity. And since methanol reforming releases carbon dioxide, the environmental benefit is much less than if a way can be found to store the direct hydrogen on board, which in Iceland's case would mean complete elimination of greenhouse gas emissions. It's a difficult decision, notes Skúlason: "We must deal with the technologies we are given by the global companies."

Iceland will have to choose between two options: producing and distributing pure hydrogen and storing it on board vehicles (the "direct hydrogen" option) or producing hydrogen on board vehicles from other fuels—natural gas, methanol, ethanol, or gasoline—using a reformer (the "onboard reformer" option). In general, the automobile industry strongly favors the onboard option, using methanol and gasoline, because most existing service stations already handle these fuels. A third path, reforming natural gas at hydrogen refueling stations, is under consideration in countries like the United States that already have an extensive natural gas network, but is not practical in Iceland.

The up-front costs of direct hydrogen will be high because such a change requires a new infrastructure for transporting hydrogen, handling it at fueling stations, and storing the fuel on board as a compressed gas or liquid. According to DaimlerChrysler's Ferdinand Panik, retrofitting 30 percent of service stations in the U.S. states of New York, Massachusetts, and California for methanol distribution would cost about $400 million. Supplying hydrogen to these stations would cost about $1.4 billion.

But in terms of long-term societal benefits, direct hydrogen is the clear winner. Using hydrogen directly is more efficient because of the extra weight of the processor and lower hydrogen content of the methanol or gasoline. It is also less complex than having a reformer on board each vehicle—which adds $1,500 to the cost of a new car, takes time to warm up, and creates maintenance problems. As the vehicle population grows large enough to cover the capital costs of providing refueling facilities, the costs of direct hydrogen will become comparable to the onboard option. Once the infrastructure and vehicles are put in place, using hydrogen fuel will be more cost-effective than having cars with reformers—even excluding the environmental gains.

If Iceland, with its heavy renewable energy reliance, were to switch directly to hydrogen, the country would have *no* greenhouse emissions. And in fact, it is much easier to produce hydrogen than methanol from renewable energy through electrolysis. Thus, as renewables become more prominent around the world, a hydrogen infrastructure will emerge as the most practical option. In Iceland, rather than require that hydrogen first be used to create, and then be reformed from, methanol, the simplest approach would be to use geothermal power and hydropower, augmented by geothermal steam, to electrolyze water, creating pure hydrogen to drive cars and boats. But behind the seeming solidarity of the public/private venture, a fateful struggle may be emerging.

A FORK IN THE ROAD

In spite of the long-term economic and environmental advantages of the direct hydrogen approach, industry and government—both in Iceland and worldwide—have devoted substantially greater attention and financial support to the intermediate approach of using methanol and onboard reformers. Car companies are hesitant to mass-produce a car that cannot be easily refueled at many locations. Energy companies, similarly, are loathe to invest in pipelines and fuel stations for vehicles that have yet to hit the market. This is a classic case of what some engineers call the chicken-and-egg dilemma of creating a fueling infrastructure. But the potential public benefits—especially for addressing climate change—give governments around the world incentive to steer the private sector toward the optimal long-term solution of a hydrogen infrastructure by supporting additional research into hydrogen storage and by collaborating with industry.

In Iceland's case, producing pure hydrogen through electrolysis by hydropower is at the moment three times as expensive as importing gasoline. But the fuel cells now being readied for the transportation market are three times as efficient as an internal combustion engine. In other words, running the island's transport and fishing sectors off pure hydrogen from hydropower is becoming economically competitive with operating conventional gasoline-run cars and diesel-run boats.

Since the methanol reformers that these fuel cells will presumably use are still several years away from mass production, some scientists see the next few years as an important window of opportunity to prove the viability of direct hydrogen technology. But the history of technology is littered with examples of inferior technologies "locking out" rivals: witness VHS versus Beta in the videocassette recorder market. If methanol does gain market dominance and locks out the direct hydrogen approach, it may be decades before real hydrogen cars become widespread—a wrong turn that could take the Icelandic venture kilometers from its destination. By the time a full-blown methanol infrastructure was put in place, it would probably no longer be the preferred fuel—committing the country to a fleet of obsolete cars and causing the consortium to strand millions of *kronur* in financial assets.

Yet some outside developments are pointing in the direction of direct hydrogen. In California, where legislation requires that 10 percent of new cars sold in 2003 must produce "zero emissions," a consortium called the California Fuel Cell Partnership is planning to test out 50 fuel cell vehicles and build two hydrogen fueling stations that will pump hydrogen gas into onboard fuel tanks. Hydrogen fueling stations have already been built in Sacramento (California's capital), Dearborn, Michigan (home to Ford headquarters), and the airport at Frankfurt, Germany—the last of which expects to eventually import hydrogen from Iceland. The prospect of Iceland becoming a major hydrogen exporter, perhaps the new energy era's "Kuwait of the North," surfaces several times during my interviews—and is no doubt a good selling point for the strategy to officials inclined to think more in narrow economic terms.

Skúlason assures me that there is a "very open discussion" underway within the consortium and says "We have to take steps slowly because there might be a shift." He admits that he would prefer to see compressed hydrogen gas used, noting the advantages of having a direct hydrogen fuel infrastructure and vehicles. Shell and DaimlerChrysler themselves seem to recognize the potential competitive advantage of putting up hydrogen filling stations and reformer-free cars right from the beginning, giving them a head start in preparing for a world fueled by hydrogen. At a June 2000 conference in Washington, DC,

Shell Hydrogen CEO Don Huberts asserted that direct hydrogen was the best fuel for fuel cells and suggested that geothermal energy converted to hydrogen would be the main means for converting the Icelandic economy. DaimlerChrysler representatives have admitted that their methanol reformers are relatively expensive and large—they take up the entire back seat—and the company has recently rolled out "next-generation" prototype cars that run on liquid and compressed hydrogen—prime candidates for the Iceland strategy.

"The transition is messy," the politician Hjalmar Árnason tells me. "We have one leg in the old world, and one in the new." It's an apt metaphor, given Iceland's geography.

But the question is whether the Icelandic venture will, in rather un-Viking fashion, cautiously creep ahead—sticking to the onboard methanol approach—or, brashly set *both* feet in the new world, voyaging straight to direct hydrogen. As a world leader in utilizing renewable energy sources, if Iceland does not take the "newest" path, governments and businesses elsewhere may extract the wrong conclusion from its experiment and give short shrift to the direct hydrogen option. Skúlason nails the conundrum: "How many times will we shift? Will it be cheaper for society to pay a little more now and not have to rebuild? This argument doesn't always work with government or the consumer."

Professor Árnason is quick to note that, whichever short-term infrastructural path the country takes, "the final destination is the same:" pure hydrogen, derived from renewable energy and used directly in fuel cells. But he acknowledges that there may be significant costs in taking the gradual approach. And he agrees that the assumption on which his scenario is based—that methanol is the most economical option—is "subject to revision." The cost and efficiency of fuel cells will continue to improve, and advances in carbon nanotubes, metal hydrides, and other storage technologies are making it more feasible to store hydrogen on board. The high cost of electrolysis is likely to decline sharply with technical improvements, while other sources of hydrogen—tapping solar, wind, and tidal power, splitting water with direct sunlight, playing with the metabolism of photosynthetic algae—are on the horizon. And new climate policies or fluctuating fuel prices from volatile oil markets "would change the whole picture."

WHY ICELAND?

When he first met with his prospective joint venture partners, Bragi Árnason posed this query: "Why are you interested in coming to Iceland?" He asked the question because "we were quite surprised to learn about the strong interest of these companies in participating in a joint venture with little Iceland." Their answers shed light on some of the elements that may be useful for developing a hydrogen economy elsewhere in the world.

Without a doubt, the most critical element of getting the Iceland experiment underway has been the government's clearly stated commitment to transforming itself into a hydrogen economy within a set timeframe. A similar dynamic is at work in California, where the zero-emission mandate has forced energy and transport companies to join forces with the public sector to seriously explore hydrogen. For Dr. Árnason, the lesson is clear: a strong public commitment can attract and encourage the participation of private-sector leaders, resulting in partnerships that provide the financial and technical support needed to move toward environmental solutions. "You *must* have the politicians," he says.

In addition, companies have shown interest in the Iceland experiment because the results will be applicable around the world. While the country's hydrogen can be produced completely by renewable energy, its car and bus system and heavy reliance on petroleum—amplified by its island setting—are common characteristics of industrial nations, making the result somewhat adaptable. The island's head start in transitioning to renewable energy also makes it a good place to test out this larger shift.

Iceland may also have something more to tell us about the more general cultural building blocks that can enable the evolution of a hydrogen society. Icelanders treasure their hard-won independence, and the prospect of energy self-reliance is attractive. Hjalmar Árnason likes to emphasize his homeland's "free, open society," which he believes has maintained a political process more conducive to bold proposals and less subject to special-interest influence and partisan gridlock. He points, too, to the country's openness to new technology—to its willingness to take part in international scientific endeavors, such as global research in human genetics. He hopes Iceland will become a training ground for hydrogen scientists from around the world, cooperating internationally to convert its NATO base to hydrogen. Skúlason cites a poll of Reykjavík citizens indicating that 60 percent of the citizens were familiar with and supportive of the hydrogen strategy—though some ask about the safety of the fuel (it is as safe as gasoline), pointing to the need for public education campaigns before people will be persuaded to buy fuel cell cars.

Another important cultural factor has been what Árni Finnsson, of the Icelandic Nature Conservation, describes as his nation's relatively recent but increasing "encounter with the globalization of environmental issues." This encounter originated with the emotional whaling disputes of the 1970s and 80s, and today includes debates about persistent organic pollutants and climate change. As Icelanders seek to become more a part of global society, so too do they seek legitimacy on global issues, forcing their government to sensitize itself to emerging cross-border debates—a process that has sometimes created Iceland's political equivalent of volcanic eruptions.

Finnsson points out that, thanks to the "Kyoto dilemma," Icelandic climate policy is not terribly progressive, consisting mainly of efforts to create loopholes that would allow additional greenhouse gas emissions from its new aluminum smelters. But there is little doubt that this dilemma has also unwittingly helped encourage the hydrogen strategy by forcing the nation to explore deep changes in its energy system. In a land that, even as it becomes wired to the information age, routinely blocks new road projects due to age-old superstitions of upsetting elves and other "hidden people," it's a contradiction that somehow seems appropriate. A country that has stubbornly refused to sign the Kyoto Protocol provides the most compelling evidence to date that climate change concerns—and commitments—will increasingly drive the great hydrogen transformation.

But my favorite, if least provable, theory for "Why Iceland?" comes from the heroic ideals of its sagas. One of the recurring themes of these remarkable literary works is that a person's true value lies in renown after death, in becoming a force in the lives of later generations through one's deeds. Listening to Bragi Árnason, who is now 65, one cannot help but wonder whether this cultural concern for renown is playing a part in the saga now unfolding: how Iceland became the world's first hydrogen society, inspiring the rest of the globe to follow its lead. "Many people ask me how soon this will happen. I tell them, 'We are living at the beginning of the transition. You will see the end of it. And your children, they will live in this world.'"

QUESTIONS

1. Who first proposed the use of hydrogen as a viable energy source?
2. What percent of Iceland's energy needs is met by renewable energy sources?
3. Describe the operation of a fuel cell.
4. Discuss various features of Iceland's "hydrogen society" strategy.
5. Discuss potential impediments to the completion of the hydrogen experiment in Iceland.

Current Status of Iceland's Hydrogen Economy: Use Internet search engines (Google, Yahoo!, or AltaVista, for example) to answer the following questions:

6. What is the status of development and marketing of Ford and DiamlerChrysler fuel cell cars?
7. Where does Iceland stand today in its pursuit of becoming a "hydrogen society"? Read "Missing in Action: Iceland's Hydrogen Economy," by Freyr Sverrisson, *WorldWatch*, November/December 2006 (Available at http://www.worldwatch.org/node/4664) to answer this question.

Biomass Energy

AHMED S. KHAN

Biomass refers to any organic material—except coal, oil, natural gas, etc.—that is used for energy generation. It includes terrestrial and aquatic vegetation, agricultural and forester residues, and animal wastes. It can be regarded as the solar energy that has been collected by photosynthesis and is stored as chemical energy from the organic material. In photosynthesis, plants convert radiant energy from the sun into chemical energy by combining CO_2 and water to form carbohydrate molecules in the form of glucose or sugar. Figure 1 illustrates the bioenergy/carbon cycle. The photosynthesis process can be summarized as a chemical equation.

$$mH_2O + nCO_2 + \text{Light energy} \text{------>} C_6(H_{12}O)_m + nO_2$$

Biomass is a renewable energy source—it draws energy from the ambient environment rather than from the consumption of mineral fuels. The growth of new plants and trees replenishes the supply. Renewable energy systems are independent of fuel supply and price variations and are economically more viable. In contrast to fossil fuels, renewable energy sources are more uniformly distributed geographically. Many renewable energy technologies are presently being employed or have the potential of becoming viable: biomass power, hydropower, geothermal power, wind power, ocean wave power, ocean tidal power, solar and thermal heating power, and ocean thermal electric power.

Typical applications for biomass include industrial-process heat generation; electrical generation; generation of heat and power for residential, industrial, or agricultural purposes; and production of liquid fuel for vehicles. Table 1 lists the key biomass energy application areas and conversion processes, and Table 2 presents a summary of the biomass energy sources, conversion techniques/technologies, products, and their applications.

In the pre-industrial revolution era, wood was the main fuel used for cooking and space heating. Until the mid 1800s, wood provided 90 percent of the energy used in the United States. Even today, in many developing countries, wood supplies more than one-half of the energy consumption. In the United States in 1997, biomass provided 29 percent of renewable energy and 1.25 percent of the total energy. In 2002, biomass sources supplied about 47 percent of all renewable energy consumed in the United States.

The industrial sector remains the largest sector of biomass energy users. Seventy-seven percent of the energy produced by biomass sources is used by the industrial sector. Globally, biomass accounts for 14 percent of the world's energy needs. In the developed countries, about 7 percent of total primary energy comes from renewable sources, with hydropower accounting for 75 percent of this energy. In the developing countries, future demand for biomass will be determined by two opposing factors: (1) increase in the demand for biomass energy due to population growth and (2) reduction

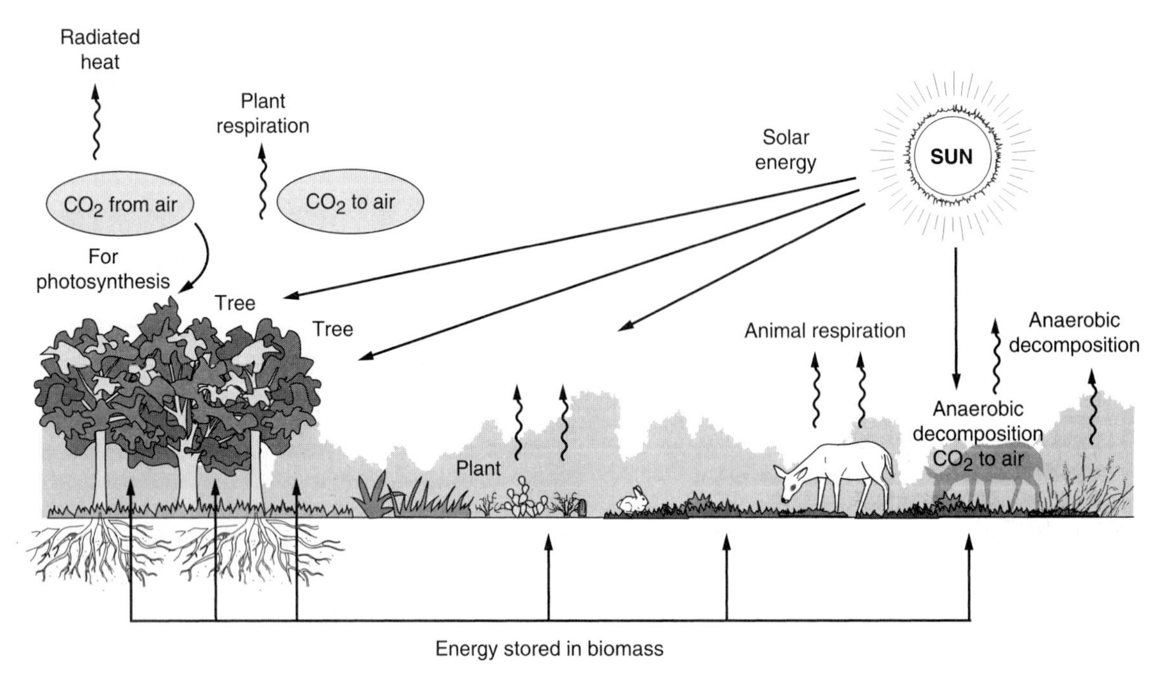

Figure 1
Bioenergy/carbon cycle.

Table 1
Biomass Energy Key Applications and Conversion Processes

Key Applications	
Biofuels	Conversion of biomass into liquid fuels for transportation
Biopower	Conversion of biomass into gaseous fuel or oil to generate electricity
Bioproducts	Conversion of biomass into chemicals for making products that generally are made from petroleum
Key Conversion Processes	
Thermochemical Conversion	
Combustion	Burning of biomass directly in a stove, furnace, or boiler
Gasification	Conversion of biomass into a gaseous fuel composed of H_2 and CO that retains most of its heating value
Pyrolysis	Thermal decomposition of biomass to produce a combination of solid, liquid, and gaseous products that are combustible
Biochemical Conversion	
Fermentation	Conversion of biomass (carbohydrates) through distillation to ethanol (C_2H_5OH) for blending it with gasoline for motor vehicle use
Anaerobic digestion	Conversion of biomass to a gaseous mixture of CO_2 and CH_4

Table 2
Biomass Energy Sources, Conversion Techniques/Technologies, Products, and Their Applications

Source	*Conversion Technique/Technology*	*Products*	*Applications*
Terrestrial Biomass	Biochemical/Ethanol production	Liquid fuels or biofuels (e.g., ethanol)	Generation of electric power for transportation sector
Forestry, agricultural and animal residues	Thermochemical/ chemical /biodiesel production/methanol production	Gaseous fuel or oil (e.g., biodiesel) /biopower	Onsite fuel supply for industrial/agricultural sector
Aquatic biomass	Photochemical	Petrochemical substitutes/bioproducts	Residential and commercial sector
		Electricity and heat	Utility sector

Source: Hunt, V. (1982). *Handbook of Energy Technology: Trends and Perspectives.* New York: Van Nostrand Reinhold Company, p 437.

in the demand of biomass energy as result of a shift to fossil fuels.

Advantages

- There is an abundance of biomass sources (e.g., crops, plants, trees, wood residues, crop waste, animal wastes).
- Biomass sources are inexpensive and widely available.
- Biomass sources are easy to store and transport.
- Biomass is a renewable source of energy.
- Biomass sources allow the disposal of biomass waste material.
- Biomass sources do not contribute to a net increase in CO_2 (provided that the new biomass growth balances the biomass used for energy generation).
- Biomass sources contain small amounts of sulfur and nitrogen and do not produce pollutants that cause acid rain.

Disadvantages

- Air emissions from the combustion of biomass sources are not always less than emissions from fossil fuels.
- Growth of new plants and trees to replenish the consumed amount of biomass requires considerable time. (It may require a few months for new crops and plants to grow and may take years and decades for trees to grow.)

REFERENCES

"Biomass Energy." Available at http://www.egov.oregon.gov/ENERGY/RENEW/BiomassHome.shtml. Accessed September 5, 2006.

Biomass Energy Basics. Available at http://www.nrel.gov/learning/re_biomass.html. Accessed September 5, 2006.

Boyle, Godfrey. (2004). *Renewable Energy: Power for a Sustainable Future,* 2nd ed. Oxford: Oxford University Press.

Fay, James, and Golomb, Dan. (2002). *Energy and the Environment.* New York: Oxford University Press.

Hunt, V. (1982). *Handbook of Energy Technology: Trends and Perspectives.* New York: Van Nostrand Reinhold Company, pp. 33, 434–446, 376–377.

"What Is Biomass?" Available at http://lsa.colorado.edu/essence/texts/biomass.htm. Accessed September 5, 2006.

QUESTIONS

1. How does light energy get stored as chemical energy in biomass materials?
2. Why is biomass energy considered to be a renewable source?
3. Discuss the future growth of biomass energy in both the developed and the developing countries of the world.

4. List advantages (+) and disadvantages (–) of biomass energy.

(+)	(–)

5. What biomass conversion techniques/technologies are used to produce the following bioproducts?
   a. biofuels (ethanol)
   b. biodiesel
   c. bioproducts

Solar Energy

AHMED S. KHAN
BARBARA A. EICHLER

The sun was the major source of energy before the Industrial Age (see Table 1). In today's electro-optics industrial age, it still remains a vast reservoir of energy that can fulfill today's and tomorrow's growing demands for energy. According to A.B. Lovins, "Solar power in general has several unique implications which do not arise from its obvious advantages. For example, it could help to redress the severe temperate and tropical zones; its diffuseness is a spur to decentralization and increased self-sufficiency of population; and as the least sophisticated major energy technology it could greatly reduce tensions resulting from the uneven distribution of fuels and from limited transfer of technology."

The sun is a continuous source of energy. The mass of the sun is 2.2×10^{27} tons. The sun converts its mass into energy at a rate of 4.2 million tons per second which is 0.02×10^{18} percent of its mass per second. At this rate, the sun is expected to radiate energy for billions of years.

Solar radiation is Earth's main income energy. Outside the earth's atmosphere, the sun provides energy at a rate of 1,353 watts/m normal to sun. Considering the diametral plane of Earth (1.27×10^{14} m), the solar input is 1.73×10^{17} watts, but only a small fraction of this reaches Earth's surface. The majority of solar energy incident on Earth is lost due to reflection by the atmosphere, scattering by clouds, water particles, and dust particles (see Figure 1). The average solar radiation reaching the Earth's surface is approximately 630 watts per meter. The energy emitted by the sun is spread over a broad band of an electromagnetic spectrum. Most of the solar radiation lies in the range of 0.3 micrometers (μm) to 1.3 μm. Solar radiation does not arrive with constant intensity during the day. The average intensity varies with cloud cover, latitude, season, and time of the day. Solar energy could be used in a number of applications (see Figure 2). In its intrinsic form, the solar radiation can be used for heating. For other applications, it needs to be converted into other forms of energy (see Figures 3, 4).

In 2003, cumulative solar PV-generating capacity rose 31 percent to 3,146 megawatts. The largest markets for solar PVs in 2003 were Japan, Germany, and California. Japan and Germany together installed 53 percent of the PV capacity produced that year, bringing Japan's cumulative PV capacity to 887 megawatts and Germany's total to 417 megawatts. Only five countries (Denmark, Germany, India, Japan, and Spain) and a half dozen U.S. states account for approximately 80 percent of the world market for wind and solar energy technologies.

Advantages

- Solar energy is a continuous source of energy.
- Solar energy is a clean source of energy.
- Solar energy is a safe source of energy.

Disadvantages

- The average solar power is not steady and depends on cloud cover, season, latitude, and time of day.

Table 1
Use of Solar Energy

B.C.	Solar heat for distillation of liquids and drying agricultural products
1913	Solar power electrical plants in Egypt
1960	Use of solar water heaters in Florida
1967	Use of silicon solar cells in Japan for isolated radio repeaters
1968	Development of large-scale mirrored solar oven; produces one megawatt of power per day
1974	Breakthrough in mass production of photovoltaic (PV) cells
1980s	Afghan refugees in Pakistan use 100,000 solar cookers for preparing food.10-MW Solar One System constructed in Barstow, California
1990s	President Clinton announces the One Million Solar Roofs program.
2000s	World PV production exceeds 700 MW. In 2004, the U.S. domestic shipment of photovoltaic cells and modules reached an all-time high (78,346). For the period from 1995 to 2004, 301,530 shipments were made. Solar energy consumption in the United States reached 0.006 quadrillion Btu.

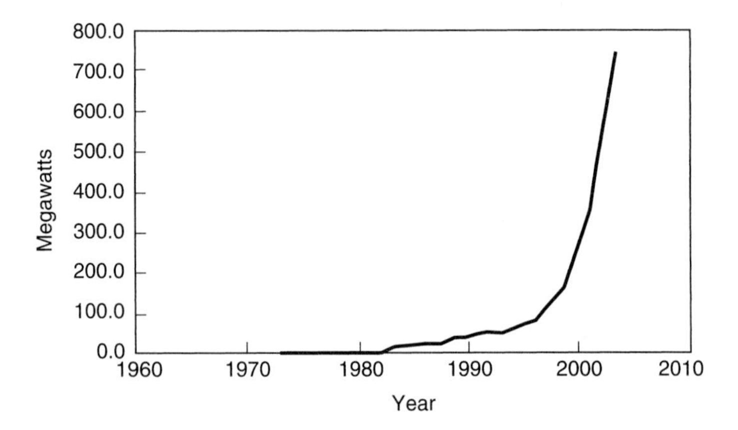

Figure 1
World photovoltaic production, 1971–2003.
Source: *Vital Signs 2002*, Worldwatch Institute, Washington, D.C.

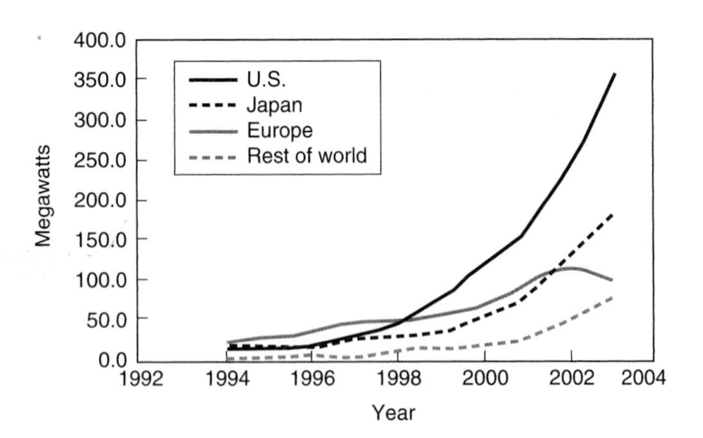

Figure 2
Photovoltaic production by country or region, 1994–2003.
Source: *Vital Signs 2002,* Worldwatch Institute, Washington, D.C.

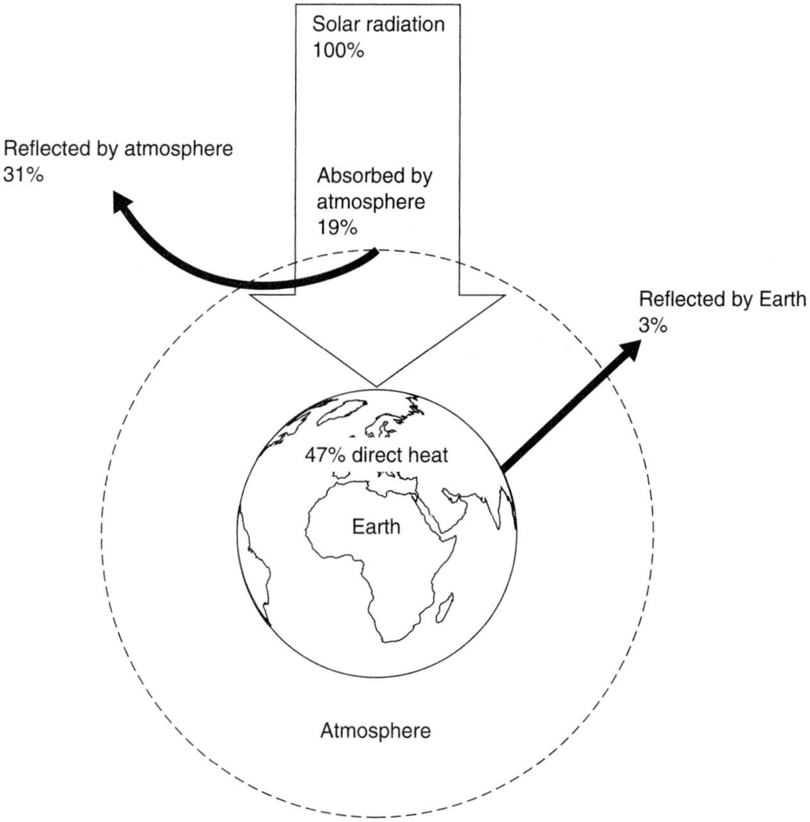

Solar radiation
100%
Reflected by atmosphere
31%
Absorbed by
atmosphere
19%
Reflected by Earth
3%
47% direct heat
Earth
Atmosphere

Figure 3
Solar radiation incident on Earth.

Solar systems
Terrestrial
Extraterrestrial
Solar collectors
Photovoltaics
Solar collectors
Photovoltaics

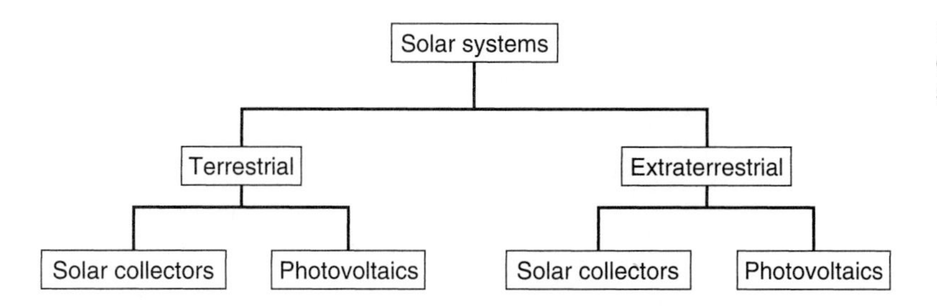

Figure 4
Classification of Solar Energy Systems.

BOX 1 Terrestrial and Extraterrestrial Systems

Terrestrial Systems

Terrestrial systems consist of flat-plate or concentrating collectors that transfer solar heat to a carrying medium, such as water. The impounded heat can be stored or converted immediately to work. Terrestrial collectors are subject to the intermittent nature of solar cycles and local climates.

Extraterrestrial Systems

The amount of solar power available for power generation in a geosynchronous orbit (35,800 km from Earth) is about 15 times that available on Earth. The idea of an orbital solar power station was presented by P. E. Glaser in 1973. According to Glaser, a 5000-MW orbital power station employs concentrators to reflect sunlight on photoelectric solar cells assembled into two large arrays, with each being 4.33 km $\times$ 5.2 km in dimension. The generated electrical power is converted into microwave power that can be transmitted toward Earth in a focused beam with the help of a 1-km diameter antenna. On the Earth, the microwave radiation is received by a 7.12-km diameter antenna. The received microwave power can be converted into direct current (DC) or 60-Hz alternating current (AC) for power distribution. The power density of microwaves varies as a function of its frequency—the higher the frequency, the higher the power density. High-frequency microwaves can cause damage to living tissue.

BOX 2 Scenario

Fort Worth, Texas: March 1, 20XX

TERASOLAR-I, the first orbital solar power generating station (see Figure 5), has completed its twentieth year of service in space. During its 20-year service, it has provided an average of 8,000 MW (45 percent of the power consumption of Texas) of power to a receiving station.

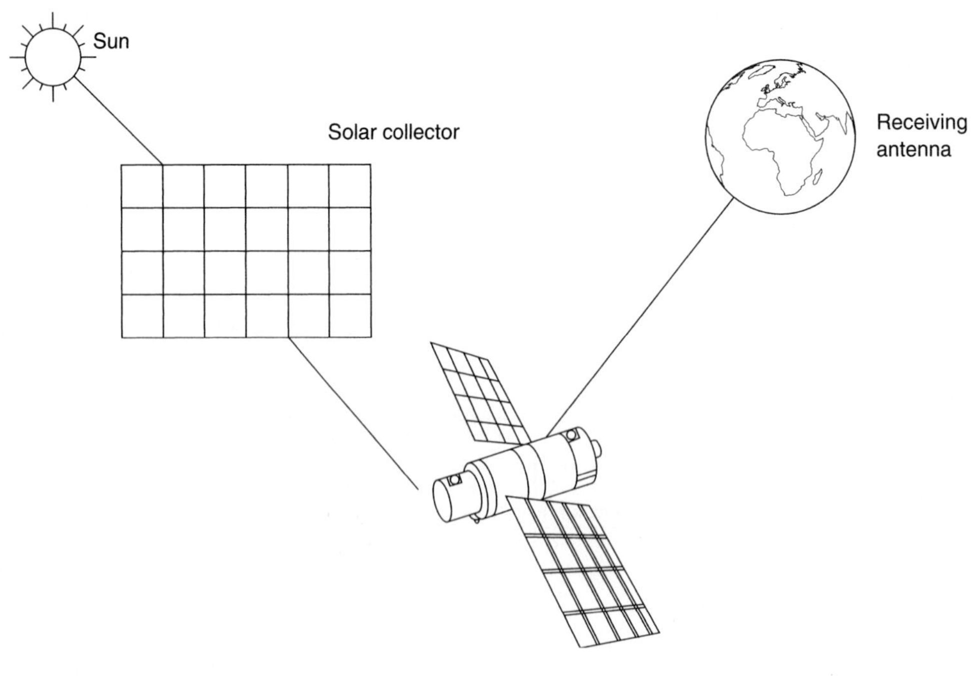

Figure 5
TERASOLAR-1, an orbital solar power–generating station.

The station transmits power at 30 GHz. Originally, the power station was designed to use a microwave frequency band of 300 MHz, but to decrease the size of the orbital transmitting antenna and Earth's receiving antenna and to cut the cost of the project, the use of a 30-GHz band was approved. A recent medical survey in Dallas reports that the number of cases of cancer has increased by 30 percent.

Response

Do you think that the radiation from the orbital solar power station is responsible for an increase in the number of cancer cases in the Fort Worth area? How could this have been avoided? Going back into the twentieth century, develop an energy strategy that could have avoided this situation.

- Solar energy is not available at night.
- Solar energy needs to be converted into other forms (e.g., electrical) for useful applications.
- Extraterrestrial solar energy will increase the heat burden of the biosphere.
- Efficiency of photovoltaic (PV) cells is low (10-20%). [See figure below.]

An array of photovoltaic (PV) solar panels.

REFERENCES

Dorf, R. (1981). *The Energy Factbook*. New York: McGraw-Hill.

Flavin, Christopher (2004). *Trends Overview: Energy.* Signposts 2004 CD-ROM. Washington; Worldwatch Institute.

Gabel, M. (1975). *Energy, Earth, and Everyone*. San Francisco: Straight Arrow Books.

Knoepfel, H. (1986). *Energy 2000: An Overview of the World's Energy Resources in the Decades to Come*. New York: Gordon and Breach Science Publishers.

World Resources 1988–89, A Report by World Resources Institute. New York: Basic Books.

QUESTIONS

1. Define the following terms.
   a. Terrestrial systems
   b. Extraterrestrial systems

2. List some advantages (+) and disadvantages (–) of solar energy.

(+)	(–)

3. Define the following terms.
   a. Photovoltaic
   b. Inverters
4. Determine the solar (photovoltaic) power–generating capacity for the following developed and developing countries.

Developed Countries	*Solar Power–Generating Capacity (MW)*	*Developing Countries*	*Solar Power–Generating Capacity (MW)*
United States		Bangladesh	
France		China	
Japan		Ethiopia	
Australia		Ghana	
Norway		India	
Sweden		Indonesia	
Switzerland		Kenya	
Italy		Malaysia	
Singapore		Pakistan	
United Kingdom		Venezuela	
Germany		Zimbabwe	

Wind Power

AHMED S. KHAN
BARBARA A. EICHLER

The atmosphere is a reservoir of solar radiation. The wind is continuously regenerated in the atmosphere as the solar radiation is converted into kinetic energy. The winds are local as well as regional. It is estimated that the average power available from shifting air masses all over the earth is 1.8×10^{15} watts. Wind power available at any location depends on its topographical features. The earth's surface offers resistance to wind, thereby decreasing its power density. In some locations, a wind power density of 500 W/m 10^2 is available at a nominal height of 25 m from the ground.

The use of wind power dates back thousands of years. It was employed for grinding grain and pumping water. Table 1 lists the history of wind power. Wind power can be used to turn vanes, blades, or propellers attached to a shaft. The revolving shaft spins the rotor of a generator, which produces electricity.

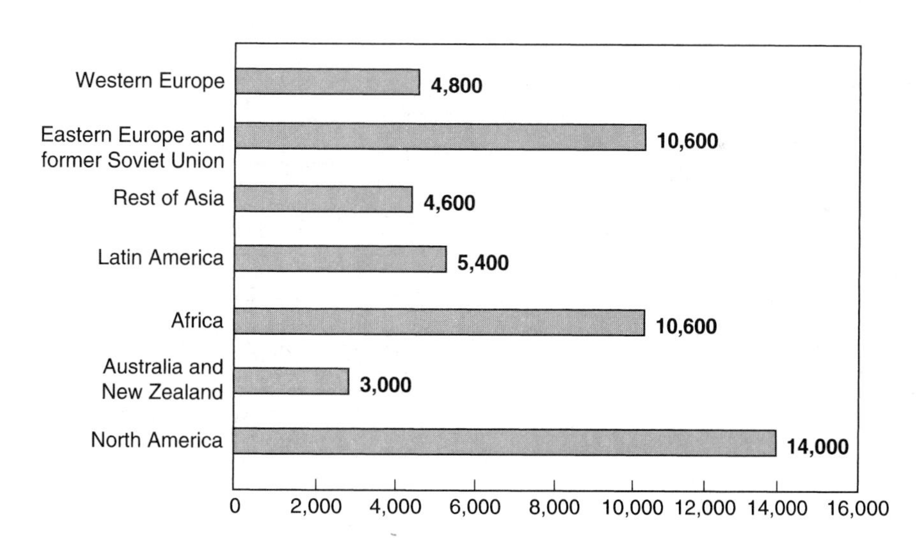

Figure 1
World land-based energy resources (TWh).
Source: *Vital Signs 2003*, Worldwatch Institute, Washington, D.C.

Table 1
History of Wind Power

1000 B.C.	Wind power of sailing ships
1850	Use of windmills in America
1894	First use of wind power for electric generation by Arctic explorer Nasen
1929	Development of electric wind turbine, 20 meters in diameter, in Bourget, France
1931	Development of 100-kW wind turbine capacity in Yalta, USSR
1941	Development of 1,250-kW capacity wind turbine in Vermont, USA
1950	Development of 10-kW Hutter wind generator consisting of 200-foot blades atop 475-foot tower
1951	Development of Thomas 6,500-kW generator, consisting of 200-foot blades atop 475-foot tower
1954	In USSR, the number of wind power plants reached 29,500, with a total capacity of 1 billion kWh.
1957	Development of 200-kW fully automated unit with three 45-foot blades mounted on a 75-foot tower, in Denmark
1960	600-kW Gedser generator designed
1980s	Advances in wind power generators
1990s	Development of efficient wind turbines, with an average turbine size of 100–200 kW
2000s	Average turbine size of 900 kW; wind power-generating capacity surpasses 40,000 MW

During the past two decades, wind energy technology has evolved to the point where it can compete with most conventional forms of power generation. In 2004, wind power–generating capacity increased by 26 percent and surpassed the 40,000-megawatt mark for the first time. New advances in turbine technology have increased the lifetime of wind turbines, improved performance, and reduced costs. The average turbine size has increased from 100–200 kilowatts (kW) in the early 1990s to more than 900 kW in 2004 (see Figures 3, 4).

The world's total land-based wind energy potential is estimated at over 53,000 terawatts-hours (TWh) per year (Grubb and Meyers; see Figure 1). Figure 2 illustrates the world wind-generating capacity from 1980–2003.

Advantages

- Wind power is a continuous source of energy.
- Wind power is a clean source of energy, with no emissions into the atmosphere.
- Wind power does not add to the thermal burden of the Earth.

Disadvantages

- For most locations, the wind power density is low.
- In most cases, wind velocity must be greater than 7 mph to be usable.
- A problem exists in the variation in the power density and duration of wind.
- Wind power may have some environmental effects, depending on the location and number of wind power plants (local climate, bird migration patterns, etc.).

REFERENCES

Dorf, R. (1981). *The Energy Factbook.* New York: McGraw-Hill.

Gabel, M. (1975). *Energy, Earth, and Everyone.* San Francisco: Straight Arrow Books.

Knoepfel, H. (1986). *Energy 2000: An Overview of the World's Energy Resources in the Decades to Come.* New York: Gordon and Breach Science Publishers.

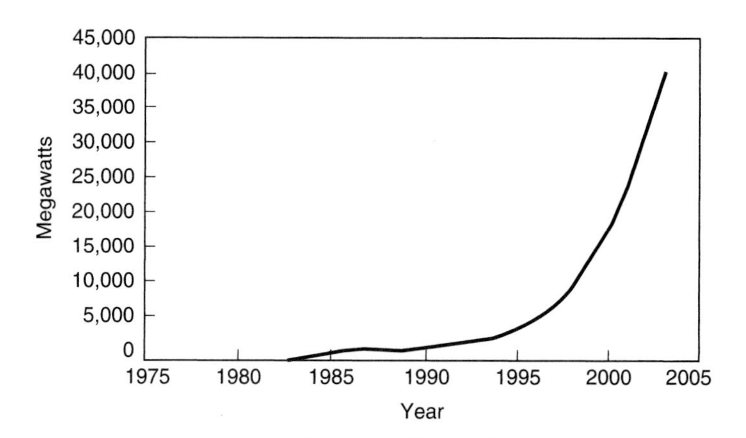

Figure 2
World wind energy-generating capacity, 1980–2003.
Source: *Vital Signs 2003*, Worldwatch Institute, Washington, D.C.

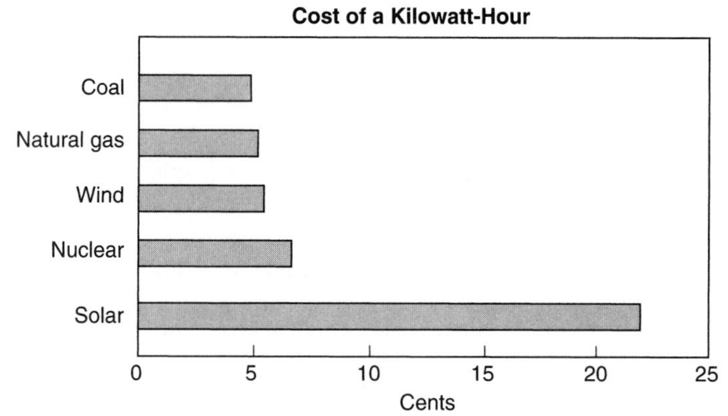

Figure 3
Cost of a Kilowatt-Hour

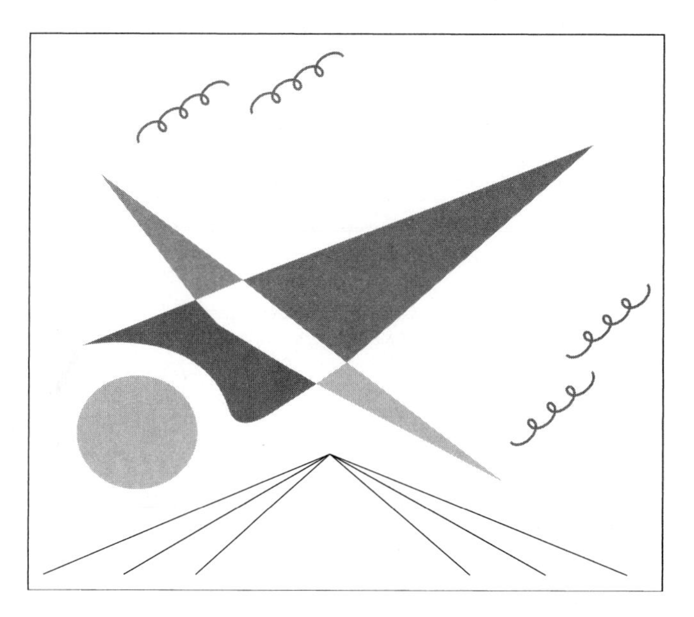

Figure 4
Wind Power: Wind power takes off.
Source: Ahmed S. Khan.

World Resources 1988–89, A Report by World Resources Institute. New York: Basic Books.

Signpost 2004 CD-ROM. Energy Trends. Washington, D.C.: Worldwatch Institute.

QUESTIONS

1. How is wind power used to generate electricity?
2. List the advantages (+) and disadvantages (–) of wind power.

(+)	(–)

3. Determine the wind power-generating capacity for the following developed and developing countries.

Developed Countries	*Wind Power–Generating Capacity (MW)*	*Developing Countries*	*Wind Power–Generating Capacity (MW)*
United States		Bangladesh	
France		China	
Japan		Ethiopia	
Australia		Ghana	
Norway		India	
Sweden		Indonesia	
Switzerland		Kenya	
Italy		Malaysia	
Singapore		Pakistan	
United Kingdom		Venezuela	
Germany		Zimbabwe	

Use an Internet search engine (Google, Yahoo!, AltaVista, etc.) to answer the following questions:

4. What percentage of the world's total electricity is produced by wind power?
5. Can wind power compete with new fossil fuel–fired plants?
6. Compare the wind-generating capacities of Asia, Europe, and North America.
7. Discuss various factors that impede the development of wind power projects in the developing world.
8. Will wind power someday replace fossil fuels?

23 Hydroelectric Power

AHMED S. KHAN
BARBARA A. EICHLER

Hydroelectric power is the conversion of the gravitational pull of the falling water of rivers and the controlled release of water reservoirs through turbine generators. Hydroelectric power (see Tables 1, 2, and 3) provides a clean and efficient means of producing electric power. It supplied 21 percent of electricity worldwide in 1986, less than coal and oil, but more than nuclear power. In 1984, the global yearly production of hydro energy amounted to about 1,700 billion kilowatt hours (kWh); an additional 550 was under construction. In 1999, hydro power provided 19 percent (2,659 trillion watt-hours [TWh]) of world's electricity supply. The estimated technically feasible global hydroelectric potential is about 14,400 TWh per year, but only 56 percent of this potential is currently considered to be economically feasible for development. Total global installed hydro-power capacity is about 692 GW, and an additional capacity of 110 GW is under construction.

According to the World Bank, 31 developing countries doubled their hydropower capacity between 1980 and 1985, much of it with small-scale projects. Small-scale hydropower generation provided almost 10 billion watts worldwide by 1983. China is the world leader in small-scale hydro power, with about 90,000 small hydropower stations supplying electricity to rural areas.

In the United States by early 1988, according to the Federal Energy Regulatory Commission (FERC), more than 2,000 hydro projects were operating. But according to the Environmental Protection Agency (EPA), there were 15,000 private hydro dams. The Unites States is the second-largest producer of hydro-power in the world; Canada is number one. Hydro-power provides about 10 percent of electricity in the United States. (see Figures 1 and 2).

Throughout history, there have been instances of dam failure and discharge of stored water, which have caused

Table 1
History of Hydro Power

B.C.	Use of water wheels
1000	Development of water-driven blast furnace
1500	Use of water-pumping works
1882	First hydroelectric power station built at Fox River in Appleton, Wisconsin, with a capacity of 25 kW
1885	First large hydroelectric power station built at Niagara Falls, New York
1936	Hoover Dam built with a capacity of 1345 MW
1971	Earthquake severely damaged the Lower San Fernando Dam north of Los Angeles
1975	Almost one million people perished in a series of hydroelectric dam failures
1984	First of eighteen 700-MW generators comes online at Itaipú Dam in Brazil
1993	Construction of The Three Gorges dam started in China
2002	World total electricity production from hydropower reaches 2,700 TWh from a capacity of 740 GW

Table 2
Power Generation Capacity of Major Dams of the World

Name of Dam/Location	*Capacity (MW)*
Itaipú Dam, Brazil/Paraguay	14,000
Guri, Venezuela	10,200
Grand Coulee Dam, Washington	6,809
Churchill Falls, Canada	5,429
Bratsk, Siberia, Russia	4,500
Terbala Dam, Pakistan	3,500
Aswan Dam, Egypt	2,100
Hoover Dam, Nevada	2,080
Mangla, Pakistan	1,000
Three Gorges Dam, China	18,200 expected completion date: 2009

Table 3
Estimated Hydro-Power Potential (TWh/year) of the World

Region	*Theoretical Potential*	*Technical Potential*
Africa	3,876	>1,888
Asia	16,443	>4,875
Europe	5,392	>2,706
Middle East	688	<218
North America	6,818	>1,668
South America	6,891	>2,792
Oceania	596	>232
Total World	>40,704	>14,379

Source: WEC Member Committees, 2000/2001: Hydropower & Dams World Atlas 2001, supplement to the International Journal on Hydropower & Dams (http://www.worldenergy.org/wec-geis/publications/reports/ser/hydro/hydro.asp)

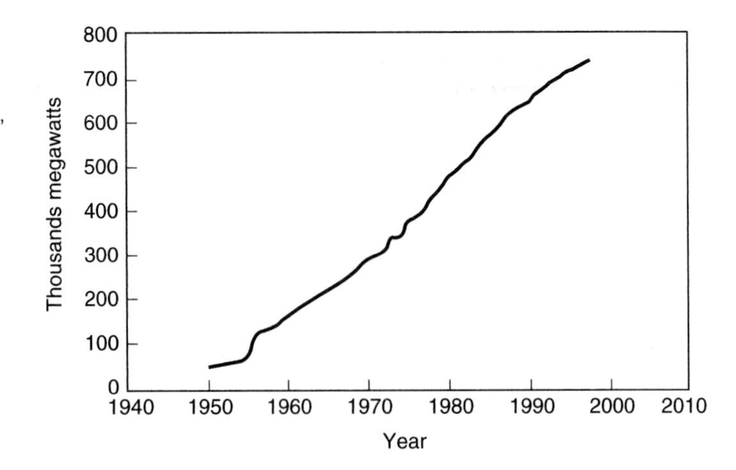

Figure 1
World hydroelectric-generating capacity, 1950–1998.
Source: *Vital Signs 2003*, Worldwatch Institute, Washington, D.C.

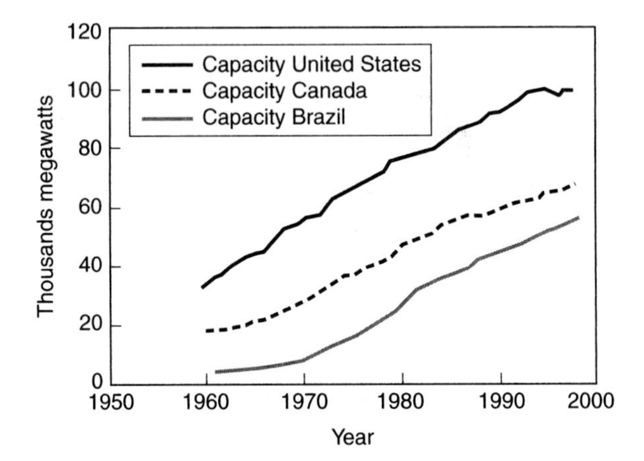

Figure 2
Hydroelectric-generating capacity in the United States, Canada, and Brazil, 1960–1998.
Source: *Vital Signs 1998*, Worldwatch Institute, Washington, D.C.

considerable loss of life and great damage to property. Advances in soil mechanics and structural engineering have revolutionized dam construction and hence increased safety aspects. However, it is estimated that about 150,000 dams around the world present a potential hazard to life or property; there have been more than 200 failures since 1900. Table 4 lists major failures resulting in major loss of life.

Table 4
Major Dam Failures

Year	*Dam*	*Country*
1626	San Ildefonso	Bolivia
1802	Puentes	Spain
1864	Bradfield	England
1889	Johnstown	U.S.
1890	Walnut Grove	U.S.
1895	Bouzey	France
1911	Austin	U.S.
1916	Bila Densa	Czechoslovakia
1917	Tigra	India
1923	Gienco	Italy
1928	St. Francis	U.S.
1935	Alla S. Zerbimo	Italy
1948	Fred Burr	U.S.
1959	Malpasset	France
1961	Kuala Lumpur	Malaya
1961	Babi Yar	Soviet Union
1963	Baldwin Hills	U.S.
1963	Vaiont	Italy
1972	Buffalo Creek	U.S.
1975	Banqiao	China
1975	Shimantan	China
1976	Teton	U.S.
1977	Kelly Barnes	U.S.
1979	Machhu II	India
1982	Lawn Lake	U.S.
1985	Val di Stava	Itay
1997	Opuha	New Zealand
2004	Camara	Brazil
2004	Big Bay	U.S.
2005	Shakidor	Pakistan
2005	Taum Sauk	U.S.

The International Commission on Large Dams (ICOLD) was formed in 1928 by six countries with the purpose of developing and exchanging dam design experience, and it has grown to 83 member countries. In 1982, ICOLD established a committee on dam safety to define common safety principles, integrate efforts, and develop guidelines, and in 1987 ICOLD published "Dam Safety Guidelines." In 2006, ICOLD has national committees from 83 countries with approximately 7,000 individual members.

Source: *McGraw-Hill Encyclopedia of Science and Technology,* 7th ed., New York: McGraw Hill, p. 20, and ICOLD history (http://www. www.icold-cigb.net).

Advantages

- Hydroelectric power produces no air, thermal, or chemical pollution in its electricity generation.
- Hydroelectric power has low production costs.
- Hydroelectric power has high efficiency (about 90 percent) converting from water to electrical energy.
- The water reservoir can provide potential flood protection for downstream currents.
- Groundwater reserves are increased by recharging from the dam's water reservoir.
- The dam's water reservoir can store large volumes of water for long periods of time; thus, downstream flow can be controlled for water quality and seasonal stream extreme conditions.

Disadvantages

- Hydroelectric power has high construction costs.
- There are limited feasible sites for dam construction.
- Electrical power production may be discontinued due to severe drought conditions.
- Dam construction causes loss of land suitable for agriculture.
- Construction of dams impacts the ecological cycles of the rivers and surrounding landscape.
- Silt accumulation and sedimentation changes flow and land drainage patterns.
- Water stored in the dam's reservoir by impounding the river is low in oxygen; therefore, the water issued from the dam is low in oxygen and affects the species of the water stream.
- Dam construction prevents upstream migration of fish.

REFERENCES

Dorf, R. (1981). *The Energy Factbook.* New York: McGraw-Hill.

Gabel, M. (1975). *Energy, Earth, and Everyone.* San Francisco: Straight Arrow Books.

Knoepfel, H. (1986). *Energy 2000: An Overview of the World's Energy Resources in the Decades to Come.* New York: Gordon and Breach Science Publishers.

McGraw-Hill Encyclopedia of Science and Technology, 7th ed. New York: McGraw-Hill.

"Survey of Energy Resources: Hydropower." Available at http://www.worldenergy.org/wec-geis/publications/reports/ser/hydro/hydro.asp.

World Resources 1988–89, a Report by World Resources Institute. New York: Basic Books.

BOX 1 China's Three Gorges Dam

Estimated completion date 2009 (expected)

Estimated cost $17 billion (government estimates) $75 billion (according to some estimates of dam opponents)

Power generation capacity 18,200 MW

China's most ambitious project since the Great Wall—the Three Gorges Dam—will displace approximately 2 million people and will swallow up cities, farms, and the canyons of the Yangtze River, called Chang Jiang (Long river) by the Chinese people. The Chinese government plans to harness the power of this great river to control its furious flooding and transform its raw power into electrical power.

The dam will stand 607 feet high and run 1.3 miles from the foreground to the far shore. The dam will be capable of generating a peak power of 18,200 megawatts using the world's largest 26 turbines of 400 tons each. When completed in 2009, the dam will be the most powerful dam ever constructed—the biggest project completed in China since the construction of the Great Wall 2000 years ago. The population of some 1,400 rural towns and villages will be resettled near the dam's reservoir in thirteen replacement cities or elsewhere at sites chosen by the Chinese government. The enormous capacity of the dam will enable China to move into the 21st century with a hydropower bang.

Benefits

- The 18,200-MW power-generation capacity is equivalent to the capacity of 18 nuclear power plants.
- The dam will help control flooding problems which historically have killed hundreds of thousands of people and have displaced more than a million people from their homes.
- The dam's massive locks will enable large vessels to navigate in the river.
- Hydropower will reduce China's consumption of fossil fuel (coal) and thus reduce carbon dioxide emissions.

Drawbacks

- The dam's reservoir will drown thousands of archaeological sites.
- Approximately 1.9 million people will be displaced. The Chinese government will pay 5,000 yuan ($600) a head for resettlement.
- The habitat for animals, such as the Chinese river dolphin, will be lost.
- The dam will prevent the distribution of rich sediment deposits, a main ingredient for the fertility of the agricultural land.

Sources: A. Zich, "China's Three Gorges," *National Geographic*, September 1997, pp. 2–33. http://www.chinaonline.com/refer/ministry-profiles/threegorgesdam.asp; http://www.cnn.com/SPECIALS/1999/China.50/asian.super power/three.gorges/.

QUESTIONS

1. Define the following terms.
   a. KWh
   b. Hydroelectric
   c. EPA
2. List some advantages (+) and disadvantages (–) of hydroelectric power using the chart on page 181.

Advantages (+)	*Disadvantages (–)*

BOX 2 Ataturk Dam

Ataturk Dam, Turkey.
Photo courtesy of Mary Jane Parmentier (May 2002).

The Turkish government's South Eastern Anatolian Project (GAP is the Turkish acronym) has as its cornerstone development project the Ataturk Dam, third-largest dam of this type in the world, created from the Euphrates River. This large hydroelectric project utilizes water power as an energy source and water for large-scale crop irrigation, providing critically needed power and water to the infrastructure of this region (one of the poorest in Turkey). However, the Turkish government is also aware that large infrastructure projects don't automatically create local development and can cause problems as well, such as the rise in mosquito-borne illnesses that have come with the big reservoir of water. GAP has implemented smaller-scale projects that target the economic and social needs of the people of the region, some of whom have been displaced by the creation of the dam. In addition to technology and infrastructure, GAP must focus as well on education, health care, housing, employment, and the environment. There are also international political issues, with down-river countries accusing Turkey of changing the flow of water and flooding patterns of the Euphrates due to the dam projects.

Source: "GAP Project." *All About Turkey*, http://www.allaboutturkey.com/gap.htm, accessed June 26, 2005.

The Electric Car Arrives—Again

SETH DUNN

It was two bicycle mechanics from Massachusetts, Charles and Frank Duryea, who rolled out the first commercially manufactured automobiles—13 of their Duryea Motor Wagons—in Detroit in June of 1896. Dubbed "horseless carriages," they ran on a noisy new invention known as the internal combustion engine—and a pungent fuel called gasoline. Soon, cars were spreading across the countryside—their costs falling rapidly, thanks to Henry Ford's assembly lines—and the world was moving toward a heavy dependence on oil.

One casualty of the internal combustion engine's triumph was the electric car, which had become quite popular in the 1890s. Proclaimed as quieter, cleaner, and simpler than the engine-driven car, the electric vehicle was widely expected to dominate the automotive market of the twentieth century. Instead, it quietly disappeared as automobile companies chose to pour billions of dollars into developing, and incrementally improving, the internal combustion engine. The electric auto, it seemed, was destined for the scrap heap of technological wrong turns.

It may have taken a century, but suddenly the electric car has returned from the dead. Its comeback has been fueled in large part by the engine-induced smog now filling urban areas like Athens, Bangkok, and Los Angeles—just as the manure piling up in the streets of America's cities a hundred years ago prodded the search for alternatives to the horse-drawn carriage. After years of false starts and heated debate, electric cars are on the road again: already an estimated 7,500 "engineless carriages" are now in use worldwide.

The most telling sign of life for this comeback invention came last December [1996] with the appearance of General Motors' long-awaited electric sports car, the EV1, in Saturn showrooms in southern California and Arizona. With its high-profile launch, the world's biggest automaker joined a rapidly growing list of companies around the world marketing, or set to market, electric cars—among them Honda, Mercedes, Peugeot, and Renault. These giants will do battle with a quickly growing army of some 250 entrepreneurial startups, each with visions of becoming the next Henry Ford.

While the electric cars on the market so far are expensive and can only travel limited distances, they are taking carmakers in new directions. With the push of government mandates soon to be overtaken by the pull of market opportunities, the drive to produce the most practical, economical electric car is quickly becoming a competitive auto race. As Michael Gage, President of CALSTART, a California electric vehicle consortium, puts it, "We are entering the tornado."

THE WHIRRING NINETIES

This time it was a maker of flying machines, Paul MacCready, who got the creaky wheels of car innovation turning. MacCready, inventor of the first human-powered aircraft to fly across the English Channel—a bicycle-like device that earned him "engineer of the century" plaudits from his contemporaries—had designed a solar-powered car, the Sunraycer, for GM. In 1987, the Sunraycer won

Reprinted with permission from Worldwatch Institute, Washington, D.C., *WorldWatch,* March /April 1997, pp. 19–25.

the first Solar Challenge race, crossing Australia on the equivalent of five gallons of gasoline. GM then asked MacCready's firm, AeroVironment, to build a concept electric car for the company.

Three years later, the resulting prototype—called the Impact—was greeted with such unexpected plaudits at a Los Angeles auto show that then-President Jack Smith brashly vowed to begin mass-producing the car immediately. In doing so, he gave this second automotive revolution a much-needed push. The actual jumpstart came later that year, when the California Air Resources Board—at the time facing worsening air pollution in Los Angeles and other cities, and greatly encouraged by GM's vow—passed the toughest auto emissions standards in the world. Most notable was the industry-shaking requirement that 2 percent of cars sold in the state by the seven major carmakers in 1998 be "zero-emission," with the share rising to 10 percent by 2003.

Auto industry lobbyists immediately turned on their own creation and—joined by the oil industry—began a bitter right to roll back the electric car mandate. At times, it seemed some of these companies had devoted more money to badmouthing zero-emission cars than to designing them. This certainly appeared to be the case for Chrysler, whose chairman—in the midst of scaling back its program—went so far as to declare, "There is absolutely no economic basis for electric vehicles in the world."

Eventually, the lobbying paid off, and California legislators lifted the 1998 mandate. But the big automakers are still required to make the more stringent 10 percent target in 2003, which means that some 800,000 zero-emissions cars should be on California's roads by 2010. This would be a giant leap from the approximately 2,300 electric cars in use in the entire country today.

ROADBLOCKS

Whether these targets will be met depends on whether prospective consumers can be helped around the immediate barriers of cost and range. At today's low-volume production levels, electric cars are more expensive to buy or lease than conventional cars with internal combustion engines. The problem is not the electric motor, a highly evolved technology used in everything from tiny dentist drills to huge freight locomotives. In fact, today's electric motors are already between four and five times as efficient as internal combustion engines.

The biggest roadblock for electric cars is, and always has been, storing the electricity needed to run them. The EV1—the product of the California mandate and a $345 million investment by GM—carries 1,175 pounds of lead-acid batteries (the same kind used to start conventional cars), but has a range of just 70 to 90 miles between recharges (the electric equivalent of refueling, recharging takes three hours, using a 220-volt inductive "paddle" at home or in public charging stations—of which California already has more than 400). In part because of the cost of the battery, the EV1s now on the market in California and Arizona lease for as much as $34,000 over 3 years.

But the energy–weight ratio of lead-acid batteries has been cut by 60 percent over the last decade, and further gains are likely down the road. Virtually all of the major carmakers are working hard to lower the cost of more advanced batteries with greater energy density—including nickel–metal-hydride models that could double the EV1's range to between 150 and 200 miles. Other alternatives in the works include nickel–cadmium and lithiumion batteries and even flywheels—mechanical batteries consisting of a rapidly spinning disk made of synthetic materials.

These new batteries are still too expensive for wide commercial use, but experts at California's Air Resources Board believe that they should be on the market soon. The next advance is expected to come from Honda, whose Formula One race car engineers are now fully devoting their work to electric cars and whose solar car has displaced GM's as the Solar Challenge champion. Honda plans to launch its EV Plus in California this May. Billed as the first family-oriented electric car, the compact four-seater will be equipped with nickel–metal-hydride batteries which give it a range of 125 miles—but at substantial cost: the batteries alone go for an eye-popping $40,000. Whether people will be willing to pay $500 a month (though this does include insurance and roadside service) to lease a car that is virtually indistinguishable from a standard economy car is uncertain; Honda expects to lease just 300 of the cars over the next three years.

Ironically, the limitations of today's batteries have made these first electric cars far more advanced than they otherwise might have been. Forced to stretch the range of bulky batteries, designers threw away the book on conventional automotive design and construction. In the search for a commercially viable electric car, automakers—for the first time in decades—turned their engineers loose on truly revolutionary concepts.

Author Michael Schnayerson, who was given inside access to GM's program, notes in his book, *The Car That Could*, how engineers struggled for eight years to fuse unconventionality and practicality, garnering 23 electronics and aerospace patents and a slew of engineering achievements. The EV1's aluminum car frame is the world's lightest; its teardrop-shaped body has aerodynamics equal to a modern fighter plane; and its brakes can regenerate, recharging the battery. With a super-efficient electric motor and low-resistance tires, it accelerates from zero to sixty miles per hour in less than nine seconds—faster than most conventional cars.

With these technological wonders, the EV1 has impressed most of those who have test-driven it. But the far greater challenge ahead for electric cars will be to make their way from the engineering track to the suburban garage. Although many buyers may be lured by the "zero-emission" label, performance, convenience, and cost are the benchmarks by which the cars must ultimately be judged.

ALTERNATE ROUTES

Automakers are taking a variety of approaches to "niche marketing" their first-generation electric cars. GM, for example, appears to be aiming at young, wealthy environmentalists—perhaps Hollywood stars and executives looking for a fast, sporty, pollution-free car that can speed through Beverly Hills. This is, after all, a group already accustomed to spending $30,000 to $40,000 for a Mercedes or Lexus. (Including tax incentives, the monthly lease rate for the EV1 falls to between $480 and $640 per month—less than the figure for luxury cars, like the Cadillac DeVille.)

So far, the strategy seems to be working. Thanks to a major ad campaign, demand for the EV1 has been stronger than GM anticipated, with about 50 cars leased out on the very first day. GM's electric car expert, Robert Purcell, believes the EV1 will assume a "second car" role for commuting (the average U.S. commute is less than 35 miles) and for short trips. Purcell notes similarities to the microwave oven, which was unpopular at first but which eventually caught on as a second oven. While production plans have been kept under wraps, Detroit's labor press estimates that at least 80 EVs were made last year [1996].

European automakers appear to be targeting a different set: green-oriented urban dwellers. The City Bee, for example, is a lightweight, fully recyclable two-seater with limited range, developed by the Norwegian consortium PIVCO. Scheduled for assembly in Europe and California later this year [1997], it will cost around $10,000 at a production volume of 10,000 and is intended only for use in the inner city and for quick rental at rail stations.

A multitude of other small companies, meanwhile, are preparing their own models for production. And the "urban car" niche has now begun to attract bigger players as well, resulting in some intriguing partnerships. German auto giant Mercedes-Benz and Swiss watchmaker Swatch have teamed up to develop a "Smart" car: a two-seater that can be built in under five hours, using plastic parts with interchangeable color schemes like those of the Swatch watches. The Smart car has proven crashworthy even though it is less than 10 feet long. Cast as a modern version of the "runabouts" seen at the turn of the century and aimed at a younger audience, it can use electric as well as other drive systems. Some 200,000 Smart cars, a sizable percentage of which could be electric, are scheduled to roll out in Europe in March 1998 at a price of between $10,000 and $13,000.

Other European automakers are taking a more conservative approach to marketing electric cars, at much lower cost, by converting conventional cars into electrics. One of the leaders is Peugeot, which has put out a car using nickel–cadmium batteries with a range of 50 miles. Having successfully tested 500 of its cars in the city of La Rochelle, the Peugeot-Citroen group believes there will be around 100,000 electric cars in Europe by 2000 and plans to grab a quarter of the market; it produced more than 4,000 in 1996.

Renault, meanwhile, sold 215 of its electric conversions in the first six months of last year. It will make 1,000 cars this year [1997], and expects to continue increasing output in 1998. On a smaller scale, Fiat and Volkswagen are selling conversions, which they are reportedly making at a rate of about one a day, in Swiss cities. These conversions have a more limited range than cars designed to be electric from the start, but they allow the automakers to enter the market and gain experience without a huge upfront investment.

Much as L.A.'s smog sped the move to electric cars, air pollution and congestion in Europe have prompted some carmakers to rethink the car's conventional ownership and role in transportation. In France, Peugeot and Citroen are involved with the design of a transit system that will allow Parisian commuters to rent electric cars over short distances under a credit card-like system. Renault plans, as part of a consortium of electric utilities, carmakers, and government agencies, to operate 50 such "multiuser" cars in one of the city's high-tech suburbs. Similar efforts are underway in Switzerland. Swiss carmaker Horlacher will soon offer a lightweight "instant taxi," while Fiat has begun to rent electric cars in Geneva.

Whatever the route to consumers, proponents note that, despite their current high price and limited range, electric cars already have a number of advantages over today's cars. Their relative noiselessness, practicality, and simplicity appeal to many drivers. Electric cars cost less to refuel and service and have many fewer parts to break down. Their owners are likely to spend less time on maintenance, and if they recharge at home, will rarely have to go to the service station. These time savings have real value in today's busy world. On a life cycle basis, then, the cost gap between cars that pollute and those that don't is not all that great, even now; with another decade of battery development and mass production, it could be closed entirely.

Air regulators in California and New England argue further that if the avoided costs of urban air pollution, acid rain, and global climate change were included, electric cars would look like a steal. Of course, a fair comparison must include the emissions from the power plants that are used to charge the batteries. Fortunately, emissions from stationary power sources are easier to control than those from vehicles.

More importantly, running cars on electricity opens up a host of new fuel options not based on oil—including renewable resources, such as wind power and solar energy. Already, some of the municipal governments promoting electric cars are erecting solar cells on the roofs of their parking garages to recharge them. The California city of Sacramento, for example, through the involvement of automakers, utilities, and local authorities, has installed more than 70 charging stations.

TOOL KITS

As the Sacramento example suggests, governments can play an important role in accelerating the transition to electric cars. Several of them, in fact, offer incentives that can lessen the electric car's initial cost. In the United States, federal and state tax credits and local rebates are available in many areas: customers in Los Angeles, for example, can use a 10 percent federal tax credit and a $5,000 rebate from the South Coast Air Quality Management District.

A 5,000-franc subsidy, meanwhile, is available to those buying electric cars in France. Paris and several other French cities add on tax credits, as do Switzerland, Austria, and Denmark. Switzerland already has about 2,000 electric cars in operation and aims to use tax incentives to help make 8 percent of its cars electric by 2010. Germany, with more than 2,400 electric cars, gives its customers federal tax exemptions and state-level subsidies. Electric cars are exempt from sales taxes (which are typically high in Europe) in Italy, Norway, Sweden, and the United Kingdom. In Japan, which has the ambitious if distant goal of producing 200,000 electric vehicles by 2000, buyers may see costs cut in half by federal tax credits, incentives, and depreciation allowances—and even further by municipal governments.

Spurred by these and other less-noticed but important policies and programs—installing recharging stations, setting up demonstration projects, funding battery research and development—an electric car industry appears to be taking shape. Consortiums like CALSTART—a network of 200 government agencies, environmental groups, and aerospace, defense, and electric power companies—are helping these groups work horizontally (in contrast to today's "vertical" auto industry) to start programs and attract funding. Electric vehicle associations in North America, Europe, and Asia report growing memberships and are holding well-attended conferences and exhibitions each year to share the latest surveys, technologies, and production plans.

Surprisingly, the production hub for cost-competitive electric cars may turn out not to be in today's auto industry powers, the United States and Japan, but in the developing countries. With the world's most polluted urban air, some of these countries are just starting to develop automobile industries and make the associated investment in oil refineries, service stations, and the like—and therefore have less vested interest in the internal combustion engine. Their industrialists recognize that if they are already making computers and televisions, electric cars should be within reach.

This notion has already taken hold in Asia, where prospective electric car manufacturers have held extensive discussions with potential manufacturing partners. Thailand, which offers tax exemptions to electric car makers, is producing electric versions of its three-wheeled "tuk-tuk" taxi. Korean car makers Daewoo and Hyundai are working to produce lightweight electric cars for sale by the end of 1997 and 1998, respectively. China, which hopes to have several carmaking plants by 2000, plans to link domestic and foreign makers. One fledgling Chinese carmaker has cut a deal with Peugeot-Citroen to produce a small electric model.

As their production picks up, electric cars' prices will fall noticeably. Comparing their price history with that of the traditional automobile, Daniel Sperling of the University of California at Davis estimates full-scale production could reduce the cost of electric cars to well below half the current level. And new technological breakthroughs will not be needed for costs to plummet, according to Tufts University's Global Development and Environment Institute. The

Institute projects price declines analogous to those that have occurred in personal computers, for example, with the costs of an electric car approaching—and with government support, falling below—those of conventional cars. And some analysts believe electric cars will be competing with gasoline-powered cars, *without* subsidies, within a decade. Table 1 lists major commercial or near-commercial cars.

Table 1
Major Commercial or Near-Commercial Electric Cars

Car and Model	*Battery*	*Range (miles)*	*Location*	*Date*
"GROUND-UP" ELECTRIC CARS				
General Motors EV1	lead–acid	70–90	US	1996
American Honda EV	nickel–metal hydride	125	US	1997
			Japan	1997
Solectria Sunrise	various	120	US	1998
SMALL ELECTRIC CARS				
PIVCO City Bee	nickel–cadmium	60–70	Europe	1997
			US	1997
Mercedes/Swatch "Smart"	various	various	Europe	1998
MODIFIED CONVENTIONAL CARS				
Peugeot 106	nickel–cadmium	50	Europe	1994
Citroen AX	nickel–cadmium	50	Europe	1994
Renault Clio	nickel–cadmium	50	Europe	1996
Fiat Panda Elettra	lead–acid	36	Europe	1996
Volkswagen CitySTROMER	lead–acid	30–54	Europe	1996
Solectria Force	various	60	US	1994

BACK TO THE FUTURE?

Once the electric car catches up to today's cars by lowering up-front costs and extending range, its continued success will depend on the ability of manufacturers to lure consumers with inexpensive, battery-powered versions of commuter, family, or sports cars. Gage of CALSTART foresees "an improved driving experience" analogous to the shift from long-playing records and cassettes to the compact disc: the former worked, but the latter is better. Carmakers may also, however, find themselves dusting off hundred-year-old advertising pitches. As *Scientific American* observed in 1896, "The electric automobile . . . has the great advantage of being silent, free from odor, simple in construction, capable of ready control, and having a considerable range of speed."

Perhaps that praise was premature, but precisely a century later, *Scientific American* has returned to the topic—and suggests that the tables have turned. Writes Sperling of UC-Davis in a recent issue: ". . . it seems certain that electric-drive technology will supplant internal-combustion engines—perhaps not quickly, uniformly, nor entirely—but inevitably. The question is when, in what form and how to manage the transition." Those words, ironically, recall a very similar statement anticipating the adoption of the internal combustion engine a century ago. After watching the Duryea brothers race their wagons one afternoon in 1895, a young inventor named Thomas Edison boldly announced, "It is only a question of time when the carriages and trucks in every large city will be run with motors."

While technological dreams do not always come true, the electric car now seems to have more than a fighting chance. The major limitations of today's models are within sight of being overcome: indeed, a host of companies are betting billions of dollars on their ability to make that happen. And they cannot but be pleasantly surprised by the initial response to the EV1: according to Saturn dealers, the waiting list of several hundred hopeful lease holders continues to grow. As any car executive will tell you, the Model T took off when its costs were cut in half and cheap gasoline became an available fuel: now, they seem to be musing, might the electric car do likewise once its prices fall and batteries improve dramatically?

But larger forces, as much as the battery, will determine how far the electric car ultimately goes. On these counts, it seems to have a good deal riding in its favor. Its biggest constituency has always been women, who have far more purchasing power today. And its advantages, curious virtues in 1897, have become serious necessities by now as the burdens of the gasoline culture grow heavier. Urban air pollution, congestion, dependence on oil imports, and global warming: these mounting societal concerns are recharging the electric car.

Still, if it is to avoid the fate of its predecessors, the EV1 and its companions must now handle the rocky road test of the market by selling, leasing, or renting. Making inroads on today's car population will not happen overnight, yet GM's model is already gracing the pages of the *The New York Times'* Automobile Section, alongside the conventional competition. This, perhaps, is the real story of the electric car's reemergence, lost amid the press conferences and television ads: that, true to the invention's own nature, its entry into the mainstream may resemble less a loud revving than a quiet hum.

REFERENCES

CALSTART. (1996, October). *Electric Vehicles: An Industry Prospectus*. Burbank, CA: CALSTART, Inc.

Flavin, C., and Lenssen, N. (1994). *Power Surge: Guide to the Coming Energy Revolution*. New York: W.W. Norton & Co.

MacKenzie, J. (1994). *The Keys to the Car*. Washington, DC: World Resources Institute.

Natural Resources Defense Council. (1996). *Green Auto Racing*. Washington, DC: NRDC.

Schiffer, M.B. (1994). *Taking Charge: The Electric Automobile in America*. Washington, DC: Smithsonian Institution Press.

Shnayerson, M. (1996). *The Car That Could: The Inside Story of GM's Revolutionary Electric Vehicle*. New York: Random House.

Sperling, D. (1995). *Future Drive: Electric Cars and Sustainable Transportation*. Washington, DC: Island Press.

Current Status of Hybrid Vehicles

Box 1 presents a timeline of the development of hybrid vehicles, and Box 2 compares the technical specifications of hybrid vehicles available in the U.S. market.

BOX 1 Hybrid Car Development: A Timeline (1993–2006)

1991	The United States Advanced Battery Consortium (USABC) lauches a major program to produce a "super" battery to get viable electric vehicles on the road. The consortium would invest more than $90 million in the nickel hydride (NiMH) battery. The NiMH battery can accept three times as many charge cycles as lead-acid, and can work efficiently in cold climate.
1991	Toyota Corporation proclaims the "Earth Charter," a document outlining goals to develop and market cars with lowest emissions possible.
1993	The Clinton administration announces a government initiative called the Partnership for a New Generation of Vehicles (PNGV). Through this program, the government will work with the American auto industry to develop a clean car that could operate at up to 80 miles per gallon.
1997	Toyota introduces the Toyota Prius in the Japanese market, two years before its original launch date, and prior to the Kyoto global warming conference held in December. First-year sales were about 18,000 vehicles. Audi introduces the hybrid Audi Duo in the European market. The Duo was not a commercial success and therefore was discontinued, prompting European carmakers to focus their R&D investment on diesels.
1997–1999	A small selection of all-electric cars from major automakers (Honda's EV Plus, GM's EV1 and S-10 electric pickup, a Ford Ranger pickup, and Toyota's RAV4 EV) were introduced in California. Despite the enthusiasm of early adopters, the electrics failed to reach beyond a few hundred drivers for each model. Within a few years, the all-electric programs were dropped.

1999	Honda releases the two-door Insight, the first hybrid car, in the U.S. market. The Insight receives EPA mileage ratings of 61 mpg city and 70 mpg highway.
2000	Toyota releases the Toyota Prius, the first hybrid four-door sedan, in the U.S. market.
2002	Honda introduces the Honda Civic Hybrid, its second commercially available hybrid gasoline-electric car.
2004	Due to increased demand for Prius, Toyota pumps up production from 36,000 to 47,000 for the U.S. market. Interested buyers wait up to six months to purchase the 2004 Prius. Ford releases the Escape Hybrid, the first American hybrid and the first SUV hybrid.
2006	All major automakers introduced Hybrid cars, SUVs/Minivans and trucks in the U.S. market. Cars Honda Accord Honda Civic Honda Insight Lexus GS 450h Toyota Camry Toyota Prius SUVs & Minivans Ford Escape Lexus RX 400h Toyota Highlander Mercury Mariner Trucks Chevrolet Silverado / GMC Sierra Dodge Ram Ford Motor Company delivers a Mercury Mariner Hybrid "Presidential Edition" to former President Bill Clinton—the first hybrid vehicle to be outfitted for presidential service.
2007–2008	Major automakers plan to release the following vehicles in the U.S. market Cars Honda Fit Hyundai Accent (2009) SUVs & Mini Vans Toyota Sienna Minivan Porsche Cayenne GMC Yukon Dodge Durango Chevrolet Tahoe

Source: http://www.hybridcars.com/history.html.

BOX 2 Comparison of Major Commercial Hybrid Vehicles Available in the U.S. Market

	Honda Accord Hybrid	*Toyota Prius*	*Honda Civic Hybrid*	*Ford Escape Hybrid (SUV)*	*Lexus RX 400h (SUV)*
Year	2006	2006	2006	2006	2006
Available Years	2005–2006	2000–2006	2006	2005+	2006
Base MSRP	Approx. $30,000	Approx. $25,000	Approx. $21,850	Approx. $29,000	Approx. $48,500
Seating	5	5	5	5	5
Mileage	30/37	60/51	50/50	2wd 36/31 4wd 33/29	31/27
Tank size	17.1 gal	11.9 gallons	13.2 gallons	15 gallons	17.2 gallons
Greenhouse Gas Emissions/15k miles	8,700 pounds	5,100 pounds	5,700 pounds	2wd 8,500 pounds 4wd 9,200 pounds	9.806 pounds
Net Power	255 horsepower	110 horsepower	110 horsepower at 6,000 rpm	155 horsepower	268 horsepower
Gas Engine horsepower	240 horsepower	76 horsepower	95 horsepower	133 horsepower	208 horsepower
Electric Motor Power	16 horsepower	67 horsepower	20 horsepower	94 horsepower	167 horsepower front; 68 horsepower rear
Gas Engine Torque	232 lb.-ft @ 5000 RPM	82 lb.-ft. @ 4200 RPM	123 lb.-ft @ 1000-2500 RPM (Net)	124 lb.-ft. @ 4250 RPM	212 lb-ft @ 4400 rpm
Electric Motor Torque	100 lb.-ft @ 840 rpm	295 lb.-ft. @ 0-1200 RPM	NA.	100 lb.-ft. @ 840 RPM	NA.
Displacement	3.0 liter	1.5 liter	1.3 liter	2.3 liter	3.3 liter V6
Voltage	144V	500 volts	158 volts	330 volts	650 volts
Drag Coefficient	.29	.26	.28	N/A	.35

Source: http://www.hybridcars.com/accord_specs.html

QUESTIONS

1. Does the electric car reduce pollution? What are some of the pro and con arguments?
2. Discuss the roadblocks that stand in the way of mass production of electric cars.
3. How do you think we can safely dispose of a high volume of used batteries that are a waste product of electric car technology? Is any ecologically safe approach being considered? If so, which?
4. Compare the marketing strategies of U.S. and European electric car manufacturers.
5. Discuss the role of the oil industry in the development and success of the electric car.
6. Using Internet search engines compare the technological features of the following hybrid-electric cars:
   a. Toyota Prius
   b. Honda Civic
   c. Ford Explorer

BOX 3 CO_2 Emissions

- The combined 1995 population of Africa, Asia, Oceania, and Central and South America, which that year had a total of 200 million motor vehicles: 4.40 billion.
- The 1995 population of the United States, which also had a total of 200 million motor vehicles: 0.27 billion.
- The amount of carbon dioxide that a car getting 27.5 miles per gallon emits over 100,000 miles: 31,752 kilograms.
- The amount of carbon dioxide that a human walking that same distance would produce: 59 kilograms.

Source: "Matters of Scale." Reprinted with Permission from Worldwatch Institute, Washington, D.C., November/December 1997, p. 39.

Box 4 One Step Forward, a Hundred Steps Back?

Reduction in carbon emissions by World Bank-funded projects measure in the thousands of tons per year . . . but the emissions generated by bank-funded projects measure in the millions of tons.

Selected Fossil Fuel Power Plants Financed in Part by the World Bank in the 1990s

Location	*Plant Description*	*Estimated Cost ($ billion)*	*Annual Carbon Emissions (million tons)*
Tuoketo, China	3,600 MW, coal fired	4	7.19
Dolana Odra, Poland	1,600 MW, coal fired	0.3	3.196
Paiton, Indonesia	1,230 MW, coal fired	1.8	2.457
Pangasaman, Philippines	1,200 MW, oil fired	1.4	2.397
Hub River, Pakistan	1,469 MW, oil fired	2.4	2.345*

*Mr. Z. A. Khan, Divisional Manager of Chemistry & Environment, National Power International, disputes these emission numbers. For details, visit the International Power Web site: www.internationalpowerplc.com.

Hub River Power Plant, Pakistan.
Photo Courtesy of Ahmed S. Khan.

Selected Global Environment Facility Energy Projects, Some Cofinanced by the World Bank

Location	*Plant Description*	*Estimated Cost ($ billion)*	*Annual Carbon Emissions (million tons)*
Leyte, Luzon, Philippines	440 MW, geothermal	1,300	872
Countrywide, Indonesia	Installation of 200,000 PV systems	118	120
Nine cities, China	Energy-efficiency upgrades of industrial boilers	101	41–68
Guadalajara and Monterrey, Mexico	Dissemination of high-efficiency lighting (1.7 million fluorescent lamps)	23	32
Tejona, Costa Rica	20 MW	31	16

Source: "Banking Against Warming," Christopher Flavin, *World Watch Magazine*, November/December 1997, pp. 32–33.

Case Study 1

Chernobyl

Ahmed S. Khan

At 1:24 a.m. on April 26, 1986, two massive explosions destroyed unit four of the Chernobyl nuclear power plant. The plant, located near the Belarus–Ukraine border on the River Pripyat, about 80 miles north of Kiev, the capital of Ukraine (a republic of the former Soviet Union), had four reactors. The plant's roof was blown off, and radioactive gases and materials were released into the atmosphere. The Soviet government did not make any public announcement about the accident. It was not until April 28, when the monitoring instruments in Sweden detected a dramatic increase in wind-borne radiation, that the Soviets acknowledged the disaster (Newton 1994). On May 14, Soviet President Mikhail Gorbachev, in his address to the Soviet people, said:

> "Good evening, comrades. All of you know that there has been an incredible misfortune—the accident at Chernobyl nuclear plant. It has painfully affected the Soviet people, and shocked the international community. For the first time, we confront the real force of nuclear energy, out of control." (CNN, 1986)

The accident at Chernobyl became the greatest human experiment with radiation exposure. According to recent studies, an estimated 50 to 250 million curies (ci) of radiation were released as a result of the accident. A curie is a measure of the intensity of radiation and is equal to 37 billion disintegrations per second. As a reference point, the Hiroshima and Nagasaki bombings released an estimated 1 million curies.

The Chernobyl disaster has rendered an area the size of New York State unsafe for human habitation. Some 200,000 residents of the area were evacuated, and clouds of radioactivity were carried all over the globe. The accident killed 31 people immediately and, within the next few days, Soviet officials admitted the deaths of 224 others. Soviet officials also prohibited doctors from diagnosing illnesses as radiation induced. The actual death toll is probably several thousand. In the next few years, the number may reach hundreds of thousands due to radiation-induced illnesses.

Since the Chernobyl accident, cancer rates in the former Soviet Union countries have risen by 300 percent, respiratory disease rates by 2,000 percent, and the number of babies born with birth defects has soared to record numbers. The estimates predict anything from 14,000 to 475,000 cancer deaths worldwide from Chernobyl. According to the World Health Organization (WHO), about 6.7 million people were exposed to radiation fallout, leading to a tenfold increase in thyroid cancer among children in affected areas. Of the 80,000 firefighters and soldiers involved in clean-up operations at the reactor in the years after the explosion incident—the so-called liquidators—30,000 are reported to have received radiation doses of more than 0.5 sievert (Sv); a radiation dose of greater than 0.5 Sv is considered a high-level radiation dose. According to statistics released by government agencies in the three former Soviet states affected (Ukraine, Belarus, and Russia), about 25,000 liquidators have died (www.chernobyl.info). However, no one will ever know with certainty the exact toll of the Chernobyl disaster.

The key event leading to the accident was the failure of a controlled experiment that went out of control. The plant ran on electricity tapped off the national power grid. If the plant failed to receive the electric power, it would have to be shut down immediately. Any failure in the shutdown process or the failure of back-up diesel generators to provide enough power to the plant would lead to the failure of water pumps that circulate the water required to cool the reactor core. In the absence of cooling water, the core temperature would rise quickly and result in the meltdown of the core. Plant operators wanted to discover what would happen if there was a power failure and steam stopped flowing in the turbines. Would the inertia of the spinning turbine blades be sufficient to produce electricity to power cooling pumps until the back-up diesel generators turned on? To find out the answer to this emergency scenario, the operators decided to conduct a controlled test simulating the emergency conditions.

During this "controlled test," a combination of major mistakes by workers led to the disaster. The major mistake was the operators' decision to shut down the reactor's emergency cooling system. At the start of the test, the reactor began to lose power. The test could only be continued if the reactor remained operational, so the operators turned off the cooling system. The reactor continued to lose power. In an attempt to increase power, the technicians removed all of the control rods, which are essential in controlling the fission reaction. This action resulted in an increased fission rate, and thus the amount of heat produced soared (Newton 1994).

As the power level in the core started to increase, the operators tried to lower the control rods. But the channels into which the rods were supposed to go had deformed due to intense heat in the core. The control rods did not drop properly and, in just four seconds, the power level in the core surged to 100 times the normal level. With intense heat, the temperatures reached 5,000 degrees Celsius and the reactor core started to melt. The molten metal in the reactor core reacted with remaining coolant water, generating hydrogen gas and more steam, to produce the first explosion that blew the reactor's 1000-ton concrete lid into the air. The details of the second explosion are not very clear. Some experts believe that it occurred due to the hydrogen gas, and others consider it the result of a pure nuclear reaction. A part of the molten core may have achieved critical mass during the meltdown, resulting in an atomic bomb-like explosion. Between the two major explosions, the roof of the reactor was destroyed. Since there was no containment shell, several tons of radioactive particles escaped into the atmosphere (Newton 1994).

The Chernobyl disaster has had both short-term and long-term effects on regional and international levels and on the future of the nuclear power industry. The radioactive cloud that rose from the ruins of the reactor has left the city of Pripyat—built for Chernobyl personnel—a radioactive ghost town. The radioactive radiation released from Chernobyl has circled the Earth several times. Over time, its radionuclides are settling to the land and sea to enter the planetary ecosystem. No one has a viable plan to clean up the ruins of the reactor while radioactive fuel is slowly leaking into the ambient ground and air.

Almost two decades after the accident, the level of radiation still remains so high in some fallout areas that no crops can be grown or consumed. The main contaminants are cesium-137 and strontium-90, having half-lives of 30 and 28 years, respectively. These isotopes, if not removed, will constitute to be a hazard for many decades.

Soviet engineers used a wide array of techniques to extinguish the flames in the reactor. First they pumped liquid nitrogen to the core to put out fires. Then they used 30 military helicopters to dump 1,800 tons of sand and 2,400 tons of lead and boron on top of the plant. These materials absorbed neutrons and thus helped extinguished the fires (www.chernobyl.info and Newton 1994).

After the fires were extinguished, workers built a huge steel and concrete sarcophagus over the damaged plant to isolate unit four of the plant and to contain the release of the radioactivity. The first shell did not prove to be very effective; by 1992, it developed cracks and began to leak radioactivity into the environment. In 1997, the Ukrainian government in collaboration with the G-8 countries, the European Union (EU), and the European Bank for Reconstruction and Development (EBRD), launched the Shelter Implementation Plan (SIP) to construct a second shell to cover the first to provide the containment of radioactivity for at least the next 100 years (www.chernobyl.info).

At the time of the accident, the number four reactor at Chernobyl was loaded with 185 tons of Uranium-235. After the core meltdown, much of the fuel had poured into the basement of the building. There existed the possibility that the remaining fuel could reach the critical mass and explode in an atomic chain reaction. In July 1991, the radioactivity readings indicated that such a reaction was under way. Fortunately, workers were able to locate the place and pumped chemicals in to stop the reaction.

The Chernobyl power plant consists of four nuclear reactors constructed between 1977 and 1983. By 1986, the units were operating at near capacity and generating 4 million kilowatts (kW) of electricity. The reactors are of the type known as RBMK, a Russian acronym meaning Large Power Boiling Reactor. In such reactors, graphite is used as the moderator element to slow down neutrons produced by fission reaction in the fuel rods, a function performed by coolant water in the majority of other types of nuclear reactors. The core of an RBMK reactor, a cylinder 46 feet in diameter and 23 feet in height, consists of stacks of nuclear fuel assemblies packed within columns of graphite blocks weighing about 2,000 tons. Each column also has a channel into which a control rod can be lowered. The rods, made of boron carbide, absorb the free neutrons and reduce the nuclear reaction when lowered into the channel.

The RBMK reactors contain a number of design traits that render them to be a high-risk operation. One problem is that the RBMK design makes no provision for a containment shell, which can retain gases and other nuclear material released during an accident in the reactor core. The containment shell is present in all nuclear reactors in the United States and most reactors in other countries. Another problem is that a loss of cooling water increases the rate of fission reaction, which leads to increased power production. This process is the reverse of water-moderated reactors, in which loss of cooling water results in decreased power production. This design flaw poses a serious threat to the safe operation of the reactor at low power levels.

The RBMK-type reactors were constructed exclusively in the former Soviet bloc countries. They were constructed because they have the ability to produce electricity and plutonium at the same time. Soviets wanted to produce enough plutonium to stay at par with the United States in the area of nuclear weapons development. After the accident, Soviet scientists were reluctant to modify the design of RBMK reactors, but they eventually modified the design to make it safer. One of the revised design features is that the control rods cannot be removed completely. At present, 40 RBMK-type reactors are still operational in eastern European and former Soviet states. In 32 countries, 440 nuclear plants are used to generate electricity (Newton 1994).

The final legacy of the Chernobyl accident is that it has destroyed the myth of so-called clean energy. What appears to be a clean source of energy are a few megawatts of cheap energy today, but in the long run, it is an immortal radioactive contamination of the environment.

How many more accidents will take place in the future? What will become of the accidents that have already occurred? How many accidents have been kept secret? Was Chernobyl a warning? Will it remain the greatest nuclear accident throughout history? Or is there another tragedy waiting to happen? Will the future cost outweigh the price of nuclear power? Are we unwisely choosing a few megawatts of cheap energy today that carries the price of an irreversible contamination of the environment?

REFERENCES

Barringer, Felicity. (1991). Chernobyl: The Danger Still Persists. *The New York Times Magazine*, April 14.

"Chernobyl Nuclear Disaster." Available at http://www.chernobyl.co.uk. Accessed September 5, 2006.

Chernobyl information. Available at http://www.Chernobyl.info. Accessed September 5, 2006.

Chernobyl 10 Years Later: A Threat to the Future. (1996). CNN Presents Chernobyl: Legacy of a Meltdown. April 4. Available at http://www.cnn.com/WORLD/9604/04/cnnp_chernobyl/index.html. Accessed September 5, 2006.

"Europe Stages Nuclear Crisis Test." (2005). *BBC News*, May 11. Available at http://news.bbc.co.uk/2/hi/europe/4535899.stm. Accessed September 5, 2006.

Edwards, Mike. (1994). Living with the Monster—Chernobyl. *National Geographic*, August.

Flavin, Christopher. (1987). *Reassessing Nuclear Power: The Fallout from Chernobyl*. Washington, D.C.: Worldwatch Institute.

Lenssen, Nicholas. (2004). Confronting Nuclear Waste. *State of World-1994*. New York: W.W. Norton & Company.

Medvedev, Zhores. (1991). *The Legacy of Chernobyl*. New York: Basic Books.

Newton, David. (1994). Chernobyl Accident. *When Technology Fails: Significant Technological Disasters, Accidents and Failures of the Twentieth Century*. Schlager, Neil (ed.). Detroit, MI: Gale Research Inc. pp. 529–535.

BOX 1 Chernobyl: Price of "Clean Energy"

Cause of accident	Human error coupled with flawed reactor design
Amount of radioactivity released	50–250 million curies of radiation (the atomic bombs in Hiroshima and Nagasaki released an estimated 1 million curies)
Number of people exposed to radiation fallout	6.7 million (BBC News, May 2005)
Number of people evacuated	200,000+
Short-term death toll	224 (according to Soviet authorities); actual numbers are close to several thousand
Long-term death toll	Estimates range from 14,000 to 475,000 cancer-related deaths

The unleashed power of the atom changed everything, save our modes of thinking.

ALBERT EINSTEIN

QUESTIONS

1. Suppose you were the Chernobyl plant director in 1986. What decisions and actions would you have taken to prevent the disaster? Complete the flowchart in Box 2 by listing your decisions and actions in chronological order.
2. Do you think that nuclear energy is a "clean" source of energy?
3. With more than 440 operational nuclear plants worldwide and around 28 nuclear plants under construction, is the chance for a nuclear accident growing larger or smaller? What might be the consequences of a major nuclear accident at local, regional, and global levels?
4. What alternatives are there to nuclear energy?
5. List short-term and long-term effects of the Chernobyl disaster.

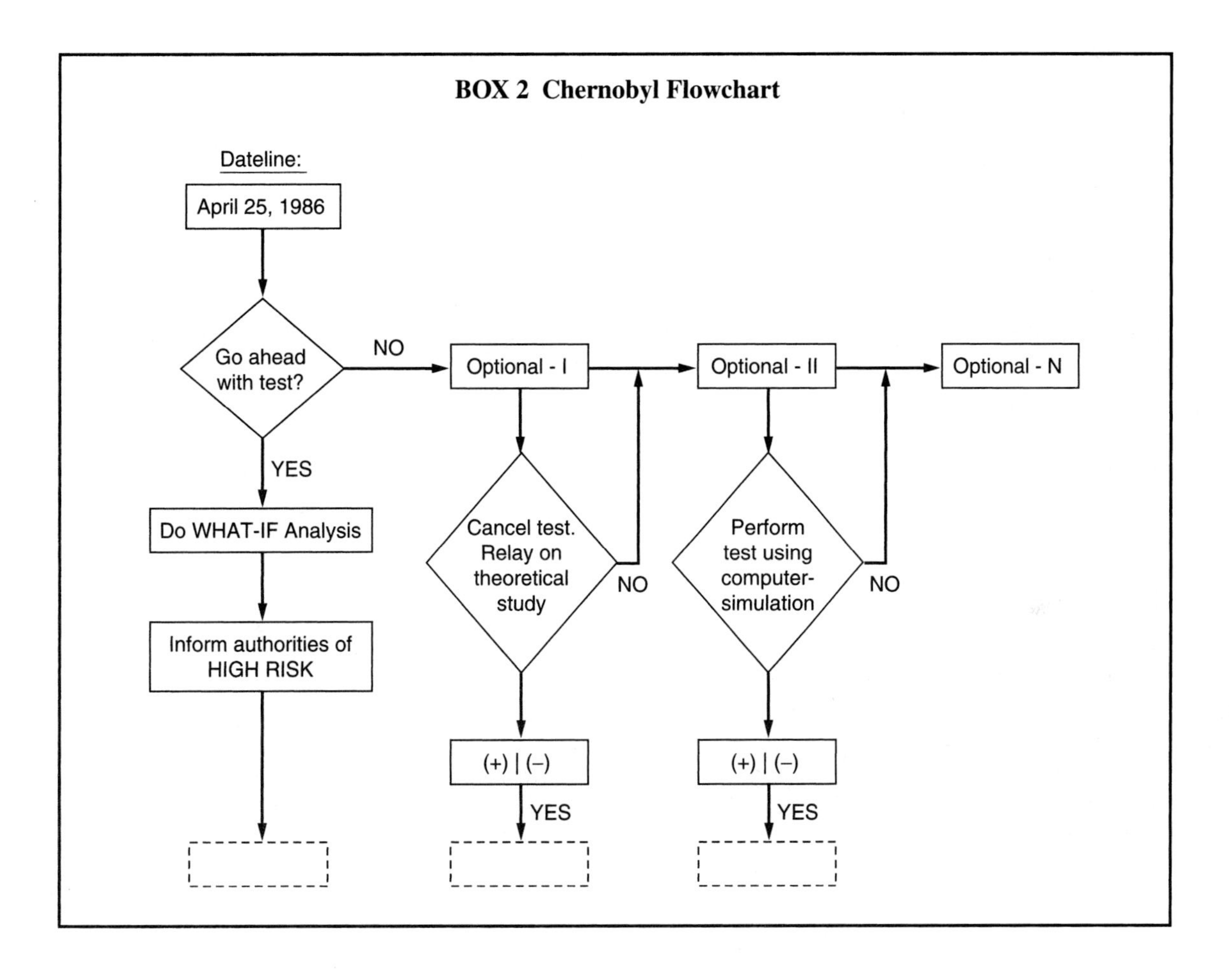
BOX 2 Chernobyl Flowchart
Dateline:
April 25, 1986
Go ahead with test?
NO
YES
Optional - I
Optional - II
Optional - N
Do WHAT-IF Analysis
Inform authorities of HIGH RISK
Cancel test. Relay on theoretical study
NO
Perform test using computer-simulation
NO
(+) | (–)
YES
(+) | (–)
YES

Case Study 2

Tasman Spirit Oil Spill

Ahmed S. Khan

A Greek oil tanker, *Tasman Spirit*, which was carrying over 67,500 tons of crude oil, ran aground near the Clifton Beach of Karachi, Pakistan (Bakhtiar & Karim 2003). Due to the Karachi Port Authority's slow response to managing this emergency, the ship broke into two halves. As a result, millions of gallons of crude oil spilled out, contaminating a 14-km long section of Clifton Beach. The beach was covered with thick layers of crude oil, and thousands of fish, turtles, and birds were killed in a short period of time.

The damaged oil tanker continued to lie aground off Karachi, and the oil slick continued to scar Karachi beaches, causing irreversible damage to ecological systems (FOE press release 2003). The sequence of events and emerging evidence suggest that it was a man-made disaster. This disaster could have been avoided through adequate contingency planning and acquisition of pollution control materials, but the Karachi Port Authority did not have any contingency plans to deal with such an emergency. The Port Authority failed to develop appropriate plans to contain an oil spill and did not acquire mechanical containment tools such as booms (fence, curtain, nonrigid inflatable), skimmers (oleophilic, suction, weir), and sorbents (natural organic, natural inorganic, and synthetic). They did not train personnel to contain and control oils spills and to do shoreline cleanup (pressure washing, raking, and mopping).

During the past 50 years, the gigantic increase in the use of fossil fuels has wreaked havoc on the earth's atmosphere in the form of carbon emissions and oil spills. Oil spills cause such havoc to ecological systems that even a very thorough cleanup and control measure can never restore these systems to their pre-spill conditions. The *Exxon Valdez* and 1991 Gulf War episodes are vivid examples of the environmental impact of oil spills.

On March 24, 1989, the single-hull oil tanker, *Exxon Valdez*, hit a rock off the coast of Prince William Sound near Alaska. More than 11 million gallons of oil spilled out within five hours. The resulting oil slick was 1,000 feet wide and four miles long. Estimates indicate that the oil spill killed more than 250,000 seabirds, 2,800 sea otters, 300 harbor seals, 250 bald eagles, 22 killer whales, and billions of salmon and herring eggs (Sierra club press release, 2004). During the 1991 Gulf War, a total of 720 oil wells were damaged in Kuwait and 294 million gallons of oil was lost (Gist 1991). Vast amounts of wildlife were contaminated including birds, fish, and animals. Many were killed. Vast numbers of birds were found dead in oil lakes. Sea turtles, shellfish, shrimp, and commercial fish species were in jeopardy.

According to conservative estimates, more than 20,000 tons of oil were spilled from the oil tanker *Tasman Spirit*. An oil spill of this magnitude initiated a cycle of irreversible damage to ecological systems off Karachi's coast. This is the first time in the history of man-made disasters that such a big oil spill occurred so close to a populated coastline. This will lead to short-term and long-term adverse health effects for thousands of people. According to recent studies, the impact of an oil spill on marine life is not directly related to the size of the spill; even a small spill in an ecologically sensitive area can have long-term adverse effects. In an oil spill, the most deadly toxins are a class of organic compounds known as polycyclic aromatic hydrocarbons (PAHs) (Lamoureux 2001). Scientists have found that PAHs and other toxic compounds have adverse effects on marine species even at very low concentrations (Poupart et al. 2005). Also, the dispersants used to control the spreading oil slick are harmful to humans. The health of the thousands of people who live close to the shoreline is jeopardized. Coastline inhabitants will need to be monitored for short-term and long-term adverse health effects associated with the exposure to PAHs and dispersants.

Developing countries need to develop and implement appropriate legislation to prevent possible contamination of land and water as a result of oil transportation by major corporations and to hold companies liable for environmental damage. To guard against potential oil spills, only those oil tankers that have safety features, such as a double hull and protective ballast tanks, should be allowed to transport crude oil.

During the past few years, too many people have become victims of natural and man-made disasters. In this regard, the time has come for the Pakistan government to form a national disaster relief organization in collaboration with private-sector and philanthropic organizations to offer help to the masses and develop contingency plans to deal with future calamities. Technological tools should be used to forewarn people of natural and man-made disasters (rains, floods, earthquakes, and oil spills) to reduce and avoid unnecessary human suffering and environmental damage.

REFERENCES

Bakhtiar, Indrees, and Askari, Hussain. (September 2003). Ship of Fools. Karachi *Herald*.

Askari, Hussain. (September 2003). All at Sea. Karachi *Herald*.

Bakhtiar, Indrees. (September 2003). "Something's Rotten in Karachi." Karachi *Herald*.

Ferrell, O., and Fraedrich, J. (1991). *Business Ethics: Ethical Decision Making and Cases*. Boston: Houghton Mifflin.

Gist, Ginger. (Spring 1991). The New Dead Sea. *Journal of Environmental Health*.

Karim, Abdul. (August 14, 2003). Karachi: Why Did Tasman Spirit Run Aground? *Dawn*. Available at http://www.dawn.com/2003/08/14/local10.htm.

McCarthy, Rory (August 15, 2003). Fears of Massive Oil Spill as Karachi Tanker Cracks. *The Guardian Unlimited*. Available at

Table P3.2
Chronology of Events

1. July 27, 2003: *Tasman Spirit*, the Greek-owned oil tanker, ran aground off the Karachi coast near Clifton Beach. Karachi Port Trust (KPT) authorities attempted to push the ship back into the shipping channel with the help of two tug boats.
2. July 31, 2003: *Endeavor II*, a larger oil tanker, fails to reach the *Tasman Spirit* to offload 67,500 tons of crude oil.
3. August 7, 2003: Lighterage operation begins with the help of *Fair Jolly*, a small ship. Poor weather coupled with the ship's limited capacity permits the transfer of 8,000 tons of oil from the *Tasman Spirit* in 48 hours.
4. August 15, 2003: More than 20,000 tons of oil leaked into the sea, contaminated Karachi beaches, and damaged the ecosystem.

Source: *Herald/Dawn*, September 2003, Karachi.

The *Tasman Spirit* oil spill caused massive contamination of Karachi's coastline.
Photo courtesy of Azhar Iqbal, editor, *Pakistan Medicine Digest.*

http://www.guardian.co.uk/international/story/0,,1019114,00.html.

"Not Again! Fishing, Mangroves, Wildlife Threatened by Karachi Oil Spill." (August 13, 2003). Friends of the Earth press release. Available at http://www.foe.co.uk/resource/press_releases/not_again_fishing_mangrove.html.

Lamoureux, Marc (2001). Polynuclear Aromatic Hydrocarbons (PAHs) Interactions with Some Model Soils and Sediments. Available at http://www.pr-ac.ca/prac_5966.html.

Poupart, N., et al (2005). Polycyclic Aromatic Hydrocarbons Monitoring in Terrestrial Plant Biota after Erika Oil Spill. VERTIMAR-2005, Symposium on Marine Accidental Oil Spills. Available at http://otvm.uvigo.es/vertimar2005/comunicaciones/.

"Risk of Oil Spill Disasters Still High, Fifteen Years after *Exxon Valdez* Tragedy." (March 24, 2004). Sierra Club press release. Available at http://www.sierraclub.org/pressroom/releases/pr2004-03-24.asp.

QUESTIONS

1. What steps need to be taken at national and international levels to prevent future oil spills?

2. What recommendations would you make for formulating a rapid response to clean an oil spill in oceans and ports? Complete the following flowchart.

3. Identify financial, technological, and manpower resources required to implement plans for cleaning oil spills.

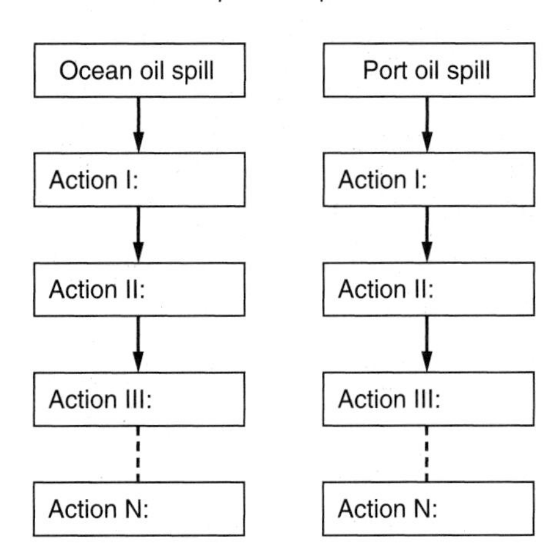

BOX 1 Water Pollution Cause Damage to the Ecosystem of Mangrove Forests near Karachi

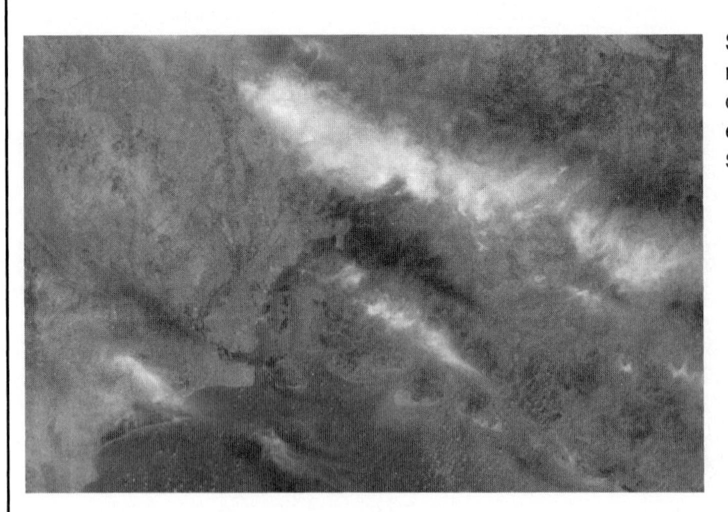

Satellite image of mangrove forests near Karachi. Water pollution has caused serious damage to the ecosystem of mangrove forests.
Source: NASA, http://eol.jsc.nasa.gov/.

SCENARIO I

Dateline: February 19, 20XX

Today the U.N. Security Council passed a resolution demanding the African republic of Bango to accept 500,000 tons of nuclear waste for permanent storage. All African and third-world countries have protested against this resolution. The government of BANGO has announced that it will not comply with the U.N. resolution. The U.N. is also considering a proposal for establishing common storage sites in Third World countries for the nuclear waste generated by developed countries.

Response

Discuss the implications of the U.N. resolution against BANGO. Is the U.N. justified in asking third-world countries to accept the nuclear waste generated by developed countries?

St. Louis, April 1, 20XX

A train carrying high-level nuclear waste from a nuclear power plant in Illinois collided with an east-bound freight train while crossing a bridge over the Mississippi River. The cars containing waste canisters were badly damaged. Due to the colossal impact of the accident, many canisters ripped open and some fell into the river.

Response

Discuss the impact of this accident on the environment. Should the transfer of high-level waste by train be allowed to continue?

SCENARIO II

Los Angeles, July 12, 20XX

The powerful earthquake (7.7 on the Richter scale) that shook the city this morning has severely damaged the core of the XXXXXX nuclear reactor. A radioactive gas cloud has escaped into the atmosphere. Efforts are being made to contain the radioactivity.

Response

Discuss how this accident could have been avoided. Are nuclear power stations better or worse than fossil fuel power-generating stations?

Dateline: January 30, 20XX

A ship containing 50,000 cubic meters of nuclear waste returned to New York after visiting South America, Africa, and Asia in search of potential dump sites. All third-world countries have refused to accept the nuclear waste, despite lucrative offers.

Today, U.S. nuclear waste volume has reached 10 million cubic meters compared with 500,000 cubic meters in the 1990s. Due to the saturation of temporary storage sites for nuclear waste and because of a lack of proper planning and development of permanent storage for nuclear waste in the last century, the world's nuclear waste volumes have reached alarming levels. The potential leakage from temporary storage sites poses the greatest threat to the environment.

Action Item/Response

Going back to 1960, draw a timeline and label it by proposing appropriate action taken in each decade that could have prevented the nuclear waste dilemma that the world faces today.

Time line ____

1950 1960 1970 1980 1990 2000 2010 2020 2030 20XX

Conclusion

During the past 50 years, the global economy has increased fivefold, world population has doubled, and world energy use has tripled. Will these trends continue in the 21st century? Increased production and use of fossil fuel could have severe local and regional impacts. Locally, air pollution takes a toll on human health. Acid precipitation and other forms of air pollution can degrade downwind habitats, especially in lakes, streams, and forests. On a global level, the increased burning of fossil fuels will result in increased emissions of greenhouse gases, which in turn could lead to global warming and other adverse climate changes.

In the new millennium, population pressures and the deterioration of the environment present a challenge to humankind: how to satisfy the ever-growing appetite of the energy-hungry genie—how to develop and use energy sources that are friendly to the environment. We face a dilemma: Energy technologies often enhance material well-being across the planet, but the continuation of current trends could lead to a degraded planet, yielding an uncertain existence for future generations.

According to the International Energy Outlook 2004 report, world energy consumption is projected to increase by 54 percent from 404 quadrillion British thermal units (Btu) in 2001 to 623 quadrillion Btu in 2025. The gross domestic product (GDP) in developing countries in Asia is expected to increase to 5.1 percent compared with 3 percent for the world as a whole. This growth could account for 40 percent of the increased energy consumption. In contrast, for the developed countries, the projected growth is about 1.2 percent for the same forecast period.

World oil demand is expected to increase by 1.9 percent annually until 2025 from 77 million barrels per day in 2001 to 121 million barrels per day in 2025. Much of the increase in oil demand is projected to occur in the United States and in developing Asia. However, if consumers in developed countries, such as the United States, continue to show their lack of concern for the environment by buying more SUVs and trucks rather than fuel-efficient vehicles, and if the people in China, India, Pakistan, and other developing nations try to emulate the SUV-buying habits of American consumers, there will be an exponential increase in the demand for oil. Oil prices may soar to more than $100 a barrel, which will cause consumers to spend a considerable amount of their earnings on energy costs.

It is predicted that, if energy consumption styles do not change, world energy consumption will increase by 50 to 60 percent. Carbon dioxide emissions will also increase by 50 to 60 percent. Therefore, as we plan for the future, we must use energy sources that enable us to sustain our environment. Appropriate energy policies at the individual, national, and international levels, along with efficient energy technologies and conservation efforts, could play an important role in achieving a balance between economic growth and a sustainable environment.

Part IV explores the impact of increased energy consumption on the ecology of our planet.

Chaash mo hum Chaash mee
Eye wants, what eye sees.
PERSIAN PROVERB

Holy Mother Earth, the trees and all nature are witnesses of your thoughts and deeds.
A WINNEBAGO WISE SAYING

The use of solar energy has not been opened up because the oil industry does not own the sun.
RALPH NADER

INTERNET EXERCISES

Use any of the Internet search engines (e.g., Google, Alta Vista, Yahoo!, Infoseek) to research information for the following questions.

1. Answer these questions pertaining to the *Exxon Valdez* oil spill.
   a. When did the oil tanker *Exxon Valdez* (shown below) run aground on Bligh Reef in Prince William Sound in Alaska?
   b. How much oil was spilled into the Sound?
   c. What was the major cause of the accident?
   d. How did the construction of the *Exxon Valdez* contribute to the accident?
   e. What are the short-term and long-term effects of the oil spill in Prince William Sound?
   f. List the 10 largest oil spills to date.
   g. Compare the Prince William Sound oil spill with the Persian Gulf War oil spill in terms of size and short-term and long-term effects.
   h. What actions should be taken by multinational oil companies, governments, and international organizations to prevent future oil spills?

The *Exxon Valdez* floats serenely in the midst of an oil spill in the waters of Prince William Sound, Alaska, flanked by another tanker.
Photo Courtesy of the U.S. Coast Guard.

2. Answer these questions pertaining to the Chernobyl nuclear power plant accident.
   a. Where is Chernobyl located?
   b. What happened on April 26, 1986, at Chernobyl?
   c. What was the cause of the accident at Chernobyl?
   d. What are the short-term and long-term effects of the nuclear disaster at Chernobyl?
3. Define the following terms.
   a. Fossil fuel
   b. Acid rain
   c. Global warming
   d. Greenhouse effect
   e. Radioactive waste
   f. GNP
   g. Per capita oil consumption
4. Determine the electricity consumption per capita for the following developed and developing countries.

Developed Countries	*Electricity Consumption per Capita (kWh)*	*Developing Countries*	*Electricity Consumption per Capita (kWh)*
United States		Bangladesh	
France		China	
Japan		Ethiopia	
Australia		Ghana	
Norway		India	
Sweden		Indonesia	
Switzerland		Kenya	
Italy		Malaysia	
Singapore		Pakistan	
United Kingdom		Venezuela	
Germany		Zimbabwe	

5. What alternative energy sources look promising for the energy demands of the twenty-first century?
6. What percentage of the world's total electricity is produced by hydroelectric power?
7. Compare the features of large dams with those of small dams.
8. Complete the following table.

Country	*Hydroelectric Power-Generating Capacity (MW)*
United States	
Mexico	
Canada	
Egypt	
China	
Brazil	
Pakistan	
India	
Russia	
Japan	
Argentina	
Venezuela	
Italy	

9. What percentage of the world's total electricity is produced by solar power?
10. Will solar power some day replace fossil fuel as a source of energy? Why or why not? Explain your answer.

11. Complete the following table.

Country	*Solar Power-Generating Capacity (MW)*
United States	
Mexico	
Canada	
Nigeria	
China	
Brazil	
Pakistan	
India	
South Africa	
Japan	
Malaysia	
Indonesiaz	
Australia	

12. Determine the wind power-generating capacity for the following developed and developing countries.

Developed Countries	*Wind Power–Generating Capacity (MW)*	*Developing Countries*	*Wind Power–Generating Capacity (MW)*
United States		Brazil	
France		China	
Japan		Nigeria	
Australia		Ghana	
Norway		India	
Sweden		Indonesia	
Switzerland		Jordan	
Italy		Malaysia	
Singapore		Pakistan	
United Kingdom		Venezuela	
Germany		Zimbabwe	

13. Using Internet search engines, find pertinent information regarding capital costs for generating electricity by using fossil fuels and renewable sources, then complete the following table and answer the following questions.
    a. Which energy technologies have a low capital cost for generating electricity?
    b. Which energy technologies have a high capital cost for generating electricity?
    c. Which energy technologies have low capacity factors?
    d. Which energy technologies have a high capital cost of electricity?
    e. Can renewable energy technologies compete with fossil fuel technologies in terms of their economic costs? Yes/No? Explain your answer.
    f. Formulate a road map, consisting of five to ten key steps, for developing national and international policies that can increase the economic competitiveness of renewable energy technologies.

Energy Technology	*Capital Cost ($/kw)*	*Capacity Factor* (%)*	*Capital Cost of Electricity (cents/kWh)*
Fossil fuel (oil)			
Fossil fuel (natural gas)			
Solar (photovoltaic)			
Biomass (wood-fueled steam plant)			
Wind			
Hydro			

* Capacity factor is the ration of average to installed power.

14. Using Internet resources, research the details of the U.S.–India nuclear energy agreement and discuss its pros/cons and global implications.

15. Visit NASA's Multimedia Gallery at http://www.nasa.gov. The site contains a large collection of photos of the Earth taken from space by various space shuttle missions depicting the impact of different types of human interaction. Answer the following questions:

    a. List the impact of the construction of dams on ecology and soil erosion in developing and developed countries.

    b. List those dams in the United States where the impact of construction on ecology and soil erosion is most evident.

TRENDS OF THE 21ST CENTURY

Chart the following trends into the 21st century and explain your opinion.

1. Do you feel that these trends are important for the twenty-first century? Indicate the degree of importance that these issues will have for the twenty-first century.

Trend	*Very Important*	*Important*	*No Opinion*	*Not Important*	*Insignificant*
Fossil fuel consumption					
Family values					
Pollution					
Humanity					
Availability of resources					
State of environment					
Stress					
Moral and ethical values					
Materialism					

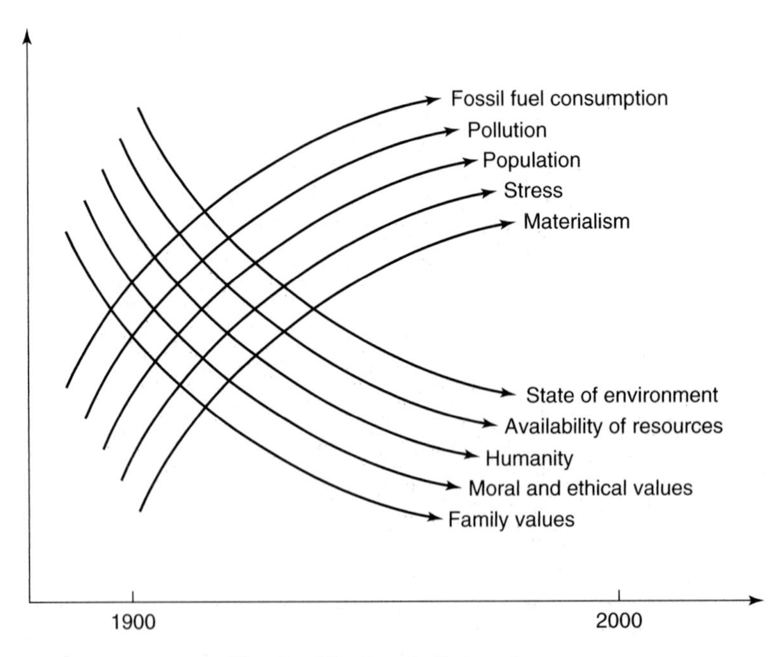

Trends of the twenty-first century.

Beyond the Bridge: Explore Future Options . . .

2. The gigantic increase in the use of fossil fuels in the twentieth century has wreaked havoc on the Earth's atmosphere. What kind of risks does increased fossil fuel consumption impose on society at personal, national, and international levels? What efforts must be made to minimize these risks? Complete the following table.

	Personal Level	*National Level*	*International Level*
Risks			
Efforts			

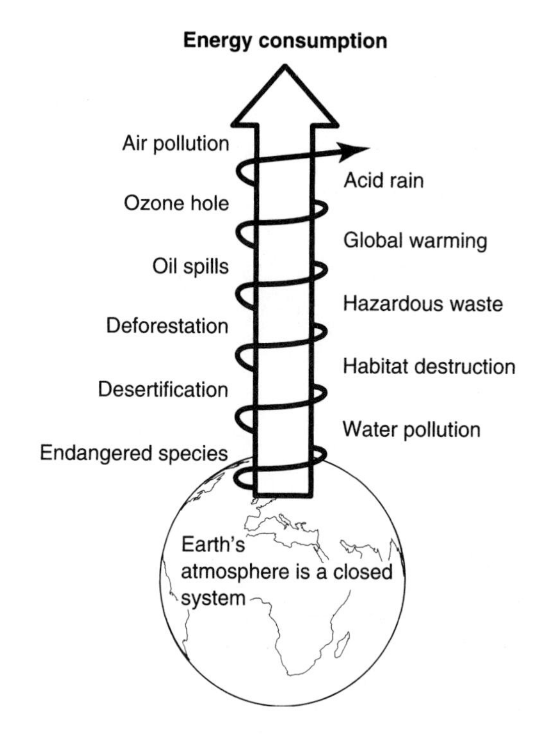

Impact of increased fossil fuel consumption.

3. What types of options and policies must be adopted at personal, national, and international levels to curb the increased consumption of fossil fuels?

	Personal Level	*National Level*	*International Level*
Options/Policies			
I			
II			
III			

4. What types of renewable and alternative energy technologies will promote economic growth for a sustainable environment in the twenty-first century?
5. Can the impact of technology in the twentieth century be reversed? Yes or No? Explain your answer.
6. World gasoline prices have reached record high levels. Discuss the economic impact of increased prices at personal, national, and international levels.

"Record high gas prices are burning holes in peoples' pockets and harming our economy. They are squeezing everyone, farmers and truckers, small businesses and families. When oil prices rise, the prices of all goods increase."
(Lane Evans)
Photo courtesy of Ahmed S. Khan

USEFUL WEB SITES

http://www.nrel.gov	National Renewable Energy Laboratory
http://www.fsec.ucf.edu	Florida Solar Energy Center
http://www.nei.org	Nuclear Energy Institute
http://www.chernobyl.info/	Chernobyl Web links
http://www.worldwatch.org	Worldwatch Institute
http://www.worldbank.org	World Bank
http://www.undp.org	United Nations Development Programme
http://www.lib.umich.edu/govdocs/statsnew.html	Statistical resources on the Web
http://www.sandia.gov/Renewable_Energy	Renewable energy sources at the Sandia National Laboratories
http://www.hooverdam.com/Gallery	Hoover Dam photo gallery
http://www.energy.gov	Department of Energy
http://tidalelectric.com/	Tidal power
http://www.bwea.com	British Wind Energy Association
http://www.awea.org	American Wind Energy Association
http://www.Bera1.org	Wind and marine renewable energy alternatives
http://www.piarc.org/en/	World Road Association
http://www.apta.com/	American Public Transportation Association
http://www.bts.gov/	Bureau of Transportation Statistics
http://www.hpva.us/	Human Powered Vehicle Association
http://www.earth-policy.org	Earth Policy Institute
http://www.transportenvironment.org	European Federation for Transportation and Environment
http://www.uic.asso.fr/	International Union of Railways
http://www.cmdl.noaa.gov/	Global Monitoring Division and atmospheric conditions
http://www.epa.gov/ozone	Ozone depletion at the U.S. Environmental Protection Agency
http://www.epa.gov/docs/acidrain/	Acid rain at the U.S. Environmental Protection Agency
http://www.ucsusa.org/warming/index.html	Energy issues from the Union of Concerned Scientists
http://www.epa.gov/globalwarming	Global warming at the U.S. Environmental Protection Agency

PART IV Ecology

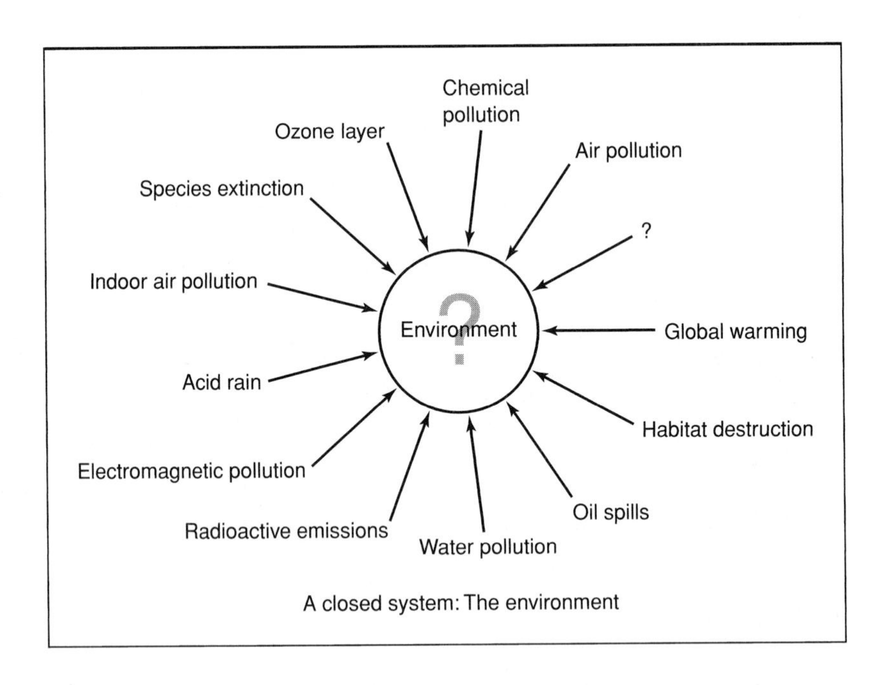

A closed system: The environment

OBJECTIVES

After reading Part IV, Ecology, *you will be able to*

1. Identify the historical background and major and minor issues of ecology.
2. Examine the various problems, urgencies, and implications of specific ecological issues.
3. Distinguish between industrialized and non-industrialized nations' ecological issues and viewpoints.
4. Analyze and use tables and graphs to promote specific and additional ecological perspectives.
5. Evaluate the implications of increasing technological and industrial wastes.
6. Evaluate the impact of rapidly increasing human resource use on the environment.
7. Develop ethical considerations with environmental issues.
8. Evaluate the strength, ethics, and potential of planned environmental approaches.
9. Appreciate the importance and contributions of individual ecological involvement and action.

INTRODUCTION

The term "environment" simply means the world that is all around us. This definition, then, includes no less than all of our natural world—our ecosystem. The study of ecology examines the mutual relationship between organisms and this natural world—the environment—and therefore analyzes the changes and effects on our entire environment due to organisms and their tools. Of course, one section or part of a book can hardly deal with such an all-embracing subject, but it can present an overview of some of the major issues of our ecosystem as man,—along with the powerful and unpredictable use of technology—changes its balance, supportive systems, and even beauty. The purpose of Part IV is to enhance awareness, thought, understanding, responsibility, strategies, and involvement for a future environment that is desirable, inhabitable, and sustainable. As this part of the text is presented, historical, social, and economic views will be discussed as well to offer a more complete understanding of the issues from differing perspectives.

Looking into the environment of the past, present, and future.
Photo courtesy of Ahmed S. Khan.

HISTORICAL BACKGROUND AND MAJOR ISSUES

The environmental movement formally began in the United States on Earth Day, April 22, 1970, when the American public began to awaken to the ecological destruction surrounding it. At that time, pollution controls did not exist for cars. People and cities dumped untreated sewage into the nation's rivers, some of which were so filled with chemical waste that they actually caught fire. Industrial cities were clouded with pungent, acid smoke. Since then, many of these problems have diminished and have been effectively addressed. The Environmental Protection Agency (EPA), established in 1970, monitors air quality around the country, and toxic emissions from smokestacks, factories, and incinerators have been sharply reduced. Mandatory pollution-control standards on automobiles have led to a drop in lead emissions, and recycling as a way of reducing solid waste has made a significant impact on the health of the environment.

However, as our use of materials continues to grow on a worldwide basis, along with an exponentially increasing population growth and level of need, new environmental strategies cannot help but involve an ecological cost of use as well as the cost of the pollution of our supportive natural resources. Based on the findings of a scientific advisory group, the EPA has ranked world environmental issues in the following manner (Wright 2004).

1. *High risk*: Global warming, species extinction, habitat destruction, ozone layer depletion, loss of biological diversity
2. *Medium risk*: Herbicides and pesticides, surface water pollution, airborne toxic substances
3. *Low risk*: Oil spills, radioactive materials, groundwater pollution
4. *Human health risk*: Indoor air pollution, outdoor air pollution, exposure of drinking water to chemicals

In 2004, The BBC News produced a six-part series titled *Planet Under Pressure* that outlined the six most pressing global environmental problems for the twenty-first century according to world environmental scientists and their research. These six most urgent global issues include (1) biodiversity, (2) water, (3) energy, (4) food, (5) pollution, and (6) climate change. (1) Concerning biodiversity, many scientists believe that the earth is currently entering the sixth great extinction in its history of plants and animals. (2) Concerning water shortages, by 2025, two-thirds of the people of the world will possibly experience acute water shortages. (3) Oil production will peak and be in short supply possibly as early as 2010. (4) Food shortages continue. An estimated one of six people in the world suffer from hunger and malnutrition complicated by poor agricultural practices that further damage the land and erode its capability to supply their food needs. (5) Chemical pollution has spread everywhere. Hazardous chemicals are now present in the bodies of all newborn babies, and approximately one of four people globally are exposed to dangerous amounts of air pollutants. (6) Climate change remains the world's greatest environmental challenge. The earth's surface temperature has risen about 1 degree within the last century, accompanied by increasing temperatures and CO_2 emissions (Kirby 2004, 2–3). Nearly three times as much carbon was released in 2004 than in 1960 (Worldwatch Institute, 2005), with the hottest years on record occurring in the last decade. It is very clear that both the environment and the management of ecology in the twenty-first century are the most urgent issues for our basic sustainability.

For an overview of and a brief introduction to these high-risk issues, a brief discussion of the major issues in each category follows.

When the environment is finally forced to file for bankruptcy because its resource base has been polluted, degraded, dissipated, and irretrievably compromised, the economy will go down with it.

Timothy Wirth, U.S. Department of State

Important pathways for a sustainable environment. Photo courtesy of B. Eichler.

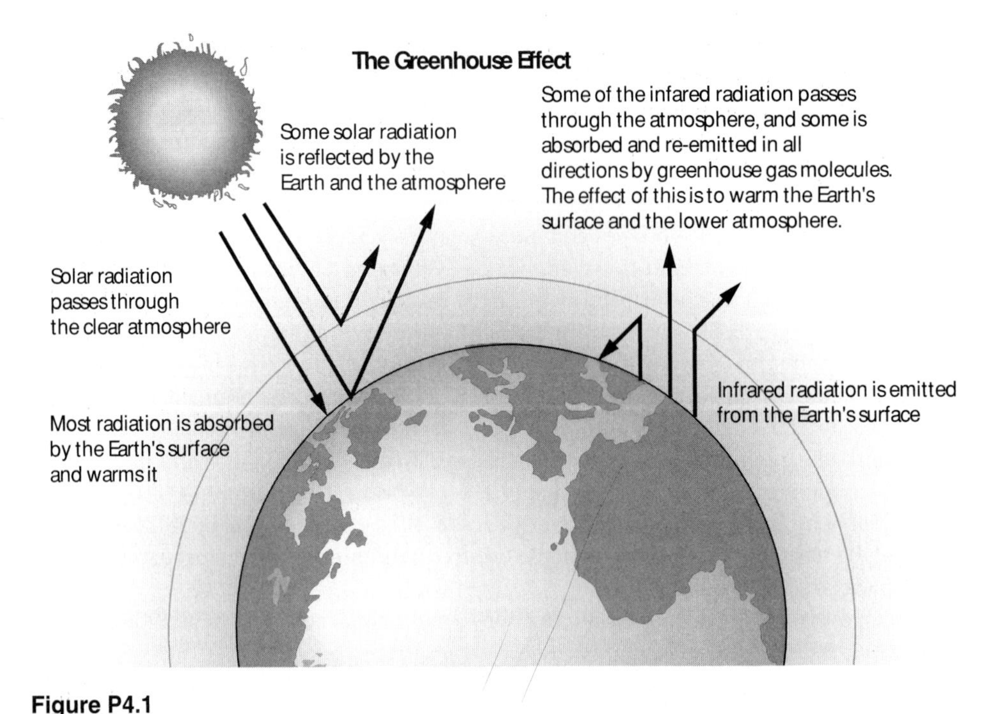
The Greenhouse Effect
Some solar radiation is reflected by the Earth and the atmosphere
Some of the infared radiation passes through the atmosphere, and some is absorbed and re-emitted in all directions by greenhouse gas molecules. The effect of this is to warm the Earth's surface and the lower atmosphere.
Solar radiation passes through the clear atmosphere
Most radiation is absorbed by the Earth's surface and warms it
Infrared radiation is emitted from the Earth's surface

Figure P4.1
Greenhouse effect.
Source: U.S. Environmental Protection Agency, 2000. "Global Warming – Climate" http://yosemite/epa.gov/oar/globalwarming.nst/content/climate.html.

INTRODUCTORY SUMMARY: HIGHEST RISK AREAS

Global Warming

Scientists have increasing evidence that the earth has begun to warm significantly over the past century—by about 1 °F—and could further warm 2.5–10 °F over the next 100 years. If this estimate is true, the effects would be devastating, considering that the earth's temperature has only risen 9 degrees since the last ice age 12,000 years ago. Such an increase in warming would cause polar ice caps to melt, thus causing a rise in the sea level worldwide that would destroy coastal areas, islands, water supplies, forests, and agriculture in many parts of the world. Global warming is caused from gases in the atmosphere that prevent sunlight from being reflected away from the earth. Usually, sunlight that reaches the surface of the earth is partly absorbed and partly reflected. The absorbed light heats the surface and is later released as infrared radiation. Gases (called "greenhouse gases") that are not released collect the heat and keep it in the atmosphere. The earth's atmosphere is only 0.03 percent CO_2, but together with other gases, it can trap 30 percent of the reflected heat and maintain the earth's average temperature at about 59 °F (Wright 2005).

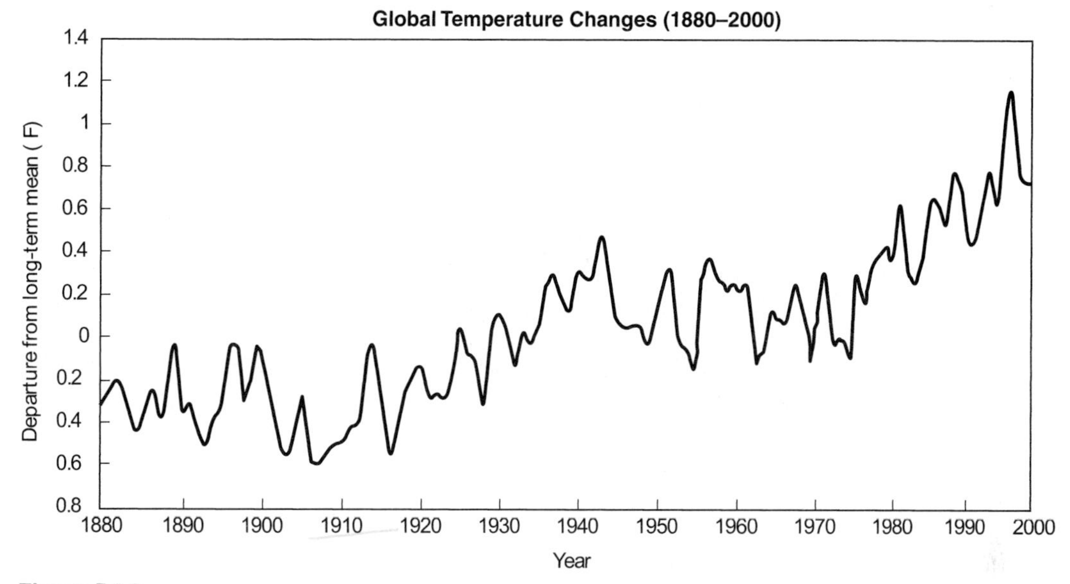

Figure P4.2
Global temperature changes (1880–2000).
Source: U.S. National Climatic Data Center, 2001.

Since the 1800s, an unprecedented amount of the four primary greenhouse gases (those that are not transparent to infrared radiation and collect heat such as carbon dioxide, chlorofluorocarbons (CFCs), methane, and nitrous oxide) have been released into the atmosphere. The worst of these gases are CFCs which, in addition to contributing to the greenhouse effect, are also destroying the stratospheric ozone that shields the earth's surface from ultraviolet rays. In 2000, the United States, Russia, Japan, and various countries in Europe agreed to a 100-percent phaseout of CFCs , and CFCs are now slowly being phased out. Concentrations of carbon dioxide have increased by 25 percent within the last century. As greenhouse gases increase in concentration, they will produce a domino effect that could cause environmental chaos in weather patterns, warming patterns, and sea level. Scientists increasingly agree that global warming is occurring as evidence continues to accumulate from plant and animal cycles, chemical analysis, historical tracking, and atmospheric and geographic data. The 1997 Kyoto Accord attempted to reduce worldwide emissions by 5 percent. However, the United States refused to sign, even though it contributes one-fourth of the world's CO_2. Please see Chapter 26, "Climate Change," for further discussion.

Species Extinction and Loss of Biological Diversity

There is serious concern that the sixth global mass extinction of plant and animal species in the history of the planet may be beginning to occur, with this extinction being as great as the mass extinction of the dinosaurs 65 million years ago. Rapid loss of forest, freshwater, and marine ecosystems are pressuring animals and plants into smaller areas and threatening their existence. This is primarily due to rapid human population as our numbers have almost tripled in half a century from 2.5 billion in 1953 to 6.2 billion in 2005. A minimum of 50,000 species alone are rendered extinct per year, which comes to about 140 extinctions each day at five to six per hour! The world's 10 to 80 million species are threatened by deforestation, loss of wetlands, and urban sprawl, as well as shifting climate and vegetation zones. Another cause for animal and plant loss is the change to monocultures in agriculture, which is the production of only one strain of crops for food that then becomes vulnerable to disease and other threats.

Tropical rain forests contain at least half of the world's species, and much of them are being decimated at extremely rapid rates. Less than half of the tropical rain forests exist today that existed in 1800; there are 3.5 billion acres today, which is down from the 7.1 billion acres found in 1800. The United States estimates that by 2020, less than 5 percent of tropical forests will remain in pristine condition. Within 15 years, about one-fifth of central Africa's forest will have disappeared, and this is a worldwide trend. The World Conservation Union's (IUCN) Red List in 2003 stated that more than 12,000 species (of the 40,000 assessed) faced some extinction risk including one bird in eight, 13 percent of the world's flowering plants, and one-quarter of all mammals (Kirby, Biodiversity, 2–3). This collapse of rain forest ecosystems will likely lead to the disappearance of up to 10 percent of the world's species within the next 25 years unless we prevent it. Please see Chapter 30 on species extinction for further discussion.

The U.S. Endangered Species Act of 1966 is the chief protector of species in the United States, and the Convention on International Trade in Endangered Species of Wild Fauna and Flora (CITES), signed in 1975 by 122 nations, continues to be one of the main forces in the preservation of species internationally. To abate the trend toward species extinction, governments around the world have set aside approximately 16.4 million square miles of protected lands in about 3,500 parks and preserves. In addition, many countries try to identify those species threatened with extinction so that they may be carefully monitored and protected.

The following are some of the selected endangered species of the world: cheetah, African chimpanzee, Dugong (sea cow), gibbon, gorilla, jaguar, leopard (three species), monkey (many species), orangutan, giant panda, rhinoceros, tiger, whale (eight species), wolf, birds (over 1,000 of 9,672 species), reptiles (many species of alligators, crocodiles, iguanas, and sea turtles), amphibians (all frogs and toads in the United States), fish (many species), plants (about 21 percent, or 4,200 of 20,000 species; as many as 750 plant species in the United States could become extinct by the first years of the twenty-first century). Refer to Tables P4.1, P4.2, and P4.3 for a more detailed breakdown of data on endangered species.

Habitat Destruction

Habitat destruction includes the destruction of wild lands that harbor the biological diversity of earth. Estimates of the total number of plant and animal species are between 10 and 80 million, of which only about 1.4 million species have been identified. The mass extinction of this biodiversity at unprecedented rates has already begun to cause the loss of 50,000 invertebrate species per year (nearly 140 each day) in the rain forests. By the year 2050, it is believed that 25 percent of the earth's species will become extinct if rain forest

Table P4.1
U.S. List of Endangered and Threatened Species

Group	*Endangered*		*Threatened*		*Total Species*[1]	*Species with Recovery Plans*
	U.S.	*Foreign*	*U.S.*	*Foreign*		
Mammals	69	251	9	17	346	55
Birds	77	175	14	6	272	78
Reptiles	14	64	22	15	115	33
Amphibians	11	8	10	1	30	14
Fishes	71	11	43	1	126	95
Clams	62	2	8	0	72	69
Snails	21	1	11	0	33	23
Insects	35	4	9	0	48	31
Arachnids	12	0	0	0	12	5
Crustaceans	18	0	3	0	21	13
Animal subtotal	390	516	129	40	1,075	416
Flowering plants	571	1	144	0	716	578
Conifers and cycads	2	0	1	2	5	2
Ferns and allies	24	0	2	0	26	26
Lichens	2	0	0	0	2	2
Plant subtotal	599	1	147	2	749	608
GRAND TOTAL	989	517	276	42	1,824	1,022

[1] Some species are classified as both endangered and threatened. The table tallies those "dual-status" species only once as endangered, except for the olive ridley sea turtle, which is dual status but tallied as a U.S. threatened species. The other dual-status species, all tallied as endangered, are (U.S.) Chinook salmon, gray wolf, green sea turtle, piping plover, roseate fern, sockeye salmon, steelhead, Steller sea lion, (non-U.S.) argali, chimpanzee, leopard, saltwater crocodile.
Source: *The World Almanac and Book of Facts*, New York: World Almanac Books, 2005.

destruction is continued. Tropical forests are losing an average of 2.8 acres per month . . . 93,000 acres/day . . . 3,800 acres/hour . . . 64 acres/minute (World Conservation Union).

According to a 1997 World Resources Institute (WRI) assessment, just one-fifth of the earth's original forests remain in large, relatively natural ecosystems known as frontier forests. An analysis by country finds that 76 countries (of the world's 193) have lost all of their frontier forests, and 11 nations are close to losing their last remaining frontier forest with fewer than 5 percent of these forests left, all of which are threatened (World Resources Institute 2006; see Table P4.4). Other habitats, such as islands, wetlands, and freshwater lakes, are also losing great proportions of their life forms due to environmental loss and degradation.

Rain forests are home to at least half of the planet's species, and they have been reduced by nearly half of their original area (from 14 percent of the earth's land to only 6 percent today), thereby losing an area about the size of the state of Washington each year

Table P4.2
Endangered and Threatened Species in the United States, 1980–2004

Year	*Endangered* Animals	Plants	Total	*Threatened* Animals	Plants	Total	Total Species
1980	174	50	224	48	9	57	281
1985	207	93	300	59	25	84	384
1990	263	179	442	93	61	154	596
1995	324	432	756	113	93	206	962
1998	357	567	924	135	135	270	1,194
1999	358	581	939	126	140	266	1,205
2000	368	593	961	129	142	271	1,232
2001	386	595	981	128	145	273	1,254
2002	387	596	983	128	147	275	1,258
2003	388	599	987	129	147	276	1,263
2004[1]	390	599	989	129	147	276	1,265

[1]As of September 22, 2004.
Source: U.S. Species Listed by Calendar Year, 1980–Present, U.S. Fish and Wildlife Service.

Table P4.3
Countries of the World with the Most Threatened Species (organized by region)

Country	*Mammals*	*Birds*	*Reptiles*	*Amphibians*	*Fishes*	*Invertebrates*[1]	*Plants*
Cameroon	43	18	4	53	39	1	355
Madagascar	48	36	18	55	72	8	277
South Africa	28	38	20	21	28	123	73
Tanzania	35	39	5	41	130	25	241
China	84	88	34	91	59	5	442
Japan	38	56	11	20	35	18	12
Russian Fed.	44	53	6	0	22	28	7
India	89	82	26	68	35	20	247
Indonesia	146	121	22	47	45	2	688
Malaysia	51	43	28	39	105	28	387
Philippines	51	74	9	48	58	17	215
Vietnam	45	42	27	18	30	0	148
Mexico	74	62	21	204	109	35	261
United States	41	79	27	53	159	303	243
Brazil	73	124	22	26	58	13	382
Ecuador	34	76	11	165	14	0	1,832
Australia	64	65	39	47	85	107	56
New Guinea	58	32	10	10	37	10	142

[1]Invertebrates include insects, crustaceans and others
Source: Red List Data, International Union for Conservation of Nature and Natural Resources, 2006.

Table P4.4
Frontier Forests Suffer a Variety of Threats

Region	% of Frontier Forest[1] under Moderate/High Threat	Percentage of Threatened Forests at Risk from Logging	Mining, Roads, and Other Infrastructure	Agricultural Clearing	Excessive Vegetation Removal	Other[2]
Africa	77	79	12	17	8	41
Asia	60	50	10	20	9	24
North & Central America	29	83	27	3	1	14
North America	87	54	17	23	29	13
Central America	26	84	27	2	0	14
South America	54	69	53	32	14	5
Russia & Europe	19	86	51	4	29	18
Europe	100	80	0	0	20	0
Russia	19	86	51	4	29	18
Oceania	76	42	25	15	38	27
WORLD	39	72	38	20	14	13

[1] Frontier forests considered under immediate threat, as a percentage of all frontier forests assessed for threat. Threatened frontier forests are places where ongoing or planned human activities, if continued over the coming decades, are likely to result in the significant loss of natural qualities associated with all or part of these areas.
[2] Includes such activities as overhunting, etc.
Source: World Resources Institute, *Earth Trends. The Environmental Information Portal*, Washington, D.C: World Resources Institute, 2006, http://earthtrends.wri.org.

due to deforestation. During the past 30 years, about one-third of the world's rain forests have disappeared. Tropical rain forests are lush habitats filled with more diversity of life than any other habitat. A typical patch of rain forest covering four square miles contains 750 species of trees, 750 species of other plants, 125 species of mammals, 400 species of birds, 100 species of reptiles, and 60 species of amphibians. This rich plant life supports food chains that include many of the world's most spectacular animals such as brightly colored frogs, eagles, monkeys, tigers, sloths, and army ants. Many rain forest species cannot survive in any other type of environment and are, like all species, biologically unique. The destruction of species that accompanies deforestation becomes a large-scale loss of species to the earth and mankind. For example, of the 3,000 plant species that have anticancer properties, 70 percent are found in tropical rain forests. Rain forests also have an impact on world climate, affecting the chemistry and ecology of the ocean, consuming huge quantities of CO_2, producing a large percentage of the world's oxygen, and slowing global warming.

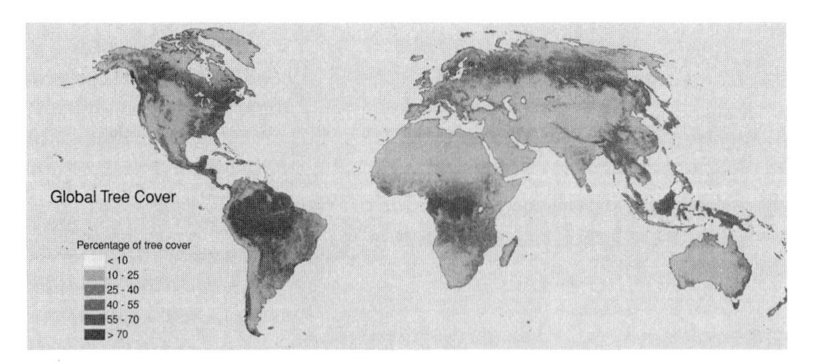

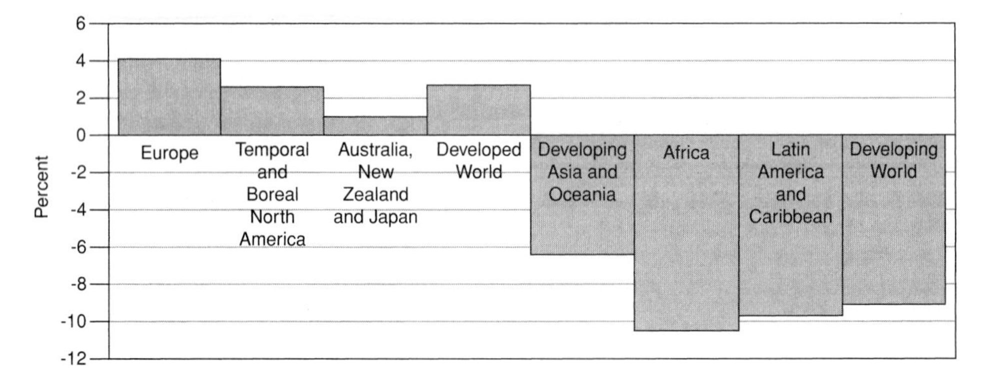

Figure P4.3

Changing extent of forests. Since the 16th century, the forests of the northern temperate zone have suffered the most extensive losses as a result of human activity. In recent years, they have begun to recover. However, these gains have been more than offset by rapid decreases in the more extensive and species-rich forests of the developing world. Many of the world's trees grow within areas that are only partially forested. These lands provide many of the goods associated with forests, especially woodfuel, species habitat, and soil protection. Such areas are particularly vulnerable to clearance, however, since they are often more accessible and less likely to be legally protected than forest areas with higher tree cover.

Source: World Resources Institute. 2000. *World Resources 2000–2001: People and Ecosysems: The Fraying Web of Life.* Washington, D.C.

The Amazon River basin, one of the largest rain forests, holds 66 percent of the earth's fresh water and produces 20 percent of the earth's oxygen. Figure P4.3 below illustrates loss and change of the forest landscape on a world scale from 1980 to 1995 with the highest percentage loss in the species-rich forests of the developing world.

Wetlands are also important environmental ecosystems that provide habitats for many species of animals and plants; a third of the endangered or threatened species in the United States live in or are dependent on them. Wetlands are the nurseries for many fish, all amphibians, and many birds and mammals. Wetlands also filter, purify, and hold water, even eliminating pesticides and other toxins by speeding up degradation by microbes. They store water, slow down its flow, and buffer floods and erosion. Coastal wetlands are spawning grounds for between 60 and 90 percent of U.S. commercial fisheries, and hunting depends entirely upon wetland habitats. Wetlands are among the most productive natural ecosystems on earth in terms of total biological mass per unit of area. Since colonial times, half of the wetlands in the United States have been destroyed, with only about 105 million acres remaining (excluding Alaska). Refer to Table P4.5 for more detailed information on the loss of state wetland acreage.

Ozone Layer

The chemical ozone (O_3) exists as smog at ground level—where there is an abundance of it, causing pollution—and in the stratosphere, where there is not enough of it, causing a very serious weakening of the ozone shield that protects us from the sun's ultraviolet rays and other types of radiation. A significant reduction of the ozone layer would lead to an increase in skin cancers and cataracts, the loss of small ocean algae and bacteria, and the breakdown of other natural systems. Chlorofluorocarbons (CFCs), chemicals manufactured since the 1930s for refrigeration, insulation, and packaging, add chlorine to the atmosphere and absorb the ozone in the stratosphere. CFCs are very stable and have lifetimes of 75–110 years. They drift into the upper atmosphere, where their chlorine components are released. The freed chlorine atoms find ozone molecules and react, creating chlorine monoxide. In a subsequent reaction, the chlorine monoxide releases its oxygen atom to form molecular oxygen, and the chlorine atom is freed again to repeat the process. With the long lifetime of chlorine, each chlorine atom can destroy about 100,000 molecules of ozone before the chain reaction ends. Ice crystals in the Antarctic and Arctic accelerate the process, producing "ozone holes" near both poles. Studies indicate that stratospheric ozone has declined by 60 percent during the worst years and has fallen 5–10 percent over the United States depending on the season.

Table P4.5
States with the Most Wetland Acreage, 1780s–1980s

State[1]	*Wetlands in 1780s*		*Wetlands in 1980s*		*Percent*
	Acres	*% of Area*	*Acres*	*% of Area*	*Lost*
Alaska	170,200,000	45.3	170,000,000	45.3	–0.1%
Florida	20,325,013	54.2	11,038,300	29.5	–46
Louisiana	16,194,500	52.1	8,784,200	28.3	–46
Minnesota	15,070,000	28.0	8,700,000	16.2	–42
Texas	15,999,700	9.4	7,612,412	4.4	-52
North Carolina	11,089,500	33.0	5,689,500	16.9	–49
Michigan	11,200,000	30.1	5,583,400	15.0	-50
Wisconsin	9,800,000	27.3	5,331,392	14.8	–46
Georgia	6,843,200	18.2	5,298,200	14.1	-23
Maine	6,460,000	30.4	5,199,200	24.5	-20

[1]Ranked by Wetlands Acreage in the 1980s.
Source: Wetlands in the United States, 1780s to 1980s, U.S. Dept. of Interior, Fish and Wildlife Service, 1990.

In 1987, the Montreal Protocol was signed and the signatory nations agreed to reduce the use of CFCs and other ozone-depleting substances. This was later amended because scientific data indicated worse damage than previously expected. In 1992, the Protocol agreement stated that it would end the production of halons (fire-extinguishing agents) by the beginning of 1994 and of CFCs by the beginning of 1996 in developed countries. As of January 1, 1996, only recycled and stockpiled CFCs were available for use in developed countries. Today, 189 countries have ratified the treaty. Due to the measures taken under the Protocol, emissions of ozone-depleting substances are already declining. Stratospheric chlorine levels have peaked and are no longer increasing based on chlorine measurements that had stopped increasing in 1997–1998. According to the most recent Scientific Assessment of Ozone Depletion, 2006 of the World Meteorological Organization and the United Nations Environmental Programme assessment, if the ban on CFCs continues and CFC use is curtailed according to the Montreal Protocol, the natural ozone production process will heal the ozone layer by the end of the 21st century. Thus far, this is a tremendous success story for international ecological cooperation and urgent environmental global action (EPA 2006).

There are many environmental issues that need to be discussed. The issues, readings, and case studies in Part IV highlight our environmental needs, causes, perspectives, problems, and successes. These readings and case studies not only address the issues themselves, but also include societal and economic responses since this is where the effects and actions

Understanding the value of our ecology.
Photo courtesy of Ahmed S. Khan.

can be seen and understood. Four of the readings provide a global analysis of the environmental consequences of increasing intense resource use by man. *The Kyoto Protocol* provides an overview of how the world is wrestling with climate changes. There are three readings in the field of ecology that deal with specialized issues in specific locations of the world that have focused outcomes and lessons to instruct humankind in our future decisions. The readings and case studies presented here were selected because they wrestle with central issues and responses that enhance awareness and decision-making for our future world. These studies also consistently underscore the vital necessity to acknowledge the increasing value and importance of the world around us—our life-sustaining world, the environment.

AN ECONOMIC VIEW OF ECOLOGY

In the United States each year, $20 billion worth of crops are pollinated by honeybees. Many of the bees are *leased* because wild bees are essentially gone from certain areas of the country. The Millennium Ecosystem Assessment, a UN-commissioned report, states that 60 percent of life-supporting functions—from fisheries to fresh water—are in decline. This report was compiled over four years by 1,300 scientists in 95 countries. It calls on societies to provide a financial accountability assessment and audit for all of the services provided by nature. The study argues that people can protect and reverse the damage in those ecosystems that they have profoundly corrupted in the last 50 years if the environment is given the appropriate value. Chief scientist of Environmental Defense, Dr. Bill Chameides, states, "The services a forest provides in clean water, watershed managements and carbon storage are worth much more than its lumber. Yet we cut down our forest for timber; it's often a poor economic choice." (*Solutions* 12)

Another major study from 13 researchers published in a May 1997 report in the journal *Nature* tried to establish the economic value of the annual worth of the earth's natural goods and services—the annual worth of our ecosystem. This value is estimated at $33 trillion and includes the earth's basic services and products—from beaches and forest lumber to the oceans' regulation of carbon dioxide. In comparison with the total ecosystem's worth, the world's annual gross national products alone total about $18 trillion.

The actual worth of our ecosystem is probably much larger than $33 trillion since the calculations were conservative, were not all inclusive, and came from the conversion of a variety of ecosystem values from more than 100 studies. For example, the relationship of shrimp harvest yields to the local wetlands was calculated to the average value of a hectare (about 2.5 acres) of wetland, which was then applied to the international size of that habitat. Recreational value came from people's reports of their willingness to pay for access to lakes, reefs, and other recreational areas. The estimates omit ecosystem services in desert and tundra areas because these areas have not been studied in terms of ecosystem values. The study also omits services in urban areas, such as green spaces in New York's Central Park.

The most valuable ecosystems per hectare turned out to be estuaries and wetlands because they have been studied the most often. Other ecosystems will probably increase in value as they also are studied and understood. This field of ecological economics is a fairly new field, being only about a decade or so old. This financial study of ecosystems is just

Table P4.6
Primary Human-Induced Pressures on Ecosystems

Ecosystem	*Pressures*	*Causes*
Agroecosystems	Conversion of farmland to urban and industrial uses Water pollution from nutrient runoff and siltation Water scarcely from irrigation Degradation of soil from erosion, shifting cultivation, or nutrient depletion Changing weather patterns	Population growth Increasing demand for food and industrial goods Urbanization Government policies subsidizing agricultural inputs (water, research, transport) and irrigation Poverty and insecure tenure Climate Change
Coastal ecosystems	Overexploitation of fisheries Conversion of wetlands and coastal habitats Water pollution from agricultural and industrial sources Fragmentation or destruction of natural tidal barriers and reefs Invasion of nonnative species Potential sea level rise	Population growth Increasing demand for food and coastal tourism Urbanization and recreational development, which is highest in coastal areas Government fishing subsidies Inadequate information about ecosystem conditions, especially for fisheries Poverty and insecure tenure Uncoordinated coastal land-use policies Climate change
Forest ecosystems	Conversion or fragmentation resulting from agricultural or urban uses Deforestation resulting in loss of biodiversity, release of stored carbon, air and water pollution Acid rain from industrial pollution Invasion of nonnative species Overextraction of water for agricultural, urban, and industrial uses	Population growth Increasing demand for timber, pulp, and other fiber Government subsidies for timber extraction and logging roads Inadequate valuation of costs of industrial air pollution Poverty and insecure tenure
Freshwater systems	Overextraction of water for agricultural, urban, and industrial uses Overexploitation of inland fisheries Building dams for irrigation, hydropower, and flood control Water pollution from agricultural, urban, and industrial uses Invasion of nonnative species	Population growth Widespread water scarcity and naturally uneven distribution of water resources Government subsidies of water use Inadequate valuation of costs of water pollution

		Poverty and insecure tenure Growing demand for hydropower
Grassland ecosystems	Conversion or fragmentation owing to agricultural or urban uses Induced grassland fires resulting in loss of biodiversity, release of stored carbon, and air pollution Soil degradation and water pollution from livestock herds Overexploitation of game animals	Population growth Increasing demand for agricultural products, especially meat Inadequate information about ecosystem conditions Poverty and insecure tenure Accessibility and ease of conversion of grasslands

Source: World Resources Institute (WRI) in collaboration with United Nations Environment Programme (UNEP), United Nations Development Programme (UNDP), and World Bank, 2000, *People and Ecosystems: The Fraying Web of Life,* Washington, D.C: World Resources Institute, 2000–2001.

Table P4.7
Primary Goods and Services Provided by Ecosystems

Ecosystem	*Goods*	*Services*
Agroecosystems	Food crops Fiber crops Crop genetic resources	Maintain limited watershed functions (infiltration, flow control, partial soil protection) Provide habitat for birds, pollinators, soil organisms important to agriculture Build soil organic matter Sequester atmospheric carbon Provide employment
Coastal ecosystems	Fish and shellfish Fishmeal (animal feed) Seaweeds (for food and industrial use) Salt Genetic resources	Moderate storm impacts (mangroves; barrier islands) Provide wildlife (marine and terrestrial) habitat Maintain biodiversity Dilute and treat wastes Provide harbors and transportation routes Provide human habitat Provide employment Provide for aesthetic enjoyment and recreation
Forest ecosystems	Timber Fuelwood Drinking and irrigation water Fodder Nontimber products (vines, bamboos, leaves, etc.)	Remove air pollutants, emit oxygen Cycle nutrients Maintain array of watershed functions (infiltration, purification, flow control, soil stabilization) Maintain biodiversity

	Food (honey, mushrooms, fruit, and other edible plants; game) Genetic resources	Sequester atmospheric carbon Moderate weather extremes and impacts Generate soil Provide employment Provide human and wildlife habitat Provide for aesthetic enjoyment and recreation
Freshwater systems	Drinking and irrigation water Fish Hydroelectricity Genetic resources	Buffer water flow (control timing and volume) Dilute and carry away wastes Cycle nutrients Maintain biodiversity Provide aquatic habitat Provide transportation corridor Provide employment Provide for aesthetic enjoyment and recreation
Grassland ecosystems	Livestock (food, game, hides, fiber) Drinking and irrigation water Genetic resources	Maintain array of watershed functions (infiltration, purification, flow control, soil stabilization) Cycle nutrients Remove air pollutants, emit oxygen Maintain biodiversity Generate soil Sequester atmospheric carbon Provide human and wildlife habitat Provide employment Provide for aesthetic enjoyment and recreation

Source: World Resources Institute (WRI) in collaboration with United Nations Environment Programme (UNEP), United Nations Development Programme (UNDP), and World Bank, 2000, *People and Ecosystems: The Fraying Web of Life*, Washington, D.C: World Resources Institute, 2000–2001.

BOX 1 The Value of Nature

What's the Annual Dollar Value of Nature?

- Wildlife watching in the United States: $85 billion
- Recreational saltwater fishing in the United States: $20 billion
- U.S. employment income generated by wildlife watching: $27.8 billion
- State and federal tax revenues from wildlife watching: $6.1 billion
- Wild bee pollinators to one coffee farm in Costa Rica: $60,000
- Natural pest control services by birds and other wildlife to U.S. farmers: $54 billion

Source: Morrison, J. (2005). How much is clean water worth? *National Wildlife*, 43 (2):27.

a starting point, according to researchers, and they acknowledge many limitations with it. However, even as a starting point, the power of the analysis is the acknowledgment of the per-hectare value of the various habitats, the assignment of value to processes and areas taken for granted, and the way in which the assignment of value will influence ecologically friendly local decisions and conservation when it comes to planning.

REFERENCES

Kirby, A. (October, 2004). Planet Under Pressure. *BBC News*. Accessed May 30, 2005. Available at http://news.bbc.co.uk/1/hi/sci/tech/3667300.stm.

Mastny, L. (2005). Vital Signs 2005. Worldwatch Institute. New York: Norton.

Mlot, C. (May 17, 1997). A Price Tag on the Planet's Ecosystems. *Science News* 151:303.

U.S. Environmental Protection Agency. 2000. http://www.epa.gov.

U.S. Environmental Protection Agency. 2004. http://www.epa.gov.

U.S. Environmental Protection Agency. (January 7, 2000). "Climate." Available at http://yosemite.epa.gov/oar/globalwarming.nsf/content/climate.html.

U.S. Environmental Protection Agency. (September 19, 2006). "EPA marks 19 years of ozone layer preservation progress." Available at http://yosemite/epa.gove/opa/admpress.nef/7c02ca8c86062aOf85257018004118a6/0656ec6.

What's Nature Done for You Lately? Scientists Take Stock. (May/June 2005). *Solutions*, p. 12.

World Conservation Union (IUCN). 2005. http://www.IUCN.org.

World Resources Institute. 1998-9. Fragmenting forests: The loss of large frontier forests. Feature article. Available at http://earthtrends.wri.org/features/view_feature.php?theme=9&fid=14.

World Resources Institute. 2002. *World Resources 2000-2001: People and Ecosystems; The Fraying Web of Life.* Washington, D.C.

Worldwatch Institute, 2005. *Vital Signs 2005; The Trends That Are Shaping Our Future.* New York: W.W. Norton and Co.

Wright, J. W. (1995). *The Universal Almanac 1996*. Kansas City: Andrews and McMeel, pp. 604-618.

Wright, J. W., ed. (2004). *The New York Times Almanac 2005*. New York: Penguin.

QUESTIONS

1. Is the term "ecosystem" the same as the term "environment?" What are some different connotations of the two terms?
2. How have attitudes toward the environment changed from the 1970s?
3. Why do you think the EPA's world environmental issues are prioritized as they are? Why are some of the issues labeled as high risk, others as medium risk, and yet others as low risk? Explain.
4. As you think about the six most urgent global issues from the Planet Under Pressure series, identify six of your ideas (one for each issue) for improving the situations. If possible, share these in class discussion to better understand the global problems and issues.
5. What causes global warming? What are the four gases mostly responsible for heat absorption of the atmosphere?
6. Why is the loss of species an issue? Is it necessary to have so many species?
7. Why is the loss of rain forests an issue? Give three significant reasons.
8. Provide three reasons why wetlands are important.
9. What do you think we can do to prevent habitat destruction? Creatively think of three workable ideas.

BOX 2 Ecological Statistics

Global Facts

- Over 40 percent of all tropical forests have been destroyed and, at a minimum, another acre is lost each second (National Resource Defense Council – NRDC).
- Each year, humankind 6–8 billion tons of carbon to the atmosphere by burning fossil fuels and destroying forests, pumping up the concentration of greenhouse gases responsible for global warming—an effect that could raise temperatures by 3–10 degrees by the year 2050 (NRDC).
- While the United States. makes up only 5 percent of the world's population, we produce 72 percent of all hazardous waste and consume 33 percent of the world's paper (NRDC).
- Worldwide, thousands of pounds of plutonium are being produced, used, and stored under conditions of inadequate security. Using current technology, only two pounds of plutonium are required to make a nuclear device (NRDC).
- The amount of unwanted fish discarded globally each year is 60 billion pounds. The percentage of the world's catch that is discarded is 25 percent (Environmental Defense).
- The percentage of China's rivers that were severely polluted in 2000 was 42 percent. In 2002, it was 71 percent (Environmental Defense).

Air Facts

- As many as 70,000 people nationwide may die prematurely from heart and lung disease aggravated by particulate air pollution (NRDC).
- More than 100 million Americans live in urban areas where the air is officially classified by the EPA as unsafe to breathe (NRDC).
- In many urban areas, children are steadily exposed to high levels of pollutants, increasing the risk of chronic lung disease, cell damage, and respiratory illnesses.
- Dioxin and other persistent pollutants that are released into the air accumulate in our waterways, wildlife, food supply, and human bloodstreams. These poisons may cause cancer and reproductive disorders in human beings and other animal species.

Water Facts

- Millions of pounds of toxic chemicals, such as lead, mercury, and pesticides, pour into our waterways each year and contaminate wildlife, seafood, and drinking water.
- One-half of our nation's lakes and one-third of our rivers are too polluted to be completely safe for swimming or fishing (NRDC).
- Raw sewage, poison runoff, and other pollution have caused more than 18,000 beach closures and pollution advisories worldwide in 2003 (Environmental Defense).
- We are losing once pristine national treasures—such as the Everglades, Lake Superior, and the Columbia River system—to toxic pollution, chemical spills, development, and diversion of freshwater flows.
- All but one species of the magnificent ocean-going salmon in the Pacific Northwest face a growing risk of extinction throughout most of their range due to habitat degradation and overfishing (NRDC).

BOX 3 Ecological Statistics

Energy Facts

- The United States is responsible for about 25 percent of the world's total energy consumption. We use 1 million gallons of oil every two minutes (NRDC).
- Energy currently wasted by U.S. cars, homes, and appliances equals more than twice the known energy reserves in Alaska and the U.S. Outer Continental Shelf.
- We could cut our nation's energy consumption in half by the year 2030 simply by using energy more efficiently and using more renewable energy sources. In the process, we would promote economic growth by saving consumers $2.3 trillion and producing one million new jobs (NRDC).
- When just 1 percent of America's 140 million car owners tune up their cars, we eliminate nearly a billion pounds of carbon dioxide—the key cause of global warming—from entering the atmosphere (NRDC).
- Each gallon of gasoline pumps out 19 pounds of CO_2. The SUV's popularity in the United States. has caused a 20-percent increase in CO_2 pollution since the early 1990s. An SUV can add 43 percent more greenhouse gas emissions than the average car (and gets worse gas mileage than Ford's model T!). U.S. vehicles release more CO_2 than all of the energy users in India. Americans (with 5 percent of the world's population) burn one-quarter of the world's oil, and 40 percent is used in passenger cars that burn 8.7 million barrels a day (Ridgley, National Wildlife).
- The percentage of America's 128 million workers who use public transportation is 4.7 percent (Environmental Defense).
- The increase in global wind power capacity in the past six years is 415 percent (Environmental Defense).

Health and Habitat Facts

- Americans are exposed to 70,000 chemicals, of which some 90 percent have never been subjected to adequate testing to determine their impact on our health (NRDC).
- The average American uses five cordless electronic products daily, up from three in 1999. In one to five years, after being recharged between 500 and 1,500 times, they become electronic waste that is full of toxic chemicals. By 2005, cell phone batteries alone will account for 32,500 tons of waste. Americans dispose of 28 million pounds annually of spent rechargeable batteries as chemical waste, of which only about 4 million are collected and disposed of correctly (Best, NRDC).

Success Eco-Facts

- Recycling is one of the best environmental success stories of the late twentieth century. Recycling, including composting, diverted 72 million tons of material away from landfills and incinerators in 2003, up from 34 million tons in 1990. By 2002, almost 9,000 curbside collection programs served roughly half of the American population. Curbside programs, along with drop-off and buy-back centers, resulted in a diversion of about 30 percent of the nation's solid waste in 2001 (EPA). Seventy-eight percent of waste tires are recycled as rubberized asphalt, highway noise barriers, flooring material, and other products or are used as an energy source in cement manufacturing (NRDC).
- According to the UN Food and Agriculture Organization, 852 million people go hungry every day, about 18 million more than during the mid-1990s. Hunger kills more than 5 million children each year or about one child every five seconds. Yet every dollar invested in reducing hunger can yield from five to over 20 times its worth in benefits. More than 30 countries in the developing world have successfully reduced hunger through planned, long-term, food-sustaining investments (Worldwatch).

BOX 3 Ecological Statistics (Continued)

References

Best, J. (2003). The afterlife of batteries. *Onearth* 25 (3). Natural Resources Defense Council.

Environmental Defense. (2004). Earth index. *Solutions.* New York: Environmental Defense.

Natural Resources Defense Council. (1995). "Ecology Facts," *25 Year Report 1970–1995*, New York: Natural Resources Defense Council.

Natural Resources Defense Council. (2004, Fall). When the rubber hits the road. *onearth.* 26 (3).

Ridgley, H. (2005, April/May). Driving down the heat. *National Wildlife* 43 (3):56-57.

United States Environmental Protection Agency (2006). Municipal solid waste. Available at http://www.epa.gov/epaoswer/non-hw/muncpl/recycle.htm.

Worldwatch Institute (2005). *Vital Signs 2005*. New York: W.W. Norton & Co.

10. What can the industrial world do to help abate the extinction of species?
11. Explain how CFCs destroy ozone. What can be done to combat CFC use? Give some specific and practical ideas.
12. Why is it a good idea to place economic value on the earth's natural habitat and goods and services? Explain.
13. Refer to the title page of Part IV. Why is the environment a "closed system"? Explain.

QUESTIONS ABOUT THE TABLES

1. Compare the information contained in Table P4.1 and Table P4.2. What common conclusion can be drawn from this material? What are some of the differences between the various pieces of information presented?
2. Summarize in a short written paragraph the overall conclusions in Table P4.2 and in Table P4.3.
3. What implications are drawn from Table P4.4. Where is most of the deforestation occurring? Give some suggestions to abate the highest deforestation areas?
4. Summarize the information contained in Table P4.5. What conclusions do you draw, and what action might you take from these conclusions?
5. Relate the findings and conclusions from Tables P4.4 and P4.5 with the "Economic View of Ecology" discussion. How much economic value is lost along with the loss of resources? Discuss your thoughts in a paragraph or two.
6. Review Tables P4.6 and P4.7. Write a one page essay of the impact to you and your thoughts on some of the issues that you are most concerned about.

Ecology Vocabulary Exercise

As you proceed through this part of the book and continue your reading and research of the issues, keep a listing of ecological terms and organizations that you would like to learn more about. Use various resources including other texts, articles, encyclopedias, and the Internet (refer to the Internet Exercise at the end of Part IV).

The Grim Payback of Greed*

ALAN DURNING

A man is rich in proportion to the things he can afford to let alone.
HENRY DAVID THOREAU

Early in the age of affluence that followed World War II, an American retailing analyst named Victor Lebow proclaimed that an enormously productive economy "demands that we make consumption our way of life . . . . We need things consumed, burned up, worn out, replaced and discarded at an ever increasing rate."

Americans have responded to Mr. Lebow's call, and much of the world has followed. The average person today is four-and-a-half times richer than were his great-grandparents at the turn of the century. That new global wealth is not evenly spread among the Earth's people, however. One billion live in unprecedented luxury; one billion live in destitution.

Overconsumption by the world's fortunate is an environmental problem unmatched in severity by anything but perhaps population growth. Their surging exploitation of resources threatens to exhaust or inalterably disfigure forests, soils, water, air, and climate.

Of course the opposite of overcompensation—poverty—is no solution either to environmental or human problems. Dispossessed peasants slash-and-burn their way into the rain forests of Latin America, and hungry nomads turn their herds out onto fragile African rangeland, reducing it to desert.

If environmental destruction results when people have either too little or too much, we are left to wonder how much is enough. What level of consumption can the Earth support? When does having more cease to add appreciably to human satisfaction?

The Consuming Society Skyrocketing consumption is the hallmark of our era. The trend is visible in statistics for almost any per capita indicator. Worldwide, since mid-century, the intake of copper, energy, meat, steel, and wood has approximately doubled; car ownership and cement consumption have quadrupled; plastic use has quintupled; aluminum consumption has grown sevenfold; and air travel has multiplied 32 times.

Moneyed regions account for the largest waves of consumption since 1950. In the United States, the world's premier consuming society, on average people today own twice as many cars, drive two-and-a-half times as far, use 21 times as much plastic and travel 25 times as far by air as did their parents in 1950. Air conditioning has spread from 15 percent of households in 1960 to 64 percent in 1987, and color televisions from 1 to 93 percent. Microwave ovens and video cassette recorders found their way into almost two-thirds of American homes during the eighties alone.

The eighties were a period of marked extravagance in the United States; not since the Roaring Twenties had conspicuous consumption been so lauded. Between 1978 and 1987, sales of Jaguar automobiles increased eightfold, and the average age of first-time fur coat buyers fell from 50 to 26. The select club of American millionaires more than

Reprinted with permission from Worldwatch Institute. Durning, A. (May–June 1991). "The Grim Payback of Greed." *International Wildlife*. Vol. 21, No. 3, pp. 36–39. (* Editor's note: This reading even though written in the 90's is still relevant and classic. It contrasts pre-1950's and post-1950's consumption. Such pre and post 50's are often used in anthropological time studies.)

doubled its membership from 600,000 to 1.5 million over the decade, while the number of American billionaires reached 58 by 1990.

Japan and Western Europe have displayed parallel trends. Per person, the Japanese of today consume more than four times as much aluminum, almost five times as much energy, and 25 times as much steel as people in Japan did in 1950. They also own four times as many cars and eat nearly twice as much meat. In 1972, one million Japanese traveled abroad; in 1990, the number was expected to top ten million. As in the United States, the eighties were a particularly consumerist decade in Japan, with sales of BMW automobiles rising tenfold.

Like the Japanese, West Europeans' consumption levels are only one notch below Americans'. Taken together, France, West Germany, and the United Kingdom almost doubled their per capita use of steel, more than doubled their intake of cement and aluminum, and tripled their paper consumption since mid-century. Just in the first half of the eighties, per capita consumption of frozen prepared meals—with their excessive packaging—rose more than 30 percent in every West European country except Finland; in Switzerland, the jump was 180 percent.

The Cost of Wealth Long before all the world's people could achieve the American dream, however, the planet would be laid to waste. Those in the wealthiest fifth of humanity are responsible for the lion's share of the damage humans have caused to common global resources. They have built more than 99 percent of the world's nuclear warheads. Their appetite for wood is a driving force behind destruction of the tropical rain forests and the resulting extinction of countless species. Over the past century, their economies have pumped out two-thirds of the greenhouse gases that threaten the Earth's climate, and each year their energy use releases perhaps three-fourths of the sulfur and nitrogen oxides that cause acid rain. Their industries generate most of the world's hazardous chemical wastes, and their air conditioners, aerosol sprays, and factories release almost 90 percent of the chlorofluorocarbons that destroy the Earth's protective ozone layer. Clearly, even 1 billion profligate consumers are too much for the Earth.

Beyond the environmental costs of acquisitiveness, some perplexing findings of social scientists throw doubt on the wisdom of high consumption as a personal and national goal: rich societies have had little success in turning consumption into fulfillment. A landmark study in 1974, for instance, revealed that Nigerians, Filipinos, Panamanians, Yugoslavians, Japanese, Israelis, and West Germans all ranked themselves near the middle of a happiness scale. Confounding any attempt to correlate affluence and happiness, poor Cubans and rich Americans were both found to be considerably happier than the norm, and citizens of India and the Dominican Republic, less so. As Oxford psychologist Michael Argyle writes, "There is very little difference in the levels of reported happiness found in rich and very poor countries."

As measured in constant dollars, the world's people have consumed as many goods and services since 1950 as all previous generations put together. Since 1940, Americans alone have used up as large a share of the Earth's mineral resources as did everyone before them combined. If the effectiveness of that consumption in providing personal fulfillment is questionable, perhaps environmental concerns can help us redefine our goals.

In Search of Sufficiency Some guidance on what the Earth can sustain emerges from an examination of current consumption patterns around the world. For three of the most ecologically important types of consumption—transportation, diet, and use of raw materials—the world's people are distributed unevenly over a vast range. Those at the bottom fall below the "too little" line, while those at the top, in what could be called the cars-meat-and-disposable class, consume too much.

About 1 billion people do most of their traveling, aside from the occasional donkey or bus ride, on foot, many of them never going more than 100 kilometers from their birthplaces. Unable to get to jobs easily, attend school, or bring their complaints before government offices, they are severely hindered by the lack of transportation options.

The massive middle class of the world, numbering some 3 billion, travels by bus and bicycle. Kilometer for kilometer, bikes are cheaper than any other vehicles, costing less than $100 new in most of the Third World and requiring no fuel.

The world's automobile class is relatively small: only 8 percent of humans, about 400 million people, own cars. Their cars are directly responsible for an estimated 13 percent of carbon dioxide emissions from fossil fuels worldwide, along with air pollution, acid rain, and a quarter-million traffic fatalities a year.

The global food consumption ladder has three rungs, as well. At the bottom, the world's 630 million poorest people are unable to provide themselves with a healthful diet, according to World Bank estimates.

On the next rung, the 3.4 billion grain-eaters of the world's middle class get enough calories and plenty of plant-based protein, giving them the most healthful basic diet of the world's people. They typically receive less than 20 percent of their calories from fat, a level low enough to protect them from the consequences of excessive dietary fat.

The top of the ladder is populated by the meat-eaters, those who obtain close to 40 percent of their calories from fat. These 1.25 billion people eat three times as much fat per person as the remaining 4 billion, mostly because they eat so much red meat. The meat class pays the price of its diet in high death rates from the so-called diseases of affluence—heart disease, stroke, and certain types of cancer.

The Earth also pays for the high-fat diet. Indirectly, the meat-eating quarter of humanity consumes nearly 40 percent of the world's grain—grain that fattens the livestock they eat. Meat production is behind a substantial share of the environmental strains induced by the present global agricultural system, from soil erosion to overpumping of underground water.

In raw material consumption, the same pattern emerges. About 1 billion rural people subsist on biomass collected from the immediate environment. Most of what they use each day—about a half-kilogram of grain, 1 kilogram of fuelwood, and fodder for their animals—could be self-replenishing renewable resources. Unfortunately, because these people are often pushed by landlessness and population growth into fragile unproductive ecosystems, their minimal needs are not always met.

These materially destitute billion are part of a larger group that lacks many of the benefits provided by modest use of nonrenewable resources—particularly durable things like radios, refrigerators, water pipes, high-quality tools, and carts with lightweight wheels and ball bearings. More than 2 billion people live in countries where per capita consumption of steel, the most basic modern material, falls below 50 kilograms a year. In those same countries, per capita energy use—a fairly good indirect indicator of overall use of materials—is lower than 20 gigajoules per year (compared to 280 gigajoules in the United States).

Roughly 1.5 billion people live in the middle class of materials' use. Providing each of them with durable goods every year uses between 50 and 150 kilograms of steel and 20–50 gigajoules of energy.

At the top of the heap is the throwaway class, which uses raw materials extravagantly. A typical resident of the industrialized fourth of the world uses 15 times as much paper, 10 times as much steel, and 12 times as much fuel as a Third World resident. The extreme case is the United States, where the average person consumes most of his own weight in basic materials each day—18 kilograms of petroleum and coal, 13 kilograms of other minerals, 12 kilograms of farm products, and 9 kilograms of forest products, or about 115 pounds total.

In the throwaway economy, packaging becomes an end in itself; disposables proliferate and durability suffers. Four percent of consumer expenditures on goods in the United States goes for packaging—$225 a year. Likewise, the Japanese use 30 million disposable single-roll cameras each year, and the British dump 2.5 billion diapers. Americans toss away 180 million razors annually, enough paper and plastic plates and cups to feed the world a picnic six times a year, and enough aluminum cans to make 6,000 DC-10 airplanes.

In transportation, diet, and use of raw materials, as consumption rises on the economic scale, so does waste—both of resources and of health. Yet despite arguments in favor of modest consumption, few people who can afford high consumption levels opt to live simply. What prompts us, then, to consume so much?

The Cultivation of Needs "The avarice of mankind is insatiable," wrote Aristotle 23 centuries ago, describing the way that, as each of our desires is satisfied, a new one seems to appear in its place. That observation, on which all of economic theory is based, provides the most obvious answer to the question of why people never seem satisfied with what they have. If our wants are insatiable, there is simply no such thing as enough.

Much confirms this view of human nature. The Roman philosopher Lucretius wrote a century before Christ: "We have lost our taste for acorns. So [too] we have abandoned those couches littered with herbage and heaped with leaves. So the wearing of wild beasts' skins has gone out of fashion .... Skins yesterday, purple and gold today—such are the baubles that embitter human life with resentment."

Nearly 2,000 years later, Russian novelist Leo Tolstoy echoed Lucretius: "Seek among men from beggar to millionaire, one who is contented with his lot, and you will not find one such in a thousand .... Today we must buy an overcoat and galoshes, tomorrow, a watch and a chain: the next day we must install ourselves in an apartment with a sofa and a bronze lamp; then we must have carpets and velvet gowns: then a house, horses and carriages, paintings and decorations."

What distinguishes modern consuming habits from those of interest to Lucretius and Tolstoy, some would say, is simply that we are much richer than our ancestors and consequently have more ruinous effects on nature. There is no doubt a great deal of truth in that view, but there is also reason to believe that certain forces in the modern world encourage people to act on their consumption desires as rarely before.

The search for social status in massive and anonymous societies, omnipresent advertising messages, a shopping culture that edges out nonconsuming alternatives, government biases favoring consumption, and the spread of the commercial market into most aspects of private life—all these things nurture the acquisitive desires that everyone has. Can we, as individuals and as citizens, act to confront these forces?

A Culture of Permanence When Moses came down from Mount Sinai, he could count the rules of ethical behavior on the fingers of his two hands. In the complex global economy of the late twentieth century, in which the simple act of turning on a light sends greenhouse gases up into the atmosphere, the rules for ecologically sustainable living run into the hundreds.

The basic value of a sustainable society, though, the ecological equivalent of the Golden Rule, is simple: each generation should meet its needs without jeopardizing the prospects of future generations to meet their own needs.

What is lacking is the thorough practical knowledge—at each level of society—of what living by that principle means. For individuals, the decision to live a life of sufficiency—to find their own answer to the question, "How much is enough?"—is to begin a highly personal process. The goal is to put consumption in its proper place among the many sources of fulfillment and to find ways of living within the means of the Earth. One great inspiration in this quest is the body of human wisdom passed down over the ages.

Materialism was denounced by all the sages, from Buddha to Muhammad. These religious founders, observed historian Arnold Toynbee, "all said with one voice that if we made material wealth our paramount aim, this would lead to disaster." The Christian Bible echoes most of human wisdom when it asks, "What shall it profit a man if he shall gain the whole world and lose his soul?"

For those people experimenting with voluntary simplicity, the goal is not ascetic self denial. What they are after is personal fulfillment; they just do not think consuming more is likely to provide it.

Still, shifting emphasis from material to nonmaterial satisfaction is hardly easy: It means trying both to curb personal appetites and to resist the tide of external forces encouraging consumption.

Many people find simpler living offers rewards all its own. They say life can become more deliberate as well as spontaneous and even gain an unadorned elegance. Others describe the way simpler technologies add unexpected qualities to life.

Realistically, however, voluntary simplicity is unlikely to gain ground rapidly against the onslaught of consumerist values. The call for a simpler life has been perennial through the history of North America, from the Puritans of Massachusetts Bay to the back-to-the-landers of the 1970s. None of these movements ever gained more than a slim minority of adherents.

It would be naive to believe that entire populations will suddenly experience a moral awakening, renouncing greed, envy, and avarice. What can be hoped for is a gradual weakening of the consumerist ethos of affluent societies. The challenge before humanity is to bring environmental matters under cultural controls, and the goal of creating a sustainable culture—a culture of permanence—is a task that will occupy several generations.

The Ultimate Fulfillment In many ways, we might be happier with less. Maybe Henry David Thoreau had it right when he scribbled in his notebook beside Walden Pond. "A man is rich in proportion to the things he can afford to let alone."

For the luckiest among us, a human lifetime on Earth encompasses perhaps a hundred trips around the Sun. The sense of fulfillment received on that journey—regardless of a person's religious faith—has to do with the timeless virtues of discipline, hope, allegiance to principle, and character. Consumption itself has little part in the playful camaraderie that inspires the young, the bonds of love and friendship that nourish adults, the golden memories that sustain the elderly. The very things that make life worth living, that give depth and bounty to human existence, are infinitely sustainable.

QUESTIONS

1. How much richer are Americans now than at the turn of the twentieth century?
2. What is another major cause (other than consumption) of severe environmental problems?
3. What two groups are most destructive to the environment? Give two examples from each of the groups.
4. From the article, name three statistics of the industrialized countries' (one-fifth of the world population) destruction of global resources. Discuss.

5. Give some evidence that refutes the idea of consumerism as an avenue for happiness.
6. Which world class of people does the least damage to the environment? Give three examples.
7. Why do the industrialized nations keep accelerating their consumerism?
8. What can be done to curb this exponential consumerism?
9. Explain briefly the major points in "The Grim Payback of Greed." What did you learn from this article?
10. Define the consuming society. Do you think that this will continue to be a problem in the next few decades? Why or why not?
11. How has wealth affected ecological issues? Does materialism create ecological concerns? How?
12. What is the ultimate fulfillment?
13. How can the payback of greed be stopped? Why or why not?

BOX 1 How Much Do We Consume?

Humans consume goods and services for many reasons: to nourish, clothe, and house ourselves, certainly. But we also consume as part of a social compact, since each community or social group has standards of dress, food, shelter, education, and entertainment that influence its patterns of consumption beyond physical survival (United Nations Development Programme (UNDP) 1998:38-45).

Consumption is a tool for human development—one that opens opportunities for a healthy and satisfying life, with adequate nutrition, employment, mobility, and education. Poverty is marked by a lack of consumption, and thus a lack of these opportunities. At the other extreme, wealth can—and often does—lead to excessive levels of material and nonmaterial consumption.

In spite of its human benefits, consumption can lead to serious pressure on ecosystems. Consumption harms ecosystems directly through overharvesting of animals or plants, mining of soil nutrients, or other forms of biological depletion. Ecosystems suffer indirectly through pollution and wastes from agriculture, industry, and energy use and also through fragmentation by roads and other infrastructure that are part of the production and transportation networks that feed consumers.

Consumption of the major commodities ecosystems produce directly—grains, meat, fish, and wood—increased substantially in the last four decades and will continue to increase as the global economy expands and world population grows. Plausible projections of consumer demand in the next few decades suggest a marked escalation of impacts on ecosystems (Matthews and Hammond 1999:5).

- Global wood consumption has increased 64 percent since 1961. More than half of the 3.4 billion m^3 of wood consumed annually is burned for fuel; the rest is used in construction and for paper and a variety of other wood products. Demand for lumber and pulp is expected to rise between 20 and 40 percent by 2010. Forest plantations produce 22 percent of all lumber, pulp, and other industrial wood; old-growth and secondary-growth forests provide the rest (Matthews and Hammond 1999:8, 31; Brown 1999:41).
- World cereal consumption has more than doubled in the last 30 years, and meat consumption has tripled since 1961 (Matthews and Hammond 1999:7). Some 34 percent of the world's grain crop is used to feed livestock raised for meat (USDA 2000). A crucial factor in the rise in grain production has been the more than fourfold increase in fertilizer use since 1961 (Matthews and Hammond 1999:14). By 2020, demand for cereals is expected to increase nearly 40 percent, and meat demand will surge nearly 60 percent (Pinstrup-Andersen et al. 1999:11).
- The global fish catch has grown more than sixfold since 1950 to 122 million metric tons in 1997. Three-fourths of the global catch is consumed directly by humans as fresh, frozen, dried, or canned fish and shellfish. The remaining 25 percent are reduced to fish meal and oil, which is used for both livestock feed and fish feed in aquaculture. Demand for fish for direct consumption is expected to grow some 20 percent by 2010 (FAO 1999:7, 82; Matthews and Hammond 1999:61).

Figure 1
World resource production.
Sources: FAO 1999; 2000.

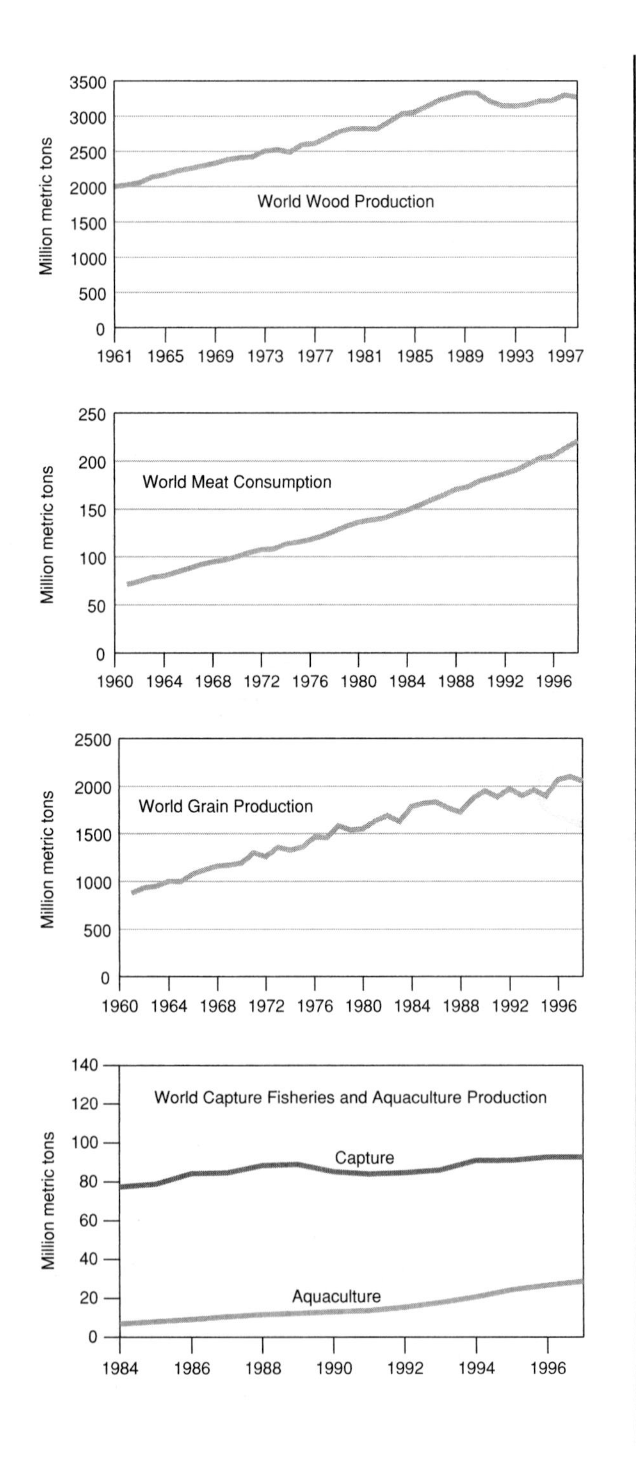

The Unequal Geography of Consumption

While consumption has risen steadily worldwide, there remains a profound disparity between consumption levels in wealthy nations and those in middle- and low-income nations.

- On average, someone living in a developed nation consumes twice as much grain, twice as much fish, three times as much meat, nine times as much paper, and eleven times as much gasoline as someone living in a developing nation (Laureti 1999:50, 55).
- Consumers in high-income countries—about 16 percent of the world's population—accounted for 80 percent of the money spent on private consumption in 1997—$14.5 trillion of the $18 trillion total. By contrast, purchases by consumers in low-income nations—the poorest 35 percent of the world's population—represented less than 2 percent of all private consumption. The money spent on private consumption worldwide (all goods and services consumed by individuals except real estate) nearly tripled between 1980 and 1997 (World Bank 1999:44, 226).

Source: World Resources Institute (WRI) in collaboration with United Nations Environment Programme (UNEP), United Nations Development Programme (UNDP), and World Bank, 2000, *People and Ecosystems: The Fraying Web of Life*, Washington, D.C: World Resources Institute, 2000-2001.

Table 1
Disparities in Consumption: Annual per Capita Consumption in Selected High-, Medium-, and Low-Income Nations

Country	*Total Value of Private Consumption* (1997)*	*Fish (kg) 1997*	*Meat (kg) (1998)*	*Cereals (kg) (1998)*	*Paper (kg) (1998)*	*Fossil Fuels (kg of oil equivalent) (1997)*	*Passenger Cars (per 1,000 people) (1996)*
United States	21,680	21.0	122.0	975.0	293.0	6,902	489.0
Singapore	16,340	34.0	77.0	159.0	168.0	7,825	120.0
Japan	15,554	66.0	42.0	334.0	239.0	3,277	373.0
Germany	15,229	13.0	87.0	496.0	205.0	3,625	500.0
Poland	5,087	12.0	73.0	696.0	54.0	2,585	209.0
Trinidad/ Tobago	4,864	12.0	28.0	237.0	41.0	6,394	94.0
Turkey	4,377	7.2	19.0	502.0	32.0	952	55.0
Indonesia	1,808	18.0	9.0	311.0	17.0	450	12.2
China	1,410	26.0	47.0	360.0	30.0	700	3.2
India	1,166	4.7	4.3	234.0	3.7	268	4.4
Bangladesh	780	11.0	3.4	250.0	1.3	67	0.5
Nigeria	692	5.8	12.0	228.0	1.9	186	6.7
Zambia	625	8.2	12.0	144.0	1.6	77	17.0

*Adjusted to reflect actual purchasing power, accounting for currency and cost-of-living differences (the "purchasing power party" approach).
Sources: Total Private Consumption (except China and India): World Bank 1999, Table 4.11; Fish: Laureti 1999, 48-55; Meat: WRI et al. 2000a: Agriculture and Food Electronic Database; Paper: WRI et al. 2000b: Data Table ERC.5; Fossil Fuels: WRI et al. 2000b: Data Table ERC.2; Passenger Cars: WRI et al. 2000b: Data Table ERC.5.

26 Climate Change*

BARBARA A. EICHLER

"No single bit of scientific evidence makes a convincing argument that global warming is having an impact on wildlife and plants, but the cumulative evidence cannot be ignored. The question is no longer 'Is global warming happening?' The question is, what are we going to do about it?"
DOUG INKLEY, National Wildlife Federation, 2005

Future hopes for a world climate agreement.
Photo courtesy of Ahmed S. Khan.

Is climate change really happening? Or is it the pronouncement of over-zealous environmentalists and media hype? Scientists worldwide are trying to assess the realistic effects of accumulating greenhouse gases and incoming data from many predicted scenarios. The following discussion examines their findings from varying viewpoints about one of the most discussed topics of the twenty-first century—what is really going on with global warming?

*Also referred to as "the greenhouse effect" or "global warming."

WARMED ONE DEGREE IN THIS PAST CENTURY

As discussed in the introduction of this "Ecology" section, according to the National Academy of Sciences, the Earth's surface temperature has risen by about 1 °F in the past century, with much of the warming occurring during the past two decades. There is accumulating strong evidence that human activities are mostly responsible for this warming, especially over the last 50 years, by altering the chemical composition of the atmosphere. Atmospheric alterations are principally caused by the buildup of greenhouse gases—carbon dioxide, methane, chloroflourocarbons (CFCs), sulfur hexaflouride, and nitrous oxide—which undisputedly trap heat. What is unknown and is currently being assessed is how the earth's climate is responding to increasing volumes of these gases and their heat-holding properties that we add each year (EPA 2000, 1).

HOW GLOBAL WARMING WORKS

The sun's energy controls the earth's weather and climate and heats the earth's surface. The earth radiates this energy back into space. Atmospheric greenhouse gases (water vapor, carbon dioxide, nitrous oxides, methane, and other gases) trap some of the outgoing energy and prevent it from returning back into space, much like a greenhouse's glass panels. Some of this trapping of the natural "greenhouse effect" is positive because it enables the temperate, average, 60-degree global temperatures that allow life on earth. (Refer to Figure P4.1 in the Introduction to Part IV.)

Problems arise, however, when the atmospheric concentration of greenhouse gases increases, which has been happening since the Industrial Revolution. Since the 1800s, atmospheric concentrations of carbon dioxide have increased nearly 35 percent, methane has doubled, and an additional 15 percent of nitrous oxides have been added. These increases strengthen the heat-trapping function of the earth's atmosphere (EPA 2000, 1). Research from the World Meteorological Organization, the U.S. Environment Programme (UNEP), and the Intergovernmental Panel on Climate change (IPCC) find that carbon dioxide, which is released into the atmosphere from fossil fuel burning, is the single most important greenhouse gas contributing to the "anthropogenic forcing of climate change" or the warming of the earth's surface (Dunn and Flavin 2002, 26). The problematic volume of CO_2 emissions from industries and motor vehicles is at 372 parts per million (PPM) (it was about 280 parts per million in the 1850s), which is the highest concentration in 420,000 years as indicated from gas analysis trapped in ancient ice (DiSilvestro 2005, 23). CO_2 emissions are increasing at the rate of approximately two PPM per year. Such increases are of special concern because they had been fairly stable over the last million years at a rate between 200 and 300 PPM. Some predict that, in ten years, CO_2 emissions will increase to 400 PPM, which will result in a 3.5-degree temperature increase (Dyer 2005, 2).

HISTORIC PATTERNS

Scientists feel that the combustion of fossil fuels (along with other human activities, such as the destruction of forests) is the primary reason for the increased concentrations of carbon dioxide. The Environmental Protection Agency (EPA) states: "Plant respiration and the decomposition of organic matter release more than 10 times the CO_2 released by human activities; but these releases have generally been in balance during the centuries leading up to the industrial revolution, with carbon dioxide absorbed by terrestrial vegetation and the oceans." (EPA 2000, 1) In the last few hundred years, human fossil fuel use from transportation, home heating, and energizing factories have dramatically increased greenhouse gas accumulations. At present in the United States, fossil fuel emissions from cars and trucks, heating, and power plants contribute 98 percent of U.S. carbon dioxide emissions, 24 percent of methane emissions, and 18 percent of nitrous oxide emissions. In addition, increased agriculture, deforestation, landfills, industrial production, and mining also contribute to increased emissions. (At the close of the twentieth century, the United States released approximately one-fourth to one-fifth of the world's total global greenhouse gases.)

In the United States, CO_2 emissions from fossil fuels keep increasing rapidly. The average American burns enough fossil fuels annually to produce 22 tons of CO_2. In 2002, the most recent data available, "the U.S. emitted a record 5.84 billion metric tons of CO_2 into the atmosphere, a 21-percent increase since 1990 and about twice the level of 1960. Petroleum products were responsible for 2.43 billion metric tons of the total, while coal and natural gas accounted for 2.09 billion and 1.27 billion metric tons. Since the start of the Industrial Revolution in 1850, the burning of fossil fuels, the primary cause of global warming, has raised the amount of carbon dioxide in the earth's atmosphere by 35 percent." (Marland 2005).

Trying to predict future emissions is complicated because of many demographic, economic, and governmental considerations. Yet by 2100, some projections estimate that world carbon dioxide concentrations could be in the range of 30–150 percent higher than today in the absence of emissions control policies (EPA, 2000, 1–2). In 2002, the World Wildlife Fund (WWF) report, *Habitats at Risk: Global Warming and Species Loss in Terrestrial Ecosystems*, examined the impact of climate change on special ecosystem regions identified as the Global 200, areas where ecological wealth is most rich, where loss will be most severely felt, and where conservation must be fought

Table 1
U.S. Greenhouse Gas Emissions from Human Activities, 1990–2002

Gas and Major Sources(s)	*1990*	*1996*	*1997*	*1998*	*1999*	*2000*	*2001*	*2002*
Carbon dioxide (CO_2)	5,002.3	5,498.5	5,577.6	5,602.5	5,676.3	5,859.0	5,731.8	5,782.4
Fossil fuel combustion	4,814.7	5,310.1	5,384.0	5,412.4	5,488.8	5,673.6	5,558.8	5,611.0
Methane (CH_4)	642.7	637	628.8	620.1	613.1	614.4	605.1	598.1
Landfills	210.0	208.8	203.4	196.6	197.8	199.3	193.2	193.0
Natural gas systems	122.0	127.4	126.1	124.5	120.9	125.7	124.9	121.8
Enteric fermentation[1]	117.9	120.5	118.3	116.7	116.6	115.7	114.3	114.4
Coal mining	81.9	63.2	62.6	62.8	58.9	56.2	55.6	52.2
Nitrous oxide (N_2O)	393.2	436.9	436.3	432.1	428.4	425.8	417.3	415.8
Agricultural soil management	262.8	288.1	293.2	294.2	292.1	289.7	288.6	287.3
Hydrofluorocarbons (HFCs), perfluorocarbons (PFCs), and sulfur hexafluoride (SF_6)[2]	90.9	114.9	121.7	135.7	134.8	139.1	129.7	138.2
Total U.S. emissions	6,129.1	6,687.3	6,764.4	6,790.5	6,852.5	7,038.3	6,883.9	6,934.6
Net U.S. emissions[3]	5,171.3	5,632.1	5,943.5	6,084.7	6,176.8	6,348.2	6,194.1	6,243.8

Note: Emissions given in terms of equivalent emissions of carbon dioxide (CO_2), using units of teragrams of carbon dioxide equivalents (Tg CO_2 Eq).
[1] Digestive process of ruminant animals, such as cattle and sheep, producing methane as a byproduct.
[2] These gases have extremely high global warming potential, and PFCs and SF_6 have long atmospheric lifetimes.
[3] Total emissions minus carbon dioxide absorbed by forests or other means.
Source: U.S. Environmental Protection Agency, "Global Warming–Climate," January 2000, http://yosemite.epa.gov/oar/globalwarming.nst/content/climate.html, accessed April 2, 2005.

the hardest. Their findings below predict what will result in 100 years if $C0_2$ concentrations are doubled (doubled concentrations would be less than the current predictions).

1. More than 80 percent of the areas tested will experience extinctions of plant and animal species because of global warming.
2. Habitat changes from global warming will be more severe at higher latitudes and altitudes than in lowland tropical areas.
3. The most unique and diverse natural ecosystems may lose more than 70 percent of the habitats that support their plant and animal species.
4. Many habitats will change approximately 10 times faster than the rapid changes during the recent postglacial period, causing extinction among species unable to adapt or migrate (Morgan 2002, 5).

EFFECTS AND CHANGING CLIMATE

There are significant patterns and effects of increased temperature during the last 100 years that lead to serious affirmation of climate change. Global mean surface temperatures have increased 0.5–1.0 °F since the late nineteenth century or 1.4 °F since 1750 (DiSilvestro 2005, 23). The 10 warmest years of the twentieth century occurred in the last 15 years of the century, with 2005 being the warmest year on record since instrumental record-taking began in the 1860s.

National Geographic assesses global warming using three major scientific classification indicators: geographic "geosigns," ecological "ecosigns," and time "timesigns." Their analysis has found accumulating and persuasive evidence in all three areas that points to the conclusion of climate warming (Appenzeller, *National Geographic* 2004, 11). The EPA states that indicators of increased climate change include the decrease of the snow cover in the

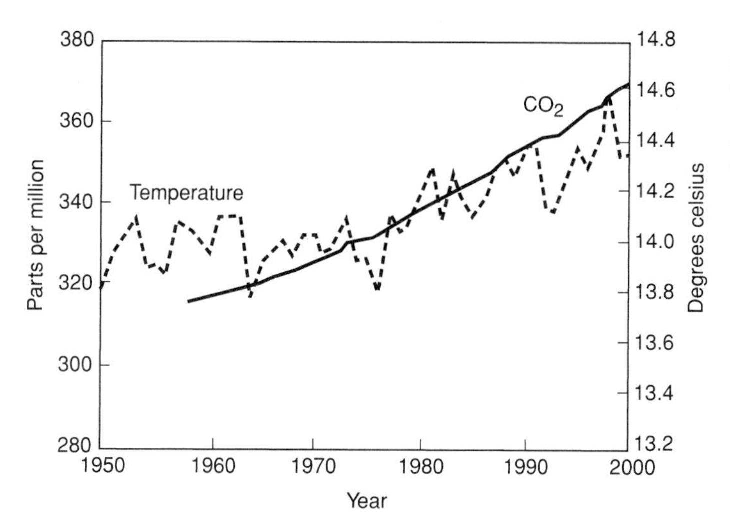

Figure 1
Atmospheric carbon dioxide concentrations and global average surface temperature, 1950–2000.

Source: Worldwatch. (2003) *Vital Signs 2003.* New York: Norton and Co.

Northern Hemisphere, the shrinking of glaciers and floating ice in the Artic Ocean, and a rising sea level of four to eight inches over the past century. In addition, more turbulent weather patterns are emerging with more extreme rainfall events in the United States and around the world (EPA 2000, 2).

Geosigns

According to the United Nation's Intergovernmental Panel on Climate Change (IPCC), composed of 1500 climatologists (and agreement of the National Academy of Sciences, the American Meteorological Society, the American Geo-physical Union, and the American Association for the Advancement of Science), an overview of the following specific phenomena have been observed (DiSilvestro 2005, 23).

Polar Change The rate of warming in the Artic has been eight times faster during the past 20 years than the previous 100 years. The artic regions are warming at nearly twice the rate of the rest of the planet, where average winter temperatures in Alaska and western Canada have increased as much as 7 degrees during the past 60 years. NASA satellites show that the area of permanent ice cover is contracting at about 9 percent per decade. Arctic summer

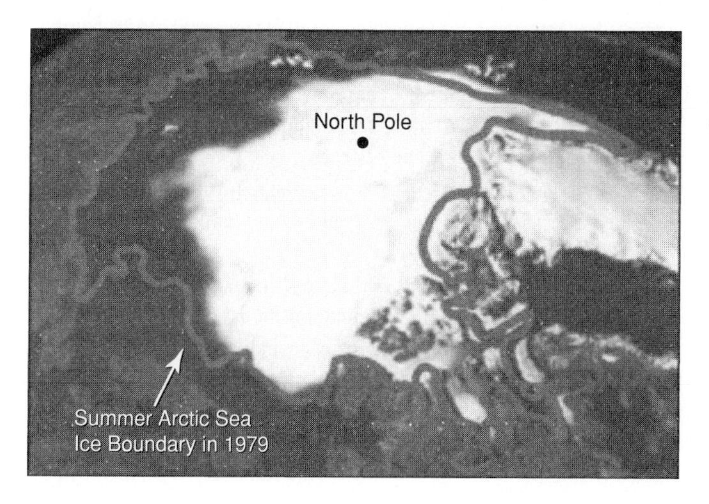

Figure 2
Summer Arctic Sea ice boundary in 1979.

Source: NASA, http://www.nrdc.org/globalWarming/qthinice.asp.

sea ice has already shrunk by 15–20 percent over the past 30 years and thinned by up to 40 percent since the 1960s. If this trend continues, summers in the Arctic could become ice-free by the end of the century. Arctic warming has caused the sea of ice around the North Pole since the 1970s to lose an area the size of Texas and Arizona combined. In Antarctica, average temperatures have increased by as much as 4.5 °F since the 1940s, which is also among the fastest rates of change in the world (DiSilvestro 2005, 23).

Glaciers and Ice Sheets Antarctica's famous 1200-acre Larsen B ice shelf, which is more than 700 feet thick, partially collapsed in March 2002, the third largest shelf to collapse since 1995. Such a disintegration of shelves that are about 12,000 years old is an unusual event. Arctic sea ice is now melting 20 percent faster than it did two decades ago, resulting in about 10 percent less ice per decade. The freezing-line in the Rockies and the Alps has risen by almost 500 feet since 1970. Some 80 percent of Kenya's Mount Kilimanjaro's snowcap has disappeared, and the 150 glaciers of Glacier National Park that existed in 1910 have been reduced to fewer than 30 (DiSilvestro 2005. 23). Other "geo-signs" include NASA's observations that the edges of Greenland's ice sheet are shrinking. In the Northern Hemisphere, spring freshwater ice breakups occur nine days earlier and autumn freezes occur 10 days later than they did 150 years ago. "From the arctic to Peru, from Switzerland to the equatorial glaciers of Irian Jaya in Indonesia, massive ice fields, monstrous glaciers, and sea ice are disappearing, fast." (Glick 2004, 14) Such overall effects of global melting patterns have played a major role in raising global sea levels between four and eight inches in the last 100 years, according to the IIPCC the EPA and other major sources.

Table 2
Selected Examples of Ice Melt Around the World

Name	*Location*	*Measured Loss*
Arctic Sea ice	Arctic Ocean	Has shrunk by some 8 percent (an area larger than Denmark, Sweden, and Norway combined) over the past 30 years, and the melting is accelerating. Has thinned by 15–20 percent overall since the late 1960s, with losses in some areas near 40 percent.
Greenland ice sheet	Greenland	Surface melt area increased 16 percent between 1979 and 2002 to a record 685,000 square kilometers. Margins are now melting by as much as 10 meters per year, 10 times faster than in 2001. Speed of flow of the largest outlet glacier has doubled since 1997 to 12.6 kilometers per year.
Glaciers	Alaska, United States	Now thinning by 1.8 meters a year on average, more than twice the annual rate observed from the 1950s to the mid-1990s. Total ice loss is estimated at 96 cubic kilometers each year.
Glacier National Park	Montana, United States	Since 1850, the number of glaciers has dropped from 150 to less than 40. The park's remaining glaciers could disappear completely in 30 years.
Central and Southern Andes	Peru	Have lost 20 percent of their 2,600 kilometers of glaciers in the past 30 years. All 18 glacier-capped mountains are now melting. At current rates, glaciers below 5,500 meters could disappear by 2015.
Patagonian Ice Fields	Argentina and Chile	Northern and southern fields have been in retreat for roughly a century. Average rate of thinning was twice as fast between 1995 and 2000 as between 1975 and 2000. Now lose 42 cubic kilometers of ice a year.

Table 2 (*Continued*)

Amundsen Sea area	West Antarctica	Glaciers discharge some 250 cubic kilometers of ice each year, nearly 60 percent more than they accumulate from inland snowfall. Thinning rates in 2002–2003 were much higher than during the 1990s.
Tibetan glaciers	China	Retreated 7.5 percent between 1850 and 1960 and a further 7 percent in the following 40 years. In the 1990s alone, the glaciers shrank by more than 4 percent.
Himalayas	Nepal	Average snow and ice cover in the East has decreased by 30 percent in the last 30 years. Within the next 35 years, the total glacial area is expected to shrink by one-fifth to 100,000 square kilometers.
Mount Kilimanjaro	Tanzania	Lost 82 percent of its ice between 1912 and 2000, shrinking from 12 square kilometers to 2.6 square kilometers. Ice could disappear completely by 2015.
Alps	Europe	Glaciers lost roughly one-third of their area and one-half of their mass between 1850 and 1980. Since 1980, a further 20–30 percent of the remaining ice has melted. The hot summer of 2003 alone accounted for a large share of this loss: Swiss glaciers retreated 3 meters in 2003 compared with 70 centimeters a year in the 1990s. Three quarters of Swiss Alpine glaciers are projected to disappear by 2050.

Source: Worldwatch (2005). *Vital Signs 2005*. New York: W.W. Norton and Co.,89.

Ocean CO_2 Absorption Oceans function as important absorption centers for CO_2 and absorb about one-third of human-generated CO_2. Surface absorption is rising at about the same rate as atmospheric CO_2, but the deeper levels at the Bermuda Atlantic Study stations are showing an altered phenomenon. In waters between 820 and 1476 feet deep, CO_2 levels are rising at nearly twice the rate as in surface waters. Nicholas Bates, principal investigator for the Bermuda Study stations, states, "It's not a belief system; it's an observable scientific fact. And it shouldn't be doing that unless something fundamental has changed in this part of the ocean." (Glick 2004, 28).

Weather Events The weather events attributed to climate change are vast. Higher global temperatures could fuel extreme weather. Some of the more specific indicators include the following information. Yearly precipitation in southern New England has increased by more than 25 percent during the last 100 years, but snowfall in northern New England since 1953 has decreased by 15 percent. Snowfall in Australia has declined by 30 percent during the past 40 years. Lakes in Pennsylvania freeze on average about 10 days later and thaw about 9 days earlier than they did 50 years ago. Severe droughts affect 30 percent of the earth's surface today compared with 10–15 percent 35 years ago (DiSilvestro 2005, 23).

Sea Level Sea temperatures have risen up to 2 °F during the past 20 years, although the link to global warming has not been established. In addition, the mean global sea level has risen as much as 7.8 inches during the past century (DiSilvestro 2005, 24).

Ecosigns

There are many indicators of species changing habitat patterns or shifting species. What worries many biologists is that certain species respond to changing climate in different ways from others who share their same habitat. When this happens, vital coordination among interdependent species can occur, such as with a plant and its pollinator. This link can be broken and could result in a loss of synchrony that not only threatens the species in the relationship, but may also disrupt the entire plant and animal communities. Biologist John Weishampel of the University of Central Florida states, "In nature, timing is everything. It's like a symphony that's ruined if one instrument comes in at the wrong time" (Tangley 2005, 28). Because the earth's climate seems to be changing rapidly, the primary concern is that warming could take place so quickly that species will not have the time to adapt and avoid extinction. The natural cycles of interdependent creatures, such as birds or animals and their diets of specific insects or flowers, may fall out of sync and thus cause populations to decline. The following list reflects a summary of some of these concerns (DiSilvestro 2005, 24).

- The tree line in Russia's Ural Mountains has moved about 500 feet higher since the early twentieth century, and in Canada's Banff National Park, spruces are growing up past their tree line by about 180 feet since 1990. Oak trees in England are leafing two weeks earlier than they did 40 years ago. In Europe, the vegetative growing season of trees, shrubs, and herbs has increased 11 days since 1960.
- Globally, plants are now blooming about 5.2 days earlier each decade, according to Stanford University. In Washington, D.C., data collected for 100 flowering species are blossoming an average of 4.5 days earlier than they did in 1970, while only 11 are flowering later. A similar study in Edmonton, Alberta, found that spring flowering is occurring eight days earlier than 60 years ago (DiSilvestro 2005, 24) In Europe, many plants are flowering about one week earlier and shedding their leaves five days later than they did 50 years ago (Montaigne 2004, 40).
- A study completed in 2002 found that 1700 species of birds and butterflies have changed their ranges to the north by four miles annually since the 1960s. Twenty bird species in the United Kingdom between 1971 and 1995 are laying their first eggs an average of nine days earlier, from an analysis of 74,000 nesting records of 65 species. The Sachem Skipper butterfly has expanded its range 420 miles northward from California into Washington in just 35 years; in 1998, the warmest year on record, the range expanded 75 miles alone (DiSilvestro 2005, 24). In Europe, a study of 35 nonmigratory butterfly species found that, in the last few decades, about two-thirds have expanded their ranges to the north by 20–150 miles (Montaigne 2004, 40).

BOX 1 Wildlife on the Hot Seat

Changes in the timing of seasonal behaviors are only some of the ways global warming has affected wildlife. Species are always on the move, with their ranges shifting for the most part northward or to higher elevations. Distributions of some species, particularly those that have small ranges or live at the edge of suitable habitat, are contracting, while a handful of others are expanding. As with all major changes, there are winners as well as losers, yet the losers include some of Earth's most vulnerable and endangered wildlife.

Beyond shifts at the species level, entire communities are being transformed as plants and animals in the same habitat respond differently to climate change. According to Stanford University biologist Terry Root, coauthor of The Wildlife Society report *Global Changes and Wildlife in North America,* "In the future, well-balanced wildlife communities as we know them will likely be torn apart."

1. **Black guillemots:** On Alaska's Cooper Island, populations of black guillemots are declining as sea ice recedes, taking with it the birds' primary prey: Arctic cod, which live under the ice.
2. **Sockeye salmon:** Extremely high water temperatures accompanied by drought-induced low water flows led to the 1998 deaths of tens of thousands of sockeye salmon in British Columbia—a preview of what's to come, say some scientists.
3. **Sooty shearwaters:** Between 1987 and 1994, the number of sooty shearwaters off the U.S. west coast decreased by 90 percent. Some scientists attribute the decline to changes in ocean temperatures and currents caused by global warming.
4. **Snails, sea stars, and other intertidal creatures:** In Monterey Bay, California, the ranges of intertidal invertebrates, such as limpets and snails, have been shifting northward for the past six decades as sea and air temperatures rise.
5. **Edith's checkerspot butterflies:** The distribution of Edith's checkerspot butterfly, which extends from the west coast of southern Canada through northern Mexico, is contracting. In the southern

portion of the butterfly's range, 80 percent of all populations have become extinct.

6. **Mexican jays:** In southern Arizona, the breeding season of the Mexican jay advanced by 10 days between 1971 and 1998. The change correlates with spring monthly temperatures, which rose 4.5 °F during the same period.
7. **American robins:** In Colorado, robins are migrating from low to high elevations where they breed two weeks earlier than they did in the late 1970s. Many birds now arrive before snow melts.
8. **American pikas:** Scientists say warming in the U.S. Great Basin has contributed to the extinction of 7 of 25 populations of American pika. Sensitive to temperature changes, these high-elevation mammals have nowhere to move.
9. **Polar bears:** On Canada's Hudson Bay, early melting sea ice has decreased the amount of time polar bears have to hunt seals, their primary prey. As a result, bears today weigh less and give birth to fewer cubs than they did 20 years ago.
10. **Red-winged blackbirds:** These migratory birds are arriving at their breeding grounds in northern Michigan 21 days earlier than they did in 1960.
11. **American lobsters:** In western Long Island Sound, large numbers of lobsters died mysteriously during September 1999. Some researchers suggest that high water temperatures wiped out the crustaceans, which were living at the southern limit of the species' range.
12. **Prothonotary warblers:** For nearly two decades, prothonotary warblers have been returning to breeding grounds in Virginia from wintering grounds in South America and the Caribbean a day earlier each year as springtime temperatures rise.
13. **American alligators:** The distribution of the American alligator, which ranges from the Carolinas south to Florida and west to Texas, appears to be shifting northward in some regions. Rising sea level may also force the freshwater reptiles inland, where they would encounter more development.
14. **Loggerhead sea turtles:** Off the Atlantic coast of Florida, threatened loggerhead sea turtles are coming ashore to nest 10 days earlier than they did in 1989. During the same period, offshore ocean temperatures have increased 1.5. °F.
15. **Coral reefs:** Off the shores of many Caribbean islands, higher water temperatures are causing corals to expel their symbiotic algae—or bleach—which can lead to coral death and damage to entire reef ecosystems.
16. **Golden toads:** In the first species extinction attributed to global warming, the amphibians have disappeared from their only known habitat, Costa Rica's Monteverde Cloud Forest.

Source: Reprinted with permission from *National Wildlife* magazine, National Wildlife Federation, April/May 2005.

- Australia's Great Barrier Reef, the world's largest (1240 miles), is home to some 400 species of coral and 1500 fish and is the ocean's parallel to a tropical rain forest. Coral starts to bleach and lose some key organisms from its tissues in waters warmer than about 85 °F. The current warming trend is damaging and bleaching the coral, causing both the coral and its resident marine life to begin to be in peril (Montaigne 2004, 45).
- In England, great white sharks and Portuguese man o' war jellyfish are being found in waters off Devon and Cornwall, waters that previously were too cold for these species. U.S. species of the red fox, rufous hummingbird, two subtropical dragonflies in Florida, and several marine species found off Monterey, California, are moving northward (DiSilvestro 2005, 24). Adelie penguins (one of only two ice-dependent polar penguin species) are declining by more than 66 percent on Antarctica islands, where numbers have plummeted in 30 years from 32,000 breeding pairs to 11,000 and are being replaced by a subantarctic species of the Gentoo penguin that prefers warmer climates. Ecologist Bill Fraser, who studies the Adelie penguin, states, "Lesson number one for me has been the realization that ecology and ecosystems can change…like that. In geologic time, it's a nanosecond." (Montaigne 2004, 39)

Timesigns

Geologists, climatologists, and other scientists are trying to analyze the earth's more rapid fluctuations when the earth switches from frozen ice age to a warmer temperature and back again and not the 100,000-year glacial ice ages of the past million years or so. They are asking questions these rapid types of changes and what they tell us about the earth's climate both now and in the future. To do this, they are using a variety of sources ranging from glacial ice, stalagmites from caverns, and tree rings to coral, dust and sand dunes, and microscopic shells to human inscriptions and people's logs and diaries.

- Corals produce annual rings like trees. Observations taken from corals are that temperatures were steady from 1850 to 1950; after 1950, temperatures rose dramatically. Geologist Greg Wiles, in data taken from reading tree rings from 585 A.D, reports that living trees seem to be experiencing stresses they haven't seen in the past thousand years (Morell 2004, 59–60).
- Ice cores from Greenland that have remained undisturbed for over 100,000 years indicate some of the best records for past temperatures, amount of precipitation, and atmospheric conditions. Results here see bursts of warming and cooling activity, with the recent climate being especially erratic.

Coral reefs and marine sediment also indicate how important the ocean circulation system is. The "North Atlantic conveyor" (the ocean current that pulls warm water from the tropics northward, thus keeping temperatures more mild in the north) is highly important to the climate of the entire planet. It lost strength and even stopped during the ice ages, which caused a series of events that led to warmer temperatures in the Southern Hemisphere and colder temperatures in the north. Differing concentrations of fresh water from weather and ice melts in the ocean alter the ocean's salinity and currents, can slow the conveyor, and can decrease the amount of warm water pulled from the tropics, thereby changing dynamics and temperatures as far south as Antarctica.

Since about 50 percent of the surface of the planet is located in the tropics—a major heat source—a global conduit such as the North Atlantic conveyor plays a larger role in driving climate change than was first realized. Greenhouse gases, especially CO_2, seem to be switching these natural circulatory and conveyor systems along with the earth's climate. Projections are that we will see a 2–5 °C warming over the next 100 years, resulting in higher minimum temperatures as well as fewer cold and frost days over all of the planet's land areas. The general conclusion is that we are now entered into the fastest climate-warming phase of the past 10,000 years, attributed primarily to the extremely high CO_2 levels that are being added to the atmosphere, mostly from fossil fuel use (Morell 2004, 58–75). Simon Tett, a United Kingdom climate specialist, states that "We'd need to get to zero emissions to stabilize the CO_2 that's already in the atmosphere. And that's not the path we, as societies, have chosen. Even if we were to stop CO_2 emissions now, we are committed to warming." (Morell 2004, 75)

Doug Inkley, the National Wildlife Federation Senior Science Advisor, states: "No single bit of scientific evidence makes a convincing argument that global warming is having an impact on wildlife and plants, but the cumulative evidence cannot be ignored. The question is no longer, 'Is global warming happening?' The question is, what are we going to do about it?" (DiSilvestro 2005, 24).

According to Lee Hannah, a biologist for Conservation International's Center for Applied Biodiversity, "Climate change is the most significant new threat for species extinctions in the twenty-first century." (Tangley 2005, 4c). A study by 18 international scientists published in 2004 in

BOX 2 Hottest Years on Record

1. 2005
2. 1998
3. 2002
4. 2003
5. 2004
6. 2001
7. 1997
8. 1990
9. 1995
10. 1999

Source: Union of Concerned Scientists. (2005). 2005 Vies for Hottest Year on Record. Available at http://www.ucsus.org/global_warming/science/recordtemp2005.html.

Nature magazine focused on 1,103 species from a variety of terrestrial ecosystems in Europe, Australia, Mexico, South Africa, and Brazil and was the first comprehensive analysis of the impact of climate change on plant and animal global habitats. By proposing a variety of increased temperature scenarios, the study predicted that between 15 and 37 percent of the targeted species would be "committed to extinction" by 2050. J. Alan Pounds of Costa Rica's Tropical Science Center states that "the threat to life on Earth is not just a problem for the future. It is part of the here and now." (Tangley 2005, 34c)

In the future, increasing concentrations of greenhouse gases will most likely accelerate the rate and effects of climate change. Predictions are that the average global surface temperature could rise 1–4 °F in the next 50 years and 2.2–10 °F in the next century, accompanied by many regional variations (EPA 2000, 2). Global precipitation will increase as evaporation increases due to the warming climate. Coastal areas will become flooded, implying disaster for the hundreds of millions of people who live in those areas and cities. Other effects include a decrease in soil moisture in many regions, an increase in intense rain storms, an elevation of the sea level along most U.S. coasts of approximately two feet, and the loss of plant and animal habitats. Predictions for specific areas remain more difficult to forecast than overall global effects and trends (EPA 2000, 2).

SOLUTIONS

There are many possible paths to take to counter global warming and initiate solutions. Countries with the largest CO_2 emissions especially need to provide leadership and participation in counter-emission activities. These countries include the United States, with carbon emissions that are more than double those of China, which is the second world leader in CO_2 emissions, followed by Russia in third place (see Table 3). World response to climate change can be effective if it involves national and international policy strategies that either limit or reduce greenhouse gas emissions,

Table 3
Fifteen Major Nations Producing Carbon Dioxide Emissions, Ranked by 2002 Totals, 1980–2002

Country	*1980*	*1985*	*1990*	*1995*	*2000*	*2001*	*2002*
United States	1,296.59	1,250.41	1,366.60	1,442.32	1,587.10	1,557.96	1,568.02
China	394.01	507.58	616.89	787.72	822.85	866.11	906.11
Russia[1]	837.21	958.58	1,037.47	433.26	420.89	423.54	415.16
Japan	261.18	246.13	269.89	298.63	322.53	322.27	321.67
India	82.67	120.41	161.80	236.48	271.67	275.49	279.87
Germany[2]	208.63	188.81	192.75	238.73	229.93	234.62	228.62
Canada	125.58	119.36	130.05	134.77	157.34	157.29	161.37
United Kingdom	168.16	160.98	163.66	152.60	151.48	155.52	150.77
Korea, South	35.16	44.69	63.82	109.28	114.96	117.60	122.88
Italy	103.31	102.56	113.24	118.45	122.43	122.50	122.36
Australia	54.67	61.49	72.37	79.59	97.78	106.68	111.92
France	136.02	108.56	102.00	100.69	112.09	112.36	111.08
Ukraine[3]	NA	NA	NA	121.80	97.64	102.56	105.89
South Africa	64.09	81.65	80.84	93.85	104.43	103.08	102.99
Mexico	64.67	74.62	83.96	86.91	102.50	100.08	98.87
WORLD TOTAL	5,082.65	5,353.00	5,901.28	6,029.22	6,515.83	6,607.66	6,690.73

[1] Numbers for 1980–1990 are for the former Soviet Union.
[2] Numbers for 1980–1990 are for the former West Germany.
[3] Included in Russia prior to 1995.
Source: U.S. Department of Energy; World Almanac Education Group. The World Almanac and Book of Facts – 2005. New York.

such as the initiatives of the Kyoto Protocol. Such policy suggestions might include "carbon/energy taxes, tradeable permits, removal of subsidies to carbon energy sources, provision of subsidies and tax incentives for carbon-free sources, refund systems, technology or performance standards, energy mix requirements, product bans, voluntary agreements and investment in research and development." (Dunn and Flavin 2002, 33)

The United States has the opportunity to make a significant difference by taking responsibility for its high emissions, living standards, and high energy consumption. U.S. citizens emit 21 tons of heat-trapping gases per capita (twice as much as western Europeans, who enjoy the same standard of living), typically due to gas-guzzling vehicles and inefficient home heating. Transportation vehicles in the United States are responsible for about one-fourth of the country's annual CO_2 emissions. Although the United States has only 4 percent of the world's population, it is responsible for one-fourth of all global carbon dioxide emissions, which remain in the atmosphere for decades.

Americans can participate in the solution by maximizing the efficiency of their homes and using renewable energy. Additionally, governmental action at both the national and international levels would be necessary to require power plants and factories to reduce emissions, raise fuel efficiency for automobiles and other vehicles, stop subsidies for the fossil fuel industries, and invest in energy-efficient technologies and renewable energy (Tolme' 2005, 42). The following sections present suggested solutions and policies for combating global warming.

Individuals DO Make a Difference

"According to the Union of Concerned Scientists, if every U.S. family replaced one regular incandescent bulb with a compact fluorescent, the U.S. could decrease CO_2 emissions by more than 90 billion pounds annually—equivalent to taking 7.5 million cars off the road." (Tolme' 2005, 43) The American Council for an Energy Efficient Economy states that if every U.S. household used the most efficient appliances, the nation could save $15 billion in energy costs and prevent 175 million tons of greenhouse gas emissions. Buying high-efficiency appliances not only reduces energy use, but sends a powerful message to manufacturers that people want such products. By demanding renewable energy, consumers influence the market and reduce both the emissions of greenhouse gases and their energy expenses (Tolme' 2005, 43).

Cars and Trucks

Purchasing a more fuel-efficient car is significant because even modest improvements in fuel economy can make a difference. For example, driving a truck that gets 16 mpg instead of one that gets 14 mpg can save 134 gallons of gas per year. Since each gallon of gasoline pumps out from19 to 25 pounds of CO_2, more than 2,000 pounds of CO_2 will not be released into the atmosphere. SUVs and pickup trucks have become a national passion and are being bought nationally every four seconds. Their popularity has caused a 20-percent increase in CO_2 pollution since the early 1990s. An SUV can add 43 percent more greenhouse gas emissions than the average car, and it gets worse mileage than Ford's Model T! Technology does exist to raise the average fuel economy for all vehicles from the current 24 mpg to 40 mpg within the next 10 years without significantly changing the appearance or performance of the vehicles. Raising overall fuel economy by 10 percent could cost as little as $465 a vehicle according to a 2001 National Academy of Sciences panel.

There are some interesting car emission facts. Vehicles in the United States release more CO_2 pollution than the entire country of India emits from all of its uses of energy combined. Americans also burn one-fourth of the world's oil, with 40 percent coming from passenger vehicles that burn 8.7 million barrels a day. New cars and trucks today have approximately the same gas efficiency that they did in 1982—about 25 miles per gallon. Hybrids, however, constitute a huge improvement by averaging 40–50s mpg. States Jim Kliesch of the American Council for an Energy-Efficient Economy: "That we are still producing vehicles as inefficient as they were 20 years ago is pathetic. Compounding this problem, we are driving more miles than ever before." (Ridgley 2005, 56–57)

Green Buildings

Building homes and offices that are energy efficient can begin a significant greenhouse gas reduction standard. The average U.S. home produces nearly 10,000 pounds of greenhouse gas emissions per year, and office buildings are even more profligate energy guzzlers. Energy conservation approaches for office buildings include Green

Building Ordinances that require all buildings to be certified by the U.S. Green Building Council Leadership in Energy and Environmental Design program (LEED). Buildings are then rated according to their "greenness." Such approaches for energy efficiency include overall building design, rooftop solar panels, "embedded energy" construction materials that reduce the amount of energy used in manufacturing materials and transport, natural lighting utilization, special employee programs to encourage energy conservation, and many more. For homes, an Energy Star program labels both new and existing homes as energy efficient, which results in a higher resale value. Using this market preference, it is hoped that green homes will become the rule rather the exception (Tolme' 2005, 42).

Other Solutions

1. Develop a national energy strategy to promote wildlife conservation and combat global warming, such as that found in the recommendation of the National Wildlife Federation's report, *Fueling Environmental Progress*. Their recommendations would require power plants and factories to reduce their emissions to year 2000 levels by 2010, would raise fuel efficiency for automobiles (which account for one-third of the nation's CO_2 emissions), would require investment in energy-efficient technologies and renewable energies, and would eliminate subsidies for the fossil fuel industry (Schweiger 2005, 20).
2. Set national and international standards and policies that either limit or reduce greenhouse gas emissions as suggested above. Such policies could include "carbon/energy taxes, tradeable permits, removal of subsidies to carbon energy sources, provision of subsidies and tax incentives for carbon-free sources, refund systems, technology or performance standards, energy mix requirements, product bans, voluntary agreements, and investment in research and development" (Dunn and Flavin 2002, 3).
3. Use more wind power both locally and nationally, which has the greatest growth potential. Since the 1980s, the price of wind power has dropped 40 percent and is one of the cheapest energy sources. Wind power constitutes only about 1 percent of the electricity market, but the figure could easily grow to 6 percent by 2020 (Tolme' 2005, 43).
4. Buy renewable energy from utilities. By spending a little extra money, about half of electricity consumers in the United States have this opportunity. Yet, most consumers know nothing about this option. More than 500 utility companies in 34 states offer green pricing programs (Tolme' 2005, 430).
5. Turn your home into a power plant with solar and wind power systems. With state rebate programs and energy reciprocity laws, homeowners can feed power back into the utilities grid that they are no longer using and watch their electricity bill shrink (Tolme' 2005, 43).
6. Help your community and state to implement Climate Action Plans that meet the goals of the Kyoto Protocol even though the federal government is dragging its feet. Previously enacted examples include the following: Oregon set limits on CO_2 emissions from new power plants in 1997. New York's Public Service Commission requires 25 percent of the state's electricity to come from renewable resources by 2013. California has decided that only cars with lower CO_2 emissions can be sold in the state. In addition, cities and towns such as Seattle, Portland, Salt Lake City, Chicago, Los Angeles, San Francisco, and many others are implementing programs to cut emissions and are instituting various other climate protection programs. In all, more than 150 communities across the country and 600 worldwide have joined the Cities for Climate Protection program (www.iclei.org/us/ccp). Director Abby Young says, "At first we had to solicit communities; now we have cities coming to us. Until we get some leadership at the federal level, communities and states must keep the pressure on." (Tolme' 2005, 45)

Many people see the need for positive action as soon as possible to control the increasing climate change problem before it accelerates into difficult and unsolvable global consequences. In answer to skeptics who claim that reducing dependence on fossil fuels would wreck the economy, many instead predict that investments in renewables and efficiencies would spur an economic revolution similar to the birth of the Internet and high-tech boom of the past two decades. Economist Michelle Manion of the Union of Concerned Scientists states: "We have done a good job refuting the science skeptics [who dismiss the reality of global warming]. Now we need to do a better job refuting the economic skeptics." (Tolme' 2005, 46)

BOX 3 Reduce Your Own Greenhouse Gas Emissions

Individual choices can have an impact on global climate change. Reducing your family's heat-trapping emissions does not mean forgoing modern conveniences; it means making smart choices and using energy-efficient products, which may require an additional investment up front, but often pay you back in energy savings within a couple of years.

Since Americans' per capita emissions of heat-trapping gases is 5.6 tons—more than double the amount of western Europeans—we can all make choices that will greatly reduce our families' global warming impact.

1. **The car you drive: the most important personal climate decision.** When you buy your next car, look for the one with the best fuel economy in its class. Each gallon of gas you use releases about 25 pounds of heat-trapping carbon dioxide (CO_2) into the atmosphere. Better gas mileage not only reduces global warming, but will also save you thousands of dollars at the pump over the life of the vehicle. Compare the fuel economy of the cars you're considering and look for new technologies, like hybrid engines.
2. **Choose clean power.** More than half the electricity in the United States comes from polluting coal-fired power plants. And power plants are the single largest source of heat-trapping gas. None of us can live without electricity, but in some states, you can switch to electricity companies that provide 50–100 percent renewable energy. (For more information, go to www.Green-e.org.)
3. **Look for Energy Star.** When it comes time to replace appliances, look for the Energy Star label on new appliances (refrigerators, freezers, furnaces, air conditioners, and water heaters use the most energy). These items may cost a bit more initially, but the energy savings will pay back the extra investment within a couple of years. Household energy savings really can make a difference: if each household in the United States replaced its existing appliances with the most efficient models available, we would save $15 billion in energy costs and eliminate 175 million tons of heat-trapping gases.
4. **Unplug a freezer.** One of the quickest ways to reduce your global warming impact is to unplug the extra refrigerator or freezer you rarely use (except when you need it for holidays and parties). This can reduce the typical family's carbon dioxide emissions by nearly 10 percent.
5. **Get a home energy audit.** Take advantage of the free home energy audits offered by many utilities. Simple measures, such as installing a programmable thermostat to replace your old dial unit or sealing and insulating heating and cooling ducts, can each reduce a typical family's carbon dioxide emissions by about 5 percent.
6. **Light bulbs matter.** If every household in the United States replaced one regular light bulb with an energy-saving model, we could reduce global warming pollution by more than 90 billion pounds over the life of the bulbs, the same as taking 6.3 million cars off the road. So, replace your incandescent bulbs with more efficient compact fluorescents, which now come in all shapes and sizes. You'll be doing your share to cut back on heat-trapping pollution and you'll save money on your electric bills and light bulbs.
7. **Think before you drive.** If you own more than one vehicle, use the less fuel-efficient one only when you can fill it with passengers. Driving a full minivan may be kinder to the environment than two midsize cars. Whenever possible, join a carpool or take mass transit.
8. **Buy good wood.** When buying wood products, check for labels that indicate the source of the timber. Supporting forests that are managed in a sustainable fashion makes sense for biodiversity, and it may make sense for the climate too. Forests that are well managed are more likely to store carbon effectively because more trees are left standing and carbon-storing soils are less disturbed.
9. **Plant a tree.** You can also make a difference in your own backyard. Get a group in your neighborhood together and contact your local arborist or urban forester about planting trees on private property and public land. In addition to storing carbon,

trees planted in and around urban areas and residences can provide much-needed shade in the summer, reducing energy bills and fossil fuel use.

10. **Let policymakers know you are concerned about global warming.** Our elected officials and business leaders need to hear from concerned citizens.

Source: Courtesy of Union of Concerned Scientists, www.ucsusa.org/global_warming.

BOX 4 What's Your Role in Global Warming?

Americans make many choices every day that increase their collective output of carbon dioxide, the planet's leading climate-changing gas. Take the following quiz to see how you rank as a carbon dioxide emitter on the climate-change scale.

1. Do you use energy-efficient light bulbs, such as compact fluorescent bulbs (CFLs), throughout your home or apartment?

   If yes, add 1 point to the emission meter.
   If no, add 4 points to the emission meter.

Fact: Where electricity is produced from coal-fired power plants, each CFL used prevents 1,300 pounds of carbon dioxide and 20 pounds of sulfur dioxide from being spewed into the atmosphere every year.

2. Do you turn off lights when not using them?

   If yes, add 1 point to the emission meter.
   If no, add 3 points to the emission meter.

3. Do you buy and install appliances, such as refrigerators, freezers, washers and dryers, or dishwashers, that bear the federal Energy Star stamp designating them as the most energy-efficient products in their class?

   If yes, add 2 points to the emission meter.
   If no, add 4 points to the emission meter.

Fact: Home appliances account for 30 percent of electricity use in industrial countries and contribute 12 percent of their total greenhouse gas emissions.

4. Where does your electricity come from? If you don't know, you can find out by going to www.green-e.org and clicking on "Your Electricity Choices" or by calling your local electricity supplier.

   - coal-fired plant — add 6 points
   - oil-fired plant — add 5 points
   - gas-fired plant — add 4 points
   - garbage — add 3 points
   - nuclear — add 2 points
   - low-impact hydro — add 1 point
   - wind power — add 0 points
   - solar (panels and photovoltaic cells) — add 0 points
   - geothermal — add 1 point
   - biomass — add 1 point
   - combination of green power — add 1 point

5. Does your computer wear an Energy Star sticker?

   If yes, add 2 points to the emission meter.
   If no, add 4 points to the emission meter.
   I have no computer, add 0 points.

Fact: Buying green power (renewable energy) as the sole power source for the average U.S. home for 1 year prevents carbon dioxide from being emitted; it is equivalent to planting two acres of trees, removing a car from the road or not driving 12,000 miles.

6. What kind of car/truck do you drive?

   - compact car with highest fuel-efficiency standards — add 2 points
   - SUV — add 5 points
   - light truck — add 6 points
   - full-sized car meeting minimum efficiency standards — add 3 points
   - hybrid electric car — add 1 point
   - fuel cell or hydrogen car — add 1 point
   - I don't own a vehicle — add 0 points

Fact: Cars and light trucks in the United States account for 40 percent of the nation's oil use and contribute as much to climate change as the entire Japanese economy.

7. What kind of clothes do you usually wear?
    - all cotton add 1 point
    - cotton/polyester blend add 3 points
    - wool add 0 points

Fact: The production process for a cotton T-shirt blended with polyester uses petrochemicals and releases roughly 10 times the shirt's weight in carbon dioxide.

How Do You Stack Up?

The lower the emission meter, the more energy efficient and environmentally sound your lifestyle and consumer choices are.

- If your emission meter is under 12 points, congratulations, you are a genuine energy saver and carbon-emission eliminator!
- If your emission meter is between 13 and 17 points, you are on your way to being a genuine energy conservationist and emission eliminator.
- If your emission meter is between 18 and 25, you are an average consumer and carbon emitter.
- If your emission meter is over 25, you are a human smokestack! Cut back on your energy use.

Source: Reprinted with permission from *National Wildlife* magazine, National Wildlife Federation, April/May 2005.

REFERENCES

Appenzeller, T., Dimick, D. (2004). Signs from the Earth. *National Geographic* September 10-11.

DiSilvestro, R. (2005). The Proof Is in the Science. *National Wildlife* 43 April/May (3):22–24.

Dunn, S., and Flavin, C. (2002). Moving the Climate Change Agenda Forward. *State of the World, 2002*. New York: Norton.

Dyer, G. (2005). "Nuclear Power Is the Least-Bad Option to Fix Energy Woes." *The Salt Lake Tribune*, May 23. Accessed June 6, 2005. Available at http://www.sltrib.com/opinion/ci_2754455.

Environment Canada. *Comparison of Global Emissions*. Accessed May 9, 2005. Available at http://www.ec.gc.ca/pdb/ghg/global_emissions_e.ctm.

Glick, D. (2004). Geosigns; The Big Thaw. *National Geographic* 206 September (3):14–33.

Intergovernmental Panel on Climate Change. (2001). "*Climate Change 2001*." IPCC. Accessed May 9, 2005. Available at http://www.grida.no/climate/ipcc_tar/wg1/052.htm.

Malcolm, J., Liv, C., Miller, L., Allmutt, T. Hansen, L. (2002). Habitats at Risk: Global Warming and Species Loss in Terrestial Ecosystems. Switzerland, Gland: WWF-World Wide Fund for Nature. Available at http://assets.panda.org/downloads/habitatsatriskfull.pdf.

Marland, G. (2005). Carbon Dioxide Emissions by Fuel Type. *Audubon* 107 July/August (4):12.

Montaigne, F. (2004). Ecosigns; No Room to Run. *National Geographic* 206 September (3):34–55.

Morell, V. (2004). Timesigns: Now What? *National Geographic* 206 September (3):56–75.

Morgan, J. (2002). Habitats at Risk: Global Warming and Species Loss in Terrestrial Ecosystems. *World Wildlife Climate Change Campaign*, February.

National Resources Defense Council. "*Global Warming*." Accessed May 9, 2005. Available at http://nrdc.org/global-Warming/qthinice.asp.

Ridgley, H. (2005). Driving Down the Heat. *National Wildlife* 43 April/May (3):56–57.

Schweiger, L. (2005). Looking North, Seeing Our Future. *National Wildlife* 43 April/May (3):18–20.

Tangley, L. (2005). Out of Sync. *National Wildlife* 43 April/May (3):28–34.

Tolme', P. (2005). It's the Emissions, Stupid. *National Wildlife* 43 April/May (3):40–46.

Union of Concerned Scientists. (2006). 2005 Vies for Hottest Year on Record. Available http://www.ucsusa.org/global_warming/science/recordtemp2005.html.

U.S. Environmental Protection Agency. (January 2000). "*Global Warming–Climate*." Accessed April 2, 2005. Available at http://yosemite.epa.gov/oar/globalwarming.nst/content/climate.html.

QUESTIONS

1. How much has the earth's temperature risen in the past two decades? What do most scientists feel is the cause of this recent warming?
2. What is the most controversial and unknown aspect of climate change?

3. Which greenhouse gas contributes most to the warming of the atmosphere and the earth's surface?
4. Explain how global warming works.
5. How much does the United States contribute to total global greenhouse gases? Why does it contribute so much?
6. Name three convincing geosigns that reflect significant climate change. Explain why they are important and significant indicators.
7. Identify three convincing ecosigns that reflect significant climate change. Explain why they are important and significant indicators.
8. Identify three convincing timesigns that reflect significant climate change. Explain why they are important and significant indicators.
9. Identify three solutions to global warming that make the most sense to you. How would they be most effective on local and global levels? Write a few paragraphs explaining your scenario, as well as its effectiveness and possible obstacles.

The Kyoto Protocol

BARBARA A. EICHLER

BACKGROUND

The Kyoto Protocol initially began as the United Nations Framework Convention on Climate Change Treaty (UNFCCC) at the 1992 Earth Summit in Rio de Janeiro and was officially enacted in March 1994. The UNFCCC Treaty established the increasingly important objectives of (1) world cooperation (2) towards the priority of stabilizing atmospheric concentrations of greenhouse gases at levels that will avoid "dangerous anthropogenic interference with global climate" and still allow economic development to proceed. Three principles were the foundation for the Treaty:

1. Scientific uncertainty must not be used to avoid precautionary action.
2. Nations must have "common but differentiated responsibilities."
3. Industrial nations with the greatest historical contributions to climate change must take the lead in addressing the problem. (Dunn and Flavin 2002, 27).

The Treaty at that time committed 181 nations and the European Union to the goals of addressing climate change, focusing on effects, and reporting their actions. It committed signatory industrial and transitional countries to report their climate policies and greenhouse gas inventories. It also developed the voluntary goals of returning greenhouse emissions to 1990 levels by the year 2000 and an approach for providing financial and technical support to nonindustrial nations to help achieve these goals (Dunn and Flavin 2002, 27).

By 1995, however, the UNFCCC signatories concluded that not enough progress was being made and launched a tighter initiative for a legally binding protocol. This developed into the Kyoto Protocol, which was adopted in December, 1997. The Protocol collectively committed 38 industrial and former Eastern Bloc nations (called Annex I nations) to reduce their greenhouse gas emissions between 2008and 2012 by 5.2 percent below 1990 levels (Dunn and Flavin 2002). The Kyoto Protocol required the endorsement of at least 55 industrial nations that together account for at least 55 percent of 1990 global greenhouse gas emissions (Prugh 2005, 21).

The Kyoto Agreement focused on reduction of carbon-rich gases—mainly the byproduct of burning oil, gas, and coal—that scientists believe could dramatically change weather patterns. The Accord allowed for other "flexibility mechanisms" that help to reduce total emissions levels by "trading of emission permits, the use of forests and other carbons 'sinks,' and the earning of credits through a Clean Development Mechanism or joint implementation projects (carbon-saving initiatives that take place in developing or Annex B [Annex I] nations)." (Dunn and Flavin 2002, 27.) Developing countries were required to continue with their existing commitments to monitor and deal with their emissions and were included in the Treaty, but they were excluded from emission quotas on economic grounds.

In further negotiations in Bonn, Germany, in March 2001, the United States withdrew from the negotiating process. President Bush stated that the restriction imposed by the Treaty would harm the U.S. economy and was not binding enough to the developing nations (BBC News 2001). The White House said that nations such as China and India should also have emission targets and that signing the Kyoto Protocol could cost the United States more than 5 million jobs. From the Protocol's provisions, the United States would have to cut emissions by 7 percent from 1990 levels (Bloomberg 2005, 2–3). However, without the United States, 178 nations still reached progressive agreement on principal points, even though there were many compromises on emissions trading, sinks, and compliance that allowed more flexibility in attaining the Kyoto emission levels (Dunn and Flavin 2002, 27).

Additional Evidence on Climate Change by the IPCC

Also in 2000–2001, the Intergovernmental Panel on Climate Change (IPCC), comprised of more than 1,500 scientists from approximately 100 countries, published *Climate Change 2001,* an encompassing and ambitious study that brought strong evidence and concern about climate change to the Kyoto Protocol climate negotiations. The IPCC announced increasing climate change concern and accumulating evidence that "unless greenhouse gas levels are stabilized, Earth's average surface temperature will rise by up to 5.8 °C by the end of the century. …if unchecked, CO_2 levels in the air will be between 650 and 970 parts per million (ppm). To stabilize CO_2 at 450 ppm (twice the pre-Industrial Revolution level), which would limit global warming to about 2 °C, total global greenhouse gas emissions must be cut by 60–80 percent of today's emissions within 50 years at the latest." (Dixon 2004, 2). If the polar ice caps remain constant, climate change by greenhouse gases could raise sea levels between 20 centimeters and 1 meter by 2100. There will be more severe and extensive flooding, storms, and droughts. The poorest countries would be the most affected and least able to adapt (Dixon 2004, 2).

KYOTO GOES INTO EFFECT!

Twelve years after the initial efforts of the UNFCCC in Rio de Janiero in 1992 and eight years after the beginning of the Kyoto Accord, Russia became the final country of the 35 Annex I industrial nations (which includes the European Union [15 countries]) needed for ratification that would account for 55 percent of emissions. As of February 16, 2005, the Kyoto Protocol, the first international effort to slow down global warming, went into effect despite years of battling about its difficult economics and other associated requirements. It was ratified by 141 nations, including most industrialized countries except Australia and, most notably, the United States, which emits the most CO_2 (about 17 percent, which is about twice the amount of China and four times the amount of Russia). See Figure 1 and Table 1.

Specific Developments

The Kyoto Protocol, as of February 2005, is realized as a global enforcement among signatories and represents

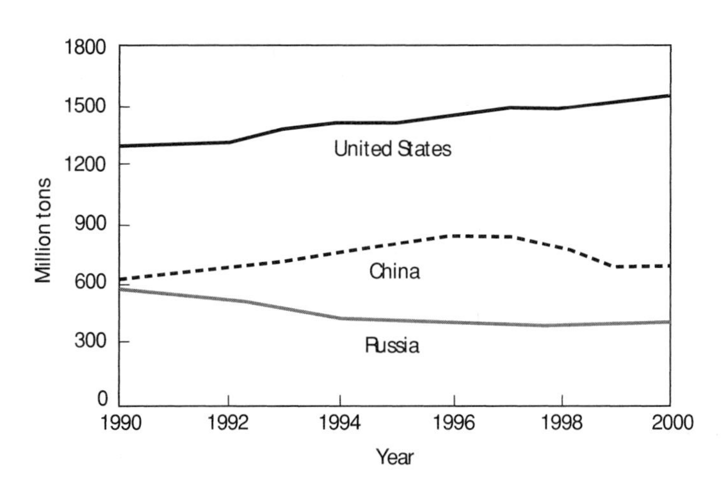

Figure 1
Carbon emissions in the United States, China, and Russia, 1990–2000.
Source: S. Dunn and C. Flavin, (2002). Moving the Climate Change Agenda Forward, *State of the World 2002*, New York: Norton, p. 35.

Table 1
Kyoto Emission Targets, First Commitment Period (2008–2012)

Country/Region	*Target 1990–2008/12*[1]	*Actual Emisions 1990–2008*[2] *(percent)*
United States	–7	+18.1
European Union	–8	–1.4
Japan	–6	+10.7
Canada	–6	+12.8
Australia	+8	+28.8
Russia	0	–30.7
All Annex B countries	–5.2	–1.7

[1] Basket of six greenhouse gases.
[2] Carbon only.
Source: S. Dunn and C. Flavin. (2002). Moving the Climate Change Agenda Forward, *State of the World 2002*, New York: Norton, p. 35.

important progress on an urgent global problem. Both the World Wildlife Fund (WWF) and Greenpeace stated that a 50-percent global drop in emissions from 1990 levels by 2050 is needed to stop climate change from becoming dangerous; however, they feel that Kyoto is the first step toward tackling the problem. WWF European Director of Climate and Energy Stephan Singer stated: "This shows that the majority of the world can work together to tackle one of the biggest global challenges, climate change. . . We need to see increased and strengthened caps for industrialized nations and we need to broaden the participation of developing countries such as China." (Bloomberg 2005, 1)

Kyoto requires signatories to reduce emissions with a range of national targets that vary by country between 2008 and 2012. Iceland and Norway have caps set at a higher level from 1990 emissions. European Union nations have individual requirements, but have a collective goal of reducing emissions by 8 percent from 1990 levels. Germany—Europe's biggest emitter—has an 8-percent emissions cut requirement, but has given itself a national target of reducing greenhouse gases by 21 percent by 2012, the highest reduction goal of all industrialized nations. It has already made progress by cutting emissions by about 18 percent from 1990 levels. Japan's requirement to cut CO_2 emissions by 6 percent from 1990 levels was a "difficult" target, according to Prime Minister Junichiro Kozumi. "Japan has increased its greenhouse emissions by 8 percent since 1990 and that means we have to cut down 14 percent against 1990 levels....That is a tough target for Japan, " stated Masaaki Nakajima, a spokesman for the Greenpeace environmental group (Bloomberg 2005).

The intense debate among nations practically left the Protocol close to being scrapped, and in the end, it will have only a small effect. The Treaty will only lead to a 2-percent cut in global greenhouse gas emissions, which is short of the 50–80 percent that experts state is necessary to avert significant climate change. Another problem is that the Kyoto Protocol expires in 2012. However, supporters consider the Treaty to be essential in initiating the process and important in learning how to develop international environmental treaties and systems.

German's Environment Minister Juergen Trittin stated: "The Kyoto Protocol is an urgently needed first step on a long path to reach climate stability. Industrialized nations have to assume more ambitious targets... We have to include the U.S. again in the international process of climate protection. The world's biggest emitter has to assume responsibility." (Bloomberg 2005, 4)

THE ROLE OF THE UNITED STATES

After 2001 and the incongruity of the U.S. pullout from the Kyoto Accord (despite the fact that they produce about one-third of the world's $C0_2$ emissions), further developments, progress, and political implications continue with the Kyoto Accord both with and without the input of the United States. Most of the nations of the industrialized world (31 of 34 except Australia, Monaco, and the United States) and many from the developing world (141 total nations) have elected to adopt the Kyoto Accord, despite its rejection by Washington. The Kyoto Accord has become somewhat "watered down" during the years of negotiation, in which it "has reduced the average cut in greenhouse gas emissions required by the year 2012 from 5.2 percent below 1990 levels and has incorporated a number of negotiation positions . . . such as crediting nations for maintaining large forests to serve as 'carbon sinks' to soak up the offending gas." (Karon 2005, 2) These negotiation positions are in response to the demands of industrialized countries, such as Japan and Canada, even though they were originally initiated by the Clinton administration. Yet the Kyoto Protocol is generally being heralded as an historic breakthrough by the nations of the world and environmentalist activist groups, for the Treaty's signatories still collectively produce more than twice as much greenhouse gas as the United States (Karon 2005).

Many believe that the real significance of the Kyoto Accord is found not only in the momentum it generated to deal with the earth's climate, but also that it survived the withdrawal of the United States. The Treaty demonstrates the determination of nations of the industrialized world to negotiate a binding agreement in which the United States no longer automatically holds the leadership role among Western nations. Some believe that President Bush gambled by withdrawing from the negotiations without offering a feasible alternative so as to force the international community to seek an agreement more favorable to the United States and its high fossil fuel- and gas-consumptive economy. Yet almost every country stayed with the Protocol and proceeded by simply moving forward without America. As a result, there is a "collective sense of achievement among the overwhelming majority of the world's industrialized and developing nations at the fact that they fashioned an epic international consensus on global warming despite the objections of the one nation that still aspires to global leadership." (Karon 2001, 5)

McCain-Lieberman Bill

With the withdrawal of the United States from the Kyoto Protocol, the McCain-Lieberman "Climate Stewardship Act of 2003" proposed by Senators John McCain (R-AZ) and Joseph Lieberman (D-CT) suggests an alternative proposal for addressing climate change in the United States. This Bill requires domestic, mandatory, and economy-wide emission reductions not only in carbon dioxide, but also in sulfur dioxide, nitrogen oxides, and mercury. The Bill also involves a climate research program, an emissions registry, and a trading program. The legislation would establish a limit on greenhouse gas emissions beginning on

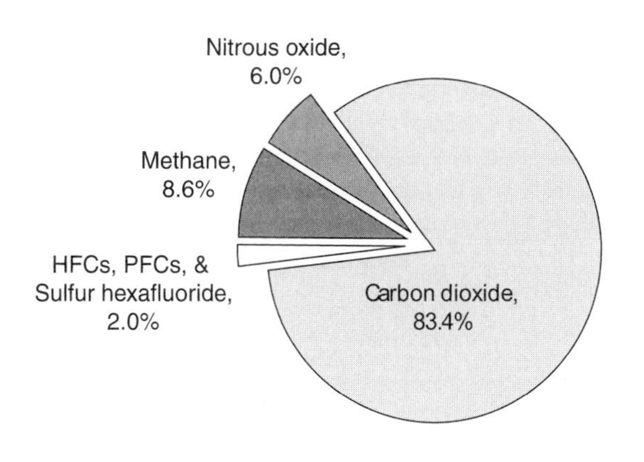

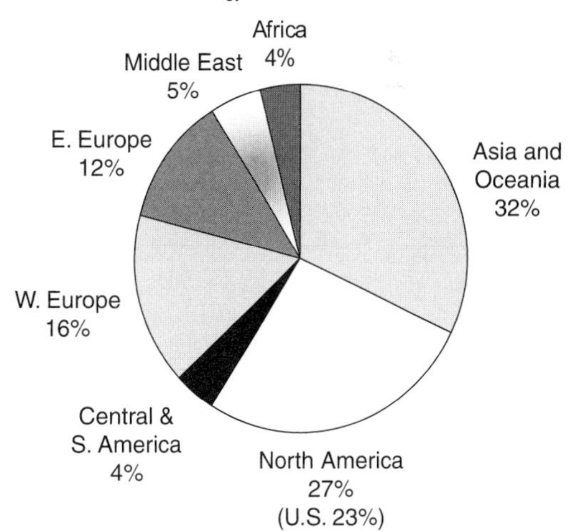

Figure 2

Emission statistics and average global temperatures.

Sources: U.S. Environmental Protection Agency, U.S. Energy Information Administration, National Oceanic and Atmospheric Administration

BOX 1 Carbon Emissions Trading

Carbon emissions trading is increasing and entails the trading of permits that allow countries to emit carbon dioxide and other greenhouse gases, which are calculated in tons of carbon dioxide equivalent (tCO_2e). This is an option that enables countries to meet Kyoto Protocol levels to reduce emissions and slow global warming. According to sources, "107 million metric tons of carbon dioxide equivalent (tCO_2e) have been exchanged through projects in 2004, a 38% increase relative to 2003 ($78mtCO_2e$) (http://carbonfinance.org/docs/CarbonMarketStudy2005.pdf).

Sources: Original source: http://carbonfinance.org/docs/CarbonMarketStudy2005.pdf. Wikipedia 2006, "Carbon Emissions Trading," http://en.wikipedia.org/wiki/Carbon_emmissions_trading, accessed May 28, 2006.

Table 2

Ratifications of Industrialized Countries and Their 1990 Share Percentages

As of August 2006, according to the United Nations Framework Convention on Climate Change (UNFCCC) 165 countries and other entities have ratified the Kyoto Protocol, which took effect February 16, 2005. For the Kyoto Protocol to enter into force, two specific conditions had to be met: (1) a minimum of 55 industrial nations that together account for at least (2) 55 percent of global greenhouse gas emissions in 1990 had to endorse the treaty. The conditions have been satisfied. The table below indicates signatory countries with CO_2 emission requirements that contribute to the emissions total.

Industrialized Country (in chronological order of ratification)	*Global Share of 1990 CO_2 emissions*
2001 – Romania and Czech Republic	2.48
May 23 – Iceland	.02
May 30 – Norway	.26
Slovakia	.42
May 31 – European Union (15 members)	24.23
June 4 – Japan	8.55
July 5 – Latvia	0.17
August 15 – Bulgaria	.60
August 21 – Hungary	.52
October 14 – Estonia	.28
2002 – December 13 – Poland	3.02
December 17 – Canada	3.33
December 19 – New Zealand	.19
2003 – Switzerland	.32
2004 – November – Russia	17.40
TOTAL	61.79

Source: Climate Action Network Europe. Table. Percentage Shares of Annex 1 Countries of 1990 Emissions. Available http://www.climnet.org/EUenergy/ratification/1990sharestable.htm.

Table 3
Total Carbon Dioxide Emissions and Their 1990 Global Shares of the 34 Industrialized Countries (Annex I) for the Purposes of Article 25 of the Kyoto Protocol*

Party	*Emissions (Gg)*	*Percentage*
Australia (not a signatory)	288,965	2.1
Austria	59,200	0.4
Belgium	113,405	0.8
Bulgaria	81,990	0.6
Canada	457,441	3.3
Czech Republic	169,514	1.2
Denmark	52,100	0.4
Estonia	37,797	0.3
Finland	53,900	0.4
France	366,536	2.7
Germany	1,012,443	7.4
Greece	82,100	0.6
Hungary	71,673	0.5
Iceland	2,172	0.0
Ireland	30,719	0.2
Italy	428,941	3.1
Japan	1,173,360	8.5
Latvia	22,976	0.2
Liechtenstein	208	0.0
Luxembourg	11,343	0.1
Monaco (not a signatory)	71	0.0
Netherlands	167,600	1.2
New Zealand	25,530	0.2
Norway	35,533	0.3
Poland	414,930	3.0
Portugal	42,930	0.3
Romania	171,103	1.2
Russian Federation	2,388,720	17.4
Slovakia	58,278	0.4
Spain	260,654	1.9
Sweden	61,256	0.4
Switzerland	43,600	0.3
United Kingdom	584,078	4.3
United States of America (not a signatory)	4,957,022	36.1
TOTAL	13,728,306	100.0

Gg = Greenhouse gases
*31 out of the world's 34 industrialized (Annex 1) nations have ratified.
Source: Climate Action Network Europe. Table. Percentage Shares of Annex 1 Countries of 1990 Emissions. Available at
http://www.climnet.org/EUenergy/ratification/1990sharestable.htm.

January 1, 2010. During the first six years of the program (2010–2016), annual greenhouse gas emissions would be reduced to the amount released in 2000. In following years, the limit would be allocated to 1990 emission levels (Pizer and Kopp 2003, 1–2).

In analyzing the legislation, "the initial reductions equal 14% of forecast [world] emissions levels of 6.2 billion tons in 2010... By 2020, reductions will equal 39% of forecast [world] emissions levels of 7.5 billion tons. Compared with the Kyoto Protocol, which would have required reductions of 2 billion metric tons in 2010, the McCain-Lieberman Act is relatively modest. Compared with calls by the Bush administration for an 18-percent improvement in greenhouse gas intensity—or an approximate 350-million metric ton reduction in 2012—McCain-Lieberman is relatively aggressive." (Pizer 2003, 3). The McCain-Lieberman Act creates a single trading approach that covers more than 70 percent of all U.S. carbon dioxide and industrial greenhouse gas emissions and is one of the most cost-effective proposals to date (Pizer and Kopp 2003). The Bill, although not yet passed, does represent a significant step forward in U.S. national conscience and accountability for its high greenhouse gas emissions and global responsibility.

REFERENCES

BBC News. (2001). "Europe Backs Kyoto Accord." March 31. Accessed May 13, 2005. Available at http://new.bbc.co.uk/1/hi/world/europe/1252556.stm.

Bloomberg. (2005). "Kyoto Accord Comes into Force to Cut Gas Emissions." *Climate Art—Climate Change Portal*. Accessed May 13, 2005. Available at http://www.climateart:org/articles/readers.asp?linkid=39091.

CAN-EUROPE. (2005). "Who Is Still Missing?" *Climate Action Network—Europe*. Accessed May 16, 2005. Available at http://www.climnet.org/aboutcnc.htm.

Climate Action Network Europe. Table. Percentage Shares of Annex 1 Countries of 1990 Emissions. Available at http://www.climnet.org/EUenergy/ratification/1990sharestaggle.htm.

Dixon, Norm. (2004). "Global Warming: Is Kyoto Accord the Answer?" *Green Left Weekly*. Accessed May 13, 2005. Available at http://www.greenleft.org.au/back/2004/510/610p8.htm.

Dunn, S., and Flavin, C. (2002). Moving the Climate Change Agenda Forward. *State of the World 2002*. New York: Norton, pp. 24–50.

Karon, T. (2001). "When It Comes to Kyoto, the U.S. Is the 'Rogue Nation.'" *Time Online Edition,* July 24. Accessed May 13, 2005. Available at http://www.time.com/time/world/article/0,8599,168599,168701,00.html.

Pizer, W. A., and Kopp, R. J. (2003). Summary and Analysis of McCain-Lieberman—"Climate Stewardship Act of 2003." *Resources for the Future*. Accessed May 20, 2005. Available at http://www.rff.org/rff/News/Feat.../Understanding-the-MCCain-Lieberman-Stewardship-Act.cf.

Prugh, T. (2005). Russia Ratifies Kyoto Protocol. *World Watch Magazine* (January/February):18, 21.

QUESTIONS

1. What were the three principles used as the foundation of the UNFCCC treaty? Explain what you think were the reasons for those three principles.
2. What was the reason that the Kyoto Protocol developed from the UNFCCC treaty? Why do you think it might have been necessary?
3. Why did the United States withdraw from the Kyoto Protocol? Using numeric reasoning and other research, further investigate or evaluate what you think were the principal reasons for U.S. withdrawal. Do you think the United States was justified in withdrawing? Please justify your answer.
4. Do you think the IPCC report influenced the Kyoto Protocol? Research and explain your answer.
5. Why do you think Russia signed the Kyoto Protocol? Consider different scenarios.
6. Do you think the Kyoto Protocol has enough significance? What do you think should happen after 2012? Explain your answers.
7. What was the momentum for the Kyoto Accord to be ratified, even without the United States? List your answers.
8. Explain your reaction to the McCain-Lieberman Bill. Do you think it is an adequate approach for the United States in addressing their global greenhouse emissions? Support your answers.
9. Research the current status and modifications of the McCain-Lieberman Bill and other U.S. greenhouse emission bills being proposed.

Young at Risk: Dioxins and Other Hazardous Chemicals

EDITED BY BARBARA A. EICHLER

Children get 12 percent of their lifetime exposure to dioxin in their first year of life. On a daily basis, the infant is getting about 50 times the exposure an adult gets during what may be a critical developmental stage.
EPA Toxicologist Linda Birnbaum

During the eight years after an industrial dioxin (a group of chlorine-based chemicals from wastes of papermaking incineration of chlorinated plastics and other processes) pollutant explosion in Seveso, Italy, an unusual scarcity of male babies being born was noticed—twice as many girls were born as were boys—differing from the usual ratios of baby boys slightly outnumbering girls. In addition, excess cancers turned up among Seveso's adults. Clinical pathologist Polo Mocarelli theorized that the dioxin interfered with hormonal balances in developing embryos, either making normal male growth impossible or killing males. Such an effect of dioxin affecting sex ratios is well known in wildlife. For example, crossed bills in double-crested cormorants with the presence of dioxin in the Great Lakes region during the 1980s almost always occurred in females; scientists speculated that the males died before they hatched with this deformity (Monks 1997, 18).

Dioxins are a chemical byproduct of many common industrial processes; most (84 percent) is released as air pollution from waste incineration. Ironically, 53 percent of dioxin release comes from medical waste incineration due to the high level of plastic garbage accumulated by the medical industry. To add to this negative picture, the Environmental Protection Agency (EPA) says that the dioxins stored in the environment are 15–36 times that of known annual emissions, and the amount found on the ground from unidentified sources in the United States is two to five times that from identified sources (Mazza 1996, 2).

Dioxins and furans are a class of chemical compounds which are some of the most toxic chemicals ever made by humans. Both dioxins and furans (often just referred to as dioxins) have no useful purpose and are produced as the unwanted by-products of processes of industry in the manufacturing of PVC materials, incineration, pesticide production, paper bleaching with chlorine and more. They are one of the most lethal synthetic chemicals known and according to various health organizations are very potent as very small amounts can pollute and effect large amounts of animals and people.

Dioxins have already been established in animal studies as one of the most potent carcinogens; they have also been linked to human illnesses that affect almost every major body system including diabetes, bronchitis, irregular heartbeat, and nervous system and thyroid disorders. One of the most comprehensive studies of dioxins by the EPA in 1994 found that dioxins are more prevalent and dangerous in the population than previously reported. According to the EPA study, the average American has accumulated dioxins amounting to nine parts per trillion in his or her body. Studies have shown that dioxins begin to slow the action of the immune system at about seven parts per trillion, which leads to the conclusion that the dioxin insult to the body is already above the level that has been shown to cause harm (Mazza 1996, 1).

A growing body of scientific literature and observation strongly suggests that the young of most animals are far more susceptible to toxins (such as dioxins, polychlorinated biphenyls [PCBs], and other chemicals) than adults. "Children and animal young eat and breathe more for their body weights than adults do, so they get bigger proportional doses of whatever is out there," explains Herbert Needleman of the University of Pittsburgh, who pioneered studies linking lowered intelligence with early childhood exposures to lead (Monks 1997, 20). Significantly, in 1993, the National Academy of Sciences concluded that infants and children are not sufficiently protected by pesticide regulations since the risks have been calculated for adults. These various toxins not only cause cancer, but also affect the young's immune systems, brains, and reproductive organs. "It's important for us to realize that if we're seeing abnormalities in wildlife, similar mechanisms may exist in humans. We are just another species in the ecosystem; if other species are harmed, we may be too," says University of Florida zoologist Louis Guillette (Monks 1997, 20).

DEFORMED NEWBORNS

In Minamata, Japan, in the 1950s, before people understood the effects of industrial pollution, mercury discharges from a chemical plant poisoned the seafood that habitated the surrounding area and those who ate it. By the 1960s, evidence of the harm of the mercury began to mount from poisonings, for which animals died and people became sick. Fishermen noticed that seabirds were dying; feral cats that ate scavenged fish became stiff legged; cerebral palsy and mental retardation significantly increased in children, and adults were frequently ill. At the time of the Minamata poisonings, science held that the womb was a protected environment capable of filtering out harmful substances. However, in Japan, many women who ate the contaminated fish did not become ill themselves, but they gave birth to children with severe mental retardation and physical deformities. That incident changed scientists' thinking to a hypothesis that, in actuality, the womb was not protective from toxins. Instead, the fetus was sharing the mother's toxic load and, in a way, was actually protecting the mother since the fetus was absorbing some of the mercury, thus reducing the mother's exposure to it. The fetus received at least the same doses as did its mother—and the fetus was far more susceptible to toxic pollutants.

Not only are the young exposed to toxic chemicals in the womb, but mothers also unload toxins in their milk to the young. The milk from many species, ranging from beluga whales to dairy cows, has measurable concentrations of chemicals including dioxins, PCBs, and various pesticides. Children, in their first year of life, get 12 percent of their lifetime exposure to dioxin; on a daily basis during their critical developmental months, infants get about 50 times the exposure of an adult. Milk is still recommended for its antibodies, protection, and nourishment, and its benefits clearly outweigh its potential risks, but exposure of the young to toxins is increasing and is of growing concern (Monks 1997, 20).

Another study, which took place from 1987 to 1992 by the EPA, indicated strong evidence that lake trout embryos exposed to dioxin could develop a lethal syndrome called "blue sac." The yolk sac of young healthy trout is a rich golden color. During the first month or so, when the fry (baby fish) rely exclusively on the yolk for nutrients, they become vulnerable to blue sac syndrome, where fluid leaks out of the blood vessels and into the yolk sac, turning it milky and slightly blue. Many conditions can cause blue sac, but this study confirmed that certain dioxins, such as chemicals found in the Great Lakes at very small concentrations of just 60 parts per trillion (ppt), will cause 50 percent of lake trout fry to develop the blue sac disorder—making this fish the most vulnerable species known. This finding is in comparison to the same mortality rate of rainbow trout from the same chemicals at 400 ppt (Raloff 1997, 306–307).

DAMAGED IMMUNITY

Along the coast of Florida, bottlenose dolphins' firstborn calves die between the ages of three and six. In four generations of dolphins in the past 25 years, only one firstborn is known to have survived. Although the cause of the deaths is not certain, high levels of toxins exist in the fat of marine mammals. Research suggests that mother dolphins unload as much as 80 percent of their accumulation of pollutants into each of their calves, most likely through nursing. The firstborn calf receives the highest dose through the mother's accumulated toxins. The chemicals found in Florida dolphins' blubber are some of the most deadly and long-lived contaminants of the industrial age, including dioxins and PCBs (although banned, PCBs are still found in the insulation of electrical systems). The toxins are so persistent and widely distributed, moving into the food chain from the soil and water, that people and other animals continue to be exposed to them worldwide. "What we are seeing now is the impact of damage that was done over the last few decades," says biologist Randall Wells of the Chicago Zoological Society (Monks 1997, 22).

The evidence is that these and other toxic chemicals that are still being manufactured can interfere with the immune system. In 1987, about 700 bottlenose dolphins, half of the migrant Atlantic population, died and washed up along the Atlantic coastline of the United States from New

Jersey to Florida. Analysis determined that they were killed by infectious disease and that their bodies contained high levels of PCBs, dichlorodiphenyltrichloroethane (DDT), and other compounds that suppress the immune system—evidence that scientists think explains the dolphins' susceptibility to disease. If these chemicals are damaging immunity in adult dolphins, they may be doing even more harm to juveniles because mammalian immune systems are not fully functional until months or years after birth, according to immunologist Garet Lahvis of the University of Maryland School of Medicine (Monks 1997, 22).

Regarding the Inuits, an Eskimo people in the Canadian Arctic, researchers from Quebec are analyzing the relationship of unusually high rates of infectious disease among Inuit children and exposure to toxic chemicals. Even though no polluting industries operate near the region, contaminants enter the ecosystem from high-altitude winds and migrant wildlife. The contaminants accumulate in greater density with every link up the food chain as PCBs, pesticides, and other organochlorines progress from plant and fish to seal, whales, polar bears, and humans. Inuit babies in their first year of life have rates of infectious disease that are 20 times greater than those of babies in southern Quebec, and Inuit women have rates of PCBs in their breast milk that are seven times greater than those of women from the urban, industrialized south of Quebec. With the Inuits, acute ear infections are common, causing hearing loss for nearly one in four Inuit children, and the usual childhood immunizations do not work very well. In another study of the Inuits conducted in 1993, it was found that babies nursed by mothers with the highest contaminant levels in their milk were afflicted with more acute ear infections than were bottle-fed Inuit babies. The babies with the highest exposures to contaminants also produced few of the helper T cells that play an important role in eliminating bacteria and other harmful invaders. Even though there may be other factors causing these problems, data suggest that contaminants remain a significant factor. In the Netherlands, researchers have concluded that even infants with mild exposures to contaminants may experience weakened immunity; a correlation was discovered between PCB–dioxin exposure and suppressed levels of disease-fighting white blood cells that would cause immune system changes (although not extreme) that could persist throughout life or provoke autoimmune diseases.

In nonmammalian species, biologist Keith Grasman of Wright State University measured immune suppression that is mediated by T cells in young Caspian terns and herring gulls from contaminated colonies around the Great Lakes between 1992 and 1994. Grasman states that the same PCB and organochlorine pollutants found in these birds have also been measured in seals, dolphins, humans, and other species with similar T-cell immune problems. Many chicks with suppressed immune systems die before they leave the breeding grounds. Some contaminated terns do grow and migrate south, but most never return to breed (Monks 1997, 23–24).

LOWERED INTELLIGENCE

Since the metabolism of young animals is faster than that of adults and because young animals do not excrete contaminants or store them away in fat in the same manner that adults do, babies and young get continuous exposure to toxins at the time that all of their organs, including their brains, are still developing. In an adult, a blood–brain barrier guards the brain from potentially harmful chemicals in the body, but in a child, that barrier does not become fully developed until six months after birth.

The developing brains of the young of various wildlife species are also far more sensitive to toxic contaminants than are the brains of the adults of the species. As an example, in the late 1980s, great blue heron hatchlings from dioxin-contaminated colonies in Canada developed gross asymmetries and other abnormal changes in brain structures. The susceptibility of human young to such toxic effects was evidenced from an accidental PCB poisoning in Taiwan in 1979. In the Taichung province of Taiwan, more than 2,000 people were exposed to PCB-contaminated cooking oil, resulting in "Yu-Cheng," or oil disease. In the first three years after the accident, many newborns died, and others developed blotch patches of dark skin as well as fingernail and toenail deformities. As the children grew, they were mentally slower than other kids their age and displayed hyperactive and other behavioral problems. These developmental delays and IQ deficits have not gone away as the children have aged. The exposed mothers continued to deliver babies with problems as late as 1985, even though the accident occurred in 1979. As Walter Rogan of the National Institute of Environmental Health Sciences (NIEHS), who studied the case, explains it, a large portion of the PCBs that these women consumed ended up being stored in their fat, a process that happens with many toxic chemicals. During pregnancy, women metabolize a great deal of body fat, and the contaminants in the fat are also passed to their children—even years later, as in the case of Yu-Cheng. The same mechanism also applies to low-level toxic exposures from food, air, and water, for which even small amounts of pollutants—accumulated by women throughout their lives—can have lasting consequences for a child exposed to the pollutants in the womb.

A Michigan study found persistent intellectual deficits in children exposed before birth to much lower doses of PCBs than those in the case of Yu-Cheng. In 1981, two Wayne State University psychologists, Sandra and Joseph Jacobson, measured PCB levels in mothers and newborn infants. Since consumption of fatty fish from contaminated water is a major source of PCB exposure, the Jacobsons selected mothers for their study who had eaten Lake Michigan salmon or lake trout during the years before their children were born. The infants with the highest exposures grew more slowly than other babies and, at four years old, had poorer short-term memory. By the time they were 11 years old, the 30 most exposed children had average IQs six points lower than those of the least exposed group. Twenty-three percent of the high-exposure kids were two years behind in reading, while 10 percent of the least exposed group were two years behind. The Jacobsons also found that fish-free diets did not guarantee lower PCB levels since there were some very highly exposed children from mothers who did not eat the fish. The exposure might have come from other fatty foods, such as butter, cheese, beef, or pork, but there is no way of knowing the source since exposure to toxins is a societal problem. As Jacobson said, "We are all walking around with PCBs in us." (Monks 1997, 24).

SEXUAL IMPAIRMENT

Sexual development in the growing fetus may be as sensitive to toxic effects as the brain is. When certain chemicals bind to hormone receptors, they can interfere with the work of natural hormones in the development of male or

BOX 1 Dioxin's Effect on Fish Raises Questions of Effects on Higher Animals

Pollutants are considered dioxin-like if they connect to the Ah receptor in cells (a protein that reacts to these pollutants and turns genes on or off). The receptor was identified in fish in 1988 after being recognized in mammals in 1986. Since then, researchers have been examining other lower species to see how far down the evolutionary ladder this receptor exists as well as the accompanying vulnerability to dioxin. Scientists at Woods Hole Oceanographic Institution have been examining various types of animals for this Ah receptor and have so far found sharks to be the most primitive animal with the Ah receptor. (Sea lampreys have something that resembles the receptor, but this substance does not appear to bind to dioxin-like compounds.)

Scientists have also been using lower animals as a useful model of the common effects of dioxins on all animals. It has been found from a joint study between Cornell University and the University of Wisconsin-Madison that dioxin-like chemicals target the cardiovascular system of lower animals as they do in mammals. Using zebra fish, which have transparent embryos, scientists found that the pollutants slow the blood flow feeding the head and gills and also slow the heart's rate. Richard Peterson, of the University of Wisconsin-Madison, says that there appears to be a pruning of these blood vessels, which may account for the head malformations that often accompany blue sac syndrome. Dioxin-like compounds also appear to weaken blood vessels once they form, which may explain why blood vessels become leaky in blue sac.

The dioxin chemicals trigger early death in blood vessel cells caused by oxidant damage. This effect may trace to the ability of dioxins to turn on genes that increase the production of detoxifying enzymes, which then starts a process that releases oxidative compounds that do not respond to normal controls. The overproduction of oxidants can damage the vessels. Because there is no reason to suspect that this effect occurs only in fish, scientists are also searching for it in birds, reptiles, and mammals—including humans.

Source: Reprinted with permission from *Science News*, the Weekly Newsmagazine of Science, Copyright © 1996, 1997 by Science Service.
Excerpted from J. Raloff, Those Old Dioxin Blues, *Science News*, May 17, 1997;151:307.

female organs, resulting in any number of reproductive disorders. These chemicals, known as endocrine disrupters, include PCBs, dioxins, and many pesticides. The growing body of evidence suggests reason for concern about the effects of endocrine-disrupting chemicals found in the environment. Among the Yu-Cheng children of Taiwan, the boys with high PCB exposures had smaller than average penises. University of Florida biologists found the same phenomenon in alligators born in a lake poisoned by pesticides. In the highly polluted St. Lawrence River, biologists found a male beluga whale with a fully developed set of female organs in addition to the whale's male apparatus. This male carried a very high load of endocrine-disrupting contaminants in its blubber.

In South Florida, 13 of the 19 male panthers that still survive have undescended testicles. Because such males produce abnormal sperm and have low sperm counts, biologists are worried about the potential for saving the endangered animals from extinction. It is suspected that environmental endocrine-disrupting chemicals may be contributing to these cats' sexual abnormalities. The panthers are exposed to heavy doses of pesticides and toxic metals, such as methyl mercury from their diet of raccoons, which ingest the pollutants in fish. It is possible that most of the problems of the Florida panther could be attributed to pesticides. If this assertion is true, then the introduction of female Texas cougars to improve the panthers' genetic diversity may not accomplish much. If all of the panthers' habitat is contaminated, the animal may not be able to be saved.

According to a 1996 study by U.S. and European scientists, data from several countries show substantial increases since the 1950s in the number of baby boys born with undescended testicles and other sexual abnormalities. One London study found that 5.2 percent of low-birthweight boys born in the 1980s had undescended testicles compared with 1.74 percent from the 1950s. Testicular cancers nearly doubled among older teenagers in the United States between 1973 and 1992.

Many of the health effects documented in young wildlife from toxic contamination may not apply to human children, but wildlife can give indicators as to where to look for problems and answers. Deformed frogs in Minnesota may yield clues about the reasons for high rates of birth defects among the region's farm children. Links are being established everywhere between environmental toxic contaminants and the health of young wildlife and children. The warnings found in wildlife may help us to do something about environmental toxicities before we permanently contaminate and deform the animal world of which we are a representative part (Monks 1997, 25).

AN IMPRESSIVE SUCCESS STORY

Although toxic dioxins and dioxin-like compounds are pervasive and continue to increase around the world,

BOX 2 Success in Curbing Toxic Emissions

Significant Results of the Inventory of Sources of Dioxin in the U.S. (2005 External Review Draft) are listed below.

- 1987–2000: 89-percent reduction in the release of dioxin-like compounds to the circulating environment of the United States from all known sources combined.
- 2000–2005: 92-percent decline from all known sources combined.

a. In 1987 and 1995, the leading sources of U.S. dioxin emissions were municipal waste combustors.

b. Bleached chlorine pulp and paper mills were significant sources of dioxin to the aquatic environment in 1987, but a minor source in 1995 and 2000.

c. A major source of dioxin in 2000 was the uncontrolled burning of refuse in backyard burn barrels in rural areas of the United States.

Source: U.S. Environmental Protection Agency . (2005) The Inventory of Sources of Dioxin in the United States (External Draft 2005). EPA National Center for Environmental Assessment. Available at http://cfpub.epa.gov/hcea/cfm/recorddisplay.cfm?deid+132080.

BOX 3 The Industrial Tundra

1. Blubber samples from beluga whales off the north coast of Alaska contain unhealthy levels of toxaphene, an insecticide commonly used on cotton in the Caribbean and Central America.

3. Russian electrical equipment and metal refineries leak polychlorinated biphenyls into rivers, which then empty into the Svalbard region of Norway. Some polar bears there have both female and male sexual organs. PCBs are a prime suspect.

4. Global atomic testing has left small amounts of cesium-137, a radioactive isotope, in most human bodies. But some Scandinavian reindeer herders have 300 times that amount, the result of Russian nuclear tests and the 1986 Chernobyl accident.

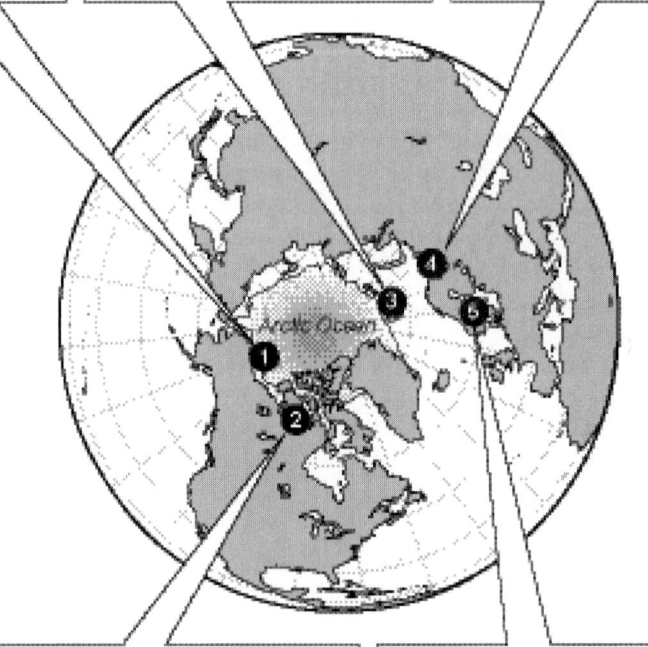

2. Umbilical cord studies reveal that one in five indigenous women throughout northern Canada have enough mercury in their blood to potentially harm their children's development.

5. Levels of polybrominated diphenyl ethers, flame retardants added to furniture padding and plastics, have increased 40-fold in the breast milk of Swedish women since 1972. Also found in moose, seals, and sperm whales, the chemicals could be toxic to infants' developing brains.

Tracking the Migration Patterns of Civilization's Waste

In the middle of the arctic winter, some 9 million square miles of ice may cover the Arctic O cean—explaining why the region has largely escaped 150 years of industrialization. That doesn't mean the Arctic is free of industrial *waste*, however. Today it is a veritable dumping ground for heavy metals and other contaminants, which travel thousands of miles from Russia, the United States, and Europe on northbound air and ocean currents. "The cold atmosphere just traps them," says Theo Colborn, director of the World Wildlife Fund's wildlife and contaminants program. From Alaska to Norway, industrial chemicals have been found in the bodies of arctic animals, people, and their offspring.

- Blubber samples from beluga whales off the north coast of Alaska contain unhealthy levels of toxaphene, an insecticide commonly used on cotton in the Caribbean and Central America.
- Umbilical cord studies reveal that one in five indigenous women throughout northern Canada have enough mercury in their blood to potentially harm their children's development.
- Russian electrical equipment and metal refineries leak PCBs into rivers, which then empty into the Svalbard region of Norway. Some polar bears there have both female and male sexual organs. PCBs are a prime suspect.
- Global atomic testing has left small amounts of cesium-137, a radioactive isotope, in most human bodies. But some Scandinavian reindeer herders have 300 times that amount—the result of Russian nuclear tests and the 1986 Chernobyl accident.
- Levels of polybrominated diphenyl ethers, flame retardants added to furniture padding and plastics, have increased 40-fold in the breast milk of Swedish women since 1972. Also found in moose, seals, and sperm whales, the chemicals could be toxic to infants' developing brains.

Source: Reprinted by permission of *OnEarth*, © 2003 by *OnEarth*, www.nrdc.org/onearth.

some success stories exist in reducing their presence and use in the environment. Dioxins are chemical contaminants that have no commercial usefulness in themselves. As stated, they are formed as byproducts of combustion and manufacturing processes from such simple things as waste incineration, backyard trash burning, herbicide use, and also herbicide and paper manufacturing (NIEHS 2002). Dioxins decompose very slowly in the environment, and their sediment is deposited on plants and ingested or absorbed by animals and aquatic life. Dioxins then concentrate up in the food chain so that animals (where it is stored in the fat) have higher concentrations than plants, water, soil, or sediments. In laboratory animals, dioxins are highly toxic, cause cancer, and alter various functions such as reproductive, developmental, and immunity capabilities. In humans, high levels lead to an increase in cancers, reproductive and developmental problems, increased heart disease, and diabetes. They also show an ability to cause birth defects, and it is possible that they can affect genes (by binding to Ah receptors that can bind to DNA and alter the gene) and produce congenital defects.

In our industrial and chemical age, there are increasing levels of dioxins and dioxin-like compounds around the world. In September 1994, The EPA described dioxin as a serious public health threat that may rival the impact that DDT had on public health in the 1960s. However, there have been significant reductions of dioxin-like compounds in the United States from 1987 to 2005. The EPA Inventory of Sources of Dioxin in the U.S. (2005 External Review Draft) shows that dioxin emissions from all quantified sources plummeted by 89 percent between 1987 and 2000 and were projected to decline by 92 percent by 2005! This can be attributed to (1) source-specific improved regulations, (2) improvements in source technologies, (3) advancement in those pollution control technologies specific to controlling dioxin discharges and releases, and (4) the voluntary actions of U.S. industries to reduce or prevent dioxin releases (EPA 2005). As dioxin emissions to the environment have declined, accordingly so have measured levels in sediments, foods, and humans tissue. These declines represent an impressive victory for industry, the EPA, and environmentalists whose efforts have made and hopefully will continue to make an impact through the use of government regulations, voluntary industry efforts, and citizen campaigns. Dioxins, however, remain in the environment in increasing levels, are easily produced, and seriously imperil our living environment and life on all levels. The problems resulting from dioxins and other highly hazardous chemicals are not yet solved, but through our efforts, they can be curtailed and hopefully controlled … even, perhaps, eliminated. (Please refer to Case Study 1 for more on Dioxins with Love Canal.)

BOX 4 Not for the Squeamish! Possible Low-Tech Solution for a High-Tech Problem

The Humble Earthworm May Hold the Key to Removing One of Our Most Deadly Environmental Toxins: PCBs

Charles Darwin admired the earthworm extravagantly. "It may be doubted," he wrote in 1881, "if there are any other animals which have played such an important part in the history of the world as these lowly organized creatures." The earthworm is a natural organic chemist, cultivator, and fertilizer. But it may have yet another talent, one that Darwin would never have discovered: toxic cleanup specialist. It turns out that PCBs—among the nastiest of modern pollutants—may be no match for the humble earthworm.

In addition to their toxicity (the Environmental Protection Agency has designated them a probable human carcinogen), polychlorinated biphenyls are notorious for two reasons. First, they "bioaccumulate." In other words, their concentration in tissue increases dramatically as they move up the food chain. Second, they are virtually indestructible.

For half a century until their use was banned in 1979, PCBs were regarded as a miracle of the industrial age. With their stability, low volatility, and extreme fire resistance, they made wonderful insulators for electrical transformers and capacitors. Their largest user was General Electric, whose plants in New York State and Pittsfield, Massachusetts, discharged massive amounts of PCBs into the Hudson and Housatonic Rivers. A series of consent decrees in the 1990s obliged G.E. to spend hundreds of millions of dollars to clean up the mess it had left behind. But the settlements left a large question unanswered: How was this to be accomplished?

There are two basic alternatives for disposing of PCBs. One is incineration, the other dumping in hazardous-waste landfills. Both methods are expensive and controversial. At the G.E. facility in Pittsfield, for example, one of the largest landfills used for PCB-contaminated soil is Hill 78, which looms high above the Allendale Elementary School.

Tim Gray, who was trained as a chemist and biologist before becoming executive director of the Housatonic River Initiative, has been battling G.E. for years. Sites like Hill 78 keep him awake at night.

So when Gray learned of the work of a Slovak scientist named Oto Sova, he was excited. Sova was using the Californian earthworm, *Eisenia foetida*—or ground-up earthworm enzymes, to be precise, in a solution called Enzymmix—to remediate soils contaminated with petroleum derivatives or, most ambitiously, toxins such as phenols, cresols, DDT, and even PCBs. Better yet, Enzymmix promised to work in situ, either by being sprayed directly on contaminated surfaces or by being injected into the ground through plastic tubes.

Sova's remedy has been successfully tested at gas stations, on airport runways, and at other contaminated sites in five European countries. "I didn't know if this was for real or if it was too good to be true," Gray says. So he went to Europe to observe the enzymes in action. He was impressed enough with what he saw to take soil samples from G.E.'s Pittsfield site to be tested at the State University of New York at Albany and at HydroTechnologies, a private environmental laboratory in New Milford, Connecticut.

HydroTechnologies' lab director, Lawrence Paetsch, tested two different soil samples, with PCB levels of 250 and 450 parts per million (ppm), respectively. (The EPA considers anything above 50 ppm to be hazardous.) Paetsch says that a single application of the enzymes "showed a significant reduction in PCB concentration: on average, 62 percent of the original PCB content was removed." A second round of tests showed 76 percent removal.

If Sova's earthworm cocktail is to be approved for use in the United States, the next step is for it to be tested by the EPA's Superfund Innovative Technologies Evaluation Program (SITE). Tim Gray is optimistic. "When I went to Europe," he says, "I was about 1 percent convinced. Now I'd say I'm 80 percent there."

Source: Reprinted by permission of *OnEarth*, © 2003 by *OnEarth*, www.nrdc.org/onearth.

Table 1
Superfund Hazardous Waste Sites, 1981–2004

Date	*Number of Sites*	*Date*	*Number of Sites*
Oct. 23, 1981[1]	115	Feb. 23, 1994	1,190
Dec. 30, 1982[1]	418	Sept. 29, 1995	1,232
Sept. 8, 1983	406	June 17, 1996	1,227
Oct. 15, 1984	538	April, 1997	1,208
June 10, 1986	703	Feb., 1998	1,197
July 22, 1987	802	July 22, 1999	1,226
Oct. 4, 1989	981	July 31, 2000	1,236
Feb. 21, 1990	1,081	June 14, 2001	1,236
Feb. 11, 1991	1,189	March, 2002	1,223
Feb., 1992	1,183	Sept., 2003	1,233
May 10, 1993	1,201	Sept. 23, 2004	1,244

[1] Proposed sites only. Final sites not calculated until release of first National Priorities list in 1983.
Source: Environmental Protection Agency, National Priorities List, http://www.epa.gov/superfunds/sites.

Table 2
Hazardous Waste Sites in the United States, 2004

	Total Proposed		*Total Final*				*Total Proposed*		*Total Final*	
State/Territory	*Gen*	*Fed*	*Gen*	*Fed*	*Total*	*State/Territory*	*Gen*	*Fed*	*Gen*	*Fed*
Alabama	2	0	10	3	15	Nevada	0	0	1	0
Alaska	0	0	1	5	6	New Hampshire	1	0	18	1
Arizona	0	0	7	2	9	New Jersey	3	0	104	8
Arkansas	0	0	11	0	11	New Mexico	1	0	11	1
California	3	0	72	24	99	New York	2	0	86	4
Colorado	2	0	13	3	18	North Carolina	1	0	27	2
Connecticut	1	0	14	1	16	North Dakota	0	0	0	0
Delaware	0	0	13	1	14	Ohio	6	2	26	3
District of Columbia	0	0	0	1	1	Oklahoma	1	0	9	1
Florida	1	0	45	6	52	Oregon	0	0	9	2
Georgia	1	0	12	2	15	Pennsylvania	4	0	87	6
Hawaii	0	0	1	2	3	Rhode Island	0	0	10	2
Idaho	3	0	4	2	9	South Carolina	1	0	23	2
Illinois	5	1	36	4	46	South Dakota	0	0	1	1
Indiana	1	0	29	0	30	Tennessee	1	1	9	3
Iowa	1	0	12	1	14	Texas	2	0	38	4
Kansas	1	1	9	1	12	Utah	4	0	11	4

Table 2 (*Continued*)

	Total Proposed		*Total Final*				*Total Proposed*		*Total Final*	
State/Territory	*Gen*	*Fed*	*Gen*	*Fed*	*Total*	*State/Territory*	*Gen*	*Fed*	*Gen*	*Fed*
Kentucky	0	0	13	1	14	Vermont	1	0	10	0
Louisiana	3	0	12	1	16	Virginia	0	0	19	11
Maine	0	0	9	3	12	Washington	0	0	33	14
Maryland	1	0	9	9	19	West Virginia	0	0	7	2
Massachusetts	1	0	24	7	32	Wisconsin	1	0	38	0
Michigan	1	1	67	0	69	Wyoming	0	0	1	1
Minnesota	0	0	22	2	24	Guam	0	0	1	1
Mississippi	2	0	3	0	5	Puerto Rico	1	1	10	0
Missouri	0	0	23	3	26	Virgin Islands	0	0	2	0
Montana	1	0	14	0	15					
Nebraska	1	0	10	1	12	Total	61	7	1,086	158

Gen = general Superfund sites; Fed = federal facility sites.
Source: U.S. Environmental Protection Agency, *National Priorities List,* Sept. 2004.

Table 3
World Hazardous Waste
The following table contains industrial and hazardous waste generation in member states of the Organisation for Economic Co-operation and Development (OECD), eastern Europe, and the rest of the world in the late 1980s.

Region	*Hazardous and Special Wastes (million metric tons per year)*	*Industrial Wastes (million metric tons per year)*
World	338	2100
OECD	303	1430
North America	278*	821
Europe	24	272
Pacific	<1	333
Eastern Europe	19	520
Rest of the world	16	180

*The value of the U.S. (275 million metric tons per year) that was used to derive the regional total for North America includes liquid wastes that are classified as hazardous.
Source: Encyclopedia of Global Change, 2002. New York: Oxford University Press.

Table 4
Toxic Release Inventory, United States, 2001–2002
This table documents releases of toxic chemicals into the environment by manner of release and industry sector; pollutant transfers are entered by the destination of transfer. Totals below may not add because of rounding.

Pollutant Releases	*2001 (mil lb)*	*2002 (mil lb)*	*Top Industries, Total Releases*	*2001 (%)*	*2002 (%)*
Air releases	1,657	1,632	Metal mining	46	26
Surface water discharges	230	230	Electric utilities	17	23
Underground injection	216	222	Primary metals	9	16
On-site land releases	2,953	2,195	Chemicals	9	12
Off-site releases	558	514	Hazardous waste/solvent recovery	4	4
TOTAL on- and off-site releases	5,612	4,792	Paper	3	4
Pollutant transfers			All others	12	15
To recycling	1,695	1,986	*Top Carcinogens*, Air/Water/Land Releases	(mil lb)	(mil lb)
To energy recovery	840	804	Styrene	50	50
To treatment	280	276	Formaldehyde	21	19
To publicly owned treatment works	340	347	Acetaldehyde	13	14
Other transfers	2	1	Dichloromethane	22	12
Off-site to disposal	600	596	Trichloroethylene	9	8
TOTAL	3,757	4,010	Ethylbenzene	8	7

Note: This information does not indicate whether (or to what degree) the public has been exposed to toxic chemicals.
Source: U.S. Environmental Protection Agency.

Table 5
Common Hazardous Chemicals Requiring National Response Center Notification (partial list)

Substance	*Remarks*	*Amount to Be Reported if Released*
Acetic acid	Vinegar is generally 2 percent acetic acid, but acetic acidis used in many manufacturing processes. Toxic as vapor at 10 ppm in air.	5,000 lbs. (2,270 kg)
Acetone	Toxic chemical (1,000 ppm in air) used in large amounts as solvent for resins and fats.	5,000 lbs. (2,270 kg)
Aluminum sulfate	Sometimes used in dyeing or in foam fire extinguishers.	1,000 lbs. (454 kg)
Ammonia	Use as a fertilizer does not need to be reported; however, it is toxic and extremely hazardous; emergency planning required if 500 lbs. possessed.	100 lbs. (45.4 kg)
Benzene	Used in drugs, dyes, explosives, plastics, detergents, and paint remover; can cause cancer; toxic.	10 lbs. (4.54 kg)
Chlorine	Widely used to disinfect water, the gas is toxic at concentrations of 1 ppm in air; extremely hazardous; requires emergency planning if 100 lbs. is possessed.	10 lbs. (4.54 kg)

Table 5 (*Continued*)

Substance	*Remarks*	*Amount to Be Reported if Released*
Cumene	Additive for high-octane fuels; toxic to skin at 50 ppm in air.	5,000 lbs. (2,270 kg)
Cyclohexane	Petroleum derivative.	1,000 lbs. (454 kg)
Ethylbenzene	Toxic at 100 ppm in air.	1,000 lbs. (454 kg)
Ethylene dichloride	Additive to gasoline that combines with lead to make "ethyl" gasoline; also used in making plastics.	100 lbs. (45.4 kg)
Ethylene oxide	Widely used in making plastics; toxic at 50 ppm in air; extremely hazardous; requires emergency planning if 1,000 lbs. possessed.	1–10 lbs. (0.454–4.54 kg)
Formaldehyde	Used in wood substitutes and plastics; toxic and may cause cancer; extremely hazardous; requires emergency planning if 500 lbs. possessed.	100 lbs. (45.4 kg)
Hydrochloric acid	Used in petroleum, manufacturing, and metals industries; toxic; as the gas hydrogen chloride, it is extremely hazardous; requires emergency planning if 500 lbs. possessed.	5,000 lbs. (2,270 kg)
Methanol	Commonly called wood alcohol; used as antifreeze, solvent, and starting material for other compounds; toxic.	5,000 lbs. (2,270 kg)
Nitric acid	Used in preparing fertilizers and explosives; toxic and extremely hazardous; requires emergency planning if 1,000 lbs. possessed.	1,000 lbs. (454 kg)
Phenol	Used in making plastics; vapor is toxic to skin at 5 ppm in air; extremely hazardous; requires emergency planning if 500 lbs. possessed and further planning if 10,000 lbs. possessed.	1,000 lbs. (454 kg)
Phosphoric acid	Used as a flavoring agent, in pharmaceuticals, and in manufacturing fertilizers; toxic.	5,000 lbs. (2,270 kg)
Sodium hydroxide	Commonly known as lye or caustic soda; toxic.	1,000 lbs. (454 kg)
Styrene	Used in manufacture of styrene plastics and artificial rubber; toxic.	1,000 lbs. (454 kg)
Sulfuric acid	Most common chemical used in U.S; toxic and extremely hazardous; emergency planning required if 1,000 lbs. possessed.	1,000 lbs. (454 kg)
Toluene	Used in making explosives, drugs, and dyes; toxic.	1,000 lbs. (454 kg)
Vinyl chloride	Used to make plastics and aerosols; causes cancer; toxic.	1 lb. (0.454 kg)
Xylene	Used to make other compounds; toxic.	1,000 lbs. (454 kg)

ppm = parts per million.
Source: U.S. Environmental Protection Agency, *Title 3 List of Lists*, January 1990. (Note: Although this list is from 1990 – it remains a valuable guideline of common chemicals and their dangerous release amounts.)

REFERENCES

Environmental Protection Agency. (2005). The Inventory of Sources of Dioxin in the United States (External Review Draft 2005). EPA National Center for Environmental Assessment. Available at http://cfpub.epa.gov/ncea/cfm/recorddisplay.cfm?deid=132080.

Mazza, P. (1996). "Love Canal Is Everywhere: The Pervasive Threat of Dioxin." February 21. Available at http://www.tnews.com/test /dioxin.html.

Monks, V. (1997). Children at Risk. *National Wildlife* 5 June/July (4):18–27.

National Institute of Environmental Health Sciences (NIEHS). (2002) Dioxin Research at the National Institute of Environmental Health Sciences. Accessed June 11, 2005. Available at http://www.niehs.nih.gov/oc/factsheets/dioxin.htm.

Raloff, J. (1997). Those Old Dioxin Blues. *Science News* 151 (17 May):306–307.

QUESTIONS

1. Why are the young of most animals more susceptible to toxins than adults?
2. Why do you think infants and children are not sufficiently protected by pesticide regulations? How do you think we can use pesticide regulations to better protect infants and children? Explain some specific ideas.
3. What does the case study in Minamata, Japan, teach us about the effects of toxic pollution on fetuses? How are infants doubly exposed to pollutants?
4. What does the study regarding blue sac syndrome in lake trout embryos teach us about dioxins?
5. Give three examples of damaged immunity caused by dioxin and PCB contamination.
6. Give two examples of lowered intelligence caused by toxic pollution. Why are the young particularly susceptible to toxins affecting intelligence? Do you think that the examples in the article can support the claim that such toxins affect intelligence in humans? Explain.
7. Give two examples of sexual impairment caused by toxic pollution. Do you think that the examples in the article can support the claim that such toxins affect sexual impairment in animals and humans? Explain.
8. What is the reason that animals should be so closely watched for their reactions to chemical toxins? Why is the issue of the young such a special case?
9. How can evidence of the effects of chemical toxins, especially on the young, be made more apparent to determine clearer relationships and, hence, approaches to prevention? Outline appropriate studies that could be carried out to this end.
10. How can such chemical toxic pollution be prevented? What worldwide plan would realistically prevent such pollution and casual dumping?

QUESTIONS ABOUT THE TABLES

1. Examine Tables 1, 2, and 3. Draw numeric and locale conclusions. What do you conclude from this data on hazardous waste sites?
2. Read the list of chemicals in Table 5. How often do you use any of these chemicals? How do you dispose of them? Were you aware of their classification as hazardous before examining Table 5?
3. Are you aware of the health hazards of the chemicals listed in Table 5? Please list the health hazards of which you are aware for any of the chemicals mentioned. Share this knowledge in discussion.
4. Discuss approaches to chemical dumping and chemical disposal.
5. Research superfund hazardous waste site dumps. What progress is being made? How much money is being spent?
6. How should we dispose of hazardous wastes? Discuss some plausible methods.

Fisheries: Exploiting the Ocean—What Will Be Left?

EDITED BY BARBARA A. EICHLER

Developing countries hire private companies to conduct surveillance and enforcement of new fisheries laws. These companies are zealous: One has proposed to watch over fisheries from a blimp, which could descend to launch a patrol boat.

MICHAEL PARFIT

Technology has applied its increasing power to the fish of the ocean . . . and there is now trouble at sea. "There are too many fishermen and not enough fish" (Parfit 1995, 9). Fifty years of rapidly improving fishing technology has created an immensely powerful industrial fleet—37,000 freezer trawlers that catch and process a ton or more of fish an hour, manned by about a million people worldwide. This fleet contrasts with small-boat fishermen, who probably number about 12 million, but who catch only about half the world's fish.

The problem is that fish stocks are being damaged by pollution, by destruction of wetlands that serve as nurseries and provide food, by the waste of unprofitable fish (called "bycatch"), and, most of all, by overfishing. These practices have caused the collapse of some fish stocks and the fishing of many important groups of fish beyond sustainable capacity. As of 2002 (the most recent data available), the world's fishers harvested 133 million tons of fish and shellfish from streams, oceans, and other bodies of water. "This record catch was about 2 percent more than in 2001 and nearly seven times the global harvest in 1950. Over the same period, the amount of fish harvested per person tripled to 21 kilograms per year." (See Figures 1 and 2.) Fishers from the developing world catch about three of four wild fish (by weight), and they also eat most of the world's fish, although they consume less per capita: "The developing world consumes an average of 14.2 kilograms a year compared with 24 kilograms in the industrial world. For nearly 1 billion people, mostly in Asia, fish supply 30 percent of protein: worldwide, the figure is just 6 percent (Halweil 2005, 26).

The global cost of this fishing intensity will be felt with future shortages of certain types of fish and in sustainable yields. Although technology has helped to quadruple the world's catch of seafood since 1950, a nearly empty basket is typical for what a lone fisherman has to show for hours of work—a complaint heard around the world. Morocco and most small coastal or island countries are extremely worried about overharvesting and want to reduce fish quotas taken in their waters by foreign fleets. One Moroccan official asserts, "People using traditional techniques will not survive. The waters are being emptied by industrial fishing." One marine scientist says, "We've come to our reckoning, the next ten years are going to be very painful, full of upheaval for everyone connected to the sea." (Parfit 1995, 2–11).

According to the U.N. Food and Agriculture Organization (FAO), almost two-thirds of the world's 200 commercially important distinct fish populations are either exploited or fished to the edge, and another 10 percent have been harvested so heavily that it will take years for fish populations to recover. "In 2004, marine scientists estimated that industrial fleets have fished out at least 90 percent of all large ocean predators—tuna, marlin, swordfish, sharks, and flounder—in just the past 50 years!" (Halweil 2005, 26). The FAO states that there is an urgent need for the development of effective measures to reduce and control fishing capacity and effort. The visual images of overfishing are very apparent when

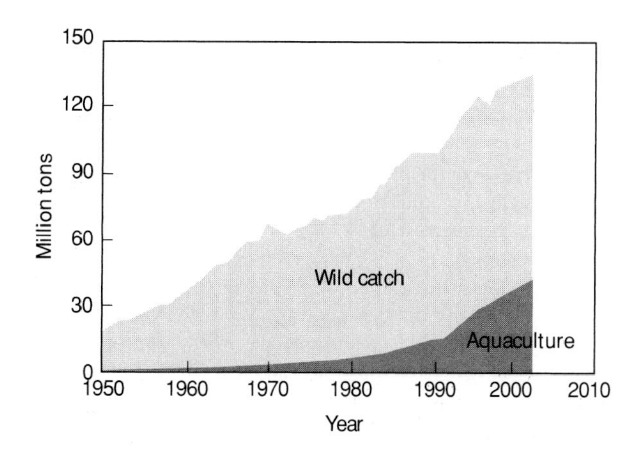

Figure 1
World fish harvest, 1950–2002.
Source:Worldwatch Institute. (2005). *Vital Signs.* New York: W.W. Norton & Co.

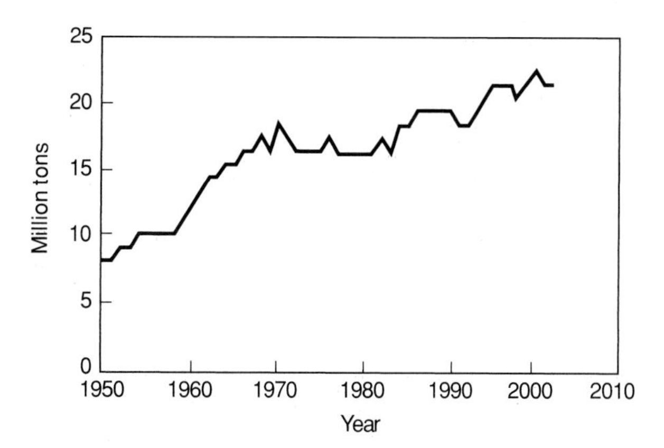

Figure 2
World fish harvest per person, 1950–2002.
Source: Worldwatch Institute. (2005). Vital Signs. New York: W.W. Norton & Co

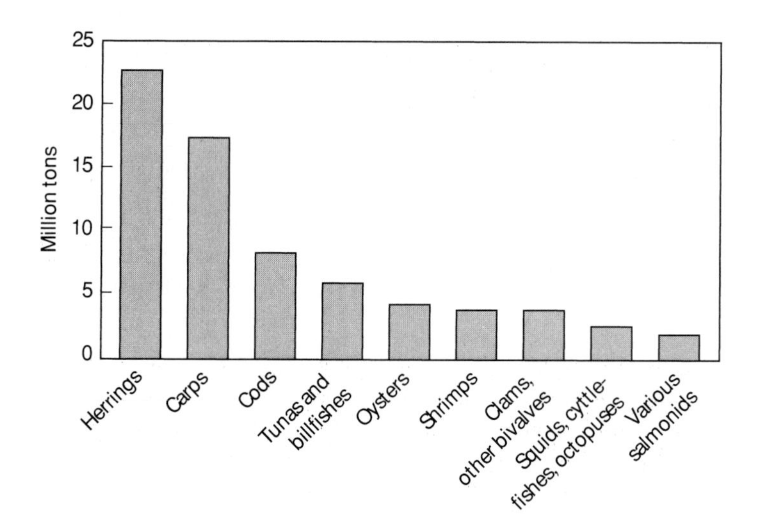

Figure 3
Top fish species harvested worldwide, 2002.

- trawlers line up abreast to sweep the life from the sea with nets so large that each could drag up a dozen jumbo jets;
- vessels called "long-liners" trail out thousands of baited hooks on lines stretching 80 miles across the ocean; and
- Japan's squid fleet is so enormous that its lights to draw squid to its nets in the north Pacific can be seen by orbiting astronauts.

Within two decades, giant vessels have combined the latest technology, including satellite navigation, spotter planes, fish-scanning sonar, and lightweight nets, to capture the "limitless" bounty of the seas into finite resources. Big trawling vessels can pull up at least 20,000 pounds of fish in 20 minutes (Hanley 1997, 1–2).

The problem is not easily resolvable, and it could grow into a catastrophe. Fishing is a 70 billion-dollar-a-year industry with strong roots in national pride, profits, and age-old traditions of freedom. As governments struggle to solve the problems at sea, they inevitably create laws that challenge traditional freedoms. Throughout the world's seas, fishing vessels are attacked in competitive territorial battles and ownership of waters. The following are common examples.

- In Patagonia, an Argentinian gunboat chases and fires on a vessel from Taiwan; the crew is rescued, but the trawler sinks.
- In the North Atlantic, the Stern trawler REX is arrested west of Scotland for trespassing in British waters. REX is officially multinational to evade fishing laws and is owned by Icelanders, registered in Cyprus, and crewed by fishermen from the Faroe Islands.
- In the South Atlantic, a patrol boat from the Falkland Islands chases a Taiwanese squid boat 4,364 nautical miles from home waters, all the way past South Africa. The boat gets away.

An increasing aggressiveness of the law on the free oceans began after World War II such that, by the 1970s, most nations pushed their territorial control from 12 to 200 nautical miles offshore to grab valuable fishing grounds, thereby pushing the boats of other nations (the "distant-water fleets") far out to sea. Now, 200 nautical miles is not enough. Many fish roam from national to international waters, where they are taken by intense fishing methods outside of any nation's control. Therefore, those nations whose roaming salmon, cod, or pollock are caught before returning to their home waters fight with those nations that intercept them. Increasingly, more countries are taking the fight to sea. In Canada, the Great Banks of Newfoundland were in danger of fishing collapse, necessitating that Canada shut down its own fishery there, thus putting about 40,000 people out of work. However, distant-water trawlers from Spain, Portugal, and other nations continued to fish just outside the 200-mile limit, making Canada very angry and distrustful. Typical strong offensive responses from Canada or other nations in this situation included strategies of using spy planes to record any suspicious ship's activities, shooting at poachers, implementing blockades, employing patrol boats, and making arrests.

Such fishing-rights arguments have made the oceans a sort of oceanic Wild West, with one state acting as the self-appointed lawmaker, as well as sheriff and judge. In actuality, except for the past few decades, most of the sea was indeed like the Old West: free, wild, unregulated, and a place of opportunity for any brave enough to venture forth. But declining populations of fish and feuding fishermen, along with the power of technology, are proving that even the limits of the grandest piece of the planet—our seas—have been reached (Parfit 1995, 10–20).

HIGH-TECH FISHING VESSELS AND METHODS

The number of fishing vessels has doubled since 1970 to more than 1 million. Not only has the number doubled, but also the types of fishing vessels used have radically changed. Now in use are super-efficient floating ship factories where all fish from within the ship's enormous fishing path are processed onboard and where all products are used—nothing is wasted. A typical new factory ship is 376 feet long and operates both as a mothership for fleets of smaller vessels and also catches its own fish, processing and packaging fish at sea as it catches them. The typical "technological kit" of such ships includes a variety of gigantic nets (some large enough to swallow the Statue of Liberty), highly powerful lights that act as bait, literally miles of fishing lines that dangle lures or baited hooks, sonar arrays and computers to locate and track the fish, and sensors on the net. The fish are not so much strained from the sea by the nets, but are instead herded by the net's cables and winch-like leading edges. A typical factory ship is capable of processing more than 600 metric tons of pollock a day into surimi, the protein paste used in imitation seafood products. All processing occurs onboard, including the tasks of processing into various products, quick freezing, and packaging. To keep morale high among the 125 crew members of such ships, living decks on the ships are

Table 1
World fish catch and aquaculture, 1950–2002.

Year	*Catch (million tons)*	*Aquaculture (million tons)*
1950	19	0.6
1955	27	1.2
1960	34	1.7
1965	48	2.0
1970	63	2.6
1971	63	2.7
1972	59	3.0
1973	59	3.1
1974	62	3.6
1975	62	3.6
1976	65	3.7
1977	64	4.1
1978	66	4.2
1979	66	4.3
1980	67	4.7
1981	69	5.2
1982	71	5.7
1983	71	6.2
1984	77	6.9
1985	78	8.0
1986	84	9.2
1987	84	10.6
1988	88	11.7
1989	88	12.3
1990	85	13.1
1991	84	13.7
1992	85	15.4
1993	87	17.8
1994	92	20.8
1995	92	24.4
1996	94	26.7
1997	94	28.7
1998	88	30.6
1999	94	33.4
2000	96	35.5
2001	93	37.8
2002	93	39.8

Source: Worldwatch. (2005) *Vital Signs 2005.*.New York: Norton & Co.

separated from work decks and typically can include well decorated cafeterias, gymnasiums, bathrooms with Japanese soaking tubs, and televisions in most of the crew's cabins (Parfit 1995, 16–17).

Other new technologies could help restore depleted fisheries and habitats or could also exploit them. Multibeam sonar systems can send dozens and even hundreds of beams of sound simultaneously to the ocean bottom instead of sending only a single beam of sound, as is found in most boats. This sonar system allows mappers to construct 3-D maps of the seafloor with incredible detail to identify schools of fish and differences in habitats that attract certain fish. Such multibeam sonar systems could be used to zone the ocean floors to protect sensitive habitats and monitor fish populations (Woodard 2003).

AQUACULTURE BEGINS TO AUGMENT WILD FISHING

Due to the astounding depletion of natural fish stocks, almost all growth in the global catch is coming from aquaculture or farmed fish. Aquaculture harvests have doubled in the last decade to 39.8 million tons and accounts for approximately 30 percent of the global fish harvest. It is projected that, by 2020, aquaculture could produce nearly half of all fish caught. China raises 70 percent of the world's aquaculture harvest, which already accounts for about two-thirds of total fish production. Some species, such as salmon, are farm harvested more than they are wild harvested (Halweil 2005, 26).

The Diminishing Catch Nearly 40 percent of the world's oceans have been locked up by territorial claims and exclusive fishing zones. The 200-nautical-mile coastal boundaries have not stopped the overexploitation of species. Rich nations buy into poor countries' waters. High-tech fleets grab fish migrating outside protected coastal zones, and the fish just keep getting scarcer and smaller. After peaking in 1989 and despite a temporary rise in the Pacific Ocean catch, the tonnage of seafood harvested worldwide has reached a foreboding plateau. Meanwhile, demand continues to grow, with an annual per-capita consumption of 66.6 kilograms (147 pounds). Japan has the world's biggest appetite; China is gaining rapidly in bulk catches. Please refer to Table 2. Some of the species being caught include the following.

Salmon: Hydropower, pollution, and logging have diminished salmon-spawning streams worldwide. The North Pacific waters remain the healthiest waters for salmon.

Shrimp: About one-fourth of all shrimp consumed are raised in man-made ponds. Viruses recently devastated farmed stocks in China, Ecuador, and Texas.

Herring, sardines, anchovies: These small open-ocean fish are used mainly for industrial purposes, such as the production of fish meal and fertilizer (although they remain a delicacy in many countries). Herring has been heavily fished in both the Atlantic and Pacific Oceans. Anchovies are caught mainly off the Pacific coast of South America, where populations fluctuate dramatically depending on the El Niño phenomenon.

Cod, pollock, haddock: These fish are the mainstays of human consumption and have been heavily exploited. The cod fishery in the northwest Atlantic recently collapsed and, in Newfoundland, it completely disappeared. Pollock is still plentiful, and much of it is processed into fish sticks and fast food. Cod, haddock, and flounder stocks have collapsed in New England.

Jacks, sauries, capelins: Used both for eating and industrial purposes, these fish are prey for larger, more favored fish, such as cod or tuna.

Redfish, seabreams, roughies: When more easily accessible fish stocks run out, fishermen turn to alternative species. Redfish, a restaurant basic, have been fished commercially since the 1930s. Orange roughy fishing began only in the 1970s. The danger is that little is known about the biology of these alternative food fish. Some scientists think that orange roughies may live longer than humans. If this is true, then orange roughies will need ample time to replenish from intense fishing.

Tuna, billfish: These fish are the glamour fish of the industry and command luxury prices. A two-bite portion of sushi made from prime meat from the bluefin tuna can cost $75 in Japan. Large fish are hunted with harpoons and spotter planes (Parfit 1995, 12–13).

Our seas have changed, and the unlimited bounty of the sea is endangered. This does not mean that the sea is ruined, however. It is more like a forest than a mine in that it will keep producing as long as we do not plunder it without restraint, and that is just what we do not know how to do yet. The Dust Bowl did not kill American agriculture; it just changed it into a big industry—highly regulated and tidy. Fishing may go the same route. Fish farming has produced the only productive growth in recent years and will continue to grow, as will the regulation of the sea itself. We

Table 2

Commercial Catch of Fish, Crustaceans, and Mollusks for 20 Leading Countries, 1997–2002[1] (thousands of metric tons; live weight; ranked for 2002)

Country	*2002*	*2001*	*2000*	*1999*	*1998*	*1997*	*Country*	*2002*	*2001*	*2000*	*1999*	*1998*	*1997*
China	27,767	26,050	24,581	22,790	20,795	19,316	Phillipines	443	435	394	353	313	327
India	2,192	2,120	1,942	2,135	1,908	1,864	Egypt	376	343	340	226	139	86
Indonesia	914	864	789	749	630	623	Taiwan	330	297	244	248	240	258
Japan	828	802	763	759	767	807	S. Korea	297	294	293	304	327	392
Bangladesh	787	713	657	593	575	486	Spain	264	313	312	321	315	239
Thailand	645	724	738	694	595	540	France	250	252	267	265	268	287
Norway	554	511	491	476	411	368	Brazil	246	208	177	141	104	88
Chile	546	566	392	274	293	272	Italy	184	218	214	207	206	191
Vietnam	519	519	511	467	413	405	U.K.	179	171	152	155	137	130
U.S.[2]	497	479	456	479	445	438	Canada	172	153	128	113	91	82

[1] Includes aquaculture.

[2] Includes weight of clam, oyster, scallop, and other mollusk shells. This weight is not included in U.S. landings statistics.

Source: U.N. Food and Agriculture Organization (FAO).

BOX 1 The Seas . . . Now and Then . . .

When my father began diving in the Mediterranean Sea in the early forties, the water was clean. Great beds of seagrass and algae thrived there, along with dense schools of fish and rich invertebrate fauna. Gorgonians were abundant, and so were huge groupers and spiny lobsters. It was the rich sea-floor community the world was seeing in the early Cousteau films.

Since those days, the Mediterranean coast has become densely populated. Industries, hotels, and homes line the coast. Sewage and other wastes stream into the sea. Sadly, the same waters where the first Aqua-lung divers discovered the sea's beauty and diversity are today biologically impoverished. And this scene—where urban development meets the water—is spreading rapidly around the world today.

Jean-Michel Cousteau (1990)

The very survival of the human species depends upon the maintenance of an ocean clean and alive, spreading all around the world. The ocean is our planet's life belt.

Marine Explorer Jacques-Yves Cousteau (1980)

will still have fish, but not the fish or traditions or industries that we have previously known.

SOME FACTS

- The world fishing fleet has doubled in size over the past two decades and now includes 37,000 "industrial" vessels of more than 100 tons. U.S. and other government subsidies encouraged this growth.
- The ocean catch has exploded from 18 million tons in 1950 to 93 millions tons in recent years. Fish stocks began to decline in the 1990s, and production could no longer increase at the same rates. If the same fishing practices continue, the average marine catch could decline by 10 million tons a year.
- In 2004, marine scientists estimated that industrial fleets have fished out at least 90 percent of all large ocean predators—tuna, marlin, swordfish, sharks, cod, halibut, skates and flounder—in just the past 50 years (Halweil 2005, 26).
- The impact on seafood prices has been moderate due to aquaculture of new fish, shrimp, and scallop farms in China, Thailand, and elsewhere.
- Fishing jobs directly or indirectly employ about 200 million people in the world. There are an estimated 3 million fishing vessels worldwide made up of mostly small vessels in developing countries. The 3 million vessels include the supertrawlers of 37,000 ships of more than 100 tons. These ships caught 90.7 million tons of marine and inland fish in 1995. Aquaculture produced 21.3 million tons in 1995 (Hanley 1997, 1–2).

THE MAGNUSON FISHERY CONSERVATION AND MANAGEMENT ACT

Enactment of the Magnuson Fishery Conservation and Management Act (MFCMA) in 1976 extended fishery jurisdiction to 200 nautical miles offshore and established the current federal fishery management structure. This act, however, did little to preserve fisheries—its intended objective. By banishing foreign competition from U.S. coastal waters and offering generous subsidies to U.S. fishers, it has encouraged thousands of new U.S. boats and, in most cases, imposed no restrictions on the amount of fish an individual fisher could take. As a result of this "open access" policy, U.S. fisheries have become packed with competing boats in an intense, dangerous race for any fish left in the water. In 1992, eight lives were lost in a frenzied race for halibut in which the entire season was compressed into a 48-hour time period. MFCMA also intensifies problems by letting fishery managers use economic and social factors to modify scientists' estimates of levels for sustainable catches. Then, because of monetary and social pressure, the managers reset the allowable catch to be high.

The Environmental Defense Fund (EDF) would like Congress to reauthorize a Magnuson Act that defines and prohibits overfishing and requires the allowable catch to be ecologically sustainable rather than based on short-term economic demands. One suggested scheme would assign each fisher a tradable share of the total allowable catch. Such shares, called Individual Transferable Quotas (ITQs), could be used or sold, giving fishers an equitable base and vested interest in the fishery's future. Through market

BOX 2 What Can We Do About Bycatch?

How much fish do fishermen throw away that is caught in the nets of large-scale fishing operations? It is estimated that, in 2000, U.S. fishermen discarded 2.3 billion pounds of sea life (called bycatch), roughly 25 percent of the catch. Species are scooped up accidentally and some, such as the barndoor skate, are near extinction. The bycatch is typically thrown back into the ocean in poor shape, where most of the fish die. It is of great concern that the fish ecosystem will be impacted by waste of such high numbers. The following list provides some of the species involved and solutions to the problem.

- *Red snapper*: The Gulf of Mexico's shrimping industry has one of the worst bycatch records. For every pound of shrimp caught, up to 10 pounds of other species will be netted, such as juvenile members of the overfished Gulf red snapper. Bycatch Reduction Devices let fish escape through a small hole and can cut bycatch by 50 percent.
- *Waved albatross*: As many as 12,000 hooks can be attached to a single longline. When a fisherman unfurls it into the sea, birds such as the albatross fly in to snag the bait. Too often, they also snag a hook. Simple devices, such as bird-scaring lines—essentially streamers that flap in the wind—are just $260 per pair and can reduce seabird bycatch by 92 percent.
- *Leatherback turtles*: Gillnets—a wall of net up to 1,200 feet long—are used along the coast of California to catch sharks and swordfish. The nets can also ensnare the endangered leatherback turtle. Large areas of ocean are now closed to gillnets during the turtle's migratory season.
- *Blue shark*: A common deepwater species, the blue shark is also routinely caught on lines set to catch tuna and swordfish. A 1996 study by the National Marine Fisheries Service noted that 100,000 blue sharks per year were caught by Hawaiian longliners—more than the numbers of targeted species.
- *Yellowtail flounder*: In southern New England and around Cape Cod, hundreds of tons of yellowtail flounder are discarded every year by the fisheries that catch scallop, haddock, cod, and shrimp. The flounder population in the region is just 2 percent of what is considered healthy.

Source: Reprinted by permission of OnEarth, excerpted from "Lost Haul" in "Saving Maine," by C. Woodard, © 2005 by *OnEarth*, www.nrdc.org/onearth.

forces, ITQs would reduce fishing by ending the race for fish. As fish populations again increase, ITQs would increase in value, giving fishers a strong incentive to help fish populations recover. The EDF is advocating reforms to reduce overfishing (*EDF Letter* 1994, 1–2).

The MFCMA was reauthorized and signed by President Clinton on October 11, 1996. The reauthorization focuses on the issues of habitat degradation, overfishing, bycatch (discards and other fish), and funding; for many people, it did not do enough to strengthen sustainable yield approaches. Provisions for ITQs remain controversial. The strength, effects, and direction of the MFCMA reauthorization will be a central ecological issue that affects us all and that we should watch carefully (Buck 1996).

REFERENCES

Buck, E. H. (1996). 95036: Magnuson Fishery Conservation and Management Act reauthorized. *Congressional Research Service Issue Brief,* December 4, pp. 1–14. Available at http://www.cnie.org/nle/mar-3.html.

EDF Letter. Vol. XXV, No. 1, pp. 1–2. Available at http://www.edf.org/pubs/EDF-Letter/1994/Jan/c_overfish.html.

Halweil, B. Aquaculture Pushes Fish Harvest Higher. *Vital Signs 2005*. Worldwatch Institute. New York: Norton.

Hanley, C. J. (1997). "Fishermen Endanger 'Limitless' Seas." *Chicago Tribune*, July 21, pp. 1–2.

New Bill Seeks to Reverse Crashes of Fish Populations. (1994). *EDF Letter*, Vol. XXV, No. 4 (July), pp. 1, 3.

Parfit, M. (1995). Diminishing Returns, Exploiting the Ocean's Bounty. (1995). *National Geographic*, Vol. 188, No. 5 (November), pp. 2–37.

Woodard, C. (2003). Saving Maine. *OnEarth* 25(2):14–22.

QUESTIONS

1. What has caused the depletion of worldwide fish stocks?
2. In what way has the increase of technology contributed to the depletion of the world's fish stocks?

3. Explain why nations are using warlike tactics to protect their national fishing waters.
4. Why are the oceans of the past compared with the Wild West? What has changed?
5. What is the role of floating fish factories in the depletion of fish stocks? What do you think can be a sustainable approach for the future with the existence and use of these fish factory trawlers?
6. Devise a plan for future world fisheries that accounts for sustainable yields, the use of high technology, and national fishing rights. Contrast that plan with current practices.
7. Explain the cause and effect of the 200-mile offshore coastal waters fishing jurisdiction.
8. What is the future of the small independent fisherman? Explain.
9. What was the result of the Magnuson Fishery Conservation and Management Act in 1976 on world fisheries? Explain both the positive and negative results.
10. How could the Magnuson Act be rewritten to deal with some of the problems of world fisheries? Research the new reauthorization of the 1996 Magnuson Act. What are its strengths and weaknesses?
11. Explain the concept of Individual Transferable Quotas (ITQs).
12. Consult Figure 3 and Table 2 for fishing data patterns for discussion.
13. Research open-sea and fishery areas. In what ocean is the largest catch found?
14. Refer to Figure 3. Why do you think that total fish production has peaked? Which species do you think are fished the most? Which species of fish do you feel we still have abundant supplies of?

QUESTIONS ABOUT THE TABLES

1. According to Table 3, where are the biggest threats to the seas? How does this result compare with other threats such as environmental pollution and population concerns. Explain your answer.
2. What does Table 2 tell you about the health of the seas? Select one of the threats and outline a plan to help abate some of the damage.
3. Examine Table 2. What countries have increased or decreased their yields during the five-year period of 1997-2002? Explain these behaviors.

Table 3
Dying Seas

Sea Threat	*Baltic*	*Bering*	*Black*	*Caspian*	*Mediterranean*	*South China*	*Yellow*
Overfishing	H	H	H	H	H	H	H
Eutrophication	M	N	M	L	L	M	M
Dams	L	N	H	H	M-H	M	M
Organo-chlorines	H	M	M	L	M	M	M
Heavy metals	M	L	H	M	M-H	H	H
Oil drilling	L	L	H	H	M	H	H
Population (millions)	80	1.4	165	45	360	517	250
Area (thousands of sq. km)	370	420	2,292	371	2,500	3,685	404

H = high, M = medium, L = low, N = negligible.
Source: Dying Seas, *World Watch* Magazine, January/February 1995. Reprinted with permission of the Worldwatch Institute.

What Will Happen to the Endangered Species Act?

EDITED BY BARBARA A. EICHLER

... half of the recorded extinctions of mammals over the past 2,000 years have occurred in the most recent 50-year period.

"Nothing is more priceless and more worthy of preservation than the rich array of animal life with which our country has been blessed.

PRESIDENT NIXON, on signing the Endangered Species Act, December 28, 1973

The Endangered Species Act (ESA) still stands as a monument after 30 plus years as an environmental powerhouse that places species conservation over development concerns. It has earned the highest respect for its many notable successes and the simple yet powerful nature of its law. When the law was enacted in 1973, there were 109 species listed as endangered; today there are about 1,500 endangered species, both U.S. and foreign. The bald eagle, brown pelican, and peregrine falcon were all in danger of extinction from thinning eggshells due mostly to dichlorodiphenyltrichloroethane (DDT) and other pesticide pollution. The American alligator was threatened from overexploitation and loss of habitat. Since 1973, however, it appears that species disappearance is accelerating. An Interior Department Assistant Secretary told the House Merchant Marine and Fisheries Committee that "the truth in this is apparent when one realizes that half of the recorded extinctions of mammals over the past 2,000 years have occurred in the most recent 50-year period." (Curtis and Davison 2005, 1).

The ESA is a short, effective law and its simplicity has proven visionary and far-reaching. "The Act establishes a broad goal and expansive definitions of what it protects. It then sets up a process to determine whether an animal or plant species is either in danger of becoming extinct (endangered) or likely to become endangered (threatened). Those species found to be threatened or endangered are added to the list of species that are protected under the ESA. Once a species is listed, two key types of protection follow: it is illegal for anyone to kill or harm an individual animal or plant of a listed species, and all federal agencies are required to ensure that they don't fund or do anything that would be likely to put the continued existence of a listed species in jeopardy or adversely modify habitat critical to that species." (Curtis and Davison 2005, 2).

This broad purpose of the ESA was visionary by not only focusing on the conservation of threatened and endangered plant and animal species, but also on the conservation of these species' environments and ecosystems. This ecological understanding of the relationship between

BOX 1 Definitons

- *Endangered species*: A species that could become extinct in the future.
- *Threatened species*: A species that could become endangered soon.
- *Recovery point*: The time at which the measures provided by the Endangered Species Act are no longer necessary.

habitat and species has provided strength and success for the ESA. Habitat control also remains one of its central political hotspots and battlefronts. Increased grazing, logging, mining, and other environmentally dependent businesses in the United States threaten the existence of many animal and plant species. The conservation efforts of the ESA present a serious (and at times highly emotional) conflict in many affected areas to local economic operations and can result in unemployment and loss of private property rights. Opponents believe that the ESA ignores the economic considerations and interests of the specific location, while supporters believe that the ESA ensures and preserves the species and a variable gene pool. The debate then becomes the following: Can the Endangered Species Act accommodate increasing human demands and decreasing biodiversity?

In 1973, President Richard Nixon signed the Endangered Species Act into law; the ESA's main purpose is to maintain a list of endangered or threatened species. Based on the assumptions that each life-form may prove invaluable in ways we cannot yet measure and that each is entitled to exist, the act gave the federal government sweeping powers to prevent extinction. No commitment of this magnitude to other life-forms had ever been made before. This Act provides guidance to two agencies: the Fish and Wildlife Service and the National Marine Fisheries Service, whose main responsibilities are to stop further endangerment of species. The services then strive to protect wildlife and remove them from the list when they have gained recovery. The decision to remove them from the list is to be based solely on scientific data rather than on economic and political aspects. In 1973, the U.S. list contained 109 names; in 1995, the total included more than 900 (1,400 including foreign species), with 3,700 officially recognized candidates waiting to be listed (Chadwick 1995, 7–9). In 2004, the U.S. list contained over 1,265 names.

Conjecture regarding possible removal of the ESA brings much reaction from both critics and supporters, businessmen and environmentalists. Of course, businessmen who depend on the land for revenue would not mind seeing the removal of the ESA altogether, and other opponents of the act feel that the ESA is poorly written and simply stands in the way of economically beneficial activities. On the other side, supporters stand on the foundation that one cannot put a price on the value of a species. However, there are many additional problems—aside from economic issues—with the way that the ESA is written and managed.

The ESA categories of threatened, endangered, and secure are not well defined and allow for different views of interpretation. Since the ESA enables the government to prioritize the future of a species above any human disturbance that may cause a decline of that species, any vague interpretations allow groups to twist ESA categories to their own needs or to define data differently. The resulting ambiguity yields conflicting reports of population statuses, inaccurate information, and ineffective—and even destructive—action taken on a species population. The ESA lacks population guidelines for each species so that it can be accurately categorized and protected. Without that clear frame of reference, the protection of a species becomes difficult and arbitrary while human impacts are steadily increasing.

Another problem is that the ESA is used as a last-chance approach. After a species qualifies for the ESA due to decreased numbers, the situation is already extremely dire. In these cases, the act is dealing with the last population of a species, and action must be taken immediately to curb extinction. Such desperation forces the government to devise drastic strategies to save the species; consequently, a more balanced approach involving the needs of local human populations and economies are not considered. Such last-minute recovery plans magnify tensions and are not nearly as effective when the Act must salvage a species that is nearly lost. If the Act could practice planned "preventive medicine" instead of "reactive medicine," resources would be more strategically used, productivity would be increased, and both humans and animal species would experience decreased tension.

An additional problem is that the ESA does not adequately take into account the amount of habitat required for a species' numbers to increase. The ESA does not make provisions for species that need larger tracts of land or for the construction of habitat bridges that link smaller tracts of land together to allow cross-movement. Instead, the ESA protects islands of habitat surrounded by urban influence, allowing very little chance of survival for nomadic species such as the fox, wolf, grizzly bear, lynx, cougar, and many others. The larger nomadic predators need very large tracts of land to search for food. The ESA does make adequate attempts at preserving national parks as large areas of habitat, but national parks are not usually biodiverse ecosystems that meet the food needs of these species; they are more appealing for their aesthetic beauty. Typically, when large predators roam from their small habitats in search of food, they are shot when entering an agricultural or populated area. Conflict then arises because the ESA prohibits shooting an endangered species, but humans feel betrayed by the government for simply defending their land, lives, and resources (Endangered Species Act 1997, 1).

Other problems exist in processing and interpreting the list of species on the ESA. Some deletions from the list exaggerate the rate of success of the ESA. In 1993, the Burneau host spring snail was removed from the list after a federal judge ruled that it was listed erroneously; this species was one of eight others removed from a total of 21 species because of listing errors. Another example is that of the Rydberg milvetch (a member of the pea family), which existed only in Southwestern Utah in 1905 and is believed to be extinct. However, in 1980, taxonomists decided that a dozen populations of a close relative (the plataua) should be counted as Rydberg milvetches, and the species was removed from the list even though the plant was not the same one.

Species that have been delisted because their status improved also need correct interpretation. The Arctic peregrine falcon was struck from the list in October 1995, but its improvement probably had more to do with the reduction of pesticides than changes in hunting laws and protection of habitat. Also, the smallness of recovery numbers may be a basic reflection of the dangers faced by the species at the time of the listing rather than the ineffectiveness of recovery tactics. The belief is that, allowing for data errors, extinctions, and other extraneous factors, success stories under the ESA can be misinterpreted and the effectiveness of the Act cannot be realistically appraised (*Science* 1995).

According to most sources, the ESA does fundamentally work, but the Act would prove more effective if changes were made. The ESA has many success stories: the Palau dove, Palau flycatcher, Palau owl, Atlantic brown pelican, and gray whale. An increase in the number of red wolves within the continental United States is due mostly to anti-shooting laws as set forth by the ESA. The ESA, in conjunction with the banning of DDT, has increased many populations of birds, including the bald eagle. The ESA has also funded the recovery of some severely endangered species, such as the California condor. According to research numbers, the ESA does help to slow rates of extinction; however, it also has been suggested that many species go extinct while waiting for ESA recognition (Endangered Species Act 1997, 2).

When first written in 1973, the ESA appeared to be the answer for increased environmental concern and pressure for diminishing species. There remain some fundamental problems, however, that are now prompting views that species recovery planning and implementation processes are not working very well. These problems also agitate involved parties by pitting business interests against environmentalists and federal agencies.

Some ESA classic case studies include the Northern spotted owl, snail darter, red-cockaded woodpecker, and gnatcatcher.

Northern Spotted Owl

***Habitat*: Northwest old-growth forests**

***Status*: Threatened**

The government declared 8 million acres to be off limits to chainsaws, saving the environment as well as the owl, but it took away thousands of jobs and millions of dollars in timber sales in a depressed area. Controversy surrounded the importance of economics versus endangered species. The result was that new regulations for national forest use were created that emphasized conservation and not consumption. Management of both federal and state forest timberlands shifted from lumber production to environmental protection. The result is that cities around national forests have new environmentally friendly industries moving into the region, with the emergence of strong growth opportunities (Mitchell 1997).

Snail Darter

***Habitat:* Little Tennessee River**

***Status*: Endangered**

Work during the middle of construction of the Tellico Dam in the Little Tennessee River was halted due to concern over the future of the snail darter (a tiny fish). Criticisms of the ESA began due to the costs of salvaging a species versus completing the half-finished multimillion-dollar dam. Supporters of the ESA argued that the snail darter existed only in the Little Tennessee River, but other populations have since been found. The conclusion was that the fish were transplanted and the dam was completed.

Red-Cockaded Woodpecker

***Habitat*: Southeast old-growth pine forests**

***Status:* Endangered**

This woodpecker inhabits the old-growth pine forest ecosystem and passes its nest from one generation to the next. The loss of 88 million acres of pine forest due to development threatens the woodpecker as well as many other potentially endangered species. The conclusion of this case is as yet unresolved.

Gnatcatcher

Habitat: Sage shrub ecosystem of California

Status: Endangered

The gnatcatcher blocked the future construction of a development in southern California. Supporters said that the planned site of the development was the only remaining location for the species, while opponents said that preservation attempts were prohibiting the construction of a development with high property values. To break the impasse, Secretary of the Interior Bruce Babbitt changed the classification of the species to "threatened" rather than "endangered" to allow continuing "controlled development" to the area. In return, the developing companies had to set aside a certain percentage of sage shrub habitat to ensure the recovery and future of the gnatcatcher.

The ESA has continued to evolve, especially in the area of private land conservation. There is much support for regulatory approaches toward private land as well as many other additions to be incorporated into the law. The ESA has been amended twice in the last 25 years, but it remains the subject of nearly continuous changes in its regulations, especially in the last 10 years, because it is the only law with firm standards on core issues that are not based solely on procedural approaches. The Act forces change and conservation that is often highly contested and controversial. It is ironic that the ESA continues to be the focus for reform (even though amended twice) while most other laws have not been reexamined in decades (Curtis and Davison 2005, 6–8).

In the next few years, efforts will continue to weaken the ESA through legislation and administrative changes. One of the main attacks will likely be a call to base ESA decisions on "sound science." At present, the ESA already requires decisions to be made from the best scientific research, but the "sound science" call is a political effort to weaken the ESA to require absolute proof of significant harm to a species before economic restrictions can be enforced. Such absolute proof will rarely be possible to demonstrate fully and scientifically between habitat modification and species survival. The ESA was developed 30 years ago on the principle of the "institutionalization of caution," which set the standard that the ESA should err on the side of the species when uncertainty exists. Another possible attack on the Act's standards could be the blurring of the distinction between hatchery and wild salmon, which could pose a significant threat to wild salmon (Curtis and Davison 2005, 8).

In 2005, Congress, landowners, and industry tried to limit the ESA's jurisdiction. The threat to the power of the ESA was more significant than previous attempts, and the Bush administration gathered much support for the position that the government goes too far to protect threatened species and thereby curtails people's ability to utilize their own land. In 2005 a bill passed the house that would weaken the powers of the ESA in favor of more power for property rights. The bill would eliminate federally mapped "critical habitat" for endangered species and threatened species and allow states not to be restricted by some of the ESA federal laws. In 2006, the bill awaits the Senate for its decision. The political threats and strong offensives continue, but none have yet succeeded in passing. Lawsuits have been abundant in the past 30 years, especially in the West, where much land has been classified as critical habitat for species, thus subjecting it to federal controls that limit construction, logging, fishing, and other activities. These litigations and conflicts are becoming more intense as the needs of newly recognized endangered species interfere more profoundly with the demands of area developments. Even without Congressional revision, the federal agencies involved in enforcing the ESA seem to have a more compromising stance toward raising the request for additional scientific proof and considering the economic impact of ESA decisions (Barringer 2005, 1–7).

The effectiveness of the ESA ultimately depends upon the resolve and commitment of federal agencies to maintain standards and enforce its provisions. Curtis and Davison explain the dilemma as follows.

> The Supreme Court has said the ESA represents "the most comprehensive legislation for the preservation of endangered species ever enacted by any nation." But in the end, our commitment to species conservation must go beyond the ESA. Precious few species have been recovered, and the ESA has often succeeded in maintaining species and their habitat at only threshold levels. Often underfunded, with the agencies that administer the ESA and the Act itself often under attack, species conservation cannot be carried on the back of one law.... We need, instead, to devote more effort toward ensuring that the laws governing the management of our public lands and our nation's waters make the ESA a rarely used, last resort rather than the only barrier to loss of the nation's biological heritage" (Curtis and Davison 2005, 9).

Table 1
Some Endangered Animal Species

Common Name	*Scientific Name*	*Range*
Albatross, Amsterdam	*Diomedia amsterdamensis*	Amsterdam Island, Indian Ocean
Antelope, giant sable	*Hippotragus niger variani*	Angola
Armadillo, giant	*Pridontes maximus*	Venezuela, Guyana to Argentina
Babirusa	*Babyrousa babyrussa*	Indonesia
Bandicoot, desert	*Perameles eremiana*	Australia
Bat, gray	*Myotis grisescens*	Central, southeastern U.S.
Bear, brown (grizzly)	*Ursus arctos horribilis*	Palearctic
Bison, wood	*Bison bison athabascae*	Canada, northwestern U.S.
Bobcat, Mexican	*Felis rufus escuinapae*	Central Mexico
Caiman, black	*Melanosuchus niger*	Amazon basin
Camel, Bactrian	*Camelus bactrianus*	Mongolia, China
Caribou, woodland	*Rangifer tarandus caribou*	Canada, northwestern U.S.
Cheetah	*Acinonyx jubatus*	Africa to India
Chimpanzee, pygmy	*Pan paniscus*	Congo (formerly Zaire)
Condor, California	*Gymnogyps californianus*	U.S. (AZ, CA, OR), Mexico (Baja California)
Crane, whooping	*Grus americana*	Canada, Mexico, U.S. (Rocky Mts. to Carolinas)
Crocodile, American	*Crocodylus acutus*	U.S. (FL), Mexico, Caribbean Sea, Central and South America
Deer, Columbian white-tailed	*Odocoileus virginianus leucurus*	U.S. (OR, WA)
Dolphin, Chinese River	*Lipotes vexillifer*	China
Dugong	*Dugong dugon*	East Africa to southern Japan
Elephant, Asian	*Elephas maximus*	South-central and southeastern Asia
Fox, northern swift	*Vulpes velox hebes*	Canada
Frog, Goliath	*Conraua goliath*	Cameroon, Equatorial Guinea, Gabon
Gorilla	*Gorilla gorilla*	Central and West Africa
Hartebeest, Tora	*Alcelaphus buselaphus tora*	Egypt, Ethiopia, Sudan
Hawk, Hawaiian	*Buteo solitarius*	U.S. (HI)
Hyena, brown	*Hyaena brunnea*	Southern Africa
Impala, black-faced	*Aepyceros melampus petersi*	Angola, Namibia
Kangaroo, Tasmanian forester	*Macropus giganteus tasmaniensis*	Australia
Leopard	*Panthera pardus*	Africa, Asia
Lion, Asiatic	*Panthera leo persica*	Turkey to India
Manatee, West Indian	*Trichechus manatus*	Southeastern U.S., Caribbean Sea, Mexico
Monkey, spider	*Ateles geoffroyi frontatus*	Costa Rica, Nicaragua
Ocelot	*Felis pardalis*	U.S. (AZ, TX) to Central and South America
Orangutan	*Pongo pygmaeus*	Borneo, Sumatra
Ostrich, West African	*Struthio camelus spatzi*	West Sahara
Otter, marine	*Lutra feline*	Peru south to Straits of Magellan

Panda, giant	*Ailuropoda melanoleuca*	China
Panther, Florida	*Felis concolor coryi*	U.S. (FL)
Parakeet, golden	*Aratinga guarouba*	Brazil
Parrot, imperial	*Amazona imperialis*	West Indies (Dominica)
Penguin, Galapagos	*Spheniscus mendiculus*	Ecuador (Galapagos Islands)
Puma, eastern	*Felis concolor couguar*	Eastern North America (presumed extinct in wild)
Python, Indian	*Python molurus molurus*	Sri Lanka, India
Rat-kangaroo, brush-tailed	*Bettongia penicillata*	Australia
Rhinoceros, black	*Diceros bicornis*	Sub-Saharan Africa
Rhinoceros, northern white	*Ceratotherium simum cottoni*	Congo, Sudan, Uganda, Central African Republic
Salamander, Chinese giant	*Andrias davidianus*	Western China
Sea lion, Steller	*Eumetopias jubatus*	Alaska, Russia
Sheep, bighorn	*Ovis canadensis*	California
Squirrel, Carolina northern flying	*Glaucomys sabrinus coloratus*	U.S. (NC, TN)
Tiger	*Panthera tigris*	Asia
Tortoise, Galapagos	*Geochelone elephantopus*	Ecuador (Galapagos Islands)
Turtle, Plymouth red-bellied	*Pseudemys rubriventris bangsi*	U.S. (MA)
Whale, gray	*Eschrichtius robustus*	North Pacific Ocean
Whale, humpback	*Megaptera novaeangliae*	Oceania
Wolf, red	*Canis rufus*	U.S. (FL, NC, SC)
Woodpecker, ivory-billed	*Campephilus principalis*	Cuba
Yak, wild	*Bos grunniens mutus*	China (Tibet), India
Zebra, mountain	*Equus zebra zebra*	South Africa

Source: Fish and Wildlife Service, U.S. Department of the Interior.
Source: *The World Almanac and Book of Facts—2005.* New York

REFERENCES

Barringer, F. (2005). "Endangered Species Act Faces Broad New Challenges." *New York Times*, June 26, pp. 1–7. Accessed July 1, 2005. Available at http://www.nytimes.com/2005/06/26/politics/26species.html.

Chadwick, D. H. (March 1995). Dead or Alive: The Endangered Species Act. *National Geographic* (March): 2–41.

Curtis, J., and Davison, B. (2005). "The Endangered Species Act: Thirty Years on the Ark. *Open Spaces Quarterly* 5(3):1–9. Accessed June 14, 2005. Available at http://www.open-spaces.com/article-v5n3-davison.php.

Environmental Protection Agency. (2006). Endangered Species Act. Available at http://www.epa.gov/Region 5/defs/htm/esa.htm.

Haynes, V. D. (1997). U.S. Wants to Amend Endangered Species Law. *Chicago Tribune*, July 27, Section 1, p. 4.

Is Endangered Species Act in Danger? (1995). *Science* (3 March):267.

Mitchell, J. G. (1997). In the Line of Fire: Our National Forests. *Scientific American* 191(3):57–58.

NOAA Fisheries Office of Protected Resources. (2006). Endangered Species Act. Available at http://www.nmfs.noaa.gov.pr/laws/esa/.

FURTHER READING

Graham Jr., F. (1994). Winged Victory. *Audubon* (July/August): 36–49.

Horton, T. (1995). The Endangered Species Act: Too Tough, Too Weak, or Too Late? *Audubon* (March/April):68.

Kohm, K. (1991). *Balance on the Brink of Extinction.* Washington, D.C.: Island Press.

Korn, P. (1992). The Case for Preservation. *The Nation* (30 March):414.

Mann, C. C., and Plummer, M. L. (1995). California vs. Gnatcatcher. *Audubon* (January/February):40–48.

National Research Council. (1993). *Science and the Endangered Species Act.* National Research Publishing Company.

Rauber, P. (1996). An End to Evolution. *Sierra* (January/February): 28.

QUESTIONS

1. Do you think that the Endangered Species Act is needed? Do you think that this legislation was too strong or too weak in 1973? If you were to change the law now, how would you change it?
2. Where do you stand in your political view on the environment? Could you be regarded as a critic or supporter of the ESA? Explain your reasons in detail. Should the ESA be eliminated? If so, what should take its place? If not, why not?
3. Name some significant successes of the ESA. Support your list of successes with the role and contribution of the ESA for these successes.
4. Name some of the problems of the ESA, and give some suggestions to correct those problems.
5. Research some of the classic ESA cases (e.g., bald eagle, gnatcatcher, Northern spotted owl, California condor, milvetch), and present arguments for and against ESA involvement with and protection of these species.
6. Do you feel that the ESA works? Give arguments for and against your conclusions.
7. Why do you think the ESA is increasingly under attack by the government and Congress? Do you think such attacks are appropriate?
8. Why is the "sound science" approach a threat to the power of the ESA?
9. Explain in the last paragraph what Curtin and Davison mean when they say "species conservation cannot be carried on the back of one law." What needs to be done in this context?
10. Research the species found on the ESA list. Does your research support or refute the utilization of the ESA?
11. What is your overall reaction to the issue of the ESA? Give your views regarding the environmental, ethical, economical, and political aspects of this issue.

The balance of nature and species is a needed priority of the highest concern.
Photo courtesy of B. Eichler.

Air Poisons Around the World

EDITED BY BARBARA A. EICHLER

There are studies showing that on days when there are more particulates in the air, more people die. You can't get more basic that that.... This country has a law called the Clean Air Net, which says that Americans should be able to breathe without harm.

JOHN H. ADAMS, Executive Director, Natural Resources Defense Council

Air pollution in its many forms—sulfur dioxide, ozone, fine particles, carbon monoxide, and nitrogen oxide—is causing a deathly fog around the world. "From Hong Kong to Mexico City, the toll taken in terms of human death and disease caused or aggravated by air pollution almost certainly is measured in the millions," says Dr. Alfred Munzer, past president of the American Lung Association and respiratory specialist.

In Japan, various respiratory illnesses suffered by thousands of citizens living in heavily polluted areas of Japan were deemed indisputably to be caused by sulfur dioxide. Within the next year, the Japanese government set up a tracking program and medical reimbursements for certified victims of air pollution ranging from bronchitis to asthma. Even though the program monitored only sulfur dioxide and was limited to just a few industrial areas, it still certified more than 90,000 air pollution victims before it was stopped in 1988 because of pressure from polluters. Since no other nation has routinely collected this kind of data on air pollution and human health, this number serves as an indicator of a worldwide unprecedented problem.

Sulfur dioxide, along with fine particles, result in newly formed acidic compounds that claim an unbelievable number of victims in all parts of the world. When weather conditions trapped sulfur dioxide over the Mae Moh region of Thailand in 1992, 4,000 residents required medical treatment, cattle died with blistered hides, and crops withered. In the 1990's in Poland, medical tracking of army inductees revealed four times more asthma and three times more bronchitis in areas that were polluted by sulfur dioxide. In Krakow, men who lived in the city's most polluted areas were tracked for 13 years; the study found losses of lung function similar to smoking.

Ozone, the dominant chemical in smog, is an invisible toxic gas that is the result of reactions of unburned gasoline with other pollutants. Ozone scars tissue, burns eyes, and promotes coughing, wheezing, and rapid, painful breathing. Within minutes of entering the lungs, ozone burns through the cell walls. The immune system tries to defend the lungs, but they are stunned by the ozone. Cellular fluid then seeps into the lungs, and breathing becomes rapid and painful. Ozone destroys ciliated cells in the nose and airways, which the body replaces with thick-walled, abnormal squamous cells. With time, the lungs stiffen and the ability to breathe decreases. Children raised in ozone-polluted areas have unusually small lungs, and adults lose up to 75 percent of their lung capacity. The massive cell death caused by ozone triggers precancerous physiological responses.

Mexico City's air, which contains some of the world's worst ozone pollution, reaches unhealthful ozone levels about 98 percent of the time. Healthy men newly exposed to Mexico City's air developed precancerous cell alterations in nasal and airway passages. Mexico City residents are generally in the second of three stages of cancer in which the third stage is the production of cancer.

In Los Angeles, one in four fatal accident victims aged 14 to 25 had severe lung lesions of the sort caused by ozone, which is a destructive, irreversible disease in young people. Los Angeles residents exposed to ozone had double the risk of cancer compared with residents of cleaner cities.

Particulates are a catch-all term for everything from road dust to soot to mixtures of pollutants—solids as well as liquids, microscopic as well as larger grains that vary in the environment. Scientists believe that fine particles small enough to lodge deep into the lungs are the most dangerous. Fine particles result from the burning of coal, oil, and gasoline; they also result from the atmospheric change of oxide from sulfur and nitrogen into sulfates and nitrates.

Bangkok police work in gritty air pollution thick with air particulates from motorcycles and thick clouds of burnt diesel fuels from buses and trucks. The effects on these officers include abnormal lung function tests and spots on their lungs as well as general congestion. Roughly one of every nine Bangkok residents has a respiratory ailment caused by air pollution.

The evidence that particulates kill is "absolutely complete" according to University of British Columbia's Dr. David Bates, a foremost air pollution and health expert. Studies from around the world and within the United States confirm that, as particulate pollution rises, so do sickness and death.

Carbon monoxide is invisible and oxides of nitrogen are nearly invisible; therefore, it is impossible to tell when the air contains dangerous levels of both. These gases are produced when gasoline and other fuels are burned incompletely. Roughly 90 percent of urban carbon monoxide occurs from motor vehicle tailpipes, especially in cities such as London. When inhaled, carbon monoxide starves the body of oxygen, which then causes dizziness and unconsciousness. Oxides of nitrogen form the reddish-brown layer that can be seen from an airplane one mile up. Nitrous oxides cause oxygen and nitrogen in the air to combine, further causing air and ozone pollution. Like ozone, oxides of nitrogen destroy organic matter, such as human tissue, and make organisms more susceptible to bacterial infections and lung cancer.

Some of the worst air pollution occurred in the "Black Triangle," where the Czech Republic meets Poland and the former East Germany during the post war era from the 1950's to the late 1980's. (Please refer to the case study at the end of this part, "Europe's Black Triangle Turns Green.") More sulfur per square meter fell there than on any other place in Europe. In Poland, high lead levels found in the soil as a result of air pollution were unprecedented elsewhere in the Western world. According to experts, the "Black Triangle" was among the most polluted locations in the world. Respiratory disease in children up to 14 years was epidemic. In this triangle, black smoke clung to everything. The World Health Organization concluded that approximately 15 percent of infant mortality and 50 percent of postneonatal respiratory mortality in the Czech Republic were connected to air pollution. According to studies, children living in cities with air pollution lagged as much as 11 months behind in bone growth compared with those children breathing cleaner air. In Poland, lead-laden air pollution was so severe that it caused lead poisoning and intelligence loss; some experts estimated that 10 to 15 percent of the nation's citizens had been affected.

Governments that are forced to choose between protecting the public and shielding industry from regulation too often choose to sacrifice their men, women, and children. People around the world require the basic necessity of breathing clean air. Governments must prioritize their efforts toward improving the health of their citizens and utilizing technology for the benefit of their citizens as well as their economies; if not, their economic decisions may become as short-lived as their suffering people.

U.S. HISTORICAL AND SOCIETAL PROGRESS REGARDING AIR POLLUTION

After a lawsuit was initiated by the Natural Resources Defense Council in 1973, the Environmental Protection Agency (EPA) created a five-year program to gradually reduce the lead content in gasoline. At that time, tetra-ethyl lead (TEL) was routinely added to lower-grade gasoline to create more efficient burning and prevent gasoline "knock" caused by uneven combustion. For decades, the nation had allowed the combustion of leaded gas to emit millions of tons of lead into the air, where it was breathed in or deposited in soil and dust. Lead, however, is toxic, even in the smallest amounts. Children with elevated blood–lead levels can suffer lowered IQs, slower neural transmission, hearing loss, and disruption of the formation of hemoglobin red blood cells; acute lead poisoning causes even greater physiological damage. By the mid 1970s, physiological links were established between lead content in gasoline and health problems in children.

After many political battles as well as opposition by special-interest groups, the EPA proposed stricter regulations by 1985 with the goal of eliminating the lead in gasoline entirely by the mid-1990s. Two factors drove this development. One was new health data demonstrating that blood–lead levels previously accepted as safe had adverse effects. The other was data by Joel Schwartz (formerly of the EPA and a professor at the Harvard School of Public Health) illustrating that, although it would cost industry about $575 million to meet the newest lead standards by 1986, that sum would be small in comparison with the $1.8 billion saved in 1986 alone from reduced needs for medical care, lower pollution emissions from catalytic converters, greater fuel efficiency, and less vehicle maintenance. In 1994, the Centers for Disease Control and Prevention (CDC) estimated that 1 million children under six years of age in the United States still had blood–lead levels exceeding the recommended thresholds, but the number had been reduced significantly from previous data. By 2000, estimates had been further reduced to about half that number, closer to about a half million (CDC, 2004).

Aside from lead, the two worst air pollution components are ozone and fine particulates. Several major air pollution disasters during the early part of the twentieth century—Meuse Valley, Belgium (1930); Donora, Pennsylvania (1948); London, England (1950)—demonstrated that high concentrations of air pollution fine particulates could kill large numbers of people. By the late 1980s, there was accumulating evidence that the particulate matter in this pollution was closely associated with death and illness (Skelton 1997, 27–29).

FINE-PARTICULATE REGULATIONS

People who are hurt the worst are those with the most vulnerable lungs—children, the elderly, and asthmatics. For people with heart or lung disease, fine particulates can cause an earlier death by one year or more. Over the past 10 years, hundreds of studies have indicated that the damage caused by these particulate pollutants is more serious than previously known and that more people die on those days when there are more particulates in the air. Fine particulates are currently the leading air pollution health threat in the country. Researchers believe that about 60,000 Americans may die annually as a result of particulate pollution—a number larger than any other form of pollution. Currently, however, less than one-third of all funds to reduce air pollution are directed toward removing fine particulates.

Airborne particles have different sources and come in different sizes. Currently, the smallest particulates regulated by the federal government have a diameter of 10 microns or less (referred to as PM-10). The largest of these particulates are dust and dirt, but the smallest of these particulates (PM-2.5) cause the greatest health threat. These very fine particles evade the body's clearance mechanism and penetrate deeply into the lungs' most sensitive areas. Additionally,

Air pollution surrounding a city. U.S. Public Health Service.

these particulates are mostly byproducts of combustion (from coal-fired power plants, industrial boilers, highway vehicles, and other pieces of machinery) that in themselves form chemicals that are health hazards. In 1971, the EPA set a general limit of particulate standard concentrations at PM-10 or above. Yet by the end of the decade, it was becoming clear that the smallest particulates were the worst health risks. In 1987, the EPA set a tougher standard limit for particles at PM-10, the first time that a standard limit had been set on fine inhalable particles. Evidence continued to mount that the lives of thousands of Americans were being shortened by several years, on average, due to exposure to particulates at levels below the PM-10 standard.

Additional studies became available that tracked the health of more than 8,000 people from six cities for 14 to 16 years. These studies from the 1980s and 1990s concluded that fine particles increased the risk of premature death for residents of the most polluted city by 26 percent. Another study that tracked 500,000 people in 151 cities came to the conclusion that people living in the most polluted areas containing fine particles had a 17-percent greater risk of mortality. In 1996, a study of 239 cities found that about 64,000 people may die prematurely from heart or lung disease annually due to particulate pollution. In 1996, the EPA finally proposed a tougher standard that included the newest guidelines pertaining to the finest, most dangerous particulates—PM-2.5. It is estimated that these new standards and the new law would annually prevent 9,000 hospital admissions, 250,000 cases of aggravated asthma, and 20,000 premature deaths every year. In 2006 the EPA revised the standards to address two categories of particle pollution: (1) fine particles which are 2.5 micrometers in diameter and smaller and inhalable coarse particles which are smaller than 10 micrometers in diameter and larger than 2.5 micrometers. This issue is better put in the following manner: As a standard for the entire and future industrial world, the United States and the world require such protection for rights to and needs for clean air.

In 1987, U.S. law required certain industries to disclose their emissions of 650 toxic chemicals. Since the disclosure began, selected industries have reduced those chemical emissions by 31 percent; however, companies still admitted to emissions of 3.8 billion pounds of toxic chemicals in 1992. According to the Washington Post (2004), the EPA's Toxic Release Inventory figures in 2002 represent the first time since 1997 that emissions increased with an increase by 5 percent more than 2001. Releases of lead increased 3.2 percent and mercury jumped by 10 percent. However, emissions of dioxin fell by 5 percent. Various environmental groups are highly concerned that there is too much under-reporting of industrial emissions and lack of tight monitoring of such toxic emissions. Phil Clapp, president of the National Environmental Trust states "This is an across-the-board increase in pollution (Washington Post, 2004)." Other environmental groups suggest that government figures understate emissions, especially carcinogens such as benzene and butadiene by as much as 400 or 500 percent.

Public disclosure has been quite effective in prompting industry to clean up its approach, but there is still a long way to go. A study in 1997 indicated that the quantitative breakdown of emissions from industry was not very accurate in terms of the analysis of location, composition, and emission rates; the study stated that more accurate methods needed to be developed that relied on actual measurements rather than calculations on paper.

As of 2005, the 1990 Clean Air Act remains the most recent version of U.S. law that regulates air standards. It was originally passed in 1970 and has federal jurisdiction under the EPA that pertains to the entire country, even though the states must create the mechanisms to enforce the Act, such as granting permits or enforcing fines. The Act establishes standards that set limits on how much of a pollutant is allowed into the air throughout the United States Even though some states have elected to set tougher laws, the Clean Air Act regulates firm basic standards for the entire country. The following list details some of the areas covered by the Clean Air Act.

- Enactment of state implementation plans that explain how each state will perform its job under the Clean Air Act
- Development of standards and infrastructure concerning interstate air pollution and its enforcement; institution of public participation programs and market approaches as well as economic incentives for reducing air pollution
- Regulation of smog and other "criteria" (based on scientific standards for protecting health, environment, and property) air pollutants. Changes were revised in 1997 to include higher standards for ground-level ozone (smog) and particulate matter.
- Maintenance of standards for mobile sources of air pollution (cars, trucks, buses, off-road vehicles, planes). Cars today produce 60–80 percent less pollution than cars in 1960, but motor vehicles still release more than 50 percent of the hazardous air pollutants.

- Reduction of acid air pollutants that cause acid rain
- Establishment and maintenance of standards that eliminate the production of chemicals that destroy the ozone layer. All significant ozone-destroying chemicals are no longer produced in the United States including chlorofluorocarbons (CFCs), halons, carbon tetrachloride, methyl chloroform, and hydro CFCs (HCFCs).
- Regulation of consumer products for release of smog-forming volatile organic compounds (VOCs) and ozone-destroying chemicals (EPA 2005).

REFERENCES

Adams, J. H. (1997). Past Time for Clean Air. *The Amicus Journal* (Spring):2.

Center for Disease Control and Prevention (CDC). (2004). Children's Blood Lead Levels in the United States. Available at http://www.cdc.gov/nceh/lead/research/kidsBLL.htm.

Eilperin, J. (2004). Toxic Emissions Rising, EPA Says. Washingtonpost.com. Available at http://www.washingtonpost.com/wp-dyn/articles/A61795-2004Jun22.html.

Environmental Protection Agency (EPA). (2006). PM Standards Revision – 2006. Available: http://www.epa.gov/oar/particlepollution/naaqsrev2006.html.

It's the Ecosystem, Stupid. (1994). 26th Environmental Quality Index. *National Wildlife* (February/March):40.

U.S. Environmental Protection Agency. (1993). The Plain English Guide to the Clean Air Act, April, 1993. Last updated January 2005. Accessed July 2, 2005. Available at http://www.epa.gov/oar/oaqps/peg_caa?pegcaain.html.

Moore, Curtis A. (1995). Poisons in the Air. *International Wildlife* 25 September/October (5):38–45.

Raloff, J. (1997). Industries Tally Air Pollution Poorly. *Science News* 151 (28 June):396.

Skelton, R. (1997). Clearing the Air. *The Amicus Journal* (Summer):27–30.

QUESTIONS

1. What compounds mentioned in this article cause air pollution? How are each of these compounds created?
2. What can be done to stop the creation of these air pollution compounds?
3. In your estimation, is such air pollution a significant health risk, or is it the necessary price and inconvenience of technology? What priority would you place on curbing these air pollution compounds?
4. What are some approaches that industrialized nations can take to curb such air pollution? How can these efforts be made effective?
5. How can industrialized nations aid Third World nations in the effort to clean up the air?
6. What were some of the deciding factors that drove the passage of stricter lead regulations in gasoline? Explain how economics and society–health arguments won this standard. What are the implications of this situation?
7. Why are fine particulates so dangerous?
8. Do you feel that there should be a PM-2.5 standard? How can such a standard be enacted? (You may wish to refer to Question #6 for some ideas.)
9. Has industrial disclosure of toxic emissions helped to reduce industrial chemical emissions? Why do you think this has happened? What problems exist with self-disclosure concerning pollution control effectiveness? What approaches would you further recommend for reducing emissions levels?
10. Refer to Tables 2 and 3. Observe the historical improvement of air toxins and air pollutants. Comment on these trends in each of the tables.
11. Refer to Tables 4 and 5. Discuss how each pollutant could realistically be minimized in industry, transportation, and domestic uses. How does your plan compare with what is being done today? How can we improve governmental and societal efforts to reduce pollutants?

Table 1
Air Pollution in Selected World Cities*

City and Country	*Particulate Matter*	*Sulfur Dioxide*	*Nitrogen Dioxide*	*City and Country*	*Particulate Matter*	*Sulfur Dioxide*	*Nitrogen Dioxide*
Accra, Ghana	31	NA	NA	Milan, Italy	36	31	248
Amsterdam, Netherlands	37	10	58	Montreal, Canada	22	10	42
Athens, Greece	50	34	64	Moscow, Russia	27	109	NA
Bangkok, Thailand	82	11	23	Mumbai (Bombay), India	79	33	39
Barcelona, Spain	43	11	43	Nairobi, Kenya	49	NA	NA
Beijing, China	106	90	122	New York, NY	23	26	79
Berlin, Germany	25	18	26	Oslo, Norway	23	8	43
Cairo, Egypt	178	69	NA	Paris, France	15	14	57
Cape Town, South Africa	15	21	72	Prague, Czech Republic	27	14	33
Caracas, Venezuela	18	33	57	Quito, Ecuador	34	22	NA
Chicago, IL	27	14	57	Rio de Janeiro, Brazil	40	129	NA
Cordoba, Argentina	52	NA	97	Rome, Italy	35	NA	NA
Delhi, India	187	24	41	Seoul, South Korea	45	44	60
Kolkata (Calcutta), India	153	49	34	Sofia, Bulgaria	83	39	122
London, UK	23	25	77	Sydney, Australia	22	28	81
Los Angeles, CA	38	9	74	Tokyo, Japan	43	18	68
Manila, Philippines	60	33	NA	Toronto, Canada	26	17	43
Mexico City, Mexico	69	74	130	Warsaw, Poland	49	16	32

NA = not available.
* Particulate data (1999): World Bank study, "The Human Cost of Air Pollution: New Estimates for Developing Countries." Sulfur dioxide and nitrogen dioxide data (1998 or earlier): Healthy Cities Air Management Information System, World Health Organization, and the World Resources Institute.
Source: *The World Almanac and Book of Facts 2005.* New York: World Almanac Education Group.

Table 2
Emissions Estimates for EPA-Monitored Pollutants, 1970–2003 (thousand short tons)

Source	*Carbon Monoxide (CO)*		*Nitrogen Oxides (NO_x)*		*Volatile Organic Compounds*		*Sulfur Dioxide (SO_2)*		*Particulate Matter (PM-10)*	
	1970	*2003*	*1970*	*2003*	*1970*	*2003*	*1970*	*2003*	*1970*	*2003*
Fuel combustion										
Electric utility	237	530	4,900	4,458	30	56	17,398	10,929	1,775	683
Industrial	770	1,377	4,325	2,775	150	170	4,568	2,227	641	317
Other	3,625	3,003	836	729	541	878	1,490	596	455	461
Chemical and allied processing	3,397	329	271	102	1,341	218	591	329	235	51
Metals processing	3,644	1,422	77	91	394	72	4,775	285	1,316	141
Petroleum and related industries	2,179	138	240	137	1,194	380	881	323	286	38
Other industrial processes	620	634	187	504	270	412	846	426	5,832	410
Solvent utilization	NA	73	NA	7	7,174	4,562	NA	2	NA	17
Storage and transport	NA	241	NA	16	1,954	1,178	NA	6	NA	86
Waste disposal and recycling	7,059	1,854	440	137	1,984	427	8	32	999	386
On-road vehicles	163,231	58,807	12,624	7,381	16,910	4,428	273	256	480	187
Non-road vehicles	11,371	24,447	2,652	4,103	1,616	2,572	278	443	164	308
Miscellaneous[1]	7,909	14,033	330	289	1,101	704	110	88	839	19,854
Total	204,043	93,706	26,805	20,492	34,659	15,429	31,218	15,848	13,023	24,942

NA = not available.
Note: Some columns may not sum due to independent rounding.
[1] Miscellaneous includes PM-10 natural sources, such as fugitive dust that arises from construction activities, mining and quarrying, and paved road resuspension. It is the leading source of PM-10 emissions.
Source: Wright, J. W. (ed.), *The New York Times Almanac*, New York: Penguin, 2005. Data from U.S. Environmental Protection Agency, August 2004, www.epa.gov/airtrends.

Table 3

Emissions of Principal Air Pollutants in the United States, 1970–2003 (thousand tons, estimated)

Source	*1970*	*1975*	*1980*	*1985*	*1990*	*1995*	*2000*	*2001*	*2002*	*2003*
Carbon monoxide	204,043	188,398	185,407	176,844	154,186	126,777	114,467	106,262	112,054	106,886
Nitrogen oxides[1]	26,883	26,377	27,079	25,757	25,529	24,956	22,598	21,549	21,102	20,728
Volatile org. compounds[1]	34,659	30,765	31,106	27,404	24,108	22,041	17,512	17,111	16,544	16,056
Particulate matter[2]	13,023	7,556	7,013	41,324	27,752	25,819	23,747	23,708	22,154	22,940
Sulfur dioxide	31,218	28,043	25,925	23,307	23,076	18,619	16,347	15,932	15,353	15,943
TOTAL[3]	309,826	281,139	276,530	294,636	254,651	218,212	194,671	184,562	187,207	182,553

[1] Ozone, a major air pollutant and the primary constituent of smog, is not emitted directly to the air, but is formed by sunlight acting on emissions of nitrogen oxides and volatile organic compounds.
[2] PM-10 = particulates 10 microns or smaller in diameter.
[3] Totals are rounded, as are components of totals.
Source: *The World Almanac and Book of Facts 2005.* New York: World Almanac Education Group. Data from U.S. Environmental Protection Agency, Office of Air Quality Planning and Standards.

Table 4
Particulate Matter (PM-10) Emissions, 1970–2003 (in thousand tons)*

Source	*1970*	*1975*	*1980*	*1985*	*1990*	*1995*	*2000*	*2001*	*2002*	*2003*
Fuel combustion, elec. util.	1,775	1,191	879	280	295	268	687	696	695	683
Industrial processes	8,310	4,267	3,433	1,199	1,199	1,132	931	967	855	957
Transportation	644	665	689	712	715	643	552	529	515	495
TOTAL	13,023	7,556	7,013	41,324	27,752	25,819	23,747	23,708	22,154	22,940

Source: U.S. Environmental Protection Agency, Office of Air Quality Planning and Standards
*PM-20 refers to particulates equal or smaller than 10 microns in diameter, and so capable of entering deep in to the respiratory tract. Totals include miscellaneous sources not determined.

Table 5
Common Air Pollutants (Criteria Air Pollutants)

Name	*Source*	*Health Effects*	*Environmental Effects*	*Property Damage*
Ozone (ground-level ozone is the principal component of smog)	Chemical reaction of pollutants; VOCs and NOx	Breathing problems, reduced lung function, asthma, irritated eyes, stuffy nose, reduced resistance to colds and other infections, may speed up aging of lung tissue	Ozone can damage plants and trees; smog can cause reduced visibility	Damages rubber, fabrics, etc.
VOCs† (volatile organic compounds); smog formers	VOCs are released from burning fuel solvents, paints, and other products used at work or at home. Cars are an important source of VOCs. VOCs include chemicals such as benzene, toluene, methylene chloride.	In addition to ozone (smog) effects, many VOCs can cause serious health problems such as cancer and other effects	In addition to ozone (smog) effects, some VOCs such as formaldehyde and ethylene may harm plants	
Nitrogen dioxide (one of the NOx): smog-forming chemical	Burning of gasoline, natural gas, coal, oil etc. Cars are an important source of NO^2.	Lung damage, illnesses of breathing passages and lungs (respiratory system)	Nitrogen dioxide is an ingredient of acid rain (acid aerosols), which can damage trees and lakes.	Acid aerosols can reduce visibility. Acid aerosols can eat away stone used on buildings, statues, monuments, etc.

Table 5 (*Continued*)

Name	*Source*	*Health Effects*	*Environmental Effects*	*Property Damage*
Carbon monoxide (CO)	Burning of gasoline, wood, natural gas, coal, oil, etc.	Reduces ability of blood to bring oxygen to body cells and tissues. CO may be particularly hazardous to people who have heart or circulatory (blood vessel) problems and people who have damaged lungs or breathing passages		
Particulate matter (PM-10); (dust, smoke, soot)	Burning of wood, diesel and other fuels; industrial plants; agriculture (plowing, burning off fields); unpaved roads	Nose and throat irritation, lung damage bronchitis, early death	Particulates are the main source of haze that reduces visibility	Ashes, soots, smokes and dusts can, dirty and discolor structures and other property, including clothes and furniture
Sulfur dioxide	Burning of coal and oil, especially high-sulfur coal from the Eastern United States; industrial processes (paper, metals)	Breathing problems, may cause permanent damage to lungs	SO_2 is an ingredient in acid rain (acid aerosols), which can damage trees and lakes. Acid aerosols can also reduce visibility.	Acid aerosols can eat away stone used in buildings, statues, monuments, etc.
Lead	Leaded gasoline (being phased out), paint (houses, cars), smelters (metal refineries); manufacture of lead storage batteries	Brain and other nervous system damage; children are at special risk. Some lead-containing chemicals cause cancer in animals. Lead causes digestive and other health problems.	Lead can harm wildlife.	

†All VOCs contain carbon (C), the basic chemical element found in living beings. Carbon-containing chemicals are called organic. Volatile chemicals escape into the air easily. Many VOCs, such as the chemicals listed in the table, are also hazardous air pollutants, which can cause very serious illnesses. EPA does not list VOCs as criteria air pollutants, but they are included in this list of pollutants because efforts to control smog target VOCs for reduction.
Source: U.S. Environmental Protection Agency.

Rain Forests May Offer New Miracle Drugs

EDITED BY BARBARA A. EICHLER

In the remote rain forests of Suriname, families have traditionally used local plants to treat common illnesses, such as fever and stomachache, and more exotic ailments, such as leishmaniasis, an Amazon parasite that produces skin lesions. Folklore knowledge not only includes which plants are specific to certain diseases, but also prescribes the methods of preparation and application of the health-restoring tinctures.

Synthetic production of drugs has been the primary focus of Western pharmaceutical companies since the 1940s. Producing drugs in the laboratory allows these companies to safeguard their investment in research and development through intellectual property legislation enacted via patents and licenses. In recent times, international drug companies, such as Bristol-Myers Squibb, are refocusing their interest on the natural laboratory that lies beneath the canopy of the rain forest, which is able to offer an albeit undocumented but nevertheless tried and perfected methodology toward drug refinement and employment. Significant discoveries include tree extracts from Indonesia help fight HIV infection, a Brazilian shrub that is used to treat diabetes, and fungi and bacteria from the Philippines that show promise in treating a variety of ailments.

The process of searching for new natural medicines is called bioprospecting. Recent technological advances in screening samples of plants, animals, and other natural substances for curative ingredients has made it cost effective to collect and analyze the large number of specimens needed to obtain viable medicines. In addition, the natural proclivity of diseases, such as malaria, to build up resistance to drugs has accelerated the need for alternative forms of treatment.

Bioprospecting accounts for 10 percent of the research budget in most large U.S. drug companies, which can amount to a significant investment for firms such as Merck & Co., which have annual R & D budgets up to $1 billion. In addition, government agencies, such as the National Institutes of Health, the National Science Foundation, and the U.S. Agency for International Development, are also turning to bioprospecting. These three agencies will spend millions of dollars searching the rain forests of Costa Rica, Peru, Suriname, and Cameroon as well as the deserts of Mexico, Chile, and Argentina.

Bioprospecting successes include the landmark Taxol, a treatment for advanced ovarian cancer found in the bark and needles of the Pacific yew, and Vinblastine, a drug used for treating cancers, including Hodgkin's disease and leukemia, that is found in the rosy periwinkle, a Madagascan plant. Two other natural substances, Michellamine B from a Cameroon vine, and Prostratin, from a Samoan rubber tree help in the fight against AIDS.

Plants are not the only source of potential life-saving drugs. On the French Guinea border, the Wayana people eat the heads of soldier ants to obtain a chemical that they believe will ward off malaria.

The search for curative compounds does not only occur in the natural environment. Medicinal plant researchers scour local markets in various parts of the world seeking folk cures that may lead to the next wonder drug. In addition, these researchers rush to record potential remedies that may be lost to the passing of generations. Traditional cures are often complex combinations of herbs and animal extracts taken together to produce the required effect. Plants that reduce the side effects of curative substances are often taken with their medicinal counterparts. Researchers fear that the knowledge of which combinations are effective may be lost as cultures are assimilated into the mainstream and the natural environment is destroyed. For example, at least 90 percent of the plants in the vanishing Brazilian Amazon have never been chemically analyzed.

Another problem is the time required to bring a discovered drug to market, which is often 10–15 years. Lab-developed drugs are frequently cheaper to refine and document for approval. It is reported that aspirin, derived from willow tree bark, would never have been approved under the modern approval process due to uncertainties about the specific mechanism that makes it effective.

Researchers are also struggling with the dilemma of how to return part of the profits to the people and countries who provided the cure. In Suriname, bioprospectors set up a Forest People's Fund, which received an initial $50,000 royalty from Bristol-Myers Squibb, and which will receive between 1and 5 percent of any royalties from marketed drugs. This program is designed to encourage the local conservation of threatened ecosystems with financial incentives.

Thus, while bioprospecting is showing increasing promise as a source of medicines and disease-fighting drugs, researchers are fighting environmental and cultural destruction in the race to discover the next wonder drug. While the search for traditional cures accelerates, no such acceleration in the process of bringing these cures to market seems imminent.

RAIN FOREST DIVERSITY AND DESTRUCTION

As stated in the introduction, the rain forests of the world need protection because they significantly affect world climate by consuming large quantities of CO_2, contributing a large percentage of the world's oxygen, and slowing global warming. Also, rain forests guard most of the planet's biodiversity (an umbrella term for the variety of ecosystems, species, and genes present). Scientists theorize that tens of millions of species exist; however, they have described only between 1.4 and 1.5 million of them—a small percentage. One-half of the identified species live in tropical forests, but scientists estimate that 90 percent of all existing species live in rain forests and would be identified if only they had ample time to study them. The destruction of rain forests is rampant at nearly 100 acres a minute (approximately the size of the state of Washington destroyed annually). While rain forests now cover only about 5 percent of the earth's land surface, they contain at least half of its plant and animal species. Over the last 30 years, about one-third of the world's rain forests have disappeared, and this rate is increasing (Jukofsky and Wille 1993, 20). At current rates, 5–10 percent of tropical forest species at a minimum will become extinct every decade.

An example of this biological richness and abundance is demonstrated in a study that found that a single hectare (about 2.5 acres) of rain forest in Peru contained 300 tree species—almost half the number of tree species native to all of North America. In another study, scientists found more than 1,300 butterfly species and 600 bird species within one five-square-kilometer patch of Peruvian rain forest. (The entire United States harbors 400 butterfly species and about 700 bird species.) In the same Peruvian jungle, Harvard entomologist Edward O. Wilson determined 43 ant species in a single tree, about the same number as exists in all of the British Isles (Rice et al. 1997, 48).

ECONOMIC VALUE

One reason why such plant and animal diversity is vital is because it is essential for creating food, medicines, and raw materials. Wild plants have the genetic resources to breed pest- and disease-resistant crops. An estimated 120 clinically useful prescription drugs originate from 95 species of plants, of which 39 grow in tropical forests. Botanists estimate that 35,000–70,000 plant species located primarily in tropical forests provide traditional remedies in countries throughout the world. Without the rain forest, these plant species and the vast array of existing and potential medicines derived from them would forever be lost (Rice et al. 1997, 48).

Dr. Robert Balick of the New York Botantical Garden and Dr. Robert Mendelsohn of Yale University have studied the potential economic value that might result from medicines derived from the flora of the rain forest. Effective new drugs are valued at an average of $94 million each to drug companies. They concluded that at least 328 drugs await discovery with a projected value of some $147 billion (Medical Herpetology 1997).

About half of the earth's 250,000 flowering plants exist in tropical forests, but less than 1 percent has been thoroughly tested for medicinal uses. Nearly one-half of all drugs prescribed in the United States have originated from plant life, with 47 medications currently on the market that are derived from the tropical forest including codeine, quinine, and curare (see Table 1) (*Science* 1995).

Table 1
Rain Forest Pharmacy: Partial List

Drug	*Derivation*	*Medical Uses*
Atropine	Belladonna	Eases asthma attacks
Cocaine	Coca	Local anesthetics
D-Tubocurarine	*Chondodendron tomentosum*	Skeletal muscle relaxant
Diosgenin	Mexican yam	Oral contraceptive
Papain	Papaya	Eases chronic diarrhea
Picrotoxin	Levant berry seeds	Eases schizophrenic seizures
Pilocarpine	Pilocarpus	Glaucoma
Quinine	Cinchona	Malaria
Reserpine	Snakeroot	Tranquilizers and treatment of hypertension
Vinblastine	Rosy periwinkle	Helps Hodgkin's disease
Vincristine	Rosy periwinkle	Helps acute leukemia

Source: Jukofsky and Wille 1993, 26.

It is hoped that the economic value of rain forest studies will emphasize the importance of research and accelerate the conservation of invaluable world resources. However, efforts to slow destruction of the world's rain forests have met with limited success. Entire plant and animal populations and species are disappearing rapidly. Aside from deforestation, other factors that may cause this loss, including acid rain, other types of pollution, and increased ultra violet-B exposure, are being investigated. It is hoped that the discovery of pharmaceutical and other products derived from the rain forest may result in increased support for saving the rain forests and their biodiversity and help create funding for continuing research (Medical Herpetology 1997).

BOX 1 Biological Diversity

With an estimated 13 million species on Earth, few people take notice of an extinction of a variety of wheat, a breed of sheep, or an insect. Yet it is the very abundance of species on Earth that helps ecosystems work at their maximum potential. Each species makes a unique contribution to life.

- Species diversity influences ecosystem stability and undergirds essential ecological services. From water purification to the cycling of carbon, a variety of plant species is essential to achieving maximum efficiency of these processes. Diversity also bolsters resilience—an ecosystem's ability to respond to pressures—offering "insurance" against climate change, drought, and other stresses.
- The genetic diversity of plants, animals, insects, and microorganisms determines agroecosystems' productivity, resistance to pests and disease, and, ultimately, food security for humans. Extractions from the genetic library are credited with annual increases in crop productivity worth about $1 billion per year; yet the trend in agroecosystems is toward the replacement of polycultures with monocultures and diverse plant seed varieties with uniform seed varieties. For example, more than 2,000 rice varieties were found in Sri Lanka in 1959, but just five major varieties in the 1980s.
- Genetic diversity is fundamental to human health. From high cholesterol to bacteria fighters, 42 percent of the world's 25 top-selling drugs in 1997 were derived from natural sources. The global market value of pharmaceuticals derived from genetic resources is estimated at $75–$150 billion. Botanical medicines like ginseng and echinacea represent an annual market of another $20–$40 billion, with about 440,000 tons of plant material

in trade, much of it originating in the developing world. Not fully captured by this commercial data is the value of plant diversity to the 75 percent of the world's population that relies on traditional medicine for primary health care.

The threat to biodiversity is growing. Among birds and mammals, rates may be 100–1,000 times what they would be without human-induced pressures—overexploitation, invasive species, pollution, global warming, habitat loss, fragmentation, and conversion (Reid and Miller 1989). Regional extinctions, particularly the loss of populations of some species in tropical forests, may be occurring 3–8 times faster than global species extinctions.

Of the estimated 250,000–270,000 species of plants in the world, only 751 are known or suspected to be extinct. But an enormous number—33,047, or 12.5 percent—are threatened on a global scale. Even that grim statistic may be an underestimate because much information about plants is incomplete, particularly in the tropics. Such localized extinctions may be just as significant as the extinction of an entire species worldwide. Most of the benefits and services provided by species working together in an ecosystem are local and regional. If a keystone species is lost in an area, a dramatic reorganization of the ecosystem can occur. For example, elephants disperse seeds, create water holes, and trample vegetation through their movements and foraging. The extinction of elephants in a piece of savanna can cause the habitat to become less diverse and open and cause water holes to silt up, which would have dramatic repercussions on other species in the region (World Resources Institute 2000).

Table 2
Origins of the Top 150 Prescription Drugs in the United States

Origin	*Total Number of Compounds*	*Natural Product*	*Semi-Synthetic*	*Synthetic*	*Percent*
Animal	27	6	21	—	23
Plant	34	9	25	—	18
Fungus	17	4	13	—	11
Bacteria	6	5	1	—	4
Marine	2	2	0	—	1
Synthetic	64	—	—	64	43
Total	150	26	60	64	100

Source: World Resources Institute (WRI) in collaboration with United Nations Environment Programme (UNEP), United Nations Development Programme (UNDP), and the World Bank, 2000. World Resources 2000-2001: People and Ecosystems: The Fraying Web of Life. Washington, DC: WRI.

REFERENCES

"Drugs and Money Abound in Rain Forests." (July 1995). *ION Science*. Available at http://w.w.winjersey.com/media/Ion-Sci/glance/news795/raindrug.html.

Goering, L. (September 12, 1995). Rain Forests May Offer New Miracle Drugs. *Chicago Tribune*, Section 1, 16.

Jukofsky, D., and Wille, C. (April/May 1993). They're Our Rain Forests Too. *National Wildlife* 31(3):18–37.

"Medical Herpetology." (1997). Available at http://www.world-corp.com/biodiversity/newsletter/two/herp.htm.

Rice, R. E., Gullison, R. E., and Reid, J. W. (April 1997). Can Sustainable Management Save Tropical Forests? *Scientific American* 276(4):44–49.

World Resources Institute. (2000). *World Resources 2000–2001: People and Ecosystems*. Washington, D.C.: World Resources Institute.

QUESTIONS

1. Should local inhabitants be rewarded for drugs discovered in their region? Can ownership of a renewable living resource (such as a plant) be treated in the same way as a nonrenewable resource (such as mineral rights)?
2. The reading mentions that it can take 10–15 years to gain approval and bring a drug to market. Is there a justification for performing drug trials on patients before full testing has been completed and the drug is proven safe and effective? If a patient has a terminal illness, what guidelines would you suggest in administering experimental medicines?
3. Researchers are concerned that potential cures will be lost to the destruction of the environment and the assimilation of folklore into popular culture. While there is much focus on saving the environment, the preservation of cultural diversity attracts little attention. What responsibility falls on Western nations in safeguarding native cultures in the countries that they influence? Conversely, what right do they have to deny modern "civilization" to less technological societies?
4. The goal of the medical community, including drug companies, is to keep people alive for the longest possible time regardless of their usefulness to society as a whole. It is also true that the earth has limited resources, and natural law dictates that exponential population growth is not sustainable. This dictates that at some point, mankind will be beset by natural and uncontrollable events (such as epidemics) that will cull the world's population back to sustainable levels. Extend this logic to produce an argument supporting euthanasia or other solutions to world over-population. What are the ramifications of such arguments?
5. Project how long the rain forests might exist given the current rate of destruction. What approaches can you think of to stop this rate of destruction?
6. What specifically are the ramifications of our increasing loss of rain forests? From this answer, derive solutions for preventing this loss.
7. How effective do you think estimations of the economic value of potential drugs will be to save rain forest diversity? How would you use this strategy of placing an economic value on rain forest resources to support rain forest conservation?
8. Research a more extensive list of rain forest medicines.
9. The reading states that plant and animal diversity is vital for creating food, medicines, and raw materials. What are some other reasons?

Case Study 1

Buried Displeasure: The Love Canal and One Person Who Made a Difference

EDITED BY BARBARA A. EICHLER

This is the story of perhaps the country's most notorious waste site battle. It ignited people's emotions, made headlines, and brought together homemakers, scientists, corporations, politicians, government officials, and activists to the environmental nightmare and corruption of industrial wastes that literally lay underneath the feet of families and growing children. It brought together the nexus of environmental responsibility, the technological age, and neighborhoods of children caught between unseen consequences and ignored responsibility.

Love Canal was the dream community of William T. Love. He imagined a village nestled in the rolling hills and orchards near the Niagara River in upstate New York. The success of the venture, begun back in 1894, depended on the construction of a canal that would connect the two branches of the Niagara River. The canal would tap the power of the rapids just before Niagara Falls and produce hydroelectric power, which would attract business and industry to the area.

But Love's dream was not to be. Economic difficulties caused many of his backers to pull out of the scheme. The only trace of Love's plan was a 3,000-foot long, 60-foot wide canal. In 1927, the area was annexed to the city of Niagara Falls. Around 1946, the city was approached by the Hooker Chemical Company, which was looking for a place to dump chemical wastes. Love Canal seemed like the perfect spot, sparsely populated, and the thick, clay-like soil provided protection against any possible leakage. Hooker began dumping chemical wastes into the canal that year and for every year until 1952. At least 150 chemicals were placed into Love Canal, including dioxin, the most toxic man-made chemical known. There was little public awareness at the time about the dangers of these chemicals and their connection with nervous system disorders, kidney problems, respiratory distress, deafness, and birth defects.

The potential for a technological disaster at Love Canal mounted during the 1950s. The post-war baby boom had created a pressing need for housing and schools. Officials in Niagara Falls began looking at Love Canal as a place to expand. In 1953, the Niagara Falls Board of Education announced its intention to acquire Love Canal for the purpose of building an elementary school in the area. Hooker, aware the land could be taken from them via eminent domain, sold the land to the city for $1, but made city officials sign a deed acknowledging that the area had been used as a chemical dump site and absolving Hooker of any deaths, loss, or damage to property once construction began.

The 99th Street School was built on the central part of the canal, and the city sold the southern part to developers. As the bulldozers began tearing away the top soil, the rain and snow gradually began to seep into the drums and containers holding the chemical wastes and, in time, a chemical mixture, leachate, began to flow out of Love Canal. The leachate that began to appear in the late 1950s produced skin burns on a few children who were playing in the area. But that was nothing compared with the complaints that began surfacing in the 1970s as the danger of Love Canal became apparent. Heavy rains during that decade brought the buried chemicals to the surface, and with them came increased reports of miscarriages, birth defects, liver abnormalities, and cancer. In 1978, over 200 families were forced from their homes after toxic fumes were detected in their basements. Two years later, the federal government and the state of New York, reacting to an Environmental Protection Agency (EPA) report, moved an additional 800 families out of the Love Canal area.

Love Canal remained a near ghost town throughout the 1980s as the EPA, Hooker Chemical, the School Board and city of Niagara Falls, as well as numerous insurance companies thrashed out the legal and possibly criminal costs of this disaster. In 1988, Love Canal residents received a $20 million settlement in addition to the $30–$40,000 each family received for their homes. Individual claims were in the $2,000–$4,000 range. By 1990, 1,000 cases were still pending.

ONE PERSON MAKES A DIFFERENCE

Lois Gibbs was a housewife living in Love Canal in 1978. She discovered an epidemic of miscarriages, birth defects, nervous system problems, and respiratory disorders across the neighborhood; she also learned that the neighborhood had been built next to a huge toxic waste dump. She became a neighborhood organizer and ultimately became responsible, as much as anyone, for bringing this toxic disaster into the public forum and causing responsible parties to pay for victim relocation.

She continued to become one of the most prominent grassroots environmental activists in the United States. She founded the Citizens Clearinghouse on Hazardous Waste (CCHW) and has written extensively on toxic hazards, the prevention of new ones, and approaches for communities with existing toxic dumps. Ms. Gibbs has also written on the effects of dioxin as both a potent carcinogen as well as a cause of illness affecting almost every major system of the body such as diabetes, chronic bronchitis, irregular heartbeat, nervous disorders, thyroid and immune system disorders, and a variety of reproductive system disorders including birth defects, miscarriages, and lowered fertility. Gibbs recommends a strong dose of democracy as a cure for toxic environmental problems. She concludes, "The job is too big for some national organization or remote coalition to achieve on our behalf . . . Our country's power is vested in the people, and the people must act." (Mazza 1996).

SOME INFORMATION TO CONSIDER

- Manufacturing wastes are being created at an accelerating pace of 6 percent per year, which means that total waste production is doubling every 12 years.
- In the first 12 years since the problems at Love Canal were discovered, American industry pumped out a total amount of toxic waste equal to all of the toxic waste created prior to Love Canal (from the years 1880–1978).
- The Niagara River has the greatest concentration of toxic dumps anywhere on the North American continent with 65 huge chemical dumps along the banks of the river. Love Canal, at 20,000 tons of toxins, is not the largest of these dumps: the Hyde Park dump contains 80,000 tons of toxins, the "S" site contains 70,000 tons, and the 102 Street Site contains 80,000 tons. All of these sites are within a few hundred yards of the river (Montague 1990).

When Lois Gibbs founded the CCHW in 1981, its main focus was to help community groups who were suffering from the effects of toxic dump sites, such as Love Canal. It has since expanded its programs to address a broad range of environmental issues including toxic waste, solid waste, air pollution, incinerators, medical waste, radioactive waste, pesticides, sewage, and industrial pollution. The CCHW is a now a more than 20-year-old nonprofit environmental organization and remains the only national organization started and led by grassroots organizers. As of 1996, the CCHW had worked with over 8,000 community-based groups nationwide (Gibbs 1996).

The CCHW has changed its name to the Center for Health, Environment and Justice, where Ms. Gibbs remains the executive director. It continues to grow into an effective, broad-based information, referral, and support resource organization to help individuals, organizations, and communities eliminate exposure to hazardous chemicals. It specializes in dioxin as well as other toxic, chemical, and hazardous wastes and provides various publications and resources for communities, community leaders, and strategists devoted to hazardous waste information and elimination.

Complete and discuss the following flowchart.

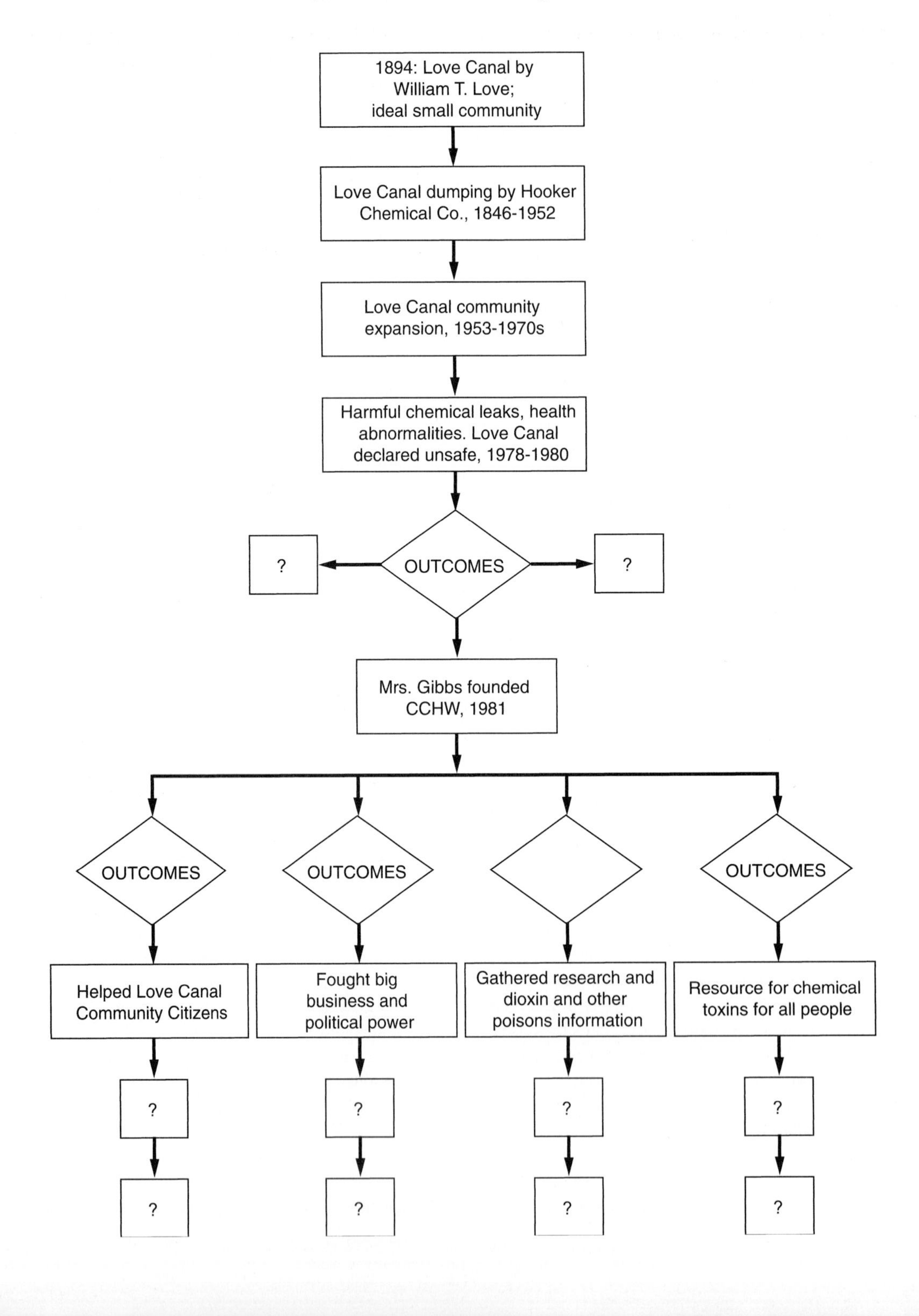

REFERENCES

Center for Health, Environment and Justice. Available at www.chej.org.

Ferell, O. C., and Fraedrich, John. (1991). Excerpted from *Business Ethics*. New York: Houghton Mifflin. Additional information provided by researchers Denise Schultz, Paula Ingerson, and Joanne Hammond.

"Lois Gibbs' Biography Page." (1996). Available at http://www.medaccess.com/newsletter/n10415/gibbiog.htm.

Mazza, P. (February 1996). "Love Canal Is Everywhere: The Pervasive Threat of Dioxin." Available at http://www.tnews.com/text/dioxin.html, pp. 1–3.

Montague, P. (June 1990). The Niagara River—Part I: How Industry Survived Love Canal. Rachel's Hazardous Waste News #186. Annapolis: Environmental Research Foundation. Available at http://xp0.rtknet.org/E3540T132, pp. 1–3.

QUESTIONS

1. List the unethical and poor environmental decisions made that caused the development and large-scale damage of the Love Canal scandal.
2. List parallel ethical and sound environmental corrective steps that could be made in this Love Canal case.
3. From the reading and your interpretation of her actions, what characteristics do you think made Lois Gibbs such a successful leader? Discuss.
4. Consult the Web site of CCHW to further research its community outreach and support in dealing with hazardous wastes and especially dioxins.

Case Study 2

Europe's Black Triangle Turns Green

BRUCE STUTZ

Who says environmentalism is nothing but bad news?

For two days it has been cold and pouring continually, but each morning the caravan of scientists rolls out from the inn on the square in the small northwestern Czech village of Horni Blatna and heads an hour north into the mountains. At the group's study site, just a few miles from the German border, the forest is full-grown Norway spruce about a hundred years old. The trees survive on the western edge of the notorious Black Triangle, the heavily industrialized region where Poland, Germany, and the Czech Republic meet. During the Communist era, this 12,000-square-mile area was one of the most polluted industrial landscapes on the face of the globe.

The group unpacks its gear—from pruners that can reach branches 30 feet off the ground to small glass lab dishes in which a single spruce needle can be cut up and preserved—and hikes into the woods. Barrett Rock, a professor of natural resources at the University of New Hampshire, a tall, ruddy-faced, white-haired Vermonter, briefs the researchers on their procedures and they set to work, some at branches, some at trunks, some at roots, like a Lilliputian surgical team operating on a giant. Their patient, however, is not one tree or a single group of trees but the forest itself. They want to know what effects the region's pollution has had on it. And then, they hope, if their measurements and instruments are sensitive enough, their analyses can be used to chart the pathology of this or any other forest. These Czech scientists and students working in the forest with their American counterparts regard its recovery in the Krusné hory—the Ore Mountains—as something of an ecological allegory, a tale in which natural fortunes reflect political ones.

I watch Rock drill into one of the tree trunks with a hollow bit. He removes a pencil-thick core more than a foot long and holds it up to show me the growth rings.

"You see how they get wider?" he asks. "I can read the changes in government in the record of the tree rings."

The changes Rock refers to began in 1948. Having just emerged from the grip of Nazi Germany, Czechoslovakia became a Soviet satellite. Moscow chose the Most basin, the sprawling plain south of the Krusné hory that is named for its major city, to be Czechoslovakia's industrial center. Known until then for its Bohemian glass, ceramics, and textile industries, the basin had a ready source of fuel—lignite—that lay close to the surface and in deposits 300 feet deep.

Lignite is a coal that never fully evolved, no longer peat but only halfway to becoming bituminous, or hard, coal. It's soft, brown, and crumbles easily. It's too delicate to be

OnEarth, Spring 2005 (27), 1, pp. 14–21.

shipped long distances, and it burns very quickly (disintegrates might be a better way to put it). Pound for pound, lignite generates less heat than hard coal and produces four times the sulfur when it is burned. What recommended lignite to Soviet managers was that it was available and it was cheap.

The lignite strip mines expanded and deepened. As industries moved in—chemical plants, refineries, steam heating and power plants—so did workers. To provide heat and electricity for the growing population, more coal was burned. When still more coal was needed, the government had no compunction about bulldozing entire villages so the strip mines could devour more of the landscape. North of the basin, where the three national borders converge, the mountains formed an angular barrier that contained the increasingly polluted air, enveloping the Most basin in ash and ozone. When the wind blew, plumes of pollution swept up into the spruce forests.

Writing in 1987, electrical engineer Eduard Vacka described life in the Black Triangle town of Teplice:

> It was one of those miserable fall days, when you wake up in the morning with a throbbing headache. Out the window, it looks like a dark sack has been thrown over the whole town, just as it had all week. "Back into this s***," you mutter under your breath as you close the door. "God, what a stench! What the hell are they putting in the air? It's unbelievable: they're waging chemical warfare against their own people."
>
> If you say you can't breathe, there are two meanings. The first is symbolic, that the mental environment is stifling, choked with lies and hypocrisy: there is no breathing room. The second meaning is more immediate, that the air itself is corrupted and you are literally choking to death.
>
> The first is a sigh of despair; the second a cry for help.

For many Czechs the Black Triangle became a symbol of Soviet callousness toward both nature and their culture. Josef Richter, a physician who heads the department of immunology at the Institute of Public Health in Usti, witnessed the region's decline. Now a stout, barrel-chested man with white hair, Richter has lived all his 70 years in the Most basin. During the Soviet years his father, for the offense of being half-Gypsy, or Roma, had been forced to work in the infamous uranium mines in nearby Jakimov. For him, as for many others, work in the mines with no protective gear was a sentence to a slow but certain death. When the younger Richter had the temerity to express a desire to attend medical school, Communist authorities told him that he, too, had to first serve a year in the mines.

Before I can ask him what it was like, he stands, lifts his shirt, and points to a scar on the right side of his chest.

"Bronchogenic carcinoma. All of us had tumors removed," he says, lowering his shirt and sitting. "I would have to walk with a uranium stone for five kilometers."

After earning his degree Richter came to the Usti hospital.

"The main chemical plant was built here during the time of Hitler," he tells me. "In the Communist era they began building energy plants. Eighty percent of Czechoslovakia's electricity was produced here, and all of it from brown coal."

The population was also changing. Following the war, Richter explains, the Czechs, with clearly justified grievances but with sometimes savage zeal, drove the German population out of Bohemia, the area known by the Germans as Sudetenland. Much of the textile and glass industries went with them. "What we got in exchange were people with little education and with no roots in the region—many Gypsies, very few with a high school education."

The remark sounds harsh in light of his own heritage, but Richter is a straightforward man.

He adds, "It was a new social-economic construct, and the impact was great."

With no attachments to the place and the land, very few of the newcomers felt they had a stake in the region; few protested the environmental havoc being wreaked around them. Those who objected did so by leaving.

Richter stayed on, and throughout the Soviet years he wrote some 150 papers on the region's pollution and health problems. But under the Communist regime there was no possibility of publishing his work outside Czechoslovakia. The rest of the world never saw his data or their damning conclusions.

I ask him what the Soviet managers said when he warned them how bad the pollution was.

"'Comrade,'" they told me, 'you must develop men so strong they can live in this pollution.'"

By the 1980s, sulfur dioxide concentrations in the month of January, when heating demands were at their height, could average more than 75 parts per billion (ppb), nearly double what the U.S. Environmental Protection

Agency (EPA) considers the highest acceptable level. During the same period the average annual nitrogen dioxide concentration was 25 ppb, slightly above acceptable levels.

But the killers, literally, were the particulate emissions—the fine particles of soot and dust, less than a quarter the diameter of a human hair—that were released into the air. Photographer Antonin Kratochvil still recalls the haze of soot that used to hang in the Most basin, the surgical masks children wore on their way to school. Among the shades of "Kafka gray," as he puts it, that characterized Czechoslovakia under Soviet rule, the Most basin was the darkest.

The EPA requires that over the course of a year, particulate emission densities not average more than 40 micrograms per cubic meter. In 1980, densities in the Black Triangle averaged more than three times that, reaching nearly 200 micrograms per cubic meter during the winter months.

In a single generation, the average Czech life expectancy had fallen to seven years lower than that of Western Europe. The infant mortality rate was 40 percent higher than the European norm. The incidence of respiratory infections was five times greater. Those who could moved away, a migration caused not by war or natural disaster but by an environment made uninhabitable.

Once released into the air, sulfur dioxide and nitrogen dioxide, which do so much damage to the human respiratory system, undergo a chemical change. With sunlight as a catalyst, they combine with atmospheric water vapor and oxygen to form sulfuric and nitric acids. In a high mountain forest such as the Kru[[scaron]]né hory, leaves soak in the acidic moisture from low clouds and mist. When it rains or snows, the acidic precipitation acidifies the soil. Nutrients in the soil get broken down and washed away. In their place substances toxic to trees, such as aluminum, are released and the trees, being the efficient hydraulic systems that they are, suck them up. Between 1972 and 1989 about half of the Kru[[scaron]]né hory forests died, 115 square miles' worth, and had to be clearcut.

Ducking into my car to find some refuge from the day's rain, Rock explains that his studies of the effects of acid rain in the northeastern United States were what brought him to the Czech Republic. As a research scientist at the government's Jet Propulsion Laboratory in Pasadena, California, Rock had considered using satellites for environmental monitoring. When he came to the University of New Hampshire in 1987, Rock began trying to assess the decline of New England forests. What he concluded immediately was that airborne pollution appeared to hang in a layer at 3,000 feet, the altitude at which the bottoms of clouds flatten out.

"What I was seeing was that the bottom of a cloud could have a pH of 2.5 while the top of the cloud could be 4 to 4.5."

This is not a trivial difference. Each unit increment on the pH scale indicates a tenfold difference in the concentration of hydrogen ions. The lower the pH, the higher the acidity. What Rock was recording was a pH 100 times lower at the base of the clouds than at their top. Pure water has a neutral pH of 7. Normal rain, because of dissolved carbon dioxide, has a pH of 5.5. Rain is considered acidic when the pH level drops to 4. At 3,000 feet Rock was measuring vinegar. He tells me that in one Vermont forest, on September 20, 1988, scientists measured a single cloud with a pH of 2.6. "For three hours it hung there and just toasted the needles."

Rock concluded that pollution damage hits first and hardest at 3,000 feet, though other researchers questioned this. "I was asked whether this 3,000-foot acid rain damage applied only in a specific location in Vermont where the pollution input was constant. I had to find a place where, at 3,000 feet, they had a range of pollution levels," depending on proximity to industrial sources and wind direction.

"So I came to the Kru[[scaron]]né hory. When I looked at the Landsat images I couldn't believe what I saw. The damage was appalling."

This was in 1989. What Rock found on the ground when he arrived in May of that year confirmed the worst.

"Retention of needles is a key indicator of health in these trees. A healthy tree may have 12 or more years of needles on its limbs. These had only two or three. The trees were skeletons with tufts of needles. When I looked at the cells of these needles I saw they were suffering plasmolysis, the inability to retain water. The cell content pulls away from the cell wall. The cells become physiologically crippled."

When Rock takes out photographs of damaged needle cells they recall for me comparisons between normal lungs and smokers' lungs. The cells' chloroplasts, which are responsible for photosynthesis, disintegrate. Acidic tannins, looking like tar, accumulate inside the cell. The cell's walls go flaccid and the needles' normally orderly interior structure disintegrates. Says Rock, "Sulfur dioxide and ozone are the only things that do this."

As it turned out, the most damaged Czech forests were those located at 3,000 feet.

In November 1989, a few months after Rock's first trip to the Krusné hory, the great change came. In what would become known as the Velvet Revolution, the shaky Communist regime fell and gave way to a new Czech democracy. The right to clean air was among the first demands of the revolution's leaders, and very quickly the new republic set out to clean up the Most basin. What was required was clear. Smokestacks needed scrubbers to remove sulfur from emissions. Inefficient energy plants had to be shut down or switched to low-sulfur, low-nitrogen fuels, such as hard coal or natural gas. The loss of these plants would also require the development of new sources of low-polluting energy. All of these changes would take money.

In 1990, the Czech Republic received financial aid from the European Union's Phare plan, under which grants were given to nations intending to join the E.U. From 1990 to 2003, the country received some 1.1 billion euros ($1.42 billion) in aid. To this was added $626 million from the World Bank. Combining these funds with technical assistance from Europe and the United States, the Czechs began dismantling the wasteful and damaging energy complex that had been imposed by the Soviet central committee four decades earlier.

At the same time the Czechs began gradually to diversify their sources of power, adding natural gas, hydropower, and a new nuclear power plant. They ended price controls on energy costs. They passed new standards limiting sulfur dioxide emissions that mirrored the EPA standard of "best available technology." The legislation also enabled them to levy fines on polluters. And they began to share energy resources with three neighboring countries—Germany, Hungary, and Poland—which were also receiving assistance from the E.U. to implement similar changes. Since 75 percent of the pollution produced in the Black Triangle areas of Germany and the Czech Republic ended up in Poland, that country immediately benefited from the changes that were made by its neighbors. Poland began treating its own emissions as well, especially those from its huge Turow power plant. Germany, meanwhile, began to retrofit its massive power plant in Boxburg, adding desulfurization technology, as well as decommissioning smaller, lignite-powered plants. To complement these efforts, the three Black Triangle states set up a joint air-monitoring system, which continuously measures major air pollutants at 42 stations.

Henry Manczyk, an energy expert who has studied the efforts of the Czech Republic, Poland, and Germany to reduce pollution, says the success of the program sets up a paradigm for other former Soviet states, with Bulgaria, Ukraine, and Belarus all likely to follow as their political circumstances change.

Mining output in the region dropped from 80 million tons of lignite in 1984 to 56 million tons in 1993 and then to 50 million tons in 1999. By 1996 the Czechs could measure a decline in sulfur dioxide emissions, from 2.5 metric tons in 1982 to less than 1.4. Rock, working by then with Czech scientists, most closely with Jana Albrechtova, a plant physiologist at Charles University in Prague, began seeing changes in the forest: The acute damage appeared to have come to a halt.

Landsat imagery of the Kru[[scaron]]né hory in 1997 showed Rock that the areas whose appearance had shocked him when he first came to the region were now experiencing the beginning of a remarkable reversal of fortunes.

During Rock's tenure at the Jet Propulsion Laboratory, NASA satellites had begun measuring the reflected infrared light from the forests below. Light in the infrared spectrum is invisible to the human eye. ("A good thing," he says, "or we'd be blinded by a walk in the woods.") Each plant's leaves reflect a specific range of infrared light. By using a spectrometer, the satellites showed what trees made up a forest and, to a lesser degree, how healthy those trees were.

Rock, however, was dissatisfied with the resolution of the satellite imagery. He believed that by doing close-up spectrometry he'd be able to tell what was going on inside the leaves. If he could see that, he might also be able to recognize when a forest was in danger of dying.

So he took his students out into the Kru[[scaron]]né hory with a spectrometer. And what the spectrometer did in fact see was what was happening at the cellular level. It was like going from photograph to CAT scan. They could measure the leaves' chlorophyll concentrations, water content, and cellular development—essentially, the health of the leaf, the tree, and the forest.

Outside in the chilling rain, Rock, Albrechtova, and other researchers and students are hustling from tree to tree, pruning branches, taking cuttings, digging soil samples, and collecting needles.

Albrechtova is an intense woman with an athlete's lithe build. In the forest she's all business. She points out that some of the trees now have seven or eight years of needles on them. The needles are dark and plump. In the young forests, planted only five or six years ago, the trees are healthy and reproducing on their own. A once-denuded hillside is damp and redolent of young pines. It's possible once again to appreciate the rain.

The environmental indicators—sulfur dioxide emissions, particulate densities—have continued to improve steadily. On graphing paper the slope of pollution decline is a slide into a ravine. Nearly as rapid has been the improving health data coming from the Most basin. While it still lags behind the rest of the country, life expectancy there has increased and deaths from respiratory illnesses have declined.

Josef Richter is sanguine about the changes, yet he makes it clear that the cultural and social damage will not be so easily fixed. There is a lack of education here, and unemployment stands at 20 percent. These conditions bring with them their own health problems, from poor diet to smoking to drug addiction. Richter adds that the cost of cleaner fuel is high. Natural gas, for instance, must be imported from Russia, and hard coal costs more than lignite. As a result, more of the population is turning to burning wood, which produces almost as much pollution as coal.

The culmination of the week comes at a meeting in the offices of the Czech forest service in the Horni Blatna town hall. Forest managers from across the country, most dressed in their dark green uniforms—collarless jackets with satin piping around the lapels that appear to have been fashioned a century ago (and then for a marching band)—sit at long folding tables. On the walls hang mounted elk antlers, ram horns, and boar hides.

Albrechtova has changed from jeans to a tailored dress suit. She is introduced, strides with her usual forcefulness to the podium, makes her formal acknowledgments to the director of the service, and, switching between English and Czech, describes the results of the present fieldwork. "The good news," she says, "is that there has been no acute damage to the needles since 1991."

She explains that she and Rock and their associates have been able to identify the chemical changes that take place in the needles of spruce trees under stress. Further, since Rock's on-site spectral analyses match up with her off-site laboratory studies, they know they can quickly find out a forest's state of health.

"The not-so-good news is that the forests are still not ecologically stable." This, because of the effects of 40 years of acid precipitation on the soil.

The foresters appear to listen carefully. They look like a tough crowd, and Albrechtova has told me that many of them long ago concluded that the only way to counteract the acidifying of the forest was aerial dusting with lime. "Some of them are very dedicated to liming," she tells me after the meeting. "But the liming may not be having the effects they think and it's very expensive. Still, now that they've been doing it for a couple of years, they're suddenly glad to hear that we found the forests have improved."

She feels she had to make the case, first, that even aside from the liming, the forest chemistry is changing and the forest improving due to cleaner air; and, second, that the techniques she and Rock have developed will allow them to monitor the forest's continued improvement. Monitoring is key, and whether the foresters believe it's being done to measure the effects of liming or just changes in pollution is, at this point, not important.

After Albrechtova speaks I look around the room and there seems to be general approval. The scientist sitting next to me says how important he considers the work that Rock and Albrechtova are doing. Rock is pleased as well, both with their findings and with the meeting. His only concern now, he says, is getting his samples of tree branches past customs when he returns home.

BOX 1 Return of the Native

From a 3,000-foot ridge atop the Czech Republic's Krusné hory, or Ore Mountains, the sight of young spruces in dense stands takes photojournalist Antonin Kratochvil by surprise. When he last visited 14 years ago, the vista was far different.

"I'm telling you," he says, "there was nothing here. These hills were bare except for the stumps of dead trees."

A Czech émigré who fled Communist Czechoslovakia in 1967, Kratochvil thinks I don't believe him.

"I have pictures to prove it," he insists as he walks off down the slope planted with 10-year-old Norway spruces, trees healthy enough to be engendering seedlings of their own.

"Come here," he calls. "Look, I told you: old stumps! That was all that was here."

Too truculent a man to keep his thinking to himself, Kratochvil tells me as we walk back to the car that the sight of what has been restored here reminds him of what is elsewhere irretrievable. For while the forests of the Krusné hory were dying, the Czech world around them was suffering prodigious sorrow. Under 40 years of Soviet rule, cultures vanished, families dissolved, and tens of thousands of lives came to ruin or death. Some, like Kratochvil, escaped and found refuge elsewhere.

"You can't understand what was lost," he tells me. "The forest can return. The people and their cultures can't."

Today, however, the air is clear, and it's only as we get close to the sprawling Chemopetrol plant that we smell anything but fall in the air. The complex is huge. I count in one quick scan of the scene a dozen concrete towers billowing steam. Along the roadside, stands of cattails are the only signs that streams meander through the property. But even the cattails are new to Kratochvil. When he was here last, there was nothing at all along the road to soften the grimy industrial landscape.

As we drive through the city of Most, Kratochvil tells me he is intent on finding a small town near Litvinov, which he photographed 14 years ago. Like many other towns in this area, it no longer appears on our road map.

"They were planning to move the town to make more room for the mine. I just want to see the little church that I shot."

During the Communist days, even when the mines swept away the people and their homes, it was customary to leave the churches standing. In the case of the old town of Most, they went so far as to move the church to a new site.

Since the map is no help, Kratochvil stops to ask passersby about the missing town. "You have to ask the old people," he says. "The young people have never even heard of it."

But even the old people have to squint to recall the place. The roads they indicate, bordering open fields and lined with abandoned pear and apple trees, lead us only to the edge of the vast open pit that is the mine—a rough-sided gouge that stretches for miles. Between the mine and the surrounding villages the land has been filled and is covered with high grasses.

"None of this was filled when I was here," says Kratochvil. The missing towns and rolling hills of fill are disorienting him.

A man walking down the road seems to know.

"Is the church there?" Kratochvil asks.

The man shakes his head and mutters something.

"What did he say?" I ask.

"They finally destroyed the church as well."

We wind our way through a small village and suddenly find ourselves driving on a road out among empty fields.

"It was here," Kratochvil says. The road ends at a gate surrounded by piles of rubble. Beyond, there is unplanted landfill, and beyond that, the mine.

I'm thinking, he's a great photographer but even he can't photograph what's no longer there. But I can see it in his face.

QUESTIONS

1. Provide three reasons why the Most Basin in the "Black Triangle" became so toxic. Discuss the political, economic, and cultural reasons why this occurred.
2. Provide three reasons why the Most Basin is on its way to recovery. Discuss the political, economic, and cultural reasons why this is occurring.
3. Discuss what this case study teaches us about environmental destruction and environmental recovery.
4. What are two special challenges that lay ahead for the Most Basin in the Czech Republic for recovery to continue and become permanent?

Complete and discuss the following flowchart.

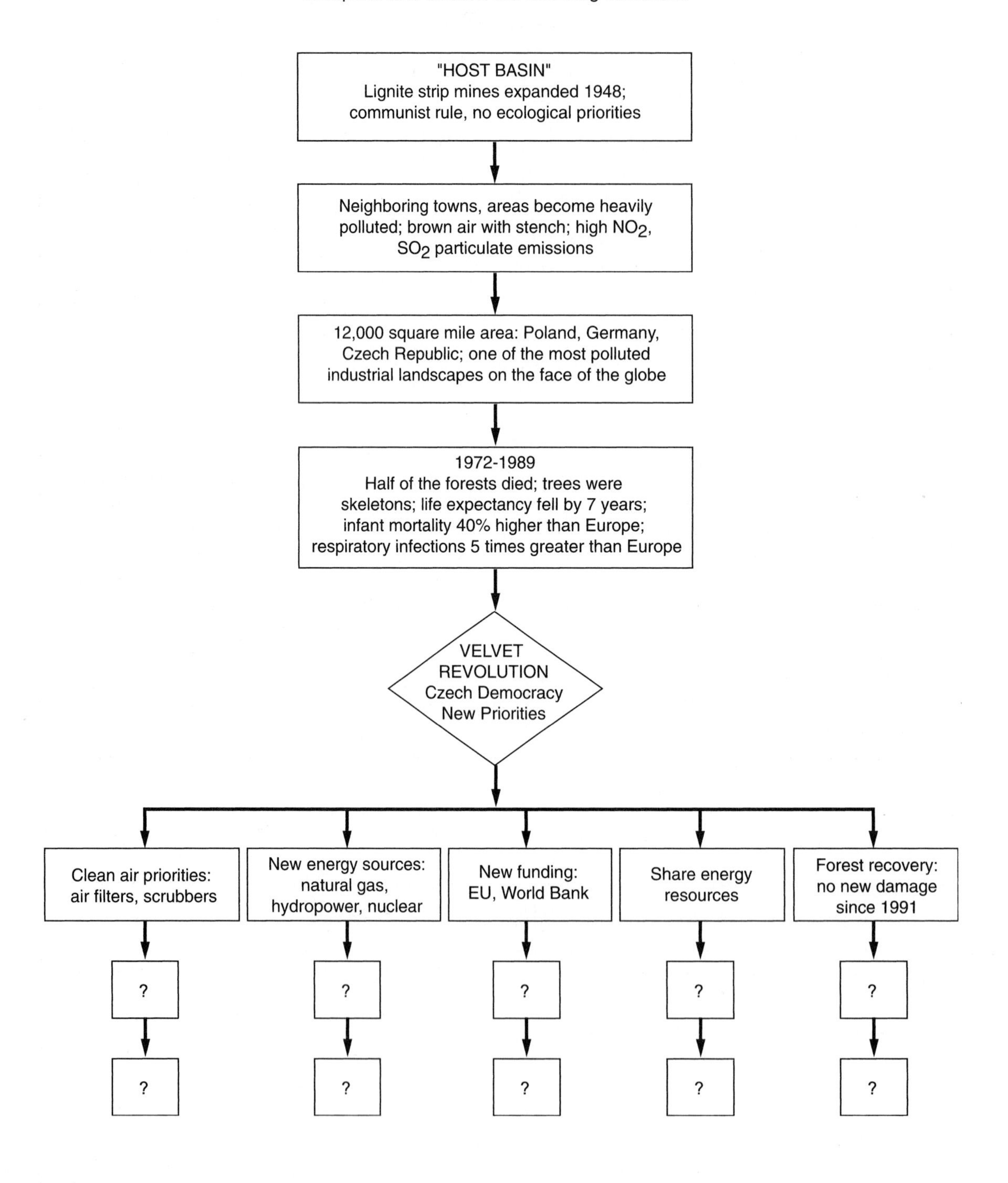

SCENARIO I

California Leads the Nation in Driving a Solution to Global Warming

California has passed legislation that approved the nation's first rule to cut global warming pollution from cars. The new standard, which takes effect for new cars and light trucks beginning in 2009, will require automakers to reduce tailpipe emissions of CO_2 and other pollutants by 30 percent with 2016 models. "With the Bush Administration and Congress refusing to take action on global warming, California is leading the way to show that Americans can help solve this problem," says Natural Resources Defense Council Policy Director Roland Hwang (Natural Resources Defense Council 2005, 1).

The new rule is highly supported in California, which has the largest auto market in the United States and buys about 1.7 million new cars and light trucks each year. Yet this new standard could also reach across the country. California was regulating tailpipe pollution (with stricter standards than the Clean Air Act) before the federal government passed the Clean Air Act; therefore, it has the authority to adopt its own motor vehicle standards. Seven states (including New York, New Jersey, and Massachusetts) use California's stricter standards than those of the federal government. Therefore, these same states and others are likely to also adopt California's new global warming pollution standards. This could triple the number of cars required to cut global warming emissions and thereby make a difference in U.S. emissions.

The technology exists to make less polluting cars, SUVs, minivans, and pickups, but many automakers are attempting to block the new standard in courts, even while the public is expressing its preference for cleaner cars with their high demand for advanced hybrid cars from Toyota, Honda, and Ford. "Automakers have fought every health and safety requirement from catalytic converters to seat belts," says Hwang. "California's plan is sensible, feasible, and legally sound. It's time for automakers to innovate, not litigate." (Natural Resources Defense Council 2005, 1).

QUESTIONS

1. Research your state's emissions and emission standards by using an almanac or other source. Calculate how much CO_2 would be saved in your state from these emissions standards.
2. Considering that the United States emits one-fourth of the world's CO_2, research and calculate how much the United States would lessen its emissions from these standards.

Source: Natural Resources Defense Council, California Drives a New Solution to Global Warming. *Nature's Voice*, January/February 2005.

SCENARIO II

Earth Day 2000: A 30-Year Report Card

On the first Earth Day in 1970, experts warned that the planet's natural systems were being dangerously destabilized by human industry. Here is how we have fared on some key fronts since then.

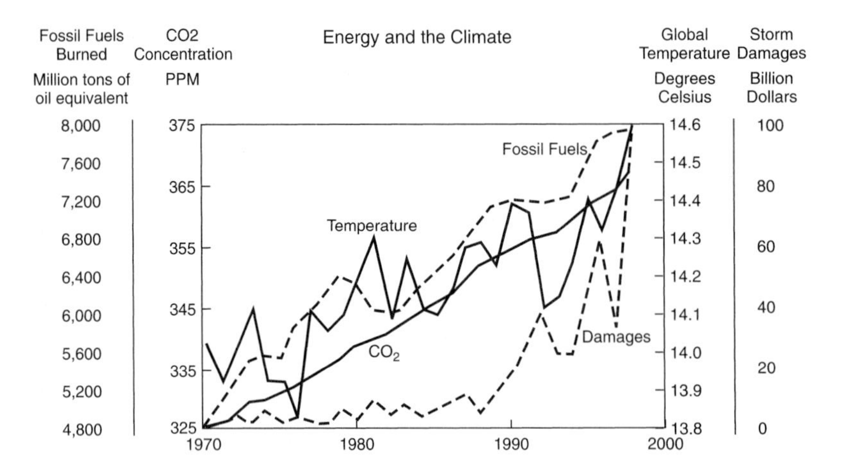

Energy and the climate.

As our growing population increased its burning of coal and oil to produce power, the carbon locked in millions of years' worth of ancient plant growth was released into the air, laying a heat-retaining blanket of carbon dioxide over the planet. Earth's temperature increased significantly. Climate scientists had predicted that this increase would disrupt weather. And indeed, annual damages from weather disasters have increased over 40-fold.

Solution: A faster shift to nonpolluting, renewable solar, wind, and hydrogen energy systems.

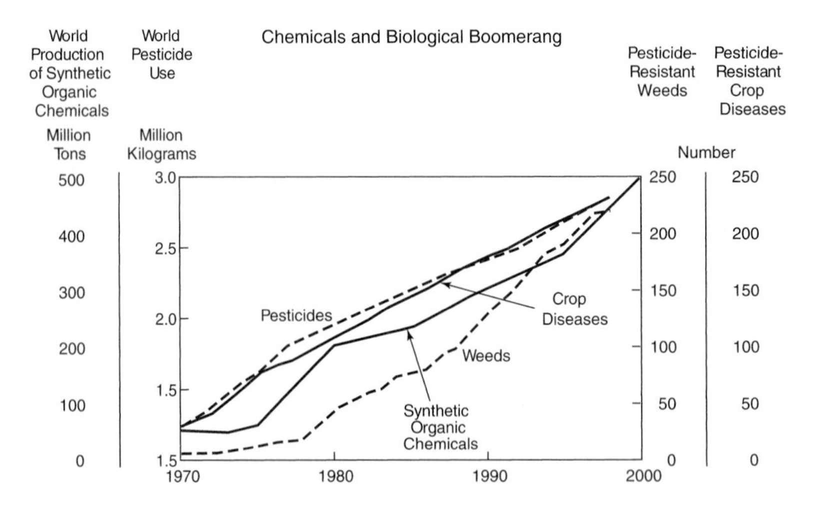

Chemicals and the biological boomerang.

Our consumption of chemicals has exploded, with about three new synthetic chemicals introduced each day. Almost nothing is known about the long-term health and environmental effects of new synthetics, so we have been ambushed again and again by belated discoveries. One of the most ominous chronic effects is that, as pesticide use has increased, so has the evolution of pesticide-resistant pests.

Solution: A large-scale shift to organic farming, a shift away from excessive consumption of synthetic chemical products, and application of the precautionary principle to the chemical industry.

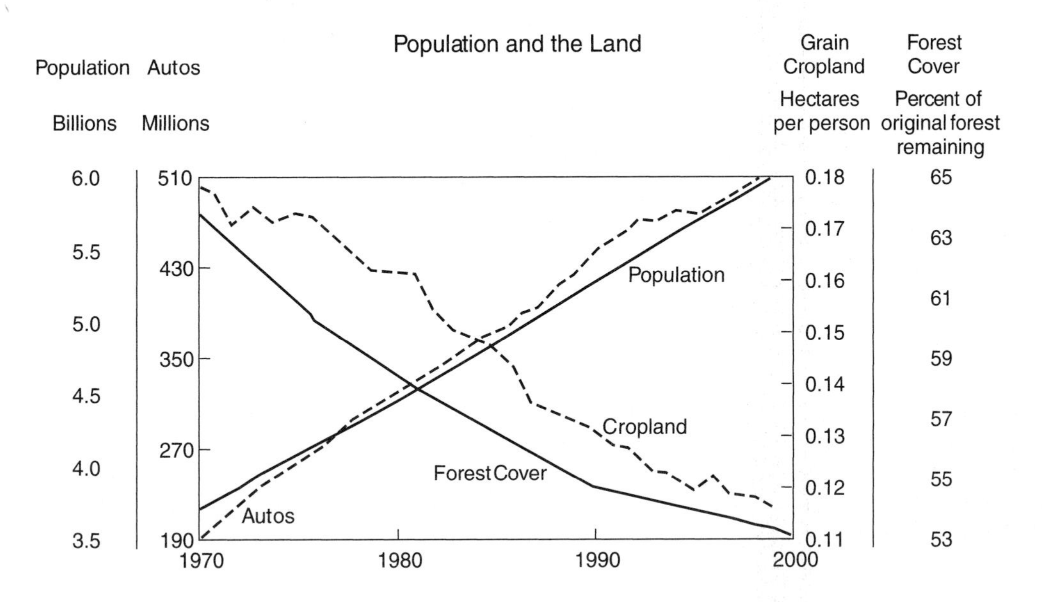

Population and the land.

Population has increased as much in the past 30 years as it did in the 100,000 years prior to the mid-twentieth century. As the number of people has grown, the amount of land used by each person—either directly or through economic demand—has also expanded. As a result of this double expansion, incursions of human activity into agricultural and forested land have accelerated.

Solution: Stabilize population, especially by improving the economic and social status of women; design cities in ways that reduce distances traveled between home, work, shopping, and school; and in urban transit systems, shift emphasis from cars to public transportation, bicycling, and walking.

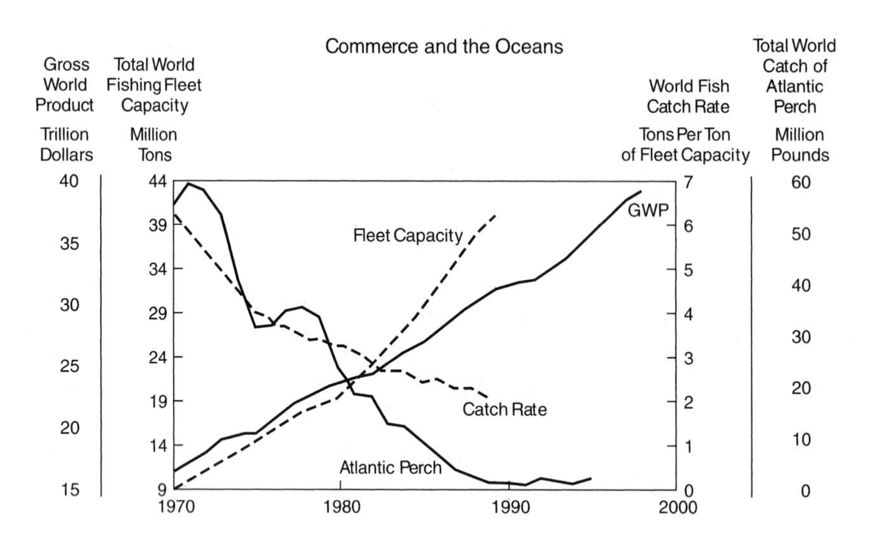

Commerce and the oceans.

The global economy has more than doubled in the past 30 years, putting pressure on most countries to increase export income. Many have tried to increase revenues by selling more ocean fish—for which there is growing demand, since the increase in crop yields no longer keeps pace with population growth. The result is that overfishing is decimating one stock after another, and the catch is getting thinner and thinner.

Solution: Stabilize population growth, stop subsidizing fishing fleets, and end the practice of feeding ocean-caught fish to farmed fish (it takes five pounds of ocean catch to produce one pound of farmed fish), which is still a very profitable and common practice.

SEVEN MOMENTS THAT HELPED DEFINE THE TRENDS OF THE PAST 30 YEARS . . .

The Car: *Mannheim, Germany 1885*

Karl Friedrich Benz takes the world's first gasoline-driven automobile out for a test drive and reaches a speed of none miles per hour. It's not yet faster than a horse, but the global infatuation with motorized speed is about to begin. Although petroleum has been around for decades and is used mainly for lighting lamps, the advent of the internal combustion engine causes a surge in demand, and the fossil-fuel age begins.

The Gusher: *Masjid-I-Salaman, Persia, May 26, 1908*

Drillers strike oil, and the rights are quickly acquired by the British government. The new enterprise, British Petroleum, turns out to be sitting atop the largest oil reservoir in the world, and thus is established a Western dominance of oil that will prevail throughout the twentieth century. That dominance will be strengthened by the establishment of the U.S.-controlled Arab-American Oil Company (Aramco) in 1933 and the Iranina coup in 1953. The resulting flow of cheap oil allows the fossil-fuel economy to dominate global industrialization.

The Golden Arch: *Oak Park, Illinois, Late 1950s*

McDonald's decides to open franchises all over the world. To establish uniform standards of production for its French fries, the company requires suppliers in each country it enters to grow its global standard potato—the Idaho russet. Other varieties, often better adapted to local conditions of soil, rainfall, temperature, and growing seasons, are displaced. The French fries policy becomes a model for the "mono-culturization" of agriculture on a global scale. It is an approach that eventually increases food supply for the expanding human population, but also opens the way to increased erosion, soil depletion, dependence on fertilizers and pesticides, nitrogen pollution of rivers and bays, and the decline of genetic diversity in the world's major food crops.

The TV: *Western Europe, 1952*

The first international standard for transmission of TV images (in lines per frame and frames per second) is established, opening the way to mass-audience broadcasts. Appetites for consumption are stimulated first in the industrial countries where TVs catch on quickly, then in the developing world where subtitled or dubbed American or European shows serve as implicit but vivid advertisements for first-world overconsumption.

The Highway: *Washington, D.C., 1956*

The U.S. Congress passes the Interstate Highway Act, authorizing construction of a national network of high-speed roads across the United States. The American penchant for traveling long distances, even in routine trips between home, work, shopping, and recreation, is greatly facilitated. Suburbanization is accelerated, natural areas are paved over, and pollution increases as major cities build beltways and open the way to "edge" cities. High mobility becomes a model for other countries, which develop their own highway systems—causing massive increases in deforestation, oil spills, air pollution, and carbon dioxide emissions.

The Backlash: *India, mid-1970s*

The Indian government, faced with surging population, adopts a policy of enforced birth control. Many men and women undergo compulsory sterilization. The policy triggers a great backlash, and the birthrate climbs instead of declining. Demographers project that by 2010, India will have passed China as the most populous country on Earth.

The Flood: *Yangtze River Basin, China, 1998*

Chinese developers clear thousands of hectares of forest to make space for the country's burgeoning population—thus setting the stage for one of the largest disasters in history. Stripping tree cover reduced the watershed's capacity to slow the flow of the surface water. Global warming increases evaporation—and thus increases rainfall. When the monsoon of 1998 comes, the heightened volume and velocity of the runoff—and unprecedented numbers of people living in the water's path—drive over 100 million people from their homes. The following year, Hurricane Mitch inundates Honduras and Belize, where similar deforestation has taken place. The disruptive impacts of climate change appear to be well under way.

AND SEVEN MOMENTS (PAST AND FUTURE) THAT COULD BE KEYS TO THE NEXT 30 YEARS . . .

Civil Society: *Uttarakhand, India, 1958*

A popular movement arises to protest government mismanagement of Himalayan forest and the operations of large timber companies engaged in what is widely regarded as a form of looting. Led mainly by women, the Chipko movement asserts the traditional rights of villagers to manage their local forests rather than submit to management by a distant bureaucracy. The Chipko movement raises the profile of nongovernmental environmental movements in India, as thousands of women stand in the way of tree-cutters. In the ensuing years, grass-roots groups proliferate and become more numerous in India than in any other country. By the 1990s, they have become a "third force" in human organization worldwide—a "civil society" that may soon be strong enough to begin to counterbalance unresponsive government and industry.

Precautionary Principle: *New York, 1962*

Rachel Carson publishes a book, *Silent Spring,* calling attention to the rising burden of chemical pollutants on the environment. As the burden continues to worsen in the following decades, it provokes discussion of a new

Precautionary Principle—the principle that the burden of proof of safety should be on those who wish to introduce a new chemical, not on those who claim to have been injured by it. In the 1990s, the principle will be invoked by members of the Intergovernmental Panel on Climate Change, a network of the world's leading climate scientists, in their argument that "uncertainty" in climate science should not be a reason to avoid preventive action on climate change.

Earth Summit: *Stockholm, Sweden, 1972*

The United Nations Conference on Human Development becomes the first global effort to place the protection of the biosphere on the official agenda of international policy and law. It will be followed by the UN Conference on Human Settlements (HABITAT) in 1976, the first World Climate Conference in 1979, and the UN Conference on Environment and Development (Earth Summit) in 1992—leading to what has become an essentially continuous process of international discussion on issues that concern transnational threats to human security.

Micropower: *Sri Lanka, about 1900*

In 100 villages, solar panels are installed on rooftops to provide low-cost electricity to homes that are not on the electric grid. Around the same time, similar installations are being made in the Dominican Republic, Zimbabwe, and other developing countries. They form the first scatterings of a movement toward the use of decentralized electric power systems, based on nonpolluting solar or wind power, that will eventually revolutionize the energy industry worldwide.

GMO-Free Food: *Western Europe, 1998*

European protesters compel transnational biotech companies to halt the rush to use genetically modified organisms (GMOs) in agriculture. Monsanto's bullish advertising campaign is scrapped, major food producers and retailers change their food-processing formulas, and Monsanto halts its program to force farmers to buy terminator seed.

The Climatic Wake-Up Call: *Somewhere on Earth, Soon*

An extreme weather event strikes a major population center head-on, with cataclysmic results. The event may be a gigantic hurricane or storm surge striking a coastal city, or it may be an inland flood inundating a heavily populated river basin. This time the disaster achieves a perceptual critical mass in the global public—an undeniable recognition that the greatest threats to human society are not those of military invasion but of environmental degradation. As a result, large-scale campaigns are undertaken to gird for—and stabilize—the future impacts of climate change.

Bioregionalism: *United States and Canadian Pacific, Early Twenty-First Century*

Along the northern Pacific coast, there is yet another clash between native peoples and the companies logging the region's remaining old-growth rain forest. Yet after decades of controversy over the management of coastal forests and waters, the native activists discover they have a constituency much broader than anything their predecessors enjoyed. From Oregon through British Columbia, they have awakened a latent bioregional awareness—a widely shared view that the region is unique, both ecologically and culturally. This awareness begins to reshape local politics so as to make it better reflect the long-term interests of the region itself. As the region thrives, people elsewhere come to believe—and act on—the principle that environmental progress often comes easier when natural regions are given precedence over political ones.

Earth Day 2000: A 20-Year Report Card, *World Watch*, March/April 2000, Worldwatch Institute, available at http://www.worldwatch.org.

What Would (Can) You Do?

Review the four graphs and their solutions on Energy and the Climate, Chemicals and the Biological Boomerang, Population and the Land, and Commerce and the Oceans. Read the two pages on "Seven Moments (and individuals) of the Past and the Future" that shaped and could shape the future. Investigate the following topics according to "what would or can you do." Relate your findings to the graphs and how they could affect trends in the future. Provide specific examples of these topics as to how you can contribute to solutions and make a possible difference. Also create your own topics and relate your actions and answers to ethical, personal, and technological guidelines.

1. Investigate home, work, and industrial energy efficiency possibilities and patterns.
2. Investigate home, work, and industrial recycling possibilities and patterns.
3. Discover initiatives for greater energy efficiency in home, work, and industrial environments.
4. Discover initiatives for increased recycling industries.
5. Discuss the types and amounts of fuel used in various cars, both individually as well as in local groups. Discuss their demographics. What are the implications?
6. Research transportation and passenger patterns, both individually and locally.
7. Investigate the creativity and development of additional products from waste technologies.
8. Investigate how to become involved with government policies.
9. Investigate resources, reading, awareness, and involvement of local energy policies.
10. Research local chemical industrial practices and local waste contaminants.
11. Investigate local fertilization, pesticide, and other community chemical applications and practices.
12. Research local and regional garbage practices such as collection, dumping, standards for garbage fills, and incineration.
13. Initiate conversations with local farmers, growers, industry, and company owners about their chemical concerns and practices.
14. Contact your local census bureau and discover local population demographics as well as various growth outbreaks of socio-economic levels, ethnic groups, and types of jobs.
15. Contact area or county map bureaus and track the type of growth of your community within the last decade. Develop a future plan of projected growth considering ecological and environmental priorities and including special needed or saved "green areas."
16. Contact local or area fish markets. Discover area fish consumption including approximate consumption (home and restaurant), types of fish, origins of catch, and amount and use of waste.
17. Develop a plan for more ecologically and environmentally sustainable approaches to fish consumption and use.
18. Create an effective approach to being ecological and ethically aware and involved.

Conclusion

As stated in the beginning of Part IV, the purpose of this part is to enhance awareness, responsibility, strategies, and involvement for a sustainable, desirable environment. This part has overviewed many major issues that cry out for more results and better solutions. Many of those solutions have begun to be implemented and have already made an important difference—but the issues contained in this part (and others) remain central and critical to our future and its quality of life. Your thoughtful awareness and involvement can and will make an important contribution. Therefore, this text is designed so that you may become aware of the issues and decisions you can make or that can be made to design the future. Your involvement and awareness will become the fulcrum for the solution. The questions, tables and figures, flowcharts illustrate this philosophy.

The issues presented here are not meant to present a negative picture of ecology, but instead are intended to bring out three different layers of central issues that need to be addressed. The three layers of the overview, the global issues, and the specific issues could be likened to the earth itself with (1) the overview being similar to the atmosphere of ecological understanding; (2) the global issues being comparable with the surface of the earth, with all the interactions of activity that take place on the surface of the globe; and (3) the specific concerns and issues being similar to the local activity that then affects global issues and understanding. The three layers comprising these issues include the following aspects:

1. The overview structural layer in the introduction discussed the history and overview of the high-risk environmental issues of habitat destruction, climate change, species extinction, global warming, and depletion of the ozone layer, as well as an emphasis on economic value.
2. The text then proceeded to the major global issues of consumerism and waste, long-term effects of pollutants on our animal and human young, dioxins—what to do with our toxic wastes—and the exploitation and lack of planning for our world fishing industries.
3. Specific studies brought certain situations to your attention regarding Love Canal, the Endangered Species Act, air pollution, rain forest pharmaceuticals, and Europe's Black Triangle.

It is hoped that these issues and experiences have given you, the reader, an urgency and understanding to the high priority of some of the major ecological issues of our time and to participate in their solutions.

Our "Ecological" Social–Economic View

Associating fiscal value with the earth's natural goods and services will begin to raise our perceptions of ecological value to its real importance and value. The estimation of our ecosystem at $33 trillion seems conservative, but it's a start in bringing to mind the replaceable and working value of the ecosystem. Pollution controls can be viewed from the perspective of dollars saved in annual health-care costs. Raw material costs should include the costs of earth materials lost and recovery techniques utilized for the environment. Costs of fish should be addressed, with monetary penalties enforced when quotas are exceeded. Rain forest conservation approaches can be partially funded through the market of its products and pharmaceuticals. And lastly, threatened and endangered species can receive the necessary funding if the extinction of species comes to be viewed as a heavy debt and monetary liability.

Not all ecological problems can be solved with a credit and debit ledger, but placing true monetary value on an interactive world ecosystem on which we all depend may help unindustrialized as well as industrialized countries to treasure and account for their resources, thus leveling the playing field. Many Third World countries retain valuable resources without receiving compensation for their use or conservation (e.g., African and Asian animals, rain forests of Central America). It makes good sense to count treasure as treasure and to acknowledge your national treasures as true global treasures.

Issues for the Future

Many of the issues discussed in this Part IV have had successful attempts at enacting solutions. Earth Summits, attended by heads of state, have addressed numerous issues with serious intent so as to make progress on issues of environment and poverty, but the solutions issuing from these summits are not necessarily working. During the part of the book that addresses the future of technology, we will discuss some of the successes, failures, and future initiatives for ecological progress in the twenty-first century.

INTERNET EXERCISES

1. Use any of the Internet search engines (e.g., Google, Yahoo) to research information about the following topics.
   a. Updates on rain forest destruction and programs to conserve rain forests
   b. Disappearance and malformation of frogs on the earth
   c. Latest research on global warming—what new evidence is appearing?
   d. Latest views on the disappearance of the ozone layer, as well as the laws attempting to protect the ozone layer and their effects
   e. Facts on improvements in conservation due to recycling
   f. Water pollution and the water we drink
   g. Additional information on Superfund
   h. Latest information on the Magnuson Fishery Conservation and Management Act
   i. Commercial and recreational fishing catches in your local area for the last decade
   j. Projects undertaken by the Citizens Clearinghouse for Hazardous Waste
   k. New revisions and implications of the Endangered Species Act
   l. Air pollution in your area
   m. Water pollution in your area
   n. Latest laws on air pollution particulates

o. People whom you admire who have made a difference ecologically
p. New medicines and products derived from flora or fauna in the rain forest
q. Garbage production in the last 10 years (locally, statewide, or nationally)
r. Ecological laws passed in the last two years (local, state, or federal)

2. Use any of the Internet search engines (e.g., Google, Yahoo) to research information about the following topics.
   a. Ecosystem, ecosphere, and synthesphere
   b. Environmental Protection Agency
   c. Biological diversity
   d. Greenhouse gases
   e. Chlorofluorocarbons and ozone
   f. Dioxins
   g. PCBs
   h. Pesticides
   i. Hazardous wastes
   j. UN Food and Agriculture Organization
   k. Individual Transferable Quotas
   l. Endangered Species Act and threatened species

USEFUL WEB SITES

http://www.chej.org	Center for Health, Environment, and Justice
http://www.edf.org	Environmental Defense Fund
http://www.rachel.org	Environmental Research Foundation
http://www.audubon.org	National Audubon Society
http://www.nationalgeographic.com	National Geographic Society
http://www.nrdc.org	National Resources Defense Council
http://www.nwf.org	National Wildlife Federation
http://www.100topenvironmentsites.com	100 Top Environment Sites
http://www.tnc.org	The Nature Conservancy
http://www.sierraclub.org	Sierra Club
http://www.worldwildlife.org	World Wildlife Fund
http://www.worldwatch.org	The Worldwatch Institute
http://www.nasa.gov	NASA
http://www.fao.org	Food and Agriculture Organization of the United Nations
http://www.redlist.org	IUCN Red List of Threatened Species
http://www.epa.gov	U.S. Environmental Protection Agency
http://www.bsri.nsu.edu	Rain Forest Research Institute
http://eobglossary.gsfc.nasa.gov	Earth Observatory

BOX 1 Dedicated to People Who Make a Difference

Planet Earth from space.
Photo courtesy of NASA Headquarters.

The registry of people who make environmental differences live in every town, county, and province of the world. What is happening notably since the 1970s is that people are becoming aware and concerned; they understand that they can make a difference and change the spread of ecological malaise. They are thinking change and enacting a system. The people mentioned in this chapter are only a few of the people that have altered ways of doing things. Magazines, newspapers, and publications of local organizations all herald individual ecological achievement, but there is still much achievement that goes unacknowledged.

The central idea here is that ecological change stems from individual conscience that has held to a commitment to improve some system or aspect that affects our earth. Ecological leadership and change begin with such individuals that care and extend energy and commitment behind their caring. From their actions, there are positive changes, new systems, and new research. This is where the hope for our future lies—with a strong, positive, ecologically minded, innovative dynamic. This box is dedicated to the many individuals who have made a difference and who will make *such* a difference.

PART
V

Population

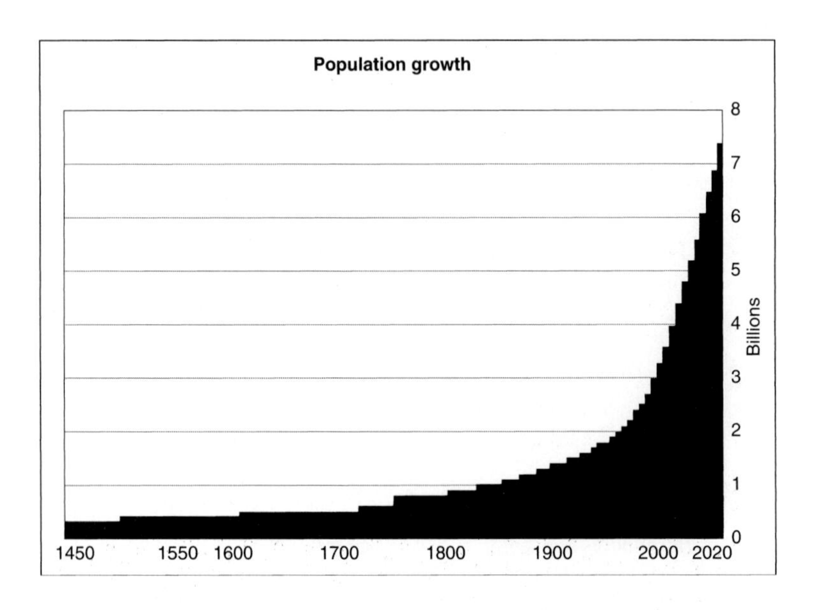
Population growth
8
7
6
5
4
3
2
1
0
Billions
1450
1550
1600
1700
1800
1900
2000
2020

38 ***Fertility Rates: The Decline Is Stalling***
Linda Starke

39 ***The Hazards of Youth***
Lisa Mastny

40 ***Earth's Carrying Capacity: Not Quite So Easy When Applied to Humans***
Edited by Barbara A. Eichler

Case Study 1: ~~Definitely~~ Probably One: A Generation Comes of Age Under China's One Child Policy
Claudia Meulenberg

Case Study 2: Japan's Unpopular Pill—An Unrealized Battle Victory
Edited by Barbara A. Eichler

OBJECTIVES

After reading Part V, Population, *you will be able to*

1. Analyze world human population growth projections and their long- and short-range implications.
2. Appraise the urgency of human growth expansion and its accompanying economic, human rights, and ecological global and local implications.
3. Evaluate population growth differences and their causes among industrialized and unindustrialized nations, along with global and local issues and their implications.
4. Analyze and use tables and graphs to develop and support specific information on population perspectives.
5. Interpret world birth control issues from varying cultural, ethical, and national perspectives.
6. Identify the roles of women's rights and education in conjunction with fertility rates.
7. Review the positive and negative impacts of various local and global strategies regarding birth control.
8. Appraise the importance and influence of effective and appropriate education, human rights, and birth control strategies.

INTRODUCTION

"Humanity is approaching a crisis point with respect to the interlocking issues of population, environment, and development."

Statement from 60 science academies led by the U.S. National Academy of Sciences and Britain's Royal Society, October 1993

Understanding world population issues: a street in Calcutta, India. Photo courtesy of United Nations/J. P. LaFonte.

"The decisions that the international community takes over the next several years, whether leading to action or inaction, will have profound implications for the quality of life for all people, including generations not yet born, and perhaps for the planet itself."

Statement made at the International Conference on Population and Development, Cairo, 1994

As you read and reflect upon these pages, you will find many issues—individual, local, national, global, religious, societal, economic, and ecological—that are all closely interrelated. Graphs in this part of the book demonstrate that we reached the first billion people in world history around 1839, our second billion 90 years later around 1930, and our third billion 30 years later in 1960. Currently, we are adding a billion people to the world's population approximately every 15 years or less, which could easily escalate our population count to the UN's upper projection of 11.9 billion by the year 2050. For sustenance, the United Nations recommends an ambitious plan to stabilize the earth's population at low to medium projections of 7.9 to 9.8 billion people by 2050. For this stabilization to occur, it is necessary to address the following issues confronting us:

- Growth projections
- Ecological implications
- Food production
- Limitation of the earth's natural resources for food production
- Specific issues raised by the ten-year progress report at the 1994 Cairo Conference
- Population control issues
- Numbers-carrying capacity in which a country becomes stabilized
- Specific case studies (China and Japan)
- Inter-dependent nature of population, poverty, women's rights and education, and the local environment

Part V's emphasis on population should help you to make informed decisions on individual, local, national, economic, ecological, and other fronts regarding this important issue. There is much to say and more to do, but it all begins with becoming informed. This is a vital assignment since only with knowledge can we hope to play a part in establishing a future population plan.

In 2005, the world's population increased by 74 million people, which is more than the population of all but the 14 largest nations. If all of these people formed a new country, it would be the fifteenth largest country in the world, just larger than Egypt and just smaller than Vietnam. As shown in Figure P5.1, from the Global Population Profile of 2002, which is the latest global report of the U.S. Census Bureau, adding people at this rate is equivalent to

- 2 1/3 people per second
- 141 people per minute
- 200,000+ people per day
- 6.2 million people per month

At this rate of growth, the world would add nearly the equivalent of the population of Western Europe (392 million persons in 2002) in five years.

For real-time population numbers, the following Internet addresses access current, second-by-second population clocks and projection numbers to give you a greater perspective

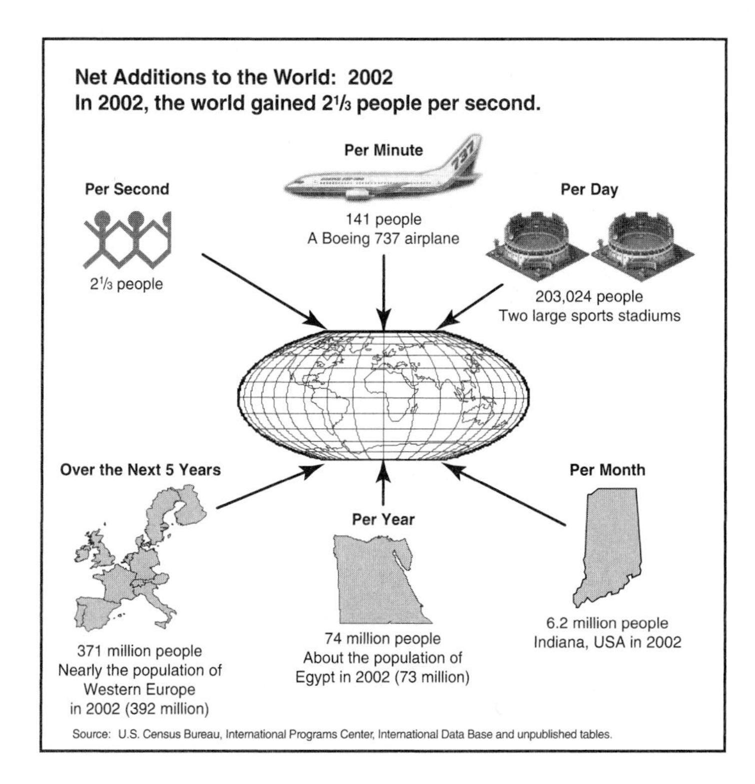

Figure P5.1
World population statistics.
Source: U.S. Census Bureau. (2002). International Programs Center, International Data Base and unpublished tables.

on our exponential patterns of population growth. You may also refer to the Useful Web Sites section at the end of Part V for more information.

National and world population

- http://www.census.gov/main/www/popclock.html
- http://www.ibiblio.org/lunarbin/worldpop
- http://www.census.gov/cgi-bin/popclock
- http://www.statistics.gov.my

Similar to other parts of this book, Part V presents factual, historical, social, and economic weavings along with the readings to provide a cross-analysis of the issues from

BOX 1 Malthus vs. Demographic Transition Theory

The rapid growth of Europe's population because of more abundant food supplies and better living conditions caused Thomas Malthus (1766–1834), an English economist, to predict uncontrolled population growth and, hence, future doom. In 1798, he wrote An Essay on the Principle of Population in which he predicted what became known as the Malthus Theorem. His principal argument was that, while population grows geometrically (from 2 people to 4 to 8 to 16, etc.), food supply increases arithmetically (from 1 unit to 2 to 3 to 4, etc.), concluding that if births proceed unchecked, the population will outgrow its food supply (Henslin 2006 374). However, a different and somewhat opposite argument exists, called the three-stage demographic transition theory, based on an interpretation of American demographer Warren Thompson in 1929. Stage I has high birth rates and high death rates while population remains stable; Stage II experiences high birth rates and low death rates while population surges; and Stage III has low birth rates and low death rates while population stabilizes and the economic quality of life improves. The theory proposes that all nations proceed through these stages in economic and societal growth before reaching stabilization (Montgomery, 2006). Consider the following historical world population patterns in Table P5.1 on the next page.

Figure P5.2
UN world projection plan.
Permission granted by the United Nations Population Fund.

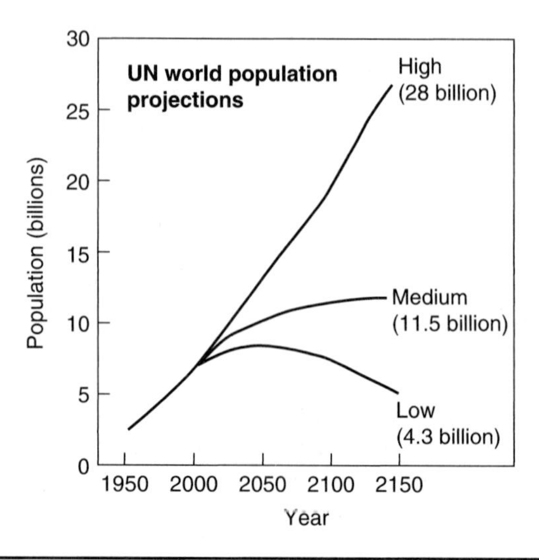

Table P5.1
Historical World Population Trends

Approximate Year	*Estimated Population*
1 million B.C.	125 thousand
8000 B.C.	5.3 million
4000 B.C.	86.5 million
1 A.D.	200 million
1000 A.D.	275 million
1650 A.D.	545 million
1750 A.D.	728 million
1839 A.D.	1 billion
1850 A.D.	1.17 billion
1900 A.D.	1.55 billion
1930 A.D.	2.0 billion
1950 A.D.	2.5 billion
1960 A.D.	3.0 billion
1970 A.D.	3.6 billion
1976 A.D.	4.0 billion
1982 A.D.	4.5 billion
1995 A.D.	5.72 billion
1999 A.D.	6.0 billion
Projected	
2025 A.D.	8.5 billion
2050 A.D.	10 billion

Sources: Pytlik 1985, 148; Wright, 2005

differing perspectives. The tables and figures that follow the introductory text present a necessary empirical base with which to realize the issues numerically before the descriptive discussions of the readings and case studies are presented. This format builds an understanding from both data-based and cross-descriptive views.

QUESTIONS ABOUT THE TABLE

1. What are the slowest periods of population gain? What are the most rapid?
2. Using these population figures and those leading to current and future growth, what do you feel are projections for world food supply, economic growth, political stability, and quality of life?
3. Apply your conclusions to the Malthus Theorem and the demographic transition theory. Do your conclusions change when you consider these viewpoints? What validity do you think these arguments hold?

BOX 2 World Population Continues to Soar

Concern for the exponentially increasing world population is becoming more widespread. Refer to Figure 2 for three different projections to 2150. One of the most important questions in the world today is whether human beings are wise enough to see what is coming. Today's global population cannot be sustained at the current lifestyle level of the United States, Europe, or Japan. Over the past few centuries, the following statements were issued about population growth:

- Around 200 years ago (1798), Englishman Thomas Malthus stated, "Population, when unchecked, increases in a geometrical ratio. Subsistence only increases in an arithmetical ratio."
- About one century later, Charles Darwin said, "Man tends to increase at a greater rate than his means of subsistence; consequently he is occasionally subjected to a severe struggle for existence."
- Nearly 100 years after that (1968), Paul Ehrlich spoke of population calamity in *The Population Bomb.*

Other Comments on the Issue of Continued Increasing Growth and the Resulting Decreasing World Stability

- Joel E. Cohen, in "How Many People Can Earth Hold?", states, "According to every plausible calculation that's ever been done, Earth could not feed even the 695 billion people that the UN projected for 2150 if present fertility rates were to continue" (Cohen 1992, 100). He quotes Princeton demographer Ansely Coale's conclusions: "Every demographer knows that we cannot continue a positive rate of increase indefinitely. The inexorable arithmetic of compound interest leads us to absurd conditions within a calculable period of time. Logically we must, and in fact we will, have a rate of growth very close to zero in the long run" (Cohen 1992, 100).
- Lester R. Brown and Hal Kane, in their book *Full House*, explain the following: "The demands of the 90 million people added each year for grain and seafood are being satisfied by reducing consumption among those already here. This is a new situation, one that puts population policy in a new light . . . [This population expansion rate] can devastate local life-support systems" (Brown and Kane 1994, 49–50).

Debate on this issue by nations and their people heatedly continues as the world's population continues to leap geometrically from 2.5 billion in 1950 to 5.6 billion in the 1990s, 6 billion in 1999. It is projected that there will be perhaps 9 billion by 2025 or 2030. Overpopulation is one of the chief obstacles to universal well-being. A 1994 Cornell University study concluded that the world can only support 2 billion people at the current standard of living now experienced by the industrialized nations of the world. In 1992, 1,600 prominent world scientists, including half of the living science Nobel laureates, presented a declaration to the world's leaders stating that the continuation of destructive human activities "may so alter the living world that it will be unable to sustain life in the manner that we know." (Brown and Kane 1994, 30).

The current global growth pattern adds about the equivalent of Egypt's population (74 million) to the world every year. The U.S. population hit the 300 million mark in October, 2006. The United States remains one of the world's most populous nations, trailing only China and India. It is the 6th largest contributor to annual world population growth behind India (21%), China (12%), Pakistan (5%), Bangladesh (4%), Nigeria (4%), and United States (4%). One major reason for this growth is that immigrations add more than 1 million people a year to the U.S. population (U.S. Census Bureau).

Questions about the Boxed Material

1. Using an almanac or the Internet, find the current population figures for the world and for the United States. Chart the new percentage-growth figures. What are your conclusions regarding the various statements quoted in this box?
2. Do you agree with the statements quoted in this box? Why or why not? Build your position on the figures of population growth.
3. Consult world population growth charts. Where are the highest percentages of population growth occurring? What are some of your conclusions about food resources in these areas?

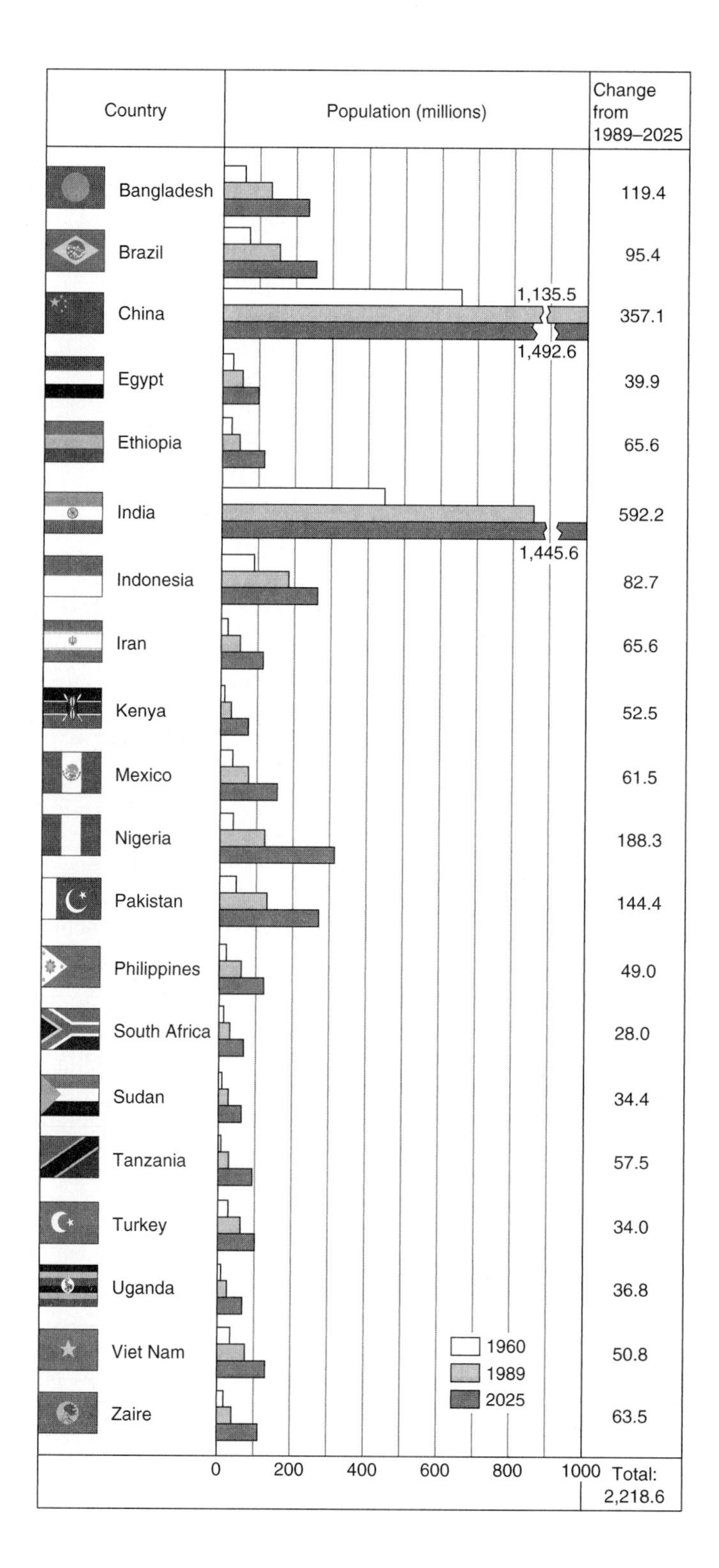

Figure P5.3
High-growth countries.
Seventy percent of the projected increase in world population by the year 2025 will occur in these 20 less developed countries.
Source: Data from the United Nations Department of International Economic and Social Affairs. Reprinted with Permission from Ian Worpole.

Table P5.2
Top Ten Most Populous Countries: 1950, 2002, and 2050*

*1950***	*2002*	*2050*
1. China	1. China	1. India
2. India	2. India	2. China
3. United States	3. United States	3. United States
4. Russia	4. Indonesia	4. Indonesia
5. Japan	5. Brazil	5. Nigeria
6. Indonesia	6. Pakistan	6. Bangladesh
7. Germany	7. Russia	7. Pakistan
8. Brazil	8. Bangladesh	8. Brazil
9. United Kingdom	9. Nigeria	9. Congo (Kinshasa)
10. Italy	10. Japan	10. Mexico
Rankings of Future or Past Top Ten Countries		
11. Bangladesh	11. Mexico	14. Russia
13. Pakistan	13. Germany	16. Japan
15. Nigeria	21. United Kingdom	24. Germany
16. Mexico	22. Italy	29. United Kingdom
32. Congo (Kinshasa)	23. Congo (Kinshasa)	35. Italy

*More developed countries/less developed countries; less developed countries dominate the list of the world's ten most populous countries.
**Current boundaries.
Source: U.S. Census Bureau (2004). *Global Population Profile 2002*, 13. Available at www.census.gov/prod/2004pubs/wp-02.pdf.

4. What will happen after 2050 in terms of population growth? What does the demographic theory propose will happen? What does the Malthus Theory propose will happen with these exponential projections? What theory do you have regarding the future of population growth?
5. What do you project population numbers to be in the year 3000 A.D.? Give a planned and supported explanation of your projection.

REFERENCES

Brown, L. R., and Kane, H. (1994). *Full House: Reassessing the Earth's Population Carrying Capacity*. New York: W.W. Norton & Company.

Cohen, J. E. (1992). How Many People Can Earth Hold? *Discover* (November), Vol. 13:114–120.

Henslin. J. (2006). Essentials of Sociology. Sixth Edition. Boston: Pearson and Allyn and Bacon.

Montgomery, K. The Demographic Transition. Accessed 10/14/06. Available http://www.uwmc.uwc.edu/geography/Demotrans/demtran.htm.

Pytlik. E., Lauda, D., Johnson, D. (1985). *Technology, Change and Society.* Worcester, Mass: Davis Publications, 148.

United Nations Population Fund. Available www.unfpa.org.

U.S. Census Bureau. Available www.census.gov.

U.S. Census Bureau (2004). Global Population Profile 2002. Available www.census.gov/prod/2004pubs/wp/wp-02.pdf.

Wright, J. (ed.) (2005). *The New York Times Almanac.* New York: Penguin Books.

POPULATION VOCABULARY EXERCISE

As you proceed through Part V and continue your reading and research of various issues associated with population topics, keep a list of terms and organizations dealing with population that you would like to learn more about. Use various resources including other texts, articles, encyclopedias, and the Internet (refer to the Internet Exercises at the end of Part V).

POPULATION NUMBERS AND STATISTICS

In 2002, 5 billion more people inhabited the globe than in 1800. According to the U.S. Census Bureau, world population hit the 6-billion mark in June of 1999. This figure is over 3.5 times the size of the earth's population at the beginning of the twentieth century and roughly double its size in 1960. Never before has the earth sustained such a large human population. The time required for global population to grow from 5 to 6 billion was shorter than the interval between any of the previous billions. As depicted in Figure P5.4, it took just 12 years for this to occur, just slightly less time than the 13 years between the fourth and fifth billion, but considerably less time than the 118 years between the first and second billion.

Current Census Bureau projections indicate that it will take increasingly longer periods of time to reach the seventh-, eighth-, and ninth-billion markers of world population. Those engaged in long-term projections suggest that the current rise in world population may peak before it reaches 10 billion (U.S. Census Bureau 2004).

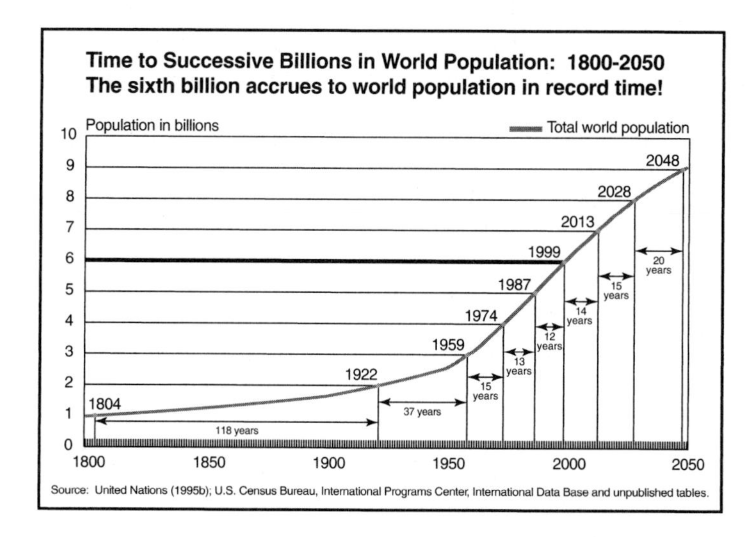

Figure P5.4
Global population profile: 2002.
Source: United Nations, U.S. Census Bureau, International Programs Center, International Data Base and unpublished tables, 1995.

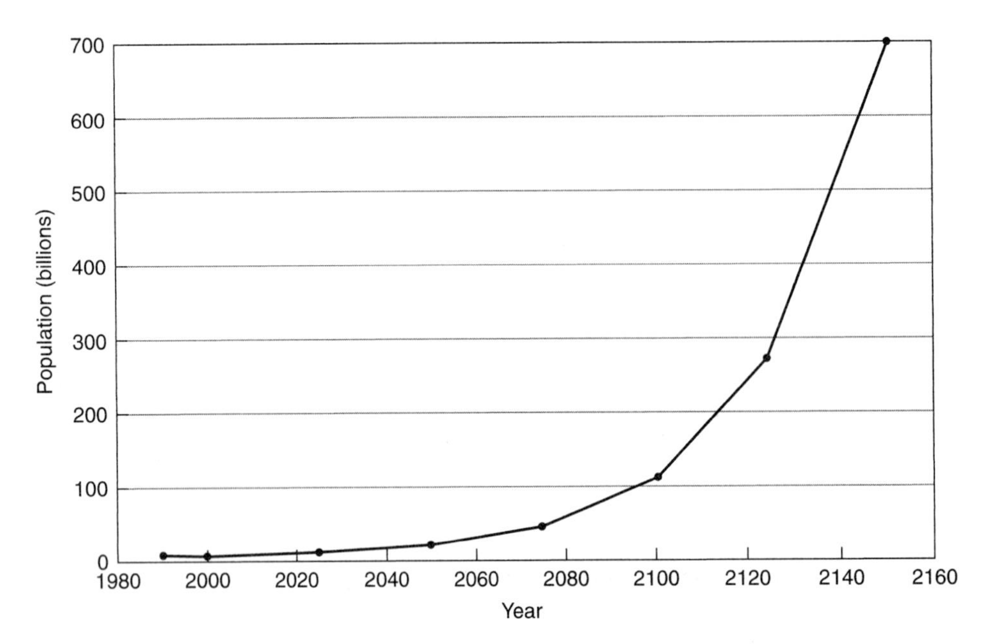

Figure P5.5
United Nations' projection of world population, assuming fertility remains constant at its 1990 levels in different regions.
Source: Original figure drawn according to data from the United Nations (1992). Permission granted by the United Nations Population Fund.

Table P5.3
World Population by Region and Development Category: 1950–2050*

Region	*1950*	*1960*	*1970*	*1980*	*1990*	*2000*	*2002*	*2010*	*2025*	*2050*
World	2,555	3,040	3,708	4,455	5,275	6,079	6,228	6,812	7,834	9,079
Less developed countries	1,749	2,129	2,705	3,374	4,132	4,887	5,030	5,588	6,582	7,836
More developed countries	807	910	1,003	1,081	1,143	1.192	1,199	1,224	1,252	1,243
Africa	227	283	361	473	627	803	839	977	1,247	1,786
Sub Saharan	183	227	291	382	508	657	687	804	1,036	1,532
North Africa	44	56	71	91	119	147	152	174	211	255
Near East	44	58	75	101	135	171	179	212	280	396
Asia	1,368	1,628	2,038	2,494	2,978	3,435	3,518	3,838	4,375	4,832
Latin America and the Caribbean	166	218	286	362	443	524	539	596	690	782

Europe and the New Independent States	572	639	702	750	788	802	803	810	814	776
Western Europe	304	326	352	367	377	390	392	398	398	373
Eastern Europe	88	99	108	117	122	121	121	120	117	104
New Independent States	180	214	242	266	289	291	290	292	300	300
North America	166	199	227	252	278	314	320	344	388	462
Oceania	12	16	19	23	27	31	32	35	40	45
Excluding China:										
World	1,982	2,374	2,868	3,446	4,109	4,786	4,918	5,436	6,350	7,627
Less developed countries	1,176	1,464	1,866	2,366	2,967	3,595	3,720	4,213	5,100	6,386
Asia	795	962	1,198	1,486	1,813	2,143	2,208	2,463	2,892	3,383
Less developed countries	711	868	1,094	1,369	1,689	2,016	2,081	2,336	2,772	3,283

*Midyear population in millions; figures may not add to totals because of rounding.
Note: Reference to China encompasses China, Hong Kong S.A.R., and Taiwan. Direct access to this table and the International Data Base is available through the Internet at www.census.gov/ipc/www.
Source: U.S. Census Bureau. (2004). Global Population Profile: 2002. Table A-3.

Table P5.4
Distribution of the World's Population, 1950–2050

Area	*1950*	*2000*	*2050[1]*
More developed regions	32.3%	19.7%	12.7%
Less developed regions	67.7	80.3	87.3
Africa	8.8	13.1	21.5
Asia	55.5	60.6	58.2
Latin America and the Caribbean	6.6	8.6	8.6
Europe	21.8	12.0	6.5
Northern America	6.8	5.2	4.7
Oceania	0.5	0.5	0.5
Total world population (millions)	2,519	6,057	9,322

[1] Based on the U.N.'s medium population projections.
Source: U.S. Bureau of the Census, *World Population Prospects: The 2000 Revision Highlights*, 2001. Located in J. W. Wright (ed.), *The New York Times Almanac*, New York: Penguin, 2005.

Table P5.5
The World's Largest Urban Areas, 1950–2015

2000 Rank, Urban Area	*Population (millions)*					*Rank*	
	1950	1970	1990	2000	2015[1]	1950	2015
1. Tokyo, Japan	11.3	23.3	32.5	34.5	36.2	2	1
2. Mexico City, Mexico	3.0	8.8	15.3	18.1	20.6	20	4
3. New York-Newark, U.S.	12.3	16.2	16.1	17.8	19.7	1	6
4. Sao Paolo, Brazil	2.3	7.6	14.8	17.1	20.0	27	5
5. Mumbai (Bombay), India	3.0	6.2	12.3	16.1	22.6	17	2
6. Calcutta, India	4.4	6.9	10.9	13.1	16.8	10	10
7. Shanghai, China	5.3	11.2	13.3	12.9	12.7	6	15
8. Buenos Aires, Argentina	5.0	8.4	11.2	12.6	14.6	8	12
9. Delhi, India	N.A.	3.5	8.2	12.4	20.9	N.A.	3
10. Los Angeles-Long Beach, U.S.	4.0	8.4	10.9	11.8	12.9	12	14
11. Osaka-Kobe, Japan	4.1	9.4	11.0	11.2	11.4	11	18
12. Jakarta, Indonesia	N.A.	3.9	7.7	11.0	17.5	N.A.	8
13. Beijing, China	3.9	8.1	10.8	10.8	11.1	13	20
14. Rio de Janeiro, Brazil	3.0	6.6	9.6	10.4	12.4	18	17
15. Cairo, Egypt	2.4	5.6	9.1	10.4	13.1	24	13
16. Dhaka, Bangladesh	N.A.	N.A.	6.5	10.2	17.9	N.A.	7
17. Moscow, Russia	5.4	7.1	9.1	10.9	10.9	5	21
18. Karachi, Pakistan	N.A.	3.9	7.1	10.0	16.2	N.A.	11
19. Metro Manila, Philippines	N.A.	3.5	8.0	9.9	12.6	N.A.	16
20. Seoul, South Korea	N.A.	5.3	10.5	9.9	9.2	N.A.	26
21. Paris, France	5.4	8.4	9.3	9.7	10.0	4	22
22. Tianjin, China	2.4	5.2	8.8	9.2	11.3	26	19
23. Istanbul, Turkey	N.A.	N.A.	6.6	8.7	11.3	N.A.	19
24. Lagos, Nigeria	N.A.	N.A.	N.A.	8.7	17.0	N.A.	9
25. Chicago, U.S.	5.0	7.1	7.4	8.3	9.4	9	24

Note: An urban area is a central city or central cities, and the surrounding urbanized areas, also called a metropolitan area.
N.A. = not available.
[1] Projected.
Source: United Nations Population Division, *World Urbanization Prospects: The 2003 Revision*, 2004. Located in J. W. Wright (ed.), *The New York Times Almanac*, New York: Penguin, 2005.

Table P5.6
World Births, Deaths, and Population Growth, 2004

Characteristic	*World*	*Developed*	*Developing*
Population	6,377,641,642	1,206,293,095	5,171,348,547
Births	129,108,390	13,216,000[1]	119,563,000[1]
Deaths	56,540,896	12,248,000[1]	41,898,000[1]
Natural increase	73,250,395	968,000[1]	77,665,000[1]
Births per 1,000 population	20.2	11.2	22.4
Deaths per 1,000 population	8.9	10.3	8.5
Rate of natural increase (percent)	1.1%	0.08%	1.4%

Source: U.S. Bureau of the Census, International Data Base, 2004; *World Population Profile, 1998.* Located in J. W. Wright (ed.), *The New York Times Almanac*, New York: Penguin, 2005.

Table P5.7
Nations with Highest and Lowest Fertility Rates, 2000–2005

Highest Rates Country	*Fertility Rate per Woman*	*Lowest Rates Country*	*Fertility Rate per Woman*
Niger	8.00	Armenia	1.10
Yemen	7.60	Latvia	1.10
Somalia	7.25	Bulgaria	1.10
Angola	7.20	Ukraine	1.10
Uganda	7.10	Spain	1.13
Mali	7.00	Slovenia	1.14
Afghanistan	6.80	Russian Federation	1.14
Burkina Faso	6.80	Czech Republic	1.16
Burundi	6.80	China	1.17
Liberia	6.80	Belarus	1.20
Ethiopia	6.75	Estonia	1.20
Congo, Democractic Republic of	6.70	Hungary	1.20
Chad	6.65	Italy	1.20

Table P5.8
Population Indicators by Region and Nation

Region/Country	*Population estimate ('000s)* 2004	2050	*Birth Rate per 1,000 2003*[1]	*Death Rate per 1,000 2003*[1]	*Life Expectancy 2000*	*Percent Urban 2003*	*Fertility Rate per Woman 2000–2005*
World total	6,377,600	8,918,700	20	9	65	48%	2.7
More developed regions	1,206,100	1,219,700	11	10	76	75	1.5
Less developed regions	5,171,500	7,699,100	23	9	63	42	2.9
Least developed countries	735,600	1,674,500	40	15	50	27	5.1
Africa	869,200	1,803,300	36	14	49	38	4.9
Asia	3,870,500	5,222,100	19	8	67	39	2.5
Europe	725,600	631,900	10	11	74	73	1.4
Latin America and the Caribbean	550,800	767,700	21	6	71	77	2.5
Northern America	328,900	447,900	14	8	77	80	2.0
Oceania	32,600	45,800	17	7	74	73	2.3
Afghanistan	24,900	69,500	41	17	43	23	6.8
Albania	3,200	3,700	15	5	74	44	2.3
Algeria	32,300	48,700	22	5	70	59	2.8
Angola	14,100	43,100	46	26	40	36	7.2
Argentina	38,900	52,800	17	8	74	90	2.4
Armenia	3,100	2,300	13	10	72	64	1.2
Australia	19,900	25,600	13	7	79	92	1.7
Austria	8,100	7,400	9	9	79	66	1.3
Azerbaijan	8,400	10,900	19	10	72	50	2.1
Bahrain	739	1,270	19	4	74	902	0.3
Bangladesh	149,700	254,600	30	9	58	24	3.5
Belarus	9,900	7,500	10	14	70	71	1.2
Belgium	10,300	10,200	11	10	80	97	1.7
Belize	261	421	31	6	71	482	2.9
Benin	6,900	15,600	43	14	51	45	5.7
Bhutan	2,300	5,300	35	14	63	9	5.0
Bolivia	9,000	15,700	26	8	64	63	3.8
Bosnia and Herzegovina	4,200	3,600	13	8	74	44	1.3
Botswana	1,800	1,400	26	31	40	52	3.7
Brazil	180,700	233,100	18	6	68	83	2.2
Brunei Darussalam	366	685	20	3	76	762	2.5
Bulgaria	7,800	5,300	10	14	71	70	1.1
Burkina Faso	13,400	42,400	45	19	46	18	6.7
Burundi	7,100	19,500	40	18	41	10	6.8
Cambodia	14,500	29,600	27	9	57	19	4.8
Cameroon	16,300	24,900	35	15	46	51	4.6
Canada	31,700	39,100	11	8	79	80	1.5
Central African Republic	3,900	6,600	36	20	40	43	4.9

Chad	8,900	25,400	47	16	45	25	6.7
Chile	16,000	21,800	16	6	76	87	2.4
China	1,313,300	1,395,200	13	7	71	39	1.8
Colombia	44,900	67,500	22	6	72	77	2.6
Congo, Dem. Rep. of	54,400	151,600	45	15	42	32	6.7
Congo, Republic of	3,800	10,600	29	14	48	54	6.3
Costa Rica	4,300	6,500	19	4	78	61	2.3
Croatia	4,400	3,600	13	11	74	59	1.7
Cuba	11,300	10,100	12	7	77	76	1.6
Czech Republic	10,200	8,600	9	11	75	74	1.2
Denmark	5,400	5,300	12	11	77	85	1.8
Djibouti	712	1,395	41	20	46	84[2]	5.8
Dominican Republic	8,900	11,900	24	7	67	59	2.7
East Timor	820	1,433	28	6	50	48[2]	3.9
Ecuador	13,200	18,700	25	5	71	62	2.8
Egypt	73,400	127,400	24	5	69	42	3.3
El Salvador	6,600	9,800	28	6	71	60	2.9
Eritrea	4,300	10,500	39	13	53	20	5.4
Estonia	1,300	700	9	13	72	69	1.2
Ethiopia	72,400	171,000	40	20	46	16	6.0
Fiji	847	969	23	6	70	52[2]	3.0
Finland	5,200	4,900	11	10	78	61	1.7
France	60,400	64,200	13	9	79	76	1.9
Gabon	1,400	2,500	37	11	57	84	4.0
Gambia	1,500	2,900	41	12	54	26	4.7
Georgia	5,100	3,500	12	15	74	52	1.4
Germany	82.500	79,100	9	10	78	88	1.4
Ghana	21,400	39,500	26	11	58	45	4.1
Greece	11,000	9,800	10	10	78	61	1.3
Guatemala	12,700	26,200	35	7	66	46	4.4
Guinea	8,600	19,600	43	16	49	35	5.8
Guinea-Bissau	1,500	4,700	38	17	45	34	7.0
Guyana	767	507	18	9	63	38[2]	2.3
Haiti	8,400	12,400	34	13	50	38	4.0
Honduras	7,100	12,600	32	6	69	46	3.7
Hong Kong	7,100	9,400	11	6	80	100	1.0
Hungary	9,800	7,600	10	13	72	65	1.2
Iceland	292	330	14	7	79	93[2]	1.9
India	1,081,200	1,531,400	23	8	64	28	3.0
Indonesia	222,600	293,800	21	6	67	46	2.4
Iran	69,800	105,500	17	6	70	67	2.3
Iraq	25,900	57,900	34	6	61	67	4.8
Ireland	4,000	5,000	14	8	77	60	2.0
Israel	6,600	10,000	19	6	79	92	2.7

Table P5.8 (*Continued*)
Population Indicators by Region and Nation

Region/Country	*Population estimate ('000s)* 2004	2050	*Birth Rate per 1,000* 2003[1]	*Death Rate per 1,000* 2003[1]	*Life Expectancy* 2000	*Percent Urban* 2003	*Fertility Rate per Woman* 2000–2005
Italy	57,300	44,900	9	10	79	67	1.2
Ivory Coast	16,900	27,600	40	18	41	45	4.7
Jamaica	2,700	3,700	17	5	81	52	2.4
Japan	127,800	109,700	10	9	82	65	1.3
Jordan	5,600	10,200	24	3	71	79	3.6
Kazakhstan	15,400	13,900	18	11	66	56	2.0
Kenya	32,400	44,000	29	16	45	39	4.0
Korea, North	22,800	25,000	18	7	63	61	2.0
Korea, South	48,000	46,400	13	6	78	80	1.4
Kuwait	2,600	4,900	22	2	77	96	2.7
Kyrgyzstan	5,200	7,200	26	9	69	34	2.6
Laos	5,800	11,400	37	12	55	21	4.8
Latvia	2,300	1,300	9	15	71	66	1.1
Lebanon	3,700	4,900	20	6	74	88	2.2
Lesotho	1,800	1,400	27	25	35	18	3.8
Liberia	3,500	9,800	45	18	41	47	6.8
Libya	5,700	9,200	27	3	73	86	3.0
Lithuania	3,400	2,500	10	13	73	67	1.3
Luxembourg	459	716	12	8	78	92[2]	1.8
Macedonia	2,100	2,200	13	8	74	60	1.9
Madagascar	17,900	46,300	42	12	54	27	5.7
Malawi	12,300	25,900	45	23	38	16	6.1
Malaysia	24,900	39,600	24	5	73	64	2.9
Maldives	328	819	37	8	68	28	5.4
Mali	13,400	46,000	48	19	49	32	7.0
Malta	396	402	13	8	78	92[2]	1.8
Mauritania	3,000	7,500	42	13	53	62	5.8
Mauritius	1,200	1,500	16	7	72	43	2.0
Melanesia	7,600	14,000	32	9	59	20	3.9
Mexico	104,900	140,200	22	5	73	76	2.5
Moldova	4,300	3,600	14	13	69	46	1.4
Mongolia							
Morocco	31,100	47,100	23	6	69	58	2.8
Mozambique	19,200	31,300	37	23	38	36	5.6
Myanmar	50,100	64,500	19	12	57	29	2.9
Namibia	2,000	2,700	34	19	44	32	4.6
Nepal							
Netherlands	16,200	17,000	12	9	78	66	1.7

New Zealand	3,900	4,500	14	8	78	86	2.0
Nicaragua	5,600	10,900	26	5	70	57	3.8
Niger	12,400	53,000	50	22	46	22	8.0
Nigeria							
Norway	4,600	4,900	12	10	79	79	1.8
Oman	2,900	6,800	38	4	72	78	5.0
Pakistan	157,300	348,700	30	9	61	34	5.1
Panama	3,200	5,100	21	6	75	57	2.7
Papua New Guinea							
Paraguay	6,000	12,100	30	5	71	57	3.8
Peru	27,600	41,100	23	6	70	74	2.9
Philippines	81,400	127,000	26	6	70	61	3.2
Poland	38,600	33,000	11	10	74	62	1.3
Portugal							
Puerto Rico	3,900	4,800	14	8	76	97	1.9
Qatar	619	874	16	4	72	92[2]	3.3
Romania	22,300	18,100	11	12	71	55	1.3
Russian Federation	142,400	101,500	10	14	67	73	1.1
Rwanda							
Saudi Arabia	24,900	54,700	37	6	72	88	4.5
Senegal	10,300	21,600	36	11	53	50	5.0
Serbia and Montenegro	10,500	9,000	14	10	73	52	1.6
Sierra Leone	5,200	10,300	44	21	34	39	6.5
Singapore							
Slovakia	5,400	4,900	10	10	74	57	1.3
Slovenia	2,000	1,600	9	10	76	51	1.1
Somalia	10,300	39,700	46	18	48	35	7.3
South Africa	45,200	40,200	19	18	48	57	2.6
Spain							
Sri Lanka	19,200	21,200	16	6	73	21	2.0
Sudan	34,300	60,100	36	10	56	39	4.4
Suriname	439	459	19	7	71	76[2]	2.1
Swaziland	1,100	900	29	21	34	24	4.5
Sweden							
Switzerland	7,200	5,800	10	8	79	68	1.4
Syria	18,200	34,200	30	5	72	50	3.3
Tajikistan	6,300	9,600	33	8	69	25	3.0
Tanzania	37,700	69,100	40	17	43	35	5.0
Thailand							
Togo	5,000	10,000	35	12	50	35	5.3
Trinidad and Tobago	1,300	1,200	13	9	71	75	1.6
Tunisia	9,900	12,900	17	5	73	64	2.0
Turkey	72,300	97,800	18	6	71	66	2.4
Turkmenistan							
Uganda	26,700	103,200	47	17	46	12	7.1

Table P5.8 (*Continued*)
Population Indicators by Region and Nation

Region/Country	*Population estimate ('000s)* *2004*	*2050*	*Birth Rate per 1,000 2003*[1]	*Death Rate per 1,000 2003*[1]	*Life Expectancy 2000*	*Percent Urban 2003*	*Fertility Rate per Woman 2000–2005*
Ukraine	48,200	31,700	10	16	68	67	1.2
United Arab Emirates	3,100	4,100	18	4	75	85	2.8
United Kingdom	59,400	66,200	11	10	78	89	1.6
United States							
Uruguay	3,400	4,100	17	9	75	93	2.3
Uzbekistan	26,500	37,800	26	8	70	37	2.4
Vanuatu	217	435	24	8	69	22[2]	4.3
Venezuela	26,200	41,700	20	5	74	88	2.7
Vietnam							
Yemen	20,700	84,400	43	9	60	26	7.0
Zambia	10,900	18,500	40	24	32	36	5.6
Zimbabwe	12,900	12,700	30	22	33	35	3.9

Note: Totals may not add because of independent rounding.
[1] Regional data is from 1998.
[2] Data from 2001.
Source: United Nations Population Fund (UNFPA); *The State of World Population 2004*, 2004; U.S. Bureau of the Census, International Data Base, 2002. Located in J. W. Wright (ed.), *The New York Times Almanac*, New York: Penguin, 2005.

According to current population data of 2005, 95 of 100 people live in the developing world. About 73 of the 100 persons in the world live in only 22 countries. About 37 people out of every 100 live in China and India. Of every 100 people in the world in 2005:

20 live in China (mainland)

17 live in India

5 live in the United States

4 live in Indonesia

3 live in Brazil

3 live in Pakistan

2 live in Bangladesh

2 live in Russia

2 live in Nigeria

2 live in Japan

2 live in Mexico

1 lives in the Philippines

1 lives in Vietnam

1 lives in Germany

1 lives in Egypt

1 lives in Ethiopia

1 lives in Iran

1 lives in Turkey

1 lives in Thailand

1 lives in France

1 lives in the United Kingdom

1 lives in Italy

Figure P5.6
Where in the world do they live?
Source: U.S. Census Bureau 2006. World Population Information. Available atwww.census.gov.

If We Shrank the Earth's Population to 100 People. . .

The following is a non-biased perspective of the world's population. The author is unknown.

If we could shrink the earth's population to a village of precisely 100 people, with all the existing human ratios remaining the same, it would look something like the following:

57 Asians
21 Europeans
14 from the Western Hemisphere, both north and south
8 Africans

52 would be female
48 would be male

70 would be non-white
30 would be white

70 would be non Christian
30 would be Christian

89 would be heterosexual
11 would be homosexual

6 people would possess 59% of the entire world's wealth and all 6 would be from the United States

80 would live in substandard housing

70 would be unable to read

50 would suffer from malnutrition

1 would be near death; 1 would be near birth

1 (yes, only 1) would have a college education

1 would own a computer

"When one considers our world from such a compressed perspective, the need for both acceptance, understanding and education becomes glaringly apparent."

Figure P5.7
Another world perspective view.

QUESTIONS

1. Using the first quote in the introduction from the U.S. National Academy of Sciences, further explain what is meant by a crisis point. Explain this term in factual, empirical terms from the information contained in the introduction.
2. Refer to Table P5.1. In the history of world population trends, where are the population acceleration points?
3. The tables in this section seem to indicate that the annual growth, birth, and death rates are slowing. Explain why population continues to rise sharply, as indicated in Figure P5.1.
4. Figure P5.2 indicates a UN World Projection Plan. Using this figure and other sources, explain what you think is a realistic projection and how it can be attained.
5. Figure P5.3 contains a projection of population growth for various countries. Compare this data with the information in other tables and figures in this section. Write a summary of your findings.
6. Box 2, "World Population Continues to Soar," states that the world can support only 2 billion people at a universal standard of living now enjoyed by industrialized nations. Interpret this statement from world projections and selected countries' population growth patterns.
7. Discuss the patterns of growth in Tables P5.3 and P5.4.
8. Examine Table P5.5. Relate this information to your findings from Question 7. What countries and areas have the most populated cities? What patterns do you find?
9. Relate Tables P5.6 and P5.7. How are births, deaths, and population growth related to fertility rates? Generalize your findings.
10. Table P5.8 provides important country-by-country profiles. Select 10 countries in various regions and compare their demographic data. What patterns do you find?
11. Consulting Table P5.8, locate the 10 countries with the highest life expectancy and the 10 countries with the lowest life expectancy. What significant patterns do you find?
12. Summarize your findings from Questions 5–11 into well-written paragraphs that strongly support your conclusions.

Interpreting Current Population Data

BARBARA A. EICHLER

Figure 1
World population, 1950–2004.
U.S. Bureau of the Census.

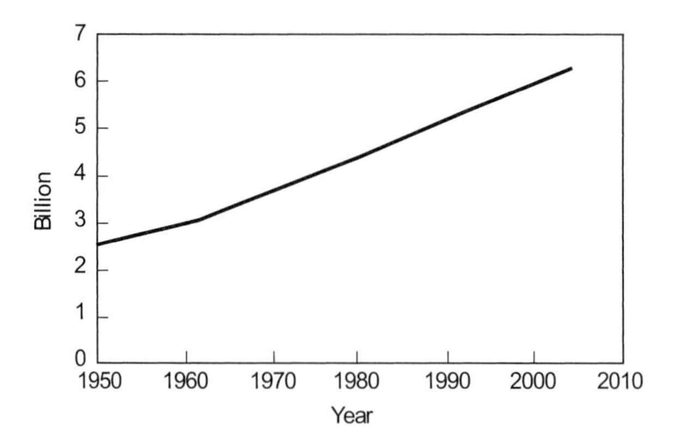

Figure 2
Annual growth rate in world population, 1950–2004.
U.S. Bureau of the Census.

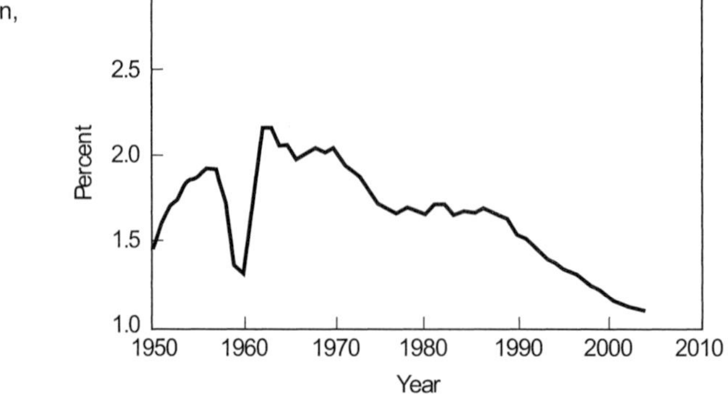

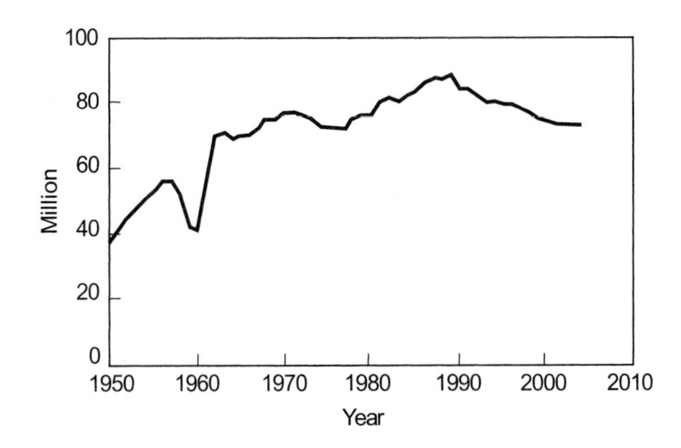

Figure 3
Annual addition to world population, 1950–2004.
U.S. Bureau of the Census.
Mid-2005 world population: 6,456,985,545
Mid-2005 United States population: 296, 749, 923

Table 1
World Population, Total and Annual Addition, 1950–2004

Year	*Total (billion)*	*Annual Addition (million)*
1950	2.56	38
1955	2.78	53
1960	3.04	41
1965	3.35	70
1970	3.71	77
1971	3.78	77
1972	3.86	76
1973	3.94	75
1974	4.01	73
1975	4.09	72
1976	4.16	72
1977	4.23	72
1978	4.30	75
1979	4.38	76
1980	4.45	76
1981	4.53	80
1982	4.61	81
1983	4.69	80
1984	4.77	82
1985	4.85	83
1986	4.93	86
1987	5.02	87
1988	5.11	87
1989	5.19	88
1990	5.28	84
1991	5.37	84

Table 1 (*Continued*)
World Population, Total and Annual Addition, 1950–2004

Year	*Total (billion)*	*Annual Addition (million)*
1992	5.45	82
1993	5.53	80
1994	5.61	80
1995	5.69	79
1996	5.77	79
1997	5.85	78
1998	5.93	77
1999	6.00	75
2000	6.08	74
2001	6.15	73
2002	6.23	73
2003	6.30	73
2004	6.37	73

Sources: Worldwatch, *Vital Signs 2005;* U.S. Census Bureau.

World population hit the 6-billion mark in June 1999 . . . five years later, it grew one-half billion more! The total population of the world's people is more than 3.5 times the population found at the beginning of the twentieth century and twice the size of the 1960 population. Within the twentieth century, the world's population twice doubled the 2-billion mark of 1922—the only century in the history of humankind to do so. The seriousness of this exponential growth strikes home when we recognize that, in the entire history of thousands of years of world civilizations, the first billion people arrived very recently in 1804 and then twice doubled themselves in the twentieth century. We are currently adding about 1 billion people to our population every 12–15 years, with a slowing of the rate to about a 20-year increment toward 2050. By 2050, the United Nations predicts that the world will add some 2.5 billion people, an amount equal to the world's total population in 1950!

GROWTH OF THE GLOBAL POPULATION

Broken down further, what does this growth mean? In 2002, slightly more than two people were being added to the world population each second (1.2 percent), producing a total of about 74 million people per year (about the size of Egypt); yet, in actuality, this represents a slowdown. From 1989–1990, approximately 87 million people were added. The slowdown in world population can be traced to the decline in fertility. In 1990, women around the world were giving birth to an average of 3.3 children (referred to as a fertility rate of 3.3). The replacement-level fertility rate is usually at 2.1 and represents the point at which each couple has only the number of births required to replace themselves in the population. By 2002, the world average fertility rate declined to 2.6, which is only one-half more of a child than needed to replace the population.

Although world fertility rates will remain above replacement level for some time, especially in sub-Saharan Africa, U.S. Census Bureau projections suggest that world averages will drop below replacement level before 2050. At present, there still remains an extremely wide variation in global child-bearing patterns, with fertility rates ranging from 1.2 in Europe to 8.0 in Niger (see Figures 4 and 5); (U.S. Census Bureau 2004).

However, declines in fertility rates do not necessarily translate into declines in the growth of population. The lack of correspondence between fertility rate declines and population growth is due to the increase in the numbers of child-bearing women in the world, which in effect increases the fertility rate (demographers refer to this phenomenon as "population momentum"). The increase in women of reproductive age was responsible for about three-fourths of global population growth in 2002 and will remain the primary

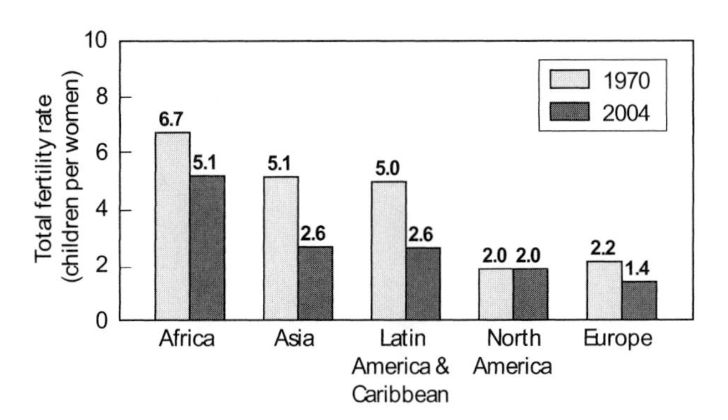

Figure 4
Childbearing trends in major world regions, 1970 and 2004.
Source: Population Reference Bureau, UN Population Division, *World Population Prospects: The 2002 Revision* (1970 data); C. Haub, *2004 World Population Data Sheet* (2004 data).

factor for global population growth until 2050 (see Figures 6 and 7; U.S. Census Bureau 2004, 3).

Ninety-six percent of the projected growth will occur in developing countries. sub-Saharan Africa (about 2.5-percent annual growth) and western Asia (2.0-percent annual growth) are the fastest-growing areas of the world; fertility rates remain high and will account for about 90 percent of the population increase by 2050. India is projected to become the most populous country, surpassing China by some 200 million people. The populations of Japan and Europe are now declining, and that decline will double by 2010 through 2015. North America continues to grow at about 1 percent annually due mainly to immigration (UNFPA 2004). Most future population growth will occur in countries that have large numbers of young people and large families. By 2050, Africa is projected to more than double to 2.3 billion people. Even though growth rates in Asia are lower, they apply to a much larger base of people—at present, half of the world's people. The 50 least developed countries of the world are expected to grow by 228 percent to 1.7 billion by 2050. The industrialized world is experiencing slower population growth and will increase their populations by only 4 percent. Some developed countries will even decline with the exception of the United States, which is averaging a 2.0 fertility ratio and 1-percent growth rate (Nierenberg 2005, 64).

Global Population Composition

Children (aged 0–14) were the largest age group in 2002, comprising 29 percent of the population. By 2050, their numbers will decline to about 20 percent, but children and adolescents still remain a large cohort impacting world age categories. In 2000, more than 100 nations contained populations in which young adults (aged 15–29) accounted

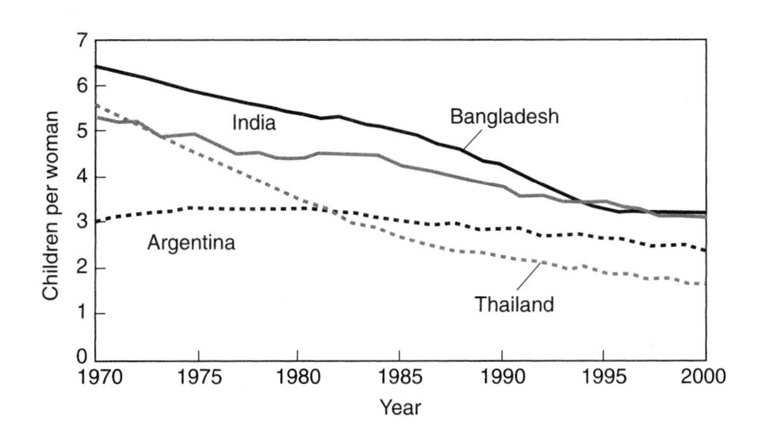

Figure 5
Different patterns of fertility decline, 1970–2000.
Sources: Registrar General of India; Instituto Nacional de Esradistica (Argentina); Population Reference Bureau, United Nations Population Division; Institute for Population and Social Research, Mahidol University, Thailand; Demographic and Health Surveys; Population Reference Bureau (PRB) estimates.

Figure 6
All world regions except Europe will continue to grow.
Source: Population Reference Bureau, 2004 World Population Data Sheet.

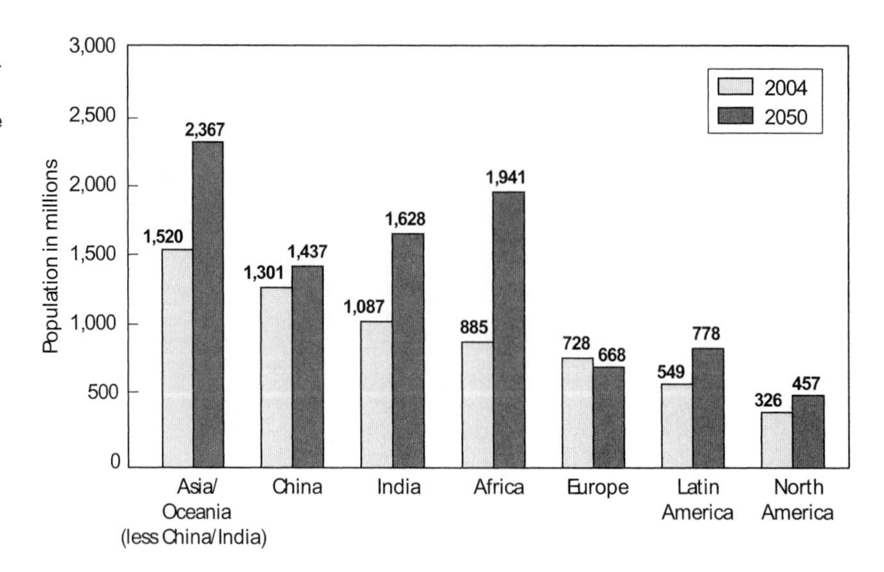

for nearly one-half of all adults. Such "youth bulges" are all found in the developing world where fertility rates are the highest. The number of women in their childbearing years (aged 15–49) is expected to increase by more than 25 percent between 2002 and 2050, but their proportion of the total population will slowly decrease (from 26 percent to 23 percent). The population of the labor force (aged 15–64)—which comprised 64 percent of the world in 2002—continues to grow, but its proportion with regard to total population stays approximately the same. Significant growth of the world's elderly (aged 65 years and older) is projected to be more than three times that of current statistics, and they will comprise about 17 percent of the population compared with 7 percent in 2002. This increase of the older population is located primarily in the developed countries (U.S. Census Bureau 2004, 3).

Figure 7
Across world regions, contrasting age structure will lead to sharply different demographic futures.
Source: Population Reference Bureau, 2004 World Population Data Sheet.

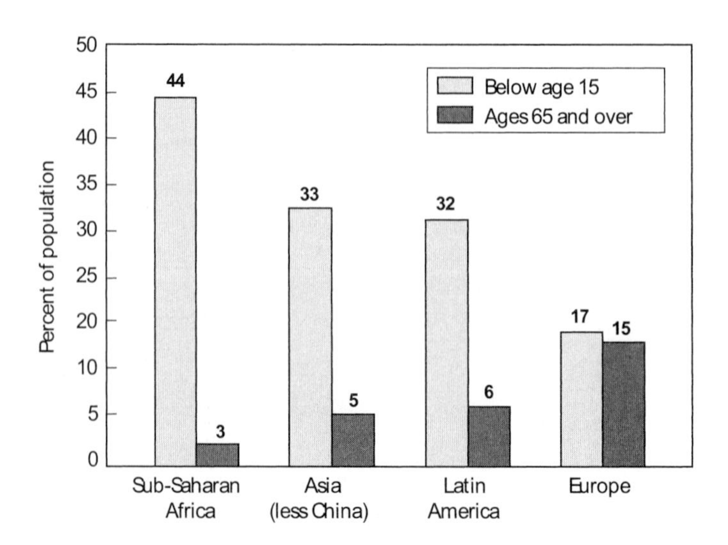

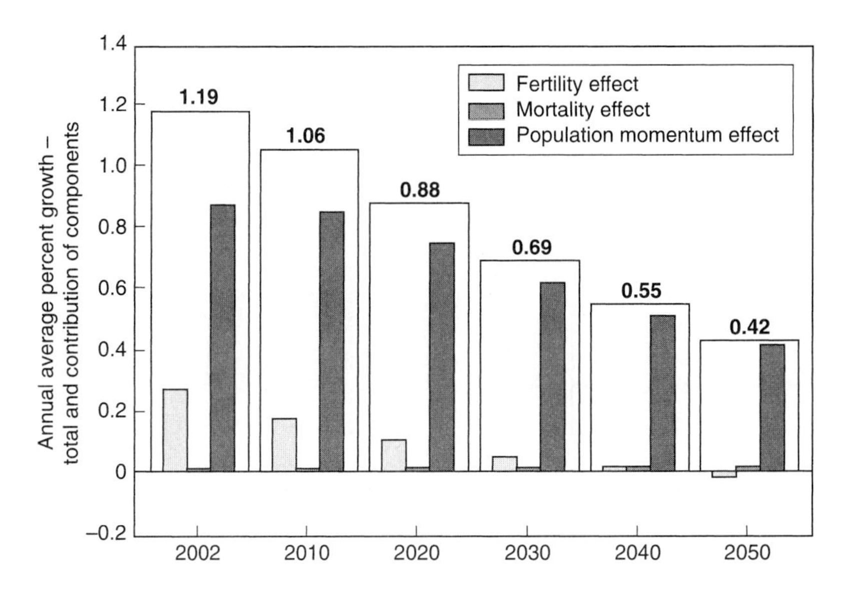

Figure 8
Components of yearly world population growth: 2002–2050. Fertility's contribution to global growth will cease by 2050.
Source: U.S. Census Bureau, International Programs Center, International Data Base and unpublished tables.

These age profiles will vary, carrying different implications per country. Western Europe has at present only 17 percent of their population under age 15, while western Africa maintains 44 percent and Asia (excluding China) has 33 percent under this age. Therefore, western Africa and Asia will experience major growth in their reproductive and productive age groups (15–49), while Europe will see a pattern of an increasingly elderly population (Ashford 2004; Population Reference Bureau) see Figures 8 9, and 10.

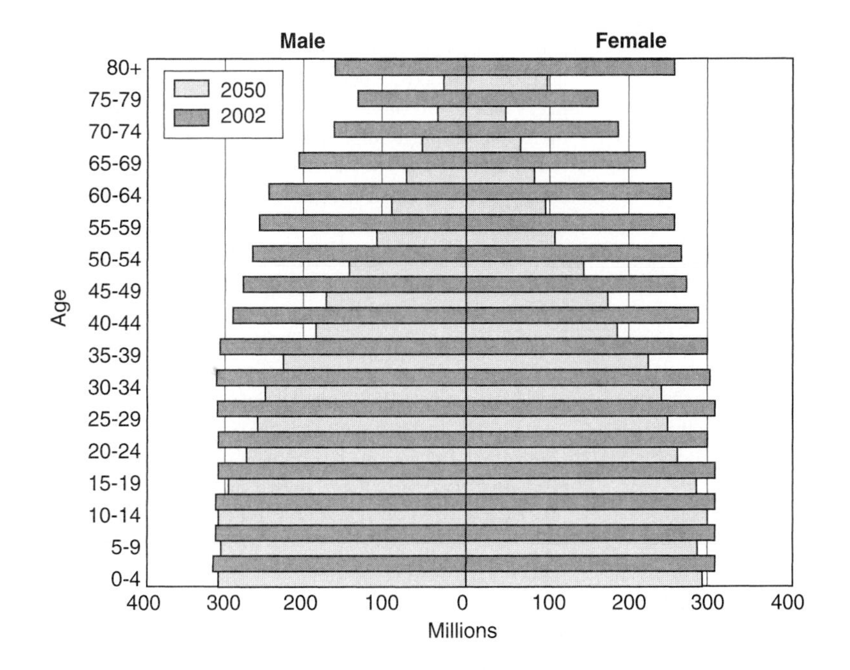

Figure 9
Age–sex structure of the world's population: 2002 and 2050.
Global population will change markedly over the next 50 years.
Source: U.S. Census Bureau, International Programs Center, International Data Base.

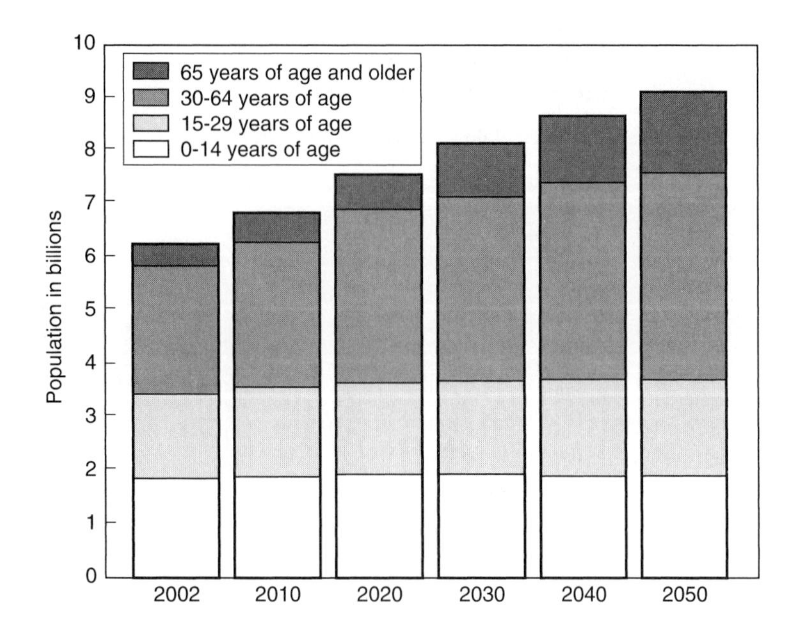

Figure 10
Population growth of specific age groups: 2002–2050.
The elderly population is projected to grow most rapidly.
Source: U.S. Census Bureau, International Programs Center, International Data Base.

CONTRACEPTIVE PREVALENCE IN THE DEVELOPING WORLD

Data indicate that contraceptive use is a primary determinant of the fertility level in a population and of differences in fertility between populations. In other words, there is a high correlation between family planning and low fertility levels. Contraceptive use is also an indicator in areas where people have access to reproductive health services. According to the UN, more than 350 million couples still lack access to adequate family planning. In addition, some 137 million women who wish to use contraception are not doing so, and 64 million other women are using less-effective methods of birth control. Contraceptive use varies greatly from region to region. In Latin American and the Caribbean, approximately 70 percent of the women are using some form of family planning, while only 21 percent of sub-Saharan African women do so. Contraceptive services are reaching more women than before, but they are not expanding quickly enough. It is projected that the demand for family planning services will increase by 40 percent by 2025 (UNFPA 2004).

The AIDS Pandemic in the Twenty-First Century

More than 20 million people have died of AIDS since the beginning of the AIDS pandemic infections of 1980. Forty million people are living with HIV—the virus that causes AIDS—and most of these people are expected to die during the next 10 years. The greatest impact of AIDS is felt in the developing countries of Africa, Asia, and Latin America, but most prominently in the sub-Saharan region of Africa. Most recent estimates from the UNAIDS and World Health Organization indicate that adult infection rates exceed 35 percent in Botswana and 10 percent in 11 other sub-Saharan African countries, constituting more than 25 million people. Because of the high rate of AIDS infection, a number of African countries will likely experience very high mortality levels that will bring the average life expectancy at birth in their regions to around 30 years—a level not seen since the early twentieth century. This is turn will result in population declines and unusual age–sex distributions.

Countries such as Thailand, Senegal, and Uganda have managed to slow the rate of the disease by prevention of mother-to-child AIDS transmission programs. Such programs provide hope that the AIDS pandemic might be controlled and curtailed (U.S. Census Bureau 2004, 4). Globally, however, the number of people living with HIV/AIDS from 2001 to 2003 has continued to increase from 35 to 38 million (Population Reference Bureau 2004). There were an estimated 5 million new HIV infections during 2003—an average of 14,000 per day—in which 40 percent were women and nearly 20 percent were children. In 2003, some 3 million people died of

Table 2
Cumulative HIV Infections and AIDS Deaths Worldwide, 1980–2004

Year	*HIV Infections (millions)*	*AIDS Deaths (million)*
1980	0.1	0.0
1981	0.3	0.0
1982	0.7	0.0
1983	1.2	0.0
1984	1.7	0.1
1985	2.4	0.2
1986	3.4	0.3
1987	4.5	0.5
1988	5.9	0.8
1989	7.8	1.2
1990	10.0	1.7
1991	12.8	2.4
1992	16.1	3.3
1993	20.1	4.7
1994	24.5	6.2
1995	29.8	8.2
1996	35.3	10.6
1997	40.9	13.2
1998	46.6	15.9
1999	52.6	18.8
2000	57.9	21.8
2001	62.9	24.8
2002	67.9	27.9
2003	72.9	30.9
2004	77.8	34.0

Source: Worldwatch, *Vital Signs 2004;* UNAIDS

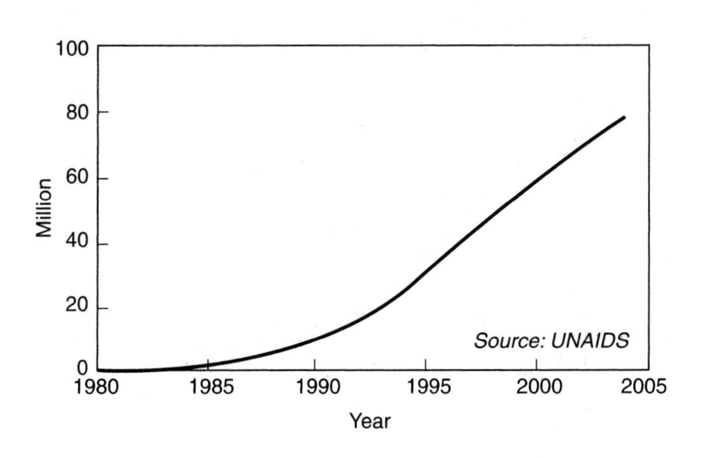

Figure 11
Estimates of cumulative HIV infections worldwide, 1980–2004.
Source: UNAIDS.

AIDS: 2.5 million adults and 500,000 children under the age of 15 (UNFPA 2004; see Figures 11, 12, and 13).

OTHER TRENDS AND IMPLICATIONS

Migration continues from rural areas of developing countries to fast-growing cities. By 2007, 50 percent of the world's population will be urban. The challenges inherent in this trend include providing urban social services (including reproductive health care), especially to poor urban areas, and also addressing the needs of rural communities (UNFPA 2004). Currently, five cities—Tokyo, Mexico City, New York, Sao Paolo, and Mumbai—have more than 15 million people. Tokyo has more than 35 million residents and is the largest city on earth (Nierenberg 2005, 64).

Along with the high resource consumption of the developed countries, population growth is contributing to stress on the global environment. Problems such as global warming, deforestation, water scarcity, and diminishing cropland make it more difficult to deal with individual needs, poverty, and gender inequality in impoverished countries. The developed countries with high resource consumption impact the earth severely by their use of materials, which places equal or compounded stress on the environment in addition to population growth. For example, the U.S. population increases by about 3 million people a year, while India's population increases by about 16 million. The United States, however, releases about 16 million tons of carbon into the atmosphere each year compared with India's approximately 5 million tons (Nierenberg 2005, 64).

Figure 12
Estimates of cumulative AIDS deaths worldwide, 1980–2004.
Source: UNAIDS.

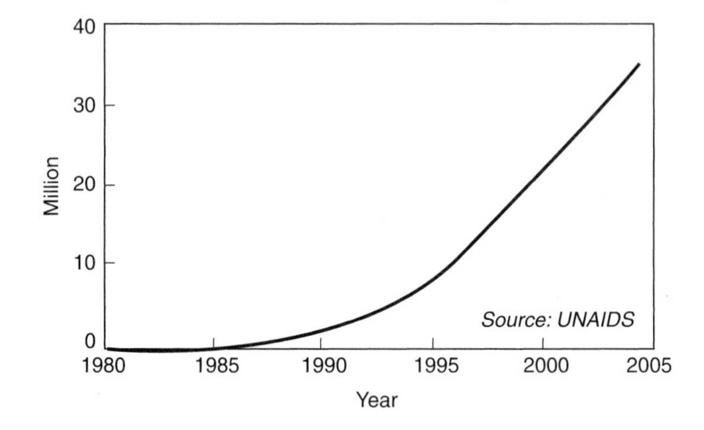

Figure 13
People living with HIV by region, 2002 and 2004.
Source: UNAIDS.

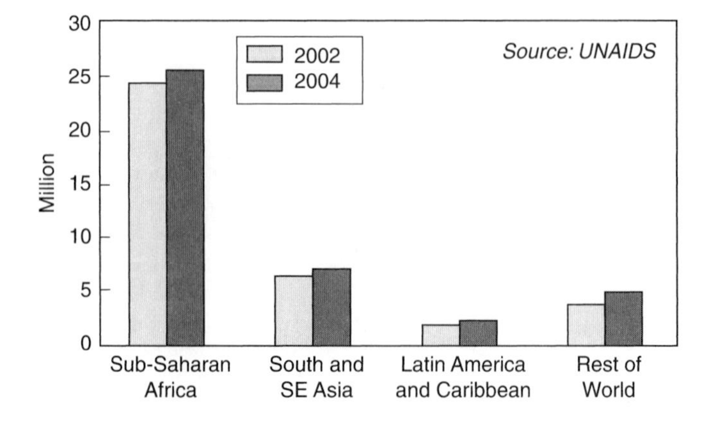

Pregnancy and childbirth complications are among the leading causes of death and illness among women of childbearing age in many developing countries. About 8 million women each year experience life-threatening pregnancy-related complications; over 529,000 die from such complications, with 99 percent of these deaths in developing countries. Many more women suffer infection or injury. In the developing world, about one-third of all pregnant women receive no health care during pregnancy; 60 percent of deliveries occur outside of health facilities, and skilled health personnel only assist in one-half of all deliveries (UNFPA 2004).

In summary, the world is growing at an unbelievably rapid pace during this century, with the specific regions of high growth being precisely those that can least support that growth. Africa, India, China, and Asia will experience unprecedented increases in population along with an accompanying stress on the resources and infrastructure needed to support that growth. Their populations will be very young, thus requiring education, job opportunities, and family planning so that these countries can sustain them. Resources such as water, food, health care, sanitation, and other basic needs will become increasingly stressed. Most people will move to cities, which will probably result in families having fewer children but also result in the requirement of health and social services as cities become more crowded. Industrialized countries will experience less or even negative growth except for the United States, which will remain about the third most populous country in the world. There will be a large increase in the percentage of people in the older population.

Resource use is a significant problem. Even though the population of developed countries will decline from about 20 percent of the world population to only about 14 percent in 2050, their resource use dominates world ecological systems, balance, and supply. New enlightened patterns of international cooperation will be required to deal with these demographics and their demands in the twenty-first century (see Figure 14).

The following box summarizes the United Nation's "World Population Prospects: The 2004 Revision" key findings and projections of historical trends and future projections for world population. Please refer to Table 3 for a detailed table of the populations of the world, Figures 15 and 16 for projection variants and projection of development groups, Table 4 for annual rate of change to 2050, Table 5 for fertility projections and Table 6 for life expectancy of major development groups and areas.

KEY FINDINGS FROM THE UN'S *WORLD POPULATION PROSPECTS: THE 2004 REVISION*

1. In July 2005, the world had 6.5 billion inhabitants—380 million more than in 2000 or a gain of 76 million annually. Despite the declining fertility levels projected over 2005–2050, the world population is expected to reach 9.1 billion according to the medium variant (see Table 3) and will still be adding 34 million persons annually by mid-century.

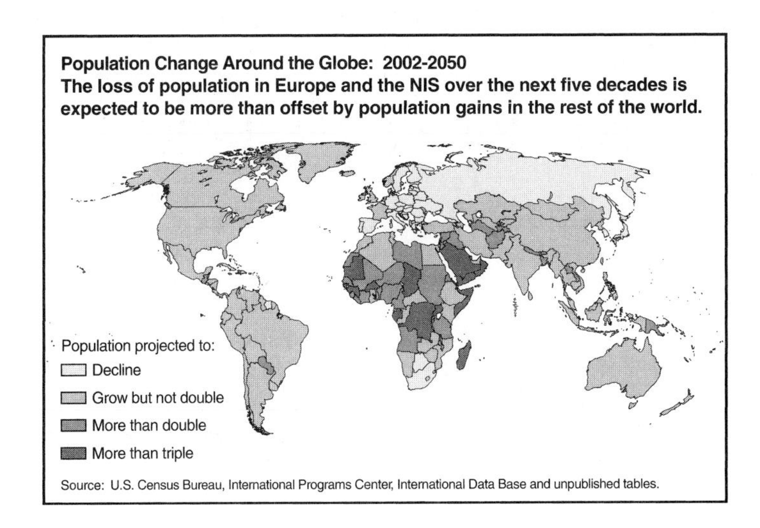
Population Change Around the Globe: 2002-2050
The loss of population in Europe and the NIS over the next five decades is expected to be more than offset by population gains in the rest of the world.
Population projected to:
Decline
Grow but not double
More than double
More than triple
Source: U.S. Census Bureau, International Programs Center, International Data Base and unpublished tables.

Figure 14
Population change around the globe: 2002–2050.
The loss of population in Europe and the newly independent states (NIS) over the next five decades is expected to be more than offset by population gains in the rest of the world.
Source: U.S. Census Bureau, International Programs Center, International Data Base and unpublished tables.

Table 3
Population of the World (Major Development Groups and Major Areas) by Projection Variants: 1950, 1975, 2005, and 2050

	Population (millions)			*Population in 2050 (millions)*			
Major area	*1950*	*1975*	*2005*	*Low*	*Medium*	*High*	*Constant*
World	2,519	4,074	6,465	7,680	9,076	10,646	11,658
More developed regions	813	1,047	1,211	1,057	1,236	1,440	1,195
Less developed regions	1,707	3,027	5,253	6,622	7,840	9,206	10,463
Least developed countries	201	356	759	1,497	1,735	1,994	2,744
Other less developed countries	1,506	2,671	4,494	5,126	6,104	7,213	7,719
Africa	224	416	906	1,666	1,937	2,228	3,100
Asia	1,396	2,395	3,905	4,388	5,217	6,161	6,487
Europe	547	676	728	557	653	764	606
Latin America and the Caribbean	167	322	561	653	783	930	957
Northern America	172	243	331	375	438	509	454
Oceania	13	21	33	41	48	55	55

Source: Population Division, Department of Economic and Social Affairs of the United Nations Secretariat, 2005; United Nations, World Population Prospects: The 2004 Revision, Highlights, New York.

2. Today, 95 percent of all population growth is absorbed by the developing world and 5 percent by the developed world. According to the medium variant, by 2050 the population of the more developed countries as a whole would be declining slowly by approximately 1 million persons a year. The developing world would be adding 35 million people annually, 22 million of whom would be absorbed by the least developed countries.

3. Future population growth is highly dependent on the path taken by future fertility. In the medium variant, fertility is projected to decline from 2.6 children per woman today to slightly over 2 children per woman

Figure 15
Population of the world by projection variants, 1950–2050.
Source: Population Division, Department of Economic and Social Affairs of the United Nations Secretariat, 2005; United Nations, *World Population Prospects: The 2004 Revision, Highlights,* New York.

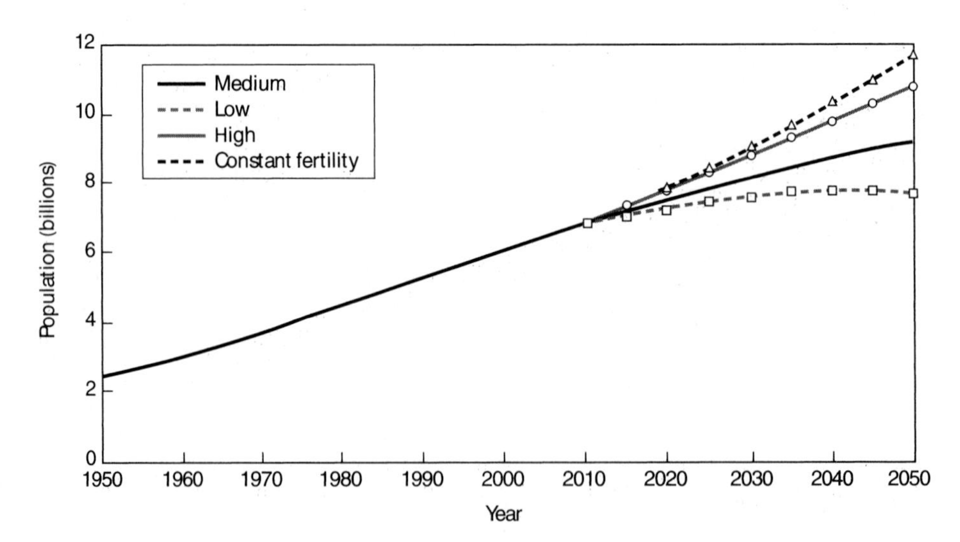

in 2050. If fertility were to remain about one-half child above the levels projected in the medium variant, world population would reach 10.6 billion by 2050. A fertility path that is one-half child below the medium variant would lead to a population of 7.6 billion by mid-century. Continued global population growth until 2050 is inevitable, even if the decline of fertility accelerates.

4. Because of its low and declining rate of growth, the population of developed countries as a whole is expected to remain virtually unchanged between 2005 and 2050 at about 1.2 billion. In contrast, the population of the 50 least developed countries is projected to more than double, passing from 0.8 billion in 2005 to 1.7 billion in 2050. Growth in the rest of the developing world is also projected to be robust, although less rapid, with its population rising from 4.5 billion to 6.1 billion between 2005 and 2050.

5. Very rapid population growth is expected to prevail in a number of developing countries, the majority of which are the least developed. Between 2005 and 2050, the population is projected to at least triple in Afghanistan, Burkina Faso, Burundi, Chad, Congo, the Democratic Republic of Congo, the Democratic Republic of Timor-Leste, Guinea-Bissau, Liberia, Mali, Niger, and Uganda.

6. The population of 51 countries or areas, including Germany, Italy, Japan, the Baltic States, and most of the successor states of the former Soviet Union, is expected to be lower in 2050 than in 2005.

7. During 2005–2050, nine countries are expected to account for one-half of the world's projected population increase: India, Pakistan, Nigeria, the Democratic Republic of Congo, Bangladesh, Uganda, the United States of America, Ethiopia, and China, listed according to the size of their contribution to population growth during that period.

8. From 2000–2005, fertility at the world level stood at 2.65 children per woman, about one-half the level it had been during 1950–1955 (five children per woman). In the medium variant, global fertility is projected to decline further to 2.05 children per woman by 2045–2050. Average world levels result from quite different trends by major development groups. In developed countries as a whole, fertility is currently 1.56 children per woman and is projected to increase slowly to 1.84 children per woman by 2045–2050. In the least developed countries, fertility is five children per woman

Table 4

Average Annual Rate of Change of the Total Population and the Population in Broad Age Groups by Major Area, 2005–2050 (medium variant)

Major area	*0–14*	*15–59*	*60+*	*80+*	*Total Population*
World	0.01	0.63	2.39	3.37	0.75
More developed regions	–0.14	–0.38	1.10	2.13	0.05
Less developed regions	0.03	0.82	2.88	4.19	0.89
Least developed countries	1.02	2.15	3.32	4.03	1.84
Other less developed countries	–0.29	0.54	2.84	4.21	0.68
Africa	0.87	2.00	3.12	3.86	1.69
Asia	–0.29	0.47	2.70	4.04	0.64
Europe	–0.36	–0.75	0.90	1.98	–0.24
Latin America and the Caribbean	–0.38	0.61	2.98	3.99	0.74
Northern America	0.23	0.37	1.67	2.30	0.62
Oceania	0.09	0.65	2.11	2.89	0.81

Source: Population Division, Department of Economic and Social Affairs of the United Nations Secretariat, 2005; United Nations, *World Population Prospects: The 2004 Revision, Highlights*, New York.

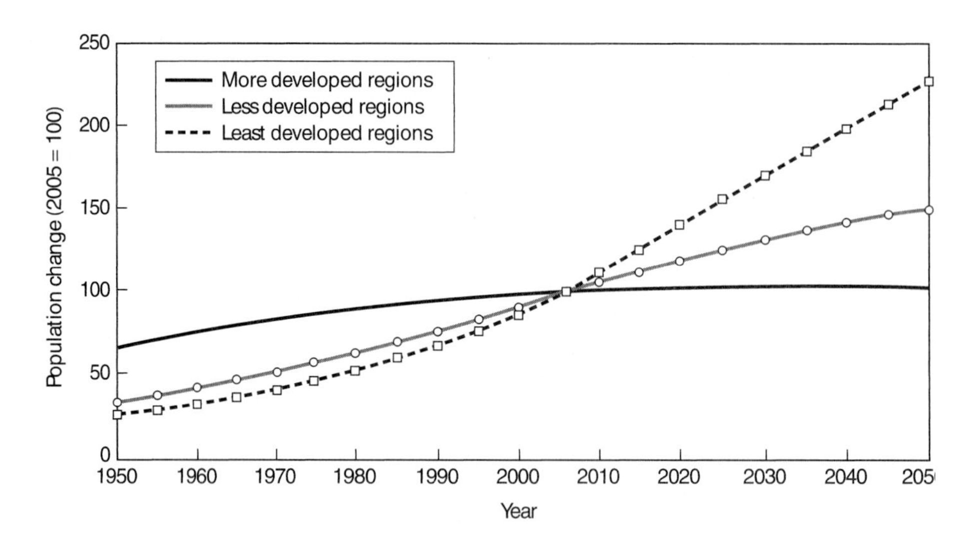

Figure 16
Population dynamics by development groups, 1950–2050.
Source: Population Division, Department of Economic and Social Affairs of the United Nations Secretariat, 2005; United Nations, *World Population Prospects: The 2004 Revision, Highlights*, New York.

and is expected to drop by about one-half to 2.57 children per woman by 2045–2050. In the rest of the developing world, fertility is already moderately low at 2.58 children per woman and is expected to decline further to 1.92 children per woman by mid-century, thus nearly converging to the fertility levels by then typical of the developed world. Realization of the projected fertility declines is contingent on access to family planning, especially in the least developed countries.

9. From 2000–2005, fertility remained above five children per woman in 35 of the 148 developing countries, 30

Table 5
Total Fertility for the World, Major Development Groups and Major Areas, 1970–1975, 2000–2005, and 2045–2050, by Projection Variants

	Total Fertility (children per woman)					
			2045–2050			
Major Area	*1970–1975*	*2000–2005*	*Low*	*Medium*	*High*	*Constant*
World	4.49	2.65	1.56	2.05	2.53	3.50
More developed regions	2.12	1.56	1.34	1.84	2.34	1.67
Less developed regions	5.44	2.90	1.59	2.07	2.56	3.69
Least developed countries	6.61	5.02	2.08	2.57	3.05	5.56
Other less developed countries	5.28	2.58	1.42	1.92	2.41	3.06
Africa	6.72	4.97	2.03	2.52	3.00	5.50
Asia	5.08	2.47	1.42	1.91	2.41	2.98
Europe	2.16	1.40	1.33	1.83	2.33	1.45
Latin America and the Caribbean	5.05	2.55	1.36	1.86	2.36	2.69
Northern America	2.01	1.99	1.35	1.85	2.35	1.99
Oceania	3.23	2.32	1.42	1.92	2.42	2.72

Source: Population Division, Department of Economic and Social Affairs of the United Nations Secretariat, 2005; United Nations, *World Population Prospects: The 2004 Revision, Highlights*, New York.

of which were least-developed countries, while the pace of decline in several countries of sub-Saharan Africa and South-Central Asia had been slower than anticipated. Overall, the countries with high fertility accounted for 10 percent of the world population. In contrast, fertility had reached below-replacement levels in 23 developing countries that accounted for 25 percent of the world population. This group included China, whose fertility during 2000–2005 was estimated at 1.7 children per woman.

10. Fertility levels in the 44 developed countries, which account for 19 percent of the world population, are currently very low. All except Albania have fertility below replacement level and 15, mostly located in Southern and Eastern Europe, have reached levels of fertility unprecedented in human history (below 1.3 children per woman). Since 1990–1995, fertility decline has been the rule among most developed countries. The few increases recorded, such as those in Belgium, France, Germany, the Netherlands, and the United States, have been small.

11. Global life expectancy at birth, which is estimated to have risen from 47 years in 1950–1955 to 65 years in 2000–2005, is expected to keep on rising to reach 75 years by 2045–2050. In more developed regions, the projected increase is from 76 years today to 82 years by mid-century. Among the least developed countries where life expectancy today is 51 years, life expectancy is expected to be 67 years by 2045–2050. Because many of these countries are highly affected by the HIV/AIDS epidemic, the projected increase in life expectancy is dependent on the implementation of effective programs to prevent and treat HIV infection. In the rest of the developing world under similar conditions, life expectancy is projected to rise from 66 years today to 76 years by mid-century.

12. Mortality in Eastern Europe has been increasing since the late 1980s. During 2000–2005, life expectancy in the region, at 67.9 years, was lower than it had been in 1960–1965 (68.6 years). The Russian Federation and the Ukraine are particularly affected by rises in mortality, resulting partly from the spread of HIV.

13. Twenty-five years into the HIV/AIDS epidemic, the impact of the disease is evident in terms of increased morbidity and mortality and slower population growth. In Southern Africa—the region with the highest HIV/AIDS prevalence of the disease—life expectancy had fallen from 62 years during 1990–1995 to 48 years during 2000–2005 and is projected to decrease further to 43 years over the next decade before a slow recovery starts. As a consequence, population growth in the region is expected to stall between 2005 and

Table 6

Life Expectancy at Birth for the World (Major Development Groups and Major Areas) 2000–2005 and 2045–2050

Major Area	*2000–2005*	*2045–2050*
World	65.4	75.1
More developed regions	75.6	82.1
Less developed regions	63.4	74.0
Least developed countries	51.0	66.5
Other less developed countries	66.1	76.3
Africa	49.1	65.4
Asia	67.3	77.2
Europe	73.7	80.6
Latin America and Caribbean	71.5	79.5
Northern America	77.6	82.7
Oceania	74.0	81.2

Source: Population Division, Department of Economic and Social Affairs of the United Nations Secretariat, 2005; United Nations, *World Population Prospects: The 2004 Revision, Highlights,* New York.

2020. In Botswana, Lesotho, and Swaziland, the population is projected to decrease as deaths outnumber births. In most other developing countries affected by the epidemic, population growth will continue to be positive because their moderate or high fertility more than counterbalances the rise in mortality.

14. The primary consequence of fertility decline, especially if combined with increases in life expectancy, is population aging, whereby the share of older persons in a population grows relative to that of younger persons. Globally, the number of persons aged 60 years and over is expected to almost triple, increasing from 672 million in 2005 to nearly 1.9 billion by 2050. Whereas six of every 10 of those older persons remain alive today in developing countries, by 2050, eight of every 10 will do so. An even more marked increase is expected in the number of the oldest segment of the population (aged 80 years and over): from 86 million in 2005 to 394 million in 2050. In developing countries, the rise will be from 42 million to 278 million, implying that by 2050, most of the oldest people will live in the developing world.

15. In developed countries, 20 percent of today's population is aged 60 years and over; by 2050, that proportion is projected to be 32 percent. The elderly population in developed countries has already surpassed the number of children (aged 0–14) and, by 2050, there will be two elderly persons for every child. In the developing world, the proportion of the population aged 60 and over is expected to rise from 8 percent in 2005 to close to 20 percent by 2050.

16. Increases in the median age—the age at which 50 percent of the population is older and 50 percent younger than that age—are indicative of population aging. Today, just 11 developed countries have a median age above 40 years. By 2050, there will be 89 countries in that group, with 45 in the developing world. Population aging, which is becoming a pervasive reality in developed countries, is also inevitable in the developing world and will occur faster in developing countries.

17. Countries where fertility remains high and has declined only moderately will experience the slowest population aging. By 2050, about one in five countries is still projected to have a median age equal to or less than 30 years. The youngest populations will be found in the least developed countries, 11 of which are projected to have median ages equal to or less than 23 years in 2050 including Afghanistan, Angola, Burundi, Chad, the Democratic Republic of Congo, Equatorial Guinea, Guinea-Bissau, Liberia, Mali, Niger, and Uganda.

18. During 2005–2050, the net number of international migrants to more developed regions is projected to be 98 million or an average of 2.2 million annually. The same number will leave the less developed regions. For the developed world, such a level of net migration will largely offset the expected excess of deaths over births during 2005–2050, which amounts to a loss of 73 million people. For the developing world, the 98 million emigrants represent scarcely less than 4 percent of expected population growth.

19. Over the 2000–2005 period, 74 countries were net receivers of migrants. In 64 of these countries, the projected net migration reinforced population growth and, in seven countries, it reversed the trend of population decline (Austria, Croatia, Germany, Greece, Italy, Slovakia, Slovenia). In three countries, the migration slowed down population decline but did not reverse it (Czech Republic, Hungary, the Russian Federation).

20. In terms of annual averages for the period 2005–2050, the major net receivers of international migrants are projected to be the United States (1.1 million annually), Germany (202,000), Canada (200,000), the United Kingdom (130,000), Italy (120,000), and Australia (100,000). The major countries of net emigration are projected to be China (–327,000 annually), Mexico (–293,000), India (–241,000), the Philippines (–180,000), Indonesia (–164,000), Pakistan (–154,000), and the Ukraine (–100,000).

REFERENCES

Ashford, L. (2004). World Population Highlights 2004. Bridge. Washington, D.C.: Population Reference Bureau.

Nierenberg, D. (2005). Population Continues Its Steady Rise. *Vital Signs 2005*. New York: Norton and Co.

Population Division, Department of Economic and Social Affairs of the United Nations Secretariat. (2005). *United Nations*

World Population Prospects: The 2004 Revisions. Highlights. New York: United Nations. Available at www.un.org/esa/population/publications/WPP2004/2004Highlights_finalrevised.pdf

Population Reference Bureau. (2004). 2004 World Population Data Sheet. Available at www.prb.org/Content/NavigationMenu/PRB/PRB_Library/Data_Sheets/Data_Sheets.htm

United Nations Population Fund (UNFPA). (2004). "The State of World Population 2004." New York: United Nations Population Fund. Accessed July 20, 2005. Available at http://www.unfpa.org/swp/2004/english/index.htm

UNAIDS. Joint United Nations programme on HIV/AIDS. Available at www.unaids.org.

U.S. Census Bureau. (2004). Global Population Profile: 2002. Washington, D.C.:

U.S. Department of Commerce, Economics and Statistic Administration and the U.S. Census Bureau.

QUESTIONS

1. Explain why the growth rate in world population is declining while world population is increasing.
2. Explain the difference between fertility rate and replacement-level fertility rate.
3. Why do declines in fertility rate not necessarily result in declines of population? Provide an example of this phenomenon.
4. Why do you think most population growth occurs in developing countries? Provide three reasons.
5. Describe what happens if countries continue to lose their population base by maintaining zero and negative growth rates.
6. In 2050, what are the implications of high percentages of youth in developing countries? What are the implications of high percentages of older people in developed countries?
7. Why do you think people in developing countries do not use family planning services or contraception? (Investigate cultural and old traditional-type rationales.)
8. Try to project what the AIDS pandemic will be like in 2050. Where will it be the worst, and where will it be more controlled? What are some solutions to control and curtail its growth?
9. What are the ramifications of high population growth for the world in 2050? Identify five issues and discuss them in class.

Putting the Bite on Planet Earth*

DON HINRICHSEN

Each year, about 90 million new people join the human race. This is roughly equivalent to adding three Canadas or another Mexico to the world annually, a rate of growth that will swell human numbers from today's 6.5 billion to about 8.5 billion by 2025.

These figures represent the fastest growth in human numbers ever recorded and raise many vital economic and environmental questions. Is our species reproducing so quickly that we are outpacing the Earth's ability to house and feed us? Is our demand for natural resources destroying the habitats that give us life? If 40 million acres of tropical forest—an area equivalent to twice the size of Austria—are being destroyed or grossly degraded every year, as satellite maps show, how will that affect us? If 27,000 species become extinct yearly because of human development, as some scientists believe, what will that mean for us? If nearly 2 billion people already lack adequate drinking water, a number likely to increase to double by 2010, how can all of us hope to survive?

The answers are hardly easy and go beyond simple demographics since population works in conjunction with other factors to determine our total impact on resources. Modern technologies and improved efficiency in the use of resources can help to stretch the availability of limited resources. Consumption levels also exert considerable impact on our resource base. Population pressures work in conjunction with these other factors to determine, to a large extent, our total impact on resources.

For example, although everyone contributes to resource waste, the world's bottom-billion poorest and top-billion richest do most of the environmental damage. Poverty compels the world's 1.2 billion bottom-most poor to misuse their environment and ravage resources, while lack of access to better technologies, credit, education, healthcare, and family planning condemns them to subsistence patterns that offer little chance for concern about their environment. This contrasts with the richest 1.3 billion, who exploit and consume disproportionate amounts of resources and generate disproportionate quantities of waste.

One example is energy consumption. Whereas the average Bangladeshi consumes commercial energy equivalent to three barrels of oil yearly, each American consumes an average of 55 barrels. Population growth in Bangladesh, one of the poorest nations, increased energy use there in 1990 by the equivalent of 8.7 million barrels, while U.S. population growth in the same year increased

Reprinted with permission from the National Wildlife Federation. D. Hinrichsen, Putting the Bite on Planet Earth, *International Wildlife*, September/October 1994.

*Editor's note: Even though this article was written in 1994, the classic nature of the article and its scope are very valuable when considering the many levels of ramifications of rapid population growth. Some of the numbers have been adjusted, but most remain in original form. Henricksen's argument even goes further as China and India are now becoming industrialized nations that are now consuming resources at ever increasing levels compared to Europe and even perhaps approaching the United State's highest levels of consumerism.

energy use by 110 million barrels. Of course, the U.S. population of 250 million in 1990 was more than twice the size of the Bangladeshi population of 113 million, but even if the consumption figures are adjusted for the difference in size, the slower-growing U.S. population still increases its energy consumption six or seven times faster yearly than does the more rapidly growing Bangladeshi population.

In the future, the effects of population growth on natural resources will vary locally because growth occurs unevenly across the globe. Over the course of the 1990s, the Third World's population ballooned by more than 900 million, while the population of the developed world added a mere 56 million. Asia, with 3.4 billion people in the 1990's, had 3.7 billion at the turn of the century; Africa's population will increase from 700 million to 867 million; and Latin America's from 470 million to 538 million. In the year 2000, the Third World's total population is nearly 5 billion; only 1.3 billion people will reside in industrialized countries.

The United Nations estimates that world population will near 11.2 billion by 2100. However, this figure is based on the assumption that growth rates will drop. If present rates continue, world population will stand at 10 billion by 2030 and 40 billion by 2110.

The United Nations Population Fund estimates that, to achieve the 11.2-billion projection, the number of couples using family planning services—such as modern contraceptives—in the developing world which is approximately 567 million in 2000 needs to double to 1.2 billion by 2025. In sub-Saharan Africa, this means a 10-fold increase by 2025 in the number of people who use family planning. If these measures do not succeed , human population growth could blast the 11.2-billion figure clear out of the ball park.

Perhaps the most ominous aspect of today's unprecedented growth is its persistence despite falling annual population growth rates everywhere except in parts of Africa, the Middle East, and South Asia. Annual global population growth stands at 1.3 percent, down from 2 percent in the early 1970s. Similarly, the total fertility rate (the average number of children a woman is likely to have) has dropped from a global average of six only three decades ago to about 2.8 today.

Population continues to grow because of tremendous demographic momentum. China's annual growth rate, for example, is only .6 percent. However, the country's huge population base—1.3 billion people—translates this relatively small rate of growth into a net increase in China's population of around 15 million yearly. Clearly, any attempt to slow population growth is a decades-long process affected by advances in medicine, extended life spans, and reduced infant, child, and maternal mortality.

The following pages survey the effects of human population growth on a wide range of natural resources.

PLANTS AND ANIMALS: THE SHRINKING ARK

Biologists have catalogued 1.7 million species and cannot even estimate how many species remain to be documented. The total could be 5 million, 30 million, or even more. Yet, we are driving thousands of species yearly to extinction through thoughtless destruction of habitat.

A survey conducted in Australia, Asia, and the Americas by the International Union for Conservation of Nature and Natural Resources—The World Conservation Union (IUCN) found that loss of living space affected 76 percent of all mammal species. Expansion of settlements threatened 56 percent of mammal species, while expansion of ranching affected 33 percent. Logging and plantations affected 26 percent.

The IUCN has declared human population growth the number one cause of extinctions. The 10 nations with the worst habitat destruction house an average of 189 people per square kilometer (250 acres), while the 10 that retain the most original habitat stand at only 29 people per square kilometer.

Future population growth poses a serious threat to wildlife habitat. Every new person needs space for housing, food, travel, work, and other needs. Human needs vary widely from place to place, but a UN survey found that the average person requires about 0.056 hectares (a hectare is a standard unit of land measurement equal to about 2.47 acres) of nonfarm land for daily living. To this must be added land for food production. This varies with land quality and available technologies, but each newborn person probably will need at least 0.2 hectares of cropland unless food production per acre increases in the years ahead. This will require the conversion of more and more wild land into cropland. In East Asia, for example, the amount of irrigated, high-yield cropland per person is near the 0.2-hectare limit.

UN consultant and author Paul Harrison estimates, very conservatively, that each new person will need at least a quarter of a hectare. Thus, every billion people that we add to the planet in the years ahead will require 250 million hectares more of agricultural land. Most of this land will have to come from what is currently wildlife habitat. The UN's projected population of 11.2 billion by 2100 would require creation of roughly 20 million square kilometers (8 million sq. mi.) of new cropland—equivalent to more than 80 percent of all forest and woodland in developing countries today.

Conversion of natural habitat for human use can even reduce the value of remaining wild areas for wildlife. When development chops wild lands into fragments, native species often decline simply because the small remnants do not meet their biological needs. For example, studies of U.S. forest birds indicate that species that prefer to nest in forest interiors are more subject to predation and lay fewer eggs when habitat fragmentation forces them to nest along forest edges. A study in southern California indicated that most canyons lose about half of native bird species dependent on chaparral habitat within 20 to 40 years after the canyons become isolated by development, even though the chaparral brush remains. Biologist William Newmark's 1987 study of 14 Canadian and U.S. national parks showed that 13 of the parks had lost some of their mammal species, at least in part because the animals could not adapt to confinement within parks surrounded by developed land.

Habitat loss in North America and in Latin American tropics has caused declines in many bird species that migrate between those regions. The Breeding Bird Survey, a volunteer group that tabulates nesting birds each June, found that 70 percent of neotropical migrant species monitored in the eastern United States declined from 1978 to 1987. So did 69 percent of monitored neotropical migrants that nest in prairie regions. Declining species include such familiar songbirds as veeries, wood thrushes, blackpoll warblers, and rose-breasted grosbeaks. As human population growth continues to push development into wild areas, fragmentation will increase and its effects on wildlife survival will intensify.

LAND LOSS: A FOOD CRISIS

Land degradation, a global problem, is becoming acute in much of the developing world. Population pressures and inappropriate farming practices contribute to soil impoverishment and erosion, rampant deforestation, overgrazing of common lands, and misuse of agrochemicals.

Worldwide, an estimated 1.2 billion hectares, an area about the size of China and India combined, have lost much of their agricultural productivity since 1945. Every year, farmers abandon about 70,000 square kilometers (27,000 sq. mi.) of farmland because soils are too degraded for crops.

Drylands, including grasslands that provide rich pastures for livestock, have been hardest hit. Although not as extensive as once thought, desertification—the ecological destruction that turns productive land into deserts—still threatens the Middle East and parts of Africa and Asia.

Because of land degradation, large portions of the Sahel, including Burkina Faso, Chad, Mali, Mauritania, Niger, and Senegal, can no longer feed their people. Although annual fluctuations in rainfall may interrupt the trend of cropland loss, the Sahel could suffer agricultural collapse within a decade. Sahelian croplands, as presently farmed, can support a maximum of 36 million people. In 2000, the rural population exceeds that number.

Since 1961, food production has matched world population growth in all developing regions except sub-Saharan Africa. In the early 1980s, the UN Food and Agriculture Organization (FAO) predicted that more than half of all developing nations examined in its study of carrying capacity (62 out of 115) may be unable to feed their projected populations by 2000 using current farming technology. Most of the 62 countries probably will be able to feed less than half of their projected populations without expensive food imports.

As a direct result of population growth, especially in developing nations, the average amount of cropland per person is projected to decline from 0.28 hectares in 1990 to 0.17 by 2025.

Three factors will determine whether food production can equal population growth:

1. *New Croplands.* Currently, the amount of new land put into production each year may equal the amount taken out of production for various reasons, such as erosion, salt deposits, and waterlogging. Thus, the net annual gain in arable land, despite widespread habitat destruction to create it, may be zero.
2. *New Water Sources.* Agricultural demand for water doubled between 1970 and 2000. Already more than 70 percent of water withdrawals from rivers, underground reservoirs, and other sources go to crop irrigation.
3. *Agrochemical Use.* Pesticides and fertilizers are boosting crop yields. However, in many areas, agrochemicals are too expensive to use, while in other areas they are overused to prop up falling yields. Agrochemicals can pose health hazards, creating another expense for developing nations.

FOREST: THE VANISHING WORLD

The quest for more crop and grazing land has sealed the fate of much of the world's tropical forests. Between 1971 and 1986, arable land expanded by 59 million hectares, while forests shrank by at least 125 million hectares. However, consultant Harrison estimates that during the same period, land used for settlements, roads, industries, office buildings, and other development expanded by more than 30 million hectares as a result of growth in urban centers, reducing the amount of arable land in surrounding areas. Consequently, the amount of natural habitat wiped out to produce the 59-million-hectare net in arable land may have exceeded 100 million hectares.

When both agricultural and nonagricultural needs are taken into account, human population growth may be responsible for as much as 80 percent of the loss of forest cover worldwide. Asia produces the highest rate of loss, 1.2 percent a year. Latin America loses 0.9 percent yearly and Africa 0.8 percent.

If current trends continue, most tropical forests will soon be destroyed or damaged beyond recovery. Of the 76 countries that presently encompass tropical forests, only four—Brazil, Guyana, Papua New Guinea, and Zaire—are likely to retain major undamaged tracts by 2010, less than a few years away.

Population pressure contributes to deforestation not only because of increased demand for cropland and living space, but also because of increased demand for fuelwood, on which half of the world's people depend for heating and cooking. The majority of sub-Saharan Africa's population is dependent on fuelwood: 82 percent of all Nigerians, 70 percent of Kenyans, 80 percent of all Malagasies, 74 percent of Ghanaians, 93 percent of Ethiopians, 90 percent of Somalians, and 81 percent of Sudanese.

In 1990, 100 million Third World residents lacked sufficient fuelwood to meet minimum daily energy requirements, and close to 1.3 billion were consuming wood faster than forest growth could replenish it. On average, consumption outpaces supply by 30 percent in sub-Saharan Africa as a whole, by 70 percent in the Sudan and India, by 150 percent in Ethiopia, and by 200 percent in Niger. If present trends continue, FAO predicts, another 1 billion people will be faced with critical fuelwood shortages into the 21st Century. Already, growing rings of desolation—land denuded for fuelwood or building materials—surround many African cities, such as Ouagadou-gou in Burkina Faso, Niamey in Niger, and Dakar in Senegal. In 2000, the World Bank estimates, half to three-quarters of all West Africa's fuelwood consumption will be burned in towns and cities.

According to the World Bank, remedying the fuelwood shortage will require planting 55 million hectares—an area nearly twice the size of Italy—with fast-growing trees at a rate of 2.7 million hectares a year, five times the 1994 rate of 555,000 hectares.

TROUBLED OCEANS: DISAPPEARING RESOURCES

Population and development pressures have been mounting in coastal areas worldwide for the past 30 years, triggering widespread resource degradation. Coastal fisheries are overexploited in much of Asia, Africa, and parts of Latin America. In some cases—as in the Philippines, Indonesia, Malaysia, China, Japan, India, the west coast of South America, the Mediterranean, and the Caribbean—economically important fisheries have collapsed or are in severe decline. "Nearly all Asian waters within 15 kilometers of land are considered overfished," says Ed Gomez, director of the Marine Science Institute at the University of the Philippines in Manila.

Overfishing is not the sole cause of these declines. Mangroves and coral reefs—critical nurseries for many marine species and among the most productive of all ecosystems—are being plundered in the name of development.

In 1990, a UN advisory panel, the Group of Experts on the Scientific Aspects of Marine Pollution (GESAMP), reported that coastal pollution worldwide has grown worse over the decade of the 1980s. Experts pointed to an overload of nutrients—mainly nitrogen and phosphorus from untreated or partially treated sewage, agricultural runoff, and erosion—as the most serious coastal pollution problem. Human activities may be responsible for as much as 35 million metric tons of nitrogen and up to 3.75 million metric tons of phosphorus flowing into coastal waters every year. Even such huge amounts could be dissolved in the open ocean, but most of the pollution stays in shallow coastal waters where it causes massive algal blooms and depletes oxygen levels, harming marine life near the shores.

Although the world still possesses an estimated 240,000 square kilometers (93,000 sq. mi.) of mangrove swamps—coastal forests that serve as breeding grounds and nurseries for many commercially important fish and shellfish species—this represents only about half the original amount. Clear-cutting for timber, fuelwood, and wood chips; conversion to fish and shellfish ponds; and expansion of urban areas and croplands have claimed millions of hectares globally. For example, of the Philippine's original mangrove area—estimated at 500,000 to 1 million hectares—only 100,000 hectares remain; 80 to 90 percent are gone.

Some 600,000 square kilometers (230,000 sq. mi.) of coral reefs survive in the world's tropical seas. Unfortunately, these species-rich ecosystems are suffering widespread decline. Clive Wilkinson, a coral reef specialist working at the Australian Institute of Marine Science, estimates that fully 10 percent of the world's reefs have already been degraded "beyond recognition." Thirty percent are in critical condition and will be lost completely in the first part of the 21st Century, while another 30 percent are threatened and will be lost in 20 to 40 years. Only 30 percent, located away from human development or otherwise too remote to be exploited, are in stable condition.

Throughout much of the world, coastal zones are overdeveloped, overcrowded, and overexploited. About two-thirds of the world's population—some 3.6 billion people—live along coasts or within 150 kilometers (100 mi.) of one. Within three decades, 75 percent, or 6.4 billion, will reside in coastal areas—nearly a billion more people than the current global population.

In the United States, 54 percent of all Americans live in 772 coastal counties adjacent to marine coasts or the Great Lakes. Between 1960 and 1990, coastal population density increased from 275 to nearly 400 people per square kilometer. By 2025, nearly 75 percent of all Americans will live in coastal counties, with population density doubling in areas such as southern California and Florida.

Similarly, nearly 780 million of China's 1.3 billion people—almost 67 percent—live in 14 southeast and coastal provinces and two coastal municipalities, Shanghai and Tianjin. Along much of China's coastline, population densities average more than 600 per square kilometer. In Shanghai, they exceed 2,000 per square kilometer. During the past few years, as many as 100 million Chinese have moved from poorer provinces in central and western regions to coastal areas in search of better economic opportunities. More ominously, population growth is expected to accelerate in the nation's 14 newly created economic free zones and five special economic zones, all of them coastal.

WATER: DISTRIBUTION WOES

Nearly 75 percent of the world's freshwater is locked in glaciers and icecaps, with virtually all the rest underground. Only about 0.01 percent of the world's total water is easily available for human use. Even this tiny amount would be sufficient to meet all the world's needs if it were distributed evenly. However, the world is divided into water "haves" and "have nots." In the Middle East, north Asia, northwestern Mexico, most of Africa, much of the western United States, parts of Chile and Argentina, and nearly all of Australia, people need more water than can be sustainably supplied.

As the world's human population increases, the amount of water per person decreases. The United Nations Educational, Scientific and Cultural Organization (UNESCO) estimates that the amount of freshwater available per person has shrunk from more than 33,000 cubic meters (1.2 million cu. ft.) per year in 1850 to only 8,500 cubic meters (300,000 cu. ft.) presently. Of course, this is a crude, general figure. But because of population growth alone, water demand in more than half the world's countries by 2000 is twice what it was as recently as 1971.

Already some 2 billion people in 80 countries must live with water constraints for all or part of the year. In the first part of the 21st Century, Egypt will have only two-thirds as much water for each of its inhabitants as it has today, and Kenya only half as much. By then, six of East Africa's seven nations and all five nations on the south rim of the Mediterranean will face severe shortages. In 1990, 20 nations suffered water scarcity, with less than 1,000 cubic meters (35,000 cu. ft.) of water per person, according to a study by Population Action International. Another eight experienced occasional water stress. The 28 nations represent 333 million people. By 2025, some 48 nations will suffer shortages, involving some 3 billion residents, according to the study.

China—although not listed as water short because of the heavy amount of rain that falls in its southern region—has, nevertheless, exceeded its sustainable water resources. According to Qu Geping, China's Environment Minister, the country can supply water sustainably to only 650 million people, not the current population of 1.3 billion. In other words, China is supporting twice as many people as its water resources can reasonably sustain without drawing down groundwater supplies and overusing surface waters.

FOSSIL FUELS: ENERGY BREAKDOWN

Human society runs on energy, principally fossil fuels such as oil, gas, and coal. These three account for 90 percent of global commercial energy production. Nuclear power, hydro-electricity, and other sustainable resources provide the rest.

The industrialized nations, with less than a quarter of the world's people, burn about 70 percent of all fossil fuels. The United States alone consumes about a quarter of the world's commercial energy, and the former Soviet Union about a fifth. In terms of per capita consumption patterns, Canada burns more fuel than any other nation—the equivalent of 9 metric tons of oil per person—followed by Norway at 8.9 metric tons of oil per person and the United States at 7.3. By contrast, developing nations on average use the equivalent of only about half a metric ton of oil per person yearly.

Known oil reserves should meet current levels of consumption for another 41 years, up from an estimated 31 years in 1970 thanks to better energy efficiency and conservation measures, along with new oil fields brought into production. Natural-gas reserves should meet current demand for 60 more years, up from 38 years in 1970. Coal reserves should be good for another 200 years.

But our addiction to fossil fuel has resulted in chronic, sometimes catastrophic, pollution of the atmosphere, in some cases far beyond what natural systems or man-made structures can tolerate. A noxious atmospheric cocktail of chemical pollutants is primarily responsible for the death and decline of thousands of hectares of European forests. Acid rain—caused by a combination of nitrogen and sulfur dioxides released from fossil-fuel combustion—has eaten away at priceless monuments and buildings throughout Europe and North America, causing billions of dollars in damages.

Urban air contains a hazardous mix of pollutants—everything from sulfur dioxide and reactive hydrocarbons to heavy metals and organic compounds. Smog alerts are now commonplace in many cities with heavy traffic. In Mexico City, for example, smog levels exceed World Health Organization standards on most days. Breathing the city's air is said to be as damaging as smoking two packs of cigarettes a day, and half the city's children are born with enough lead in their blood to hinder their development.

The only way to stretch fossil fuel reserves and reduce pollution levels is to conserve energy and use it much more efficiently than we do now. Some progress has been made, but the benefits of energy conservation have been realized in only a few industrialized countries.

Recent history has shown what can happen. In the decade following the first oil shock, per capita energy consumption fell by 5 percent in the member states of the Organization for Economic Cooperation and Development (OECD)—consisting of the industrialized countries of Western Europe and North America, plus Japan, Australia, and New Zealand—while their per capita gross domestic product grew by a third.

Buildings in the OECD countries use a quarter less energy now than they did before 1973, while the energy efficiency of industry has improved by a third. Worldwide, cars now get 25 percent more kilometers per gallon than they did in 1973. In all, increased efficiency since 1973 has saved the industrialized nations $250 billion in energy costs.

Even more savings could be realized through concerted efforts to conserve energy and improve efficiency. Three relatively simple, cost-effective measures could be introduced immediately: (1) making compact fluorescent lamps generally available in homes and offices; (2) tightening up building codes to require better insulation against cold and heat; and (3) requiring lean-burn engines, which get up to 80 kilometers per gallon (50 mpg), in all new compact cars. These three "technical fixes" could save billions of dollars in energy costs.

POLICY: BUILDING A FUTURE

The main population issues—urbanization, rapid growth, and uneven distribution—when linked with issues of environmental decline, pose multiple sets of problems for policymakers. The very nature of these interrelated problems makes them virtually impossible to deal with in balkanized bureaucracies accustomed to managing only one aspect of any problem. Population and resource issues require integrated, strategic management, an approach few countries are in a position to implement.

Sustainable-management strategies, designed to ensure that resources are not destroyed by overexploitation, are complicated to initiate because they require the cooperation of ministries or departments often at odds over personnel, budgets, and political clout. Most governments lack institutional mechanisms that ensure a close working relationship among competing ministries. Consequently, most sustainable-development initiatives never get beyond words on paper. "We talk about integrated resource management, but we don't do it," admits one Indian official in Delhi. "Our ministries are like fiefdoms, they seldom cooperate on anything."

Fragmented authority yields fragmented policies. Big development ministries—such as industry and commerce, transportation, agriculture, fisheries, and forestry—rarely cooperate in solving population and resource problems. Piecemeal solutions dominate, and common resources continue to deteriorate.

The world's population and resource problems offer plenty of scope for timely and incisive policy interventions that promise big returns for a relatively small investment. As little as $17 billion a year could provide contraceptives to every woman who wants them, permitting families throughout the globe to reduce births voluntarily. This approach might produce the same or better results than would government-set population targets, according to one study. Moreover, population specialists recognize that educating girls and women provides a higher rate of return than most other investments. "In fact, it may well be the single most influential investment that can be made in the developing world," says Larry Summers, a former World Bank economist.

But time is at a premium. The decision period for responding to the crises posed by rapidly growing populations, increased consumption levels, and shrinking resources will be confined, for the most part, to the first two decades of the 21st Century. If human society does not succeed in checking population growth, the future will bring widespread social and economic dislocations as resource bases collapse. Unemployment and poverty will increase, and migrations from poorer to richer nations will bring Third World stresses to the developed world.

QUESTIONS

1. How many people are added to the human race each year? Calculate the percentage of growth annually.
2. List some of the stresses on the earth's environment as a result of unprecedented population growth.
3. What are the difficulties of lowering world population rates even if a country's annual growth rate declines?
4. How much land does a person need to survive? How does this impact wildlife, plants, and land use?
5. Discuss the problem of land degradation. How much is the average amount of cropland per person projected to decline from 1990 to 2025? Discuss implications. What are possible solutions?
6. How soon will most tropical forests disappear? What are the implications?
7. What are some of the pressures on oceans? What are the ramifications?
8. How much have freshwater supplies shrunk from 1850? What are the issues here?
9. What is the effect of world reliance on fossil fuels? What are some solutions suggested in the reading as well as your own?
10. Name some practical solutions to building a future policy that can sustain world population. Aside from suggestions in the reading, also name your own possible solutions and approaches.

Can the Growing Human Population Feed Itself?

JOHN BONGAARTS

Demographers project that the world's population will double during the next half century, from 5.3 billion people in 1990 to more than 10 billion by 2050. How will the environment and humanity respond to this unprecedented growth? Expert opinion divides into two camps. Environmentalists and ecologists, whose views have widely been disseminated by the electronic and print media, regard the situation as a catastrophe in the making. They argue that in order to feed the growing population, farmers must intensify agricultural practices that already cause grave ecological damage. Our natural resources and the environment, now burdened by past population growth, will simply collapse under the weight of this future demand.

The optimists, on the other hand, comprising many economists as well as some agricultural scientists, assert that the earth can readily produce more than enough food for the expected population in 2050. They contend that technological innovation and the continued investment of human capital will deliver high standards of living to much of the globe, even if the population grows much larger than the projected 10 billion. Which point of view will hold sway? What shape might the future of our species and the environment actually take?

Many environmentalists fear that world food supply has reached a precarious state: "Human numbers are on a collision course with massive famines. . . . If humanity fails to act, nature will end the population explosion for us—in very unpleasant ways—well before 10 billion is reached," write Paul R. Ehrlich and Anne H. Ehrlich of Stanford University in their 1990 book, *The Population Explosion*. In the long run, the Ehrlichs and like-minded experts consider substantial growth in food production to be absolutely impossible. "We are feeding ourselves at the expense of our children. By definition farmers can overplow and over-pump only in the short run. For many farmers the short run is drawing to a close," states Lester R. Brown, president of the Worldwatch Institute, in a 1988 paper.

Over the past three decades, these authors point out, enormous efforts and resources have been pooled to amplify agricultural output. Indeed, the total quantity of harvested crops increased dramatically during this time. In the developing world, food production rose by an average of 117 percent in the quarter of a century between 1965 and 1990. Asia performed far better than other regions, which saw increases below average.

Because population has expanded rapidly as well, per capita food production has generally shown only modest change; in Africa, it actually declined. As a consequence, the number of undernourished people is still rising in most parts of the developing world, although that number did fall from 844 million to 786 million during the 1980s. But this decline reflects improved nutritional conditions in Asia alone. During the same period, the number of people having energy-deficient diets in Latin America, the Near East, and Africa climbed.

Many social factors can bring about conditions of hunger, but the pessimists emphasize that population pressure on

Reprinted with permission from *Scientific American*. J. Bongaarts, Can the Growing Human Population Feed Itself? Copyright © (1994, March) by *Scientific American*, Inc. All rights reserved.

fragile ecosystems plays a significant role. One specific concern is that we seem to be running short on land suitable for cultivation. If so, current efforts to bolster per capita food production by clearing more fertile land will find fewer options. Between 1850 and 1950, the amount of arable land grew quickly to accommodate both larger populations and greater demand for better diets. This expansion then slowed and, by the late 1980s, ceased altogether. In the developed world, as well as in some developing countries (especially China), the amount of land under cultivation started to decline during the 1980s. This drop is largely because spreading urban centers have engulfed fertile land or, once the land is depleted, farmers have abandoned it. Farmers have also fled from irrigated land that has become unproductive because of salt accumulation.

Moreover, environmentalists insist that soil erosion is destroying much of the land that is left. The extent of the damage is the subject of controversy. A global assessment, sponsored by the United Nations Environment Program and reported by the World Resources Institute and others, offers some perspective. The study concludes that 17 percent of the land supporting plant life worldwide has lost value over the past 45 years. The estimate includes erosion caused by water and wind, as well as chemical and physical deterioration, and ranks the degree of soil degradation from light to severe. This degradation is least prevalent in North America (5.3 percent) and most widespread in Central America (25 percent), Europe (23 percent), Africa (22 percent), and Asia (20 percent). In most of these regions, the average farmer could not gather the resources necessary to restore moderate and severely affected soil regions to full productivity. Therefore, prospects for reversing the effects of soil erosion are not good, and it is likely that this problem will worsen.

Despite the loss and degradation of fertile land, the "green revolution" has promoted per capita food production by increasing the yield per hectare. The new, high-yielding strains of grains, such as wheat and rice, have proliferated since their introduction in the 1960s, especially in Asia. To reap full advantage from these new crop varieties, however, farmers must apply abundant quantities of fertilizer and water.

Environmentalists question whether further conversion to such crops can be achieved at reasonable cost, especially in the developing world, where the gain in production is most needed. Presently, farmers in Asia, Latin America, and Africa use fertilizer sparingly, if at all, because it is too expensive or unavailable. Fertilizer use in the developed world has recently waned. The reasons for the decline are complex and may be temporary, but clearly farmers in North America and Europe have decided that increasing their already heavy application of fertilizer will not further enhance crop yields.

Unfortunately, irrigation systems, which would enable many developing countries to join in the green revolution, are often too expensive to build. In most areas, irrigation is essential for generating higher yields. It also can make arid land cultivable and protect farmers from the vulnerability inherent in natural variations in the weather. Land brought into cultivation this way could be used for growing multiple crop varieties, thereby helping food production to increase.

Such advantages have been realized since the beginning of agriculture: the earliest irrigation systems are thousands of years old. Yet only a fraction of productive land in the developing world is irrigated, and its expansion has been slower than population growth. Consequently, the amount of irrigated land per capita has been dwindling during recent decades. The trend, pessimists argue, will be hard to stop. Irrigation systems have been built in the most affordable sites, and the hope for extending them is curtailed by rising costs. Moreover, the accretion of silt in dams and reservoirs and of salt in already irrigated soil is increasingly costly to avoid or reverse.

Environmentalists Ehrlich and Ehrlich note that modern agriculture is by nature at risk wherever it is practiced. The genetic uniformity of single, high-yielding crop strains planted over large areas makes them highly productive, but also renders them particularly vulnerable to insects and disease. Current preventive tactics, such as spraying pesticides and rotating crops, are only partial solutions. Rapidly evolving pathogens pose a continuous challenge. Plant breeders must maintain a broad genetic arsenal of crops by collecting and storing natural varieties and by breeding new ones in the laboratory.

The optimists do not deny that many problems exist within the food supply system. But many of these authorities, including D. Gale Johnson, the late Herman Kahn, Walter R. Brown, L. Martel, the late Roger Revelle, Vaclav Smil, and Julian L. Simon, believe the world's food supply can dramatically be expanded. Ironically, they draw their enthusiasm from extrapolation of the very trends that so alarm those experts who expect doom. In fact, statistics show that the average daily caloric intake per capita climbed by 21 percent (from 2,063 calories to 2,495 calories) between 1965 and 1990 in the developing countries. These higher calories have generally delivered greater amounts of protein. On average, the per capita consumption of protein rose from 52 grams per day to 61 grams per day between 1965 and 1990.

According to the optimists, not only has the world food situation improved significantly in recent decades, but further growth can be brought about in various ways. A detailed assessment of climate and soil conditions in 93 developing countries (excluding China) shows that nearly three times as much land as is currently farmed, or an additional 2.1 billion hectares, could be cultivated. Regional soil estimates indicate that sub-Saharan Africa and Latin America can exploit many more stretches of unused land than can Asia, the Near East, and North Africa.

Even in regions where the amount of potentially arable land is limited, crops could be grown more times every year than is currently the case. This scenario is particularly true in the tropics and subtropics where conditions are such—relatively even temperature throughout the year and a consistent distribution of daylight hours—that more than one crop would thrive. Nearly twice as many crops are harvested every year in Asia than in Africa at present, but further increases are possible in all regions.

In addition to multicropping, higher yields per crop are attainable, especially in Africa and the Near East. Many more crops are currently harvested per hectare in the First World than elsewhere: cereal yields in North America and Europe averaged 4.2 tons per hectare compared with 2.9 in the Far East (4.2 in China), 2.1 in Latin America, 1.7 in the Near East, and only 1.0 in Africa.

Such yield improvements, the enthusiasts note, can be achieved by expanding the still limited use of high-yield crop varieties, fertilizer, and irrigation. In *World Agriculture: Toward 2000*, Nikos Alexandratos of the Food and Agriculture Organization (FAO) of the United Nations reports that only 34 percent of all seeds planted during the mid-1980s were high-yielding varieties. Statistics from the FAO show that, at present, only about one in five hectares of arable land is irrigated, and very little fertilizer is used. Pesticides are sparsely applied. Food output could drastically be increased simply by more widespread implementation of such technologies.

Aside from producing more food, many economists and agriculturists point out, consumption levels in the developing world could be boosted by wasting fewer crops, as well as by cutting storage and distribution losses. How much of an increase would these measures yield? Robert W. Kates, director of the Alan Shawn Feinstein World Hunger Program at Brown University, writes in *The Hunger Report: 1988* that humans consume only 60 percent of all harvested crops, and some 25 to 30 percent is lost before reaching individual homes. The FAO, on the other hand, estimates lower distribution losses: 6 percent for cereals, 11 percent for roots, and 5 percent for pulses. All the same, there is no doubt that improved storage and distribution systems would leave more food available for human nutrition, independent of future food production capabilities.

For optimists, the long-range trend in food prices constitutes the most convincing evidence for the correctness of their view. In 1992–93, the World Resources Institute reported that food prices dropped further than the price of most nonfuel commodities, all of which have declined in the past decade. Cereal prices in the international market fell by approximately one-third between 1980 and 1989. Huge government subsidies for agriculture in North America and western Europe, and the resulting surpluses of agricultural products, have depressed prices. Obviously, the optimists assert, the supply already exceeds the demand of a global population that has doubled since 1950.

Taken together, this evidence leads many experts to see no significant obstacles to raising levels of nutrition for world populations exceeding 10 billion people. The potential for an enormous expansion of food production exists, but its realization depends of course on sensible governmental policies, increased domestic and international trade, and large investments in infrastructure and agricultural extension. Such improvements can be achieved, the optimists believe, without incurring irreparable damage to global ecosystems.

Proponents of either of these conflicting perspectives have difficulty accepting the existence of other plausible points of view. Moreover, the polarity between the two sides of expert opinion shows that neither group can be completely correct. Finding some common ground between these seemingly irreconcilable positions is not as difficult as it at first appears if empirical issues are emphasized and important differences in value systems and political beliefs are ignored.

Both sides agree that the demand for food will swell rapidly over the next several decades. In 1990, a person living in the developing world ate on average 2,500 calories each day, taken from 4,000 gross calories of food crops made available within a household. The remaining 1,500 calories from this gross total not used to meet nutritional requirements were either lost, inedible, or used as animal feed and plant seed. Most of this food was harvested from 0.7 billion hectares of land in the developing world. The remaining 5 percent of the total food supply came from imports. To sustain this 4,000-gross-calorie diet for more than twice as many residents, or 8.7 billion people, living in the developing world by 2050, agriculture must offer 112 percent more crops. To raise the average Third World diet to 6,000 gross calories per day, slightly above the 1990 world average, food production would need to increase by 218 percent. And to bring the average Third World diet to a level comparable with that currently found in the developed world, or 10,000 gross calories per day, food production would have to surge by 430 percent.

A more generous food supply will be achieved in the future through boosting crop yields, as it has been accomplished in the past. If the harvested area in the developing world remains at 0.7 billion hectares, then each hectare must more than double its yield to maintain an already inadequate diet for the future population of the developing world. Providing a diet equivalent to a First World diet in 1990 would require that each hectare increase its yield more than six times. Such an event in the developing world must be considered virtually impossible, barring a major breakthrough in the biotechnology of food production.

Instead, farmers will no doubt plant more acres and grow more crops per year on the same land to help augment crop harvests. Extrapolation of past trends suggests that the total harvested area will increase by about 50 percent by the year 2050. Each hectare will then have to provide nearly 50 percent more tons of grain or its equivalent to keep up with current dietary levels. Improved diets could result only from much larger yields.

The technological optimists are correct in stating that overall world food production can substantially be increased over the next few decades. Current crop yields are well below their theoretical maxima, and only about 11 percent of the world's farmable land is now under cultivation. Moreover, the experience gained recently in a number of developing countries, such as China, holds important lessons on how to tap this potential elsewhere. Agricultural productivity responds to well-designed policies that assist farmers by supplying needed fertilizer and other inputs, building sound infrastructure, and providing market access. Further investments in agricultural research will spawn new technologies that will fortify agriculture in the future. The vital question then is not how to grow more food, but rather how to implement agricultural methods that may make possible a boost in food production.

A more troublesome problem is how to achieve this technological enhancement at acceptable environmental costs. It is here that the arguments of those experts who forecast a catastrophe carry considerable weight. There can be no doubt that the land now used for growing food crops is generally of better quality than unused, potentially cultivable land. Similarly, existing irrigation systems have been built on the most favorable sites. Consequently, each new measure applied to increase yields is becoming more expensive to implement, especially in the developed world and parts of the developing world such as China, where productivity is already high. In short, such constraints are raising the marginal cost of each additional ton of grain or its equivalent. This tax is even higher if one takes into account negative externalities—primarily environmental costs not reflected in the price of agricultural products.

The environmental price of what in the Ehrlichs' view amounts to "turning the earth into a giant human feedlot" could be severe. A large inflation of agriculture to provide growing populations with improved diets is likely to lead to widespread deforestation, loss of species, soil erosion and pollution from pesticides, and runoff of fertilizer as farming intensifies and new land is brought into production. Reducing or minimizing this environmental impact is possible but costly.

Given so many uncertainties, the course of future food prices is difficult to chart. At the very least, the rising marginal cost of food production will engender steeper prices on the international market than would be the case if there were no environmental constraints. Whether these higher costs can offset the historical decline in food prices remains to be seen. An upward trend in the price of food sometime in the near future is a distinct possibility. Such a hike will be mitigated by the continued development and application of new technology and by the likely recovery of agricultural production and exports in the former Soviet Union, eastern Europe, and Latin America. Also, any future price increases could be lessened by taking advantage of the underutilized agricultural resources in North America, notes Per Pinstrup-Andersen of Cornell University in his 1992 paper, "Global Perspectives for Food Production and Consumption." Rising prices will have little effect on high-income countries or on households possessing reasonable purchasing power, but the poor will suffer.

In reality, the future of global food production is neither as grim as the pessimists believe nor as rosy as the optimists claim. The most plausible outcome is that dietary intake will creep higher in most regions. Significant annual fluctuations in food availability and prices are, of course, likely; a variety of factors, including the weather, trade interruptions, and the vulnerability of monocropping to pests, can alter food supply anywhere. The expansion of agriculture will be achieved by boosting crop yields and by using existing farmland more intensively, as well as by bringing arable land into cultivation where such action proves economical. Such events will transpire more slowly than in the past, however, because of environmental constraints. In addition, the demand for food in the developed world is approaching saturation levels. In the U.S., mounting concerns about health have caused the per capita consumption of calories from animal products to drop.

Still, progress will be far from uniform. Numerous countries will struggle to overcome unsatisfactory nutrition levels. These countries fall into three main categories. Some low-income countries have little or no reserves of fertile land or water. The absence of agricultural resources is in itself not an insurmountable problem, as is demonstrated by regions, such as Hong Kong and Kuwait, that can purchase their food on the international market. But many poor countries, such as Bangladesh, cannot afford to buy food from abroad and thereby compensate for insufficient natural resources. These countries will probably rely more on food aid in the future.

Low nutrition levels are also found in many countries, such as Zaire, that do possess large reserves of potentially cultivable land and water. Government neglect of agriculture and policy failures have typically caused poor diets in such countries. A recent World Bank report describes the damaging effects of direct and indirect taxation of agricultures, controls placed on prices and market access, and overvalued currencies, which discourage exports and encourage imports. Where agricultural production has suffered from misguided government intervention (as is particularly the case in Africa), the solution—policy reform—is clear.

Food aid will be needed as well in areas rife with political instability and civil strife. The most devastating famines of the past decade, known to television viewers around the world, have occurred in regions fighting prolonged civil wars, such as Ethiopia, Somalia, and the Sudan. In many of these cases, drought was instrumental in stirring social and political disruption. The addition of violent conflict prevented the recuperation of agriculture and the distribution of food, thus turning bad but remediable situations into disasters. International military intervention, as in Somalia, provides only a short-term remedy. In the absence of sweeping political compromise, hunger and malnutrition will remain endemic in these war-torn regions.

Feeding a growing world population a diet that improves over time in quality and quantity is technologically feasible. But the economic and environmental costs incurred through bolstering food production may well prove too great for many poor countries. The course of events will depend crucially on their governments' ability to design and enforce effective policies that address the challenges posed by mounting human numbers, rising poverty, and environmental degradation. Whatever the outcome, the task ahead will be made more difficult if population growth rates cannot be reduced.

QUESTIONS

1. As the population of the world will probably exceed 10 billion by 2050, name the two schools of thought and their positions in dealing with the environment and the future of humanity. What was your view before accessing this reading?
2. How much has food production increased in the Third World between 1965 and 1990? How much has population expanded during this time period? What is the result?
3. What is the argument in the reading concerning the amount of arable land in the developed world and undeveloped world?
4. What is the problem with increased food per hectare from the "green revolution" and irrigation systems?
5. Explain the issue of genetic uniformity.
6. According to the optimists, how much additional land could be cultivated?
7. How can yield improvements be achieved according to the optimists?
8. Explain the argument of the optimists concerning waste, storage, and distribution losses of food.
9. How can the potential for food expansion be realized?
10. How much would food production have to rise in order to bring the Third World diet to the level found in the developed world?
11. What do you feel is the role of new technology in boosting this production?
12. Explain the author's view in accomplishing increases in food production. Explain your view.

A Ten-Year Progress Report of the 1994 International Conference on Population and Development in Cairo: 1994–2004

EDITED BY BARBARA A. EICHLER

Unless women can manage and control their own fertility, they cannot manage and control their own lives.
KAVAL GULHATI, *Indian family planner*

In September 1994, 179 delegations of the United Nations—assembled in Cairo, Egypt, at the International Conference on Population and Development (ICPD)—reached a bold and sweeping plan to stabilize world population. (The conference attendees included not only 179 national delegations, but also 2,500 government delegates, seven heads of state, five vice presidents, 20 deputy prime ministers, 1,200 nongovernmental organizations, and 4,000 journalists. The momentum of the population issue, as it climbed to the top of the global agenda with the sheer magnitude and percentage of the global community that the conference involved, was, in itself, an important accomplishment.) In the preparatory meetings leading up to the conference, delegates rejected the concept of the high trajectory of population growth reaching 11.9 billion people by 2050. Instead, they recommended an ambitious plan to stabilize world population between the low and medium projections of 7.9 to 9.8 billion—by 2050. Their strategy reflected a sense of urgency—that unless population growth can be slowed quickly, it will push human demands beyond the carrying capacity of many countries, causing environmental degradation, economic decline, and social disintegration (Roush 1994).

The plan called for active services to provide family planning to the estimated 120 million women in the world who wanted to limit the number of their children but lacked access to family planning services. It also addressed the underlying causes of high fertility, such as female illiteracy, and advocated universal primary school education for girls, recognizing that as female educational levels rise, fertility levels fall—a relationship that holds across all cultures. The goals of the conference were to achieve gender equality and permit women to have the right to control and manage their own lives. A major achievement of the Cairo conference was to change the world view of the population issue. From the previous view, which was concerned with demography and contraception, the agenda moved to a focus on the situation of women.

The Cairo conference highlighted the fact that real change in demographic patterns will occur only through fundamental changes in women's lives, such as increases in women's access to education, cash, and credit. Two-thirds of the world's illiterate people are women. Women are responsible for over 80 percent of food production in Africa, but they have very little access to agricultural extension, credit, land titles, or management. Studies show that, with better education and an increase of women's stature in the family and community, birth rates decline. Ninety-five percent of the growth in population will occur in the developing world, increasing from the current 5–6 billion people to 8.6–12 billion by the year 2100.

There is also a need for an increase in the amount of information available to adolescents around the world regarding reproduction. As there are currently over 1 billion teenagers on the planet and increasingly large numbers of them are unmarried and living in cities, "the young and the restless" are experiencing radical changes in values as differing values are beamed around the planet on television, raising questions that do not relate to their culture and the lives of their ancestors (Catley-Carlson, 1994).

The document produced by the conference set a plan for developing and supportive countries to invest at least

20 percent of their public expenditures in the social sector. Special focus was placed on a range of population, health, and education programs that would improve the health and status of women. It also requested donor spending on population assistance (about $1 billion) to increase from 1.4 percent of official development assistance to 4 percent. One of the prominent themes of the Cairo conference was to try to ensure that the responsibility and funding of this change would be shared by the whole world. These measures, the document states, "would result in world population growth at levels close to the United Nations' low [projection] of a global population of 7.8 billion by the year 2050" (Roush 1994).

Lori Ashford of the Population Reference Bureau states, "Cairo's Program of Action (PoA) is ambitious. It contains more than 200 recommendations within five 20-year goals in the areas of health, development and social welfare. A central feature of the PoA is the recommendation to provide comprehensive reproductive health care, which includes family planning; safe pregnancy and delivery services; abortion where legal; prevention and treatment of sexually transmitted infections (including HIV/AIDS); information and counseling on sexuality; and elimination of harmful practices against women (such as genital cutting and forced marriage).

"The Cairo PoA also defined reproductive health for the first time in an international policy document. The definition states that 'reproductive health is a state of complete physical, mental, and social well-being and not merely the absence of disease or infirmity, in all matter relating to the reproductive system.'

"The PoA also says that reproductive health care should enhance individual rights, including the 'right to decide freely and responsibly' the number and spacing of one's children, and the right to a 'satisfying and safe sex life.' This definition goes beyond traditional notions of health care as preventing illness and death, and it promotes a more holistic vision of a healthy individual" (Ashford 2005, 1–2).

The goals set in Cairo will be extremely difficult to achieve, but if the world succeeds in stabilizing human population at 8 or 9 billion, it will satisfy one of the conditions of an environmentally sustainable society. The plan recognizes the earth's natural limits and the urgency to respect those limits. Evidence of these global limits is seen in the world's declining fish catches, falling water tables, declining bird populations on every continent, rising global temperatures, and world grain stock inventories at the lowest levels in 20 years.

In attempting to answer the question of how many people the earth can support, we raise the reverse question: What exactly will limit the growth in human numbers? It appears that it is the supply of food that will determine the earth's population-carrying capacity. Three natural limits for food production are already slowing food yields: the sustainable yield of oceanic fisheries, the amount of fresh water produced by the hydrologic cycle, and the amount of fertilizer that existing crop varieties can effectively use.

Understanding and respecting the earth's limits are dependent on the urgent recognition of limiting the drain on those resources and translating that recognition into a worldwide population plan for limitation and stabilization. According to the plan developed in Cairo, this can happen best if women have an informed voice in the planning of their families. As the delegates gathered in Cairo debated the plan of action for nine days, the world's population grew by some 2.1 million people (Brown 1995).

BOX 1 The ICPD's Twenty-Year Goals, 1995–2015

- Provide universal access to a full range of safe and reliable family-planning methods and related reproductive health services.
- Reduce infant mortality rates to below 35 infant deaths per 1,000 live births and under-five mortality rates to below 45 deaths of children under age five per 1,000 live births.
- Close the gap in maternal mortality between developing and developed countries. Aim to achieve a maternal mortality rate below 60 deaths per 100,000 live births.
- Increase life expectancy at birth to more than 75 years. In countries with the highest mortality, aim to increase life expectancy at birth to more than 70 years.
- Achieve universal access to and completion of primary education. Ensure the widest and earliest possible access by girls and women to secondary and higher levels of education.

Source: Ashford 2005.

PROGRESS REPORT OF THE 1994 ICPD TEN YEARS LATER

Countries are making real progress in carrying out a bold global action plan that links poverty alleviation to women's rights and universal access to reproductive health. Ten years into the new era opened by the 1994 International Conference on Population and Development (ICPD) in Cairo, the quality and reach of family planning programs have improved, safe motherhood and HIV prevention efforts are being scaled up, and governments embrace the ICPD Program of Action as an essential blueprint for realizing development goals.

But inadequate resources, gender bias, and gaps in serving the poor and adolescents are undermining further progress as challenges mount, according to *The State of World Population 2004* report from UNFPA, the United Nations Population Fund.

The report, *The Cairo Consensus at Ten: Population, Reproductive Health and the Global Effort to End Poverty*, reviews countries' achievements and constraints in implementing the Program of Action, nearly halfway to the 2015 completion target date. It examines actions addressing the links between population and poverty, environmental protection, and migration and urbanization; discrimination against women and girls; and key reproductive health issues including access to contraception, maternal health, HIV/AIDS, and the needs of adolescents and people in emergency situations.

The plan, adopted by 179 countries in Cairo ten years ago, sought to balance the world's people with its resources, improve women's status, and ensure universal access to reproductive health care, including family planning. The starting point was the premise that population size, growth, and distribution are closely linked to development prospects (with emphasis on women's care and education) and that actions in one area reinforce actions in the other.

But the Cairo consensus gave priority to investing in people and broadening their opportunities rather than to reducing population growth. Empowering women and ensuring the rights of individual women, men, and young people—including the right to reproductive health and choice in determining when and whether to have children—were seen as key to sustained economic growth and poverty alleviation.

In a series of regional conferences marking ten years since the ICPD and in responses to a UNFPA global survey, governments around the world have strongly reaffirmed their commitment to the Program of Action. Its successful implementation, they agree, is critical to attaining the UN Millennium Development Goals (MDGs) for 2015, including ending extreme poverty and hunger, promoting gender equality and universal primary education, reducing maternal and child mortality, combating HIV/AIDS, and preserving the environment.

Nearly all the developing countries surveyed report they have incorporated population concerns in their development and poverty-reduction strategies; many have established laws and policies to protect women's and girls' rights; and many have begun to integrate reproductive health services into primary health care, improve facilities and training, and expand access. Non-governmental organizations (NGOs) are increasingly active in providing reproductive health services and in advocacy for implementing the Program of Action.

Use of modern contraception has risen from 55 percent of couples in 1994 to 61 percent in 2004. To reduce maternal deaths and injuries, increasing emphasis is being placed on attended delivery, emergency obstetric care, and referral systems. Countries have stepped up efforts to fight HIV/AIDS through prevention, treatment, care, and support. Adolescent reproductive health has become an emerging worldwide concern. And campaigns against gender-based violence are gaining broad support.

But much more must be done to ensure reproductive health and rights, including those of the world's 1.3 billion adolescents, to promote safe motherhood and to stem the spread of HIV/AIDS.

Ten Years After Cairo

- More than 350 million couples still lack access to a full range of family planning services.
- Complications of pregnancy and childbirth remain a leading cause of death and illness among women: 529,000 die each year, mostly from preventable causes.
- Five million new HIV infections occurred during 2003; women are nearly half of all infected adults, with nearly three-fifths from sub-Saharan Africa.
- While fertility is falling in many regions, world population will increase from 6.4 billion today to 8.9 billion by 2050; the 50 poorest countries will triple in size to 1.7 billion people.

The tenth anniversary of the ICPD is an opportunity for governments and the international community to renew their commitments and identify ways to overcome the remaining challenges.

Population and Poverty

Some 2.8 billion people—two in five—still struggle to survive on less than $2 a day. Poverty perpetuates and is exacerbated by poor health, gender inequality, and rapid population growth.

Policymakers have been slow to address the inequitable distribution of health information and services that help keep people poor. Richer population groups have far greater access than poorer groups to delivery by a skilled attendant, contraceptives, and other reproductive health services. Poor women give birth at earlier ages and have more children throughout their lives than wealthier women. Developing countries that have reduced fertility and mortality by investing in health and education have higher productivity, more savings, and more productive investment, resulting in faster economic growth.

Enabling people to have fewer children, if they want to, helps to stimulate development and reduce poverty, both in individual households and in societies. Smaller families have more to invest in children's education and health. Rapid population growth contributes to environmental stress, uncontrolled urbanization, and rural and urban poverty. Declining fertility reduces the proportion of dependent children relative to the working-age population, opening a one-time window of opportunity (before dependent older populations become a burden) in which countries can make investments to spur economic growth and help reduce poverty.

The ICPD's rights-based agenda for addressing the interdependence of population and poverty is therefore essential to achieving the MDGs. Most developing countries responding to the UNFPA global survey have adopted diverse strategies to address the links between population and poverty.

Footprints on the Planet

Unsustainable consumption and production patterns, coupled with rapid population growth, are taking their toll on the environment. More people are using more resources with more intensity and leaving a bigger "footprint" on the earth than ever before.

A rapidly growing global consumer class is using resources at an unprecedented rate, with an impact far greater than their numbers. Farmers, ranchers, loggers, and developers have cleared about half of the world's original forests. Three-quarters of the world's fish stocks are now fished at or beyond sustainable limits.

Half a billion people live in countries defined as water-stressed or water-scarce; by 2025, that figure is expected to be between 2.4 billion and 3.4 billion people. Fast-growing poor populations often have no other options but to exploit their local environment to meet subsistence needs for food and fuel.

Countries report taking action to address the linkages between population, poverty, and the environment.

Gender plays a strong role in how resources are used and developed. At both the ICPD and its 1999 review, the global community affirmed that greater equality between men and women is essential for sustainable development and sound management of natural resources.

Migration and Urbanization

Due to continued rural-to-urban migration, the number of people living in cities is growing twice as fast as total population growth. A majority of the world's people will be living in cities by 2007; by 2030, all regions will have urban majorities. Both megacities of 10 million or more (20 in all, 15 in developing countries) and small and medium-sized cities are growing, severely testing the limits of local infrastructure and services.

The ICPD recognized that people move within countries in response to the inequitable distribution of resources, services, and opportunities. Providing social services, including reproductive health care, in poor urban areas is an essential policy response, as is meeting the needs of underserved rural communities.

There were 175 million international migrants in the world in 2000—1 in every 35 persons—up from 79 million in 1960. Many people, including a growing number of women, are seeking employment abroad, with major impacts on both sending and receiving countries. The economic effects run in both directions The ICPD called on nations to address the root causes of migration, especially by addressing poverty. Three-fourths of countries report taking some action to address international migration; in 1994, only one-fifth had done so. Some countries have tightened borders, while others aim to better integrate migrants. Many countries favor greater coordination of migration policies, but the subject remains a sensitive one.

Women's Empowerment

Gender equality and women's empowerment were at the heart of the ICPD vision, strongly linked to reproductive health and rights. Since 1994, more than half of all countries have adopted national legislation on women's rights, ratified UN conventions, or established national commissions for women.

Progress has been uneven. Many countries have introduced laws on gender-based violence, but these are often not enforced. Only 28 countries have increased women's political participation. Fewer than half of all countries have initiatives in place to educate men about their and their partners' reproductive health. Efforts promoting women's advancement are susceptible to budget cuts. Only 42 countries were able to increase public spending on schools, and only 16 had increased the number of girls' secondary schools.

Priorities for improving women's status include eliminating gender gaps in education; increasing access to sexual and reproductive health information and services; investing in infrastructure to ease women's work burdens; reducing discrimination in employment, property ownership, and inheritance; increasing women's role in government bodies; and sharply reducing violence against women and girls.

Reproductive Health and Family Planning

Gaps in reproductive and sexual health care account for one-fifth of the worldwide burden of illness and premature death, and one-third of the illness and death among women of reproductive age. The ICPD's broad concept of reproductive health and rights, including family planning and sexual health, continues to gain support. This year, the 57th World Health Assembly recognized the Cairo consensus and adopted the World Health Organization's (WHO) first reproductive health strategy to accelerate progress toward the MDGs.

Family planning enables individuals and couples to determine the number and spacing of their children—a recognized basic human right. Some 201 million women, especially in the poorest countries, still have an unmet need for effective contraception. Meeting their needs would cost about $3.9 billion a year and prevent 23 million unplanned births, 22 million induced abortions, 142,000 pregnancy-related deaths (including 53,000 from unsafe abortions), and 1.4 million infant deaths.

Important progress has been made toward the ICPD goal of universal access to reproductive health services by 2015. Greater attention has been given to reproductive rights in laws and policies. Many countries have re-oriented services and stepped up training to improve quality, expand family planning method choices, and better meet clients' needs and desires. Efforts have been made to integrate treatment of sexually transmitted infections (STIs) with other services and to involve men more in protecting their own and their partners' reproductive health.

However, donor support for reproductive health commodities has declined over the past ten years, creating a growing gap between needs and supplies. Between 2000 and 2015, contraceptive users in developing countries are expected to increase by 40 percent.

Maternal Health

Obstetric complications are the leading cause of death for women of reproductive age in developing countries and constitute one of the world's most urgent and intractable health problems. Despite progress in a few countries, the issue has not been given high priority, and the global number of deaths per year has not changed significantly since 1994.

Poverty dramatically increases a woman's chances of dying. The lifetime risk of a woman dying in pregnancy or childbirth in West Africa is 1 in 12. In developed regions, the comparable risk is 1 in 4,000. Most maternal deaths stem from problems that are treatable but hard to detect. Reducing deaths requires expanded access to skilled attendance at delivery, emergency obstetric care for women who experience pregnancy complications, and referral and transport systems so those women can receive needed care quickly. Skilled personnel assist half of all deliveries in developing countries, but just 35 percent in South Asia and 41 percent in sub-Saharan Africa.

Millions of women survive childbirth but suffer illness and disability. One of the most devastating is obstetric fistula, an internal injury caused by obstructed labor that leaves women incontinent and often ostracized. UNFPA and others are working to prevent fistula, advocating against child marriage and promoting increased access to emergency care.

Since the ICPD, most countries have reported taking actions to promote safe motherhood. Countries are also increasing access to family planning services to reduce unintended pregnancies. At least 40 countries have initiated post-abortion care programs. Complications from unsafe abortions are a major cause of maternal death; the ICPD urged greater attention to this neglected issue.

Preventing HIV/AIDS

In just over two decades, the AIDS pandemic has claimed 20 million lives and infected 38 million people. This number could rise greatly if countries do not pursue strategies to prevent HIV. Fewer than 20 percent of people at high risk of infection have access to proven prevention interventions.

The consequences of AIDS are far-reaching. In some areas of sub-Saharan Africa, 25 percent of the workforce is HIV positive. Studies show that if 15 percent of a country's population is HIV positive, its gross domestic product will decline by 1 percent a year.

Since most HIV transmission is through sexual contact, reproductive and sexual health information and services provide a critical entry point to prevention by providing education on risks to influence sexual behavior; detecting and managing STIs; promoting the correct and consistent use of male and female condoms; and helping to prevent mother-to-child transmission. Linking prevention efforts and voluntary HIV testing and counseling with existing reproductive health services can improve outreach, reduce stigma, and save money.

The ICPD recognized that discrimination and violence make women and girls especially vulnerable, and efforts to address this gender inequality are growing. Infection rates among young African women aged 15–24 are two to three times higher than among young men. Married women are often unable to negotiate condom use even when they know their husbands have multiple partners.

Means to treat HIV have improved in the last decade, but the vast majority of those infected lack access to life-saving antiretroviral therapy. The WHO and UNAIDS aim to bring treatment to 3 million people by 2005 and reduce drug costs. Increased availability of treatment will bring more people into health facilities where they can be reached by prevention messages.

Three-fourths of countries answering the global survey reported having a national strategy on HIV/AIDS. New policies and coordination with NGOs mark positive changes in many countries, but many in at-risk groups are not being reached. Donor funding for condoms is far below what is needed, and education efforts are inadequate.

Serving Adolescents and Young People

Since 1994 and especially in the past few years, countries have made significant progress in addressing the often-sensitive issues of adolescents' reproductive health, including needs for information and services that will enable them to prevent unwanted pregnancy and infection. Young people aged 15–24 account for half of all new HIV infections—one every 14 seconds—with young women especially at risk. Worldwide, there is a trend toward later marriage; growing numbers of adolescents are sexually active before marriage, often without the knowledge or means to protect themselves. In a number of countries, early marriage and childbearing are still the norm for girls, usually to much older men. Married girls are less likely than others their age to finish school and more likely to contract HIV or another STI.

UNFPA and other organizations work in a variety of ways to teach young people about reproductive health in combination with developing life skills and job training. Reaching those who are married, those living in rural areas and poor urban settlements, and those out of school is a key priority. Ninety percent of countries report having taken steps to address adolescent reproductive health. But efforts lag far behind needs, and successful interventions need to be scaled up dramatically.

Helping Communities in Crisis

Since the ICPD, attention to the reproductive health needs of women made vulnerable by war and disaster has increased greatly. A quarter of the tens of millions of refugees in the world are women of reproductive age; one in five of these women is likely to be pregnant. Death from complications during childbirth is far too common in disaster-stricken countries where women have no access to maternal health care. Rape and gender-based violence occur more often during war, and victims of this abuse are more likely to contract an STI, including HIV/AIDS. Stable relationships and family units may break apart as disaster strikes. Young people are greatly affected by such changes, leading to a rise in teen pregnancy, unsafe sex, and abortions.

UNFPA and other humanitarian groups respond to emergencies by providing basic materials needed for safe childbirth and protection from unintended pregnancies and STIs. They also support education campaigns and camp safety measures to prevent sexual violence and provide counseling and treatment for those who have been violated.

Priorities for Action

Successful action to implement the Cairo agenda and combat poverty depends on adequate funding and effective partnerships. Since the ICPD and its five-year review, partnerships have developed between governments and a broad range of civil society organizations, including women's community groups, human rights organizations, trade unions, universities, private health providers, and parliamentarians. But resources have fallen short. Donors agreed to provide $6.1 billion a year for population and reproductive health programs by 2005, a third of total needs. In 2002 (the latest year with available figures), contributions were around $3.1 billion—only half their commitment.

Developing country domestic expenditures for the ICPD package in 2003 were about $11.7 billion. But a large proportion of this comes from a few large countries. The poorest countries depend heavily on donor funding for family planning, reproductive health, and other population-related activities. Additional funds are needed to combat the HIV/AIDS pandemic; funding in this area is increasing but still far less than needed.

Policy priorities for action in the next ten years include better integration of population into national planning; broadening programs to meet the needs of the poorest population groups; strengthening urban planning to provide services in marginal communities; investing in rural development; reforming laws and policies to end discrimination against women; and making civil society participation a routine part of institutional practices.

Delivery of comprehensive reproductive health and family planning services needs to be strengthened significantly through improvements in capacity, supply chains, and quality. Safe motherhood interventions should be scaled up and promoted. HIV/AIDS interventions must be linked more effectively to other components of reproductive health. Efforts to reach all adolescents in need of information and services, including those married and those not in school, should be increased.

Ten years on from the ICPD, the world needs its vision of human-centered development more than ever. Today's challenges—including security concerns, the continuing spread of HIV/AIDS, and persistent poverty alongside unprecedented prosperity—make it imperative to carry out the Cairo agenda so that its dream of a better future for all is realized.

REFERENCES

Ashford, L. S. (2005). "What Was Cairo? The Promise and Reality of ICPD." Population Reference Bureau. Accessed July 30, 2005. Available at http://www.prb.org/Template.cfm?Section+PRB&tem.../ContentDisplay.cfm&ContentID=1162.

Brown, L. (1995). Nature's Limits. *State of the World, 1995.* New York: W.W. Norton & Co., pp. 3–4.

Catley-Carlson, M. (1994). Interview. Cairo Conference Finds Women's Status Central to Slowing Population Growth, Increasing Food Production. 2020 News & Reviews (October). Available at htttp://www.cgiar.org./ifpri/2020/newslet/nv_1094/nv-1094f.htm.

Roush, W. (1994). Population: The View from Cairo. Science (26 August):1164. Available at http://www.mit.edu:8001/people/weroush/population.html.

UNFPA. (2004). State of World Population 2004. The Cairo Consensus at Ten: Population, Reproductive Health and the Global Effort to End Poverty. New York: United Nations Population Fund.

QUESTIONS

1. The Cairo conference plan is ambitious on many counts. Identify five areas in which this plan is especially ambitious.
2. What do you think is the response to the conference's resolutions from nations whose religious and cultural convictions do not support the equal status, education, and self-control of women? What can be done about their response?
3. Some sources have said that this conference was one of the most revolutionary of any UN conferences. Why do you think this might be so?

4. Relate the World Birth Control Fact Sheet (see Reading 37) to the UN resolutions from the Cairo conference. Given those facts on birth control, what are some of the biggest obstacles to accomplishing the resolutions? Do you think that the resolutions deal with the core of the problem? Why or why not?
5. In reading the progress report from the UN, identify five areas where significant progress has been made according to the goals of ICPD.
6. What are some of the problematic ecological footprints that are developing in the world?
7. What progress has been made toward gender equality and women's empowerment?
8. What are the greatest needs that would enable the ICPD to accomplish more of its goals?

37 World Birth Control Discussion Issues

EDITED BY BARBARA A. EICHLER

WORLD ARGUMENTS AGAINST POPULATION AND BIRTH CONTROL

1. *Progrowth position*: Rapid population growth in a particular country or region is a positive force on grounds of
   a. economic development (a larger population provides necessary economies of scale and a good labor supply);
   b. protection of presently underpopulated areas from covetous neighbors;
   c. differences in fertility among ethnic, racial, religious, or political population segments; and
   d. military and political power and the strength of a younger age structure.
2. *Revolutionary position*: Population programs are merely a way to hide or placate fundamental social and political contradictions that would lead to a just revolution and therefore could be viewed as inherently counterrevolutionary.
3. *Anticolonial and genocidal position*: The motives of highly developed countries of pushing less developed countries (LDCs) to adopt aggressive population programs are open to suspicion. These more developed countries went through a period of rapid population growth as part of their own development processes, and their current efforts to restrain population growth in the less developed countries are an attempt to keep control and power by retarding the development of these countries. Population limitation could be seen as an attempt of the rich countries to "buy development cheaply" or a racist or genocidal attempt to reduce the size of poor and largely nonwhite populations.
4. *Overconsumption by rich countries position*: Population problems are actually resource-scarcity and environmental-deterioration problems that derive from activities of the rich, highly developed countries and not from high fertility in LDCs. Even if fertility is too high in the LDCs, this is a consequence of their poverty, which, in turn, results from overconsumption of the world's scarce resources by rich countries.
5. *Agricultural and technological improvement position*: Growing population numbers can be accommodated as they have in the past by improvements in agricultural and industrial technologies. Past Malthusian predictions were incorrect, and the same is true of these new Malthusian predictions and solutions. Overpopulation is really underemployment. A humane and well-structured economy can provide employment and subsistence for all people, no matter what the size of the population.
6. *Distribution problem position*: It is not the population numbers themselves that are causing population problems, but their distribution in space. Many areas of the world or countries are underpopulated; others have

Reprinted by permission of *Foreign Affairs*, July 1974. M. Teitelbaum, Population and Development: Is a Consensus Possible? *Foreign Affairs*, Vol. 52, No. 4, Copyright 1974 by the Council on Foreign Relations, Inc.

too many people condensed into too small an area. Instead of efforts to moderate the rate of growth, governments should undertake efforts to reduce more population flows to urban areas and to distribute population on available land.

7. *High mortality and social security position*: High fertility is a response to high death and disease rates; if these levels were reduced, fertility would decline naturally. Living children are the primary way that poor people can achieve security in old age. Therefore, improvements in mortality rates and social security programs would lead to a reduction in fertility.

8. *Status of women position*: High fertility levels are perpetuated by the status and roles of women primarily as procreative agents. As long as women's economic and social status depends largely upon the number of children they bear, there is little possibility that societal fertility levels will decline.

9. *Religious doctrine positions*: The ideology of "be fruitful and multiply; God will provide" does not recognize population as a serious problem. The other argument holds that, while current rates of population growth are a serious problem, the primary instruments to deal with them—modern contraception, surgical sterilization, and abortion—are morally unacceptable; abortion is "murder," and surgical sterilization is "unnatural."

10. *Medical risk position*: Fertility reduction is not worth the medical risks of using the medical means of population programs. Oral contraceptives and intrauterine devices have measurable, if small, short-term risks and possible long-term effects. Sterilization and abortion are operative procedures, both of which have an element of risk, especially when performed outside a hospital.

11. *Progress and development position*: As the social and economic aspects of a society further develop, fertility rates decline. Most of the decline in fertility in LDCs derives from social and economical development rather than from the success of population control programs. International assistance for development is too heavily concentrated on population programs and is shortchanging the focus on general overall development.

12. *Social justice position*: Fertility will not decline and population programs will not be successful until the basic reasons for high fertility—poverty, ignorance, and fatalism—are eliminated through social policies that result in a redistribution of power and wealth among the rich and poor, both within and among nations.

WORLD ARGUMENTS SUPPORTING POPULATION AND BIRTH CONTROL

1. *Population activist position*: Unrestrained population growth is the principal cause of poverty, malnutrition, environmental disruption, and other social problems. The situation is desperate and necessitates action to restrain population growth, even if coercion is required: "Mutual coercion, mutually agreed upon." Population programs are fine as far as they go, but they are wholly insufficient in scope and strength to meet the urgency and reality of the situation.

2. *Services need position*: According to data and surveys, there is a great unmet demand for fertility control in all countries; therefore, the main problem is to provide modern fertility control to already motivated people. Some proponents also hold that the failure of some population service programs is due to inadequate fertility control technologies and that there is an urgent need for technological improvements in this area.

3. *Human rights position*: According to UN propositions, it is a fundamental human right for each person to be able to determine the size of his or her own family. Furthermore, some argue that each woman has the fundamental right to the control of her own bodily processes, a rationale that supports contraception and possibly abortion. Health is also a basic human right that population programs support directly and indirectly and includes the direct medical benefits of increased child spacing for maternal and child health and the indirect effects of reducing numbers of dangerous illegal abortions.

4. *Population control and development position*: Social and economic development programs help, but alone are not sufficient to bring about low mortality and fertility levels. Special population programs are also required. Too rapid population growth intensifies other social and economic problems. Some countries might benefit from larger populations, but would benefit more by moderate rates of growth over longer periods of time than by rapid rates of growth over shorter periods of time. Economic, social, and population programs are all interlinked.

Figure 1
Use of contraceptive methods by region.
Source: Courtesy of UN Population Fund.

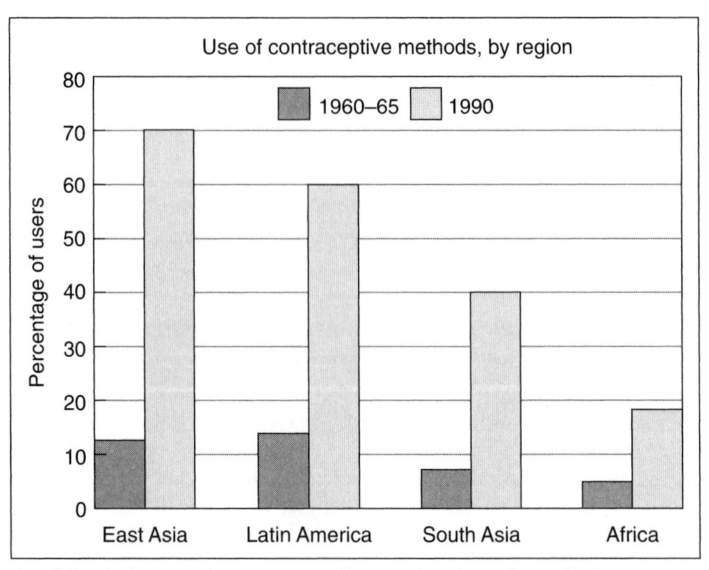

For all developing countries on average. The use of contraceptive methods has grown from less than 10% in the 1960s to more than 50% today.

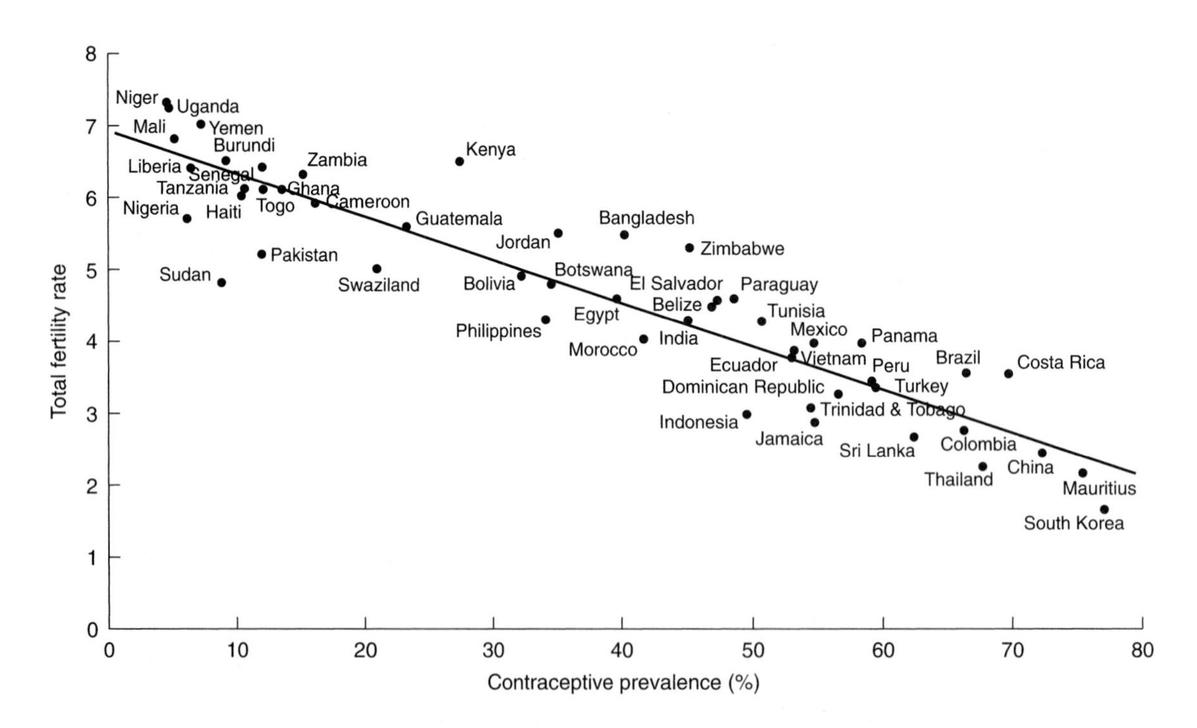

Figure 2
Relationship between contraceptive prevalence and total fertility rates in 50 countries, 1984–1992.
Source: Population Reports 1992.

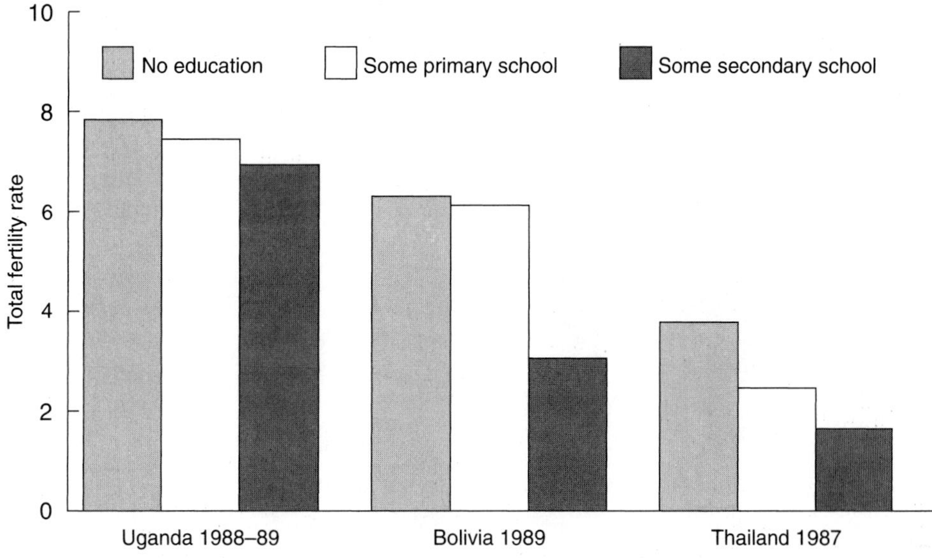

Figure 3
Total fertility rates in three countries by women's level of education.
Before fertility rates start to fall, fertility is high in all groups (Uganda). As educated women are first to reduce their fertility, differences widen among groups (Bolivia). Eventually fertility is low in all groups (Thailand).
Source: Population Reports 1992. Reprinted with permission from R. V. Short, *Contraceptive Strategies*, North American Press.

Table 1
Percentage Use of Contraceptive Methods Worldwide
Currently, 390 million couples are using modern contraceptives, about 61 percent of the total.

Female sterilization	29
Intrauterine devices	20
Oral contraceptives	14
Condoms	9
Male sterilization	8
Coitus interruptus	8
Rhythm	7
Injectables	2
Other methods	2
Barrier methods	1

Note: Methods with the lowest failure rates predominate in developing countries. Methods with the highest failure rates predominate in developed countries.
Source: Reprinted with permission from R.V. Short, *Contraceptive Strategies*, North American Press, p. 300.

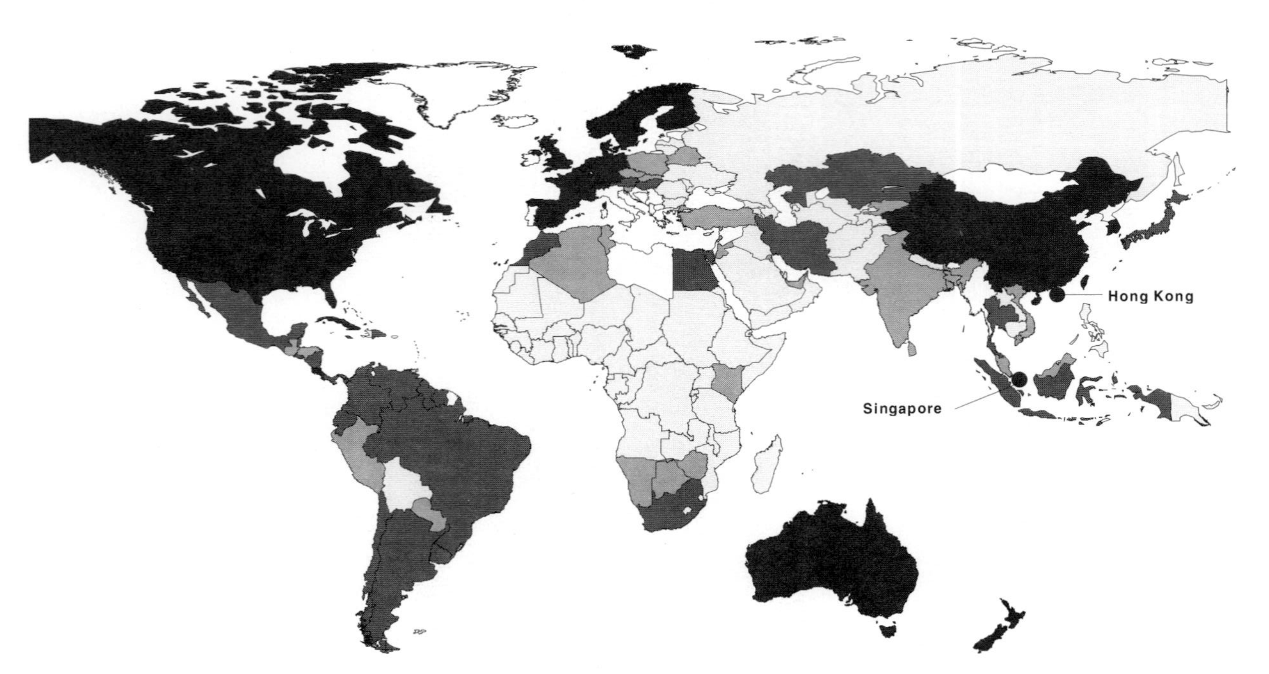

Percent of Married Women of Reproductive Age Who Use Modern Birth Control Methods*

Under 25 | 25 to 44 | 45 to 64 | 65 or more | No data

*Includes women in non-marital unions.

Rodger Doyle Copyright 1996

Figure 4
World birth control use.

Data for some countries are estimates. Data for most countries were collected in the late 1980s and early 1990s. Data apply to married women and women in nonmarital unions.

Source: Population Reference Bureau; Population Crisis International. Excerpted with permission from R. Doyle, World Birth-Control Use, *Scientific American*, Vol. 275, No. 3, September 1996, p. 34.

WORLD BIRTH CONTROL FACT SHEET

1. In the past 30 years, there has been a marked decline in world fertility rates, particularly in developing countries.
   a. Between 1960–1965, women in developing countries averaged six births over a lifetime, while in 2000–2005, they averaged 2.9 births over a lifetime.
   b. In east Asia for the past 30 years, births fell 65 percent and are now below the replacement rate of 1.6 children.
   c. In other parts of Asia, births declined by about a third, Latin America halved, and Africa only by 10 percent.
2. In the developed countries, the number of births per woman declined by about 40 percent and are now below replacement level in all these countries, including the United States.
3. Modern contraceptive methods have played a key role in lowering fertility.
   a. Among women of reproductive age who are married (or in nonmarital unions), one-half now depend on the following methods: female sterilization (most popular), male sterilization, hormonal implants (such as Norplant), injectibles (such as Depo-Provera), intrauterine devices (IUDs), birth control pills, condoms, and diaphragms.
   b. The first four methods are almost 100 percent effective, followed by IUDs, the pill, and the male condom in effectiveness. Diaphragms are among the least effective.
   c. Condoms (both the male and female type) are the only methods that provide protection against sexually transmitted diseases, such as AIDS.
   d. The percentage of women using modern contraception is as follows: 54 percent in Asia (39 percent if China is excluded), 53 percent in Latin America, 30–40 percent in the Muslim Middle East and North Africa, 48 percent in the countries of the southern tip of Africa, less than 10 percent in middle Africa, 65–75 percent in the United States and western Europe, and less than 20 percent in the former Soviet Union because of short supply.
   e. Use of contraception has risen from 55 percent in 1994 to 61 percent in 2004 (United Nations Population Fund (UNFPA), Population Issues, 2004). However according to UNFPA, in 2006 more than 350 million couples still do not have access to effective and affordable family planning services, and demand for these services will increase by 40 percent in the next 15 years.
4. Growth in birth control use and decline in fertility in developing countries is closely linked to educational opportunities for women (e.g., sub-Saharan Africa, with the highest fertility rates, has the lowest female educational levels).
   a. Increased literacy makes it easier for women to receive reliable information on contraception.
   b. The demands of education, particularly at the post-secondary level, cause women to delay marriage and childbearing.
5. Some developing countries, such as China and Cuba, are already below the replacement level of 2.1 children, mostly because of modern birth control methods.
   a. Countries such as Brazil, Indonesia, Vietnam, South Africa, Turkey, Egypt, and India should reach this level within the approximate next ten years.
   b. Pakistan and Nigeria, where few women use modern contraception, are at the other extreme and are not likely to reach the replacement rate for several decades.
   c. According to the Population Reference Bureau, Latin American and the Caribbean lead in contraceptive use among the less developed regions with about 70% of married women using some form of family planning. This compares to Asia with about 51% and sub-Saharan Africa of about 21%. (Population Reference Bureau, 2004 Data Sheet).
6. More historical and traditional methods of birth control include the rhythm method, coitus interruptus, and prolonged breastfeeding (which suppresses ovulation). Worldwide, 7 percent of all married women (or women in nonmarital unions) of reproductive age depend on these methods, which are far less reliable. The rhythm method is prevalent in Peru, and coitus interruptus is the birth control method of choice in Turkey.

Source: Except where other sources are indicated, excerpted with permission from R. Doyle, World Birth Control Use, *Scientific American*, Vol. 275, No. 3, September 1996, p. 34.

Table 2
A Day in the Life of the World

100,000,000	Acts of intercourse will take place
910,000	Conceptions will occur
150,000	Abortions will be performed
500	Mothers will die as a result of abortion
384,000	Babies will be born
1,370	Mothers will die of pregnancy-related causes
25,000	Infants in the first year of life will die
14,000	Children aged 1–4 will die
356,000	Adults will get a sexually transmitted disease
203,000	Net increase in population of the world

Source: Reprinted with permission from R. V. Short, *Contraceptive Strategies*, North American Press, p. 327.

QUESTIONS

1. What positions against population and birth control do you think have the most merit? Why?
2. What positions for population and birth control do you think have the most merit? Why?
3. What positions do you think are the most difficult to address concerning the population issue and the 1994 UN resolutions made in Cairo?
4. What do these positions for and against population and birth control tell you about other cultures and societies? What do they tell you about their priorities?
5. What plans may you have to address or change concerning the positions you identified in Questions 1 and 2?
6. Do you think that the UN resolutions adequately address the concerns of the positions against birth control?

BOX 1 Executive Summary of the World Fertility Report 2003

According to the *World Fertility Report 2003* prepared by the Population Division of the Department of Economic and Social Affairs of the United Nations Secretariat, the following are the findings of key estimates and indicators of fertility, nuptiality and contraceptive use for 192 countries. Most of their data is from the 1970's to 1990's when most of the studies were done, but more recent data is included wherever possible.

1. A major worldwide shift in the timing of marriage to older ages has occurred.
2. Both men and women are spending longer periods of their life being single.
3. Delayed marriage among young adults has not yet resulted in noticeable reductions in the percentage of persons marrying at least once over their lifetime.
4. Divorce rates have increased in most countries with data available.
5. A tremendous increase has taken place in the use of family planning. (The world average level of contraception was 38 percent in the 1970's which rose to 52 percent in the 1990's. For developing countries in the 1970's it was 27 pepcent and the 1990's it was 40 percent. By the 1990's, contraceptive prevalence was 62 percent in a quarter of all developing countries.)
6. The use of modern contraceptive methods in developing countries has generally risen. (Between the 1970's and the 1990's modern contraception in developing countries increased from 18 to 30 percent, but in a quarter of all developing countries, the use of modern contraceptive methods remains rare with levels of use remaining below 12 percent.)
7. Between 1970 and 2000, the world population experienced a major and unprecedented reduction of fertility levels, driven mostly by the decline in fertility in developing countries. (Average fertility levels in the developing world dropped from over 5.9 children per woman in the 1970's to about 3.9 children per woman in the 1990's.)
8. Whereas fertility was uniformly high in developing countries in the 1970's, the fertility levels of developing countries today vary over a wide range.
9. Fertility levels in developed countries, many of which experienced a "baby boom" during the 1950's and 1960's, have generally declined since 1970.

10. Levels of childlessness vary considerably among major areas.
11. The profound changes in fertility levels occurring since 1970 have been made possible by major behavioural transformations related to union formation, marriage and the use of contraception. (By 2001 92 percent of all Governments supported family planning programmes and the distribution of contraceptives either directly or indirectly by supporting the activities of non-governmental organizations and other associations.)

Source: United Nations. (2003) *The World Fertility Report 2003* Executive Summary. Available at ww.un.org/esa/population/publications/worldfertility/Executive_Summary.dpf.

- Family planning is rapidly increasing in the developing world. Analyze projections to see where this is happening most. Can you explain why family planning is successful in these countries?
- Considering that a quarter of developing countries rarely use contraceptive methods, what do you think are some of their major arguments against birth control and how can contraceptive use be encouraged?
- What do older age marriages and more divorces mean to population rates. Discuss your answers.

BOX 2 Population Control: A "Third World" Perspective

During recent years, many scientists and economists have expressed a concern about population growth and the scarcity of resources. World population has increased from 3.72 billion (1970) to 6.5 billion (2006) and is projected to be 11.9 billion by the year 2050. Many experts believe that the earth's population is about to surpass the planet's "carrying capacity." Contrary to this, World Bank figures show that food prices have declined dramatically to an historic low—reflecting improvements and a worldwide surplus of grain. Population growth rates in developed countries are low, whereas they are high for developing countries.

To address the question of population control, the United Nations (UN) held a conference in Cairo. The conference adopted a resolution for population control. The resolution document, called "Proposed Program of Action for the Cairo Conference," calls for adoption of various birth control methods, including abortion for population control. The majority of Third World countries with Catholic and Muslim populations believe that the UN has gone too far in making these recommendations, for they are in direct contradiction to the religious beliefs of Catholics and Muslims all over the world. The Pope has formed a Catholic–Islamic alliance to oppose UN recommendations for population control.

- Do you think the UN is justified in recommending population control strategies that contradict people's religious beliefs?
- In many Third World countries, the majority of people are farmers. Due to a lack of technology and resources, they believe in large families to help manage their farms. What population control strategy would you recommend to help Third World farmers?

Many Third World nations believe that population is an asset and that the developed countries are trying to control their future potential through the United Nations. Do you agree with this view? Why? Why not?

Fertility Rates: The Decline Is Stalling*

LINDA STARKE

During the 1970s, one of the encouraging developments in population trends was the reduction in the total fertility rate in several key countries, including the world's two largest nations—China and India. (The fertility rate measures the average number of children born to women in their childbearing years.) In China, the rate dropped precipitously, from 6.4 children per woman in 1968 to 2.2 in 1980. In India, the decline was more modest, but still significant: from 5.8 children per woman between 1966 and 1971 to 4.8 children between 1976 and 1981.

These trends helped slow the rate of world population growth from 2.1 percent between 1965 and 1970 to 1.7 percent between 1975 and 1980. At that point, however, the decline in the number of children that women were having in these two population giants stalled.

In China, despite the most aggressive and least democratic population control program in the world, the fertility rate remained around 2.5 throughout much of the 1980s as couples continued to want to marry young and to have two or more children. In India, the overzealous promotion of family planning by the ruling Congress Party through 1977 apparently backfired after the party's defeat, and progress toward lower birth rates ran out of steam.

One important lesson from these experiences is that governments must do more than just supply contraceptives; they need to lower the demand for children by making fundamental changes that improve women's lives and increase their access to and control over money, credit, and other resources.

Many countries still register fertility rates above replacement level (see Table 1), which is generally 2.1 children per woman or basically two children per couple. The total fertility rate for the world as a whole in 1991 was 3.3, ranging from 1.8 in more developed nations to 4.4 in less developed ones (excluding China). In a number of

From the Worldwatch Institute's annual report of key global indicators, *Vital Signs 1993*.

*Editor's note: Even though this reading is based on early 1990's data, the concept of fertility rates falling with the increase in population growth remains current and important to understand. World fertility rates continue to decline. In 1991, the world average fertility rate was 3.3; in 2004, it was 2.8. Even though the percentage increase in world population declined from 1970 at 2 percent to 1.3 percent in 2004, the rate is applied to a much larger population of 1.2 billion young people between the ages of 10 and 19, the largest young generation in history. At this 1.3-percent rate, the global population will double in 50 years. Because of this "youth bulge," young peoples' choices will influence the patterns and directions of the global population, but the current large global population will make these fertility choices slow to take effect. Today, growth patterns vary widely: Africa maintains a 5.1 fertility rate and 2.4-percent growth (which will double in about 23 years despite the AIDS epidemic), China has a 1.7 fertility rate and 0.6-percent growth, India has a 3.1 fertility rate and 1.7-percent growth (which will double in 41 years), Europe has a 1.4 fertility rate and -0.2-percent growth, and the United States has a 2.0 fertility rate with 0.6-percent growth (which will double in less than 70 years). It is certain that the global population in the next 50 years will be much differently distributed, incorporating at least 3 billion more people (Population Reference Bureau. (2004) *2004 World Population Data Sheet*. Available http://www.prb.org).

Table 1
World Population by Region and Development Category: 1950–2050

Region	*1950*	*1960*	*1970*	*1980*	*1990*	*2000*	*2002*	*2010*	*2025*	*2050*
World	2,555	3,040	3,708	4,455	5,275	6,079	6,228	6,812	7,834	9,079
Less developed countries	1,749	2,129	2,705	3,374	4,132	4,887	5,030	5,588	6,582	7,836
More developed countries	807	910	1,003	1,081	1,143	1,192	1,199	1,224	1,252	1,243
Africa	227	283	361	473	627	803	839	977	1,247	1,786
Sub-Saharan Africa	183	227	291	382	508	657	687	804	1,036	1,532
North Africa	44	56	71	91	119	147	152	174	211	255
Near East	44	58	75	101	135	171	179	212	280	396
Asia	1,368	1,628	2,038	2,494	2,978	3,435	3,518	3,838	4,375	4,832
Latin America and the Caribbean	166	218	286	362	443	524	539	596	690	782
Europe and the New Independent States	572	639	702	750	788	802	803	810	814	776
Western Europe	304	326	352	367	377	390	392	398	398	373
Eastern Europe	88	99	108	117	122	121	121	120	117	104
New Independent States	180	214	242	266	289	291	290	292	300	300
North America	166	199	227	252	278	314	320	344	388	462
Oceania	12	16	19	23	27	31	32	35	40	45
Excluding China:										
World	1,982	2,374	2,868	3,446	4,109	4,786	4,918	5,436	6,350	7,627
Less developed countries	1,176	1,464	1,866	2,366	2,967	3,595	3,720	4,213	5,100	6,386
Asia	795	962	1,198	1,486	1,813	2,143	2,208	2,463	2,892	3,383
Less developed countries	711	868	1,094	1,369	1,689	2,016	2,081	2,336	2,772	3,283

Denotes midyear population in millions. Figures may not add to totals because of rounding.
Note: Reference to China encompasses China, Hong Kong S.A.R., Macau S.A.R., and Taiwan.
Source: U.S. Census Bureau, International Programs Center, International Data Base (2004) *Global Population Profile 2002.* Washington, D.C. Direct access to this table and the International Data Base is available through the Internet at www.census.gov/ipc/www.

countries, such as Brazil, Egypt, Indonesia, Mexico, and Thailand, fertility rates have been dropping as they did in the 1970s in China and India. At the same time, many developing countries have not yet entered the demographic transition.

The demographic transition occurs when both birth rates and death rates in a country drop from historically high levels to low ones that translate into a stable population—one that merely replaces itself with each new generation. Traditionally, although not always, death rates have declined first, following the spread of sanitation and improved health care overall. Rapid population growth often follows this first phase of the demographic transition, as the gap between fertility and mortality rates widens for a time. Eventually, however, fertility rates fall too.

Fertility rates remain high in a number of countries. The reasons include unequal rights and opportunities for women, as well as inadequate access to birth control. Whatever the reason, the effect is the same: 67 countries, home to 17 percent of the world population, are at best in the early stages of a transition to low fertility rates. Most of them are in Africa and South Asia, and their populations are likely to double in 20 to 25 years.

This is leading to a two-tiered demographic world that is every bit as worrying as the world of economic haves and have-nots. Countries such as Nigeria and Pakistan are finding it harder to keep up with the demand for food,

Figure 1
Global fertility levels relative to replacement level: 2002–2005.
The level of global fertility is projected to drop below replacement by mid-century.
Note: Global total fertility rates were derived by calculating weighted age-specific fertility rates from country-and-age-specific births and numbers of women.
Source: U.S. Census Bureau, International Programs Center, International Data Base and unpublished tables. (2004). *Global Population Profile 2002.* Washington, D.C.

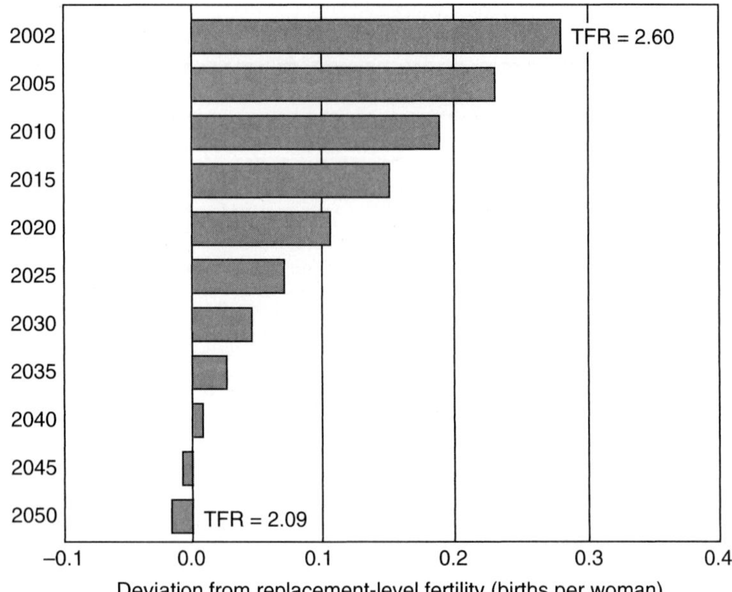

health care, jobs, housing, and education than countries that are in the middle of the demographic transition.

As Shiro Horiuchi of Rockefeller University notes, "The demographic gap seems to overlap with a growing gap in economic development." This gap has been growing for more than 20 years. In 1970, 34 percent of the world lived in countries with fertility rates below 5.5. Just five years later, thanks to the dramatic declines in India and especially China, the figure was 80 percent. It did not gain much by 1985 reaching 83 percent. More than three-fourths of the significant declines in fertility rates started in the 1965 to 1970 period.

Even when a country does reach replacement-level fertility, its population can continue growing for decades. There is a built-in momentum created by all the people who have yet to enter their childbearing years. Indeed, the decline in the world's population growth rate stalled in the 1980s in part because even in China, India, and other countries where fertility rates had been dropping, a large number of people who had been born in the 1960s reached childbearing age. So even if couples had two or three children instead of five or six, as their parents did, the population would grow substantially.

For the world as a whole, even if replacement-level fertility had been achieved in 1990, the population would continue to grow until it reached about 10 or 11 billion in 2050 because of all the young people already alive.

This built-in momentum obviously limits how quickly any country can stop population growth. Nevertheless, reaching replacement-level fertility is an all-important first step. Many of the less developed countries have not yet begun the demographic transition—nations in which invariably the government believes fertility levels are too high—could move in the right direction by providing the contraceptive and health care services that would help couples have only the number of children they desire.

For specific country by country fertility rates, please refer to Table P5.7 on page 345 in the Introduction to Part V.

QUESTIONS

1. What is the fertility rate?
2. What was the global fertility rate in 1991?
3. How many countries are in the early stages of a transition to low fertility rates? What are the characteristics of this early transition phase? What is the longer-term

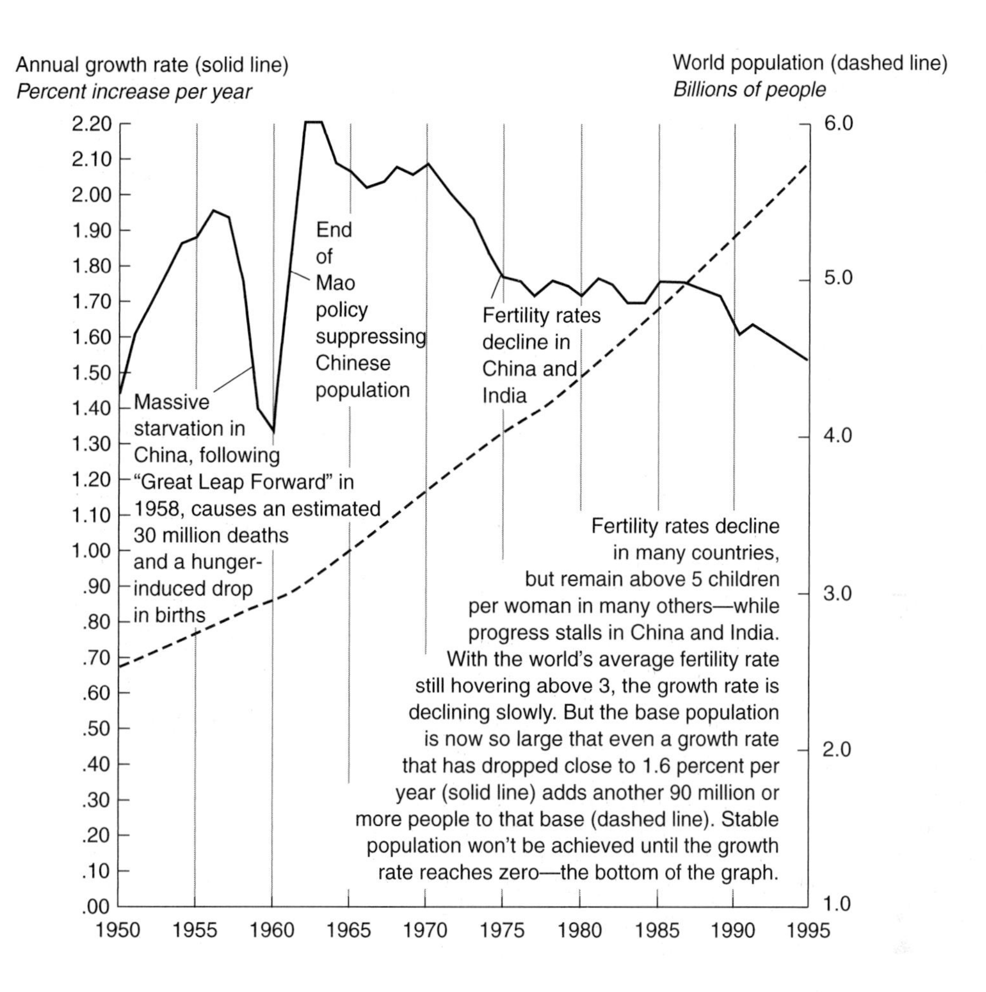
Annual growth rate (solid line)
Percent increase per year
World population (dashed line)
Billions of people
2.20
2.10
2.00
1.90
1.80
1.70
1.60
1.50
1.40
1.30
1.20
1.10
1.00
.90
.80
.70
.60
.50
.40
.30
.20
.10
.00
6.0
5.0
4.0
3.0
2.0
1.0
1950
1955
1960
1965
1970
1975
1980
1985
1990
1995
End of Mao policy suppressing Chinese population
Fertility rates decline in China and India
Massive starvation in China, following "Great Leap Forward" in 1958, causes an estimated 30 million deaths and a hunger-induced drop in births
Fertility rates decline in many countries, but remain above 5 children per woman in many others—while progress stalls in China and India. With the world's average fertility rate still hovering above 3, the growth rate is declining slowly. But the base population is now so large that even a growth rate that has dropped close to 1.6 percent per year (solid line) adds another 90 million or more people to that base (dashed line). Stable population won't be achieved until the growth rate reaches zero—the bottom of the graph.

Figure 2
World population: even as the rate of growth declines, the number of humans continues to climb.

effect of being in the early stages of a demographic transition?

4. Why is this demographic gap overlapping with an economic gap?
5. Why did the decline of the fertility rate stall in the 1980s?
6. Why, if replacement-level fertility had been reached in 1990, would the population continue to grow to 10 or 11 billion?

REFERENCES

Population Reference Bureau. (2004) *2004 World Population Data Sheet*. Washington, D.C. Available at http://www.prb.org.

Starke, L. (1993). "Fertility Rates: The Decline Is Stalling." *Vital Signs 1993*. Worldwatch Institute. New York: Norton and Company.

U.S. Census Bureau, International Programs Center, International Data Base. (2004). *Global Population Profile 2002*. Washington, D.C.

Table 2
Average Annual Rate of Growth by Region and Development Category: 1950–2050 (in percent)

Region	*1950–1960*	*1960–1970*	*1970–1980*	*1980–90*	*1990–2000*	*2000–2010*	*2010–2025*	*2025–2050*
World	1.7	2.0	1.8	1.7	1.4	1.1	0.9	0.6
Less Developed Countries	2.0	2.4	2.2	2.0	1.7	1.3	1.1	0.7
More Developed Countries	1.2	1.0	0.7	0.6	0.4	0.3	0.1	(Z)
Africa	2.2	2.4	2.7	2.8	2.5	2.0	1.6	1.4
Sub-Saharan Africa	2.1	2.5	2.7	2.8	2.6	2.0	1.7	1.6
North Africa	2.4	2.4	2.5	2.7	2.1	1.7	1.3	0.8
Near East	2.7	2.6	3.0	2.9	2.3	2.2	1.9	1.4
Asia	1.7	2.2	2.0	1.8	1.4	1.1	0.9	0.4
Latin America and the Caribbean	2.7	2.7	2.4	2.0	1.7	1.3	1.0	0.5
Europe and the New Independent States	1.1	0.9	0.7	0.5	0.2	0.1	(Z)	–0.2
Western Europe	0.7	0.8	0.4	0.3	0.3	0.2	(Z)	–0.3
Eastern Europe	1.2	0.8	0.8	0.4	–0.1	–0.1	–0.2	–0.5
New Independent States	1.7	1.3	0.9	0.8	0.1	0.1	0.2	(Z)
North America	1.8	1.3	1.1	1.0	1.2	0.9	0.8	0.7
Oceania	2.3	2.1	1.6	1.6	1.5	1.2	0.9	0.5
Excluding China:								
World	1.8	1.9	1.8	1.8	1.5	1.3	1.0	0.7
Less Developed Countries	2.2	2.4	2.4	2.3	1.9	1.6	1.3	0.9
Asia	1.9	2.2	2.2	2.0	1.7	1.4	1.1	0.6
Less Developed Countries	2.0	2.3	2.2	2.1	1.8	1.5	1.1	0.7

Z = between –0.05 percent and +0.05 percent.
Note: Reference to China encompasses China, Hong Kong S.A.R., Macau S.A.R., and Taiwan.
Source: U.S. Census Bureau, International Programs Center, International Data Base. (2004). *Global Population Profile 2002.* Washington D.C. Direct access to this table and the International Data Base is available through the Internet at www.census.gov/ipc/www.

QUESTIONS ABOUT THE TABLES AND FIGURES

1. What do you think is meant by "stabilization of population rate"?
2. Refer to Table 1. Why are the less-developed countries growing faster than the more developed countries? Compare countries for the highest growth in numbers.
3. Refer to Table 2. From your answers in question number 2, relate fast population growth to rates of growth. Discuss the differences between the numbers and the percentages.
4. Examine Figures 1 and 2. What are the differences in results? Discuss in a paragraph or two and relate to replacement levels and fertility rates.

The Hazards of Youth

LISA MASTNY

In more than 100 countries, people are getting not only more numerous, but younger. "Youth bulges," combined with economic stagnation and unemployment, can burden these countries with disproportionately high levels of violence and unrest—severely challenging their hopes for social and economic stability.

Just before dawn on April 28, a band of machete-and knife-wielding attackers launched a surprise assault on a police post in Thailand's southern province of Pattani. Failing to overrun the building, the militants fled to the nearby Krue Sae mosque, where they engaged in a three-hour shoot-out with heavily armed government security forces. Troops riddled the 16th-century red brick building with automatic weapon fire, killing more than 30 of the attackers and leaving their bodies sprawled in pools of blood.

The uprising was only one of several clashes in Thailand's restive south that day, which ended with at least 108 suspected militants dead across three provinces. It marked a severe escalation of four months of unrest in a country that had not seen such bloodshed in three decades. As news of the conflict spread, analysts attributed the tensions to rising ethnic discontent among the south's largely Muslim population, which has long complained of cultural, religious, and economic repression by the central government in Bangkok. In an address to the nation soon after the attacks, however, Prime Minister Thaksin Shinawatr pointed to another variable: the age and prospects of the combatants, most of whom were under the age of twenty. "They are poor and have little education and no jobs," he noted. "They don't have enough income and have a lot of time, so it creates a void for people to fill."

Unlike the more prosperous north, Thailand's south has lagged on several key development indicators, including demographics. Although population growth in the country overall has slowed dramatically, reaching "replacement level" at just above two children per woman by the mid-1990s, birth rates in the southern provinces remain high. Meanwhile, industrial growth in the region has stagnated, leaving few opportunities for this surging young population.

Thailand is not the only country in the world feeling the effects of a demographic imbalance. According to the United Nations, more than 100 countries worldwide had characteristic "youth bulges" in 2000, i.e., young adults ages 15 to 29 account for more than 40 percent of all adults. All of these extremely youthful countries were in the developing regions. By and large, the youth bulge is a thing of the past in North America and Europe, where the young adult share of the population is only about 25 percent of all adults.

LOSS OF OPPORTUNITY

In most cases, a youth bulge is the result of several past decades of high birth rates. It typically occurs in countries that are still in the earlier stages of their transitions to slower-growing populations: although infant and child mortality have begun to fall, birth rates still remain high, resulting in higher proportions of children surviving overall. A youth bulge usually lingers for at least two decades after fertility begins to decline, as large cohorts of children mature into young adulthood. If low fertility is maintained, however, this bulge gradually disappears.

Other demographic processes can create a youth bulge as well. Sudden drops in infant mortality or a baby boom in an industrial country can create a bulge in young

Reprinted with permission. L. Mastny, "The Hazards of Youth", *World Watch* 2004;17(5), pp. 18–21.

adults two decades later. Disproportionately high youth populations can also be present in countries where large numbers of adults emigrate or where AIDS is a major cause of premature adult death.

An excess of youth isn't necessarily a bad thing. In the United States and other industrial countries, where most young adults have been educated or technically trained, employers view young people as an asset and actively seek out their energy and ingenuity. Indeed, economists have long recognized that a large cohort of young workers can provide a demographic boost to growth in economies where the productivity, savings, and taxes of young people support smaller subpopulations of children and elderly. In Thailand, for instance, young, educated, and industrious workers—including a large proportion of young women working in the country's manufacturing and financial sectors—have contributed significantly to the growth of the country's dynamic economy.

In other circumstances, however, the predominance of young adults can be a social challenge and a political hazard. In many developing countries, labor markets have been unable to keep pace with population growth, contributing to high rates of unemployment. While unemployment tends to be high in developing countries in general, that among young adults is usually three to five times as high as overall adult rates.

Leif Ohlsson, a researcher at the University of Göteborg in Sweden, notes that young men in rural areas are often hardest hit relative to their expectations. Agriculture is the single largest source of livelihood worldwide, but many young rural men expecting to inherit land increasingly find themselves disinherited. In some cases, their fathers and grandfathers have long since divided up the family property into tiny parcels that would be unworkable if they were further divided. In other cases, the land has degenerated as a result of unsustainable practices, or larger commercial agricultural enterprises have swallowed up any remaining cropland.

Without a secure, independent living, these men find themselves unable to marry or earn the respect of their peers, contends Ohlsson. British researcher Chris Dolan has coined the expression "the proliferation of small men" in reference to the growing number of disenfranchised young men in northern Uganda who cannot fulfill their culture's expectations of a "full man." Dolan has found that such men disproportionately become alcoholic, engage in violence, or commit suicide.

Or join a militia. Insurgent organizations can offer social mobility and self-esteem, particularly in countries that are economically backward and politically repressive. During the recent civil war in Sierra Leone, young people constituted about 95 percent of the fighting forces, in part because there were few other options. Sierra Leone ranked as the world's least developed country on the United Nations' 173-nation Human Development Index in 2002, and the gross national income per capita in 2000 totaled only $140 (compared with $34,870 in the United States). An official with the Christian Children's Fund in Freetown explained of the large body of young soldiers, "They are a long-neglected cohort; they lack jobs and training, and it is easy to convince them to join the fight."

URBAN YOUTH

With few opportunities in rural areas, young people in many developing countries are increasingly forced to leave behind more traditional lifestyles and migrate to cities in search of work, education, and urban amenities. The United Nations projects that by 2007, for the first time ever, more people will be living in cities than in rural areas. This urban share could top 60 percent by 2030—with almost all of this growth projected to occur in the developing world.

With the influx of young workers and students, many urban areas are now home to significant, and potentially volatile, youth bulges. Drawing on case studies of several Asian countries, University of Hawaii political scientist Gary Fuller warns that rapidly industrializing cities and frontier areas can be spawning grounds for political unrest because thousands of young men migrate to these sites in search of livelihoods. Yet urbanization is proceeding faster than municipalities can provide infrastructure, services, and jobs. Municipal governments in the least-developed countries are often the least able to muster the human and financial resources to contend with these problems, especially when the poorest, nontaxable segment of the urban population continues to grow rapidly.

But it is not just the poor or uneducated who are discontented. "We have a large number of youth between 18 and 35 who are properly educated, but have nothing to do," lamented William Ochieng, a former government official, in Kenya's *The Daily Nation* in January 2002. Urban discord, more than the rural sort, afflicts diverse social classes, including recruits from politicized students, the angry unemployed, and the politically disaffected. Many of these, especially those from middle-class backgrounds, bring with them the skills and resources to organize and finance civil protest.

Studies show that the risks of instability among youth may increase when skilled members of elite classes are marginalized by a lack of opportunity. Yale University historian Jack Goldstone has noted that the rebellions and religious movements of the 16th and 17th centuries were led by young men of the ruling class who, upon reaching adulthood among an overly large cohort, found that their state's patronage system could not afford to reward them with the salary, land, or bureaucratic position commensurate with their class and educational achievements. Rather than allow political discontent to fester, European militarists and Ottoman expansionists induced thousands of young men of privilege to serve their interests in military campaigns and overseas colonial exploits, putting them in charge of literally millions of the unschooled from the urban and rural under-classes.

It isn't difficult to find contemporary parallels. Goldstone attributes the collapse of the Communist regime in the Soviet Union in the early 1990s in part to the mobilization of large numbers of discontented young men who were unable to put their technical educations to use due to party restrictions on entering the elite. And Samuel P. Huntington, Harvard professor and author of the controversial treatise on the "clash of civilizations," has pointed to connections between tensions in the Middle East (where 65 percent of the population is under the age of 25) and the unmet expectations of skilled youth. Many Islamic countries, he argues, used their oil earnings to train and educate large numbers of young people, but with little parallel economic growth, few have had the opportunity use their skills. Young, educated men, Huntington concludes, often face only three paths: migrate to the West, join fundamentalist organizations and political parties, or enlist in guerrilla groups and terrorist networks.

Discontented elites may in turn mobilize less-educated groups to their cause. Investigations into Thailand's recent upsurge in violence point to the possible involvement of Muslim extremist groups, who may be actively targeting young men of strong religious faith and little formal education to further their broader Islamist goals. In the town of Suso, which lost 18 men under the age of 30 to the April uprising, most of the dead had graduated from the country's privately run Islamic schools (pondoks), which are often a last resort for families that cannot afford mainstream college educations. In Pakistan, meanwhile, studies estimate that as many as 10 to 15 percent of the country's 45,000 religious schools (madrasas) have direct links to militant groups.

LOOKING AHEAD

How strong is the link between youth and conflict? In 2003, researchers with the Washington, D.C.-based group Population Action International (PAI) reviewed the data on population and past conflicts and found that countries in which young adults made up more than 40 percent of all adults were about two-and-a-half times as likely to experience an outbreak of civil conflict during the 1990s as other countries. The study identified 25 countries where a large youth bulge, coupled with high rates of urban growth and shortages of either cropland or fresh water, creates a "very high risk" of conflict. Fifteen of these countries are in sub-Saharan Africa, two are in the Middle East (Yemen and the Occupied Palestinian Territories), and the rest are in Asia or the Pacific Islands. According to Uppsala University's Conflict Database, nine of these countries experienced a civil conflict just within the first three years of this decade (2000–2002).

As evidence of this link emerges, the global security community has begun to take notice—though it's been slower to take action. In April 2002, in a written response to congressional questioning, the U.S. Central Intelligence Agency noted that "several troublesome global trends—especially the growing demographic youth bulge in developing nations whose economic systems and political ideologies are under enormous stress—will fuel the rise of more disaffected groups willing to use violence to address their perceived grievances." The CIA warned that current U.S. counter-terrorist operations might not eliminate the threat of future attacks because they fail to address the underlying causes that drive terrorists.

Large youth bulges should eventually dissipate as fertility rates continue their worldwide decline. Already, between 1990 and 2000, the number of countries where young adults account for 40 percent or more of all adults decreased by about one-sixth, primarily because of declining fertility in East Asia, the Caribbean, and Latin America. However, a more persistent group of countries in the early stages of their demographic transition—most in sub-Saharan Africa, the Middle East, South and Central Asia, and the Pacific Islands—remain as a challenge to global development and security.

Fortunately, demographics is not destiny. But the likelihood of future conflict may ultimately reflect how societies choose to deal with their demographic challenges. In its recent analysis, for instance, PAI discovered that roughly half of the very high-risk countries navigated the post-Cold War period peacefully. How? In at least some of these cases, policies were in place that provided young men with occupations and opportunities—including land reform and frontier settlement schemes, migration abroad, industrialization, and the expansion of military and internal security forces. The latter strategy, PAI suggests, probably helped repressive regimes such as North Korea, China, and Turkmenistan maintain political stability during the post-Cold War era despite large proportions of young adults.

In the short term, governments will need to tackle the underlying factors contributing to discontent among young people, including poverty and the lack of economic opportunity. And governments can address part of the risk associated with youth unemployment by investing in job creation and training, boosting access to credit, and promoting entrepreneurship.

Ultimately, however, the only way to achieve the necessary long-term changes in age structure will be through declines in fertility. Governments can facilitate fertility decline by supporting policies and programs that provide access to reproductive health services—voluntary family planning services and maternal and child health programs and counseling, including providing accurate information for young adults—and by promoting policies that increase girls' educational attainment and boost women's opportunities for employment outside the home.

For countries in the early stages of their demographic transitions, it could take nearly two decades after fertility begins to fall to observe a significant reduction in the proportion of young adults. Given the many risks of delaying the demographic transition, this only underscores the need for governments to put supportive policies into effect sooner rather than later.

QUESTIONS

1. What is the difference in the pattern of "youth bulges" in developing countries compared with North America and Europe? What are some implications of the differences in these patterns?
2. What causes "youth bulges?"
3. In rural areas in developing countries, why are young men sometimes the hardest hit with lack of opportunity? What often happens to these young men?
4. What is the situation of youth in urban areas in developing countries? Is education a deterrent to uprisings and violence?
5. Explain the relationship of high populations of young adults with violence and civil conflict.
6. What are some suggested measures that can curtail the tendency for civil conflict and violence in large young adult populations? Name at least five measures.
7. How many years will it take for the demographics of high proportions of youth and young adults to decline? What are the implications of this over the next 50 years?

Earth's Carrying Capacity: Not Quite So Easy When Applied to Humans

EDITED BY BARBARA A. EICHLER

We face a real crunch. At issue is whether we can do what is totally unprecedented—feed, house, nurture, educate, and employ as many more people in the space of four decades as are alive today.
GEOGRAPHER ROBERT W. KATES of Ellsworth, Maine

With the current worldwide population increasing by 90 million people annually and the last billion people added in just the last 12 years, people over 40 years old have actually lived through a doubling of the human population—the first time this has ever happened. This unprecedented growth evokes the concern of how many people the earth can sustain. In four decades, we again will have a doubling of the world's population—the most difficult challenge for humankind yet.

THE CONCEPT OF POPULATION CARRYING CAPACITY

Population carrying capacity is an approach that many ecologists use to mark and provide a sustainable limit philosophy that measures needed ecological policy. Economists tend to view carrying capacity with more flexibility due to technological and policy interventions so that the approach has a less fixed value.

Population carrying capacity can be defined as "the number of people that the planet can support without irreversibly reducing its capacity or ability to support people in the future." While this is a global-level definition, it applies at a national level too, albeit with many qualifications, such as for international trade, investment, and debt. Furthermore, it is a function of factors that reflect technological change, food and energy supplies, ecosystem services (such as provision of freshwater and recycling of nutrients), human capital, people's lifestyles, social institutions, political structures, and cultural constraints among many other factors, all of which interact with each other.

Two points are particularly important: (1) carrying capacity is ultimately determined by the component that yields the lowest carrying capacity and (2) human communities must learn to live off the "interest" of environmental resources rather than off their "principal." Thus, the concept of carrying capacity is closely tied in with the concept of sustainable development. There is now evidence that human numbers, with their consumption of resources plus the technologies deployed to supply that consumption, are already often exceeding the land's carrying capacity. In many parts of the world, the three principal and essential stocks of renewable resources—forests, grasslands, and fisheries—are being utilized faster than their rate of natural replenishment.

Consider a specific example—the earth's carrying capacity with respect to food production. According to the World Hunger Project, the planetary ecosystem could, with present agrotechnologies and with equal distribution of food supplies, satisfactorily support 6.5 billion people if they all lived on a vegetarian diet (the present 2006 population). If people derived 15 percent of their calories from animal products, as tends to be the case in South America, the total would decline to 3.7 billion. If they derived 25 percent of their calories from animal protein, as is the case with most people in North America, the earth could support only 2.8 billion people.

True, these calculations reflect no more than today's food production technologies. Certain observers protest that such an analysis underestimates the scope for technological expertise to continue to expand the earth's carrying capacity. We can surely hope that many advances in agrotechnologies will be discovered. But consider the population–food record over four decades. From 1950 to 1984, and thanks largely to remarkable advances in Green Revolution agriculture, there was a 2.6-fold increase in world grain output. This achievement, representing an average increase of almost 3 percent a year, raised per capita production by more than one-third. Yet from 1985 to 1989, there was next to no increase at all, even though the period saw the world's farmers investing billions of dollars to increase output (fertilizer use alone expanded by 14 percent). These big investments were supported by rising grain prices and the restoration to production of idled U.S. cropland. Crop yields had "plateaued"; it appeared that plant breeders and agronomists had exhausted the scope for technological innovation. Therefore, the 1989 harvest was hardly any higher than that of 1984. During that same period, there were an extra 440 million people to feed. While world population increased by almost 8.5 percent, grain output per person declined by nearly 7 percent.

To put the case more succinctly, every 15 seconds sees the arrival of another 35 people, and during the same 15 seconds, the planet's stock of arable land declines by one hectare. Plainly, this is not to say the first is a singularly causative factor of the second. Many other factors, notably technology, contribute to the linkage. Equally, it is not to deny that there is a strong relationship between the two factors: more people are trying to sustain themselves from less cropland.

Regrettably, there is all too little concise analysis of the concept of carrying capacity. So the evaluation of its nature, and especially of the threat of environmental overloading, must remain largely a matter of judgment. The United Nation's Population Fund of 1999 states that "Because natural conditions, technology, and consumption and distribution patterns are constantly in flux, and there is no universal agreement as to the definition of 'carrying capacity', it is unlikely that there will ever be a definitive answer" (UNFPA 1999) to what are the environmental limits to the number of people and the quality of life that the earth can support. "Most scientists who have pondered the issue have predicted that there are natural limits, but the predicted limits fall within a broad range: 4-16 billion people. What will happen as human population approaches those limits, either globally or locally, will depend on human choices—about lifestyles, environmental protection and equity"(UNFPA 1999).

But remember that in a situation of pervasive uncertainty, we have no alternative but to aim for the best choices and assessments we can muster. We cannot defer the question until such time as we have conducted enough research. After all, if we do not derive such conclusions as we can and make explicit planning decisions on their basis, there will be implicit planning decisions taken by large numbers of people who, through their daily lifestyles, are determining the outcome. In other words, decisions on the population environment nexus will be taken either by design or by default. We must make do with such information and understanding as we have at hand, however imperfect that may be (United Nations Population Fund).

EARTH'S CARRYING CAPACITY DOES NOT ACCURATELY APPLY TO HUMANS

The concept of the earth's carrying capacity works well when defining how many animals a habitat can support, but it breaks down when applied to people since the assortment of habits and environmental patterns of humans have a wide variation of resource consumption and waste generation. The typical New Yorker, for example, uses 10 to 1,000 times more resources daily than the average Chilean or Ghanaian. A second reason that the carrying capacity formula does not work well is that humans have a great ability to alter their physical environment or culture for survival. Many scientists are beginning to argue that they cannot predict how many people the earth can sustain until there exists a better way to account for the various social, cultural, and demographic habits and effects of people.

Wolfgang Lutz, a scientist in Laxenburg, Austria, has a different approach. The formula he uses does not simply account for the *number* of people; it estimates environmental impact (I) as the product of population (P) times affluence (A) times the technological efficiency (T) of a culture ($I = PAT$). The formula $I = PAT$ was first developed by scientists Paul Ehrlich and John Holdren to examine the environmental impacts of population in general. Some scientists use this formula to calculate the economic and environmental costs associated with each new birth. There was a problem with the use of this formula in calculations of the impact of CO_2 emissions, however: the formula better correlated with the number of *households* rather than individuals since there has been a trend toward the existence of more households with fewer members caused by the patterns of divorce, young adults setting up their own households, the postponement of marriage, an increase in life expectancy, and a growing sense of privacy. The trend of more, but smaller, households is most prevalent in nations that already use the most resources, and this trend could cause substantial impacts. When utilizing the *I* = *PAT* formula to derive calculations by using households rather than individuals, CO_2 emissions were 50 percent higher than those suggested by the use of individuals in the formula.

FOOD PRODUCTION AS AN EXAMPLE

Population growth impacts food production. Since 1955, nearly one-third of the world's cropland (an area larger than China and India combined) has been abandoned because its overuse has led to soil loss, depletion, or degradation. The search for cropland substitutes accounts for about 60–80 percent of the world's deforestation; however, the areas being deforested are poorly suited for agriculture and are becoming more difficult to find.

Additionally, about 87 percent of the world's freshwater is being consumed by agriculture—far more than is consumed by any other human activity. Farmers are using surface water and are mining underground aquifers. According to Population Action International, "Farmers are literally draining the continents, just as they are allowing billions of tons of topsoil to erode."

Although at present there is enough land and water to feed the earth's existing population, the per capita availability of cereal grains—the source of 80 percent of the world's food—has been declining for the last few decades. Most nations today rely on imports of surplus grains from the resources of only a few nations, which account for the most of all cereal exports, with the United States providing the largest share. If the population of the United States doubles in 60 years as projected, then U.S. food and cereal resources will instead need to be used to feed 520 million Americans. Meanwhile, if China's population continues to grow to another 500 million people and if China continues to experience the loss of cropland to erosion and unabated industrialization, its demands for imports by 2050 might be more than the world's entire annual grain exports. Lester Brown of the Worldwatch Institute says, "Who could supply grain on this scale? No one."

Considering these food demands does not, however, answer the question of how many people the earth can sustain. Joel Cohen, head of the Laboratory of Populations at Rockefeller University, states that there is no single answer. It will depend, he says, "on the number of people that are willing to wear clothes from cotton (a renewable resource) versus polyester (a nonrenewable petroleum-based product), how many will eat beef versus bean sprouts, how many will want parks versus parking lots, how many will want Jaguars with a capital J and how many will want jaguars with a small j. There is also the unknown factors and future of the genetic alteration of plants, foods and animals to increase yields, change habitat conditions and nutritional values. These choices will change in time, and so will how many people the Earth can support." If the U.S. population doubles by 2050, the share of its diet from animal products will drop by about one-half to perhaps 15 percent, and food would cost up to 50 percent of each person's paycheck. Although these percentages are typical of what Europeans spend on food now, this percentage is far more than what U.S. residents normally pay.

OTHER APPROACHES FOR A SUSTAINABLE PLAN

Another idea that addresses the problem of limited resources for a growing population is the free market economic approach. If world markets were free, the price of goods would rise as the amount of materials grew limited. Rising prices would alert entrepreneurs to the approaching scarcity and provide an entrepreneurial award for new products and creativity. Thomas Lambert of the Center of the Study of American Business feels that resource scarcity will lead to technological developments that leave everyone better off than before. "The entire history of humanity indicates that with enough economic freedom, overpopulation relative to natural resources or energy will not occur."

The problem with this free market argument, according to Thomas Homer-Dixon of the Peace and Conflict Studies Program at the University of Toronto, is that not all things are equal. In areas where young children receive inadequate nutrition, stimulation, or education, many will not mature adequately to meet the challenge of innovation. Other people face social barriers to innovation such as business corruption, banking difficulties, lack of monetary support, politics and poverty, and power group considerations. Mr. Homer-Dixon says that, as resource scarcity and population stresses rise, "the smarter we have to be socially and technically just to maintain our well-being—the more dire the problem, the more quickly we're going to have to respond." Mr. Homer-Dixon is an optimist in that he feels the institution of research in developing countries will increase the supply of ingenuity. Most of the creativity that is useful in solving scarcity and food production problems is going to be generated locally by people familiar with local geography, social relations, and resources. To stop the flow of local talent to industrialized nations where greater research opportunities exist, funding of local research needs to exist by providing scientists in the developing world with monetary and technical resources as well as computers, modems, and fax machines—a communication network to peers around the globe.

Global Research on the Environment and Agricultural Nexus (GREAN) is such a proposed organization. GREAN would set up a partnership among U.S. universities and 16 international centers that belong to the Consultative Group on International Agricultural Research. Since the demand for food in developing countries is expected to double by 2025 and triple by 2050, GREAN wants to involve the outstanding scientists of the world (including those in developing countries) in projects aimed at developing sustainable, environmentally friendly food production by and for the world's poorest people (Raloff 1996, 396–397).

Another approach is to link population to the environment more closely. The 1992 Earth Summit Conference in Rio targeted many environmental issues, but largely overlooked population. The 1994 Conference on Population in Cairo led to a comprehensive program of action, but overlooked the environment. However, the Johannesburg 2002 World Summit and the UN 2000 Millennium Project have improved in linking population, education, women's rights, poverty, and education. Establishing such a link helps to balance the intensity and sensitivity of developing countries with the highest population growth rates with the developed countries in terms of their consumptive use of earth's resources, thereby designing an entire global cooperative action plan. This type of link could measure use of global materials and resources as a means to reduce CO_2 emissions and other patterns directly related to quality of life and patterns of population. Policies on population aim to decrease fertility rates in a noncoercive fashion and focus on improving the well-being of women. This can be done by helping women avoid unwanted pregnancies, improving girls' education, aiming to achieve greater gender equality, and reducing child mortality, among other methods.

Tying environmental actions to population can develop a beneficial picture of the improvement of environmental habits and the abatement of consequences that stem from population growth in an equally noncoercive, comprehensive fashion. This process then becomes an all-encompassing and realistic method not only of reducing population, but also of improving the quality of life around the world, with all countries as equal participants in the solution (Gaffin 1996, 7).

As we approach these challenges, Mr. Cohen believes that "neither panic nor complacency is in order. Earth's capability to support people is determined both by natural constraints and by human choices. In the coming half century, we and our children are less likely to face absolute limits than difficult trade-offs" such as population size, environmental quality, and lifestyle. Many people have already been questioning the consumerism of the industrialized countries since the 1950s. Choices will start to occur when people begin to question what they are buying with their hard-earned money as well as the real long-term value of these items. For example, environmentalist mediator Paul Wilson, a lawyer in Portland, Oregon, has spent seven years simplifying his life in terms of his diet and aspirations after realizing that the products he was buying were not worth the hours of work needed to pay for them. His sense is that a transition will occur in which people and their organizations will perceive the imbalance of values in their own way until simple values predominate. Earth's carrying capacity for humans must be determined by their own well-chosen choices and strategies (Raloff 1996, 397).

REFERENCES

Gaffin, S. R., O'Neill, B. C., and Bongaarts, J. (1996). Population Growth Could Affect Global Warming. *Environmental Defense Letter* (November), Vol. 27, No. 6, p. 7.

Raloff, J. (1996). The Human Numbers Crunch: The Next Half-Century Promises Unprecedented Challenges. *Science News* (22 June), Vol. 149, No. 25, pp. 396–397.

United Nations Population Fund. Population Resource and the Environment: The Critical Challenges.

United Nations Population Fund. (1999). *The State of World Population 1999.* New York. Available http://www.unfpa.org/swp/1999/chapter2e.htm

QUESTIONS

1. Define "population carrying capacity."
2. What two points underlie carrying capacity?
3. What is the cause of the difference between the earth's carrying capacity estimates of food production for 5.5 billion people, 3.7 billion people, and 2.8 billion people, as mentioned in the reading? What number do you think will reflect a sustainable world standard of living and why? Explain.
4. What happens every 15 seconds in the way of adding and subtracting to the earth's carrying capacity?

5. What do you think of the validity of the concept of carrying capacity? Do you think that this concept can help us design our future, or will we proceed to our future by default?
6. Why doesn't the concept of carrying capacity apply easily when dealing with people?
7. What is $I = PAT$? What was the problem in using this formula? How was it better correlated?
8. How can this formula be used to project the intensity of resource use and carrying capacity? How can it be used to better predict the availability of future resources?
9. According to the reading, how are we going to resolve the inequalities of resource use?
10. What are some suggestions for helping unindustrialized countries control accelerating populations and future quality-of-life issues?
11. What is the free market approach? What is the problem with using this type of approach?
12. What type of population reduction plan can involve industrialized nations as well as unindustrialized or developing countries? Develop an outline of such a plan and present it to your class.

Case Study 1

~~Definitely~~ Probably One: A Generation Comes of Age Under China's One-Child Policy*

CLAUDIA MEULENBERG

Had China not imposed its controversial but effective one-child policy a quarter-century ago, its population today would be larger than it presently is by 300 million—roughly the whole population of the United States today, or of the entire world around the time of Genghis Khan.

The Chinese population-control policy of one child per family is 25 years old this year. A generation has come of age under the plan, which is the official expression of the Chinese quest to achieve zero population growth. China's adoption of the one-child policy has avoided some 300 million births during its tenure; without it, the Chinese population would currently be roughly 1.6 billion—the number at which the country hopes to stabilize its population around 2050. Many experts agree that it is also the maximum number that China's resources and carrying capacity can support. Standing now at a pivotal anniversary of the strategy, China is asking itself, Where to from here?

China's struggle with population has long been linked to the politics of national survival. China scholar Thomas Scharping has written that contradictory threads of historical consciousness have struggled to mold Chinese attitudes toward population issues. China possesses a "deeply ingrained notion of dynastic cycles" that casts large populations as "a symbol of prosperity, power, and the ability to cope with outside threat." At the same time, though, "historical memory has also interpreted a large population as an omen of approaching crisis and downfall." It was not until economic and development issues re-emerged as priorities in post–Mao Zedong's China that the impetus toward the one-child policy began to build rapidly. During Mao's rule, population control was often seen as inhibiting the potential of a large population, but in the years following his death, it became apparent that China's population presented itself as more of a liability than an asset. Policymakers eager to reverse the country's backwardness saw population control as necessary to ensure improved economic performance. (In 1982, China's per-capita GDP

*Editor's note: The dimensions of Chinese population are vast and difficult to comprehend. Even though China's fertility rate is below replacement level at 1.7 and its rate of growth of natural increase is 6 percent, its current population is approximately 1.3 billion people (2004), which constitutes one-fifth of the world's population. Therefore, one in five people worldwide is Chinese. In 2020, about 12 percent of China's population will be over the age of 65 compared with about 7 percent in 2004. The population is expected to peak around 2032 at 1.47 billion people (Population Reference Bureau, 2004 World Population Data Sheet.)

stood at US$218, according to the World Bank. The U.S. per-capita GDP, by way of comparison, was about $14,000.)

The campaign bore fruit when Mao's successor, Hua Guofeng, along with the State Council, including senior leaders such as Deng Xiaoping, decided on demographic targets that would curb the nation's high fertility rates. In 1979, the government announced that population growth must be lowered to a rate of natural increase of 0.5 percent per year by 1985. In fact, it took almost 20 years to reach a rate of 1 percent per year. (The overestimating was in part due to the lack of appropriate census data in 1979; it had been 15 years since the last population count and even then the numbers provided only a crude overview of the country's demography.) Nevertheless, the Chinese government knew that promoting birth-planning policies was the only way to manifest their dedication and responsibility for future generations. In 1982, a new census was taken, allowing for more detailed planning. The government then affirmed the target of 1.2 billion Chinese for the year 2000. Demographers, however, were skeptical, predicting a resurgence in fertility levels at the turn of the century.

The promotion of such ambitious population plans went hand in hand with the need for modernization. Though vast and rich in resources, China's quantitative advantages shrink when viewed from the per-capita perspective, and the heavy burden placed on its resources by China's sheer numbers dictates that population planning remain high on the national agenda. The government has also stressed the correlation between population control and the improved health and education of its citizens, as well as the ability to feed and employ them. In September 2003, the Chinese magazine *Qiushi* noted that "since population has always been at the core of sustainable development, it is precisely the growth of population and its demands that have led to the depletion of resources and the degradation of the environment. The reduction in birth rate, the changes in the population age structure, especially the improvement in the quality of the population, can effectively control and relieve the pressure on our nation's environment and resources and strengthen our nation's capability to sustain development."

THE REACH OF THE ONE-CHILD POLICY

Despite the sense of urgency, the implementation of such a large-scale family planning program proved difficult to control, especially as directives and regulations were passed on to lower levels. In 1981, the State Council's Leading Group for Birth Planning was transformed into the State Population and Family Planning Commission. This allowed for the establishment of organizational arrangements to help turn the one-child campaign into a professional state family planning mechanism. Birth-planning bureaus were set up in all counties to manage the directives handed down from the central government.

Documentation on how the policy was implemented and received by the population varies from area to area. There are accounts of heavy sanctions for non-compliance, including the doubling of health insurance and long-term income deductions as well as forced abortions and sterilizations. Peasant families offered the most significant opposition; rural families with only one daughter often insisted that they be given the right to have a second child, in hopes of producing a son. On the other hand, in some regions, married couples submitted written commitments to the birth-planning bureaus stating they would respect the one-child policy. Despite this variation, it is commonly accepted that preferential treatment in public services (education, health, and housing) was normally given to one-child families. Parents abiding by the one-child policy often obtained monthly bonuses, usually paid until the child reached the age of 14.

Especially in urban areas, it has become commonplace for couples to willingly limit themselves to one child. Cities like Shanghai have recently eased the restrictions so that divorcees who remarry may have a second child, but there, as well as in Beijing and elsewhere, a second child is considered a luxury for many middle-class couples. In addition to the cost of food and clothing, educational expenses weigh heavily: As in many other countries, parents' desire to boost their children's odds of entering the top universities dictates the best available education from the beginning—and that is not cheap. The end of free schooling in China—another recent landmark—may prove to be an even more effective tool for restricting population growth than any family planning policy. Interestingly, the *Frankfurter Allgemeine Zeitung* has reported that Chinese students who manage to obtain a university education abroad often marry foreigners and end up having more than one child; when they return to China with a foreign spouse and passport, they are exempt from the one-child policy.

There are other exceptions as well—it is rumored that couples in which both members are only children will be permitted to have two children of their own, for instance—and it is clear that during the policy's existence, it has not been applied evenhandedly to all. Chinese national minorities have consistently been subject to less restrictive birth planning. There also appears to have been a greater concentration of family planning efforts in urban centers than in rural areas. By early 1980, policy demanded that 95 percent of urban women and 90 percent of rural women be allowed only one child. In the December 1982 revision of the Chinese constitution, the commitment to population control was strengthened by including birth planning among citizens' responsibilities as well as among the tasks of lower-level civil administrators. It is a common belief among many Chinese scholars who support the one-child policy that if population is not effectively controlled, the pressures it imposes on the environment will not be relieved even if the economy grows.

MORE SERVICES, FEWER SANCTIONS

Over time, Chinese population policy appears to have evolved toward a more service-based approach consistent with the consensus developed at the 1994 International Conference on Population and Development in Cairo. According to Ru Xiao-mei of the State Population and Family Planning Commission, "We are no longer preaching population control. Instead, we are emphasizing quality of care and better meeting the needs of clients." Family planning clinics across the country are giving women and men wider access to contraceptive methods, including condoms and birth-control pills, thereby going beyond the more traditional use of intrauterine devices and/or sterilization after the birth of the first child. The Commission is also banking on the improved use of counseling to help keep fertility rates down.

Within China, one of the most prevalent criticisms of the one-child policy has been its implications for social security, particularly old-age support. One leading scholar envisions a scenario in which one grandchild must support two parents and four grandparents (the 4–2–1 constellation). This development is a grave concern for Chinese policymakers (as in other countries where aging populations stand to place a heavy burden on social security infrastructures as well as the generations now working to support them).

A related concern, especially in rural China where there is a lack of appropriate pension systems and among families whose only child is a daughter, is that it is sons who have traditionally supported parents in old age. The one-child policy and the preference for sons has also widened the ratio of males to females, raising alarms as the first children born into the one-child generation approach marriage age. The disparity is aggravated by modern ultrasound technology, which enables couples to abort female fetuses in hopes that the next pregnancy produces a son; although this practice is illegal, it remains in use. The 2000 census put the sex ratio at 117 boys to 100 girls, and according to *The Guardian* newspaper, China may have as many as 40 million single men by 2020. (There are several countries where the disparity is even greater. The UN Population Fund reports that countries such as Bahrain, Oman, Qatar, Saudi Arabia, and United Arab Emirates have male-to-female ratios ranging between 116:100 and 186:100.)

A YOUNGER GENERATION: ADAPTING TRADITION

However, the traditional Chinese preference for sons may be on the decline. Dr. Zhang Rong Zhou of the Shanghai Population Information Center has argued that the preference for boys is weakening among the younger generation, in Shanghai at least, in part because girls cost less and are easier to raise. The sex ratio in Shanghai accordingly stands at 105 boys to every 100 girls, which is the international average. Shanghai has distinguished itself over the past 25 years as one of the first urban centers to adopt the one-child policy, and it promises to be a pioneer in gradually relaxing the restrictions in the years to come. Shanghai was the first region in China to have negative fertility growth; 2000 census data indicated that the rate of natural increase was −0.9 per 1,000.

A major concern remains that, as the birth rate drops, a smaller pool of young workers will be left to support a large population of retirees. Shanghai's decision to allow divorced Chinese who remarry to have a second child is taking advantage of the central government's policy, which lets local governments decide how to apply the one-child rule. Although Shanghai has devoted much effort to implementing the one-child policy over the past 25 years, the city is now allowing qualifying couples to explore the luxury of having a second child. This is a response to rising incomes (GDP has grown about 7 percent per year over the past 20 years) and divorce rates. As noted above, however, many couples, although often better off then their parents, remain hesitant to have more than one child because of the expense.

The first generation of only children in China is approaching parenthood accustomed to a level of economic wealth and spending power—and thus often to lifestyles—that previous generations could not even have imagined. However, China also faces a rapidly aging population. In the larger scheme of things, this may be the true test of the government's ability to provide for its citizens. The fate of China's family planning strategy—in a context in which social security is no longer provided by family members alone but by a network of government and/or private services—may be decided by the tension between the cost of children and the cost of the elderly. There seems little doubt, however, that family planning will be a key element of Chinese policymaking for many years to come.

REFERENCES

Meulenberg, C. (2004). Definitely–Probably One: A Generation Comes of Age Under China's One-Child Policy. *World Watch* 17(5):31-33.

Population Reference Bureau. (2004). *2004 World Population Data Sheet*. Washington, D.C. Available: http://www.prb.org.

QUESTIONS

1. If China did not adopt its one child per family policy 25 years ago, what would the current population be? Use a population chart. How much difference exists between that number and the 2004 population?
2. Why did China adopt the one-child policy in 1979? Explain the political and social reasoning behind this action.
3. In an almanac or other resource, locate and chart the population growth of China in the last five years and their GNP. What conclusions do you draw? Discuss your findings in class.
4. How was the one-child policy encouraged and maintained within the country's policies, services, and privileges?
5. How has the one-child policy changed since the 1994 International Conference of population and Development in Cairo?
6. What are some of the current problems resulting from the one-child policy?
7. What are the future predictions for the one-child policy in China?

Please complete and discuss the following flowchart.

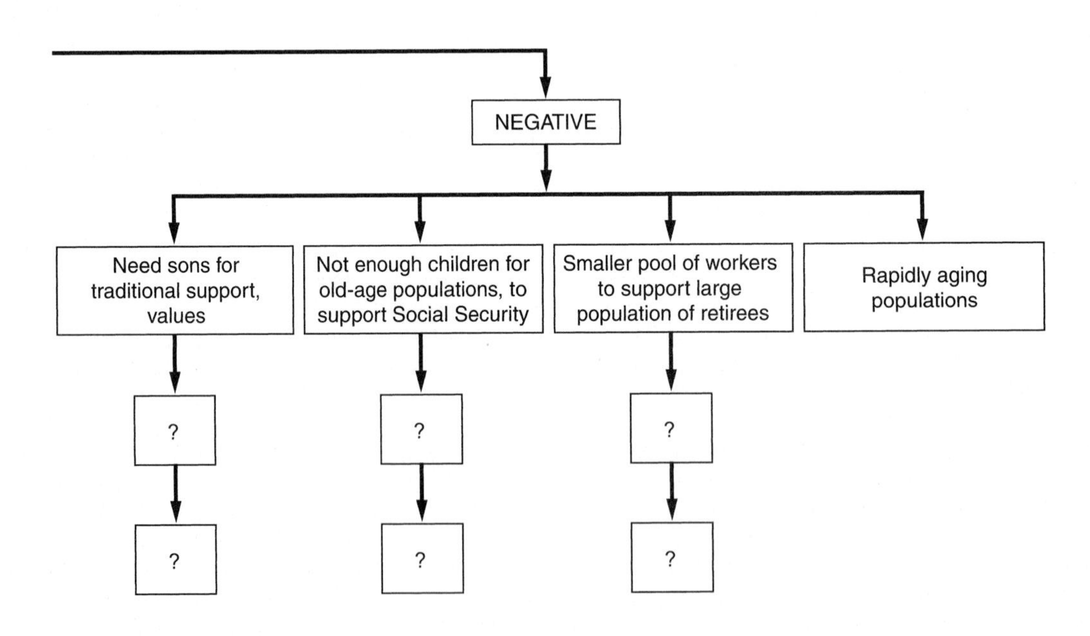

Case Study 2

Japan's Unpopular Pill—An Unrealized Battle Victory

EDITED BY BARBARA A. EICHLER

In 1999, Japan's Ministry of Health and Welfare finally announced that oral contraceptives, which had been officially banned for more than 30 years, would be approved. Even though the pill was approved, doctors said that it would not become popular quickly, as opinion polls regularly showed widespread misconceptions from women about the pill and its side effects.

After five years (2004) and their tough battle to legalize the pill, few women are using it. According to the UN, only about 1.3 percent of the females between ages 15 and 49 are using the pill compared with 15.6 percent in the United States. The continued failure of use of the pill surprises its supporters, who feel it is more effective and safer than other alternatives more widely used in Japan.

Part of the problem is the idea that contraception in Japan is not a woman's responsibility. Another reason is that the pill is a prescription drug and therefore cannot be advertised. Other problems deterring acceptance are misunderstanding and misinformation, conservatism, concerns about possible side effects, and unwillingness to move into a daily pill-taking routine. Most women are happy to use other means of contraception: condoms account for about 80 percent of the birth control market, with the rhythm method and spermicidal jelly being the next most-popular methods. Japan is doing fine without the pill. Teenage pregnancies remain low, and abortions have been decreasing from 460,000 to 330,000 in 2002. Women are also more capable of asserting their rights when asking men to use condoms, which also controls the spread of sexually transmitted diseases (Associated Press 2004).

It seems that interest in the pill has declined in Japan both because condom use controls HIV and because Japan is experiencing a rapidly aging society. The government is actually trying to persuade couples to have more babies since the country's fertility rate is very low at 1.3 and their population is shrinking (Associated Press 2004).

In the past, only women with menstrual disorders could legally use the pill in high and middling doses; however, the number of women who redirected the use of the pill from the treatment of menstrual disorders to birth control is estimated to be as high as 500,000. Many believe that Japan's abortion rate, one of the highest in the world at 400,000 per year, would be greatly reduced if pill use became more widespread.

Passage of the law allowing oral contraceptives has been a difficult road. In 1965, pill approval was denied due to fears that it would corrupt sexual morals. In 1986, the government set up a medical study group to examine the feasibility of the use of a low-dose contraceptive pill that concluded in favor of its introduction, but legislation was never forthcoming. Drug companies petitioned the government in 1990, but fears that condom use (the main form of birth control in Japan) would decline, which could furthermore lead to an increase in the spread of HIV, brought about the pill's rejection in 1992. Another possible approval was lost again in late 1995 due to publicity surrounding studies on venous thromboembolic disease and the use of combined oral contraceptives.

Many blamed the 30-year ban of the pill on a male-dominated bureaucracy that denied women contraceptive freedom. Other important factors that stopped the pill from becoming legal were the profits of condom manufacturers, profits of those who perform abortions, and anxiety about Japan's rapidly declining population. There were even debates about the polluting effects of hormone-tainted urine to the environment. Without strong feminist and antiabortion movements to keep the issue alive, complacency added to the pill not becoming legalized. In contrast, it should be noted that Japan's Ministry of Health approved Viagra, the pill that helps prevent erectile dysfunction, within six months.

"The resistance to pass the law reflects a general disregard for women and their health needs," says Michael Reich, chair, Department of Population and International Health; Iain Aitken, lecturer, Department of Maternal and Child Health Care; and Aya Gota, a former student of Harvard School of Public Health who now lives in Japan, in a December 8, 1999, article in the *Journal of the American Medical Association* (hsph.harvard.edu 2000).

In spite of the pill's passage into law, Japanese women have not rushed for pill prescriptions because, as Aitken states, "a lot of residual fears about side effects remain." Other reasons for the slow adoption, he said, could be that Japanese society as a whole resists the extensive use of pharmaceuticals. Women who use the oral contraceptive must also visit their gynecologists every three months, which they must pay for themselves; abortion costs, however, are covered by health insurance policies.

International health officials have applauded the Japanese government for finally approving the pill, but emphasize the need to increase their focus on the women's issues and health concerns indicated by this long battle for women's reproductive safety, health, and contraceptive options.

REFERENCES

Associated Press. (2004). Japanese Women Shun Birth Control Pill. Associated Press, MSNBC. Accessed August 1, 2005. Available at http://www.msnbc.msn.com/id/5726375/.

Birth Control Pill Hard to Swallow for Some Japanese. (2000). Around the School: News and Notices of the Harvard School of Public Health. February 25, 2000. Accessed August 25, 2001. Available at http://www.hsph.harvard.edu/ats/Feb25/feb25_02.html.

Gutierrez, E., and Netley, G. (1996). Japan's Ban on the Pill Seems Unlikely to Change. *The Lancet* (28 September), Vol. 346:886. Excerpted with permission.

QUESTIONS

1. Even though the ban on the pill has been lifted, why is this still an issue in Japan?
2. What does this article imply about the reproductive rights of women in Japan?
3. What does this indicate about problems of passage of laws as a case study for other countries whose laws are striving to support population control and women?
4. What might this possibly indicate about political power and power groups in other countries and the importance of political activism on population control issues?
5. Are oral contraceptives an issue for male political power groups in many countries throughout the world? Why or why not? Discuss from many perspectives including political power, women's rights, cultural, religious, historical, etc.
6. Research other countries' birth control policies and laws and discuss the implications to their population control practices, rights of women, and population growth numbers.

REFERENCE

Excerpted from McKibben, B. (1999). Taking the Pulse of the Planet. *Audubon* (November/December), 101(6).

Complete and discuss the following flowchart.

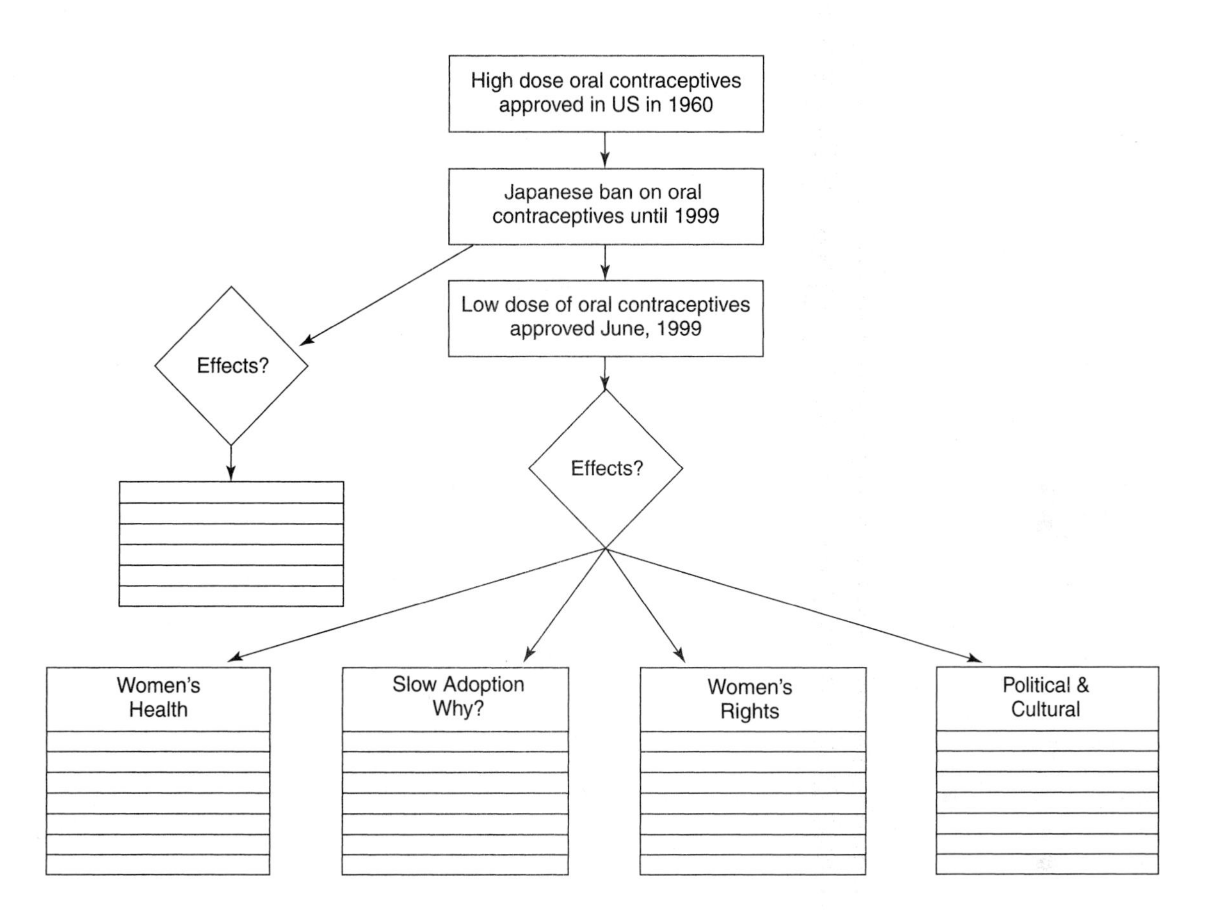

SCENARIO I: Population

The human population is growing at the rate of about 3,000 people every 20 minutes, while plant or animal species die at the same time. Directly and indirectly, rapid population growth is affecting the quality and balance of life everywhere. Billions of people face shortages of their basic needs: food, water, medical care, education, and opportunity. The highest population rates occur in the poorest places on earth. These rates and conditions challenge the Millennium Development Goals (MDGs), which were unanimously adopted by the UN in 2000.

1. Consult population charts and maps and identify where the highest areas of population growth occur.
2. Relate and research those areas to the type of ecological damage that is occurring in those areas.
3. Relate and research these areas as to their current progress for the Millennium Development Goals.
4. How can these countries remain on track with their MDG Goals? Provide three suggestions.

SCENARIO II: Population

A Conversation with Lester Brown, President of the Worldwatch Institute

Human numbers have exploded in this century from 1.5 billion in 1900 to 6.4 billion in 2004. But fertility rates are now falling worldwide. Has the population bomb been diffused? There are several answers to this question. We still are growing rapidly. There has been more growth in the world population since 1950 than during the preceding four million years since the arrival of *Homo sapiens*.

The slowing in growth is reassuring, but one must examine not simply the growth rate, but also absolute numbers. If the demand for food, energy, and other systems is more than the world can supply, the question not only becomes "How do we slow the rate of growth?" but "How do we re-establish a truly workable balance between population size and water, population size and grain, population size and seafood?" Lester Brown says, "There's no one answer, of course. How many people the earth can support depends on how those people live. If they eat like Americans, the world can grow enough grain to support only 2.5 billion of them. The planet could support 5 billion Italians—they eat low enough on the food chain." (McKibben 1999).

1. How can we deal with the extreme differences in use of resources around the world? Explain how this is tied to population growth.
2. What would be some proposals to deal with high population and inequities and disparity of food, water, education that would be acceptable to Americans, some of the highest users of resources on the planet? What would be a world sustainable approach?

Conclusion

The world is in the thrust of an unprecedented population expansion. It took hundreds of thousands of years for humans to arrive at a population level of 10 million, which occurred only 10,000 years ago. This number expanded to 100 million people about 2,000 years ago and to 2.5 billion people by 1950. Within the span of a single lifetime, it has more than doubled to 5.5 billion in 1993 and to 6.5 billion in 2006 (1 billion in 13 years). The implications of this growth are enormous: some very positive, some extremely threatening, and certainly all needing planning, awareness, a world community that works together, and wise choices. It has been stated in Part V that only a few decades remain to determine the future path of population growth and all of the repercussions that this path brings with it. Part V alludes to some of the problems and solutions—from discussions of the effects of this population growth to discussions of the earth's natural limits; from the presentation of the 10-year progress report proceedings of the 1994 Cairo Conference to the information contained in case studies of Japan and China to a presentation of details of possible plans for alleviating population problems. The problems and solutions are complex, but it is hoped that Part V has introduced the issues and made the reader aware of and involved with the important task of finding and implementing some of the best courses and considerations of action.

Progress is already being made in many countries around the world as evidenced by the reduction of birth rates, higher use of contraceptives, an increase in education, and the provision of more medical aid. However, there is still much to be accomplished when dealing with the key determinants of rapid population growth such as poverty, high childhood mortality, low status and education of women, low adult literacy, and various other negative cultural influences on progress. Our common goal is to improve the quality of life for all people and future generations, thus protecting human social, economic, and personal well-being. This goal involves the participation of all of the people on earth to use science and technology wisely, find substitutes for wasteful practices, educate others about protecting the earth's resources, and actively protect the natural environment. What happens in Africa or Asia or Puerto Rico affects us all . . . two and one-third more of us every second.

OUR POPULATION: A SOCIAL–ECONOMIC VIEW

Environmental degradation is generally the result of attempts to secure improved standards of food, clothing, shelter, and comfort for growing numbers of people. The impact of the threat to the ecosystem is linked both to human population size and resource use per person. The issues, therefore, of resource use, waste production, and environmental degradation are accelerated by factors of population growth and further exacerbated by consumptive habits, certain technological developments, social habits, and poor resource management.

As our population continues to increase, the chance of the occurrence of irreversible damaging changes also increases. There are already indicators of severe environmental stress caused by the growing loss of biodiversity, increase of greenhouse gases, deforestation, ozone depletion, acid rain, loss of topsoil, and world shortages of water, food, and fuelwood. The developed countries, with 85 percent of the gross world product but only 23 percent of the population, account for the greatest use of these resources (Graham-Smith 1993).

As developing countries seek to adopt standards of living and patterns of resource utilization from the developed world, the people of the world are further intensifying their already unsustainable demands on the biosphere. For the future, it is important that economic value be placed on the use of the earth's resources in a manner similar to that proposed by Green Net National Production or the net national product. Some equitable measures must be placed on the value of those resources that can sustain our quality of life. Productive activities and models to create a sustainable pattern of use of our natural resources must happen globally, coordinating the efforts of both the developed world and the developing world alike. The placement of monetary values on natural resources would point us in a positive direction toward conservation, thereby equalizing national population imbalances and providing for intelligent resource use.

ISSUES OF THE FUTURE

As we attend to the challenges of increasing our food supplies, limiting population growth, and educating societies to twenty-first-century imperatives, some progress has been made. Some of these new approaches are alluded to in Part IX, which deals with technologies of the future. These methods will receive more attention and refinement as we begin to tackle the population problem with the issues of the twenty-first century.

REFERENCE

Graham-Smith, F. (1993). *Population Summit. Population—The Complex Reality.* Cambridge, England: Cambridge University Press.

INTERNET EXERCISES

1. Use any of the Internet search engines (e.g., Google, Yahoo,) to research the following topics.
    a. Various population time frames ("popclocks") mentioned in Part V. Locate other resources that describe population issues. Compare the various pieces of data and determine the reasons for any discrepancies.
    b. Malthus and his predictions
    c. Three-stage demographic theory
    d. Lester R. Brown and his work with world food projections
    e. Other world demographic charts not provided in Part V
    f. Future predictions of world population growth

g. Work of the United Nations Population Fund
h. Work and predictions of the International Union for Conservation of Nature and Natural Resources (IUCN)
i. Work of the Group of Experts on the Scientific Aspects of Marine Pollution (GEASMP)
j. Figures available from the United Nations Educational, Scientific and Cultural Organization (UNESCO) on population growth, food supply, and education
k. Work of the Population Council in New York City
l. Work and predictions of Anne H. Ehrlich and Paul R. Ehrlich
m. The Club of Rome's Project on the Predicament of Mankind
n. Additional information on the 1994 International Conference on Population and Development in Cairo, Egypt
o. Any country in the world not discussed in Part V. Research its population growth, and find information on its cultural attitudes about population growth as well as its socioeconomic positions.
p. Additional information about the pros and cons of birth control
q. Latest developments in birth control, as well as its relationship to various religions and countries
r. Carrying capacity of the earth
s. Global Research on the Environment and Agricultural Nexus (GREAN)
t. I PAT
u. Status and education of women and birth control
v. Abortion rates throughout the world in various countries and their cultural views on abortion
w. Poverty and population rates around the world
x. Green Net National Production (NNP)

2. Use any of the Internet search engines (e.g., Google, Yahoo) to research the following topics.
   a. Birth rate
   b. Death rate
   c. Stabilization of population
   d. Demography
   e. Developed countries
   f. Developing countries
   g. Growth rate
   h. Infant mortality rate
   i. Projections
   j. Vital events
   k. Vital rates
   l. Fertility rate
   m. Exponential growth

n. Sink (the process of a system)
o. Birth control devices and approaches
p. Carrying capacity
q. Environmental degradation
r. Sustainability

USEFUL WEB SITES

http://www.ibiblio.org/lonarbin/worldpop	World population estimates
http://www.census.gov/main/www/popclockw	World population estimates—U.S. Census Bureau
http://popindex.princeton.edu	Extensive population database
http://www.cpc.unc.edu	Population Research Institute
http://www.overpopulation.com/faq/	Population Issues and Statistics
http://www.unfpa.org	United Nations Population Fund
http://www.prb.org	Population Reference Bureau
http://www.statistics.gov.my/English.popclock/world.htm	U.S. government's population statistics of the world.

Boundless humanity.
Photo courtesy of the United Nations.

PART
VI

War, Politics, and Technology

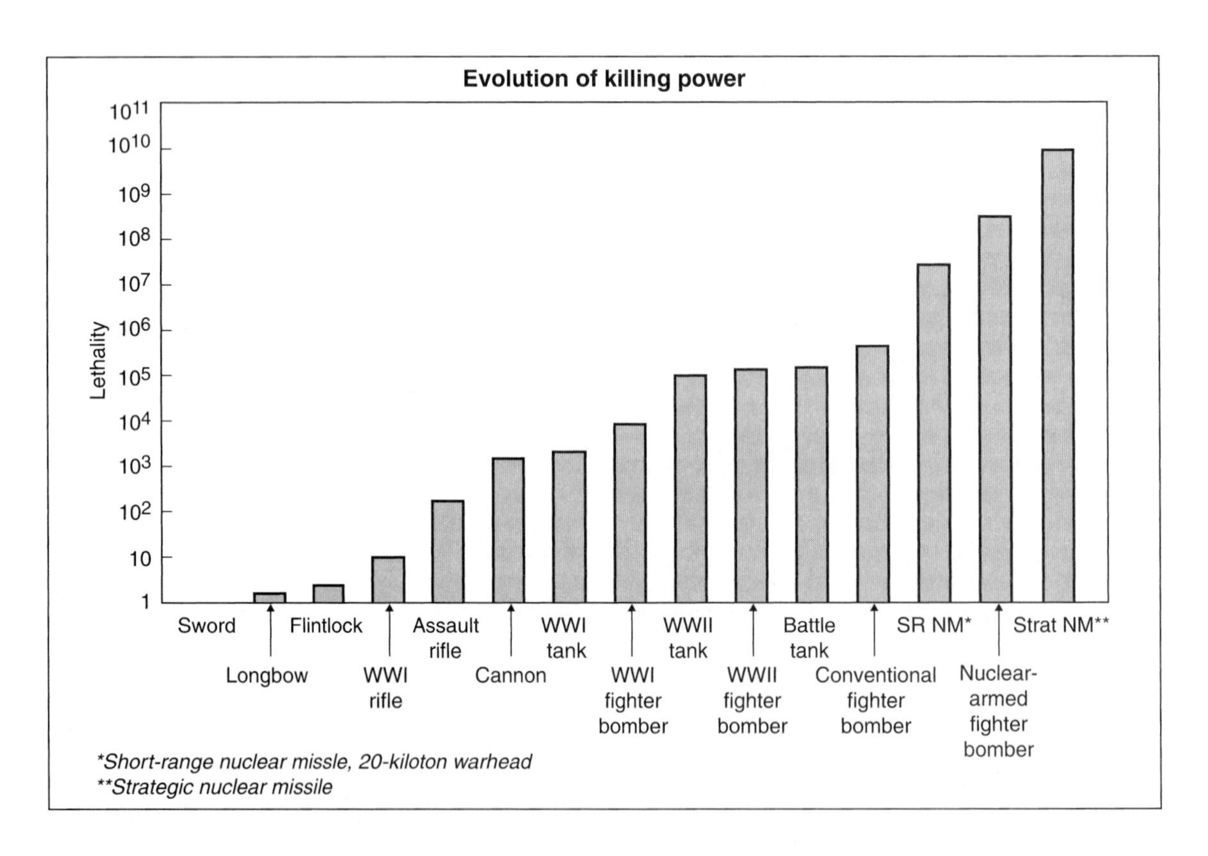
Evolution of killing power
Lethality
10^{11}
10^{10}
10^9
10^8
10^7
10^6
10^5
10^4
10^3
10^2
10
1
Sword
Longbow
Flintlock
WWI rifle
Assault rifle
Cannon
WWI tank
WWI fighter bomber
WWII tank
WWII fighter bomber
Battle tank
Conventional fighter bomber
SR NM*
Nuclear-armed fighter bomber
Strat NM**
*Short-range nuclear missle, 20-kiloton warhead
**Strategic nuclear missile

OBJECTIVES

After reading Part VI, War, Politics, and Technology, *you will be able to*

1. Define Dominant Client, Dominant Regulator, and Monopoly Maker.
2. Distinguish between a Dominant Client, a Dominant Regulator, and a Monopoly Maker.
3. Identify the impact made by technology on society in time of war.
4. Formulate opinions about the role of technology in warfare.
5. Assess the state of the relationship between technology and society both in times of war and peace.

INTRODUCTION

A World War I Tank. Tanks like this British M-1 tank first appeared in the early stages of World War I. They were slow, dangerous for their occupants, and fell victim more often to mechanical difficulties than to enemy fire. However, they changed the way in which wars would be fought in the future.

Courtesy of the U.S. Signal Corps.

It was probably a clenched fist that ended the first recorded argument. Soon the fist was replaced by a stick, and the stick was modified to accommodate a rock, which was further modified to sport a sharpened blade which would then be thrown. In time, all of the above would be rendered useless, as the fist, the stick, the rock, the club, and the spear were replaced by the sword, the bow, the flintlock, more sophisticated firearms, artillery, bombs of all shapes and sizes, airplanes, missiles, and so on. As the weapons evolved, so did the scale and scope of the conflicts in which they were used. Disputes ranged from basic disagreements over boundary lines to clashes involving regions, nations, and continents, until the entire world was occasionally consumed by conflict.

The change in scope also precipitated changes in costs. Wars could no longer be spur-of-the-moment affairs, underwritten by a quick withdrawal from a nation's petty cash drawer. Government budgets would have to reflect and plan for the possibility, if not the probability, of war. This created a ripple effect up and down society, and those ripples are still evident today. Leaders realized long ago that weaponry was a necessity and that the more they had and the more lethal they were, the better. But buying security by investing in ever-changing, state-of-the-art weapons technology presented new problems for the leaders of old, as well as their descendants. One of the first things they realized was that sophisticated weaponry was all but useless in the hands of an untrained mob pressed into temporary service. Enter the trained, professional army, followed shortly or perhaps accompanied by industries dedicated to supplying the army with tools of the trade. It seemed that, as the rise of the nation-state was occurring, the world was also being shaped by another institution: the military-industrial complex. This institution would influence what technologies would be used in war, how much they would cost, and how soon they could be used. Before long the trinity would be complete: a union of those who wielded power, those who wielded the sword, and those who wielded the forge—meaning in effect an alliance between political, military, and industrial powers. No longer just an exercise in courage, bravery, or national pride, wars became exercises in administration and execution because of their cost, which could deprive civilian populations of needed services and entailed sacrifices in their execution. also.

Part VI is about the relationship between war, technology, and politics. As you read along, you will see that Part VI makes several points.

1. As technology dominated warfare, the waging of war became as methodical and calculating as it was passionate.
2. As technologies became more expensive, civilian populations were sometimes called upon to sacrifice. Some nations enjoyed the luxury of guns and butter, while some were forced to choose between guns or butter.
3. The acquisition of military technology comes about in much the same was as non-lethal technologies. There are institutions which recognize the need for military technologies, institutions that provide those technologies, and institutions that make sure the technologies do everything promised.
4. As technologies became more commonplace on the battlefield, the human element in conflict was called into question. In effect, was the use of technology a sensitizing or desensitizing factor? Are human casualties necessary in high-tech wars? Can military

objectives be reached just as effectively by using a missile rather than a man? And if human casualties are drastically reduced by the use of weapons technologies, will political leaders be more or less inclined to take their nations to war?

5. Technology continues to develop even after the need for such evolution apparently stops. The ethical issue of weapons development to keep up with competing nations needs to be addressed.

Part VI also offers a general overview of how today's military (primarily that of the United States) gets the technology it needs. It also looks at some examples of military technologies and their development over the years and then concludes with a look at innovations on the horizon that may change the face of war.

More War in a Moment, but First, a Word from Our Sponsors

JOHN MORELLO

Some people think that military technology, like its civilian counterparts, is a result of an autonomous, self-directed force. That assumption is far from the actual case. Regardless of what technology is under consideration, they are all propelled by people and institutions in charge of their management and who benefit from their growth. Simply stated, technologies are shaped by the forces around them. Technologies need support from groups and organizations that invest resources in developing and managing them. Ron Westrum, in his book *Technology and Society: The Shaping of People and Things*, gave institutions which provide such a service a name: sponsors. Sponsors are institutions that help to develop technology and decide how and when society will use it. From initial conception to final use, most technologies are shaped by these sponsors and reflect their influence. Sponsors come from various places: government, private industry, the legal community, consumers, and even educational institutions. You may never have thought about this, but the school you attend is a sponsor of technology, and the product of that sponsorship is none other than yourself. That is not to say that each of these types of sponsors plays a role in the development of military technologies, but enough of them are present throughout the process to make them all worth investigating.

Westrum has identified three classes of sponsors: Dominant Clients, Dominant Regulators, and Monopoly Makers. Dominant Clients are those customers that, given their size and special needs, exert a major influence in the shaping of technology. This kind of relationship is evident between manufacturers of military products and their customers—the various branches of the military. The F-117 Stealth aircraft is a perfect example. The Air Force, or Dominant Client in this case, announced it was looking for an aircraft capable of making itself invisible to radar. It also mentioned other features it thought necessary to the success of the aircraft. Whatever aerospace corporation could produce such a plane would win the contract. In the Stealth case, the contractor happened to be Lockheed, which has made many F-117s over the last two decades.

Westrum also identified a class of sponsor called Dominant Regulators. Dominant Regulators are a different kind of sponsor, but they still play a role in the relationship between technology and the military. One can count among the ranks of Dominant Regulators government institutions such as the Food and Drug Administration (FDA), which ensures that the food you eat and the medicines you take are safe. The FDA can order foods and drugs off the market if it feels they are unsafe for consumers. In the case of the F-117, or any military technology for that matter, the Dominant Client is often the Dominant Regulator as well. Before the Air Force took delivery of the F-117, it was put through its paces by Air Force test pilots. They reported their findings to their superiors, and if something wasn't right, Lockheed was told to fix it. If for any reason the Air Force decided the F-117 did not meet its requirements, it could reject the aircraft, thereby exerting its influence as a Dominant Regulator.

The efforts of Dominant Regulators are determined by the amount of access consumers have to technology. The FDA prowls the marketplace in defense of millions of consumers. The military, when performing its role as a Dominant Regulator, is not always as vigilant. It's not often you'll read about the military denying itself the technologies it feels it needs, but sometimes it's called upon to exercise some restraint. You will read more about the military's struggles to balance restraint with readiness in Chapter 44 about the F-22, the replacement for the F-117. Sometimes, when restraint isn't exercised, the consequences can be deadly. Several years ago, the Marine Corps wanted an aircraft capable of taking off in short spaces. The Osprey, as it was called, was designed to carry more troops than a helicopter as well as their equipment, insert them in the battlefield, and take off vertically, if necessary. But the project was plagued by technical problems, and during a test flight, an Osprey crashed, killing a number of Marines onboard. Only later was it learned that the Marine Corps knew about the problems but pressed on with development. In effect, it had ignored its responsibility as a Dominant Regulator in favor of its desires as a Dominant Client. Years ago in Vietnam, soldiers reported that their M-16 rifles jammed in combat. The original form of the M-16, the Armalite AR-15, had been developed to fulfill the need for a high-velocity, small-caliber weapon that would be ideal for jungle warfare. However, the Army lobbied for the M-16, claiming that the AR-15 was inaccurate and did not have the desired range. The lobby effort was successful; the M-16 became the weapon issued to most combat personnel in Vietnam.

The third and final sponsor class that helps produce technology for the military is called the Monopoly Makers. Simply stated, because a corporation holds such a high percentage of the market share in a given product, any time it introduces a new technology, it can be reasonably sure that the new product will be used by customers already using its products. While the case of AT&T and IBM seem fairly obvious, the world of suppliers of military technology is a little different. No one but Lockheed can produce the technology needed for the F-117. Subsequently, when the military needs the tools to make their forces invisible to the enemy, their options are limited. Therefore, when the military does shop for new technologies, it's going to be working from an extremely short list of vendors, vendors who know that their product is going to be used by a large number of military personnel. If there's a wrinkle to this pattern, it's usually found when various branches of the military can't close ranks on a single technology to deal with a common situation. For example, the Navy, Marines, and Air Force all use aircraft, but have from time to time chosen three different vendors to meet their needs rather than join forces to get a single, all-purpose aircraft. This divergence has produced a certain amount of overlap in the military community, but with more aggressive oversight, it is gradually becoming a thing of the past.

Over the years, the sponsorship of technology by the military has followed the same rate of growth as that of its civilian counterparts. Its Dominant Clients tell producers of technology what it wants the technology to do, what it should look like, and what it should cost. Dominant Regulators ostensibly should control the conditions under which the technology will be used, as well as ensuring that it meets all the requirements imposed prior to production. Finally, Monopoly Makers maintain a fair amount of proprietary control over the technology and should impose some degree of uniformity. There are, however, some notable exceptions between the military and its civilian counterparts. Only until recently have cost, efficiency, and standardization been adopted as criteria in the military procurement process. In addition, one standard that has been unique to the military process has been the search for technologies with an ever-increasing potential for destruction. You see, in the military, the need to be one step ahead of the competition has less to do with market share and more to do with superiority and survival.

REFERENCES

Easterbrook, Greg. (1984). The Airplane That Doesn't Cost Enough. *Atlantic Monthly,* August.

Fallows, James. (1981). *National Defense*. New York: Random House.

McNaughter, T. J. (1984). The M-16 Controversies. *Military Organization and Weapons Acquisition*. New York: Praeger.

Westrum, Ron. (1991). *Technology and Societies: The Shaping of People and Things*. Belmont, California: Wadsworth Publishing.

QUESTIONS

1. Discuss how the U.S. Air Force performed its role as a Dominant Client in its effort to produce the F-117.
2. Discuss how it might be possible for a Dominant Client to serve as a Dominant Regulator simultaneously.
3. What role did Lockheed play in the F-117 project?

Tanks: The Evolution of Mechanized Warfare

JOHN MORELLO

If inventions are truly nonobvious improvements over existing practice, then the tank is probably a good reminder of just how the process works. The tanks seen today on television news programs bear little or no resemblance to the tanks of old. In fact, the first tank wasn't really a tank at all.

American engineer E. J. Pennington's late nineteenth-century design looked more a like a boat than anything else, with bathtub sides, guns mounted front and rear, and perched on a four wheeled carriage. At the time, it was an impressive picture, but that's all it was. Pennington couldn't win financial support from the U.S. government to turn his sketches into reality. So it was up to British engineer F. R. Simms to bring the creation to life. In 1899, he introduced the "War Car," a motorcycle powered by a four-stroke Daimler engine with a machine gun mounted behind an armor shield. A year later, Simms showed off a more deluxe version of the War Car, encased in armor, bristling with cannon and machine guns, mounted on four steel-tired wheels, and powered by a 16-horsepower Daimler engine. The age of armored, mechanized warfare had officially begun.

Yet despite some intriguing developments, mobile warfare continued to be executed using horsepower that required saddles and oats. World War I, however, changed all that. The "war to end all wars" was also one in which military leaders and their industrial colleagues first glimpsed the potential of mechanized warfare. Along the way, some unusual military-industrial alliances were struck as corporations experimenting with automobile development became interested in tank development. For example, the British Army joined forces with Rolls Royce, while the German Army signed on with Daimler Benz.

In the first few months of the war, wheeled armored cars enjoyed some success, but their limitations in the combat conditions of the new type of war soon became obvious to all. Trench warfare, the machine gun, poison gas, barbed wire, and especially the mud of no-man's land made them useless. Although many of these obstacles would be overcome in time, it was the lack of mobility that made wheeled armored cars obsolete. Tracked vehicles would have to be produced (Reid 1976).

That's when the British Navy stepped in and, with the help of the Royal Engineers, produced the next generation of armored vehicles. The result was, at that point, an unimaginable contraption. It was a machine of geometric proportion with steel tracks on each side, coated with armor and loaded with cannon and machine guns. At first, it was called "Centipede," then "Big Willie," and finally "Mother." The tank was born. In September 1916, British M-1 tanks went into battle. Weighing in at 28 tons, the tanks carried a crew of eight, half of whom fired the two six-pound cannons and four machine guns. They spanned the 12-foot enemy trenches without mishap, negotiated the barbed wire and shell holes, and terrified the Germans, who quickly surrendered. Tracked vehicles would now be a new and necessary weapon. By the end of the war, not only Britain, but also France, Germany, and the United States were using tanks. The United States bought hundreds of French-built Renaults and placed many of them under the command of Lt. Colonel George S. Patton, Jr. Patton, a former cavalryman, proved the tank's value not only as a supporter of advancing ground troops, but also as an independent assault weapon.

But whatever momentum tanks may have gained during the war was quickly lost once the shooting stopped. Most nations, tired of war and the financial burdens it imposed, officially halted further research and development. Both France and Britain put off new development, despite protests from military thinkers. Things were no better in the United States, where the upper echelons of the military structure also discounted the technology's importance. The experience of J. Walter Christie was a good case in demonstrating how reluctant the United States was in embracing technology that would have a major impact on modern warfare. Christie had studied the tanks of World War I and created a more modern fighting machine. Among his innovations were a suspension system that allowed the tanks to be more mobile in all types of terrain—as well as making the ride easier on the men inside, a more powerful engine, and a bigger gun mounted on a rotating turret.

But the military hierarchy wasn't interested, and junior officers who persisted in promoting the tank were warned that such views could prove harmful to their careers. When Dwight Eisenhower, then an Army major, wrote an article urging the Army to take a greater interest in mechanized warfare, he was severely rebuked for challenging military doctrine. Even Patton, who had recognized the value of the tank early on, transferred back to the cavalry. Christie, desperately in need of cash, sold his design to the Soviet Union, which used it to develop the T-34, one of the best tanks built during World War II (Fleming 1995). The T-34 would become the first in a new series of Main Battle Tanks (MBTs). MBTs would combine mobility, lethality, and survivability and were destined to be the wave of the future in terms of tank development. The T-34's construction, simplicity of operation, and firepower of its 76-mm gun proved to be up to the challenge posed by Germany's tanks when that country invaded the Soviet Union in 1941.

German military thinkers had discovered early on the value of armor, especially if it could be combined with infantry. Even though the Versailles Treaty limited the size of its armed forces, Germany still experimented with the idea of using tanks captured from Britain in World War I. Tanks, supported by infantry, could go deeper into enemy lines, wreaking even more havoc. And, in addition to attaching troops to tanks, the Germans took the technology a step further. They created entire divisions of tanks, with motorized infantry and artillery in supporting roles. It would take several years of fighting and dying before the Allied powers developed tank technology to challenge the German's Tiger and Mark series of tanks, each sporting heavier armor and greater firepower than the American and British Shermans, Churchills, and Cromwells (Frankland 1989).

The experience of those American and British tankers facing and surviving superior German technology in World War II contributed heavily to the campaign in those two countries to achieve tank superiority in the post-war world. Yes, air power could stop tanks. But what if bad weather grounded the air force? In the final analysis, went the conventional wisdom, tanks could only be stopped by better tanks (Fleming 1995). The Cold War, which pitted the Soviet Union against the United States on a variety of levels, included the race to build a better tank.

However, like before, U.S. tank development in the post-war years reflected neither advances in the available technology or progress being made by other nations. The wars that America fought in places such as Korea and Vietnam featured terrain not usually associated with extensive mechanized warfare. Additionally, air power, despite its earlier limitations, had become the key piece in the U.S. defense puzzle. Therefore, armored technology took a back seat to air-power technology. The result was a progression of U.S. tanks, the "M" series, that were big, slow, under-equipped in terms of firepower, and fueled by gasoline, making them instant fireballs if hit in the right spot. The M-60 was the last of the series and was described by General James H. Polk, writing in a 1972 issue of *Army Magazine*, as "an inferior tank, part of a tired old second-rate series."

Other countries were not so reluctant to invest in tank technology. The Soviet Union took their T-34, which had performed so well during World War II, and spun it off into the T-54, 55, 62, and 72. The "T" series, despite its cramped quarters that occasionally exposed crew members to dangerous moving parts, were formidable vehicles, with well-sloped armor, low silhouettes, and a weapons system that produced deadly accurate results. And, if imitation is the sincerest form of flattery, consider this: the Israeli Army, which in its wars with its Arab neighbors scored brilliant victories using armor, thought so much of the Russian-built T-62 that it incorporated captured T-62s into its own armored divisions.

But just as tank technology was developing, so too were efforts to stop them. A whole new generation of antitank weapons has been appearing over the years, featuring sophisticated armor-piercing warheads. The new developments had some tank advocates wondering if their vehicles were destined to be museum pieces instead of battlefield pieces. Developers of tank defense technology had to look at the problem from a whole new angle—literally. British researchers discovered that angling the layers of armor on a tank could deflect anti-tank rounds. The discovery breathed new life into tank development and new enthusiasm among American engineers about their chances for a state-of-the-art vehicle.

That enthusiasm was ultimately manifested in the M1A1 Abrams tank. Named after General Creighton Abrams, a former Army Chief of Staff and a veteran of tank operations during World War II, the M1 was a tank that combined mobility, firepower, and the highest level of survivability for its crew. Although the first M1s were delivered in the 1980s, it was not until the Gulf War in 1991 that engineers would learn whether or not their design would live up to expectations. They were not disappointed. As the M1s rolled into battle, they went equipped to deal with a variety of challenges which Saddam Hussein and his army threw at them. The tanks were fitted with an internal cooling system that also protected the crew against nuclear, chemical, and biological warfare. They cruised along smoothly at 41 mph, supported by torsion bar suspension, rotary shock absorbers, and extra-long tracks. The M1 was also under 10 feet tall, making them difficult for enemy gunners to target them. The four-person crew used a thermal imaging system to find and track their targets. The 120-mm main gun had a range of 2,500 meters and could be fired on the run with accuracy. The air-cooled engine ran on a number of fuels including diesel, was quieter, and produced less exhaust. And despite concerns that such a sensitive engine might be prone to breakdowns, especially with all the sand and dust in the area, those breakdowns were generally avoided by regularly changing the filters.

The tank has finally come of age in an age of high technology. Starting with the days of the Mark 1 for the British, the Shermans and the "M" series for the United States, and the "T" series for the Soviet Union, mechanized warfare has taken on a new look both inside and outside. The vehicles are faster, more agile, more durable, and far deadlier. The crews are better trained as well. And, despite those who might criticize the high-tech features of the new generation of tanks as well as their price tags (M1s cost about $4 million each), there is little question as to whether any weapon other than a fully equipped tank with a well trained crew can consistently stop a similar vehicle. The nearly 90 years of technological trial and error have finally produced tanks that can truly be said to be up to the task.

REFERENCES

Fleming, T. (1995). Tanks. *Invention & Technology*. Vol 10, No. 3 (Winter).

Frankland, N. (1989). *The Encyclopedia of Twentieth Century Warfare*. New York: Crown Publishers.

Reid, W. (1976). The Lore of Arms; A Concise History of Weaponry. New York: Facts on File.

QUESTIONS

1. The tank made its battlefield debut during World War I. Discuss how the tank was a combination of the efforts of inventors, Dominant Clients, and Monopoly Makers.
2. What was J. Walter Christie's contribution to the modern tank? Would his efforts classify him as an inventor?
3. How does the M1A1 Abrams tank reflect the changing nature of warfare today?

Extreme Makeover: The New U.S. Army: Leaner, But Meaner?

JOHN MORELLO

Nothing says power quite like a tank. And in this case, nothing is more tank-like than the Army's M1A1 Abrams. Weighing in at 70 tons, it's the biggest tank the Army has ever produced. It's built low to the ground to make it harder to hit and jammed with electronics to allow the crew to look for targets while making sure they don't become one themselves. Blessed with impressive speed, a suspension system that allows it to travel swiftly and smoothly, and able to lob 75-pound high explosive shells with deadly accuracy, the M1A1 Abrams, in the words of one Army officer, dominates land warfare. Too bad it's becoming obsolete.

It's not so much that the tank itself has become obsolete as the wars it was built to fight have gone out of style. There's no question that Operation Iraqi Freedom and before that, Operation Desert Storm, proved the value of tanks like the Abrams. Cruising across the desert, knocking out enemy vehicles as it went, and coming to a halt atop a sand dune so that rugged-looking tankers could pop out of their hatches and survey their accomplishments made the Abrams a legend and good television to boot. But those wars are all in the past now, and as the Army prepares for the future, it's looking like the Abrams won't be going along for the ride.

There were signs during the Cold War that the U.S. Army in general and the Abrams in particular had gotten sluggish. Before the tanks could even hope to take on their Soviet counterparts, an army of engineers had to be deployed in front of them, strengthening bridges and widening roads. But the new wars the Army thinks it will be fighting will be different. They will involve mostly urban conflict, where the battlefield could be a village square, a narrow street in some bombed-out city, or even a dark alley. In that environment, stealth and speed may be more important than size and strength. Expecting a tank weighing as much as 28 Chevrolet Suburbans to navigate in that environment isn't realistic. And expecting the Army to continue to depend on such technology in the wars of tomorrow is just as unrealistic. Tomorrow's soldiers are expected to be more than just warriors. They'll also be expected to be peacekeepers, separate rival factions, and be the first wave of an American humanitarian relief program. Trying to do all of that in the shadow of tanks the size of an Abrams may send the wrong message. So the Army, with some degree of fanfare and perhaps an even greater degree of anxiety, has begun to put those war horses out to pasture in search of something smaller and swifter to ride into battle.

Enter the Interim Armored Vehicle or IAV. It's the first step in a journey of technological evolution the Army hopes will lead it to a final solution to its armor crisis. It comes in a variety of styles. Some will carry missiles, while others will transport troops. Still others will be equipped to evacuate wounded soldiers, support advancing troops, and deal with nuclear, biological, or chemical warfare. An IAV weighs 20 tons and is nearly 70 percent lighter than an Abrams, which makes it easier to fit into a C-130 cargo plane, and even easier to be crippled by something as small as a rocket-propelled grenade. That last bit of information has a lot of armored personnel worried. In the belly of an Abrams, crew members felt indestructible; in the belly of an IAV, not so much. But if the Army is to meet its own goal of being able to deploy a full brigade anywhere in the world in 96 hours, capable of accomplishing multiple objectives, then it may be time to revise the "lean, mean, fighting machine" image into something looking smart as well as lean.

So how does the Army pare down the size of its tanks and still stay safe and sharp? Better intelligence-gathering technologies. The Army thinks good intelligence can be useful, if not more useful than a fleet of tanks with all the armor plate in the world. Knowing what's going to happen before it happens means, for one thing, a massive savings in deployment costs. In today's wars, when tankers button up and rumble into action, they have to seize and hold huge amounts of territory, establishing an almost unsustainable defensive perimeter to guard against enemy counterattacks. It also requires a massive logistical effort to keep supplies coming. By using new intelligence-gathering technology, they can seize strategic areas, identify and destroy anti-armor units the enemy might deploy, and preserve logistics and support for another day.

But how do they get the information? Some of it will be gathered from above, courtesy of satellites and unmanned aerial vehicles. The rest will come from newly formed scout units whose job will be to approach the enemy and send digital pictures of the deployment back to headquarters. Knowing where the enemy is before you go into battle may level the playing field for the new generation of tanks and the men fighting in them. It will also help them lay in wait to spring an ambush rather than confront enemy armor head on. Or it may just keep them out of harm's way altogether, allowing them to shadow enemy tanks, relay their positions to the rear, and let missiles take them out.

Right now, the Army has converted one of its armored brigades to this new way of waging war. It plans to do the same with six to seven more. The big, lumbering Abrams are going, and with them a heightened chance of survivability for the crew and the tools to punish an attacker. In their place will be the lighter IAV, slim, maneuverable, and vulnerable. To keep the IAVs and the men inside them safe, the Army is going to count on a lot of technology, which is still in development. Flying drones will gather target data and download it to an unmanned rocket launcher, which will then open fire. But the shift in hardware may be easier than the shift in personnel attitudes. There is still considerable concern about the safety of the men and the effectiveness of the new technology. It seems as if, before the Army tries to defeat its next enemy, it must first win over its own troops.

REFERENCE

Newman, Richard. (2001). After the Tank. *U.S. News and World Report,* September 18, 2001.

QUESTIONS

1. How has the changing nature of modern warfare affected tank technology?
2. What are some of the obvious concerns regarding use of IAVs on the battlefield?
3. Discuss how the Army will use intelligence-gathering technologies to compensate for the loss of traditional armored vehicles.

The F-22: The Plane Even Congress Couldn't Shoot Down

JOHN MORELLO

44

When Congress passed a funding bill for the construction of the F-22 Raptor, the event was followed by a collective sigh of relief from officials of both Lockheed Martin and the U.S. Air Force. Those two institutions joined forces over a decade ago with one goal in mind: to create a new generation of fighter aircraft to protect and defend U.S. interests in the twenty-first century. Powered by two Pratt and Whitney F-119 PW 100 engines, the F-22 can reach the Mach-2 range, allowing it to cruise at speeds today's fighters can only briefly maintain. It will also be the first agile, supersonic fighter to be stealthy, more so even than its cousin, the F-117 Stealth fighter-bomber. In addition, the F-22 will sport a lethal weapons system, including Advance Medium-Range Air-to-Air Missiles (AMRAAM), monitored by the latest computer technology and topped off by a helmet-mounted display to supply pilots with target data wherever they look (Sweetman 1995). It even comes with space for two crew members. It's all very impressive, very complicated, and very expensive. The F-22 could end up costing the Pentagon and ultimately the American taxpayer $100 million apiece, making it one of the costliest weapons systems in the Defense Department's history (Callahan 1992).

And fiscal problems aren't the only issues worrying Pentagon officials. Because the F-22 is loaded with these state-of-the-art features, will anyone be able to fly it to its fullest potential? Take the advance stealth characteristics, for example. It gives the pilot plenty of flexibility, but at the same time a lot to think about. He must weigh the opportunities to attack against the need to remain hidden. Also, depending on the distance, different radar systems can lock on to the F-22; the plane is also more visible to radar from the side or tail than from head on. What that means is the pilot must now worry if he is at greater risk trying to avoid danger rather than meet it face to face. The decision whether to fight or flee has to be made instantaneously amidst the crushing physical G-load strains of air combat, where planes can be shot down by other planes or by SAM (Surface-to-Air-Missiles) sites from miles away. It requires so much concentration that engineers worry pilots might not have the "right stuff" mentally to operate the plane at peak efficiency.

Another issue troubling Defense Department think-tankers is just whom this plane will be used against. One answer has been the Soviet Union. The Pentagon has argued that the United States must maintain technological superiority against the possible revival of Soviet power. The chances of that seem unlikely, counter the critics. With its economy in shambles and its political structure still shaky, Russia for now does not appear to be a threat for the United States. Consequently, the Congressional Budget Office (C.B.O.), which oversees government expenditures, concluded that conditions inside Russia made the need for weapons such as the F-22 unnecessary (Callahan 1992). The C.B.O. recommended the project be scrapped and existing fighter aircraft be upgraded to meet America's future security needs.

Hold on, cautioned Pentagon officials. Despite its disintegration in 1991, the "Evil Empire," as former President Ronald Reagan called it, lives on thanks to exportable military technology. Two systems that could tip the balance in any confrontation involving the United States are the

Sukhoi Su-35, an advanced and deadly jet fighter, and the Vympel R-77 air-to-air missile. The R-77 is now acknowledged to have a longer range and greater agility than the Air Force's AMRAAM (Sweetman 1995). And now the R-73E is being tested, which is said to have even more enhanced steering capabilities. A pilot carrying the R-73E can rely on a helmet-mounted sighting device, which allows him to simply look at an enemy and press a button. The instant the missile is fired, its infrared seeker locks on to its target and veers sharply toward it, allowing little or no time for evasive action (Sweetman 1995). F-22 supporters say that, because of these up-and-coming weapons systems, their plane is needed more than ever. But critics counter by saying that, while the Russians deserve credit for producing their share of sophisticated Cold War hardware, it's not just production but maintenance that also counts. A lack of spare parts and trained personnel to keep the equipment running, in their opinion, makes Russia a less than credible threat.

But the changing geo-political circumstances haven't deterred the F-22's supporters. They point to the growing military power of China, as well as the potential problems posed by North Korea. But do any of them really constitute a threat to American interests, thereby justifying a fleet of planes which, when completed in 2012, could cost more than $98 billion (Callahan, 1992)? Chinese-American relations can be rocky from time to time, but resorting to armed conflict would be unlikely. That's not to say there hasn't been a degree of bickering the last few years, especially over human rights, business practices, and China's sale of weapons to Third World countries. But the growing economic relationship between the two countries would seem to preclude any hostilities. North Korea might seem like a better prospect for trouble. But if the country collapses or improves its relationship with the outside world, it too would deprive F-22 supporters of an arena to show what the plane can do. For awhile Iraq figured as a possible place where use of the F-22 might serve to justify its existence. The 1991 Gulf War did prove that air power would be a primary instrument for winning future wars. But a closer look at the Gulf War prompts many skeptics to suggest that high-tech doesn't always carry the day. It is far from clear, claim some analysts, that Third World societies, armed with modern weapons, present the extreme danger that Pentagon officials suggest in attempting to justify annual defense expenditures at Cold War levels. In 1991, for example, Iraq had one of the finest integrated air defense systems in the developing world, including 16,000 surface-to-air missiles and 7,000 anti-aircraft guns. It also had an advanced air force that included 900 combat planes, including some of the best French and Soviet fighters available. But when the war came, the equipment proved nearly worthless. Iraq's air defenses were destroyed; when aircraft did come up to intercept American and other Coalition Force attackers, they proved to be no match for them (Callahan 1992). And, although the Gulf War showed that highly sophisticated air power elements can speed victory and save lives, it shouldn't be an invitation for a Pentagon spending spree.

Critics argue that it's not completely clear that the new big-ticket items were any more responsible for the success of Desert Storm than some of the older, cheaper, and equally capable weapons already available. Much of the success against Iraqi tanks was due in part to the use of the F-111 bomber, a relic from the late 1960s, as well as the cheapest aircraft in the Air Force, the A-10 Thunderbolt, which has been in operation since 1976. The A-10 was an especially effective tool in destroying tanks because of its relatively slow speed as well as its fuel capacity, which allowed it to remain over the target area much longer than faster and more fuel-consuming aircraft. Additionally, the A-10 was retrofitted with modern weapons technologies that increased its striking power. The Gulf War proved that, given the weakness of Third World states, low-tech weapons might still be adequate in handling many missions on the modern battlefield. Equally important might be the fact that, in the absence of high technologies, knowing how to get the most out of existing technologies might also spell the difference between victory and defeat.

Despite the success of the more modest military technologies in the Gulf War, the intent seems to be to forge ahead with the expensive and questionable F-22. It would seem as if the project, bereft of any practical justification, is being driven by domestic political concerns, namely the company building it, the branch of the military who wants it, and those political interest groups who stand to profit from its construction. That says little for any genuine concern for global security and even less for the many other critical needs that the money could address.

REFERENCES

Callahan, David. (1992). The F-22: An Exercise in Overkill. *Technology Review* (August/September), Vol. 95, No. 6:42–49.

Sweetman, Bill. (1995). Beyond Visual Range: Flying the F-22. *Popular Science* (August), Vol. 247, No. 2:44–49.

QUESTIONS

1. What weapons system does the F-22 have that makes it different from the F-117?
2. How much is the F-22 expected to cost?
3. What potential enemies does the Pentagon think this plane will be used against?
4. What planes could be used in place of the F-22 that might be as effective, yet cheaper?
5. Comment on the following statement: "Potential enemies are far behind us on weapons development. Why do we have to develop new technology when the current technology still works?"

The Human Face of War

JOHN MORELLO

Imagine for a moment that you're leading a squad of soldiers on patrol through a bombed-out neighborhood. Your mission is to clear the area of enemy snipers. So far, the patrol has been routine. But that's about to change. You haven't found any snipers yet, but one's found you. Up on a rooftop, concealed behind piles of rubble, he sets the crosshairs on his scope and draws a bead on your head. He fires. You go down. Your squad opens fires, shooting at everything and hitting nothing, while at the same time trying to call in a helicopter to evacuate you. Unsure of the threat, air strikes are called in to pulverize the neighborhood. In the confusion, the sniper sneaks away, takes up another position, and ambushes another patrol later in the day.

Now imagine this scenario. You're leading the same squad on the same patrol. You've once again come into the crosshairs of that same sniper. He looks up for a moment, takes a deep breath, and settles in behind his scope. But you're not there anymore. While he wasn't looking, your black uniform has changed to dirty brown, blending perfectly with the debris in the street around you. The sniper can't see you anymore. But you can see him, thanks to the thermal imaging system displayed on the visor of your helmet. You raise your arm, point in his direction, and whisper "fire on target." With a whoosh, one of four 15-milimeter projectiles explodes from the pod on your wrist. Tiny heat-seeking sensors have locked on to the sniper who is now on the run, trying to take evasive action. No good. The missile reaches him and detonates, spraying deadly shrapnel 30 meters in every direction. The soldier of the past has just met the soldier of the future in a brief and deadly encounter.

THE FUTURE IS NOW

Making the second encounter less of a dream and more of a reality has been the mission of the U.S. military. The Army and the Marines have been deeply involved in these projects, acting as both clients and regulators to the efforts of Monopoly Makers such as DuPont, Motorola, Honeywell, and the Massachusetts Institute of Technology (MIT). The Army has been the driving force behind what's come to be known as the Land Warrior project, which wants to use technology to help soldiers in three ways: to make them more lethal, to improve their chances of survival, and to give them all the tools necessary to make the first two possible. In many cases, the technology is already available and only needs some modifications.

The weapons system is a good example. The standard-issue M-16 rifle, manufactured by Colt Industries, will be retrofitted with a thermal imaging system and a video camera. Thermal imaging will help locate potential targets at night and even when concealed behind cover. The technology may help eliminate the old adage, "You can't hit what you can't see," from the warrior's vocabulary. The video camera serves two purposes. When connected to the communications system located on the infantryman's backpack, live pictures can be sent from the front to headquarters in the rear. Bullets may win battles, but information wins wars. The live feed can help commanders adjust strategy on the fly and possibly intervene before troops accidentally open up on noncombatants or friendly forces. When the video camera is used in conjunction with the soldier's weapon, it will allow him/her to fire from a concealed position, exposing at most their arms and hands.

Exposure to hostile fire is an occupational hazard in the military, and the risk may never be eliminated. But Land Warrior technology may help reduce the risk. Helmets will be lighter yet more protective, utilizing new materials. Uniforms will provide greater protection, especially from small arms fire at close range. This is an important feature since the future of warfare, according to experts, indicates that it's trending toward close-quarter conflict in which combatants may use less firepower, depending on mobility, to maneuver close enough to a target to get off a shot that can't miss. DuPont manufacturing's Kevlar and Nomex fabrics, already known for their protective qualities, are being improved to further withstand heat and shredding, especially from projectiles fired at close range. In time, say DuPont's scientists, these same materials will contain high-tech fibers to detect toxins, deliver medicine, or change the uniform's color, allowing the wearer to escape detection.

But what may be most crucial to protect the warriors of the future is not firepower throw-weight, but know-weight. If knowledge is indeed power, then the power of knowledge soon to be at the disposal of the infantryman will truly be awesome. The helmet the soldier will wear will do more than protect. It will also inform. A visor will be installed, giving the soldier real-time information readouts on everything from troop disposition and targets of opportunity to text message updates on the status of fire support or medevac requests. That same visor display will hook up to the weapon, allowing the soldier to see his target without being seen. All of that intelligence comes courtesy of an attached computer that will someday be built into the soldier's uniform. Each squad leader will carry one. The keyboard will be in the front, activated by a mouse possibly woven into the shirt. The processor will be placed on the squad leader's back, just above the radio and Global Positioning System (GPS) unit. Everything that's carried will be supported on a backpack, built to conform to the body to reduce fatigue and promote easier movement. For now, the wires needed to operate the computer and the radio will be found not on the backpack, but inside it. The frame will be hollow yet virtually indestructible, taking advantage of race car technology. The wires will run through the frame protecting them from possible damage.

Of course, with any new technology, problems are bound to arise, and the Land Warrior project is no different. Battery power is a major concern. The hunt is on to find batteries light enough to be carried for extended periods of time, yet powerful enough to do the job of running the thermal imaging and video systems as well as the computer and radio. Weight is another consideration. In addition to the computer, the radio, the batteries, and the weapon, the soldier of tomorrow will also need to carry spare ammunition, food, water, and first aid supplies. The current generation of soldiers can be seen lugging packs weighing more than one hundred pounds. Engineers are trying to reduce that. There are also durability issues to consider. The typical laptop or GPS gets nowhere near the amount of abuse dished out by soldiers on maneuvers or in actual combat. The equipment needs to be as tough as the people carrying it, and so far that hasn't happened. Finally, there is the cost issue. The proposed ensemble now costs about $70,000 each. To provide it to everyone in the armed forces would probably carry a multi-billion dollar price tag. As with all technologies, prices usually come down over time. But until that happens, this uniform constitutes one expensive set of threads, especially when no one's sure just how long it can last on the battlefield.

But the possibilities hinted at by the Land Warrior technology system—lethality, survivability, and superior access to information—could be a dream come true for the military. With these features, the number of men and women in uniform, on the ground, and in harm's way could be reduced. One soldier could conceivably have the firepower of 10, with a good if not better chance of success and survival. That would be good news for the politicians who send them there, and the families who had to watch them leave. But that's just the beginning. With this technology, it might be possible to revamp and even reduce the entire size of the military, cutting the bureaucratic foul-ups and logistical hassles along the way. The result could be a truly lean, mean, and rapidly deployable military, where less is more and the chances of survival could be greater than ever before. That itself could be good news for the politicians who order the military into action and the families who have to wait for them to return.

REFERENCES

Brinkerhoff, David. "DuPont, M.I.T. Look to Equip Future Soldier." Accessed April 24, 2003. Available at www.yahoo.com.

"Military Analysis Network." Accessed March 10, 2003. Available at www.fas.org/man/dod-101/sys/land/land-warrior.com.

Regan, Michael. "Future Soldiers to Have Massive Network." Accessed May 31, 2003. Available at www.channels.netscape.com.

Vizard, Frank. "The Future of Combat." Accessed April 24, 2003. Available at www.scientificamerican.com.

QUESTIONS

1. What is the intention of the Land Warrior project?
2. What technological obstacles still need to be overcome in order for the Land Warrior project to become viable?
3. Many private vendors have provided technological assistance to the Land Warrior project. Discuss how the Army, in its role as a Dominant Client, might relate to these vendors.

Case Study 1

Agent Orange

JOHN MORELLO

Agent Orange was an herbicide used by the United States during the Vietnam War to deprive Viet Cong and North Vietnamese troops of food, shelter, and camouflage. The U.S. Army expressed initial interest in the product at the end of World War II, but did not begin testing Agent Orange until 1959. The tests proved successful, and the military concluded it had the perfect defoliation tool.

Actually, the Army had a virtual rainbow of defoliants at its disposal including Agents White, Purple, Pink, Blue, and Green. But only Agent Orange apparently had enough 2-4 D, and 2,4,5-T (also known as dioxin) to be effective enough to satisfy Army requirements. After meeting with President John Kennedy in 1960, the use of Agent Orange was permitted in Vietnam, but only as part of a civilian-run counterinsurgency program. Deployment of the chemical was not a popular decision, and it drew warnings from both administration and military officials that the United States was opening itself to charges of waging chemical warfare against the Vietnamese. So in 1961, when the United States began spraying Agent Orange over Vietnam, it was done under the careful scrutiny of White House officials. However, as the war intensified, Army officials convinced Defense Secretary Robert McNamara to give them control of the program. With the Army in charge, Agent Orange applications were conducted throughout South Vietnam.

Between 1961 and 1971, over 100 million pounds of Agent Orange were sprayed over South Vietnam. U.S. Air Force C-123 cargo planes, outfitted with 1,000-gallon tanks and bars of spray nozzles attached to the undersides of their wings, crisscrossed the country, saturating 300-acre target areas in about four minutes. In a matter of weeks, all of the plant life in the area had turned brown and died. Presumably, the defoliant was doing just what it was intended to do: deny Viet Cong and North Vietnamese the food and shelter needed to wage war. But the haphazard and indiscriminate application of Agent Orange also destroyed the crops of South Vietnamese peasants and devastated about half of the country's timberlands.

But the human toll was even more devastating. Not only were Viet Cong and North Vietnamese troops exposed, but so were South Vietnamese peasants, soldiers, and American forces. GIs absorbed Agent Orange, breathing it in as they patrolled under the planes as they conducted spraying operations, getting it on their skin while at bases where Agent Orange was stored, or drinking or bathing with water in areas where defoliation had taken place. In time, news about the dangerous side effects of Agent Orange began to surface. In June, 1969, A South Vietnamese newspaper reported that women exposed to Agent Orange were giving birth to deformed babies. That same year, an independent study conducted by the National Institute of Health reached the same conclusion. In April, 1970, the Defense Department suspended the use of Agent Orange and stopped the use of all other herbicides by January, 1971.

But it wasn't the last of the Agent Orange controversy. Years after the war, doctors and medical researchers were trying to understand why so many Vietnam-era veterans were beset with health problems ranging from skin rashes to breathing dysfunctions, various types of cancers, and birth defects in their children. The conclusion reached by medical professionals seemed to indicate that all these maladies stemmed from exposure to Agent Orange and in particular the ingredient dioxin. A 1978 CBS television documentary about Agent Orange raised even more suspicions. Vietnam veterans wanted answers, action, and compensation—three things the military, chemical companies, and the Veterans Administration were reluctant to provide. Gradually, the news about Agent Orange began to leak out.

1. Agent Orange contained dioxin, something very few people knew. Dow Chemical, a leading producer of dioxin, contacted other Vietnam War civilian contractors to warn them of the health dangers of repeated exposure. However, rather than inform the U.S. Army of the problem, Dow Chemical and the other contractors chose only to recommend that the military reduce the level of exposure among its personnel. (See the flowchart on page 445.)
2. The directions for the use of Agent Orange called for it to be diluted. Instead, the Army applied it in its full strength, ignoring not only the manufacturer's own recommendations, but its own rules as well. In addition, Army documents indicated that some areas of South Vietnam were sprayed as many as 25 times in just a few months. This dramatically increased the exposure to anyone within those sprayed areas.
3. While the military promised to stay away from civilians, resettle non-combatants, and resupply food anywhere the spraying destroyed crops, the promises were never kept.

In 1985, veterans affected by Agent Orange filed a class action suit against the Veterans Administration, an action that resulted in an out-of-court settlement. That settlement included a $180-million fund established by chemical companies involved to assist veterans with legitimate claims as well as the families of veterans who died from Agent Orange exposure. In 1993, the National Academy of Science's Institute of Medicine announced that dioxin had been linked to Non-Hodgkin's Lymphoma and a skin-blistering condition called Porphyria Cutanea Tarda. Yet despite all this, medical problems among Vietnam veterans as well as their spouses and children, continue to be reported, and Agent Orange is often considered to be a likely cause.

REFERENCES

Moss, George Donelson. (1998). *Vietnam; an American Ordeal.* Upper Saddle River, NJ: Prentice Hall.

Olson, James. (1987). *Dictionary of the Vietnam War.* New York: Parsons Press.

Schlager, Neil. (1994). *When Technology Fails.* Detroit: Gale Research.

Complete and discuss the following flowchart.

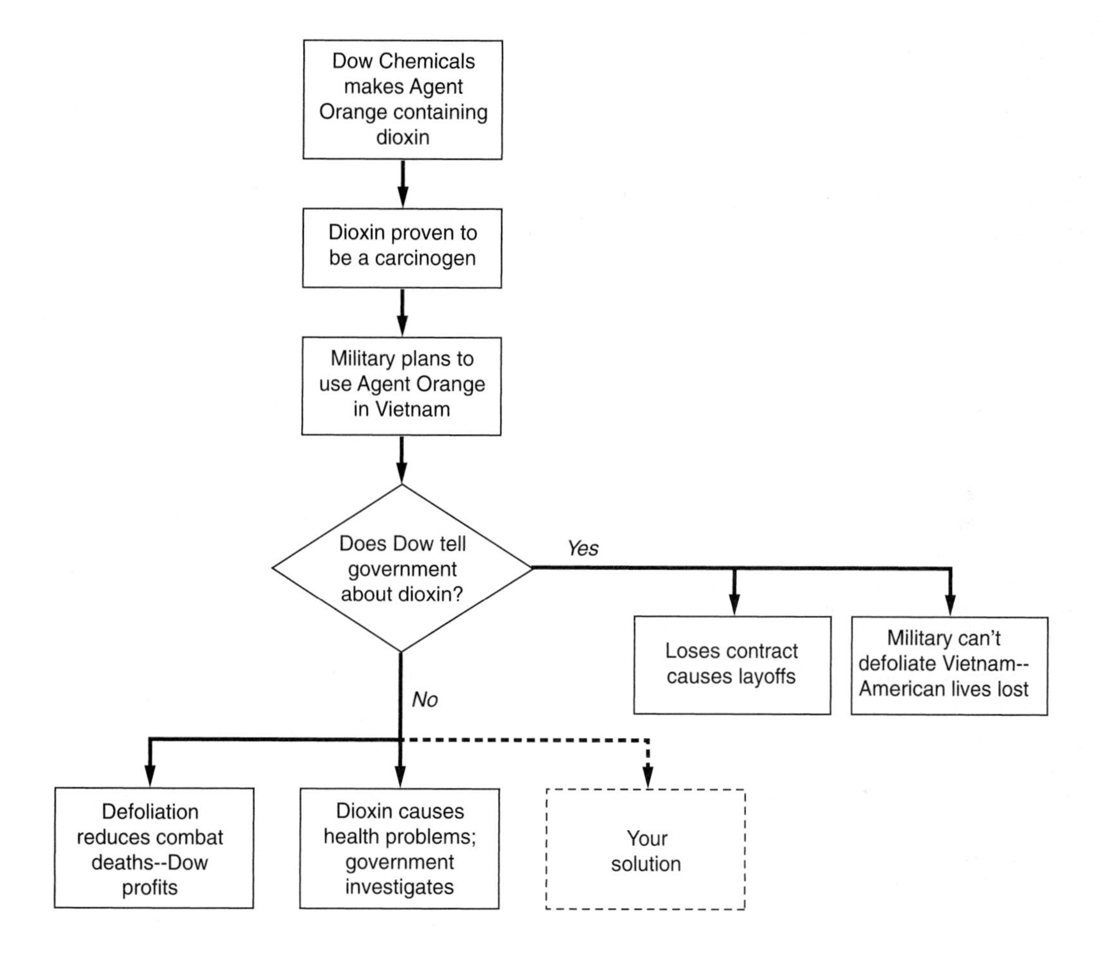

Case Study 2
The M-16 Rifle

John Morello

When technologies fail to perform the way we either want or expect them to, it's very convenient to blame the malfunction on the user and overlook the fact that the malfunction may have been caused by other factors. Sometimes design—the way something is made—is at fault, literally building a problem into the technology before the user can get his or her hands on it. And when poor design is compounded with questionable use, the result can sometimes be disastrous. The M-16 rifle might be a good example of such a technological disaster.

For years, the U.S. Army, a Dominant Client and Dominant Regulator when it came to weapons procurement and promotion, insisted that any weapon put in the hands of its soldiers should emphasize accuracy. Consequently, it took a dim view of a new weapon, the M-16. Eugene Stoner developed the M-16. He was an engineer working for the Armalite Corporation. Armalite, a Monopoly Maker in the area of weapons development, created the weapon at the request of the Army's Infantry Board, a StandardSetting Institution, which was looking for a high-velocity, small-caliber semi- or fully automatic weapon for its troops. The gun Stoner gave them was different from most weapons from conception to execution. Instead of utilizing traditional methods of assembly—namely the machining and casting of metal parts, the M-16s were stamped, pressed, and forged. And in place of wood, usually employed as a rifle's stock, plastic was used. The inside of the weapon was just as revolutionary as the outside. Like most automatic rifles, the M-16 relied on gas to propel the shells out of the barrel. Where the M-16 differed with other weapons was that, instead of using a piston to make the weapon operate, the gas was merely fed through a tube into the bolt carrier. Eliminating the piston and the wood stock helped shorten and lighten the weapon. At 39 inches in length and weighing in at six pounds, five ounces, the Infantry Board had what it thought was the perfect weapon.

But another Standard Setting Institution within the Army, the Ordinance Board, didn't like the M-16 because, in its view, the weapon was inaccurate, lacked range, and had reduced stopping power. A shell leaving the barrel of an M-16 had a muzzle velocity of about 3,250 feet per second. But after that, it was all downhill. The shell rapidly lost speed, reducing its range to 400 yards, half of what the Army considered satisfactory. The shell itself was cause for concern. The 5.56-mm cartridge was considerably lighter than the 7.62-mm slug used by America's allies or even its enemies. Any object that crossed its path, including leaves and twigs, diverted the shell's trajectory (Olson 1987). All of these factors led the Army to conclude that the M-16 should not be placed in the hands of its soldiers.

That's when Robert McNamara intervened. McNamara was the Secretary of Defense in 1962, but more importantly for the purposes of this case, he was about to pressure the Army to accept the M-16. McNamara could see down the road and knew that the next place American soldiers would be deployed would be Vietnam. There, in the jungles, surrounded by vegetation that in some cases shut out the light from the sun, accuracy and range wasn't what the soldier would need from his rifle, but rather a high rate of fire. The M-16, with its design that allowed the bullet to travel in a straight line from the moment it entered the chamber to the time it left the barrel, meant a soldier could keep his weapon on his objective, even in automatic mode, allowing him to spray his target.

Despite McNamara's influence, the U.S. Army was still reluctant to adopt the M-16 as a standard weapon until 1967, relying instead on the M-14. And for awhile, it looked like their decision was proving to be a wise one. Early use of the M-16, especially in field and more importantly in combat conditions, revealed flaws in the technology that proved potentially fatal. For starters, the gas-operated pistonless system that made the M-16 a much lighter weapon tended to jam, as gas in the chamber sometimes caused the weapon to foul up. This could only be avoided by constant cleaning. Also, soldiers who used the weapon sometimes took two of the ammunition magazines and taped them together (Olson 1987). This allowed them to pull out an empty magazine, turn it over, insert a full clip, and resume firing. But occasionally, a bullet would get caught up inside the magazine, and the rifle would jam. In both cases, lives were put at risk and sometimes lost when soldiers discovered the M-16 would not function in combat. Kept clean and fired with single clips, the M-16 proved to be an effective weapon. In 1967, the Army bowed to McNamara's wishes and adopted the M-16. (See the flowchart on page 448.) The Army had been reassured that the mechanical problems had been resolved, quite possibly because another Monopoly Maker, Colt Industries, took over production. And users were reminded that timely maintenance and proper use would guarantee successful results.

REFERENCE

Olson, James, ed. (1987). *Dictionary of the Vietnam War.* New York: Bedrick Books.

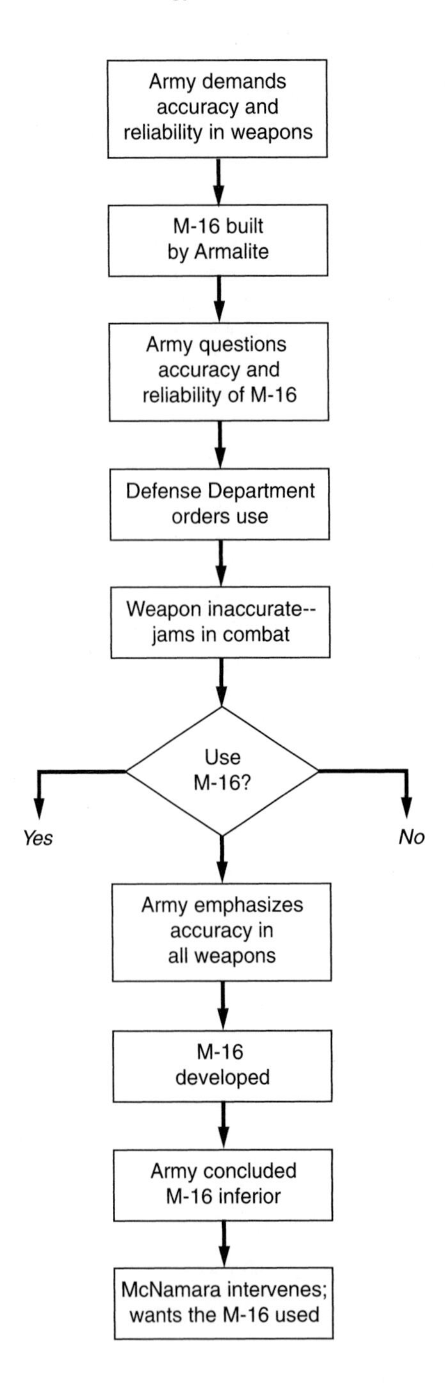
Army demands accuracy and reliability in weapons
M-16 built by Armalite
Army questions accuracy and reliability of M-16
Defense Department orders use
Weapon inaccurate--jams in combat
Use M-16?
Yes
No
Army emphasizes accuracy in all weapons
M-16 developed
Army concluded M-16 inferior
McNamara intervenes; wants the M-16 used

SCENARIO I: Friendly Fire

A weapons control officer supervising a field artillery battery in the Iraqi desert notices unusual activity on his radar screen. A column of armored vehicles appears to be headed toward him, struggling against the gale-force winds of a sandstorm. Attempts to raise the column on the radio have proven fruitless, yielding only garbled responses but no clear identification. The officer asks for air support to achieve visual confirmation, but is told all aircraft have been grounded due to the very same sandstorm. If the column is made up of friendly forces, then there's nothing to worry about. But if it isn't, they will quickly be in a position to launch an attack on allied forces. If the officer gives the command to fire, he may be shelling friendly forces; if he hesitates, his own men could be killed in an attack.

What should he do?

SCENARIO II: Corner Cutting

You are a building inspector working for the U.S. Navy in 1962. In a response to an alarming buildup of Soviet naval power, the U.S. Navy has accelerated its nuclear submarine program. Work on the Navy's last sub, the *Thresher*, is proceeding at a breakneck pace. Time, it appears, is of the essence. However, as a building inspector charged with overseeing the quality of the project, you're disturbed by what you see. Many of the boat's hydraulic valves have their open and close indicators reversed. And a disturbing number of the *Thresher's* pipe joints were pressure-tested and found to be below standards. Still, the Navy insists on having the boat ready for service as soon as possible. To report your findings could very well mean delays and possibly problems for you if you intend to continue working for the Navy. On the other hand, to keep silent could mean endangering the crew.

What should you do?

Conclusion

KINDER, GENTLER WARS?

In this part, we have examined the impact of technology on the making of war. The readings have given us a look at the relationship between the providers of military technologies, Monopoly Makers, and the regulators and consumers of those technologies, the Dominant Regulators and the Dominant Clients. In more understandable terms, we have considered the relationship between the Lockheed and Dow Chemical companies of the world and the institutions using their products, namely the military, from the top brass down to the buck private.

This part also has given the reader a shorthand history lesson on a couple of military technologies that have made their mark throughout time. Along the way, several issues were raised, at least indirectly. First, the cost, complexity, and consequences of new weapons should mean that nations deliberate more about the time and place of when to go to war. Sadly, that doesn't always happen. If anything, the growing sophistication of military technologies and the perceived reduction in human exposure to risk in combat gives the impression that wars could be waged without risk to human life. That impression may have even persuaded nations to opt for war in order to achieve national goals, believing the technology would protect soldiers from harm and their leaders from negative public opinion. But even the best of technologies can't guarantee that. War, whether it long or short, is nasty business, and someone always ends up getting hurt. And the fallout can be nuclear, chemical, biological, or even political. So, despite claims to the otherwise, nations should think long and hard about going to war.

The second issue indirectly raised in the rush for more modern and expensive military technologies is the cost it imposes on the domestic front. Although some military technologies, such as Global Positioning Systems (GPS), for example, have made the transition from lethal to leisure use, and has paid for itself along the way, it is probably an exception to the rule. Many military technologies are expensive and generally unfit for civilian consumption in their current form. So the involuntary investment citizens make through taxes provides little return, at least in the short term. In a guns and butter economy, it's possible for citizens to tolerate the cost of weapons development, as long as social services continue to be provided without interruption. They are obviously better off than are residents of guns *or* butter economies. But regardless of the vitality of its economy, no nation's patience or tolerance of denial is inexhaustible. National economies can groan and sag under the weight of defense budgets, and the procurers and providers of weapons technologies, the military-industrial complex, continues to run up the bill, blissfully unaware of the potential damage they may be doing to the public's domestic welfare.

The last issue is the human issue. War is always going to involve the human factor, yet the technologies seem to reassure us that no one really gets hurt. We know that not to be the case. Wars will always involve the human touch, augmented and supplemented by technology, and that needs to always be kept in mind, no matter how many technological bells and whistles go into the mix.

INTERNET EXERCISES

1. Use any Internet search engine (e.g., Google, Yahoo) to research the following topics.
   a. Diagram and specifications of the F-22
   b. Diagram and specifications of the A-10
   c. Amount of money the United States spent last year on defense
2. Use any Internet search engine (e.g., Google, Yahoo) to research the following topics.
   a. Diagram and specifications of the M1-A1 tank
   b. Diagram, with specifications, if available, of any World War II tank. How does that tank compare with the M1-A1 in terms of speed, crew size, armament, and weight?

USEFUL WEB SITES

http://www.mailer.fsu.edu/~akirk/tanks/ww1/WW1.html	Discusses the role of tanks during the World War I
http://www.battletanks.com	Good coverage of tanks and tank tactics during World War II
http://www.tankmuseum.org	Interactive tour of a tank museum
http://www.fas.org/man/dod-101/sys/land/mil.htm	The M1A1, America's largest Main Battle Tank
http://www.geocities.com/Pentagon/Bunker3017/air.html	Basic information on the history of warplanes
http://www.fas.org/man/dod-101/sys/ac/f-22.htm	History of the F-22 Raptor
http://www.aeroweb.lucia.it/~agretech/RAFAQ/six5th_5html	The F-22's rival, Russia's S-37
http://www.fas.org/man/dod-101/sys/land/land-warrior.htm	In-depth discussion of how the infantry will fight in the twenty-first century

Health and Technology

A DNA Double Helix Near Genetic Material.

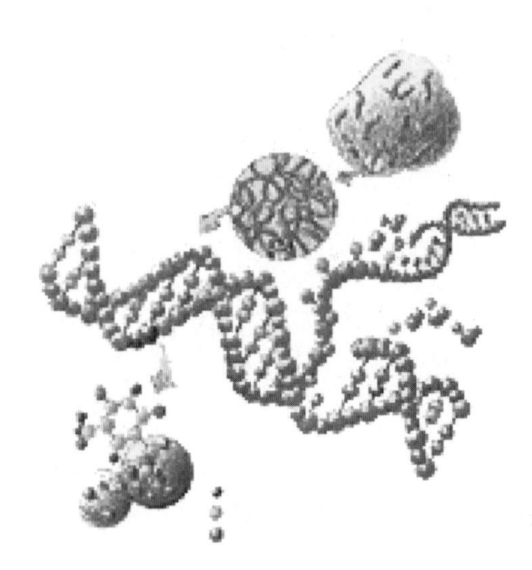

OBJECTIVES

After reading Part VII, Health and Technology, *you will be able to*

1. Identify ways to take antibiotics correctly to prevent antibiotic resistance.
2. Recall information from the Human Engineering Timeline in the first reading in this part, "Antibiotic Resistance."
3. Compare and discriminate between the benefits and harmful effects of biotech innovations.
4. Explain what is meant by "making well people 'better.'"
5. Assess the value in developing cognitive enhancers to improve memory.
6. Compare and contrast the pros and cons of genetic testing.
7. Identify the ways that a serious illness of a family member can have a financial and psychological impact on the whole family.
8. Illustrate why AIDS is a disease that the world cannot afford.
9. Address the value of telemedicine in country areas.

INTRODUCTION

You have probably heard the saying, "If you have your health, you have everything." Most would agree that our health is vital to the quality of our lives. Part VII explores the ethical considerations that doctors and patients must examine when trying to create a life of health. The interplay between health and medicines, the foods you eat, and the medical interventions that you use are explored within the framework of bioethical considerations.

In 1853, James Watson and Frances Crick determined "the double helix" structure of DNA and, in 1998, Dr. James A. Thomson isolated human embryonic stem cells that have the potential to develop into almost any kind of tissue. These amazing developments in human engineering continue to change the way we think about life, illness, and ways to treat diseases. "Making Well People 'Better'" discusses the economics of medicine and reminds the reader that pharmaceutical companies earn more money making well people better than curing people who are sick. This section also explains how the definition of ethical drugs has changed and presents examples that are easy to understand.

"The Politics of Life and Death: Global Responses to HIV and AIDS" discusses the cultural challenges faced when trying to teach teens to use condoms or drug addicts to use clean needles. Worldwide, more than 41 million children will have lost one or both parents by 2010. Now that AIDS "rivals tuberculosis as the world's most deadly infectious disease," it remains extremely important that we all learn how to prevent acquiring or spreading this fatal disease.

Are you aware that plants can be modified to contain animal genes? Experiments have been conducted in which a gene from an African frog was inserted into rhododendrons to prevent root rot. Did you eat genetically modified food today? Most likely you did because many of the grains and vegetables that you eat have been modified genetically. Human intervention in food creation dates back 8,000 years when farmers planted, grew, and harvested the hardiest foods, thereby refusing to select the weaker strains.

Modern medical technology has increased the average life span of people and, in many cases, the quality of life. Yet, it is important to analyze and study all sides of bioethical questions relating to medical care. Part VII merely touches the surface of the issues, but it should provide enough information for you to view medical care differently from this day forward.

Antibiotic Resistance

LINDA STEVENS HJORTH

"Before antibiotics were available, isolation was the only way to prevent the spread of infectious diseases. Children with tuberculosis were isolated on ferry boats during the 1920s in New York City's harbor. With the indiscriminate and reckless use of antibiotics in recent times, more and more bacteria are becoming resistant to drugs, including the microorganism that causes tuberculosis. If the trend continues, isolation may once again become necessary."
AMABILE-CUEVAS ET AL., *1995*

Have you ever gone to the doctor hoping to get an antibiotic to get rid of that hacking cough, sore throat, runny nose, and fever? Was your doctor's response something like this: "Go home, drink plenty of liquids, and rest"? This cautious response by doctors is warranted because they are correctly concerned that the desire to quickly eliminate your ills through the use of antibiotics can be a dangerous practice.

If a patient expects to be "cured" or become free from symptoms, a doctor may find it difficult to refuse that patient and not provide antibiotics; the goal of most doctors is to treat, heal, cure, and relieve pain. The doctor may feel that the patient will not be healed, and the patient may feel frustrated because they believe their discomfort should dissipate quickly if given a little pill.

The reality is that doctors must be cautious when prescribing antibiotics so that their patients will not develop antibiotic resistance (microbial resistance). According to the Food and Drug Administration (FDA), microbial resistance can be caused by

1. Administering antibiotics to patients in larger doses than recommended by healthcare and federal organizations.
2. Patients who do not finish the entire bottle of medication prescribed to them. When patients do not finish their entire prescription, the bacterial strain becomes stronger and more resistant to antibiotics.
3. Administering antibiotics for viral infections. Viral infections do not respond to antibiotics.
4. Administering antibiotic drugs to animals that are eaten by humans. These drugs are used to prevent the animal from contracting diseases and to increase production. Potentially, humans who eat the animal that has been treated with antibiotics could be resistant to medications used to treat human illness (FDA 2002).

Some turkey and chicken products contain campylobacter, which increases the risk of human infection from a bacteria that medications on the market will not easily kill. According to the Centers for Disease Control and Prevention, campylobacter is the most common bacterial cause of diarrheal illness in the United States; more than 2 million people are affected by it every year. Fever, diarrhea, and abdominal cramps can be produced in humans who eat chicken that contains campylobacter, and it can be life threatening for those with weakened immune systems (Bren 2001).

Anyone can experience antibiotic resistance. The predicament is that antibiotic resistance makes it difficult to administer cures for diseases. For example, parents have learned that after repeatedly using amoxicillin for their children's sore throats and earaches, the antibiotic loses its effectiveness and stronger antibiotics are needed. The struggle lies within a heartfelt fear: if my eight-year-old already needs a stronger antibiotic, what can my child expect when he needs antibiotic treatments in 25 years?

The FDA states that "about 70 percent of the bacteria that cause infections in hospitals are resistant to at least one of the drugs most commonly used to treat infections." (FDA 2002) It seems that the problem of antibiotic resistance continues to challenge the patients who suffer as well as the doctors who simply want to care for those patients. What can be done about this problem? The National Center for Infectious Diseases has created a strategic plan to address the problems of antibiotic resistance.

- "To assess trends in drug use and understand the relationship between drug use and the creation of infections that are antibiotic resistant."
- "To develop and evaluate new lab tests that can improve the accuracy and timeliness of antimicrobial resistance detection in clinical settings."
- "To find ways to decrease the emergence and spread of drug resistance by educating the patients about the 'right' way to use medications."
- "To evaluate a vaccine that could be used in preventing drug-resistant infections." (CDC 2001)

A heightened awareness is needed among physicians, pharmacists, pharmaceutical companies, and patients so that antibiotic resistance can be reduced. Antibiotics should only be used when necessary, but this will only happen when everyone becomes educated about the ramifications of antibiotic misuse. It is hoped that this reading will cause you to question your own need for antibiotics as well as question the need for the prescription in the first place. Pass the word about antibiotic resistance to others. By educating others, you are reducing the likelihood of the development of a medical problem that could have been prevented.

REFERENCES

Bren, Linda. (2001). Antibotic Resistance from Down on the Chicken Farm. *FDA Consumer Magazine*, January/February.

Centers for Disease Control and Prevention (CDC). (2001). Emerging Infectious Diseases: A Strategy for the 21st Century. *Target Area Booklet: Addressing the Problem of Antimicrobial Resistance*. Available at http://www.cdc.gov/ncidod/emergplan/antiresist/.

Cuevas, Amabile, Cardenas-Garcia, Maura, and Ludgar, Mauricio. (1994). Antibiotic Resistance. *American Scientist* 83:320–329.

U.S. Food and Drug Administration (FDA). (2002). "Antibiotic Resistance." Available at http://www.fda.gov/oc/opacom/hottopics/anit_resist.html.

QUESTIONS

1. How would you define antibiotic resistance?
2. What causes antibiotic resistance?
3. Describe a conversation that you might have with your doctor to express your educated concerns about antibiotic resistance.
4. What is the government doing to stop antibiotic resistance?

BOX 1 Human Engineering Timeline

1953

James Watson and Francis Crick determine the "double helix" structure of DNA. This discovery is a major breakthrough in the study of genetics and reinforces the idea that an organism's DNA is the primary and dominant determinant of its inherited traits.

1973

Stanley Cohen and Herbert Boyer create a transgenic organism using recombinant DNA technology, which allows the manipulation and transfer of pieces of DNA from one species to another.

1976

The first genetic engineering/biotech company, Genentech, is founded by Boyer and Robert Swanson. It is the beginning of the commercial use of genetic engineering technology, an industry which, by 2002, was generating revenues of $25 billion a year in the United States alone. Within two years, scientists at Genentech spliced the human gene for insulin production into *E. coli* bacteria, which then synthesizes human insulin.

1978

Louise Brown, the first "test-tube baby" (in vitro baby) is born in England, demonstrating the feasibility of growing embryos outside of the womb. In vitro fertilization is done by putting sperm and an egg together in a lab dish, where chemicals facilitate fertilization, and then implanting the embryo into a woman's uterus.

1980

The U.S. Supreme Court rules that genetically engineered microorganisms can be patented *(Diamond v. Chakrabarty),* setting a precedent for patents on life-forms.

1983

Kary Mullis devises the Polymerase Chain Reaction (PCR) technique, which rapidly replicates DNA sequences. This process of gene amplification makes gene mapping and forensics easier and cheaper.

1990

The Human Genome Project is begun by an international consortium of scientists, with most of the funding coming from the U.S. National Institutes of Health and the Wellcome Trust, a medical philanthropic organization based in London.

1996

A sheep named Dolly, the first mammal to be cloned from adult cells, is born at Scotland's Roslin Institute. Previously, cloning had only been carried out with embryo cells.

1998

Dr. James A. Thomson (University of Wisconsin) and colleagues are the first to isolate human embryonic stem cells, which have the potential to develop into almost any type of tissue. This innovation opens up the possibility of harvesting stem cells for use in treating human diseases.

2000

In June, scientists at both Celera Genomics (a private company formed in 1998) and the publicly funded Human Genome Project announce that they have completed a draft of the human genome. The announcement evokes hopes about medical advancements based on understanding of the genome, as well as controversy about the issue of public access to the information.

2001

In February, scientists at Celera Genomics and the Human Genome Project report that the number of human genes is probably about 30,000, only about twice as many as the number of genes in a fruit fly and far less than the long-standing textbook estimate of 100,000.

2002

As reports circulate that some scientists may have already begun to implant cloned embryos in women, the UN begins work on a global ban on cloning.

Source: Compiled by Vanessa Larson, "Beyond Cloning," *World Watch* magazine, July/August 2002. Reprinted with permission from Worldwatch Institute, Washington, D.C.

Making Well People "Better"

PAT MOONEY

The strategy of the biotech firms is to use sympathy for the sick to get genetic modification techniques approved, then go for the real profits—selling traits to people who aren't particularly sick.

When heads of state gathered for the Earth Summit in Rio de Janeiro 10 years ago, biotechnology was the buzzword miracle cure for world hunger and disease. A decade later, biotech has brought the poor no closer to the dinner table or better health. The reason is obvious: as ever, the poor are no one's market. Not that progress in biopharmaceuticals has lagged; advances in mapping the human genome have spawned new opportunities, and the prospects for human cloning and stem cell therapies have made headlines. However, the companies involved are actually pursuing more strategic agendas. Reproductive cloning might never be more than a niche market that the industry is happy to leave to quacks. The real money is in human performance enhancement drugs (call them "HyPEs"). And whether the focus is on pharmaceuticals developed the old-fashioned way or those that are linked via research or function to biotechnologies, they employ the same self-serving strategies.

HEALTHY MARKETS

The pharmaceutical industry has always suffered from a seemingly incurable marketing problem. Its customers are sick, and sick people are unreliable. If they die or get well, they stop buying drugs. If they remain sick, they tend to become unemployable. Unemployable sick people either can't afford drugs or (worse) they elicit sympathy and threaten prices. In the mid-1970s, pharmaceutical companies saw that the solution to the uncertainty of an ill clientele was to develop drugs for well people, who not only remain employed but never get "better." Best of all, well customers don't create sympathy and threaten price margins and profits. Now, biotechnology and the map of the human genome are making the task of creating new drugs for well people much easier.

Although the birth of biotech a quarter-century ago inspired the drive for a brave new market in well-people products, the industry has always been open to the opportunities. Morphine was purified from opium at the outset of the nineteenth century and first commercialized by Merck in Germany in 1827. Bayer was an early proponent of amphetamines and brought the world two blockbuster commercial winners, aspirin and heroin. In 1892, a Parke-Davis publication for doctors provided 240 pages of documentation extolling coca and cocaine, its two leading products; only three of the 240 pages discussed the drugs' unfortunate side effects.[1] Following World War II, the industry routinely blended barbiturates with amphetamines in diet drugs in order to encourage consumers to stay on the regime (and keep buying).[2] Sandoz (now Novartis) invented LSD, though the company was horrified by its abuses.[3]

The industry's view of "recreational" drugs has always been ambiguous. The annual global pharmaceutical market is worth roughly $300 billion, and the illicit narcotics market, valued at $400 billion in 1995,[4] is hugely inviting. New HyPE drugs could allow the industry to claim a share of this market by offering a battery of well-people products without the stigma society attaches to addictive drugs.

Reprinted with permission from the Worldwatch Institute, Washington, D.C.

DRUG ETHICS

Originally, "ethical drugs" were defined as drugs advertised only to doctors and pharmacists, but not to potential patients. Now the industry is advertising on television in the United States and elsewhere and has gone so far as to blend Internet advertising and medical research studies on websites targeting doctors. The ethical obfuscation is exemplified by the television ads that quietly have transformed Viagra from a drug to combat erectile dysfunction into an aphrodisiac.

The industry's selective ethical concern for the sick is also clear. For example, of the 1,223 drugs brought to market between 1975 and 1996, only 13 targeted the deadly tropical diseases that afflict millions of the world's poor, and just four of those drugs came from the private sector.[5] The nature of private pharmaceutical companies' commitment to patients was underscored in a 1993 study by the federal Office of Technology Assessment showing that 97 percent of the 348 ethical drugs brought to market by the 25 leading U.S. drug companies between 1981 and 1988 were copies of existing medications. Of the 3 percent offering genuine therapeutic advances, 70 percent resulted from public research. More than half had to be eventually withdrawn from sale due to unanticipated side effects.[6]

WORKING HyPE-OTHESIS

Making "well" people "better" could have significant benefits for employers. Try as we will to automate every kind of work, people are likely to remain the most versatile and efficient tool of production for many jobs. But we do have our defects, and the pharmaceutical industry is working on developing performance enhancement drugs to turn workers into superhumans. Employers (and governments) are lining up to try the new drugs. Here are some examples of recent genome-inspired innovations and some old drugs being given new, augmented lives through genetic research.

8 days a week:

Cephalon Inc. has developed a drug called Provigil for the treatment of narcolepsy (a neurological disease that causes irrepressible sleep attacks). Because Provigil is not an amphetamine, it is attracting attention as a possible alertness aid for healthy people.

Rhythm and blues:

Northwestern University has patented the circadian rhythm gene. The circadian clock regulates 24-hour rhythms in physiological systems. The patent covers the gene's uses for sleep-related problems, jet lag, alertness, stress response, diet, and sexual function and could be exploited to enhance mood in intensive care units.

Stringed-out quartets:

A "beta-blocker" drug meant for treatment of congestive cardiac failure is best known as "the musicians' underground drug" because of its effect on musical performance. (The drug blocks stage fright.) Twenty-seven percent of symphony orchestra musicians take beta-blockers.[7] A drug therapy capable of blocking anxiety would have major workplace applications.

Company genes:

In 2001, a U.S. railroad agreed under threat of a lawsuit to stop genetic testing of employees. The company had required employees claiming carpal tunnel injuries to submit to blood tests, which included searching for a genetic cause for the syndrome. Also last year, an 18-year-old Australian with a family history of Huntington's disease was told by a government official that he would be hired only if he submitted to a genetic test demonstrating that he did not have the Huntington's gene.

Our new understanding of genomics and the neurosciences is also making possible a generation of HyPE medicines that could be used in more sinister ways, e.g., to control dissent. Mood-altering drugs that dispel discontent might be individually prescribed, pressed upon workers, or even hosed into crowds. Enhancement technologies could also become disabling technologies in military or police hands. Those refusing to take HyPEs could be punished by their teachers, employers, or governments because they are refusing to maximize their potential. And if it is possible to "enhance" an infantryman's performance with a drug that turns off the brain's fear mechanism, for example, then it is also possible to switch on irrational fear in the enemy. Drugs that target hearing, memory, or alertness could be mirrored by drugs that weaken those qualities.

SMARTIES

Scientists call drugs being developed to improve memory "cognitive enhancers" or "nootropics." Consumers know them as "smart drugs" or "smarties." The market for smart drugs is already vast. Nootropics used to alleviate dementia in Alzheimer's disease victims were worth $94.5 million in

1995. The illicit market is unknown. A quick Internet search brings up dozens of companies specializing in the sale of nootropics not approved by the Food and Drug Administration.

Pharmaceutical companies are using human genomic data in their race to meet the growing demand for nootropic therapies. Ignorance of drug interactions has many worried about the long-term effects of such therapies. The excitement over using genomics to improve memory and intelligence spiked when a Princeton scientist inserted an extra copy of the gene for a particular brain receptor into a mouse. The mouse out-performed other mice on intelligence tests, and the research was hailed as a step toward decreased dementia and increased memory. However, the mouse's increased intelligence seems to have come at the cost of chronic pain.[8]

OPTIONAL EQUIPMENT

Brain Viagra?

In 1995, Cold Spring Harbor Laboratory created a fruit fly with an apparently photographic memory. The lab then partnered with Hoffman-La Roche to see if the human mind could be similarly modified. Roche Pharmaceuticals later announced a breakthrough in learning and memory that could lead to treatment for cognitive deficit diseases such as Alzheimer's, depression, schizophrenia, or aging. Several drugs are readily available and widely used as memory enhancers, though they are not proven, tested, or approved for such uses.

Trauma tamers:

After demonstrating that the fruit fly's ability to learn could also be abolished by subtle genetic alterations, Cold Spring Harbor researchers launched Helicon Therapeutics Inc. to make drugs aimed at different brain molecules. They see lucrative markets in products for boosting failing memory and medicines for blocking trauma recollection.

Learning too much?

Scientists have genetically engineered mice with enhanced memory that persists until researchers use genetic trait control technology to switch off a key memory-governing enzyme.

Social IQ:

Those who exhibit "antisocial" behavior could be subjected to genetic therapies to "cure" them of conditions such as depression, obsessive behavior, and hyperactivity. Even shyness is now being treated with the drug Seratox, originally developed as an anti-depressant. It is believed that a gene inherited from the father might act to fine-tune a part of the brain involved in social abilities.

HyPES: HOPE FOR THE POOR?

The choice between developing drugs to make ill people well or well people better is best manifested in the enormous corporate investment in diet-related medicine. Research on new forms of proteins, and on old woes like obesity and diabetes, suggests that it may be possible to develop drugs that could help people utilize food and energy more effectively.

It's clear, however, that the world's roughly 820 million malnourished poor are suffering most from a political failure to have their basic needs and human rights met by a world that is richer in food than in justice. Drug companies could at least collaborate with plant breeders to develop nutriceuticals that would enable the poor to make better use of the food they have. Instead, the pharmaceutical industry is hard at work at developing drugs that allow people to eat gluttonously without getting fat. With obesity a major health problem in industrialized countries, companies are in hot pursuit of "uncoupling protein" (UCP) molecules that interfere with the conversion of food calories into metabolic energy and release them instead as waste heat. Of course the logical solution is to eat less and exercise more. But there is a multi-billion-dollar market waiting for any pharmaceutical company that can turn UCP molecules into drugs that let people stuff their faces without losing their figures.

The poor are not entirely excluded from the search for the glutton genie. Some hunter-gatherer societies have had their own harsh encounter with obesity when they have been pushed into sedentary occupations and environments. Rising obesity has led to a rising incidence of diabetes. Under the pretext of treating it, some companies have struck deals with tropical island peoples to access their genes and identify those that aggravate obesity. Others are roving among indigenous communities in North America, studying diabetes. An estimated 15 percent of aboriginal peoples in America are pre-diabetic compared to less than 8 percent in the "white" population. However, the goal of this research is not to develop drugs that will block full-blown diabetes among the 105,000 U.S. pre-diabetic aboriginals, but to target the 11.4 million pre-diabetic white Americans.[9] But since the incidence of diabetes is correlated with rising obesity, the real goal is a magic elixir that converts indulgence into a virtue (or at least into something that is not a fashion *faux pas*). In this work, the poor are a tool, not a target.

FROM HyPE TO HEALTH

If we continue to rely upon the world's giant pharmaceutical corporations to determine research goals, our societies will remain unhealthy and become unhealthily dependent. We need to strengthen socially oriented public research and public health initiatives and, simultaneously, eliminate the patent incentive that distorts medical innovation and dictates profiteering. Until we dispel the myth that the biotech and pharmaceutical industries are working on our behalf, the prognosis is poor.

ENDNOTES

1. David T. Courtwright, *Forces of Habit: Drugs and the Making of the Modern World* (Cambridge: Harvard University Press, 2001), 86.
2. Courtwright, 105.
3. Courtwright, 89.
4. United Nations Development Program, *Human Development Report 2001–Making New Technologies Work for Human Development* (New York and Oxford: UNDP/Oxford University Press, 2001), 13.
5. UNDP, 3.
6. Pat Roy Mooney, "The Parts of Life–Agricultural Biodiversity, Indigenous Knowledge and the Role of the Third System," *Development Dialogue: A Journal of International Development Cooperation,* 1996: 1-2: 82.
7. Karla Harby, et al. "Beta Blockers and Performance Anxiety in Musicians." A Report by the beta blocker study committee of FLUTE, March 17, 1997.
8. Deborah L. Stull, "Better Mouse Memory Comes at a Price," *The Scientist* 15(7), April 2, 2001, 21.
9. Sarah Lueck, "U.S. Says 16 Million Have 'Pre-Diabetes'," *Wall Street Journal,* March 28, 2002, B8.

QUESTIONS

1. Why do pharmacological companies want to *make well people better?* Why don't they want to spend money on making sick people well? Correlate one ethical theory to this premise.
2. Explain with examples how the definition of the term *ethical drugs* has changed.
3. Do you see any complications or social ramifications of patenting a drug that changes the circadian rhythms in the body?
4. Why do musicians like to use beta-blockers? How could this drug help managers in high-stress jobs? SHOULD this medication be used for this purpose?

BOX 1 Views from Around the World

Ethiopia

Formally, the human genetic engineering project is expected to identify our genetic peculiarities so that our ailment particularities can be precisely targeted. But, as an African whose ancestors suffered for 500 years being targeted for slavery and being colonized, and whose natural resources are now being plundered, I find it difficult to expect peculiarities to be used positively. When I recall that the North has apologized to the Jews for the Holocaust and even through the Pope to the Arabs for the Crusades, and that only in 2001 the North refused to apologize to Africans in Africa and the Diaspora for slavery and colonialism, I find it difficult to feel so positive. Given this, do I expect the human genome project to make life easier for the sufferer of sickle cell anemia, or killing easier for the white supremacist who is now a major political force in the North? I leave you to guess the answer.

Berehan Gebre Egziabher
General Manager
Environmental Protection Agency, Ethiopia

South Africa

While for privileged people it may seem that the balance in the use of power flowing from scientific knowledge and technological achievements has been in favor of beneficence, different perceptions prevail among those who have been marginalized. Close links between science, technology, the military, money, and those with global power, and the use of power and secrecy to protect privilege, have undermined confidence that there is any significant concern for the future of the people of Africa.

Soloman Benatar, M.D.
South Africa

United States

Given the history of mankind, it is extremely unlikely that we will see the posthumans as equal in rights and dignity to us, or that they will see us as equals. Instead, it is most likely either that we will see them as a threat to us and thus seek to imprison or simply kill them before they kill us, [or that] the posthuman will come to see us (the garden variety human) as an inferior subspecies without human rights, to be enslaved or slaughtered preemptively. It is this potential for genocide based on genetic difference, which I have termed "genetic genocide," that makes species-altering genetic engineering a potential weapon of mass destruction, and makes the unaccountable genetic engineer a potential bioterrorist.

George J. Annas, *Chair*
Department of Health Law, Bioethics, and Human Rights, Boston University School of Public Health

Malaysia

Potential abuse of technology related to reproductive cloning of human beings not only raises moral, religious, and ethical concerns but also poses risks [of] developmental and bodily abnormalities to humans.

Hasmy Agam
Malaysian Ambassador, United Nations

India

The final goal of reproductive engineering appears to be the manufacture of a human being to suit exact specifications of physical attributes, class, caste, color, and sex. Who will decide these specifications? We have already seen how sex determination has resulted in the elimination of many female fetuses. The powerless in any society will get more disempowered with the growth of such reproductive technologies.

Sadhana Arya, ***Nivedita Menon***, *and* ***Jinee Lokaneeta***
Saheli Women's Resource Centre, Delhi University

North America

Human genetic manipulation that affects indigenous peoples is an act of war on our children.

Dave Pratt, *Dakota tribe*

United Kingdom

All the developments in and around human genomics stem from the mechanistic paradigm that still dominates western science and the global society at large.... The irony is that contemporary western science across the disciplines is rediscovering how nature is organic, dynamic, and interconnected. There are no linear causal chains linking genes and the characteristics of organisms, let alone the human condition. The discredited paradigm is perpetrated by a scientific establishment consciously or unconsciously serving the corporate agenda, and making even the most unethical applications seem compelling.

Mae-Wan Ho
Institute of Science in Society, London, U.K.

China

The main potential harm of genetic engineering is associated with artificial horizontal gene transfer experimentation. Horizontal gene transfer occurs commonly in nature. Genes can be exchanged between different bio-species. But the frequency of these natural transfers is limited by the defense systems, i.e. immune systems, of each bio-species. The immune system serves to prevent invasion by harmful foreign genes, viruses, and so forth, so that the bio-species can maintain its characteristic traits and normal metabolism. The GE method of horizontal gene transfer works by penetrating or weakening the immune system and using virulent genes as delivery vehicles. That is, the gene to be transferred is combined with a virulent gene to effect penetration. This method allows harmful virulent genes, especially those with resistance to antibiotics, to become widespread in nature.... If such virulent genes combine with the genes of harmful viruses to form new viruses, it will be disastrous for humankind.

Yifei Zhu
Hangzhou, Zhejiang Province, China

Environmental NGOs

Together with proposed techniques of inheritable gene modification, the use of cloning for reproduction would irrevocably turn human beings into artifacts. It would bring to an end the human species that evolved over the millennia through natural evolution, and set us on a new, uncontrollable trajectory of manipulation, design, and control.

Brent Blackwelder, *President, Friends of the Earth*

Mark DuBois, *International Coordinator, Earth Day 2000*

Randy Hayes, *President, Rainforest Action Network*

Robert F. Kennedy Jr., *President, Waterkeeper Alliance*

John A. Knox, *Executive Director, Earth Island Institute*

Robert K. Musil, *Executive Director, Physicians for Social Responsibility*

John Passacantando, *Executive Director, Greenpeace USA*

Michele Perrault, *International Vice President, Sierra Club*

Mark Ritchie, *President, Institute for Agriculture and Trade Policy*

Source: "Beyond Cloning," *World Watch* magazine, July/August 2002. Reprinted with permission from WorldWatch Institute, Washington, D.C.

BOX 2 Biopirates and the Poor

The promise to cure disease through human genetic engineering has moved faster on Wall Street and in the media than in basic scientific knowledge of how genes work and how genetic manipulation affects whole organisms as well as their relationships with other organisms. Within a few weeks, the "alphabet" of the "Book of Life" shrank from 100,000 to 30,000; this is just one indicator of the ocean of ignorance in which the island of human genetic engineering is floating.

The three major concerns arising from human genetic engineering are biopiracy, the transformation of socially defined traits into biologically defined ones, and the issue of privacy.

Across the world, indigenous communities are outraged at biopiracy of genes and genetic material. The recent case of collection of blood samples from the Naga tribe in northeast India is just another example of gene piracy at the human level. Such piracy can even happen in the heart of rich industrial society, as shown by the case in which University of California scientists patented the genes of a cancer patient, John Moore, without his knowledge.

What is called a deficiency—mental, physical, or other—is socially defined. For example, the perverse world order of globalization dictated by commerce, greed, and profits regularly treats women, children, and poor people as inferiors. Without strong democracy and true transparency, this kind of discrimination can be used to justify human genetic manipulation, manifested in eugenics programs.

Human genetic engineering also raises major issues about the erosion of privacy and handing people's control over their own destiny to others, such as insurance companies, pharmaceutical companies, and police states, which could combine to share genetic data without the consent and participation of the persons concerned.

Source: "Beyond Cloning," *World Watch* magazine, July/August 2002. Reprinted with permission from Worldwatch Institute, Washington, D.C.

The Politics of Life and Death: Global Responses to HIV and AIDS

MARY CARON

If effectively fighting HIV means openly getting condoms to teenagers or clean needles to addicts, or candidly discussing the prevalence of prostitution in their communities, many politicians would rather avoid the subject altogether—even if it means allowing an epidemic to flourish. Where leaders have lifted their heads from the sand, however, millions of lives have been saved.

Both Rajesh and his wife—who prefers not to give her name for fear of being ostracized by neighbors in their Bombay community—are infected with HIV. With the help of money quietly contributed by relatives, they are among the few families in India who can pay for life-prolonging anti-viral drugs—but only for Rajesh. Other couples in India are finding themselves in a similar situation. "It is the woman who is stepping back" so her husband can get treatment, said Subhash Hira, director of Bombay's AIDS Research and Control Center in a recent *Associated Press* story. "She thinks of herself as dispensable."

In Zimbabwe, where 200 people are dying every day from AIDS, life insurance premiums have quadrupled to keep up with rising costs. It would take roughly two years for the average Zimbabwean to pay for one month of treatment at U.S. rates.

In the United States, of course, incomes are much higher—about 74 times those in India and 46 times those in Zimbabwe. Yet, even here, nearly half of HIV-positive patients in a recent national study had annual incomes less than $10,000, whereas the annual costs of their care and treatment came to about $20,000. Two of every three patients in the study had either no insurance or only public health insurance—which may not adequately cover their needs.

And then there is the Central African Republic, where I worked as a Peace Corps volunteer a few years ago. The five-bedroom house where I lived was owned by my neighbor, Victor, who rented it to me while living in a more modest mud-brick house next door. He used the income from rent, plus whatever his eldest daughter could earn selling food in the market, to support ten people. One of them was a little girl of five or six who used to come over and sing me songs. I didn't realize that Victor was her uncle, not her father, until someone explained that her mother and father had died following "a long illness." There are many households like Victor's. Worldwide, more than 41 million children will have lost one or both parents by 2010, mostly as a result of AIDS. The grandparents or other surviving family members, who often have their own difficulties making ends meet, may find themselves taking care of up to a dozen children.

Clearly, AIDS is a disease that the world cannot afford. And yet, the relentless spread of the virus forces us to confront painful life-and-death choices about allocating resources. Communities, nations, and international donors are all struggling to care for a growing number of the sick; to invest in prevention that can avert millions of future infections; to fund research that can yield life-prolonging treatments; and ultimately to develop a vaccine. To do all of these things at once seems an almost impossible task. But experience in the field demonstrates that there are reasons for hope, even in relatively poor countries where HIV is already a serious problem.

While other diseases target children or the elderly, HIV often strikes otherwise strong and healthy people—those most likely to be taking care of children and contributing to the economy. And it does so in a way that has repeatedly caught societies unprepared. HIV does not kill within a

Mary Caron is press director for the Worldwatch Institute. Permission granted by Worldwatch, *World Watch* May/June 1999, pp. 30–38. www.worldwatch.org.

matter of days or weeks, like other infectious diseases; rather, it gives death a kind of "rain check." The asymptomatic period may last 10 years or longer in a country like the United States, though the infection can progress to AIDS in as little as two to three years in a country like Zimbabwe or India, where the percentage of people who can get full treatment and care is much smaller. An infected person may be ill off and on for years, requiring extended care from family or community members. And HIV, while relatively slow to develop in the body, can spread rapidly within a population. About 75 percent of HIV transmission worldwide is through unprotected sex. The rest occurs mainly through sharing of unsterilized needles, through childbirth or breastfeeding from an infected mother to her child, and from the use of infected blood in transfusions.

AIDS now rivals tuberculosis as the world's most deadly infectious disease. Every day last year, 16,000 people were infected with HIV—11 people per minute. Women now account for 43 percent of all adults with HIV/AIDS. And about half of all new infections are in 15- to 24-year-olds. Since AIDS was first recognized in 1981, more than 47 million people have become infected and nearly 14 million have died. The epidemic has taken its heaviest toll in Africa, which has just 10 percent of the world's population but 68 percent of the HIV/AIDS cases—most of them in the sub-Saharan region. In some southern African countries, one in four adults is HIV positive. Thus, the world's poorest countries are staggering under the burden of the world's most unaffordable disease.

As HIV wears down the body's defenses and the infected person becomes increasingly ill from opportunistic infections, the costs of providing care and treatment mount. Worldwide, about 63 percent of the $18.4 billion spent on HIV/AIDS in 1993 went to care, according to a 1996 study by Harvard researchers Daniel Tarantola and the late Jonathan Mann.* Another 23 percent was spent on research and just 14 percent on prevention. Moreover, only 8 percent of global spending took place in "low economy" countries of the developing world, yet more than 95 percent of all HIV-infected people live in developing countries.

Preventing an HIV infection costs much less than caring for an infected individual. And the benefit of prevention is compounded since preventing one person from getting HIV keeps that person from spreading it to others. If a man has sex with three different women in a year, shielding that man from infection also shields his three partners and any children they may have.

However, global resolve to protect the uninfected may be overwhelmed by the challenge of providing for the staggering number already infected. AIDS already rivals the horror of the smallpox epidemic which decimated Native American populations in the 16th century and the Black Death which wiped out a quarter of Europe's population in the 14th century. If one of every four adults is already infected in Botswana and Zimbabwe, what hope is there for those countries and their neighbors? Many people may now be under the impression that Africa is a continent essentially lost to AIDS and that the rest of the developing world may soon follow.

Look more closely, however, and two signs make it clear that the situation is far from hopeless.

First, more than half of the populations of developing countries—about 2.7 billion people—live in areas where HIV is still low, even among high-risk groups. Another third live in areas where the epidemic is still concentrated in one or more high-risk groups, yet "HIV prevalence"—the proportion of a population infected at a given time—is still below 5 percent in the general population. Even in hard-hit Africa, there are at least a few countries—such as Benin, Senegal, Ghana, and Guinea—where adult infections are still under 3 percent. These areas of relatively low HIV prevalence present a one-time opportunity—and one that will not last long—for policy makers to implement solid strategies for keeping HIV at bay.

Second, even in places where the epidemic has taken hold, campaigns to stop further escalation have proven successful in both the early and the later stages of an epidemic. It's instructive to see how that was done in each case—first in Thailand, where an effective campaign was able to ward off an incipient epidemic and to keep prevalence relatively low in the general population, and then in Uganda, where a high percentage of the population was already infected.

By taking action relatively early, Thailand was able to keep HIV infections from spiraling out of control in the general population. In early 1988, officials were alarmed by reports from an ongoing survey at a Bangkok hospital showing that infections among drug addicts who use needles, or "injecting drug users" (IDUs), had jumped from 1 percent to 30 percent in the preceding 6 months. In response, the Thai Ministry of Public Health set up a system to collect data on HIV infection at selected sites

*Mann and his wife, Mary Lou Clements-Mann, died in the crash of Swissair flight 111 last September. Mann founded and headed the World Health Organization's Global Program on AIDS.

throughout the country. These "sentinel surveys," as they are called, revealed even more alarming news. By mid-1989, HIV was present in all 14 provinces surveyed. In the northern city of Chiang Mai, 44 percent of the prostitutes were infected. And HIV was also found in some pregnant women, who are considered representative of the general population.

Concerned about the possibility of a general epidemic, the Thai government then conducted a national survey to identify behaviors that might be driving the spread of the virus. What it found was that more than one-fourth of the country's men were having sex with prostitutes, both before and outside marriage. In 1991, Prime Minister Anand Panyarachun assumed personal leadership of the National AIDS Committee and aggressively escalated the government's response. Official spending on HIV/AIDS was pushed from $2.6 million in 1990 to $80 million in 1996.

The Thai effort mobilized sectors of the population ranging from prostitutes to teachers to monks. In the commercial sex industry, which accounts for an estimated 14 percent of Thailand's GDP, brothel owners and employees now require every male customer to use a condom. Government STD clinics hand out about 60 million free condoms a year, and encourage their use. Several monasteries in northern Thailand are providing counseling services for HIV-infected people and helping them find employment. Schools are teaching children how to reduce sexual risk-taking.

Within three years after this heightened response got underway, there were signs that it might be working. A second national behavior survey showed that between 1990 and 1993, the percentage of 15–49-year-old men reporting sex outside of marriage had dropped from 28 percent to 15 percent. Among men who continued to engage prostitutes, the percentage reporting that they always used a condom doubled. Condom sales rose and sexually transmitted disease declined throughout the country. HIV infection also declined. Annual testing of 21-year-old military conscripts, which had found 0.5 percent of them infected in 1989, showed prevalence peaking at 3.7 percent in mid-1993 (reflecting a predictable lag between risky behavior and evidence of infection), then declining to 1.9 percent in 1997. Similarly, testing of pregnant women in all 76 provinces found HIV infection at 0.5 percent in 1990, increasing to 2.4 percent in 1995, then declining to 1.7 percent in 1997.

The health and social costs of this disease still inflict a heavy burden on the Thai economy, and the costs continue to grow. The Asian financial crisis, which began in Thailand in 1997, has forced cuts in the national AIDS budget and has put greater strains on affected families. And Thailand will have to remain vigilant to keep prevalence low. While condom use has increased in the country as a whole, it remains much lower in rural areas, among people with limited education, and among those who engage in casual sex. A 1995 survey also showed that many drug users were reverting to sharing needles. Nonetheless, by acting quickly and aggressively, Thailand may have averted a full-blown HIV epidemic.

Uganda, unlike Thailand, launched its prevention campaign at a time when a high percentage of the population had already been infected. By 1999, in a population of less than 21 million, 1.8 million Ugandans have already died and 900,000 more have HIV. Moreover, Uganda has far less financial capability than does Thailand, with a per-capita GNP of just $300 compared with Thailand's $2,960. Yet Uganda's success in bringing down high HIV prevalence provides evidence that fighting HIV is not impossible, even when the situation at the outset looks dire. When Yoweri Museveni became president in 1986, HIV was already a serious problem. Museveni quickly implemented a national plan, enlisting both government agencies and non-governmental organizations (NGOs) to join the fight. Uganda established the first center in sub-Saharan Africa where people could go for voluntary and anonymous HIV testing and counseling.

Like the Thai campaign, the Uganda one succeeded by mobilizing a wide spectrum of groups. A student heading home after class at Makerere University in Kampala, for example, may well get an update of the latest information on avoiding HIV—courtesy of her "boda boda" bicycle taxi driver, who has been trained by the Community Action for AIDS Prevention project. Or if you live in the Mpigi district, the local Muslim spiritual leader, or Imam, may stop by for a discussion of AIDS and Islam. Trained by the Family AIDS Education and Prevention Through Imams project of the Islamic Medical Association of Uganda, some 850 of these leaders have taken HIV prevention messages directly to the homes of more than 100,000 families throughout the country.

The Ugandan government has conducted regular surveys of sexual behavior, and these studies show signs of substantial change from 1989 to 1995. The share of 15–19-year-olds who report never having had sex has increased from 26 to 46 percent for girls and from 31 to 56 percent for boys. The share of people reporting that they had used a condom at least once rose from 15 to 55 percent for men and from 6 to 39 percent for women.

HIV prevalence has also dropped, most notably among young people between the ages of 13 and 24. Between 1991 and 1996, the percentage of pregnant women testing positive for HIV in some urban areas dropped by one half, from about 30 percent to 15 percent.

So it is apparently possible to keep HIV in check. But it's not easy, whether the problem is caught in its early stages, as it was in Thailand, or has become full-blown, as it was in Uganda. It's difficult to change people's behavior, especially when it means challenging highly sensitive—and very personal—questions about sex, prostitution, infidelity, and drug dependence. "We have to stop thinking that HIV/AIDS is only a health problem. It is a development problem," says the World Bank's HIV/AIDS coordinator Debrework Zewdie. To stop it will take "a commitment from governments in developed and developing countries. Zoom-in-and-zoom-out programs are not going to work; we have to build local capacity."

There is no single formula for building that capacity, although the most inno]vative and appropriate solutions often come from within communities. Wherever initiatives are taken, however, there are some basic policy principles that seem to apply.

- **Early and aggressive action:** Worldwide, we already spend nearly $5 on HIV/AIDS treatment and care for every $1 spent on prevention. Implementing prevention measures before even the first case of AIDS is reported can reduce that ratio and thereby greatly reduce the overall costs of care. On the other hand, if governments avoid thinking about prevention until AIDS cases start to burden the health system, the epidemic may already have invaded large parts of the population. Yet because symptoms of AIDS typically do not show up until several years after infection, the threat may be largely invisible until large numbers of people have been doomed.
- **Communities:** Mobilizing business, religious, and civic leaders can galvanize broad support for raising public awareness of the risks of HIV and reducing the stigmatization of those infected. In Zimbabwe, for example, the Commercial Farmers Union recruited and sponsored farm owners to participate in a Family Health International-supported program that trained more than 2 million farm employees and family members in a nationwide HIV/AIDS prevention effort.
- **Political leadership:** In both Thailand and Uganda, HIV/AIDS prevention moved from simply a public health concern to a national priority. Prevention campaigns can succeed when political leaders put them at the top of the national agenda, use their public platform to encourage safer behavior, ask communities and NGOs to join the fight, and work to change laws that prohibit such effective tools of prevention as condom advertising and needle purchases.
- **Data collection and dissemination:** HIV is a stealthy attacker that can infiltrate an unsuspecting community and spread rapidly. It is therefore important to collect infection data from health clinics and to assess behavior trends. By publicizing the results of its sentinel and behavior surveys, Thailand made its population aware of the extent of risk in the country.
- **Low-cost, high-quality condoms:** Mr. Lover Man, a human-size condom mascot, can now be seen cruising the streets, attending soccer games—and, of course, passing out condoms—in several South African cities. In Portland, Oregon, teenagers have been given discreet access to protection via 25-cent condom vending machines in public rest rooms. Using new variants of old marketing techniques, organizations like Population Services International (PSI) have dramatically increased the worldwide distribution of HIV prevention information and reliable low-cost condoms. In Zaire, PSI "social marketing" programs helped condom sales to rise from 900,000 in 1988 to 18.3 million in 1991, averting an estimated 7,200 cases of HIV.
- **Targeting interventions to high-risk groups:** HIV usually gains a foothold in one or more groups whose behavior puts them at higher risk: prostitutes, IDUs, people with another sexually transmitted disease, young military recruits, migrant workers, truck drivers, or homosexual men. The virus can spread rapidly within the group and, once established, can move to those at lower risk of infection through people who act as a bridge between high- and low-risk groups—for example, men who have visited prostitutes and then bring the disease home to their wives.

A World Bank report, *Confronting AIDS,* suggests that countries can keep HIV at bay by targeting these high-risk groups with HIV prevention. It is important to note, though, that such efforts, if not managed with particular care, can trigger unintended public reactions. Singling out particular groups may inadvertently raise perceptions that HIV is a problem only for "those" people. Some public health experts have also noted that programs to give prostitutes a regular monthly course of antibiotics, for example, may reduce STDs but are harmful to overall health. And when HIV is present in the general population, questions of equitable distribution of resources arise as well.

Preventing HIV infection in someone with a high rate of partner change can avert many more future infections than preventing infection in a person with low-risk behavior, says the World Bank report. For example, compare two prevention programs. The first, in Nairobi, Kenya, provided free condoms and STD treatment to 500 prostitutes, of whom 400 were infected. Each of the women had an average of four partners per day. Under the program, condom use rose from 10 to 80 percent. A calculation based on the estimated rate of transmission, number of partners, condom effectiveness, and secondary infections shows that this program averted an estimated 10,200 new cases of HIV infections each year among the prostitutes, their customers, and their customers' wives. If the same program had instead targeted a group of 500 men, who had an average of four partners per year, 88 new cases of HIV would have been prevented. The second program would have saved fewer than 1 percent as many people as the first.

When IDUs share needles contaminated with blood, HIV can sweep through their population even more rapidly than it does among prostitutes because the risk of transmission per contact is higher. In January 1995, HIV prevalence among such drug users in the Ukraine was under 2 percent. Eleven months later, it had shot up to 57 percent. As of December 1997, 66 percent of HIV infections in China and 75 percent in Kaliningrad, Russia resulted from shared needles. Half of all new HIV infections in the United States occur among intravenous drug users, even though less than half of 1 percent of the U.S. population injects drugs frequently. And as with prostitution, HIV can spread from this high-risk group to the population at large.

Needle exchange programs aim to reduce the transmission of blood-borne infections, including HIV, by providing sterile syringes in exchange for used, potentially contaminated syringes. After the U.S. state of Connecticut made needles available from a pharmacy without a prescription, the percentage of IDUs who share needles dropped from 71 percent to 15 percent in three years. A review of studies conducted between 1984 and 1994 showed that HIV prevalence among IDUs increased by 5.9 percent per year in 52 cities that did not have needle exchange programs, but declined by 5.8 percent per year in 29 cities that did.

The experience of the past two decades has given us a set of policies that are proven to work, at least at mobilizing communities to keep HIV in check. Such policies should be in place in every country in the world. Yet, proven policies aren't always enough. Even when faced with the specter of an ever more devastating human and economic toll, people in positions of political power too often ignore—or thwart—the most effective HIV-fighting strategies. If confronting AIDS means talking about such potentially explosive topics as distribution of condoms to teenagers, or of needles to addicts, or the prevalence of prostitution in their communities, many politicians would rather avoid the subject altogether.

In Kenya, where tourism brings in more money than exports of tea, coffee, or fruit, officials—perhaps leery of scaring off tourists—declared the country AIDS free, even when studies among Kenyan prostitutes showed 60 percent of them to be HIV infected. The government did not admit the scope of the epidemic until late in 1997. By then, more than a million Kenyans were infected. The country is belatedly taking steps to implement some HIV-fighting programs, such as an awareness campaign for students. Muslim and Catholic religious leaders, however, object to sex education in schools, saying it would corrupt students' morals. By now, the number of infected Kenyans has passed 1.6 million—about 12 percent of the adult population.

Refusal to pay serious attention has been a common failing in these battles, in which the invasion is so stealthy and the victims are often socially marginalized. Even in Thailand, where the government eventually roused itself to lead an aggressive anti-HIV campaign, there was an initial period of denial in the late 1980s when infections were burgeoning among prostitutes in the Northern provinces, particularly in the Chiang Mai area. Given the large role of commercial sex in the Thai economy, officials may at first have been more concerned about the possible loss of tourism dollars than about the risk of an epidemic. Fortunately, they did not continue to ignore the problem.

In the United States, where about half of all new HIV infections are spread through shared needles or to sexual partners of IDUs, the government bans the use of federal funds for needle exchange programs. Last April, after carefully reviewing research on needle exchange programs, U.S. president Bill Clinton declared that these programs curb AIDS without promoting increased illegal drug use. Yet, in the same announcement, he declined to lift an existing ban on federal funding for needle exchange, which applies to all domestic and overseas programs. Senator Paul Coverdell of Georgia introduced a bill that would prevent the ban from ever being lifted, and Representative Todd Tiahrt of Kansas authored a provision in the federal budget that bans the use of federal and city funding for needle exchange in the nation's capital.

Politicians do not want to appear "soft on drugs" by helping drug users, who are often perceived as a criminal element—and who some cynically believe would die of drug overdoses anyway, even if they didn't die of AIDS. Even the more self-interested argument that preventing HIV among drug users could prevent it from spreading to the general population is largely ignored.

Similarly, U.S. officials were slow to act when HIV was first recognized in the early 1980s among homosexual men. Condemnation of the gay community was widespread, and some people went so far as to suggest AIDS was a heavenly retribution for worldly sins (i.e., homosexual sex). Fortunately for the U.S. population as a whole, as well as for those segments most at risk, members of the gay community launched their own aggressive and highly organized campaign to prevent HIV. Between the 1980s and the 1990s, AIDS was turned from a marginalized problem of "those people" to a high-profile national health threat. And while about half of those infected are still not in ongoing care, prevalence has been kept low.

In less politically or economically stable countries than the United States, however, leaders are sometimes overwhelmed by social and economic upheaval that may fatally distract them from the threat of HIV. As apartheid was ending in South Africa, for example, an influx of commercial trade and migrant workers from neighboring countries opened up a kind of viral superhighway for the epidemic. Legislators, grappling with the momentous political and social changes at hand, failed to foresee that these changes might also bring deadly consequences, and no proper prevention strategy was put in place. Forced displacement of black people under apartheid and the deploying of workers far from their families had also led to higher rates of extramarital sex and prostitution. Today, more than 3 million South Africans—one of every eight adults—have HIV. In a country of just 43 million, 1,500 people are infected every day.

The political and social climate in South Africa has been slow to change. The government has been accused of stifling non-governmental action with bureaucratic restrictions. Social stigmatization runs very high. A woman who had just publicly declared her HIV-positive status as a means of helping others to fight discrimination was beaten to death just after Christmas last year by a mob of neighbors who stoned her, kicked her, and beat her with sticks.

After a long period of rarely addressing the issue, departing President Nelson Mandela declared in March that "the time for such silence is now long past. The time has come to teach our children to have safe sex, to have one partner, to use a condom."

The painful lessons learned in South Africa and other AIDS-ravaged countries can now be brought to bear on the world's two largest countries, where the future health of a large portion of humanity lies at stake. The choices that Chinese and Indian leaders make about fighting HIV in the next few years will affect the course of the epidemic for one-third of the world's people. India and China both have relatively low HIV prevalence, but alarming signs of increasing infections among some groups coupled with known risk factors make both countries precariously susceptible. If HIV prevalence in China and India were to reach the levels now seen in some southern African countries, up to 300 million people would be infected. The magnitude of the impact—on economic productivity, social and political stability, psychological health, and the human spirit worldwide—is almost unimaginable.

In India, at present, less than 1 percent of adults are infected. Still, with an adult population of almost 500 million, that comes to 4 million HIV positive people—in absolute numbers, more than any other nation. Prevalence is highest among prostitutes, truckers, and IDUs, and there are signs that HIV is also gaining a foothold in the general population. A study conducted between 1993 and 1996 in the city of Pune, south of Bombay, showed that close to 14 percent of the city's monogamous married women had been infected.

In Bombay, by now, more than 50 percent of the city's 50,000 "sex workers" are HIV-positive as compared with just 1.6 percent in 1988. Prevalence has also jumped into the double digits among prostitutes in the cities of Pune, Vellore, and Chennai (Madras). By 1993, about 70 percent of the 15,000 IDUs in India's Manipur state, located near the "Golden Triangle" of Myanmar and China, were HIV-positive. And more recently, a random survey in Tamil Nadu indicated that some 500,000 of that state's 25 million people are now infected. The epidemic has also spread among people who live and work along the major north-south truck corridor. In short, the evidence suggests that India's AIDS situation is on the verge of exploding if the country's leaders don't mobilize quickly enough to stop it. Moreover, in a country with 16 major languages, more than 1,600 dialects, and six major religions, such mobilization will require exceptionally skillful coordination and organization.

The Indian government has made a commitment to fighting HIV and is working with donors to coordinate prevention and care efforts. The question now is whether it can mobilize quickly enough. Last December, Prime Minister Atal Behari Vajpayee declared HIV and AIDS to be the country's most serious public health challenge. With financial assistance from the World Bank, the government is implementing a National AIDS Control Program. It aims to give autonomy and financial support to the country's 25 states in order to upgrade their health service delivery infrastructures and carry out HIV prevention and care targeted to high-risk groups. The state of Tamil Nadu already has a system in place to give financial and technical support to NGOs and has set a precedent for an effective decentralized anti-HIV campaign.

Greater reason for hope, though, lies in India's active local communities and a thriving network of NGOs. Following on the Gandhian legacy of grassroots resistance to British colonialism, local groups are emerging throughout India to tackle HIV. In 1992, for example, representatives from SANGRAM, a rural women's group in Maharashta, went to a local red light district and began passing out condoms, telling prostitutes, "This will save your life and mine." Some prostitutes, resentful of mainstream disdain for them, did not appreciate outsiders coming in to tell them what to do. "In the beginning, it was difficult; they even threw stones at us," said SANGRAM General Secretary Meena Seshu. Eventually, though, a small group of prostitutes took over the condom distribution and began educating their peers on how to avoid STDs and HIV. Since then, some 4,000 prostitutes in seven districts have formed their own collective, called the Veshya AIDS Muquabla Parishad (VAMP). The women attend training sessions on personal health, sexuality, STDs and superstition, negotiating condom use with clients, and how to be counselors for the infected and their families. Seshu notes that, in addition to lowering STDs and pregnancies, the collective has given the women strength to tackle difficult issues that might previously have been neglected. Whereas their health needs were often overlooked in the past, for example, the prostitutes are now demanding that doctors examine them and treat STDs properly. Organizations like SANGRAM and VAMP are gaining strength in several regions of India, and as they grow they are using their programs as a basis for advocating improved AIDS-prevention policies throughout the country.

In China, as far as we know, there is not yet a large-scale HIV epidemic. However, the potential for an enormous epidemic is becoming evident. China—shades of South Africa—is relaxing once-stringent economic constraints and opening previously closed doors to the outside world. These economic policy shifts are driving rapid social change and may also be paving the way for HIV.

Once confined to foreign visitors and small groups of injecting drug users in Yunnan province, HIV has entered a phase of "fast growth" throughout the country according to a recent report by the Chinese Ministry of Health. Left unchecked, HIV infections could exceed 10 million by 2010. The World Health Organization's most recent estimate puts the number infected in China at 600,000.

China eradicated open prostitution in 1949. Since the 1980s, however, commercial sex has resurfaced and seems to be growing. Girls, lured by money in China's burgeoning cities, are moving from rural areas and are often drawn into prostitution. Economic expansion is also increasing the number of migrant workers, who may now represent up to 15 percent of the total labor force. Often young, unmarried, or living away from their spouses, migrants may be more likely to have casual sex or sex with prostitutes, greatly increasing their risk of infection. And here, as elsewhere, the sharing of needles by drug addicts spreads HIV even faster than prostitution. Among injecting drug users in Yunnan province, almost 86 percent are now infected.

The Chinese government apparently recognizes the magnitude of threat to its more than 1.2 billion people. A national program for HIV/AIDS control has been approved by the State Council, China's highest governing body. Hypodermic syringes and needles are available for sale at all pharmacies throughout the country. More than a billion condoms were produced by Chinese manufacturers in 1998 and distributed by the State Family Planning Commission. At a time when other ministries were facing stiff cutbacks in staff and resources, the National Center for HIV Prevention and Control was set up to study the epidemiology of HIV, to develop health education, and to conduct clinical work for pharmaceuticals. The Chinese Railways Administration distributed AIDS prevention information to its staff of 6 million workers and among railway passengers, many of them migrants.

China's past performance with public health management offers additional reason to hope the country can keep HIV in check. China has a unique history of bringing about swift social changes to improve health. As part of the "barefoot doctors" program in the 1970s, village representatives throughout the country were given basic public health training. Their efforts to provide basic health care and convey preventive health messages to their communities brought significant declines in infectious disease and child mortality in China. Today, China's health indicators are much closer to those of industrialized nations than to those of the developing world.

An emphasis on preventing HIV infection is essential to stemming this global health catastrophe. But even though behavior changes can dramatically reduce the spread of infection, they will never eradicate HIV. And while scientific advances have greatly improved treatment, no drug therapy has yet been able to fully rid the human body of the virus. Furthermore, anti-viral treatment is out of reach for all but a small fraction of the more than 33 million infected.

Ultimately, successful containment and eventual eradication of HIV will require a safe, effective, and affordable vaccine. Many scientists think we can eventually develop such a vaccine—despite some significant hurdles. HIV is very efficient at making copies of itself, a replication that leads to disease despite a vigorous immune response. It also mutates rapidly and has produced many different strains of itself, which means an effective vaccine would have to be able to recognize and fight off each nuance of the virus. Nevertheless, candidate vaccines have already been able to stimulate some immune response in human volunteers and seem to be safe.

Even under the most optimistic scenarios, however, the development of an effective vaccine will take years—and the risks to humanity will continue to escalate if not powerfully addressed. In the last decade or so, 25 experimental vaccines have been tested in studies involving small numbers of volunteers but only one has advanced to larger scale "efficacy trials." "Unless there's a major breakthrough," says Dr. Seth Berkley of the International AIDS Vaccine Initiative, "it's unlikely we'll have a vaccine within the next decade."

Meanwhile, even the testing poses formidable challenges. For example, some standard preparation strategies used for other vaccines cannot be used for fear that a weakened form of the live virus or a whole killed virus might cause HIV infection in the person vaccinated. Even after the basic research on safety and effectiveness has been conducted, private pharmaceutical companies will still need to develop a commercial product—a process that takes an average of 10 years and costs at least $150–250 million. Because HIV has hit developing countries hardest, a vaccine that offers any real hope of eradication will need to be inexpensive, easy to transport and administer, require few if any follow-up inoculations, and protect against any strain or route of transmission of the virus.

Getting adequate funding, too, has been an uphill battle. Five years ago, the National Institutes of Health (NIH) decided not to fund large-scale efficacy trials of any leading AIDS vaccines—at the time causing a serious setback to the research. This year, NIH increased vaccine research funding by 79 percent. But if it takes over 10 years to develop a vaccine as Dr. Berkley expects, it could be several decades before the vaccine allows us to put HIV on the road to eradication.

Sitting on the edge of a big wooden chair in my living room, Amélia—the young niece of Victor, next door—used to sing hymns, waiting for lunchtime, while I did Saturday cleaning. She was a skinny kid so I liked giving her a good meal. But she got thinner and thinner. And then, her grandmother had to tie an old towel around her head to hide the open sores. But she couldn't cover up Amé's nose, which had somehow melted away. My neighbor came one day to tell me that Amé was no longer allowed to eat at my house for fear she would infect me.

Amé died about a year later. I could tell by the look in her eyes that she knew what was happening to her. I didn't do anything to take her pain away. I didn't go over to my neighbors and insist that they let her come back. I didn't explain to them that Amé couldn't possibly infect me by eating from my dishes. I was immobilized—perhaps because I didn't know how to answer the unspoken questions of a little girl who didn't understand why this was happening to her.

This article has been based on numbers, but for every number that adds up incrementally to give us the startling statistics we now have, there is a human tragedy. Facing up to the scale of the suffering that lies behind those rising numbers can become an overwhelming challenge—one that we'd rather not think about. Yet, if we don't think about it clearly enough to find a way of stopping HIV in its tracks, a great many more communities, on every continent, will lose those men and women who are in the prime of life, as Amélia's parents were. And then they will lose their Amélias.

QUESTIONS

1. Summarize the main impact that AIDS has had on adults and children around the world.
2. List and explain several statistics in this article that seem very surprising to you. Explain your surprise.
3. What are two signs that indicate the AIDS epidemic is not hopeless.
4. How did the Thai government deal with AIDS in their country? Provide at least three examples.
5. What are basic policy principles that seem to keep HIV in check?

Telemedicine: The Health System of Tomorrow

THOMAS BLANTON
DAVID C. BALCH

The place: North Carolina's Central Prison in Raleigh.

The problem: Inmate No. 35271 (not his real number) has an unusual skin rash that the prison doctor can't diagnose. The prison administration does not want to be sued for inadequate medical care for its prisoners, but at the same time, it does not want to run the security risk of transporting a dangerous felon to an outside specialist. Bringing the specialist to the prison would also present problems.

The solution: A telemedicine link with East Carolina University School of Medicine. In a special telemedicine booth a hundred miles away, a dermatologist sits in front of a video console and interviews the patient while directing the prison nurse where to point the tiny dermatology camera. He diagnoses the condition and prescribes a course of treatment.

The result: The prisoner's condition is treated, the prison doesn't get sued, and the state saves potentially thousands of dollars.

A far-flung view of the future? Not at all. Telemedicine is already in practice in settings such as this and may become nearly universal in the not-too-distant future.

CREATING PORTABLE HEALTH CARE

The phrase "health care reform" has been used in recent political years to describe changes in the way health care is financed. Even though health-care-financing reforms failed to make it through the last Congress, health care is changing in the United States. The real reform is developing in the way health care is delivered.

As technology has developed over the years, medical diagnoses have become more accurate and treatments more effective. But the x-ray, magnetic resonance imaging machines, and other tools are not portable. Treatment and therapeutic facilities have been centralized in hospitals and rehabilitation centers so that patients have to travel distances and wait in doctors' offices to be seen. As a result, doctors don't get to see patients on a regular basis and end up treating symptoms and diseases rather than people.

Proposed solutions have been varied, from holistic health to formerly fringe ideas such as herbal medicine and massage therapy, which are gaining in popularity. But technology may soon allow the house call to return to mainstream medicine, with data being transported rather than people.

The idea of telemedicine has been around for almost as long as there has been science fiction and has been in actual practice since the late 1950s. But the recent upsurge in medical costs, combined with advances in technology, is now making telemedicine a widespread reality.

TELEMEDICINE IN PRACTICE

At present, these programs tend to serve rural areas where population is sparse and income low. Areas such as western Texas, eastern North Carolina, rural Georgia, and West Virginia are the most active in using telemedicine. A study

Originally appeared in September–October 1995 issue of *The Futurist*. Used with permission from the World Future Society, 7910 Woodmont Avenue, Suite 450, Bethesda, Maryland 20814. 301/686-8274. (http://www.wfs.org)

by the consulting firm of Arthur D. Little, Inc., showed that healthcare costs could be reduced by as much as $36 billion a year if health information technologies such as telemedicine were widely used.

In 1989, the U.S. Public Health Service gave partial funding to Texas Tech's MedNet program, which linked health-care practitioners in hospitals and clinics from 37 rural communities in western Texas. In its first year, this project reported a net savings of 14% to 22% in the cost of health-care delivery. The agencies involved reported a decrease in salaries and ambulance costs along with an increase in earned revenue.

One important issue in developing telemedicine has been states' costs in providing health care in prisons. California alone spends $380 million per year on prison health care. With the popularity of laws such as "Three Strikes and You're Out," prison populations will increase. So will the cost of prison health care.

In 1990, a prison inmate sued the state of Florida over inadequate health care and won a $1-million settlement. That prompted Florida to begin prison telemedicine on a pilot basis.

The next year, North Carolina's largest prison—Central Prison in Raleigh—contracted with the East Carolina University School of Medicine to provide telemedicine services. Since that time, physicians from the medical school have provided more than 350 consultations with the prison. In 1994, the school built the first "telemedicine suite" in the nation—four consultation booths specially designed for telemedicine.

In addition to the prison, two hospitals in rural eastern North Carolina have consultation links with the East Carolina medical school. The Medical College of Georgia, too, is building a network of 60 sites around the state for telemedicine consultations.

Just as cable television began as a way of serving rural areas outside of broadcasters' range, telemedicine is beginning as a way of reaching out to medically underserved areas. As the possibilities of cable television offerings became better known, metropolitan areas became its heaviest users. Similarly, as physicians and patients see the possibilities in telemedicine, it should spread more widely and into urbanized areas.

Patients recovering from heart attacks can already put on a headset, connect the electrocardiogram (ECG) wires to their chests, and ride their stationary bikes. The patients stay at home and communicate by telephone with the hospital. The same phone wires that carry their conversation also carry the ECG information to the medical technician on the other end. Up to five patients at a time can be served this way, and they are in contact with each other as well as the hospital.

A pilot program for this type of cardiac rehabilitation took place at St. Vincent Charity Hospitals in Cleveland, Ohio, during the late 1970s and worked so well that it became a model for other areas, including Nashville, Tennessee, and Greenville, North Carolina. Doctors at East Carolina University are setting aside a cable channel so that the cardiac rehab patients can have visual contact with the hospital staff. As technology advances, there may eventually be two-way video interaction between the patient and the hospital.

HEALTH CARE AT HOME AND WORK

As the technology advances, telemedicine will spread out over the information superhighway to link rural hospitals with smaller medical centers; then local doctors' offices will come on-line; and, finally, telemedicine will reach into private homes.

QUESTIONS

1. Why is telemedicine gaining popularity and becoming more of a reality?
2. How much in annual healthcare costs could be saved if technologies such as telemedicine are used?
3. Why is telemedicine an especially attractive option for use in prisons?
4. Have telemedicine sites proved to be successful? Explain.
5. What are other possible applications of telemedicine?
6. If you lived in a rural area hundreds of miles from a hospital or doctor, would you be in favor of the use of telemedicine? Explain your position.
7. Explain the pros, cons, and implications of using telemedicine as a future approach to medicine.

Case Study 1
Nadeem

Ahmed S. Khan

It was a bright sunny morning. Students at New Public School were excited as Parent's Day was just around the corner, and they wanted to exhibit their academic and extracurricular activities during the Parent's Day program. Students returned to their classrooms after the morning assembly. Nadeem, a ninth-grade student, walked into his classroom. His teacher, Mrs. Usman, asked Nadeem how he was doing. Nadeem replied that he was fine and very excited about the upcoming Parent's Day program at the school. He had written a special song and would sing it for the parents.

Mrs. Usman started teaching Algebra. Nadeem was a good student. After the short lecture, Mrs. Usman gave the class a short quiz. All of the students started to take quiz. About five minutes into the quiz, Mrs. Usman noticed that Nadeem had collapsed in his chair. Mrs. Usman was aware of Nadeem's history of suffering from seizures. With the help of other students, she placed Nadeem on ground, and they all held his hands and legs and tried to comfort him. During the seizure, his muscles became rigid; a one- to two-minute period of violent, rhythmic convulsions followed. During the seizure, Nadeem would become unconscious. The seizures could occur randomly and would last two to three minutes. After the seizure, Nadeem would experience a headache and feel drowsy and confused. The frequency of his seizures was unpredictable. Sometimes he would have two to five seizures a day, and then there were times when he would not have any seizures for a few days or even weeks.

Whenever Nadeem had a seizure, Mrs. Usman would call Nadeem's mother, who would come to school or send one of Nadeem's siblings or friends to bring Nadeem home. Once Nadeem got home, he would take a nap for a few hours and then wake up and resume his activities in a normal manner.

According to the Mayo Clinic, approximately 10 percent of Americans will experience a single seizure episode during their lifetime. Only about 3 percent of these people will go on to develop epilepsy. More than 2 million Americans have epilepsy.

Nadeem started to have seizures when he was very young. In 1966, when Nadeem was about six months old, he had to be hospitalized for treatment of pneumonia. At the hospital, he was kept in the intensive care unit. One night during his stay at the hospital, his condition became very unstable. As there were no senior doctors present on duty, Nadeem's mother requested the attending intern to call senior doctors to get advice regarding Nadeem's condition. But the attending intern did not follow the advice, and in an effort to control Nadeem's condition, he administered an adult dose of antibiotics to a six-month-old baby. After a few days at the hospital, Nadeem's condition improved and his parents took him home. However, they observed a week later that Nadeem started to experience convulsive seizures: his body would shiver, he would become unconscious, and white foam would come out of his mouth. His parents took him to the hospital, where doctors performed various tests, but they could not find the cause of Nadeem's seizures. His doctor treated him with anticonvulsants.

Over time, as Nadeem grew from a baby to a young boy to a teenager, he continued to have seizures. Nadeem was a good student and was active in extracurricular activities both at school and at home. His parents made all possible efforts to find the root cause of Nadeem's seizures as well as their treatment, using modern medicine and alternative medicine. Yet their efforts were to no avail, as Nadeem continued to suffer from seizures. It became a cumbersome task for Nadeem's parents and siblings to take care of him.

One evening in 1979, Nadeem was chatting with his friends outside his home after returning from school. He suddenly collapsed and fell on the ground, in the process hitting his head against the ground and becoming unconscious. His friend carried him inside the house, and his family members put him in his bed.

They thought it was just another episode of seizures and that he would be fine after taking his post-seizure nap. However, he kept on sleeping. The next morning, his mother tried to wake him up, but he did not respond to her calls. The family members got worried and took Nadeem to the hospital. The doctors admitted Nadeem to the hospital and determined that he was in a state of coma. He remained in the coma for three months. During the coma, he continued having recurrent, repeated convulsions. He was treated as an inpatient at a local hospital with electroconvulsive therapy (ECT), which involves electrical stimulation of the brain. After three months, Nadeem regained consciousness but suffered severe short-term memory loss. He started displaying signs of dementia and was diagnosed with mental retardation. Since his head injury in 1979, it has been very challenging for Nadeem's parents to take care of him.

Today, Nadeem is a 39-year-old man. He is overweight (268 lbs.) for his age. He suffers from dementia. He lives with his parents. Most of the time, he stays at home; because of his memory loss, he cannot work. Nadeem exhibits a paucity of knowledge, and he talks very fast and in a slurred manner. He spends his day sitting idly or watching television. Nadeem's parents have a limited retirement income, so they cannot send Nadeem to a structured psychosocial rehabilitation day program. He requires constant supervision at home. Nadeem has limited self-help skills and is very dependent on his parents, who are in their late seventies and are getting weaker day by day due to old age and multiple ailments. Taking care of Nadeem brings extra burdens on them in their old age. Nadeem's parents are very concerned about who will take of him after they are gone.

QUESTIONS

1. What factors led to the present state of Nadeem's mental disorder?
2. Who is responsible for Nadeem's brain impairment?
3. What options would you recommend to Nadeem's parents for his long-term care and treatment?
4. Using Internet search tools, complete the following tables.

(a) Describe the function of the following brain structures.

Structure	*Function*
Cerebrum	
Cerebellum	
Limbic system	
Brain stem	

(b) Compare the advantages and limitations of the following brain-imaging techniques.

Imaging Techniques	*Advantages*	*Limitations*
CT (roentgen-ray computed tomography)		
MRI (magnetic resonance imaging)		
SPECT/PET (single photon/positron emission computed tomography)		

5. Visit the following Web site: http://www.med.harvard.edu/AANLIB/cases/caseNN1/mr1-dg1/015.htm. Observe the MRI brain scan images and describe what happens to the structure of the brain due to various brain disorders/diseases listed in the following table.

Brain Disorder/Disease	*Changes in Brain Structure*
Cerebrovascular disease (stroke)	
Neoplastic disease (brain tumor)	
Degenerative disease	
Inflammatory or infectious disease	

Complete and discuss the following flowchart.

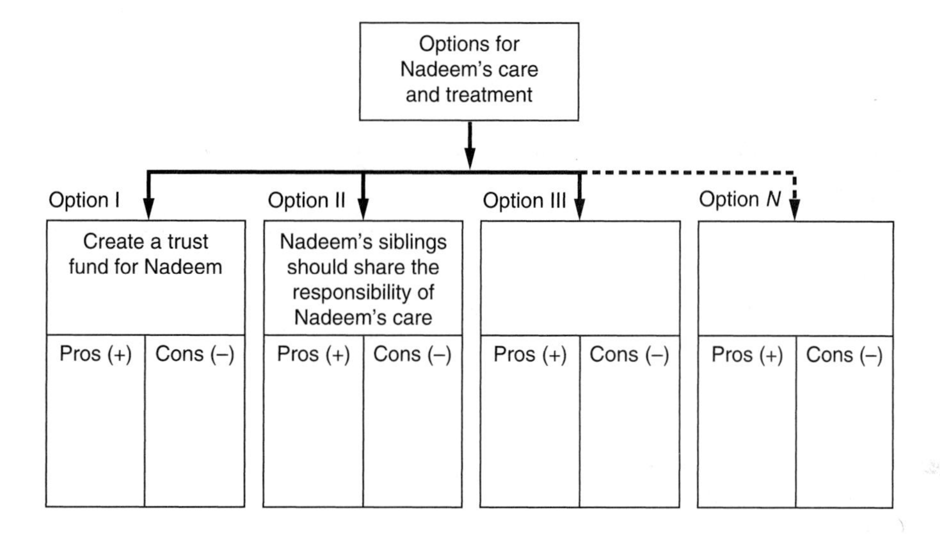

Case Study 2
Ed

AHMED S. KHAN

Last month, Ed celebrated his 73rd birthday. Ed was a professor of electrical engineering at a local university. Ed enjoyed teaching. After his retirement at age 65, he continued to teach on an adjunct basis. Three years ago, Ed had to stop teaching because his wife, Jill, was diagnosed with Alzheimer's disease (AD). The disease is named after Dr. Alois Alzheimer, a German doctor, who in 1906 noticed changes in the brain of a woman who died of an unusual brain illness. The Alzheimer's disease is a form of dementia, a brain disorder that severely affects a person's ability to carry out daily tasks and activities. Alzheimer's disease affects the parts of the brain that are responsible for thought, memory, and language functions. The disease is more common in older people, and its risk increases with age. According to the Alzheimer's Disease Education and Referral Center, about 4.5 million people suffer from AD in the United States.

Ed devotes much of his time taking care of Jill, whose memory has deteriorated rapidly during the past year, and now she is completely dependent on him. Ed and Jill have two children, one son and one daughter, who live out of town. Ed does not want to place Jill in a senior citizen home because he has heard horror stories about the mistreatment of the elderly at those places.

Every day, Ed makes sure to wake up before his wife so that he can keep an eye on her actions and activities. Sometimes, Jill gets out of the home and starts to wander in the neighborhood. Ed has had to change the door locks so that she cannot get out. Sometimes, Jill starts to set the table as if she is expecting guests for dinner. There are other times when Jill starts to pack a suitcase as if she is getting ready for a vacation. Jill's activities add up to extra work for Ed, who suffers from high blood pressure and diabetes.

Two weeks ago, Ed had to go the emergency room at the hospital, as he could not retain any food in his stomach due to vomiting. Doctors informed Ed that they would keep him in the hospital for a few days so that they could perform various medical tests to determine the nature of his ailment. Ed did not want to stay in the hospital, but he did not have any choice. He called his neighbor, Mary, a retired schoolteacher, and asked her to take care of Jill for a few days until he got out of the hospital. Mary, who suffers from arthritis, agreed to help. Ed also tried to call his son, Mark, to seek his help. However, Ed could not reach Mark because he was out of the country. Ed also called his daughter, Beverly, and asked her to come for a few days, but Beverly told him that she could not come because of her business commitments.

On the fourth day of his stay at the hospital, nephrologists informed Ed that he was suffering from kidney failure (end-stage renal disease), and they discussed with him possible treatment options (see Tables P7.1 P7.2). In end-stage renal disease, the kidneys function at less than 10 percent of normal capacity, and their reduced functionality cannot sustain life. Therefore, people with end-stage renal disease require dialysis or a transplant to stay alive. If the transplant is not possible due to a patient's poor health, then dialysis is the patient's only choice.

Nephrologists also informed Ed that they would request the radiologists at the hospital to place a temporary catheter in his neck to establish a vascular access port to

Table P7.1
Treatment Options for End-Stage Renal Disease (kidney failure)

Modality	*Treatment Setting*	*Typical Treatment Duration*	*Comments*
Hemodialysis (HD)	Dialysis center or at a hospital	Four-hour dialysis session three times a week	Most commonly used modality; removes extra fluids, chemicals, and wastes from bloodstream by filtering blood through an artificial kidney (dialyzer); requires a vascular access in the form of a fistula, graft, or by using a catheter (see Table P7.2)
Peritoneal dialysis (PD)	Home	Daily	Uses the vast network of tiny blood vessels in the patient's abdomen (peritoneal cavity) to filter blood instead of using an external dialyzer; a small, flexible tube (catheter) is implanted into the patient's abdomen and then a dialysis solution is infused into and drained out of the patient's abdomen to remove waste and excess fluid; patients should be physically capable of operating the PD machine or know someone who will operate it for them
Transplant	Surgery at a hospital	Kidney recipient is usually observed in the hospital for about one week; subsequent close follow-up in the transplant clinic and frequent monitoring of labwork; patients require life-long treatment with medications that suppress their immune response (immunosuppressive therapy) to prevent rejection of the transplanted kidney	Over 9,000 kidney transplants performed each year; availability of a donor kidney may take a few years; a donor kidney (obtained from a living related/unrelated/deceased donor) is transplanted through a surgical procedure with risks of bleeding and infection; offers the best outlook for patients with end-stage kidney disease; organ survival rates vary from 90–80% for 1–3 years and from 50 percent for 10–15 years

Table P7.2
Vascular Access Methods

A vascular access allows blood to safely leave a patient's body and travel to the dialyzer (artificial kidney) and then return to patient's body. A vascular access port is usually placed in the patient's arm, leg, or neck. A vascular access port is established using three access methods: fistula, graft, and catheter.

Vascular Access Method	*Description*	*Comments*
Fistula	An arterio-venous fistula (AVF) is made with a patient's arm or leg by surgical procedure in which a vein is connected to a nearby artery, causing the vein to grow larger and stronger because stronger arterial blood flows through it; takes typically six weeks for the fistula to get large enough to deliver an appropriate blood flow for hemodialysis	Has longer life than other access methods; probability of infection is lower than other access methods; requires fewer follow-up visits to the hospital to keep it working; may not work for all patients if they have smaller veins, bone deformity, etc.
Graft	An arterio-venous graft (AVG) is a small, soft tube placed under the skin by surgical procedure to connect one side of the tube to an artery and the other end to a vein; typically takes three weeks for a graft to be used for hemodialysis	Typically lasts for 3–5 years; probability of infection is higher than fistulas; probability of clotting is higher than fistulas
Catheter	A catheter is a small, Y-shaped plastic tube placed into a large vein in the neck; one end of the catheter is outside the body and the other end is placed in the heart; no wait period; catheters can be used for dialysis right after their placement	Probability of clotting is higher; can lead to poor removal of toxic products and fluid during dialysis; have highest risk of infection (catheter infections can lead to serious infections in the body)

perform dialysis for the short-term treatment. However, for the long-term treatment, Ed would have to choose a treatment modality. Ed is very worried and confused about selecting the right treatment option that will enable him to continue to take care of his wife. Before he can be discharged from the hospital, he needs to inform his doctors regarding his treatment modality option so that they can chart a long-term course of treatment for him.

BOX 1 Organ Transplant Statistics

- More than 80,000 men, women, and children currently await life-saving transplants.
- In 2001, 24,092 organ transplants and nearly 900,000 tissue transplants were performed.
- Every 13 minutes, another name is added to the national transplant waiting list.
- An average of 17 people per day die due to a lack of available organs for transplants.

Source: www.shareyourlife.org.

REFERENCES

Alzheimer's Disease Education and Referral Center. Available at http://www.

Kidney Failure. Available at http://www.mayoclinic.com.

Knotek, Bobbie.(2005). Choosing Your Hemodialysis Vascular Access. *National Kidney Foundation (NKF) Family Focus.* Volume 14, No. 1:6–7.

Organ Transplant Statistics. Available at http://www.shareyourlife.org.

QUESTION

1. Which option would you recommend that Ed pursue? Would your recommendation enable Ed to continue to take care of his wife? Discuss the advantages and limitations of your recommendation.

Complete and discuss the following flowchart.

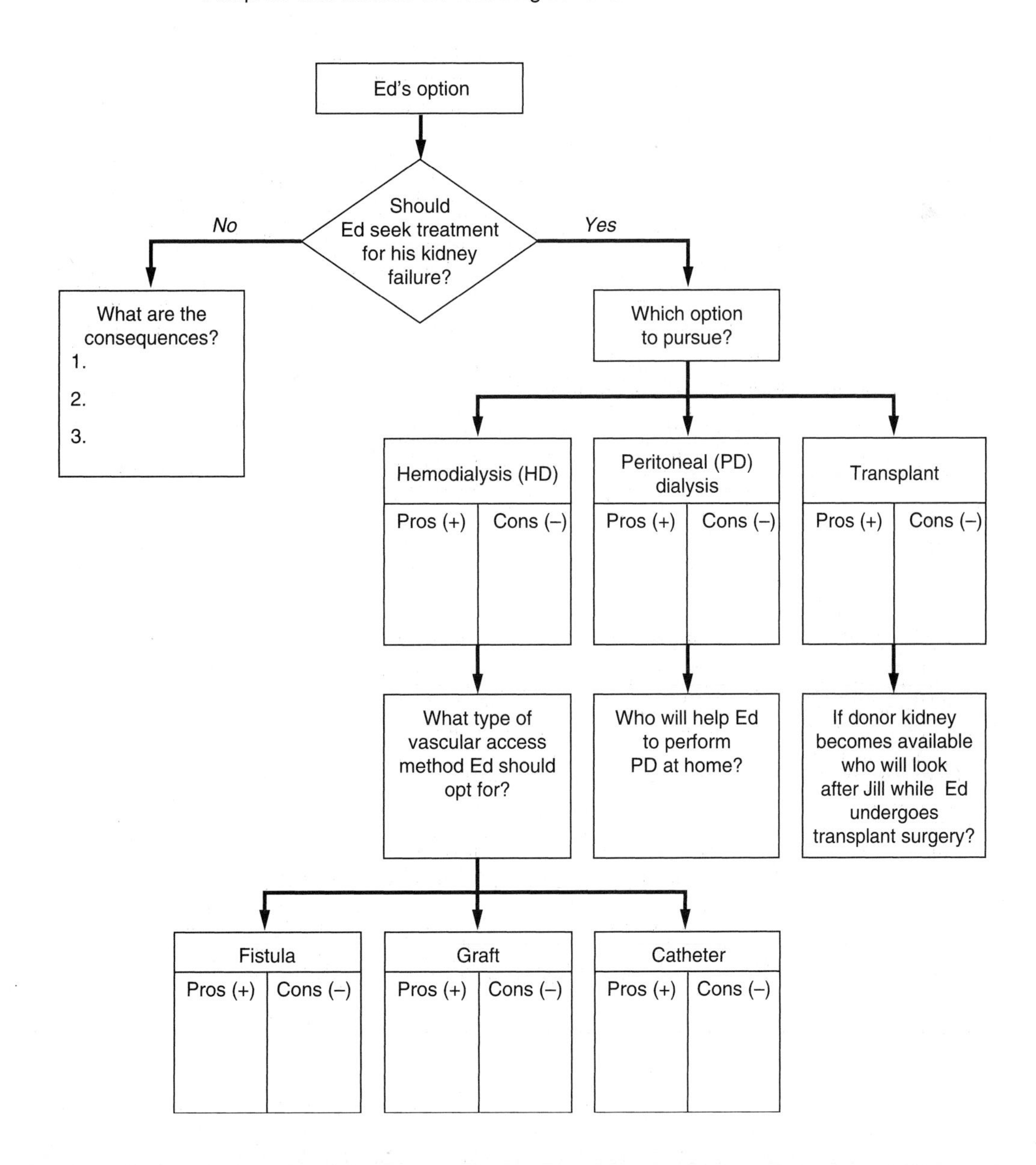

SCENARIO I: Gene Therapy

Consider the following hypothetical news broadcast.
January 20XX, New York, NY
Excerpted from the NBC six o'clock news

Researchers reported today that they can replace genes that cause disease by using artificial human chromosomes. This treatment will "fix" genes that are not working right and will replace missing genes. The concern is, however, that there could be serious medical risks. There is still hope that artificial chromosomes will be sold soon.

Discussion

How can artificial human chromosomes replace genes? What are the medical risks of such replacement? What ethical issues relate to the sale of artificial human chromosomes? Can artificial chromosomes be used to treat disorders caused by a single gene?

SCENARIO II: Xenotransplants

Consider the following scene.
October 10, 20XX Chicago, IL

John's kidney has become diseased. He has been on a donor list for months, waiting impatiently for a new kidney. The outlook for the transplant or for returning to a healthy lifestyle seems grim. However, John does not want to give up hope. As he is waiting in the examining room for the doctor, he reviews two detailed reports about xenotransplants and quickly scribbles a list of questions.

John is sitting on the examining table as the doctor walks in. Before the doctor can say a word, John says, "Dr. Michaels, I have been reading reports on xenotransplants, and I want one."

"John, let's discuss this idea of yours a little further," Dr. Michaels replies.

Discussion

Select a year for this scenario to occur (20??). Based on the time period that you choose, answer the following questions: What kind of animal should be used for John's kidney transplant? What complications might occur? How would researchers prevent animal viruses from being transferred to John? Could animals be kept in pathogen-free environments to prevent virus transfer? Should there be concern for the possibility that animals carry as-yet-unknown viruses that are fatal to humans? If you were John, what five questions would you want Dr. Michaels to answer for you (not including the previously listed questions)?

Conclusion

According to the Centers for Disease Control and Prevention (CDC), "the median age of the world's population is increasing because of a decline in fertility and a 20-year increase in the average life span during the second half of the 20th century." In 2003, the American life expectancy was 77.6 years; in 2002, it was 77.2 years. In comparison, the average life span in America in 1900 was 47.3 years; in 1950, this number increased to 68.2 years.

The National Center for Health Statistics states that age-adjusted death rates declined for eight of the 15 leading causes of death in 2002–2003. The two major causes of death in America are heart disease and cancer. According to the CDC, there was 3.6-percent decrease in heart disease in 2002 and a 2.2-percent decrease in cancer. There was also a 4.6-percent decline in strokes, 3.1-percent decline in flu/pneumonia, and 2.1-percent decline in liver disease and cirrhosis. The death rate for HIV declined 4.1 percent. In a speech delivered by Professor Carin Smith, president of the ILEP Medico-Social Commission from the University of Aberdeen, UK, "There has been a 85% reduced (rate of Leprosy) in registered cases in over 70 countries reaching the target of reducing the prevalence of leprosy to less than 1 in 10,000." (NCHS Fact Sheet 2005)

There are many reasons for this life span increase: awareness of patient responsibility in making healthy choices, public health interventions, medical treatment options, new life-saving technologies, and a decline in some diseases. Some life span increases are correlated to topics presented in Part II. The purpose of Part VII is to challenge you to think in different ways about the pros and cons of medical technology so that you begin to make informed decisions for yourself and your loved ones. We are living longer, but positive and negative quality-of-life issues continue to arise when discussing antibiotic resistance, global responses to HIV and AIDS, genetic modifications of food, the use of telemedicine, and life-sustaining technologies. It is impossible to read Part VII without wondering about the impact of medical technology on each individual's lifestyle and medical choices.

Part VIII will discuss the effects of technology on Third World countries. It will provide a clearer understanding of the role that technology plays in the development of these countries.

REFERENCES

Centers for Disease Control and Prevention. (2003). "Public Health and Aging—U.S. and World Wide." February 14, 2003/52(06);101–106. Available at http://www.cdc.gov/mmwr/preview/mmwrhtml/mm5206a2.htm.

Centers for Disease Control and Prevention. National Center for Health Statistics. (2005). "2005 Fact Sheet. Life Expectancy Hits Record High." Available at http://www.cdc.gov/nchs/pressroom/05facts/lifeexpectancy.htm.

Health, United States. Table 27: Life Expectancy at Birth at 65 Years of Age, and at 75 Years of Age, According to Race and Sex: United States, Selected Years 1900–2003. Available atftp://fxlstp.cdc.gov/pub/Health_Statistics/NCHS/Publications/HealthUS/hus05table.

Smith, Cairn. (2000). "Leprosy: Past, Present and Future." University of Aberdeen. Available at http://www.aifo.it/english/resources/online/books/leprosy/Presentation%20Cairn%20Smith%20Rome%20Jan%2000.pdf#search=%22leprosy%20reduced%20by%20%25%20in%20last%2010%20years%22.

INTERNET EXERCISES

Using any Internet search engine (e.g., Alta Vista, Yahoo, Google) to answer the following questions.

1. Internet questions
   a. What constraints and rules must researchers follow today when working with animals and humans? What is "informed consent?"
   b. What are the legal ramifications of doctors using clients for research without obtaining informed consent?
   c. Find information on current research on the misuse of antibiotics. What are the current medical concerns about patients who fail to finish antibiotic prescriptions or who use others' antibiotics without a doctor's prescription?
   d. What is the current status of the Human Genome Project?
   e. What is the current trend in the field of organ transplants? In what geographical locations are organ transplants performed most often? Least often? What age groups usually receive the most organ transplants?
   f. Define the following terms:

      Colonization

      Tissue engineering

      Bioengineered tissue

      Cell transplant

      Euthanasia

      Biotechnology
2. Xenotransplants questions
   a. What is a xenotransplant?
   b. Describe the first animal-to-human transplant. When was it? Who did the surgery? What kinds of animals were used? Did the recipient live? Did the animals live?
   c. Why is it difficult to keep patients who have had xenotransplants alive? Does the body usually reject the transplanted organ? What effect do immunosuppressive drugs have on the human body?
   d. Why do doctors not want to use baboons for xenotransplants? Why, medically, might baboons be the best animal to use for xenotransplants?
   e. Ethically, why might pigs be better to use for xenotransplants than baboons?
3. Tissue-engineering questions
   a. What parts of the body can be re-created through tissue engineering? Can a heart valve, liver, uterus, bone, skin, eye, or brain be re-created? It is doubtful that eyes could be grown in cultures, but some researchers hope to create artificial vision systems instead. What research can you find on each topic?

b. Some biotech companies are waiting to gain approval from the Food and Drug Administration for skin substitutes that would help patients suffering from burns and serious wounds. Until recently, doctors used cadaver tissue to replace dead tissue, but the cadaver tissue is sometimes rejected by the patients' immune system. By growing skin in the lab, the need for cadaver skin is eliminated. What research can you find on this technological phenomenon?

c. Tissue engineering today may start with neonatal foreskin; cow tendons may be used as replacement connective tissues. Try to find more details on this type of tissue engineering.

USEFUL WEB SITES

http://www. Cbsnews.com/stories/2000/03/02/48 hours/ main137321.html	Silent Killers: Is Your Building Sick?
http://www.ama-assn.org/amednews/site/facts03htm	American Medical News
http://www.iatp.am/economics/migr/facts.htm	Fifty Facts from the World Health Organization
http://www.er.doe.gov/production/ober/life/ HELSRD_top.html	Office of Biological and Environmental Research
http://www.ornl.gov/sci/techresources/ Human_Genorm/elsi/gmfoo.shtml	Overview of genetically altered foods, crops, and organisms
http://www.fda.gov/cber/xap/htm	USFDA xenotransplantation action plan
http://www.nlm.nig.gov/medlineplus/ aidslivingwithaids.html	Living with AIDS
http://www.transweb.org	Information about organ transplants

Technology and the Third World

The first world . . . the third world . . . the rich . . . the poor . . . where are the boundaries . . .? One Planet . . . One People. Photo courtesy of NASA Headquarters, http://visibleearth.nasa.gov/images/.

OBJECTIVES

After reading Part VIII, Technology and the Third World, *you will be able to*

1. Recognize the historical background and major issues of the developing world.
2. Distinguish between the developed and the developing world's problems and viewpoints.
3. Examine the problems, urgencies, and implications of the lack of technology in the developing world.
4. Analyze and use tables and graphs to promote a better understanding of the techno-economic gap between the First and the Third World.
5. Realize the magnitude of the intrinsic and extrinsic problems of the developing world.
6. Determine appropriate pathways and policies for the sustainable economic growth in the Third World.
7. Evaluate the importance of the use of technology in solving the economic and environmental problems of the developing world.

INTRODUCTION

All human beings are in truth akin; All in creation share one origin. When fate allots a member pangs and pain. No ease for other members then remains. If, unperturbed, another's grief canst scan, Thou are not worth of the name of man.

SAADI

The advances in new and emerging information technologies have transformed the world into a global village. The people of this global village find themselves to be close as they witness world events unfolding on television screens and computer monitors, and yet they still remain oceans apart in terms of their economic conditions and living standards. The gap between the world's poor and its rich is wide and has been getting even wider; both within and among countries. The following facts illustrate the dilemma:

- According to the 1999 Human Development Report issued by the United Nations Development Program (UNDP), the gap between the richest and poorest country, in terms of GDP per capita, was about 3 to 1 in 1820, 35 to 1 in 1950, and 72 to 1 in 1992.
- The 1999 UNDP report reveals that the assets of the three richest individuals in the world are now equal to the gross national products of the 48 least-developed countries of the world. In 1960, the richest 20 percent of the world's population controlled 70 percent of global income. By 1993, they controlled 85 percent, and the share of the poorest 20 percent had decreased from 2.3 to 1.4 percent.
- The world's poorest countries owe more than $520 billion in debt.
- Of the $23 trillion gross world product (GWP) in 1993, $18 trillion was generated in the industrial world and only $5 trillion circulated in the developing world.
- The gap in per-capita income between the industrial and developing worlds tripled over the past three decades—from $5,700 in 1960 to $15,400 in 1993.
- According to the 2003 World Bank report, the income inequality in Latin America had worsened with the richest one tenth of the population earning 48% of its total income, while the poorest tenth earns only 1.6%.
- Economic conditions and growth have been failing during the last 15 years in about 100 countries, which are home to one-quarter of the world's population. The income of the 1.5 billion people living in these countries was lower in the 1990s than in earlier decades.
- One-fifth of the world's population—those people living in developed nations—now accounts for 70 percent of the world's energy use.
- More than 80 percent of the world's population resides in developing countries.
- According to the former president of the World Bank, James Wolfensohm, the number of people living in acute poverty could grow from 3 billion to 5 billion in the next 30 years if world leaders do not act now to prevent the trend.

Many developing nations gained independence from the colonial powers after the World War II. Yet their dreams and aspirations of economic growth and prosperity have yet to materialize. People in developing countries wonder whether the colonial era has really ended or whether it has merely been replaced by an economic system in which international financial institutions and multinational corporations rule developing nations. Part VIII presents an array of issues and challenges facing the Third World today. The readings in Part VIII will enable readers to gain insight into the problems of the developing world and their potential solutions. The people of the developed world have entered the twenty-first century and a new millennium, but millions of people in many developing countries are still struggling to enter the eighteenth, nineteenth, and twentieth centuries due to their lagging technological level.

BOX 1 A Poverty-Free World

"Can we really create a poverty-free world? A world without third-class or fourth-class citizens, a world without a hungry, liberate barefoot underclass?... Yes, we can, in the same way we can create 'sovereign' states, or 'democratic' political systems, or 'free market economies.' A poverty-free world would not be perfect, but it would be the best approximation of the ideal. We have created a slavery-free world, a polio-free world, an apartheid-free world. Creating a poverty-free world would be greater than all these accomplishments, while at the same time reinforcing them. This would be a world that we could all be proud to live in."

Dr. Muhammad Yunus, Nobel Laureate

Source: Banker to the Poor, Autobiography of Muhammad Yunus, The University Press Limited, Dhaka, Bangladesh, pp. 288–289

One Planet: Many Worlds

AHMED S. KHAN

The Earth is one but the world is not. We all depend on one biosphere for sustaining our lives. Yet each community, each country, strives for survival and prosperity with little regard for its impact on others.
WORLD COMMISSION ON ENVIRONMENT AND DEVELOPMENT (WCED), Our Common Future

The great technological accomplishments of the twentieth century in the areas of telecommunications, computers, energy, agriculture, materials, medicine, genetic engineering, and defense have transformed the world and brought people closer, yet millions of people worldwide still go to bed hungry at night. These technological advances have enabled us to design advanced early warning systems to warn against missile attacks, but we have failed to develop an advanced early warning system to warn against and prevent global famine or the spread of disease. Thanks to state-of-the-art technologies, we are able to design spaceships to explore life on other planets, yet we have failed to preserve life on planet Earth. Millions of children worldwide continue to die due to malnutrition, disease, and poverty.

In this information age, thanks to the Internet, the world has been transformed into a global village. But in this global village, the economic gulf between the developed and the developing countries, commonly known as the Third World, is wide and growing. This economic disparity is one of the most challenging issues in world affairs.

In 1952, when the French demographer, Alfred Sauvey, coined the phrase "Third World," he observed that the aspirations of the nations gaining independence from the colonial powers were similar to those of the Third Estate in France at the time of the French Revolution. Like the French commoners who, in the later part of the eighteenth century, struggled against the clergy and the elite, these nations wanted to become the masters of their own destinies (Grimm 1990).

The usage of the phrase "Third World" led to coinage of the phrases "First World" and "Second World." The "First World" represents mainly economically developed Western countries. The "Second World" was meant to represent the communist countries of Eastern Europe during the Cold War era, but the term never really caught on. Today in the post-Cold War era, these countries are referred to as former Soviet-bloc countries. Some of these countries are economically worse than Third World countries, but they are never called Third World countries. Today, the phrase "Third World" is exclusively used to describe a group of nations that are struggling to feed, house, clothe, and educate their people while restrained by poverty, climate, oppression, war, the aftereffects of colonialism, and the economic bondage of the IMF (International Monetary Fund) and the World Bank. The phrase "Third World" is also used in a discriminatory manner. According to William J. Grimm, associate editor of *Maryknoll Magazine*, the Third World is "any nonwhite country whose economy is as bad as or worse than Poland's. No matter how desperate the economic situation may be in eastern Europe or on parts of the Iberian Peninsula, those places are never spoken of as being the Third World, which is a world of brown-, black-, or yellow-skinned people. No white nation is described with the term, though the country may be economically worse off than, for example, Thailand. When whites ruled Rhodesia, it was not the Third World. Now that it is black-ruled Zimbabwe, it is." (Grimm 1990).

Many African and Asian nations gained independence from the colonial powers after the second World War. But their dreams and aspirations of economic growth and prosperity have yet to materialize. People in the developing countries wonder if the colonial era has really ended or if it has merely been replaced by an economic system in which international financial institutions and the multinational corporations rule over developing nations.

As the developed world continues to make tremendous progress in technological domains in the new millennium, people in many rural areas of developing countries, due to their lagging technological level, are still struggling to enter the eighteenth, nineteenth, and twentieth centuries.

Development in the Third World is inhibited due to a number of extrinsic and intrinsic factors. The extrinsic factors include political interference by superpowers during the Cold War era, political and economic instability caused by international financial institutions, and economic exploitation of resources by multinational companies. The intrinsic factors are illiteracy, poverty, social injustice, corruption, lack of resources, population pressures, and power-hungry political, feudal, and military elite.

The developed nations have portrayed themselves to the Third World as role models of development and modernization while simultaneously producing a disproportionate share of pollutants emitted into the atmosphere through the burning of fossil fuels relative to the size of their population compared with the population of the Third World. Third World nations confront numerous environmental problems of their own, with water, soil, and air pollution being among the most common. Other problems include desertification, deforestation, extinction of animal and plant species, international trade in wildlife, export of toxic waste by developed countries to the Third World, and disposal of garbage and human waste (Myers 1990; Mennerick and Mehrangiz 1991).

Table 1
Development Indicators for the Five Most Populous Low-Income Countries

Parameter	*China*	*India*	*Pakistan*	*Bangladesh*	*Nigeria*
Population (2003) in billions	1.3	1.0	0.148	0.138	0.136
Prevalence of HIV, total (% of population aged 15–49)	0.1	0.9	0.1		5.4
Infant mortality rate per 1,000 births	130	63	74	46	98
Fertility rate (1995)	1.9	2.9	4.5	2.9	5.6
Energy use in (oil eq.) kg	959.5	513.3	454.1	154.8	718.3
Carbon dioxide emission per capita	2.3	1.1	0.7	0.2	0.3
Adult literacy rate (2003)					
% Male	91.7	67.7	59.8*	50.3	74.4
% Female	77.0	44.4	30.6*	31.4	59.4
2003 annual growth rate (GDP) %	9.3	8.6	5.1	5.3	10.7
Exports of goods & services (% of GDP)	34.3	14.5	20.5	14.2	50.0
Imports of goods & services (% of GDP)	31.8	16.0	20.4	20.0	40.9
External debt (short term) $ billion (2003)	73	4.7	1.2	0.617	3.4
Debt service (% of exports of goods and services, 2003)	7.3	18.1	16	6	6.9
Fixed lines & mobile telephones (per 1,000 people, 2003)	423.8	71	44.2	15.6	32.5
Personal computers (per 1,000 people, 2002)	27.6	7.2	4.3	7.8	7.1
Internet users (per 1,000 people)	63.2	17.5	10.3	1.8	6.1

GDP = gross domestic product.
Sources: World Bank, World Development Indicators (http://www.worldbank.org). **CIA World Factbook*, 2004.

Efforts to industrialize by Third World countries are often hindered by developed countries in the name of reducing pollution. How would the developed countries respond if the power relationship were reversed? How would they respond if Third World nations had the economic and technological power to exert pressure on these countries to reduce, for example, the use of fossil fuels and curb industrial processes to reduce pollution?

Despite the fact that average incomes for both the developed and the developing countries have increased, the gap between them has widened during recent years. Since 1960, the developed countries have become richer by an average of $5,400 in gross national product (GNP) per capita, while the developing countries have advanced by only $800. Latin American citizens are, on average, poorer than they were in 1970. Countries such as Burundi, Somalia, Burkina Faso, Chad, Bangladesh, and Sri Lanka have the lowest per-capita GNP (less than $610), whereas in most developed countries, it is more than $14,000 (Mansbach, 1994; Gardner 1995).

The debt of developing countries is estimated to have grown to over $1.8 trillion in 1994, up from $1.77 trillion in 1993. The worst debt is that of sub-Saharan Africa, excluding South Africa. Collectively, the region's debt amounts to $180 billion, three times the 1980 total, and is 10 percent higher than the region's entire output of goods and services. Many developing nations have committed one-third to one-half of their foreign currency earnings to pay the interest and principal, and it is not enough. In Latin America, all of the major debtors, especially Brazil, Argentina, Mexico, Venezuela, Peru, and Chile, are in default and have had to reschedule their repayments (Gardener 1995).

The developing countries also lag behind in areas such as health and education. In the developing countries, the infant mortality rate declined from 109 per 1,000 live births in 1970 to 71 per 1,000 in 1991. Yet, these rates are still much higher than in the developed countries. In 1970, the infant mortality rate in developed countries was 20 per 1,000 live births and declined to about eight per 1,000 by 1991. Life expectancy rose from an average of about 50 years in 1965 to slightly more than 62 in 1991 in developing countries, but remained far behind the figure for most developed countries (Mansbach 1994). The story is the same for adult literacy rates, which improved in poor countries but remained far behind those of developed countries. Today, the developed countries have set the membership standards for the First and Third Worlds. Basically, these standards are economic (Grimm 1990). Many people in the Third World believe that they are not allowed to attain these standards because of "economic and technological imperialism" imposed by Western nations. According to this belief, the First World does not allow economic growth and transfer of technology to reach the developing countries because it may lead to the elimination of jobs and affect the standard of living in developed countries.

Third World countries have found the global economic order imposed on them by the First World intolerable and have called for a new economic order. Policymakers in developed countries fear that unrest over unmet economic problems may lead the poor to "desperate politics" against the rich. People in the Third World counter that the developed countries are still practicing invisible colonialism via international economic institutions.

To create a more equitable international economic system, the developing countries in 1961 formed the "nonaligned movement" (NAM). Efforts by the NAM led to the passage of resolutions 3201 and 3202 on May 1, 1974, at the special session of the United Nations General Assembly. These resolutions defined principles and a program to improve international economic relations between the developing and the developed countries. These resolutions were also outlined in the "Charter of Economic Rights and Duties of States," passed by the General Assembly in December 1974. These documents specified six major areas of international economic reform that were needed to resolve conflict between the First World and the Third World. The demands were as follows (Mansbach 1994):

1. Transnational corporations are to be regulated.
2. Technology should be transferred from rich to poor.
3. The trading order should be reformed to assist the development of poor countries.
4. Poor countries' debt should be canceled or renegotiated.
5. Economic aid from rich to poor countries should be increased.
6. Voting procedures in international economic institutions should be revised to provide poor countries with greater influence.

The First World's response to these demands was mostly negative. Third World countries seek economic and technological help from First World countries to become

Table 2
Development Indicators for Low-, Middle- and High-Income Countries

Parameter	*Low-Income Countries*	*Middle-Income Countries*	*High-Income Countries*	*World*
Population in billions (2003)	2.3	3.0	0.972	6.3
Energy use (oil eq.) kg (2002)	493.3	1337.6	5394.7	1699.2
Carbon dioxide emission per capita (metric tons, 1999)	0.8	3.2	12.3	3.8
Infant mortality rate per 1,000 live births (2003)	79.8	29.8	5.4	56.8
Fertility rate (2003)	3.7	2.1	1.6	2.6
Adult literacy rate (1999)				
% Female	67.8	88.9		79.3
% Male	48.5	82.0		70.6
Annual growth rate (GDP, 2003	6.9	4.9	2.2	2.8
Exports of goods & services (% of GDP, 2003)	21.2	32.7	22.3	23.9
Imports of goods & services (% of GDP, 2003)	23.6	29.7	22.7	23.7
External debt $ billion (2003)	33.5	368.6		
External debt (% of exports of goods & services)	15	17.8		
Fixed lines & mobile telephones (per 1,000 people, 2003)	55.7	402.5	1267.5	405.7
Personal computers (per 1,000 people, 2002)	6.9	42.9	466.5	100.8
Internet users (per 1,000 people)	16.2	82.0	364.2	149.9

Source: World Bank, World Development Indicators (http://www.worldbank.org).

Table 3
Electricity Consumption, GNI, Personal Computers, and Literacy Statistics for Selected Developed and Developing Countries

Countries	*Electricity Consumption per capita (kWh)*	*GNI per capita ($ US)*	*Personal Computers (per 1,000 people)*	*Literacy Female (% of population) 2001*	*Literacy Male (% of population) 2001*
Afghanistan					
Algeria	662	1,720	8	58	77
Australia	9,663	19,580	565		
Austria	6,838	23,970	369		
Bangladesh	100	380	3	31	50
Belgium	7,592	22,960	241		
Bolivia	419	920	23	81	93
Brazil	1,776	2,860	75		
Canada	15,613	22,610	487		
Chad		210	2	36	53
Chile	2,617	4,340	119	96	96
Colombia	817	1,810	49	92	92
Comoros		380	6	49	63

Table 3 (*Continued*)

Countries	*Electricity Consumption per capita (kWh)*	*GNI per capita ($ US)*	*Personal Computers (per 1,000 people)*	*Literacy Female (% of population) 2001*	*Literacy Male (% of population) 2001*
Denmark	6,024	29,880	577		
Djibouti		850	15		
Egypt, Arab Republic	1,073	1,470	17		
El Salvador	595	2,110	25	77	82
Finland	15,326	23,990	442		
France	6,606	22,180	347		
Gabon	804	3,000	19		
Germany	6,046	22,860	431		
Ghana	297	270	4	65	81
Haiti	36	440		49	53
India	380	470	7		
Indonesia	411	720	12	83	92
Israel	5,857	16,090	243	93	97
Japan	7,718	33,660	382		
Kenya	120	350	6	77	89
Lao PDR		320	3	54	77
Malaysia	2,832	3,550	147		
Mexico	1,660	5,950	82		
Nepal	64	230	4	25	61
Netherlands	6,179	23,520	467		
New Zealand	8,832	13,450	414		
Niger		180	1	9	24
Norway	23,855	38,990	528		
Oman	3,177	7,830	35	64	81
Pakistan	363	490			
Peru	723	2,020	43	86	95
Saudi Arabia	5,275	8,470	130	68	84
Singapore	7,039	21,180	622		
Spain	5,048	14,610	196	97	99
Sudan	74	400	6	48	70
Turkey	1,458	2,510	45		
Uganda		240	3	58	78
United Kingdom	5,618	25,560	406		
United States	12,183	35,430	659		
Vietnam	374	430	10		
Yemen, Rep.	152	490	7	27	68
Zimbabwe	831		52		93

GNI = gross national income, PDR = People's democratic republic
Source: *CIA World Factbook 1995–96*. World Development Indicators database, http://devdata.worldbank.org.

Table 4
State of the Environment and Technology in Selected Third World Countries

Country	*Population 2003*	*Literacy, Adult Female (% of Population) 2002*	*Literacy, Adult Male (% of Population) 2002*	*GNI per capita 2002*	*Electricity Consumption kW 2002*	*Personal Computers (per 1,000 people) 2002*	*Internet Users (per 1,000 people) 2002*	*Environment (Current Issues)*
Afghanistan								Soil degradation, overgrazing, deforestation, and desertification
Algeria	31,832,612	60	78	1,720	662	8	16	Soil erosion from overgrazing; desertification; dumping of untreated sewage, petroleum refining wastes, and other industrial effluents is leading to pollution of rivers and coastal waters; the Mediterranean Sea, in particular, is becoming polluted from oil wastes, soil erosion, and fertilizer runoff; limited supply of potable water
Angola	13,522,112			680	109	2	3	Population pressures contributing to overuse of pastures and subsequent soil erosion; desertification; deforestation of tropical rain forest attributable to the international demand for tropical timber and domestic use as a fuel; deforestation, contributing to loss of biodiversity; soil erosion, contributing to water pollution and siltation of rivers and dams; scarcity of potable water
Argentina	36,771,840	97	97	4,220	2,024	82	112	Erosion from inadequate flood control and improper land use practices; irrigated soil degradation; desertification; air pollution in Buenos Aires and other major cities; water pollution in urban

Table 4 (*Continued*)

Country	*Population 2003*	*Literacy, Adult Female (% of Population) 2002*	*Literacy, Adult Male (% of Population) 2002*	*GNI per capita 2002*	*Electricity Consumption kW 2002*	*Personal Computers (per 1,000 people) 2002*	*Internet Users (per 1,000 people) 2002*	*Environment (Current Issues)*
								areas; rivers becoming polluted due to increased pesticide and fertilizer use
Bangladesh	138,066,368	31	50	380	100	3	2	Many people are landless and forced to live on and cultivate flood-prone land; limited access to potable water; waterborne diseases prevalent; water pollution, especially of fishing areas, results from the use of commercial pesticides; intermittent water shortages because of failing water tables in the northern and central parts of the country; soil degradation; deforestation; severe overpopulation
Bolivia	8,814,158			920	419	23	32	Deforestation, contributing to loss of biodiversity; overgrazing; soil erosion; desertification; industrial pollution of water supplies used for drinking and irrigation
Brazil	176,596,256			2,860	1,776	75	82	Deforestation in Amazon Basin; air and water pollution in Rio de Janeiro, Sao Paulo, and several other large cities; land degradation and water pollution caused by improper mining activities
Burkina Faso	12,109,229			250		2	2	Recent droughts and desertification, severely affecting agricultural activities, population distribution, and the economy; overgrazing; soil degradation.

Burma								Deforestation
Cambodia	13,403,644	59	81	290		2	2	Deforestation, resulting in habitat loss and declining biodiversity (in particular, the destruction of mangrove swamps threatens natural fisheries)
Cameroon	16,087,472			560	161	6	4	Waterborne diseases are prevalent; deforestation; overgrazing; desertification; poaching
Chad	8,581,741	38	55	210		2	2	Desertification
Chile	15,774,000	96	96	4,340	2,617	119	238	Air pollution from industrial and vehicle emissions; water pollution from untreated sewage; deforestation, contributing to loss of biodiversity; soil erosion; desertification
China	1,288,400,000			970	987	28	46	Air pollution from the overwhelming use of coal as fuel, producing acid rain that is damaging forests; water pollution from industrial effluents; many people do not have access to safe drinking water; less than 10% of sewage receives treatment; deforestation; estimated loss of one-third of agricultural land since 1957 to soil erosion and economic development; desertification
Colombia	44,584,000	92	92	1,810	817	49	46	Deforestation; soil damage from overuse of pesticides
Congo Dem. Rep.	53,153,360			90	43			Poaching threatens wildlife populations; water pollution; deforestation
Congo Dem. Rep.	3,757,263	77	89	610	82	4		Air pollution from vehicle emissions; water pollution from the dumping of raw sewage; deforestation
Costa Rica	4,004,680	96	96	4,070	1,611	197	193	Deforestation, largely the result of land clearing for cattle ranching; soil erosion

Table 4 (*Continued*)

Country	*Population 2003*	*Literacy, Adult Female (% of Population) 2002*	*Literacy, Adult Male (% of Population) 2002*	*GNI per capita 2002*	*Electricity Consumption kW 2002*	*Personal Computers (per 1,000 people) 2002*	*Internet Users (per 1,000 people) 2002*	*Environment (Current Issues)*
Cuba	11,326,000	97	97		1,094	32		Overhunting threatens wildlife populations; deforestation
Djibouti	705,480			850		15	7	Desertification
Dominican Republic	8,738,639	84	84	2,360	853			Water shortages; soil eroding into the sea damages coral reefs; deforestation
Ecuador								Deforestation; soil erosion; desertification; water pollution
Egypt, Arab Rep.	67,559,040			1,470	1,073	17		Agricultural land being lost to urbanization and windblown sands; increasing soil salinization below Aswan High Dam; desertification; oil pollution threatening coral reefs, beaches, and marine habitats; other water pollution from agricultural pesticides, untreated sewage, and industrial effluents; water scarcity away from Nile, which is the only perennial water source; rapid growth in population, overstraining natural resources
El Salvador	6,533,215	77	82	2,110	595	25	46	Deforestation, soil erosion; water pollution; contamination of soil from disposal of toxic waste
Eritrea	4,389,500			190		3	2	Famine; deforestation; soil erosion; overgrazing; loss of infrastructure from civil warfare.

Ethiopia	68,613,472	34	49	100	25	1	1	Deforestation; overgrazing; soil erosion; desertification; famine
Gabon	1,344,433			3,000	804	19	19	Deforestation; poaching
Gambia, The	1,420,895			270		14		Deforestation; desertification
Ghana	20,669,260	66	82	270	297	4	8	Deforestation; overgrazing; soil erosion; poaching and habitat destruction threatens wildlife populations; water pollution; limited supply of safe drinking water
Guatemala	12,307,091	62	77	1,750	361	14	33	Deforestation; soil erosion; water pollution
Guinea-Bissau	1,489,209			130			11	Deforestation; soil erosion; overgrazing
Guinea	7,908,905			410		5	5	Deforestation; inadequate supplies of safe drinking water; desertification; soil contamination and erosion
Guyana	768,888			860		27	142	Water pollution from sewage, agricultural, and industrial chemicals; deforestation
Haiti	8,439,799	50	54	440	36		10	Deforestation; soil erosion
Honduras	6,968,512			910	537	14	25	Urban population expanding; deforestation results from logging and the clearing of land for agricultural purposes; further land degradation and soil erosion hastened by uncontrolled development and improper land use practices, such as farming of marginal lands; mining activities polluting Lago de Yojoa (the country's largest source of freshwater) with heavy metals as well as several rivers and steams
India	1,064,398,592			470	380	7	16	Deforestation; soil erosion; overgrazing; desertification; air pollution from industrial and vehicle emissions; water pollution from raw sewage and runoff of agricultural pesticides; huge and rapidly

Table 4 (*Continued*)

Country	*Population 2003*	*Literacy, Adult Female (% of Population) 2002*	*Literacy, Adult Male (% of Population) 2002*	*GNI per capita 2002*	*Electricity Consumption kW 2002*	*Personal Computers (per 1,000 people) 2002*	*Internet Users (per 1,000 people) 2002*	*Environment (Current Issues)*
								growing population is overstraining natural resources
Indonesia	214,674,160	83	92	720	411	12	21	Deforestation; water pollution from industrial wastes and sewage; air pollution in urban areas
Iran, Islamic Rep.	66,392,020	70	84	1,740	1,677	75		Air pollution, especially in urban areas, from vehicle emissions, refinery operations, and industry; deforestation; overgrazing; oil pollution in Persian Gulf; shortage of drinking water
Iraq	24,699,542				1,213	8	1	Government water control projects drain inhabited marsh areas, drying up or diverting the streams and rivers that support a sizable population of Shia Muslims, who have inhabited these areas for thousands of years; the destruction of natural habitat also poses serious threats to wildlife populations; damage to water treatment and sewage facilities during the Gulf War; inadequate supplies of potable water; development of Tigris–Euphrates Rivers system contingent upon agreements with upstream riparians (Syria, Turkey); air and water pollution; soil degradation (salinization) and erosion; desertification

Jamaica	2,642,628	91	84	2,900	2,406	54	228	Deforestation; water pollution
Jordan	5,307,895	86	96	1,760	1,317	38	58	Lack of adequate natural water resources; deforestation; overgrazing; soil erosion; desertification
Kenya	31,915,850	79	90	350	120	6	13	Water pollution from urban and industrial wastes; degradation of water quality from increased use of pesticides and fertilizers; deforestation; soil erosion; desertification; poaching
Korea, Rep.	47,911,728			11,280	6,171	556		Localized air pollution attributable to inadequate industrial controls
Lao PDR	5,659,834	55	77	320		3		Deforestation; soil erosion
Lebanon	4,497,669			3,900	1,951	81	117	Deforestation; soil erosion
Liberia	3,373,542	39	72	140				West Africa's largest tribal tropical rain forest, subject to deforestation; soil erosion; loss of biodiversity
Madagascar	16,893,904			230		4	3	Soil erosion results from deforestation and overgrazing; desertification; surface water contaminated with untreated sewage and other organic wastes; several species of flora and fauna unique to the island are endangered
Malawi	10,962,012	49	76	160		1	3	Deforestation; land degradation; water pollution from agricultural runoff, sewage, and industrial wastes; siltation of spawning grounds endangers fish population
Maldives	293,080	97	97	2,150		71	53	Depletion of freshwater aquifers threatens water supplies
Mali	11,651,502			240		1	2	Deforestation; soil erosion; desertification; inadequate supplies of safe drinking water; poaching
Mauritania	2,847,869	31	51	400		11	4	Overgrazing, deforestation, and soil erosion aggravated by drought are

Table 4 (*Continued*)

Country	*Population 2003*	*Literacy, Adult Female (% of Population) 2002*	*Literacy, Adult Male (% of Population) 2002*	*GNI per capita 2002*	*Electricity Consumption kW 2002*	*Personal Computers (per 1,000 people) 2002*	*Internet Users (per 1,000 people) 2002*	*Environment (Current Issues)*
								contributing to desertification; water scarcity away from the Senegal River, which is the only perennial river
Mauritius	1,222,188			3,860		116	103	Water pollution
Mexico	102,290,976			5,950	1,660	82	98	Natural water resources scarce and polluted in north, inaccessible and poor quality in center and extreme southeast; untreated sewage and industrial effluents polluting rivers in urban areas; deforestation; widespread erosion; desertification; serious air pollution in the national capital and urban centers along U.S.–Mexico border
Mongolia	2,479,568			450		28	21	Limited water resources; policies of the former communist regime promoting rapid urbanization and industrial growth have raised concerns about their negative impact on the environment; the burning of soft coal and the concentration of factories in Ulaanbaatar have severely polluted the air; deforestation, overgrazing, and the conversion of virgin land to agricultural production have increased soil erosion from wind and rain; desertification
Morocco	30,112,644	38	63	1,170	475	24	24	Land degradation/desertification (soil erosion resulting from farming of

								marginal overgrazing, and destruction of vegetation); water supplies contaminated by untreated sewage; siltation of reservoirs; oil pollution of coastal waters
Mozambique	18,791,420	31	62	200	341	4	3	Civil strife in the hinterlands has resulted in increased migration to urban and coastal areas, with adverse environmental consequences; desertification; pollution of surface and coastal waters
Namibia	2,014,546	83	84	1,650		71	27	Very limited natural water resources; desertification
Nepal	24,659,962	26	62	230	64	4	3	The almost total dependence on wood for fuel and cutting down trees to expand agricultural land without replanting has resulted in widespread deforestation; soil erosion; water pollution (the use of contaminated water presents human health risks)
Nicaragua	5,480,000			730	279	28	17	Deforestation; soil erosion; water pollution
Niger	11,762,251	9	25	180		1	1	Overgrazing; soil erosion; deforestation; desertification; wildlife populations (such as elephant, hippopotamus, and lion) threatened because of poaching and habitat destruction
Nigeria	136,460,976	59	74	300	68	7	3	Soil degradation; rapid deforestation; desertification; recent droughts in north severely affecting marginal agricultural activities
Pakistan	148,438,768			490	363		10	Water pollution from untreated sewage, industrial wastes, and agricultural runoff; water scarcity; a majority of the population does not have access to safe drinking water; deforestation; soil erosion; desertification

Table 4 (*Continued*)

Country	*Population 2003*	*Literacy, Adult Female (% of Population) 2002*	*Literacy, Adult Male (% of Population) 2002*	*GNI per capita 2002*	*Electricity Consumption kW 2002*	*Personal Computers (per 1,000 people) 2002*	*Internet Users (per 1,000 people) 2002*	*Environment (Current Issues)*
Panama	2,984,022	92	93	3,970	1,375	38	62	Water pollution from agricultural runoff threatens fishery resources; deforestation of tropical rain forest; land degradation
Paraguay	5,643,097			1,170	842	35	17	Deforestation; water pollution; inadequate means for waste disposal present health hazards for many urban residents
Peru	27,148,000	80	91	2,020	723	43	90	Deforestation; overgrazing; soil erosion; desertification; air pollution in Lima
Philippines	81,502,616			1,030	459	28	44	Deforestation; soil erosion; water pollution; air pollution in Manila
Rwanda	8,395,000	63	75	230			3	Deforestation; overgrazing; soil exhaustion; soil erosion
Senegal	10,239,848	30	49	460	135	20	10	Wildlife populations threatened by poaching; deforestation; overgrazing; soil erosion; desertification
Sierra Leone	5,336,568			140			2	Rapid population growth pressuring the environment; overharvesting of timber, expansion of cattle grazing, and slash-and-burn agriculture have resulted in deforestation and soil exhaustion; civil war depleting natural resources
Somalia	9,625,918						9	Use of contaminated water contributes to health problems; deforestation; overgrazing; soil erosion; desertification
Sri Lanka	19,231,760	90	95	850	297	13	11	Deforestation; soil erosion; wildlife

								population threatened by poaching; coastal degradation from mining activities and increased pollution; freshwater resources being polluted by industrial wastes and sewage runoff
Sudan	33,545,726	49	71	400	74	6	3	Contaminated water supplies present human health risks; wildlife populations threatened by excessive hunting; soil erosion; desertification
Suriname	438,104			1,930			42	NA
Syrian Arab Republic	17,384,492	74	91	1,090	1,000	19		Deforestation; overgrazing; soil erosion; desertification; water pollution from dumping of untreated sewage and wastes from petroleum refining; lack of safe drinking water
Tanzania	35,888,960	69	85	280	62	4	2	Soil degradation; deforestation; desertification; destruction of coral reefs threatens marine habitats; recent droughts affected marginal agriculture
Thailand	62,014,216			2,000	1,626	40	78	Air pollution increasing from vehicle emission; water pollution from organic and factory wastes; desertification; wildlife populations threatened by illegal hunting
Tunisia	9,895,201	63	83	1,990	1,019	31	52	Toxic and hazardous waste disposal is ineffective and presents human health risks; water pollution from untreated sewage; water scarcity; deforestation; over-grazing; soil erosion; desertification
Turkey	70,712,000			2,510	1,458	45	62	Water pollution from dumping of chemicals and detergents; air pollution; deforestation
Uganda	25,280,000	59	79	240		3	4	Draining of wetlands for agricultural use; deforestation; overgrazing; soil erosion

Table 4 (*Continued*)

Country	*Population 2003*	*Literacy, Adult Female (% of Population) 2002*	*Literacy, Adult Male (% of Population) 2002*	*GNI per capita 2002*	*Electricity Consumption kW 2002*	*Personal Computers (per 1,000 people) 2002*	*Internet Users (per 1,000 people) 2002*	*Environment (Current Issues)*
Uruguay	3,380,177	98	98	4,370	1,834			NA
Venezuela, RB	25,674,000	93	94	4,090	2,472	61		Sewage pollution of Lao de Valencia; oil and urban pollution of Lago de Maracaibo; deforestation; soil degradation; urban and industrial pollution, especially along the Caribbean coast
Vietnam	81,314,240			430	374	10		Deforestation; soil degradation; water pollution and overfishing threatening marine life populations
Yemen, Rep.	19,173,160	29	69	490	152	7		Scarcity of natural freshwater resources (shortages of potable water); overgrazing; soil erosion; desertification
Zambia	10,402,959	74	86	340	583	7	5	Poaching seriously threatens rhinoceros and elephant populations; deforestation; soil erosion; desertification
Zimbabwe	13,101,754	86	94		831	52	43	Deforestation; soil erosion; land degradation; air and water pollution

Source: *CIA World Factbook 1995–96*. World Development Indicators database, http://devdata.worldbank.org.

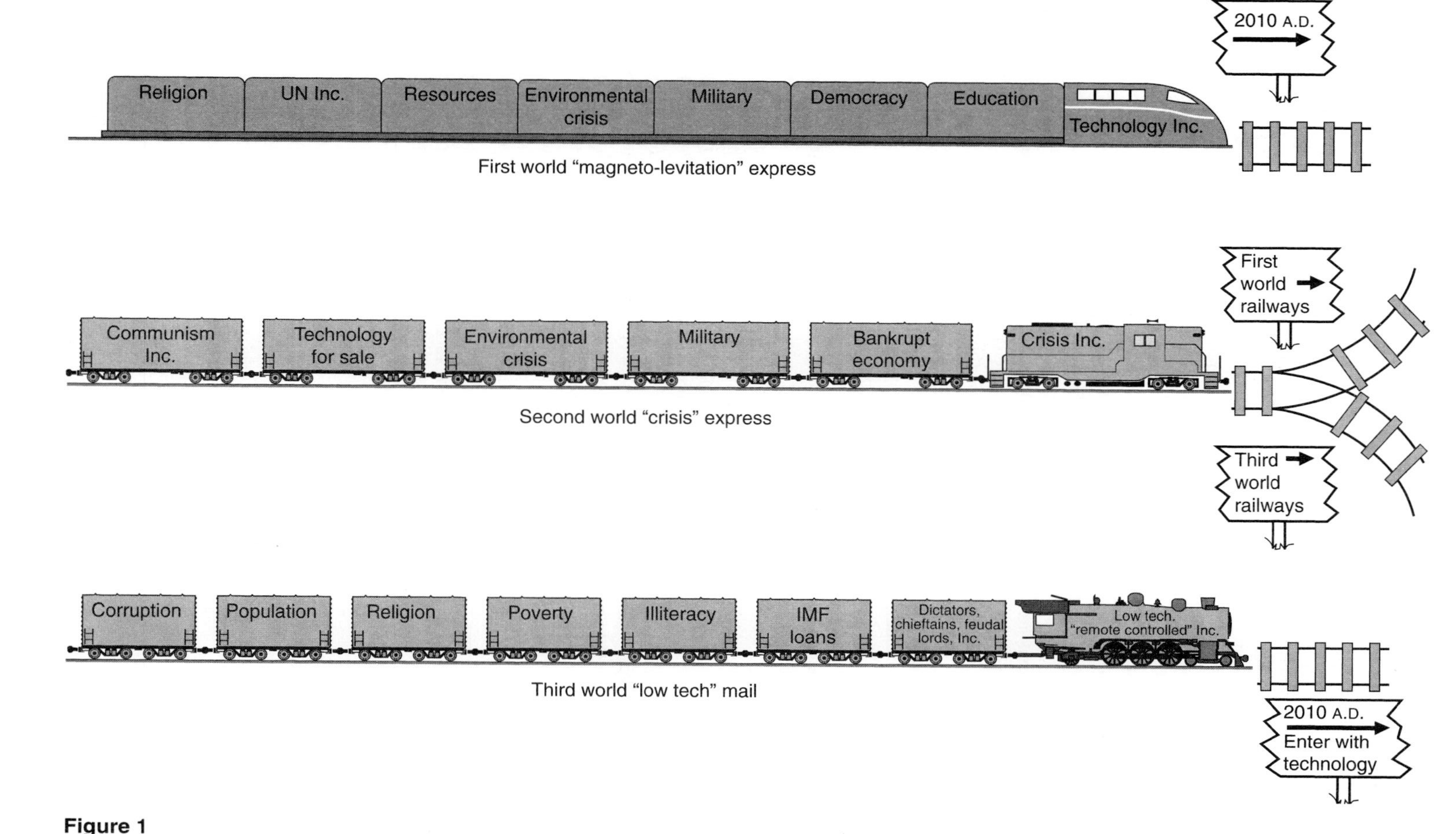

Figure 1
Comparison of the First World, Second World, and Third World: Worldwide Railways, Inc.
Source: Tasneem Khan.

economically self-sufficient. On the other hand, First World countries can learn a great deal from Third-World countries. People in developing countries take pride in family, history, cultural, and religious values, and they consider these values to be more important than economic fulfillment and greed for materialism. For the near future, the developing world will continue to struggle for economic assistance. With the end of the Cold War and with the former Soviet-bloc countries competing for economic assistance from the West, the future for the developing world remains bleak.

Positive change in the Third World will occur only when both the intrinsic and extrinsic factors inhibiting the development of Third World nations are addressed. The effects of extrinsic factors (i.e., political and economic instability caused by international financial institutions and the economic exploitation of resources by multinational companies) can be rectified only when a great change takes place in the modes of thought of the policy-makers of the developed world. Those leaders must begin to believe that all people are created equal, no one is superior to another, and treating people with equality and justice is the key to solving man-made economic, political, and environmental dilemmas.

To address the intrinsic factors (i.e., illiteracy, poverty, social injustice, corruption, lack of resources, population pressures, and power-hungry political and military elite), the key is education and appropriate use of technology. Education is a great equalizer for changing socioeconomic conditions. It could provide solutions to many of the problems of developing countries and set them free from economic bondage, autocratic rule, poverty, and disease. The developing nations should give a higher priority to educating their people. The innovative use of telecommunications technologies can provide endless opportunities in education (e.g., distance learning), medicine (e.g., telemedicine), and business (e.g., electronic trade) in developing countries. These applications could enable developing countries to increase teledensity and literacy, improve per-capita income, and narrow the technoeconomic gap with developed countries. In the process of developing economic infrastructures, the Third World countries, through proper planning and appropriate policies, could also minimize the huge cost of industrialization that the developed world has paid in terms of damage to the environment. If the developing countries fail to enhance their intellectual capital and educate their populations, it will be only a matter of time before they fall victim to a new era of technoslavery.

A number of world organizations, including the World Economic Forum (http://www.weforum.org) and the United Nations (http://www.un.org) have embarked on ambitious initiatives to change the state of the people in the developing world (see Table 5), but given the allocation of minuscule amounts of resources and the lack of political will for implementing these initiatives, the future of the poor of the world remains bleak.

The terms "First World," "Third World," "less developed countries" (LDC), and "least developed countries" (LLDC) may have significant meanings in an economic context. However, if one looks at earth from space, it is, at this point, the only living planet in the universe, without any boundaries or barriers between the First and Third Worlds. The people on earth are members of one big family with a common past and a common destiny. Problems such as ozone depletion, global warming, soil erosion, deforestation, poverty, homelessness, illiteracy, and obstruction of world peace are universal problems and cannot be solved as long as the artificial distinction between the First World and Third World exists. All people are created equal. No one is superior to others on the basis of color, race, or wealth. Treating people with equality and justice is the key to solving man-made global economic, political, and environmental crises.

Table 5
United Nations Millennium Development Goals (MDG)

1. Eradicate extreme poverty and hunger	* Reduce by half the proportion of people living on less than a dollar a day * Reduce by half the proportion of people who suffer from hunger
2. Achieve universal primary education	* Ensure that all boys and girls complete a full course of primary schooling

3. Promote gender equality and empower women	* Eliminate gender disparity in primary and secondary education preferably by 2005 and at all levels by 2015
4. Reduce child mortality	* Reduce by two-thirds the mortality rate among children under five
5. Improve maternal health	* Reduce by three-quarters the maternal mortality ratio
6. Combat HIV/AIDS, malaria, and other diseases	* Halt and begin to reverse the spread of HIV/AIDS
	* Halt and begin to reverse the incidence of malaria and other major diseases
7. Ensure environmental sustainability	* Integrate the principles of sustainable development into country policies and programs; reverse loss of environmental resources
	* Reduce by half the proportion of people without sustainable access to safe drinking water
	* Achieve significant improvement in the lives of at least 100 million slum dwellers by 2020
8. Develop a global partnership for development	* Develop further an open trading and financial system that is rule-based, predictable, and non-discriminatory; includes a commitment to good governance, development and poverty reduction—nationally and internationally
	* Address the least developed countries' special needs. This includes tariff- and quota-free access for their exports, enhanced debt relief for heavily indebted poor countries, cancellation of official bilateral debt, and more generous official development assistance for countries committed to poverty reduction
	* Address the special needs of landlocked and small island developing States
	* Deal comprehensively with developing countries' debt problems through national and international measures to make debt sustainable in the long term
	* In cooperation with the developing countries, develop decent and productive work for youth
	* In cooperation with pharmaceutical companies, provide access to affordable essential drugs in developing countries
	* In cooperation with the private sector, make available the benefits of new technologies—especially information and communications technologies

Source: UN Millennium Development Goals (MDG), http://www.un.org/mellenniumgoals/.

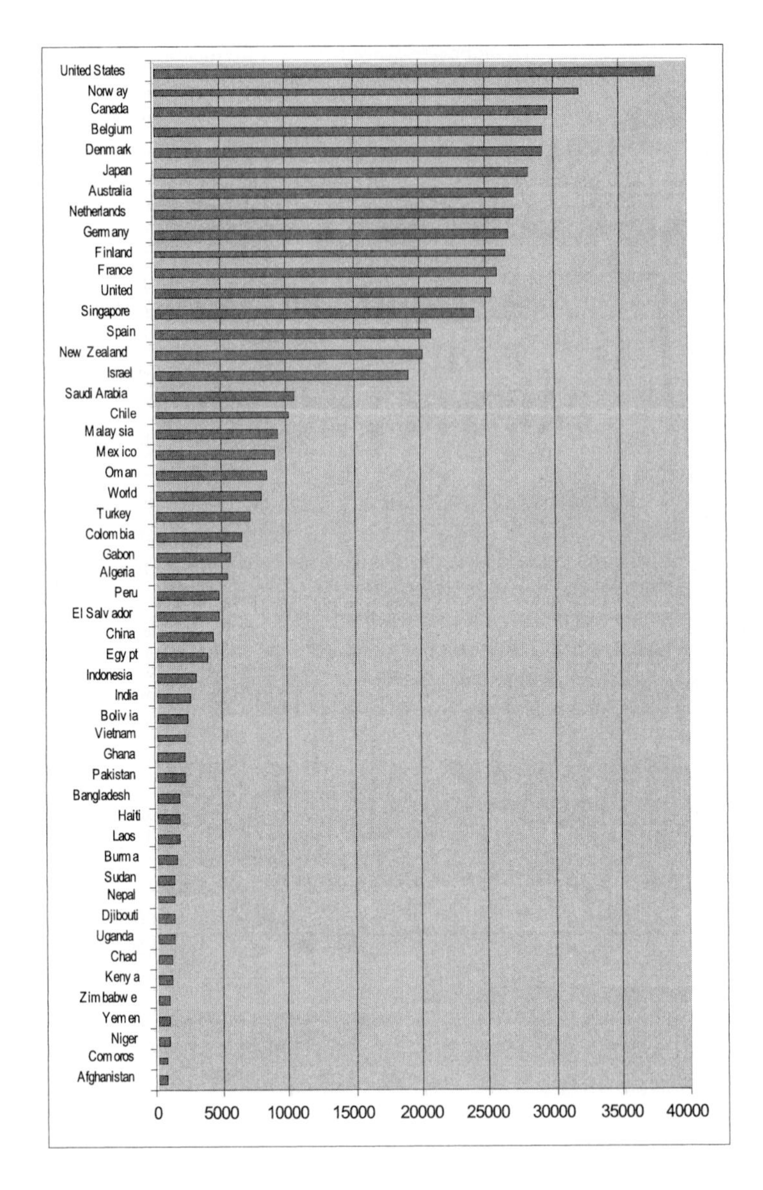

Figure 2
GDP per capita ($US) for selected developed and developing countries.
Source: www.worldbank.com.

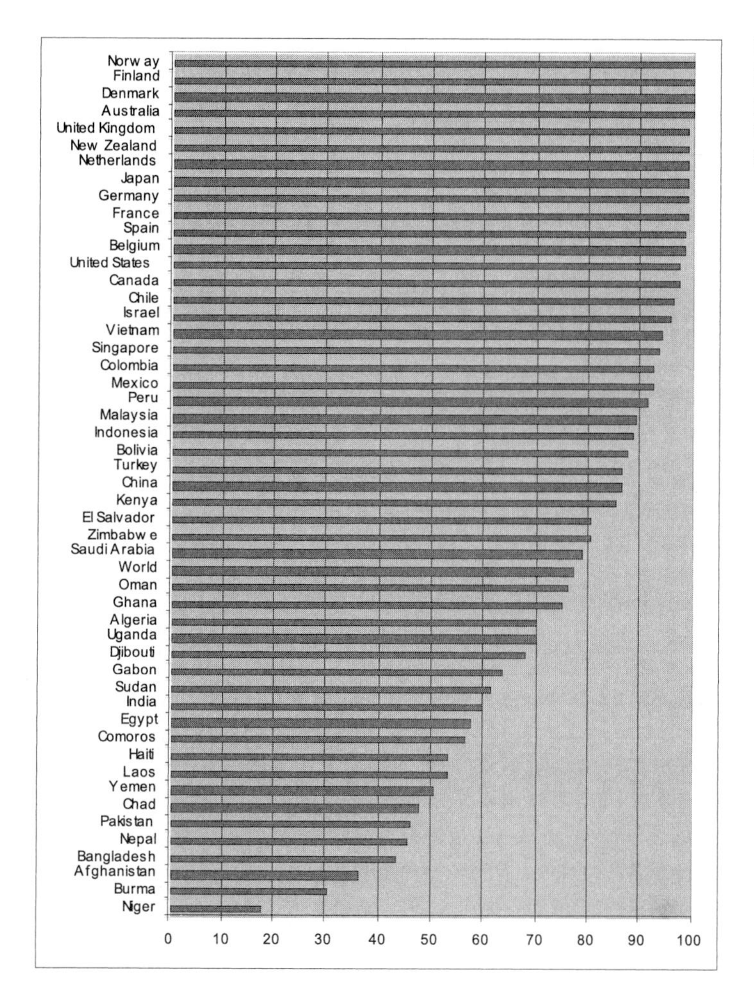

Figure 3
Literacy as a percentage of population for selected developed and developing countries.
Source: www.worldbank.com.

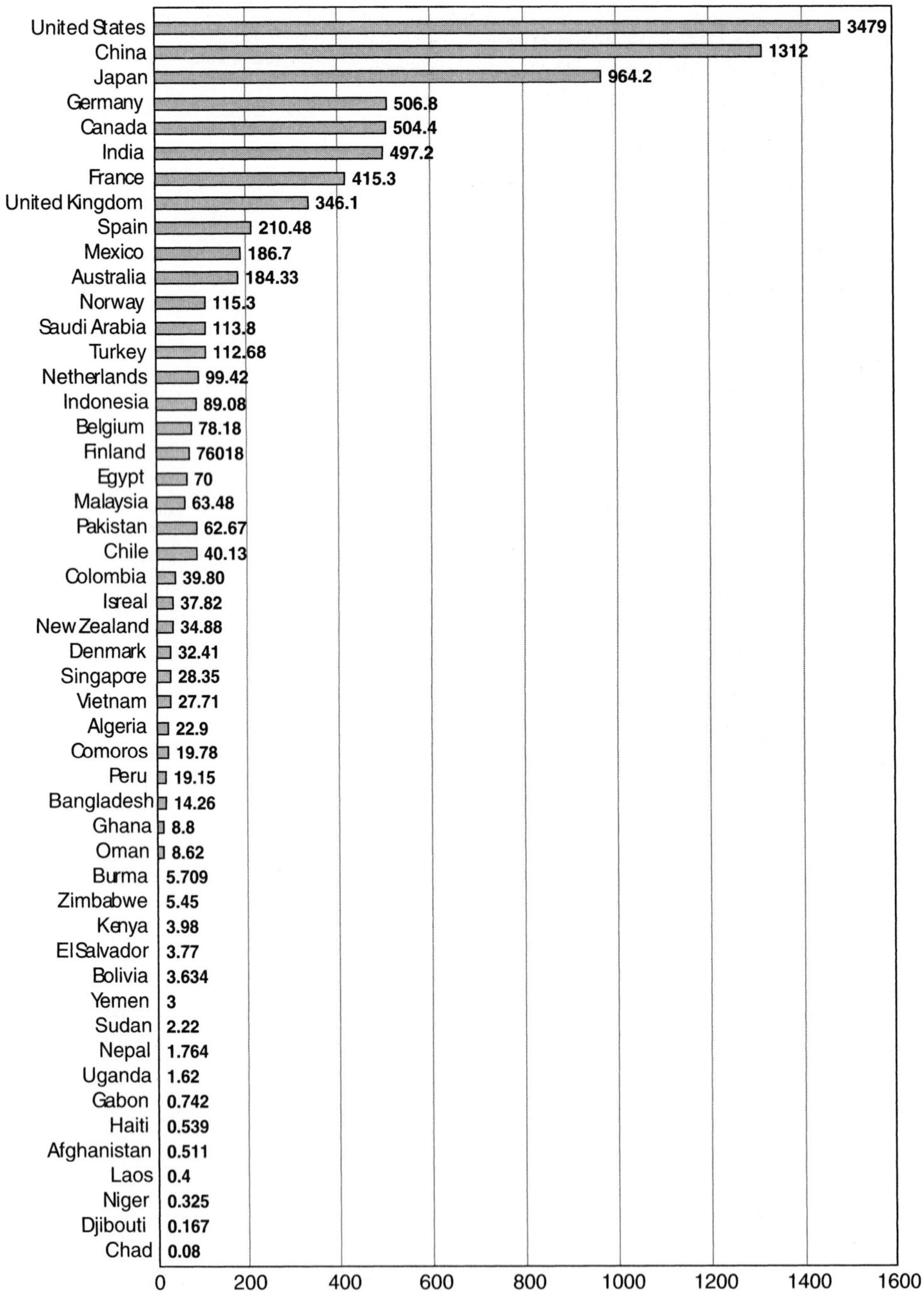

Country	Value
United States	3479
China	1312
Japan	964.2
Germany	506.8
Canada	504.4
India	497.2
France	415.3
United Kingdom	346.1
Spain	210.48
Mexico	186.7
Australia	184.33
Norway	115.3
Saudi Arabia	113.8
Turkey	112.68
Netherlands	99.42
Indonesia	89.08
Belgium	78.18
Finland	76018
Egypt	70
Malaysia	63.48
Pakistan	62.67
Chile	40.13
Colombia	39.80
Isreal	37.82
New Zealand	34.88
Denmark	32.41
Singapore	28.35
Vietnam	27.71
Algeria	22.9
Comoros	19.78
Peru	19.15
Bangladesh	14.26
Ghana	8.8
Oman	8.62
Burma	5.709
Zimbabwe	5.45
Kenya	3.98
El Salvador	3.77
Bolivia	3.634
Yemen	3
Sudan	2.22
Nepal	1.764
Uganda	1.62
Gabon	0.742
Haiti	0.539
Afghanistan	0.511
Laos	0.4
Niger	0.325
Djibouti	0.167
Chad	0.08

0 200 400 600 800 1000 1200 1400 1600

Figure 4

Electricity consumption in billions kWh for selected developed and developing countries.

Source: www.worldbank.com.

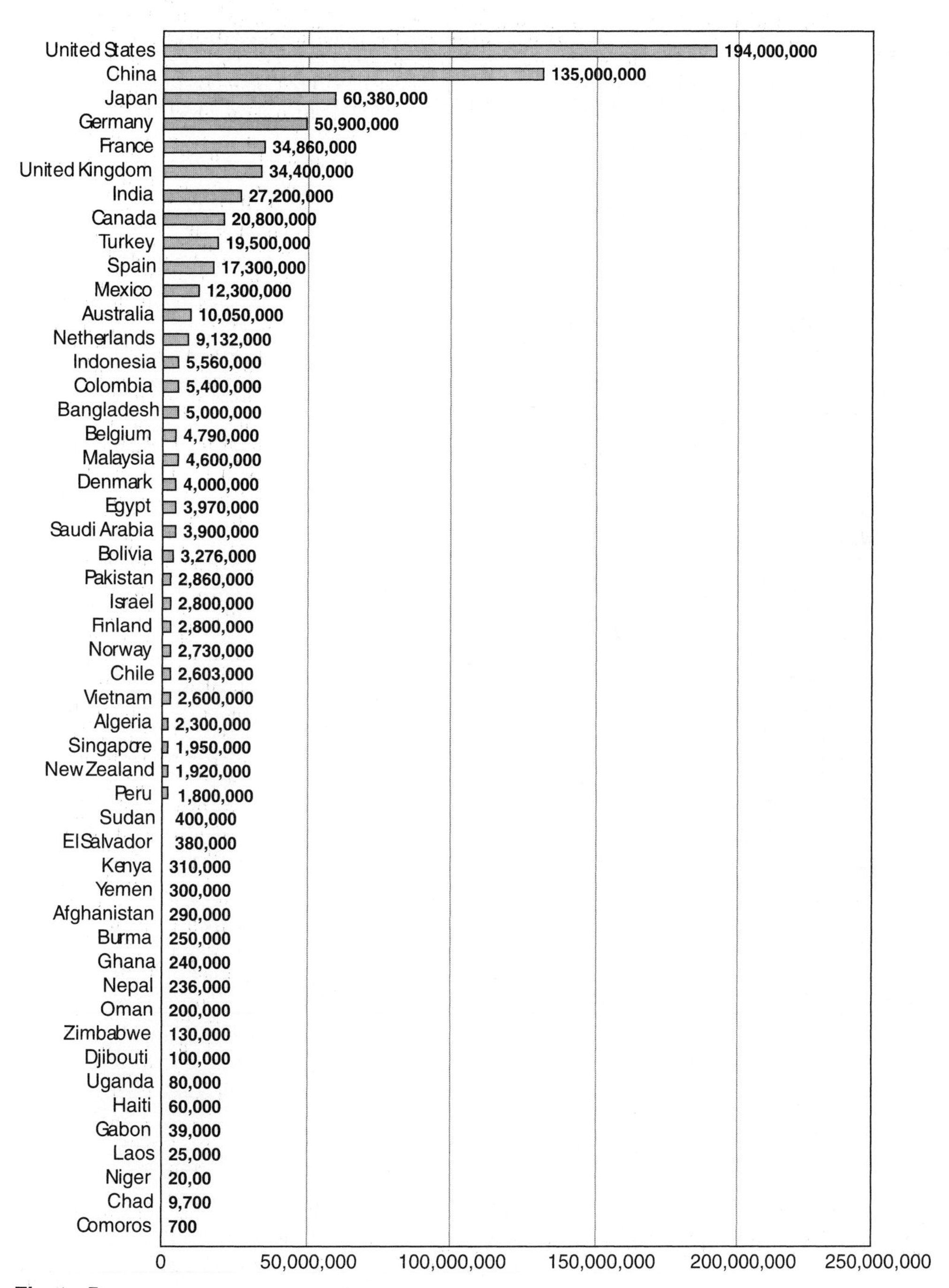

Figure 5

Number of telephone lines (millions) for selected developed and developing countries.

Source: www.worldbank.com.

REFERENCES

CIA World Factbook. (1995). Washington, D.C.: Brassey's.

Gardner, G. (1995). Third World Debt Is Still Growing. *World Watch,* , January/February, pp. 37–38.

Grimm, W. (1990). The Third World. *America: National Catholic Weekly,* , May 5, pp. 449–451.

Mansbach, R. (1994). *The Global Puzzle: Issues and Actors in World Politics.* Boston: Houghton Mifflin Company.

Mennerick, L., and Mehrangiz, N. (1991). Third World Environmental Health, Social, Technological and Economic Policy Issues. *Journal of Environmental Health,* Spring, pp. 24–29.

Myers, N. (1990). *The GAIA Atlas of Future Worlds.* New York: Doubleday.

Worldbank IED. (1997). *World Development Indicators.* September 1997. Available at http://www.worldbank.org/html/iecdd/products.htm.

Worldwatch Institute. SignPost 2005 CD-ROM. Worldwatch Institute, Washington, D.C.

Try being poor for a day or two
and find in poverty double riches.

RUMI

BOX 1 Modern Man

Love fled, Mind stung him like a snake; he could not
Force it to vision's will.
He tracked the orbits of the stars, yet could not
Travel his own thoughts' world;
Entangled in the labyrinth of his science
Lost count of good and ill;
Took captive the sun's ray, and yet no sunrise
On life's thick night unfurled.

IQBAL, The poet of east

Source: Translated by V. G. Kiernan, *Poems from Iqbal,* Karachi: Oxford University Press, 2004. p. 186.

BOX 2 Life in the Philippines

Mount Pinatubo, Island of Luzon

An outdoor laundry

Tricycle: A common mode of transportation

Shacks in metro Manila

Manila skyline
Photos courtesy of
Nickos Lambros and
Elaine Chew Lambros

Is technology bringing people together or breaking them apart?

BOX 3 The World Is Not Flat

"The advanced Western countries completed their escape from poverty to relative wealth during the nineteenth and twentieth centuries. There was no sudden change in their economic output; but only a continuation of year-to-year growth at a rate that somewhat exceeded the rate of population growth."

Nathan Rosenberg and L.E. Birdzell, Jr.

"The gap in the world between the haves and the have-nots is becoming a gulf. Will the so-called Southerners tolerate the absurd imbalances when, to some extent (through inequitable trade terms, for example), they are poor simply because the Northerners are rich when many impoverished people among the have-not nations receive less protein per week than a domestic cat in the North?"

"Intolerable Imbalances," The GAIA Atlas of Future Worlds, Norman Myers

"Sub-Saharan Africa's debt is now higher than its entire output of goods and services. Collectively, the region's debt amounts to $180 billion, three times the 1980 total, and is 10% higher than its entire output of goods and services. Debt service payment comes to $10 billion annually, about four times what the region spends on health and education combined. The burden is choking off economic development over much of the continent."

Worldwatch Institute's Annual Report, Vital Signs, 1995

"The ending of colonialism does not automatically inaugurate an era of peace and prosperity for liberated peoples. It could equally be the prelude to fraternal wars and new inhumanities. . . . It is my considered opinion that the third world war has already begun—in the Third World. The new war is likely to be a cumulation of little wars rather than one big war. Though the form is new, it is basically a war between the great powers and one fought for the realization of their ambitions and the promotion of their national interests. However, this is not obvious because in this new world war, the great powers are invisible."

Sinathamby Rajaratnam, Singapore's Senior Minister and Veteran Delegate to the Nonaligned Movement (NAM)

QUESTIONS

1. Define the following terms:
   a. First World
   b. Second World
   c. Third World
   d. LDC
   e. LLDC

2. Many people in the Third World believe that they are the victims of technological imperialism imposed by the First World. The developed countries do not allow economic growth and transfer of technology to developing countries because it may lead to the elimination of jobs and affect the standard of living in the developed countries. Do you agree with this opinion? Explain your answer.

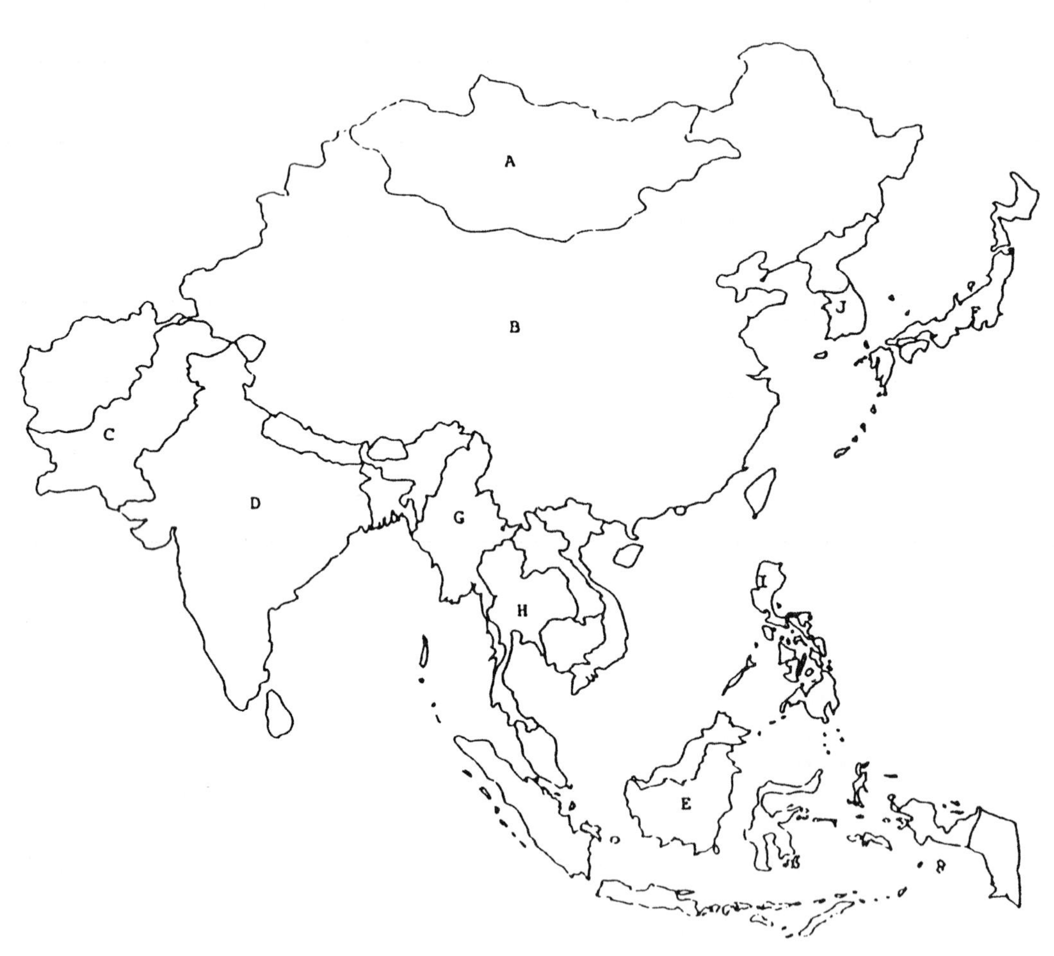

Figure 6
Asia

3. Identify the countries shown on the map of Asia in Figure 6 and define the state of technology in each country.

Country	*Name*	*State of Technology (developed/developing)*	*National Language*
A			
B			
C			
D			
E			
F			
G			
H			
I			
J			

Figure 7
Africa

4. Identify the countries shown on the map of Africa in Figure 7 and define the state of technology in each country.

Country	*Name*	*State of Technology (developed/developing)*	*National Language*
A			
B			
C			
D			
E			
F			
G			
H			
I			
J			

Figure 8
South America

5. Identify the countries shown on the map of South America in Figure 8 and define the state of technology in each country.

Country	*Name*	*State of Technology (developed/developing)*	*National Language*
A			
B			
C			
D			
E			
F			
G			
H			
I			
J			

6. Project: Comparisons of the major religions

Phase I: Using the Internet and other scholarly sources, complete the following table:

Religion	*Concept of God*	*Main Tenets/Beliefs*	*Obligations Toward God*	*Obligations Toward Fellow Human Beings*	*Accountability of One's Deeds*	*Life After Death*
Judaism						
Christianity						
Islam						
Buddhism						
Hinduism						

Phase II: Interview at least two followers of religions listed in the above table and verify the accuracy of the information collected.

Phase III: Compare and contrast the similarities between these religions.

7. Book Project:

Phase I: Read the following books:

a. *The World Is Flat: A Brief History of the Twenty-First Century* by Thomas Friedman
b. *Globalization and Its Discontents* by Joseph Stiglitz

Phase II: Write a book report by comparing the main points of thesis presented by both authors about social and technological implications of globalization. Is the world becoming flat? Is the gap between rich and poor growing or shrinking in developing world?

Use Internet resources to answer the following questions:

8. China and India are experiencing high economic growth rates. In which of the two countries, the middle class is becoming more prosperous? In which country the middle class enjoys social equality? Compare economic growth in China with India by completing the following table:

Economic Factor (2006)	*Units*	*China*	*India*	*China to India Ratio*
GDP (PPP)				
Per capita GDP growth				
Share of manufacturing in GDP				
Per capita GNP (PPP)				
Life expectancy				
Female adult literacy rate				
Under 5 mortality				
Under 5 malnutrition				
Electricity production				
Goods hauled (Railways)				
Container traffic (ports)				
Air freight				
Merchandise exports				
FDI inflow				
Foreign exchange reserves				

Complete and discuss the following flowchart.

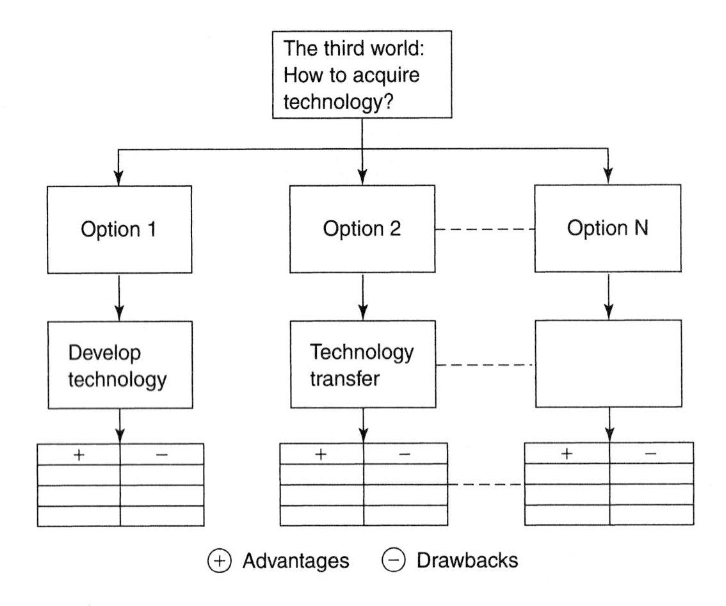

BOX 4 Richer

Number of billionaires (people with the net worth of a thousand millionaires) in the world in 1989	157
Number of billionaires just five years later in 1994	358
Number of the world's richest people whose collective wealth adds up to $762 billion	358
Number of the world's poorest people whose combined income adds up to $762 billion	2,400,000,000
Portion of global income going to the richest fifth of the population	83 percent
Portion of global income going to the poorest fifth	1 percent
Number of billionaires in Mexico in 1988	2
Number of billionaires in Mexico 1996	24
Combined income of the poorest 17 million Mexicans last year	$6,600,000,000
Wealth of the richest *single* Mexican	$6,600,000,000
Ratio of income of the richest fifth to the poorest fifth of the U.S. population in 1970	4 to 1
The same ratio in 1993	13 to 1
Combined total R&D expenditures of General Motors, Ford, Hitachi, Siemens, Matsushita, IBM, and Daimler-Benz Corporations in 1993	$31.8 billion

Military R&D expenditures of the United States in the same year	$43.5 billion
GDP of Israel in 1992	$69.8 billion
Sales of Exxon in 1992	$103.5 billion
GDP of Egypt in 1992	$33.6 billion
Sales of Philip Morris in 1992	$50.2 billion

GDP = Gross domestic product, R&D = research and development.
Source: United Nations Research Institute for Social Development, *States of Disarray: The Social Effects of Globalization*, London: UNRISD, 1995. Reprinted with Permission from Worldwatch Institute (*World Watch* July/August 1996).

BOX 5 Who Dominates the World?

Number of people employed by the United Nations	53,589
Number of people employed at Disney World and Disneyland	50,000
Total budget of the United Nations in 1995–96 (two-year budget)	$18.2 billion
Revenue of a single U.S. arms manufacturer (Lockhead Martin) in 1995	$19.4 billion
UN peacekeeping expenditures in 1995	$3.6 billion
World military spending in 1995	$767.0 billion
Number of UN peacekeepers for every 150,000 people in the world	1
Number of soldiers in national armies for every 150,000 people in the world	650
U.S. contribution to the UN budget, per capita	$7
Norwegian contribution to the UN budget, per capita	$65
Number of U.S. troops serving in UN peacekeeping operations in 1994	965
Number of U.S. troops serving in international missions under U.S. command in 1994	86,451
Cost of the 1992 Earth Summit	$10 million
Cost of the 1994 Paris Air Show and Weapons Exhibition (U.S. portion)	$12 million

Sources: U.S. Arms Control and Disarmament Agency, *World Military Expenditures and Arms Transfers 1995*; Michael Renner, "Peacekeeping Expenditures Level Off," in Lester R. Brown et al., *Vital Signs* 1996; Michael Renner, *Remaking U.N. Peacekeeping: U.S. Policy and Real Reform*, Briefing Paper 17 (Washington, D.C.: National Commission for Economic Conversion and Disarmament, 1995); U.N. Department of Public Information. Compiled by Michael Renner. Reprinted with permission from the Worldwatch Institute, Washington, D.C. (*World Watch* November/December 1996).

51

Income Gap Widens

HAL KANE

The gap in income among the people of the world has been widening. In 1960, according to United Nations statisticians, the richest 20 percent of the world's people received 30 times more income than the poorest 20 percent. By 1991, they were getting 61 times more. While the poorest one-fifth in 1960 received a meager 2.3 percent of world income, by 1991 that revenue share had fallen to 1.4 percent. The income share of the richest fifth, meanwhile, rose from 70 percent to 85 percent.

These disparities prevail both among countries and within them, and the large gap between individuals worldwide reflects the combination of both of those splits. Almost four-fifths of all people live in the developing world, where incomes are only a fraction of those in industrial countries. In turn, within countries in both categories, gaps in income between citizens can be even wider. The widest income gap reported within a country is in Botswana, where during the 1980s the richest 20 percent of society received over 47 times more income than the poorest 20 percent (see Table 1). Brazil was second, with a ratio of 32 to 1. In Guatemala and Panama, the ratio stood at 30 to 1.

The rapidly growing economies of East Asia have had income patterns similar to those of Western Europe and North America, with the richest one-fifth often earning 5 to 10 times more than the poorest fifth. In South Asia, India, Bangladesh, and Pakistan have had relatively even distributions of income, with the richest 20 percent getting only four to five times more than the poorest quintile. Some countries that have had military conflicts apparently based in part on inequities among citizens nevertheless have relatively even income distributions.

The split between countries and people can be seen in the marketplace. The value of luxury goods sales worldwide—high-fashion clothing and top-of-the-line autos, for example—exceeds the gross national products of two-thirds of the world's countries. The world's average income, roughly $4,000 a year, is well below the U.S. poverty line.

The poorest fifth of the world accounted for 0.9 percent of world trade, 1.1 percent of global domestic investment, 0.9 percent of global domestic savings, and just 0.2 percent of global commercial credit at the beginning of the 1990s. Each of those shares declined between 1960 and 1990.

These disparities are reflected in the consumption of many resources. At the start of this decade (1990), industrial countries, home to roughly a fifth of the world's population, accounted for about 86 percent of the consumption of aluminum and chemicals, 81 percent of the paper, 80 percent of the iron and steel, and three-quarters of the timber and energy. Since then, economic growth in developing countries has probably reduced these percentages. China's economy, for example, is more than 50 percent larger now than it was in 1990, and developing countries have passed industrial ones in fertilizer consumption.

Uneven income distribution is shaping some of the most important trends in the world today. It raises crime rates, for example. And it drives migration. People have long responded to economic disparities by following a path from poor regions to richer ones, as tens of millions

Reprinted with permission from Worldwatch Institute
Worldwatch March/April 1996 www.worldwatch.org.

Table 1
Income Distribution, Selected Countries, Late 1980s

Country	*Ratio of Richest One-Fifth to Poorest*
Industrial Countries	
United States	9
Switzerland	9
Israel	7
Italy	6
Germany	6
Sweden	5
Japan	4
Poland	4
Hungary	3
Developing Countries	
Botswana	47
Brazil	32
Guatemala	30
Panama	30
Kenya	23
Chile	17
Mexico	14
Malaysia	12
Peru	10
Thailand	8
Philippines	7
China	6
Ghana	6
Ethiopia	5
Indonesia	5
Pakistan	5
India	5
Bangladesh	4
Rwanda	4

Source: UN Development Programs, Human Development Report 1994 (New York: Oxford University Press, 1994).

of workers chase higher wages and better opportunities. Some 1.6 million Asians and Middle Easterners were working in Kuwait and Saudi Arabia before they fled war in 1991, and at least 2.5 million Mexicans live in the United States.

The same is true within countries: rising disparities of income are adding to the growth of cities through rural-to-urban migration. Latin America, with some of the highest disparities of income among its citizens, is also the most urbanized region of the developing world—not entirely by coincidence. Since 1950, city dwellers there have risen from 42 percent of the population to 73 percent.

For many years, China had one of the most equal distributions of income in the world. But now that is changing, as incomes in its southern provinces and special economic zones soar while those in rural areas rise much more slowly. Also not coincidentally, the Chinese National Academy of Social Sciences forecasts that by 2010, half the population will live in cities compared with 28 percent today and only 10 percent in the early 1980s.

In the early 1990s, developing world economies, especially in East Asia, have grown faster than the economies of the industrial countries. This has the potential to shrink disparities of income, if poorer countries continue to catch up. Yet even if the gaps among countries narrow, the gaps between *people* may not because economic growth is distributed so unevenly within nations. Despite the recent restoration of economic growth in Latin America, for example, U.N. economists say that no progress is expected in reducing poverty, which is even likely to increase slightly.

Meanwhile, in some regions almost no one has been getting richer. The per capita income of most sub-Saharan African nations actually fell during the 1980s. After the latest negotiations of the General Agreement on Tariffs and Trade (GATT) were completed, *The Wall Street Journal* reported that "even GATT's most energetic backers say that in one part of the world, the trade accord may do more harm than good: sub-Saharan Africa," the poorest geographic region. There, an estimated one-third of all college graduates have left the continent. That loss of talented people, due in large part to poverty and a lack of opportunities in Africa, will make it even more difficult for the continent to advance.

The economic growth that has the potential to close income gaps among peoples in the developing world is instead becoming a splitting off, with some parts of societies joining the industrial world while others remain behind. Singapore, Hong Kong, and Taiwan have begun to look like wealthy industrial countries, for example. Now parts of China are following, as are the wealthier segments of Latin American society and of Southeast Asian countries. This is good news for members of the middle-income countries and for the world. But it may do little to help the poorest fifth of humanity.

REFERENCES

CIA World Factbook (1996). Brassey's. Washington, D.C.

Girardet, E. (1985). Afghanistan—The Soviet War. New York, St. Martin's Press.

MacFarquhar, E. (1989, February 13). World Report. U.S. News & World Report. pp. 33–36.

Nixon, R. (1990, February 6). Afghanistan: Our Job Isn't Done. Chicago Tribune.

Yousuf, M. and Adkin, M. (1992). The Bear Trap—Afghanistan's Untold Story. London, Leo Copper.

QUESTIONS

1. Define the following terms.
   a. GNP
   b. GDP
   c. GATT
2. Describe the factors that are responsible for the wide gap between the per-capita GNP of the developed and the developing countries. How could this gap be reduced?

BOX 1 2005 External Debt Snapshot

Region/Group	*2005 External Debt* (Billions of Dollars)	*2005 Debt Servicing* (Billions of Dollars)
Other Emerging Markets and Developing Countries (Total)	3012.3	597.8
Africa	289.4	34.7
Central & Eastern Europe	604.7	121.4
Commonwealth of Independent States	334.0	106.1
Developing Asia	808.3	107.5
Middle East	221.8	28.2
Western Hemisphere	754.1	200.3

Source: 2006 World Economic Outlook, Statistical Appendix, External Debt and Debt Servicing, Table 37, available at: http://www.imf.org/external/pubs/ft/weo/2006/02/pdf/statappx.pdf.

BOX 2 A Fate Worse Than Debt

Amount owed by the world's 47 poorest and most indebted nations: **$422 billion**

Amount of money spent by Western industrialized nations on weapons and soldiers every 12 months: **$422**

Amount raised by Live Aid in 1985 to combat famine in Ethiopia: **$200 billion**

Amount that all African countries pay back on foreign debts every week: **$200 billion**

Amount the United Nations estimates is needed annually to curb the African AIDS epidemic through education, prevention, and care: **$15.0 billion**

Amount African nations pay to service their debts each year (amount paid just in interest): **$13.5 billion**

Annual income per person in Zaire: **$110**

Amount each resident of Zaire would have to raise to pay off the country's debt to foreign creditors: **$236**

Percent of the Zambian budget allocated for foreign debt repayment in 1997: **40%**

Percent of the Zambian budget allocated for basic social services, including vaccinations and education: **7%**

Percent of debt owed by the world's most heavily indebted nations that the World Bank and International

Monetary Fund can afford to cancel without jeopardizing their ability to function, according to the London-based accounting firm Vellacott and Chaney: **100%** percent

Percent of the debt that they have actually agreed to cancel: **33%**

Profits made by Exxon in 2002: **$16.9 billion**

Total debt burden of Benin, Burundi, Chad, Guinea Bissau, São Tomé, Togo, Rwanda, Central African Republic, Sierra Leone, Mali, Somalia, and Niger: **$16.9 billion**

Source: Matters of Scale, *World Watch*, July/August 2002.

Charity

"Cash handouts might sustain you for a few months, at the end of which your problems remain."

NELSON MANDELA

Colonialism

"Through force, fraud and violence, the people of North, East, West, Central and Southern Africa were relieved of their political and economic power and forced to pay allegiance to foreign monarchs."

NELSON MANDELA

BOX 3 Joseph Stiglitz, Winner of 2001 Nobel Prize in Economics, on the Institutional Problems of the IMF and the World Bank

"Voting rights at the IMF and the World Bank are not democratically allocated. The Europeans always choose the head of IMF while the Americans choose the head of the World bank. Not very democratic, is it? And these organizations can be run by anyone. Regardless of their qualifications. So the people selected for these jobs often don't have much experience dealing with the very problems that these institutions were created to deal with. In the case of the current president (Paul Wolfowitz)—probably no appointment could have had less support from the rest of the world."

Source: "The World is Not Flat, Q&A: Joseph Stiglitz," Alex Kingsbury, *U.S. News & World Report,* September 18, 2006, p. 28.

BOX 4 Grameen Bank's Micro Credit Program: A Viable Solution to Alleviate Poverty

Dr. Yunus started an experimental micro credit enterprise in 1977 and by 1983 the Grameen Bank was officially established. The Grameen Bank started to provide credit to the poorest of the poor in rural Bangladesh without any collateral. The credit proved to be a cost effective weapon to eliminate poverty by acting as a catalyst for improving socio-economic conditions. As of May 2006, Grameen Bank has provided loans to 6.67 million borrowers, 97 percent of whom are women. With 2247 branches, Grameen Bank provides services in 72,096 villages, covering more than 86 percent of the total villages in Bangladesh. Today Grameen bank is a $2.5 billion enterprise and its micro credit model has been replicated in more than 50 countries around the world.

Source: http://www.grameen-info.org.

BOX 5 Bangladesh: Natural and Man-Made Disasters

Global Warming: Impact of Sea-Level Rise

A rise in sea level is probably the most widely recognized consequence of global warming because, in a warmer climate, the oceans will expand when heated, and polar ice caps in Greenland and Antarctica may melt. Scientists calculate that expansion effects alone could raise sea levels 20–40 centimeters if the average temperature rises 1.5–4.5 degrees Celsius. A temperature rise in the middle of this range might increase the sea level by 80 centimeters. Such a rise could inundate low-lying areas, destroy marshes and swamps, erode shoreline, worsen coastal flooding, and increase the salinity of rivers, bays, and aquifers. Nearly one-third of the human population lives within 60 kilometers of a coast, and thus many reside on land that would be lost. In Bangladesh, the situation would be particularly severe. Bangladesh would lose 12–28% of its area, which currently houses 9–27% of its population. Floods could penetrate further inland, leaving the nation vulnerable to the type of storm that killed 300,000 people in the early 1970s.

Source: World Resources. (1988–89): An Assessment of the Resources Base that Supports the Global Economy: A Report By the World Resources Institute for Environment and Development. New York, Basic Books.

Farakka Barrage: A Threat to the Economy and the Environment

The Farakka barrage was constructed across the Ganges River at Farakka, 11 miles upstream from the border of Bangladesh. The 70-foot-high and 7365.5-foot-long barrage, with 109 bays, was built to divert 40,000 cubic feet per second (cusec) of water from the Ganges into the Bhagirathi-Hooghly River during the low-flow period. The Indian government's objective for building the Farakka barrage was the preservation and maintenance of the port of Calcutta. The Farakka barrage has triggered a series of adverse impacts in Bangladesh:

- Low water flows in the Ganges and its tributaries
- Profound effects on hydrology and river morphology
- Loss in agricultural production
- Increase in surface and groundwater salinity
- Adverse impact on navigation and fisheries
- Desertification

Cyclones

Cyclones coming from the Bay of Bengal claim thousands of lives and cause major economic damage every year. In November of 1970, cyclones and tidal waves killed 200,000 people, and 100,000 were reported as missing (http://wwwinfoplease.com/ipa/). The cyclone of May 1996, the strongest cyclone in a century, killed 139,000 people and damaged or destroyed more than a million homes (http://193.67.176.1/<climate/database/records/).

Poisoned Waters: Bangladesh Desperately Seeking Solutions

David H. Kinley III
Zabed Hossain

Bangladesh has both too much water and not enough of it. On the one hand, this poor and densely packed nation—130 million people in an area the size of New York state—is laced with the great Ganges and Jamuna Rivers and countless lesser streams. Rainfall totals about 80 inches a year. The country is largely flat, and immense tracts of floodplain become lakes during the monsoon season. Water is nothing if not abundant.

Finding water that is safe to drink is another story, however. It has long been a constant challenge for millions, especially the isolated rural poor. Now, drinking water is the villain in what CBS television once called "the greatest poisoning in human history."

In Nilkanda village, in the Sonargaon subdistrict about two hours from the capital, Dhaka, housewife Monwara Begum tells how her tragedy began to unfold. "Hand pumps helped us to avoid the diseases in the pond," she says, referring to the contamination of surface waters by human and animal waste. "But after drinking from the hand pump over many years, my husband fell ill with arsenic poisoning. We use a filter system now for all we drink, but I'm not convinced it is safe."

"More than 60 percent of the wells in this subdistrict are contaminated with arsenic and unsafe to drink from," explains Sayed Ershad, a development worker who has spent the last several years grappling with the disaster. "Many people still drink the poison water from the wells. The alternatives cost them time and money, and people here face extreme poverty."

Across the village, a thin and listless middle-aged man sits quietly in his ramshackle bamboo and thatch home. His skin is discolored and his hands and feet are pocked with callous-like growths, telltale signs of arsenicosis. "He continues to drink from the contaminated well," says Ershad. "He doesn't use a filter because he's convinced he doesn't have many more days to live."

THE BEST OF INTENTIONS

Bangladesh's high population density and lack of sanitation infrastructure keeps surface waters perpetually contaminated, and waterborne diarrheal diseases have long been a leading cause of widespread illness and premature infant death. In response, the government began installing shallow tubewells (sealed pipes extending down into the groundwater and equipped with simple hand pumps) when Bangladesh was still East Pakistan. Following the independence struggle and subsequent famine in 1971, international aid agencies (UNICEF, the World Bank, the UN Development Programme) and private interests joined the effort. Since then, several million tubewells have been sunk into the shallow water table, and hand pumps have become an icon of a better life for the rural poor. World Health Organization (WHO) reports suggest that the tubewells helped slash infant and child mortality by half over the last 40 years.

The discovery of high concentrations of naturally occurring arsenic in the groundwater is thus a bitter irony.

Reprinted with permission from Worldwatch Institute. *World Watch* January/February 2003. www.worldwatch.org.

Heavy-metal contamination was not even considered in Bangladesh until evidence of arsenicosis began to emerge in the neighboring Indian state of West Bengal in the late 1980s. Arsenic-contaminated wells were first confirmed in Bangladesh in 1993, and it wasn't until 2000 that the first comprehensive program of well testing was completed, when the British Geological Survey (BGS) surveyed a sample of about 3,500 wells nationwide.

The results were shocking. Of the BGS-estimated 6 to 11 million shallow tubewells in Bangladesh (those less than 150 meters deep), at least 1.5 million are heavily contaminated, with concentrations exceeding the national drinking-water standard of 50 parts per billion. Some 35 million people are believed to be exposed beyond the national standard, and 57 million are exposed to arsenic concentrations above the WHO standard of 10 parts per billion.

Local patterns of contamination and resulting sickness are more difficult to circumscribe. Neighboring villages and even households may be consuming well water with vastly different levels of contamination. Some villages with high levels of contamination do not show much evidence of widespread arsenicosis, which generally reveals itself first as dark spots on the skin and nodules on the palms and soles of the feet. Over 5 to 10 years or more, these symptoms become more pronounced. In many cases, internal organs, including the liver, kidneys, and lungs, are also affected. Strong evidence links arsenic poisoning with cancer, but it remains difficult to ascertain how heavy and prolonged the exposure must be to trigger the disease.

The potential human toll in Bangladesh is thus uncertain, because some cancers take as long as 20 years to emerge. According to a recent report from the WHO, in parts of southern Bangladesh where arsenic concentrations are very high, one in ten adult deaths could be due to some form of arsenic-induced cancer of internal organs stemming from long-term exposure. These risks fall heavily on the rural poor, who are rendered more vulnerable to illness by malnutrition and the large volumes of water they drink. They may also ingest additional arsenic by eating rice irrigated with poisoned water and then boiled in it. (The relatively few city dwellers get their water from largely arsenic-free deep aquifers.)

GETTING TO THE BOTTOM OF THE PROBLEM?

With almost half the country's population under threat, it is crucial to understand the root causes and patterns of contamination. But there are other dimensions to the problem besides the geographic distribution of arsenic. For instance, "we know it's not the depth of the well alone that determines its safety now or in the future," says M. Khaliquzzaman, an environmental scientist working with the World Bank. Moreover, "tapping a deep well may supply sufficient water for a village, but is it safe and sustainable for a city of a million? That's what we still don't know." And in any case, it is still necessary to find ways to quickly, cheaply, and repeatedly test water quality in millions of specific locations.

One major initiative for addressing the arsenic problem is the Bangladesh Arsenic Mitigation Water Supply Project (BAMWSP), run by the government and backed by over $35 million from the World Bank and the Swiss Agency for International Development. Launched in 1998, BAMWSP sought to provide water supply relief for large numbers of rural people and to enhance scientific analysis of the scope of contamination. The project was deliberately set up to be independent of existing government institutions, but its work has nevertheless been delayed by bureaucratic mismanagement, lack of coordination with other water sector efforts, and insufficient scientific and technical information.

"Finding solutions to the contamination that can be implemented at the community level has been a complex process," explains the former BAMWSP director, Abdul Quader Choudhary. "It took nearly 30 years to get universal coverage of drinking water supplies using hand pumps. Now we've identified mitigation technologies that can work and are affordable by the poor. But we've still got a long way to go in solving the problem in a systematic way." Indeed, four years after launching its project with much public relations fanfare, the World Bank may be ending its participation in the BAMWSP. If so, nearly 80 percent of the total project budget would be left unspent.

TECHNOLOGY GAP

Politics and bureaucracies aside, at least one part of the overall problem remains technical: determining arsenic concentrations in water at the village level. Test kits produced by the U.S. drug giant Merck began to be used to measure village well contamination in the mid-1990s, but these kits only provide a rough, and sometimes misleading, measure of contamination levels. "Testing and analysis of arsenic contamination is technically complex, difficult, and expensive," explains Dr. Abul Hasnat Milton, chief of the arsenic unit of the NGO Forum for Drinking Water Supply and Sanitation. With assistance from the Danish

Embassy, Dr. Milton's water quality testing laboratory in Dhaka has become one of the country's most sophisticated and has analyzed more than 25,000 samples for the government and aid organizations.

Unfortunately, that standard of excellence is difficult to reproduce everywhere. "There is still a great need to improve arsenic testing," says Mr. Khaliquzzaman. "There are 26 laboratories doing tests across the country, but only a third of them are capable of delivering results of an acceptable standard." To address this obstacle, the WHO and the International Atomic Energy Agency (IAEA) have been providing vital technical help through a laboratory quality assurance program over the past two years (2004–2005).

Compounding uncertainties in the measurement of arsenic levels is a lack of basic knowledge about the movement of groundwater, and the location and mobilization of arsenic in water supplies. "The geology and hydrology of Bangladesh are very complicated due to the nature of its underground structures," explains M. Nazrul Islam, director general of the Geological Survey of Bangladesh. "Try to imagine multiple layers of Himalayan sediments deposited over tens of millions of years by shifting rivers, tides, and floods. The sediment layer is up to 20 kilometers thick near the Bay of Bengal. Aquifer movements within these layers remain poorly understood."

According to S.K.M. Abdullah, who chairs a national expert committee advising the government, "the Bengal Delta is more complex than the Mississippi, the Rhine, or the Senegal River deltas. It's really a composite of three deltas in one. We know that water older than 20,000 years is largely uncontaminated with arsenic, but you can't just drill down to a certain depth and assume it is arsenic-free."

SCIENCE, DEEP AND WIDE

Bangladesh clearly needs more and better science for identifying water supply solutions. "What's really called for is analysis of deep aquifers on a country-wide basis," says Mr. Khaliquzzaman. Because of its potential for quickly and accurately enhancing this knowledge, "isotopic analysis can play a critical role in understanding and addressing the arsenic contamination problem." This technique takes advantage of the fact that most elements are mixtures of isotopes distinguished by differing numbers of neutrons in the atoms' nuclei. The oxygen in water, for instance, is about 98 percent O^{16} (atoms with 16 neutrons each), but trace amounts are in the form of O^{17} and O^{18}. The isotope mixture varies in known ways, allowing water to be "fingerprinted" and tracked through the hydrological cycle. Isotopic analysis can be used to determine the movement of groundwater, where an aquifer is being recharged, how it connects and mixes with other groundwater bodies, and how vulnerable it is to contamination.

Since 1999, IAEA has been working with the World Bank's team to deploy isotope hydrology techniques in constructing a complete model of groundwater and aquifer dynamics and arsenic mobilization. The investigators are trying to understand whether deep aquifers will remain arsenic-free over the long term if they are developed as alternative sources and how other deep aquifers may have been contaminated through mixing of deep and shallow reservoirs.

"Until very recently, the hydrogeological characterization in Bangladesh was being conducted through multiple institutions and agencies using primarily non-isotopic techniques," explains Pradeep Aggarwal, head of isotope hydrology for the IAEA in Vienna, Austria. "The integration of isotope techniques has provided the required information rapidly and at a much lower cost than previously possible. Now we are expanding the application of isotopic techniques countrywide." In particular, Dr. Allen Welch of the U.S. Geological Survey is using isotopic and conventional tools to investigate deep aquifer samples in the most heavily affected regions of the country. Meanwhile, the people of Bangladesh desperately need safe drinking water. The earliest response, initiated by the United Nations Development Programme in the mid-1990s, was an emergency initiative covering 500 of the most heavily affected villages. It screened all water sources for arsenic contamination and all the villagers for arsenicosis and conducted a community awareness program.

More recently, UNICEF has supported the country's largest development NGO, the Bangladesh Rural Advancement Committee (BRAC), and seven other NGOs in evaluating more than 161,000 tubewells supplying several million people. Contaminated wells and pumps have been marked with danger-signifying red paint. BRAC has also been field-testing low-cost water treatment systems and supply alternatives in Sonargaon and eight other subdistricts. These include arsenic filtration devices, rooftop rainwater harvesting, hand-dug ring wells, and deep tubewells. Each technology offers partial relief. Filtration systems, for example, lose their effectiveness over time as the filtering mechanism is saturated, and the captured arsenic then must be suitably disposed of without environmental contamination. Deep wells that have been tested and found safe offer an alternative source, although the associated piped water distribution systems are quite expensive. Catching the rain offers some promise during the wet season, but even simple capture and storage systems are unaffordable by many villagers.

The efforts of BRAC, UNDP, and other bilateral, private, and international aid organizations are slowly mitigating this public-health calamity. But far more extensive and effective aid will be required to find a lasting solution to the poisoned water and bring an end to the widespread suffering. "Arsenicosis is hard to diagnose and there is no known cure or treatment," says Han Heijnen, a WHO environmental health advisor in Dhaka. "We know that continuous exposure is a sure cause of early death. Arsenic poisoning at the levels we're seeing in Bangladesh will take 10 or 20 years off a person's life, without any doubt."

Moreover, arsenic-contaminated water is not limited to Bangladesh, and the scientific work now under way—if it is allowed to continue—could be a critical investment for both Bangladesh and the Asia-Pacific region. "We are only now uncovering the true extent of arsenic poisoning in India, China, and other Asian countries," says Heijnen. "The contamination could reach 100 million people in Asia—more than the numbers affected by HIV-AIDS."

REFERENCES

British Geological Survey, "Arsenic Contamination of Groundwater in Bangladesh," BGS Technical Report WC/00/19, Vol. 1, Summary, February 2001.

Bangladesh Rural Advancement Committee, "Combating a Deadly Menace: Early Experiences With a Community-Based Arsenic Contamination Mitigation Project in Bangladesh," August 2000.

"Bangladesh's Arsenic Poisoning: Who Is To Blame?" UNESCO *Courier*, Paris, January 2001.

West Bengal & Bangladesh Arsenic Crisis Information Centre. Website: www.bicn.com/acic/

QUESTIONS

1. What are the causes of contamination of surface water in Bangladesh?
2. Develop strategies to stop contamination of surface water in Bangladesh.
3. What are the technological issues regarding reliability and accuracy of tests for measuring arsenic contamination?
4. Propose appropriate use of technology to prevent arsenicosis in Bangladesh.

BOX 1 Grameen (of the Village) Bank

"Grameen is committed to social objective—eliminating poverty, providing education, health-care employment opportunities, achieving gender equality by promoting the empowerment of women, ensuring the well-being of the elderly. Grameen dreams about a poverty-free, dole-free world."

Dr. Muhammad Yunus, Nobel Laureate, Founder of the Grameen Bank

Source: Banker to the Poor, Autobiography of Muhammad Yunus, The University Press Limited, Dhaka, Bangladesh, p. 218.

China's Challenge to the United States and to the Earth

LESTER R. BROWN
CHRISTOPHER FLAVIN

During the 1990s, China has emerged as an economic superpower, boasting the world's second largest economy. It is now challenging not only U.S. economic leadership, but the earth's environmental limits.

Using purchasing power parity to measure output, China's 1995 GNP of just over $3 trillion exceeded Japan's $2.6 trillion and trailed only the U.S. output of $6.7 trillion. If the Chinese economy continues to double every eight years, the pace it has maintained since 1980, it will overtake the United States by 2010, becoming the world's largest economy.

Over the last four years (1992–1996), the Chinese economy has grown by 10 to 14 percent per year. As its population of 1.2 billion people moves into modern houses; buys cars, refrigerators, and televisions; and shifts to a meat-based diet, the entire world will feel the effects. Already, China's rapidly rising CO_2 emissions account for one-tenth of the global total.

In recent decades, many observers noted that the United States, with less than 5 percent of the world's population, was consuming a third or more of its resources. But this is no longer true. In several areas, China has overtaken the United States. For example, China now consumes more grain and red meat, uses more fertilizer, and produces more steel than the United States.

Since China has 4.6 times as many people as the United States, its per capita demands on the earth's resources are still far less. To cite an extreme example, the average American consumes 25 times as much oil as the average Chinese citizen does.

Even with its still modest per capita consumption, China is already paying a high environmental price for its booming economy. Its heavy reliance on coal, for example, has led to air pollution nearly as bad as that once found in eastern Europe. As a result, respiratory disease has become epidemic in China, and crop yields are suffering.

As China, with its much larger population, attempts to replicate the consumer economy pioneered in the United States, it becomes clear that the U.S. model is not environmentally sustainable. Ironically, it may be China that finally forces the United States to come to terms with the environmental unsustainability of its own economic system. If China were to consume as much grain and oil per person as the United States does, prices of both commodities would go off the top of the charts. Carbon dioxide emissions would soar, leading to unprecedented climate instability. Together, these trends would undermine the future of the entire world.

The bottom line is that China, with its vast population, simply will not be able to follow for long any of the development paths blazed to date. It will be forced to chart a new course. The country that invented paper and gunpowder now has the opportunity to leapfrog the West and show how to build an environmentally sustainable economy. If it does, China could become a shining example for the rest of the world to admire and emulate. If it fails, we still all pay the price.

Reprinted with permission from Worldwatch Institute. *World Watch* September/October 1996. www.worldwatch.org.

GRAIN HARVEST: CHINA

The United States, long the world's leading grain producer, was overtaken by China in 1983. Over subsequent years, the lead changed hands several times, but since 1986, China's grain harvest has usually exceeded that of the United States. It is also much more stable because China has 2.5 times as much irrigated land as the United States. U.S. production of corn, which dominates U.S. agriculture, is largely rainfed and, affected by both heat and drought, it fluctuates widely from year to year. In consumption, the gap is far wider. China now consumes 365 million tons per year and has become the world's second largest grain importer, whereas the United States consumes 200 million tons and remains the world's leading exporter.

FERTILIZER USE: CHINA

U.S. fertilizer use climbed rapidly from mid-century onward. By 1980, it exceeded 20 million tons. Then it leveled off, averaging less in the mid-1990s than in the early 1980s. Meanwhile, in China, the 1978 economic reforms in agriculture led to a meteoric climb in fertilizer use. In 1986, China overtook the United States to become the world leader. In 1995, Chinese farmers used 28 million tons of fertilizer, while U.S. farmers used just under 20 million tons. In the United States, and now increasingly in China, fertilizer use is constrained by the physiological capacity of crop varieties to effectively absorb more nutrients.

Figure 1
Grain production, United States and China, 1950–95.

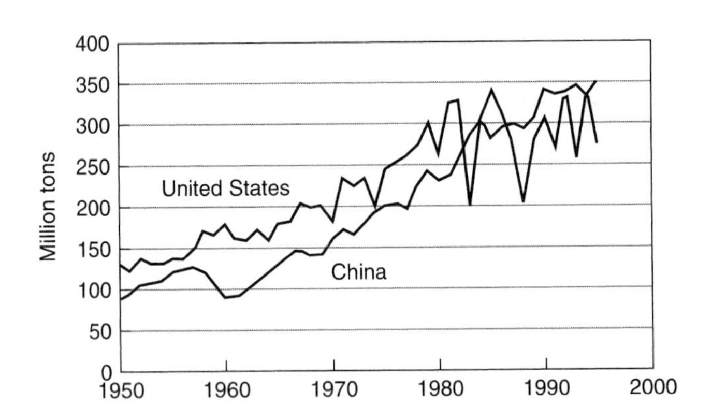

Figure 2
Fertilizer use in the United States and China, 1950–95.

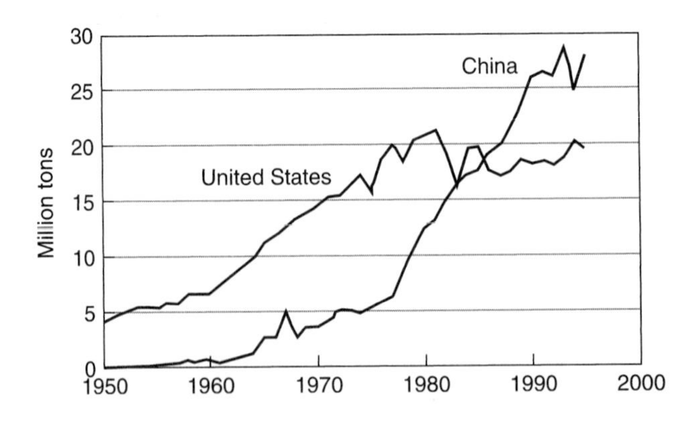

CONSUMPTION OF RED MEAT: CHINA

The consumption of red meat, particularly beef in the form of steak and hamburgers, has become a defining component of the U.S. lifestyle, but China's total consumption of red meat has eclipsed that of the United States. For all red meat combined—beef, pork, and mutton—China now consumes 42 million tons per year compared with only 20 million tons in the United States. China's pork consumption of 30 kilograms per person matches the U.S. intake of 31 kilograms, but its beef consumption lags far behind—4 kilograms to 45 kilograms. If China were to close the beef gap, its people would eat an additional 49 million tons each year. Produced in feedlots, this would take some 343 million tons of grain—roughly as much as the entire U.S. grain harvest.

AUTOMOBILE PRODUCTION: UNITED STATES

In the production of automobiles, the United States dwarfed China by some 6.6 million to 239,000 in 1995. U.S. output is not likely to increase much in the future since most of the automobiles made now are used for replacement rather than for expanding the fleet. China, by contrast, plans to boost production to 3 million per year by the end of the decade, building a fleet of 22 million automobiles by the year 2010. If China's ownership of automobiles were to reach 1 for every 2 people, as in the United States, its fleet of 600 million cars would exceed the 1995 *world* fleet of 480 million cars.

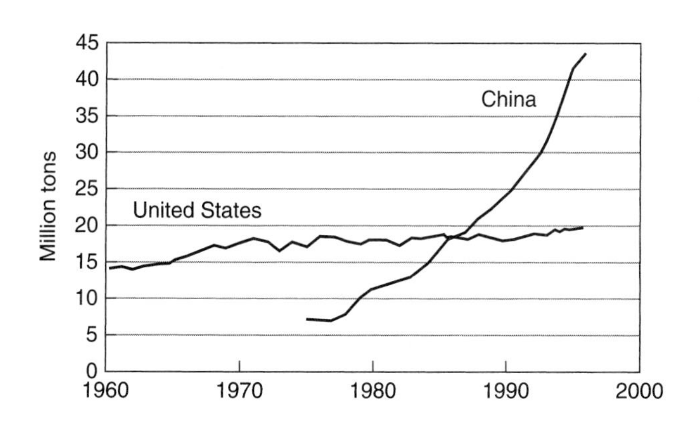

Figure 3
Red meat consumption, United States and China, 1960–96.

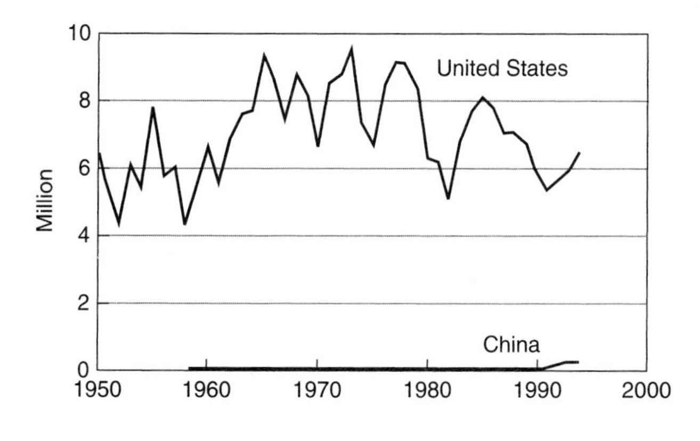

Figure 4
Automobile production, United States and China, 1950–95.

BICYCLE PRODUCTION: CHINA

If a crowded world is compelled to move toward a less polluting, less land-consuming mode of personal transport, then the bicycle may well be the transport vehicle of the future. For this shift, China is well positioned. Its bicycle production has averaged over 40 million a year in recent years compared with less than 8 million a year in the United States. This is perhaps the only major indicator for which the ratio of production between the countries reflects the ratio of population size. In global terms, China accounts for nearly two-fifths of world production of 110 million bicycles annually in recent years.

STEEL OUTPUT: CHINA

In the industrial world of an earlier era, steel production was perhaps the best single indicator of industrial progress. China has recently caught up with the United States in this industry, with both countries turning out 93 million tons in 1995. The big difference is that most of China's steel production is from iron ore. The United States, a more mature industrial society, now gets *roughly half* of its total steel from the reprocessing of scrap metal, a more energy-efficient, less polluting means of production. If China proceeds to develop an automobile-centered transportation system as currently planned, its steel needs will soar far beyond those of the United States.

Figure 5
Bicycle production, United States and China, 1970–94.

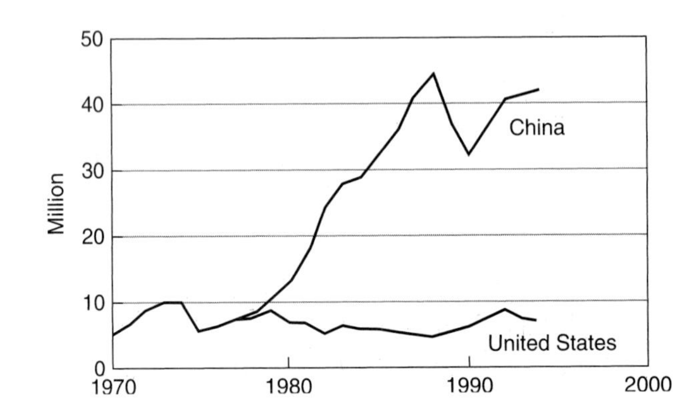

Figure 6
Steel production, United States and China, 1977–95.

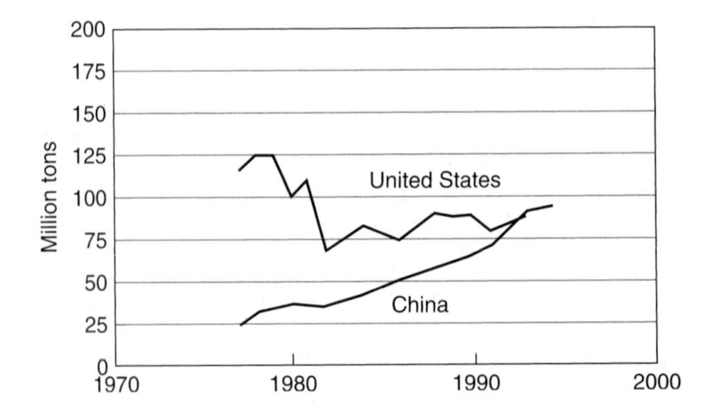

OIL USE: UNITED STATES

Despite its becoming economy, China still consumes only 3.3 million barrels of oil each day, one-fifth the 17 million consumed daily in the United States. U.S. oil use per person is a striking 25 times that in China. With its limited oil reserves, China relies on coal for 75 percent of its energy, whereas the United States relies on coal for just 22 percent. But as China becomes more dependent on automobiles and trucks, its oil use is climbing. Already, it has gone from exporting 500,000 barrels of oil per day in 1990 to *importing* 300,000 barrels per day in 1995. If China were one day to use as much oil per person as the United States does, it would need 80 million barrels daily—more than the whole world now produces or is ever projected to produce.

CARBON EMISSIONS: UNITED STATES

Carbon emissions from fossil-fuel burning totaled 1.394 million tons in the United States in 1995, 73 percent higher than the 807 million tons emitted by China. Fossil fuel burning releases carbon dioxide into the atmosphere—the main gas leading to greenhouse warming. Since 1990, U.S. carbon emissions have grown at roughly 1 percent per year while China's grew at 5 percent annually as use of coal and oil surged. Even so, the United States still emits eight times as much carbon per person as China. If China develops the sort of energy-intensive industries and lifestyles found in the United States, it would further destabilize the world's atmosphere.

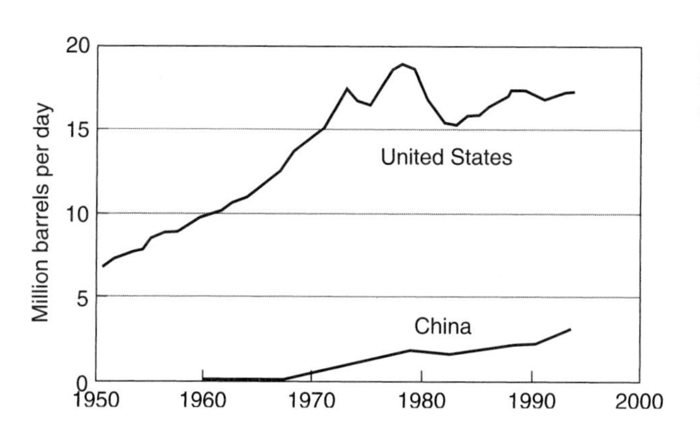

Figure 7
Oil consumption, United States and China, 1950–94.

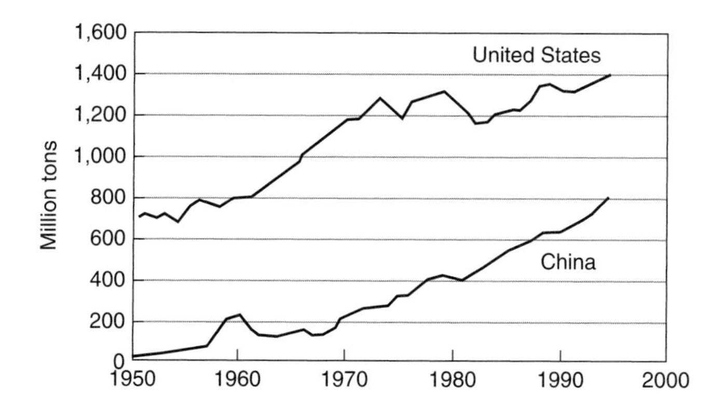

Figure 8
Carbon emission from fossil fuel burning, United States and China, 1950–95.

COMPUTER POWER: UNITED STATES

If steel production is a key indicator of progress in an industrial society, computer use is a key indicator in the information economy of the late twentieth century. In this area, there is no contest: the United States has one computer for every three people; China has one for every 1,000 people. With 74 million computers out of a worldwide total of 173 million, the United States leads the world by a wide margin in the computerization of its economy. China, which has just 1.2 million computers, lags far behind not only the United States but much of the rest of the world as well.

Figure 9
Number of computers, United States and China, 1993.

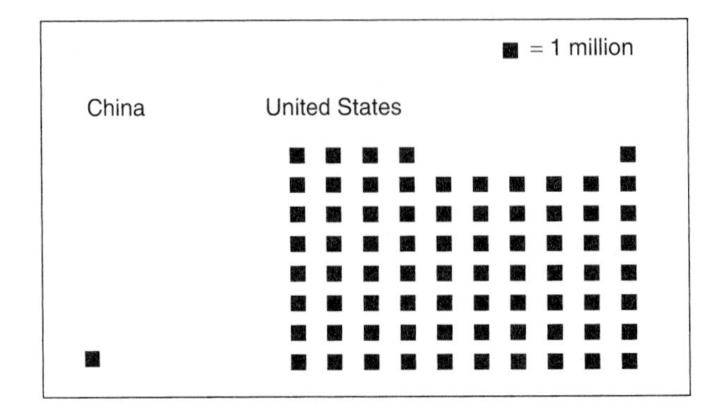

QUESTIONS

1. Compare China with the United States in terms of the following aspects: population, amounts of carbon emissions, automobile production, and oil consumption.
2. Do you think that China, despite its large population, will emerge as an industrialized nation? Why or why not? Explain your answer.
3. Consider the following table summarizing the U.S. Balance of payments (2002). The U.S. trade deficit with China is growing. The U.S. exports to China are growing, but the exports are only one-fifth of the imports from China. What is the best way to reduce the U.S. trade deficit with China?

Table 1
U.S. Balance of Payments, 2002 (in $ billions)

Current Account	*(billions of dollars)*
Exports of goods and services	+972
Imports of goods and services	-1407
Net investment income	-12
Net transfers	-56
Current account balance	-503
Capital account	
Foreign investment in the U.S.	+630
U.S. investment abroad	-156
Statistical discrepancy	29
Capital account balance	503

Source: *Survey of Current Business*, May 2003; *Economic Indicators*, May 2003.

Table 2
Fact Box: China vs. U.S.

	China 2002	*U.S. 2002*
CO_2 emissions (metric tons per capita)	2.53	20.05
Electric power consumption (Kwh per capita)	987	12,183
Energy use (kg of oil equivalent per capita)	960	7,943
Exports of goods and services (% of GDP)	29	10
Fertility rate, total (births per woman)	2	2
Fixed line and mobile phone subscribers (per 1,000)	328	1,134
GDP (current US$)	1,270,999,941,120	10,428,999,532,544
GDP growth (annual %)	8	2
GNI per capita, Atlas method (current US$)	970	35,430
High-technology exports (% of manufactured exports)	23	32
Internet users (per 1,000 people)	46	551
Person computers (per 1,000 people)	28	659
Population growth (annual %)	1	1
Population, total	1,280,400,000	288,368,992
Trade in goods (% of GDP)	49	18

Source: World Development Indicators database, http://www.worldbank.com.

BOX 1 China: By the Numbers

- 1.3 billion people (China's population density is 4.5 times that of the U.S.)
- 9.6 million square kilometers
- 135 people per square kilometer
- GDP = 7262 U.S. $ billion (China's Gross Domestic Product is second in the world)
- 174 cities with more than 1 million people
- 15% of land area is arable land
- 23.65 million people in rural areas still cannot afford daily food and clothing
- 40.67 million rural residents earn between 683 Yuan (U.S. $85) to 944 Yuan per person a year (Xinhuanews, www.chinadaily.com.cn).

"It doesn't matter if a cat is black or white, as long as it catches mice."

DENG XIAOPING

"China is a sleeping giant. When it awakens, the earth will shake."

NAPOLEON BONAPARTE

"China and the United States must, when possible, work together rather than against each other."

RICHARD NIXON

Table 3
U.S. and China: A Comparison of Intakes and Outcomes

	U.S.	*China*
Intakes		
Per capita beef consumption	43.2 kilogram	4.6 kilogram
Percent of men who smoked (2003)	28	59
Number of cigarettes smoked (2004)	379 billion	1.77 trillion
Per capita consumption of sugar (estimate)	30.4 kilogram	8.8 kilogram
Outcomes		
Deaths from heart disease per 100,000 population (2002)	1,768	543
Deaths from stroke per 100,000 population (2002)	563	1,276
Deaths from lung cancer per 100,000 males (2002)	49	37
Percent of adults (>20 years) with diabetes	8.8	2.4
Percent of adults who are obese	31	<5
Per capita spending on health care (1997)	$3,983	$143
Life expectancy at birth (years)	77.43	71.96

Source: Matters of Scale, *World Watch*, July/August 2005.

BOX 2 Confucius Said . . .

The mind of the superior man is concerned with righteousness; the mind of the inferior man is concerned with personal gain.

To error and not to reform, this indeed is an error.

In nature, men are nearly alike; in practice, they get to be wide apart.

The Scholar who cherishes comfort instead of the higher truths of virtue is not fit to be deemed a scholar.

To know what you know, and know what you do not know—this is knowledge.

The Superior man is satisfied and composed; the Inferior man is always full of emotional distress.

Source: Edward Ho and Nick Lambros (editors), *Confucius Said*, Chicago: Blue River Press, 2002.

China in transition. A painting by Hu Yichuan (1910–2000), an internationally acclaimed Chinese artist.
Photo courtesy of the Hu Yichuan family.

Shanghai in the nineteenth century. A painting by Hu Yichuan (1910–2000), an internationally acclaimed Chinese artist.
Photo courtesy of Hu Yichuan family.

Satellite image of eastern China. Beijing is not visible due to atmospheric pollution.
Source: http://visibleearth.nasa.gov/images/1036/S1999324040624_md.jpg.

BOX 3 India: Taj Mahal Threatened by Air Pollution

The Taj Mahal, one of the seven wonders of the world, located in Agra, Delhi, was built by the Mughal Emperor Shahjahan in memory of his beloved wife Mumtaz Mahal. Uncontrolled emissions from factories and plants around Taj Mahal are said to be damaging its marble facade. Agra's air has become so badly polluted that the Taj Mahal—a 350-year-old monument—often disappears behind the murk and smog. Environmentalists claim that clouds of pollution contain sulphur dioxide and nitrogen dioxide that are eating away the monument, turning the brilliant white marble into an ugly shade of gray. The environmentalists are demanding the area around the monument to be designated a total pollution-free zone. This would mean relocating all of the area's 2,000 potentially polluting industries, such as cast iron foundries, glassworks, and a state-owned oil refinery, to areas located farther from the Taj Mahal. The Taj Mahal is the biggest tourist attraction in the country and the source of much-needed foreign exchange. The Indian government has recently announced plans to clean up the environmental pollution that is slowly destroying the monument. Plans call for industries that burn coal or oil to switch to natural gas or liquefied petroleum gas. People living in the city will be required to cut down on their use of kerosene and firewood. Traffic pollution will be curbed with incentives for truck drivers to use diesel with low sulphur content. Recently, the Indian Supreme Court ordered the closing of 14 industries near the Taj Mahal that had failed to install antipollution devices. It also ordered 190 industries along the Ganges River to take steps to control their effluents.

Source: Impact International, April 1995.

Aerosol pollution over northern India and Bangladesh.
Source: http://visibleearth.nasa.gov/images/

BOX 4 Ibn Batutta and Zheng He: The Great Explorers of the Pre-Modern Times

IBN BATTUTA (1304–1377) is a celebrated traveler of premodern times. Born in Fez, Morocco, he spent his life traveling from North Africa to China, Southeast Asia and lands in between. He started his travels when he was 20 years old by going to Mecca for Hajj (pilgrimage). After completing Hajj, he continued his travels. Over the next 29 years, he traveled about 75,000 miles, visiting a large number of regions such as China, Sumatra, Ceylon, Arabia, Syria, Egypt, East Africa, and Timbuktu (equivalent to 44 present-day countries).

Following the advice of the Sultan of Morocco, Abu Inan Faris, some years after his return, Ibn Battuta dictated an account of his travels and observations to a scholar named Ibn Juzayy. The title of his accounts is "Tuhfat al-Nuzzar fi Ghara'ib al-Amsar wa-'Aja'ib al-Asfar, or A Gift to Those Who Contemplate the Wonders of Cities and the Marvels of Traveling." His accounts of what he saw and learned provided the literature of travel with some of the most objective, perceptive, and sophisticated observations ever made by a traveler. Ibn Battuta's sea voyages and references to shipping reveal the state of maritime activities of the Red Sea, Black Sea, Arabian Sea, Indian Ocean, and Chinese waters. His accounts stand, along with Marco Polo's, as the most ambitious and informative work of the travel literature of the Middle Ages.

ZHENG HE (1371–1433) [original name in Chinese, Ma He (Muhammed Ali) son of Ma Hajji (Muhammed Haji)] was a Chinese navigator of central Asian descent. As a special envoy of the Ming dynasty's Emperor Zhu Di, he commanded a fleet of as many as 200 galleons, hundreds of smaller vessels, and 28,000 crewmembers, which sailed from China to as far as East Africa. During the period of 1405–1433, he completed seven voyages. Starting from Nanjing, he sailed to more than 30 islands and coastal nations in Asia and Africa (present-day countries of Vietnam, Cambodia, Indonesia, Malaysia, Bangladesh, Sri Lanka, India, Maldives, Iran, Oman, Yemen, Saudi Arabia, Somalia, Kenya, and Tanzania). His first voyage in 1405 was almost a century before Christopher Columbus's arrival in Americas and Vasco de Gama's in India. His voyages opened a maritime "silk road" which started the East-West commercial and cultural exchanges and revealed the prestige of the Ming dynasty to other nations.

A 1:100 model of Zheng He's "Treasure Boat," donated to the Museum of Science and Industry, Chicago, on August 6, 2005 by the Zheng He Foundation, Chicago, to commemorate the 600th anniversary of Zheng He's voyages.. The model was constructed by Quanzhaou Maritime Museum and Chinese Ancient Ship Modeling Center of Fujian Province, China. The original treasure boat is believed to be 125 meters long and 51 meters wide, with a maximum loading capacity of 7,000 tons and a total water displacement of 14,800 tons. Photo: Courtesy of Ahmed S. Khan

REFERENCES

Chugtai, A. S. *Ibn Battuta—The Great Traveler.* Accessed August 1, 2005. Available at http://www.ummah.net/history/scholars/ibn_battuta/.

Dunn, Ross E. (1986). *The Adventures of Ibn Battuta.* Berkeley, CA: University of California Press.

Turner, Howard R. (1997). *Science in Medieval Islam: An Illustrated Introduction.* New Delhi: Oxford University Press,p. 120.

Viviano, Frank. (2005). China's Great Armada. *National Geographic,* August, pp. 28–53.

Complete and discuss the following flowchart.

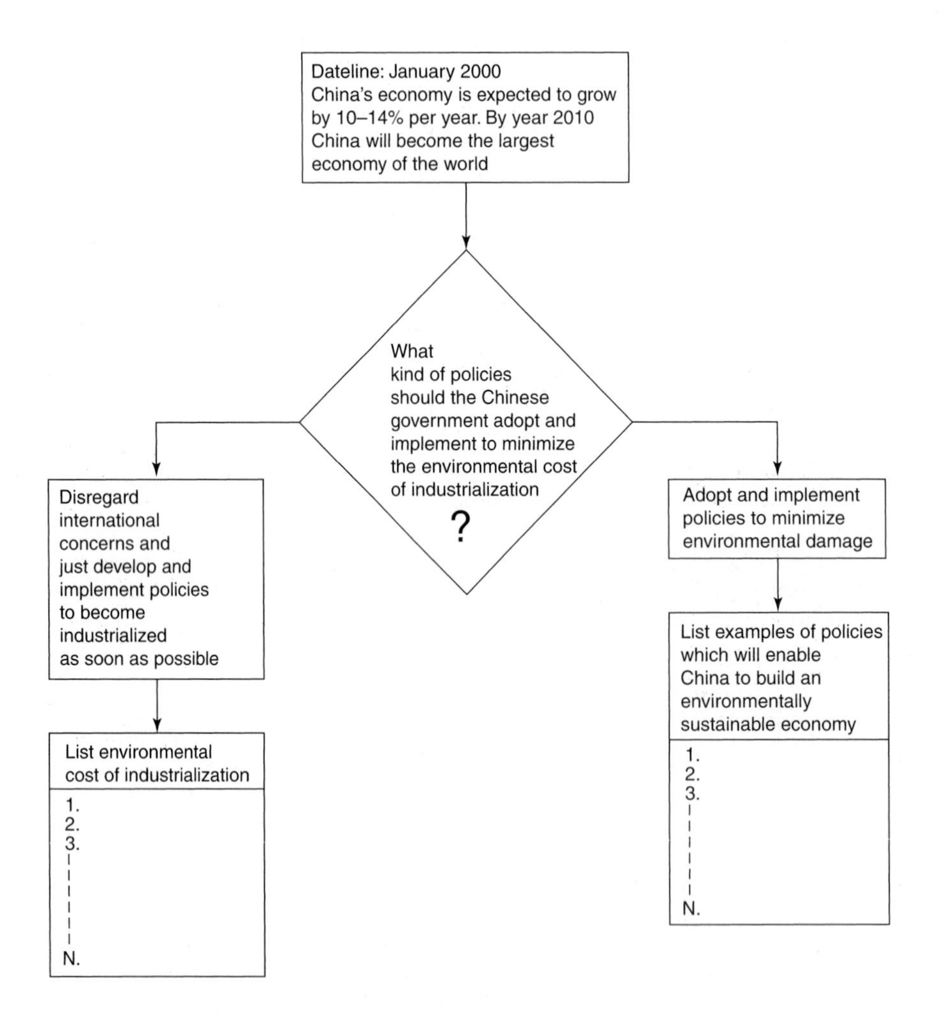

N30 WTO Showdown

PAUL HAWKEN

When I was able to open my eyes, I saw lying next to me a young man, 19, maybe 20 at the oldest. He was in shock, twitching and shivering uncontrollably from being tear-gassed and pepper-sprayed at close range. His burned eyes were tightly closed, and he was panting irregularly. Then he passed out. He went from excruciating pain to unconsciousness on a sidewalk wet from the water that a medic had poured over him to flush his eyes.

More than 700 organizations and between 40,000 and 60,000 people took part in the protests against the WTO's Third Ministerial on November 30 (2000). These groups and citizens sense a cascading loss of human and labor rights in the world. Seattle was not the beginning but simply the most striking expression of citizens struggling against a worldwide corporate-financed oligarchy—in effect, a plutocracy. Oligarchy and plutocracy often are used to describe "other" countries where a small group of wealthy people rule, but not the "First World"—the United States, Japan, Germany, or Canada.

The World Trade Organization, however, is trying to cement into place that corporate plutocracy. Already, the world's top 200 companies have twice the assets of 80 percent of the world's people. Global corporations represent a new empire whether they admit it or not. With massive amounts of capital at their disposal, any of which can be used to influence politicians and the public as and when deemed necessary, all democratic institutions are diminished and at risk. Corporate free market policies subvert culture, democracy, and community, a true tyranny. The American Revolution occurred because of crown-chartered corporate abuse, a "remote tyranny" in Thomas Jefferson's words. To see Seattle as a singular event, as did most of the media, is to look at the battles of Concord and Lexington as meaningless skirmishes.

But the mainstream media, consistently problematic in their coverage of any type of protest, had an even more difficult time understanding and covering both the issues and activists in Seattle. No charismatic leader led. No religious figure engaged in direct action. No movie stars starred. There was no alpha group. The Ruckus Society, Rainforest Action Network, Global Exchange, and hundreds more were there, coordinated primarily by cell phones, e-mails, and the Direct Action Network. They were up against the Seattle Police Department, the Secret Service, and the FBI—to say nothing of the media coverage and the WTO itself.

Thomas Friedman, *The New York Times* columnist and author of an encomium to globalization entitled *The Lexus and the Olive Tree,* angrily wrote that the demonstrators were "a Noah's ark of flat-earth advocates, protectionist trade unions and yuppies looking for their 1960s fix."

Not so. They were organized, educated, and determined. They were human rights activists, labor activists, indigenous people, people of faith, steel workers, and farmers. They were forest activists, environmentalists, social justice workers, students, and teachers. And they

---

Reprinted with permission from Paul Hawken's book titled *Another World is Possible,* (2003, Viking). Paul Hawken, co-author of *Natural Capitalism* and author of *The Ecology of Commerce* and *Another World is Possible* (2003, Viking). He can be reached at Natural Capital Institute, 3B Gate Five Road, Sausalito, CA 94965.

wanted the World Trade Organization to listen. They were speaking on behalf of a world that has not been made better by globalization. Income disparity is growing rapidly. The difference between the top and bottom quintiles has doubled in the past 30 years. Eighty-six percent of the world's goods go to the top 20 percent; the bottom fifth get 1 percent. The apologists for globalization cannot support their contention that open borders, reduced tariffs, and forced trade benefit the poorest 3 billion people in the world.

Globalization does, however, create the concentrations of capital seen in northern financial and industrial centers—indeed, the wealth in Seattle itself. Since the people promoting globalized free trade policies live in those cities, it is natural that they should be biased.

Despite Friedman's invective about "the circus in Seattle," the demonstrators and activists who showed up there were not against trade. They do demand proof that shows when and how trade—as the WTO constructs it—benefits workers and the environment in developing nations, as well as workers at home. Since that proof has yet to be offered, the protesters came to Seattle to hold the WTO accountable.

THIS IS WHAT DEMOCRACY LOOKS LIKE

On the morning of November 30th, I walked toward the Convention Center, the site of the planned Ministerial, with Randy Hayes, the founder of Rainforest Action Network. As soon as we turned the corner on First Avenue and Pike Street, we could hear drums, chants, sirens, roars. At Fifth, police stopped us. We could go no farther without credentials. Ahead of us were thousands of protesters. Beyond them was a large cordon of gas-masked and riot-shielded police, an armored personnel carrier, and fire trucks. On one corner was Niketown. On the other, the Sheraton Hotel, through which there was a passage to the Convention Center.

The cordon of police in front of us tried to prevent more protesters from joining those who blocked the entrances to the Convention Center. Randy was a credentialed WTO delegate, which means he could join the proceedings as an observer. He showed his pass to the officer, who thought it looked like me. The officer joked with us, kidded Randy about having my credential, and then winked and let us both through. The police were still relaxed at that point. Ahead of us crowds were milling and moving. Anarchists were there, maybe 40 in all, dressed in black pants, black bandanas, black balaclavas, and jackboots, one of two groups identifiable by costume. The other was a group of 300 children who had dressed brightly as turtles in the Sierra Club march the day before.

The costumes were part of a serious complaint against the WTO. When the United States attempted to block imports of shrimp caught in the same nets that capture and drown 150,000 sea turtles each year, the WTO called the block "arbitrary and unjustified." Thus far in every environmental dispute that has come before the WTO, its three-judge panels, which deliberate in secret, have ruled for business and against the environment. The panel members are selected from lawyers and officials who are not educated in biology, the environment, social issues, or anthropology.

Opening ceremonies for the World Trade Organization's Third Ministerial were to have been held that Tuesday morning at the Paramount Theater near the Convention Center. Police had ringed the theater with Metro buses touching bumper to bumper. The protesters surrounded the outside of that steel circle. Only a few hundred of the 5,000 delegates made it inside, as police were unable to provide safe corridors for members and ambassadors. The theater was virtually empty when U.S. trade representative and meeting co-chair Charlene Barshevsky was to have delivered the opening keynote. Instead, she was captive in her hotel room a block from the meeting site. WTO executive director Michael Moore was said to have been apoplectic.

Inside the Paramount, Mayor Paul Schell stood despondently near the stage. Since no scheduled speakers were present, Kevin Danaher, Medea Benjamin, and Juliet Beck from Global Exchange went to the lectern and offered to begin a dialogue in the meantime. The WTO had not been able to come to a pre-meeting consensus on the draft agenda. The NGO community, however, had drafted a consensus agreement about globalization—and the three thought this would be a good time to present it, even if the hall had only a desultory number of delegates. Although the three were credentialed WTO delegates, the sound system was quickly turned off and the police arm-locked and handcuffed them. Medea's wrist was sprained. All were dragged off stage and arrested.

The arrests mirrored how the WTO has operated since its birth in 1995. Listening to people is not its strong point. WTO rules run roughshod over local laws and regulations. It relentlessly pursues the elimination of any restriction on the free flow of trade including local, national, or international laws that distinguish between products based on how they are made, by whom, or what happens during production.

The WTO is thus eliminating the ability of countries and regions to set standards, to express values, or to determine what they do or don't support. Child labor, prison labor, forced labor, substandard wages and working conditions cannot be used as a basis to discriminate against goods. Nor can a country's human rights record, environmental destruction, habitat loss, toxic waste production, or the presence of transgenic materials or synthetic hormones be used as the basis to screen or stop goods from entering a country. Under WTO rules, the Sullivan Principles and the boycott of South Africa would not have existed. If the world could vote on the WTO, would it pass? Not one country of the 135 member-states of the WTO has held a plebiscite to see if its people support the WTO mandate. The people trying to meet in the Green Rooms at the Seattle Convention Center were not elected. Even Michael Moore was not elected.

While Global Exchange was temporarily silenced, the main organizer of the downtown protests, the Direct Action Network (DAN), was executing a plan that was working brilliantly outside the Convention Center. The plan was simple: insert groups of trained nonviolent activists into key points downtown, making it impossible for delegates to move. DAN had hoped that 1,500 people would show up. Close to 10,000 did. The 2,000 people who began the march to the Convention Center at 7 a.m. from Victor Steinbrueck Park and Seattle Central Community College were composed of affinity groups and clusters whose responsibility was to block key intersections and entrances. Participants had trained for many weeks in some cases, for many hours in others. Each affinity group had its own mission and was self-organized. The streets around the Convention Center were divided into 13 sections, and individual groups and clusters were responsible for holding these sections. There were also "flying groups" that moved at will from section to section, backing up groups under attack as needed. The groups were further divided into those willing to be arrested and those who were not.

All decisions prior to the demonstrations were reached by consensus. Minority views were heeded and included. The one thing all agreed to was that there would be no violence—physical or verbal—no weapons, no drugs or alcohol. Throughout most of the day, using a variety of techniques, groups held intersections and key areas downtown. As protesters were beaten, gassed, clubbed, and pushed back, a new group would replace them. There were no charismatic leaders barking orders. There was no command chain. There was no one in charge. Police said that they were not prepared for the level of violence, but, as one protester later commented, what they were unprepared for was a network of nonviolent protesters totally committed to one task—shutting down the WTO.

THE VICTORY THAT WASN'T

Meanwhile, Moore and Barshevsky's frustration was growing by the minute. Their anger and disappointment was shared by Madeleine Albright, the Clinton advance team, and, back in Washington, by chief of staff John Podesta. This was to have been a celebration, a victory, one of the crowning achievements to showcase the Clinton administration, the moment when it would consolidate its centrist free trade policies, allowing the Democrats to show multinational corporations that they could deliver the goods.

This was to have been Barshevsky's moment, an event that would give her the inside track to become Secretary of Commerce in the Gore Administration. This was to have been Michael Moore's moment, reviving what had been a mediocre political ascendancy in New Zealand. To say nothing of Monsanto's moment. If the as-yet unapproved draft agenda were ever ratified, the Europeans could no longer block or demand labeling on genetically modified crops without being slapped with punitive lawsuits and tariffs. The draft also contained provisions that would allow all water in the world to be privatized. It would allow corporations patent protection on all forms of life, even genetic material in cultural use for thousands of years. Farmers who have spent thousands of years growing crops in a valley in India could, within a decade, be required to pay for their water. They could also find that they would have to purchase seeds containing genetic traits their ancestors developed from companies that have engineered the seeds not to reproduce unless the farmer annually buys expensive chemicals to restore seed viability. If this happens, the CEOs of Novartis and Enron, two of the companies creating the seeds and privatizing the water, will have more money. What will Indian farmers have?

But the perfect moment for Barshevsky, Moore and Monsanto didn't arrive. The meeting couldn't start. Demonstrators were everywhere. Private security guards locked down the hotels. The downtown stores were shut. Hundreds of delegates were on the street trying to get into the Convention Center. No one could help them. For WTO delegates accustomed to an ordered corporate or governmental world, it was a calamity.

Up Pike toward Seventh and to Randy's and my right on Sixth, protesters faced armored cars, horses, and police in full riot gear. In between, demonstrators ringed the Sheraton to prevent an alternative entry to the Convention Center. At one point, police guarding the steps to the lobby pummeled and broke through a crowd of protesters to let eight delegates in. On Sixth Street, Sergeant Richard Goldstein asked demonstrators seated on the street in front of the police line "to cooperate" and move back 40 feet. No one understood why, but that hardly mattered. No one was going to move. He announced that "chemical irritants" would be used if they did not leave.

The police were anonymous. No facial expressions, no face. You could not see their eyes. They were masked Hollywood caricatures burdened with 60 to 70 pounds of weaponry. These were not the men and women of the 6th precinct. They were the Gang Squads and the SWAT teams of the Tactical Operations Divisions, closer in training to soldiers from the School of the Americas than local cops on the beat. Behind them and around were special forces from the FBI, the Secret Service, even the CIA.

The police were almost motionless. They were equipped with U.S. military standard M40A1 double-canister gas masks; uncalibrated, semi-automatic, high-velocity Autocockers loaded with solid plastic shot; Monadnock disposable plastic cuffs; Nomex slash-resistant gloves; Commando boots; Centurion tactical leg guards; combat harnesses; DK5-H pivot-and-lock riot face shields; black Monadnock P24 polycarbonate riot batons with Trum Bull stop side handles; No. 2 continuous-discharge CS (orchochlorobenzylidenemalononitrile) chemical grenades; M651 CN (chloroacetophenone) pyrotechnic grenades; T16 Flameless OC Expulsion Grenades; DTCA rubber bullet grenades (Stingers); M-203 (40 mm) grenade launchers; First Defense MK-46 Oleoresin Capsicum (OC) aerosol tanks with hose and wands; .60 caliber rubber ball impact munitions; lightweight tactical Kevlar composite ballistic helmets; combat butt packs; .30 cal. 30-round magazine pouches; and Kevlar body armor. None of the police had visible badges or forms of identification.

The demonstrators seated in front of the black-clad ranks were equipped with hooded jackets for protection against rain and chemicals. They carried toothpaste and baking powder for protection of their skin, and wet cotton cloths impregnated with vinegar to cover their mouths and noses after a tear gas release. In their backpacks were bottled water and food for the day ahead.

Ten Koreans came around the corner carrying a 10-foot banner protesting genetically modified foods. They were impeccable in white robes, sashes, and headbands. One was a priest. They played flutes and drums and marched straight toward the police and behind the seated demonstrators. Everyone cheered at the sight and chanted, "The whole world is watching." The sun broke through the gauzy clouds. It was a beautiful day. Over cell phones, we could hear the cheers coming from the labor rally at the football stadium. The air was still and quiet.

At 10 a.m., the police fired the first seven canisters of tear gas into the crowd. The whitish clouds wafted slowly down the street. The seated protesters were overwhelmed, yet most did not budge. Police poured over them. Then came the truncheons, and the rubber bullets.

I was with a couple of hundred people who had ringed the hotel, arms locked. We watched as long as we could until the tear gas slowly enveloped us. We were several hundred feet from Sgt. Goldstein's 40-foot "cooperation" zone. Police pushed and truncheoned their way through and behind us. We covered our faces with rags and cloth, snatching glimpses of the people being clubbed in the street before shutting our eyes.

The gas was a fog through which people moved in slow, strange dances of shock and pain and resistance. Tear gas is a misnomer. Think about feeling asphyxiated and blinded. Breathing becomes labored. Vision is blurred. The mind is disoriented. The nose and throat burn. It's not a gas, it's a drug. Gas-masked police hit, pushed, and speared us with the butt ends of their batons. We all sat down, hunched over, and locked arms more tightly. By then, the tear gas was so strong our eyes couldn't open. One by one, our heads were jerked back from the rear, and pepper was sprayed directly into each eye. It was very professional. Like hair spray from a stylist. Sssst. Sssst.

Pepper spray is derived from food-grade cayenne peppers. The spray used in Seattle is the strongest available, with a 1.5 to 2.0 million Scoville heat unit rating. One to three Scoville units are when your tongue can first detect hotness. (The habanero, usually considered the hottest pepper in the world, is rated around 300,000 Scoville units.) This description was written by a police officer who sells pepper spray on his website. It is about his first experience being sprayed during a training exercise:

> "It felt as if two red-hot pieces of steel were grinding into my eyes, as if someone was blowing a red-hot cutting torch into my face. I fell to the ground just like all the others and started to rub my eyes even though I knew better not too. The heat from the pepper spray was overwhelming. I could not resist trying to rub it off of my face. The pepper spray caused my eyes to shut very quickly. The only way I

could open them was by prying them open with my fingers. Everything that we had been taught about pepper spray had turned out to be true. And everything that our instructor had told us that we would do, even though we knew not to do it, we still did. Pepper spray turned out to be more than I had bargained for."

As I tried to find my way down Sixth Avenue after the tear gas and pepper spray, I couldn't see. The person who found and guided me was Anita Roddick, the founder of the Body Shop, and probably the only CEO in the world who wanted to be on the streets of Seattle helping people that day.

When your eyes fail, your ears take over. I could hear acutely. What I heard was anger, dismay, shock. For many people, including the police, this was their first direct action. Demonstrators who had taken nonviolent training were astonished at the police brutality. The demonstrators were students, their professors, clergy, lawyers, and medical personnel. They held signs against Burma and violence. They dressed as butterflies.

The Seattle Police had made a decision not to arrest people on the first day of the protests (a decision that was reversed for the rest of the week). Throughout the day, the affinity groups created through Direct Action stayed together. Tear gas, rubber bullets, and pepper spray were used so frequently that by late afternoon, supplies ran low. What seemed like an afternoon lull or standoff was because police had used up all their stores. Officers combed surrounding counties for tear gas, sprays, concussion grenades, and munitions. As police restocked, the word came down from the White House to secure downtown Seattle or the WTO meeting would be called off. By late afternoon, the mayor and police chief announced a 7 p.m. curfew and "no protest" zones, and declared the city under civil emergency. The police were fatigued and frustrated. Over the next seven hours and into the night, the police turned downtown Seattle into Beirut.

That morning, it was the police commanders who were out of control, ordering the gassing and pepper spraying and shooting of people protesting nonviolently. By evening, it was the individual police who were out of control. Anger erupted, protesters were kneed and kicked in the groin, and police used their thumbs to grind the eyes of pepper-spray victims. A few demonstrators danced on burning dumpsters that were ignited by pyrotechnic tear-gas grenades (the same ones used in Waco).

Protesters were defiant. Tear gas canisters were thrown back as fast as they were launched. Drum corps marched using empty 5-gallon water bottles for instruments. Despite their steadily dwindling number, maybe 1,500 by evening, a hardy number of protesters held their ground, seated in front of heavily armed police, hands raised in peace signs, submitting to tear gas, pepper spray, and riot batons. As they retreated to the medics, new groups replaced them.

Every channel covered the police riots live. On TV, the police looked absurd, frantic, and mean. Passing Metro buses filled with passengers were gassed. Police were pepper spraying residents and bystanders. The mayor went on TV that night to say that, as a protester from the '60s, he never could have imagined what he was going to do next: call in the National Guard.

LAWLESSNESS

This is what I remember about the violence. There was almost none until police attacked demonstrators that Tuesday in Seattle. Michael Meacher, environment minister of the United Kingdom, said afterward, "What we hadn't reckoned with was the Seattle Police Department, who single-handedly managed to turn a peaceful protest into a riot." There was no police restraint, despite what Mayor Paul Schell kept proudly assuring television viewers all day. Instead, there were rubber bullets, which Schell kept denying all day. In the end, more copy and video was given to broken windows than broken teeth.

During that day, the anarchist black blocs were in full view. Numbering about one hundred, they could have been arrested at any time but the police were so weighed down by their own equipment, they literally couldn't run. Both the police and the Direct Action Network had mutually apprised each other for months prior to the WTO about the anarchists' intentions. The Eugene Police had volunteered information and specific techniques to handle the black blocs but had been rebuffed by the Seattle Police. It was widely known they would be there and that they had property damage in mind. To the credit of the mayor, the police chief, and the Seattle press, distinctions were consistently made between the protesters and the anarchists (later joined by local vandals as the night wore on). But the anarchists were not primitivists, nor were they all from Eugene. They were well organized, and they had a plan.

The black blocs came with tools (crow-bars, hammers, acid-filled eggs) and hit lists. They knew they were going after Fidelity Investments but not Charles Schwab. Starbucks but not Tully's. The GAP but not REI. Fidelity Investments because they are large investors in Occidental

Petroleum, the oil company most responsible for the violence against the U'wa tribe in Columbia. Starbucks because of their non-support of fair-traded coffee. The GAP because of the Fisher family's purchase of Northern California forests. They targeted multinational corporations that they see as benefiting from repression, exploitation of workers, and low wages. According to one anarchist group, the ACME collective: "Most of us have been studying the effects of the global economy, genetic engineering, resource extraction, transportation, labor practices, elimination of indigenous autonomy, animal rights, and human rights, and we've been doing activism on these issues for many years. We are neither ill-informed nor inexperienced." They don't believe we live in a democracy, do believe that property damage (windows and tagging primarily) is a legitimate form of protest, and that it is not violent unless it harms or causes pain to a person. For the black blocs, breaking windows is intended to break the spells cast by corporate hegemony, an attempt to shatter the smooth exterior facade that covers corporate crime and violence. That's what they did. And what the media did is what I just did in the last two paragraphs: focus inordinately on the tiniest sliver of the 40–60,000 marchers and demonstrators.

It's not inapt to compare the pointed lawlessness of the anarchists with the carefully considered ability of the WTO to flout laws of sovereign nations. When "The Final Act Embodying the Results of the Uruguay Round of Multilateral Trade Negotiations" was enacted April 15th, 1994, in Marrakech, it was recorded as a 550-page agreement that was then sent to Congress for passage. Ralph Nader offered to donate $10,000 to any charity of a congressman's choice if any of them signed an affidavit saying they had read it and could answer several questions about it. Only one congressman—Senator Hank Brown, a Colorado Republican—took him up on it. After reading the document, Brown changed his opinion and voted against the Agreement.

There were no public hearings, dialogues, or education. What passed is an Agreement that gives the WTO the ability to overrule or undermine international conventions, acts, treaties, and agreements. The WTO directly violates "The Universal Declaration of Human Rights" adopted by member nations of the United Nations, not to mention Agenda 21. (The proposed draft agenda presented in Seattle went further in that it would require Multilateral Agreements on the Environment such as the Montreal Protocol, the Convention on Biological Diversity, and the Kyoto Protocol to be in alignment and subordinate to WTO trade polices.) The final Marrakech Agreement contained provisions that most of the delegates, even the heads-of-country delegations, were not aware of, statutes that were drafted by sub-groups of bureaucrats and lawyers, some of whom represented transnational corporations.

The police mandate to clear downtown was achieved by 9 p.m. Tuesday night. But police, some of whom were fresh recruits from outlying towns, didn't want to stop there. They chased demonstrators into neighborhoods where the distinctions between protesters and citizens vanished. The police began attacking bystanders, residents, and commuters. They had lost control. When President Clinton sped from Boeing Airfield to the Westin Hotel at 1:30 a.m. Wednesday, his limousine entered a police-ringed city of broken glass, helicopters, and boarded windows. He was too late. The mandate for the WTO had vanished sometime that afternoon.

MEDIA MYTHS AND LEGENDS

The next morning and over the next days, a surprised press corps went to work and spun webs. They vented thinly veiled anger in their columns, and pointed guilt-mongering fingers at brash, misguided white kids. They created myths, told fables. What a majority of media projected onto the marchers and activists, in an often-contradictory manner, was that the protesters are afraid of a world without walls; that they want the WTO to have even more rules; that anarchists led by John Zerzan from Eugene ran rampant; that they blame the WTO for the world's problems; that they are opposed to global integration; that they are against trade; that they are ignorant and insensitive to the world's poor; that they want to tell other people how to live. The list is long and tendentious. Outstanding coverage came from Amy Goodman's Democracy Now on Pacifica radio and The Nation.

Patricia King, one of two *Newsweek* reporters in Seattle, called me from her hotel room at the Four Seasons and wanted to know if this was the '60s redux.

No, I told her. The '60s were primarily an American event; the protests against the WTO are international.

Who are the leaders? she wanted to know.

There are no leaders in the traditional sense. But there are thought leaders, I said.

Who are they? she asked.

I began to name some: Martin Khor and Vandana Shiva of the Third World Network in Asia, Walden Bello of Focus on the Global South, Maude Barlow of the Council of Canadians, Tony Clarke of Polaris Institute, Jerry

Mander of the International Forum on Globalization, Susan George of the Transnational Institute, David Korten of the People-Centered Development Forum, John Cavanagh of the Institute for Policy Studies, Lori Wallach of Public Citizen, Mark Ritchie of the Institute For Agriculture and Trade Policy, Anuradha Mittal of the Institute for Food & Development Policy, Helena Norberg-Hodge of the International Society for Ecology and Culture, Owens Wiwa of the Movement for the Survival of the Ogoni People, Chakravarthi Raghavan of the Third World Network in Geneva, Debra Harry of the Indigenous Peoples Coalition Against Biopiracy, José Bové of the Confederation Paysanne Europèenne, Tetteh Hormoku of the Third World Network in Africa, Randy Hayes of Rainforest Action Network . . .

Stop, stop, she said. I can't use these names in my article. Why not? Because Americans have never heard of them. Instead, *Newsweek* editors put the picture of the Unabomber, Theodore Kaczynksi, in the article because he had, at one time, purchased some of John Zerzan's writings.

Some of the mainstream media also assigned blame to the protesters for the meeting's outcome. But ultimately, it was not on the streets that the WTO broke down. It was inside. It was a heated and rancorous Ministerial, and the meeting ended in a stalemate, with African, Caribbean, and some Asian countries refusing to support a draft agenda that had been negotiated behind closed doors without their participation. With that much contention inside and out, one can rightly ask whether the correct question is being posed. The question, as propounded by corporations, is how to make trade rules more uniform. The proper question, it seems to me, is how do we make trade rules more differentiated so that different cultures, cities, peoples, places, and countries benefit the most. Arnold Toynbee wrote that "Civilizations in decline are consistently characterized by a tendency toward standardization and uniformity. Conversely, during the growth stage of civilization, the tendency is toward differentiation and diversity."

Those who marched and protested opposed the tyrannies of globalization, uniformity, and corporatization, but they did not necessarily oppose internationalization of trade. Economist Herman Daly has long made the distinction between the two. Internationalization means trade between nations. Globalization refers to a system where there are uniform rules for the entire world, a world in which capital and goods move at will without the rule of individual nations. Nations, for all their faults, set trade standards. Those who are willing to meet those standards can do business with them. Do nations abuse this? Always and constantly, the US being the worst offender. But nations do provide, where democracies prevail, a means for people to set their own policy, to influence decisions, and determine their future. Globalization supplants the nation, the state, the region, and the village. While eliminating nationalism is indeed a good idea, the elimination of sovereignty is not.

GLOBALIZATION'S WINNERS & LOSERS

One recent example of the power of the WTO is Chiquita Brands International, a $2 billion dollar corporation that recently made a large donation to the Democratic Party. Coincidentally, the United States filed a complaint with the WTO against the European Union because European import policies favored bananas coming from small Caribbean growers instead of the banana conglomerates. The Europeans freely admitted their bias and policy: they restricted imports from large multinational companies in Central America (plantations whose lands were secured by US military force during the past century) and favored small family farmers from former colonies who used fewer chemicals. It seemed like a decent thing to do, and everyone thought the bananas tasted better. For the banana giants, this was untenable. The United States prevailed in this WTO-arbitrated case. So who won and who lost? Did the Central American employees at Chiquita Brands win? Ask the hundreds of workers in Honduras who were made infertile by the use of dibromochloropropane on the banana plantations. Ask the mothers whose children have birth defects from pesticide poisoning. Did the shareholders of Chiquita win? At the end of 1999, Chiquita Brands was losing money because it was selling bananas at below cost to muscle its way into the European market. Its stock was at a 13-year low, the shareholders were angry, the company was up for sale, but the prices of bananas in Europe are really cheap. Who lost? Caribbean farmers who could formerly make a living and send their kids to school can no longer do so because of low prices and demand.

Globalization leads to the concentration of wealth inside such large multinational corporations as Time-Warner, Microsoft, GE, Exxon, and Wal-Mart. These giants can obliterate social capital and local equity and create cultural homogeneity in their wake. Countries as different as Mongolia, Bhutan, and Uganda will have no choice but to allow Blockbuster, Burger King, and Pizza Hut to operate within their borders. Under WTO, even decisions made by local communities to refuse McDonald's entry (as did Martha's Vineyard) could be overruled. The as-yet unapproved draft agenda calls for WTO member governments to open up their procurement process to multinational corporations. No longer could local governments buy preferentially from local vendors. The WTO could force governments to privatize healthcare and allow foreign companies to bid on delivering national health programs. The draft agenda could privatize and commodify education and could ban cultural restrictions on entertainment, advertising, or commercialism as trade barriers. Globalization kills self-reliance, since smaller local businesses can rarely compete with highly capitalized firms who seek market share instead of profits. Thus, developing regions may become more subservient to distant companies, with more of their income exported rather than re-spent locally.

On the weekend prior to the WTO meeting, the International Forum on Globalization (IFG) held a two-day teach-in at Benaroya Hall in downtown Seattle on just such questions of how countries can maintain autonomy in the face of globalization. Chaired by IFG president Jerry Mander, more than 2,500 people from around the world attended. A similar number were turned away. It was the hottest ticket in town (but somehow that ticket did not get into the hands of pundits and columnists). It was an extravagant display of research, intelligence, and concern, expressed by scholars, diplomats, writers, academics, fishermen, scientists, farmers, geneticists, businesspeople, and lawyers. Prior to the teach-in, non-governmental organizations, institutes, public interest law firms, farmers' organizations, unions, and councils had been issuing papers, communiqués, press releases, books, and pamphlets for years. They were almost entirely ignored by the WTO.

A CLASH OF CHRONOLOGIES

But something else was happening in Seattle underneath the debates and protests. In Stewart Brand's new book, *The Clock of the Long Now—Time and Responsibility,* he discusses what makes a civilization resilient and adaptive. Scientists have studied the same question about ecosystems. How does a system, be it cultural or natural, manage change, absorb shocks, and survive, especially when change is rapid and accelerating? The answer has much to do with time, both our use of it and our respect for it. Biological diversity in ecosystems buffers against sudden shifts because different organisms and elements fluctuate at different time scales. Flowers, fungi, spiders, trees, laterite, and foxes all have different rates of change and response. Some respond quickly, others slowly, so that the system, when subjected to stress, can move, sway, and give and then return and restore.

The WTO was a clash of chronologies or time frames, at least three, probably more. The dominant time frame was commercial. Businesses are quick, welcome innovation in general, and have a bias for change. They need to grow more quickly than ever before. They are punished, pummeled and bankrupted if they do not. With worldwide capital mobility, companies and investments are rewarded or penalized instantly by a network of technocrats and money managers who move $2 trillion a day seeking the highest return on capital. The Internet, greed, global communications, and high-speed transportation are all making businesses move faster than before.

The second time frame is culture. It moves more slowly. Cultural revolutions are resisted by deeper, historical beliefs. The first institution to blossom under *perestroika* was the Russian Orthodox Church. I walked into a church near Boris Pasternak's dacha in 1989 and heard priests and *babushkas* reciting the litany with perfect recall as if 72 years of repression had never happened. Culture provides the slow template of change within which family, community, and religion prosper. Culture provides identity and, in a fast-changing world of displacement and rootlessness, becomes ever more important. In between culture and business is governance, faster than culture, slower than commerce.

At the heart, the third and slowest chronology is Earth, nature, the web of life. As ephemeral as it may seem, it is the slowest clock ticking, always there, responding to long, ancient evolutionary cycles that are beyond civilization.

These three chronologies often conflict. As Stewart Brand points out, business unchecked becomes crime. Look at Russia. Look at Microsoft. Look at history. What makes life worthy and allows civilizations to endure are all the things that have "bad" payback under commercial rules: infrastructure, universities, temples, poetry, choirs, literature, language, museums, terraced fields, long marriages, line dancing, and art. Most everything we hold valuable is slow to develop, slow to learn, and slow to change. Commerce requires the governance of politics, art, culture, and nature, to slow it down, to make it heedful, to make it pay attention to people and place. It has never done this on its own. The extirpation of languages, cultures, forests, and fisheries is occurring worldwide in the name of speeding up business. Business itself is stressed out of its mind by rapid change. The rate of change is unnerving to all, even to those who are supposedly benefiting. To those who are not, it is devastating.

What marched in the streets of Seattle? Slower time strode into the WTO. Ancient identity emerged. The cloaks of the forgotten paraded on the backs of our children.

What appeared in Seattle were the details, dramas, stories, peoples, and puppet creatures that had been ignored by the bankers, diplomats, and the rich. Corporate leaders believe they have discovered a treasure of immeasurable value, a trove so great that surely we will all benefit. It is the treasure of unimpeded commerce flowing everywhere as fast as is possible. But in Seattle, quick time met slow time. The turtles, farmers, workers, and priests weren't invited and don't need to be because they are the shadow world that cannot be overlooked, that will tail and haunt the WTO, and all its successors, for as long as it exists. They will be there even if they meet in totalitarian countries where free speech is criminalized. They will be there in dreams of delegates high in the Four Seasons Hotel. They will haunt the public relations flacks who solemnly insist that putting the genes of scorpions into our food is a good thing. What gathered around the Convention Center and hotels was everything the WTO left behind.

In the Inuit tradition, there is a story of a fisherman who trolls an inlet. When a heavy pull on the fisherman's line drags his kayak to sea, he thinks he has caught the "big one," a fish so large he can eat for weeks, a fish so fat that he will prosper ever after, a fish so amazing that the whole village will wonder at his prowess. As he imagines his fame and coming ease, what he reels up is Skeleton Woman, a woman flung from a cliff and buried long ago, a fish-eaten carcass resting at the bottom of the sea that is now entangled in his line. Skeleton Woman is so snarled in his fishing line that she is dragged behind the fisherman wherever he goes. She is pulled across the water, over the beach, and into his house where he collapses in terror. In the retelling of this story by Clarissa Pinkola Estes, the fisherman has brought up a woman who represents life and death, a specter who reminds us that with every beginning there is an ending, for all that is taken, something must be given in return, that the earth is cyclical and requires respect. The fisherman, feeling pity for her, slowly disentangles her, straightens her bony carcass, and finally falls asleep. During the night, Skeleton Woman scratches and crawls her way across the floor, drinks the tears of the dreaming fisherman, and grows anew her flesh and heart and body. This myth applies to business as much as it does to a fisherman. The apologists for the WTO want more-engineered food, sleeker planes, computers everywhere, golf courses that are preternaturally green. They see no limits; they know of no downside. But Life always comes with Death, with a tab, a reckoning. They are each other's consorts, inseparable and fast. These expansive dreams of the world's future wealth were met with perfect symmetry by Bill Gates III, the co-chair of the Seattle Host Committee, the world's richest man. But Skeleton Woman also showed up in Seattle, the uninvited guest, and the illusion of wealth, the imaginings of unfettered growth and expansion, became small and barren in the eyes of the world. Dancing, drumming, ululating, marching in black with a symbolic coffin for the world, she wove through the sulfurous rainy streets of the night. She couldn't be killed or destroyed, no matter how much gas or pepper spray or how many rubber bullets were used. She kept coming back and sitting in front of the police and raised her hands in the peace sign, and was kicked and trod upon, and it didn't make any difference. Skeleton Woman told corporate delegates and rich nations that they could not have the world. It is not for sale. The illusions of world domination have to die, as do all illusions. Skeleton Woman was there to say that if business is going to trade with the world, it has to recognize and honor the world, her life, and her people. Skeleton Woman was telling the WTO that it has to grow up and be brave enough to listen, strong enough to yield, courageous enough to give. Skeleton Woman has been brought up from the depths. She has regained her eyes, voice, and spirit. She is about in the world and her dreams are different. She believes that the right to self-sufficiency is a human right; she imagines a world where the means to kill people is not a business but a crime, where families do not starve, where fathers can work, where children are never sold, where women cannot be impoverished because they choose to be mothers and not whores. She cannot see in any dream a time where a man holds a patent to a living seed, or animals are factories, or people are enslaved by money, or water belongs to a stockholder. Hers are deep dreams from slow time. She is patient. She will not be quiet or flung to sea anytime soon.

QUESTIONS

1. Using Internet resources, define the following terms:
   a. WTO
   b. IMF
   c. World Bank

2. What is meant by oligarchy and corporate plutocracy?
3. What are the threats of globalization from the point of view of:
   a. People of the developed countries
   b. People of the developing countries
   c. Small businesses
   d. Environmental groups
4. What is your opinion about the N30 WTO showdown? Do you support Paul Hawken's stand against globalization? Why or why not? Explain your answer.

BOX 1 Intelligence and Real Perception

Intelligence is the shadow of objective Truth.
How could the shadow vie with sunshine?

RUMI

Source: E. P. Idries Shah, *The Way of Sufi*, New York: Dutton, 1970, p. 105.

BOX 2 Globalization

"...The problem for both (*IMF and World Bank*) is that economic globalization has outpaced political globalization. Governments used to ensure that capitalism was tempered and that development helped people across society. Now, we are more interdependent and need collective action on a variety of things, yet we have yet to create the political structure that allow that to be done in a democratic way."

Joseph Stiglitz, Nobel Laureate, and author of *Globalization and its Discontents*

Source: "The World is Not Flat, Q & A: Joseph Stiglitz," by Alex Kingsbury, *U.S. News and World Report,* September 18, 2006.

Pakistan: Karachi's Informal "Recycling Network"

AHMED S. KHAN

Karachi, with an estimated population of 12 million in 2005, is Pakistan's largest city. The city has been growing at more than 5 percent per annum in recent years and is expected to reach 16.2 million inhabitants by the year 2015 (2002 UN report on population). Karachi was the world's twenty-fifth largest city in 1985; sixteenth largest in 2001; and it is projected to be the tenth largest by 2015.

Karachi's high rate of population growth, coupled with scarce resources and inadequate planning, have created a number of environmental problems such as water pollution, air pollution, and waste disposal.

According to Karachi Municipal Corporation (KMC), the city generates 7,000 tons of domestic waste daily. The city has a nonmechanized system of waste disposal that is able to remove only about fifty percent of the solid waste that is generated daily, the rest remains at collection points and on dump sites (www.tvo.org and Shah 1993). The waste is removed manually by the city's 9,000 municipal sweepers and then transported by a fleet of 150 refuse vans to a single dumping site located 30 kilometers from Karachi, where it is disposed of by burning. The remaining garbage lies uncollected in city alleys and is processed by the informal "recycling network." This informal network consists of scavengers, middle men commonly known as "Kabaria," and the dealer. The scavengers hunt for recyclable garbage, in community dumpsters and city streets, and sell it to a middle man (Karabia), who sells the material to the big dealer, and finally the dealer sells recyclable waste to recycling industries.

Unlike the West, where the process of recycling has been initiated in response to environmental concerns, Pakistan's recycling industry is motivated purely by economic motives. The informal recycling network of scavengers, middlemen, dealers, and recycling units constitutes an amazing and not widely known phenomenon that performs an essential service under very adverse conditions. This informal recycling effort not only helps KMC in waste removal, but also provides a source of inexpensive raw material to local small industry and sustenance to thousands of inhabitants (Ahmad 1993).

According to a report by National Training and Consultancy Services (NTSC) for the United Nations Center for Human Settlements (HABITAT), about 40 percent of Karachi's solid waste is recycled in the informal sector, and about 2 percent of the city's population is engaged full-time in the recycling industry. Considering the other participants, such as reprocessing workers, housewives, servants, entrepreneurs, and craftsmen, it seems that almost half of Karachi's population is participating in recycling (Shah 1993).

It is estimated that there are about 25,000 scavengers operating in Karachi. Almost 95 percent of the scavengers are Afghan refugees (who fled to Pakistan after the Soviet invasion of Afghanistan in 1979), with the remainder being Burmese or Bengali. Although no accurate statistics are available, it is estimated that these scavengers reduce the work of government agencies by as much as two-thirds, saving the KMC up to 200 million rupees annually in garbage collection and disposal costs. Karachi's informal recycling network has created one of the largest recycling industries in Asia. Paper, glass, metal, plastics, even bones, are all recycled (see Table 1) and hence provide an inexpensive

Table 1
Karachi's Informal "Recycling Network"

Waste Category	*Efforts of "Recycling Network"*
Bones	About one dozen bone-processing industries have been set up in Karachi, each with a processing capacity of 5–10 tons per day. About 80 percent or 30,000 tons of bones find their way to the far eastern markets, particularly Japan. Bones are a vital raw material in the manufacture of photographic film, and Japan's $20 billion photographic film industry makes it the largest market for this Pakistani export.
Iron	Iron, especially scrap recycling, has been known to build fortunes. The leading tycoons of Indo-Pakistan, from *Mians* of Lahore to the *Tatas* of Bombay, are basically scrap dealers and scrap recyclers. According to conservative estimates, about 3,000 tons of iron scrap is generated in Karachi daily.
Glass	About 80 percent of the country's demand for glass is also met through recycling locally generated glass waste, saving Pakistan millions of rupees in imports. It is estimated that about 80,000 tons of glass scrap is produced each year in Karachi alone from both industrial and domestic sources.
Plastic	Research conducted by the NED University's Environment Department and the Pakistan Council for Scientific and Industrial Research (PCSIR) reveals that more than 1,000 tons of plastic waste is generated daily in Karachi. The material includes discarded polythene bags and the waste from hundreds of plastic factories manufacturing PVC pipes and household items. The scale of plastic recycling can be gauged from the fact that there are about 8,000 polythene bag–manufacturing units in Karachi alone.
Paper	Several tons of paper is recycled into paper and cardboard daily. The recycling process involves the low-tech process of dumping large quantities of paper scrap in a pool of water and later sifting it through a sieve.

source of raw materials for domestic and overseas industrial needs (Ahmad 1993).

Karachi's informal recycling network has emerged as an indigenous industry of miraculous proportions. About 60 to 70 percent of Karachi's reusable waste is recycled, generating employment for over 2 percent of the city's population and simultaneously saving energy and cutting down on pollution. But the industry has its pitfalls, too. More than anything else, unregulated recycling practices pose health hazards, especially for those engaged in the separation of waste and reprocessing. Laborers melt batteries to extract lead, electroplated utensils to extract aluminum, and photographic films to extract silver without using any masks. Laborers in the bone industry breathe in bone powder and hence suffer from respiratory disease. In plastic recycling units, laborers breathe toxic emissions (Shah 1993).

The existence of the vast informal recycling network suggests that, contrary to popular notions about developing

Table 2
Daily Garbage Generation per Person for Selected Cities

City	*Daily lbs. of Garbage/Person*
New York	4
Paris	2.4
Hamburg	1.9
Rome	1.5
Cairo	1.1
Karachi	1.4
Calcutta	1.1
Hong Kong	1.9
Jakarta	1.3
Singapore	1.9
Tokyo	3

Shah, an Afghan refugee, collects 50–60 kilograms of paper waste daily at Karachi's Clifton Beach and earns about Rs. 90 ($2) by selling it to Kabaria.
Photo courtesy of Ahmed S. Khan.

countries, Pakistan is a highly waste-conscious country where a staggeringly large proportion of garbage is recycled (Ahmad 1993).

REFERENCES

Ahmed, F. (1993). The Waste Merchants. *The Herald (Karachi)*, July, pp. 107–112.

Carson, W. (1990). *The Global Ecology Handbook: What You Can Do about the Environmental Crisis.* Boston: Tomorrow Coalition, Boston Press.

Population Growth and Policies in Mega-Cities: Karachi, Population Policy Paper No. 13. (1988). New York: United Nations.

Shah, N. (1993). Our Wonderful World of Waste. *Newsline (Karachi)*, June, pp. 71–77.

Waste Busters—Pakistan (July 2000). Available at: http://www.tve.org/ho/doc.cfm?aid=640 retrieved on November 1, 2006.

QUESTIONS

1. Discuss the pros and cons of Karachi's informal "recycling network." Use the chart on the next page to list the pros and cons.
2. What steps need to be taken by the city, provincial and federal government agencies of Pakistan to make the processes of waste disposal and recycling safer and more efficient?

Karachi's Informal "Recycling Network"	
(+) Pros	*(–) Cons*

Karachi's Informal "Recycling Network"

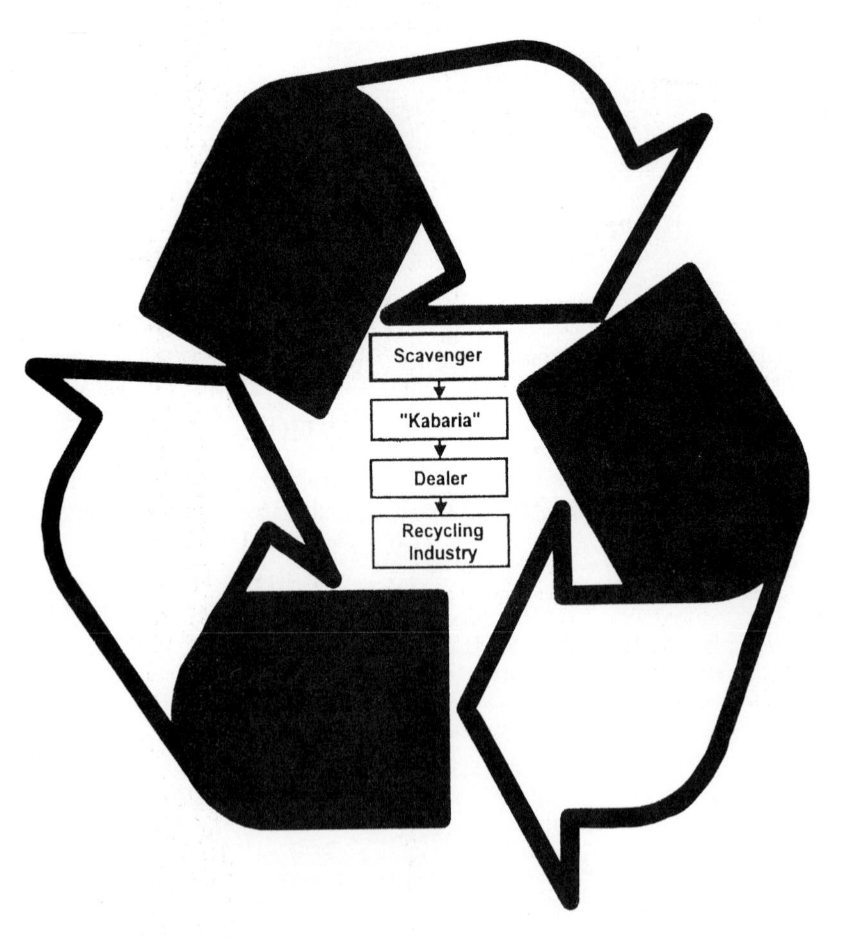

The Digital Divide in Ecuador

Mary Jane C. Parmentier*

The digital divide refers to the gap between those societies, and portions of the population within societies, that possess and can utilize new information and communications technologies—in particular the personal computer and access to the Internet—and those that do not have access and most likely lack the knowledge to utilize the technologies as well. This divide exists globally, or between nations and whole regions, as well as within all nations, including the United States and other developed countries. This distinction between the technological "haves" and the technological "have nots," however, is dramatically illustrated by the small Andean nation of Ecuador.

The United Nations Development Program (UNDP), which assesses the state of human development around the world, concluded in 2005 that Latin America in general had made some important strides in recent years, but that the alleviation of poverty and its manifestations, such as high infant mortality rates, continue to be priority issues for the region. Ecuador in particular is classified as an *intermediate developed country*, with living standards below those of Costa Rica, Chile, and Cuba. However, in the capital city of Quito, technology companies, such as Microsoft and Hewlett Packard, have offices, offering consulting services and products. Educated and upper-class citizens in the larger cities, including Quito, Guayaquil, and Ambato, have access to computers and cellular technology and utilize these tools in business and their personal lives, just as people do in developed countries. Yet the persistent levels of poverty in Ecuador—which in 2004 the UNDP reported had a yearly per capita income of $3,580 per year—coupled with the relative lack of social mobility (the ability to change one's socioeconomic status) have contributed to a widening technological gap between these elite and the rest of the population, many of whom have never touched a computer or perhaps never even made a phone call. This digital divide is "widening" because, as the "technological classes" acquire ever newer innovations in information and communications technology, those outside of the information society are left further behind. And with personal computers and the Internet accessible in 2004 to only around 4 people out of 100, even middle-income people have a hard time connecting to the information highway since the cost for service is on par with the United States, but salaries are far lower.

In Ecuador, this is evident in small towns and rural areas, particularly in indigenous Indian regions such as the small town of Salasaca near Ambato in the lower Andes, where schools lack even older technologies, such as copy machines and reliable electrical service. One of the most significant underlying causes of the technology gap is actually the energy gap. Without electricity, computers and other technologies cannot run unless an alternative method of energy can be implemented. One school in Salasaca received a donated solar panel from a university in the United States, and it is to be used to generate power to run one personal computer, the first that most of the students have ever seen, for several hours a day. A second donated panel powers a water irrigation system for the school garden, an integral

*Mary Jane C. Parmentier, Ph.D. teaches in the Global Technology and Development Program, College of Technology and Applied Sciences, Arizona State University.

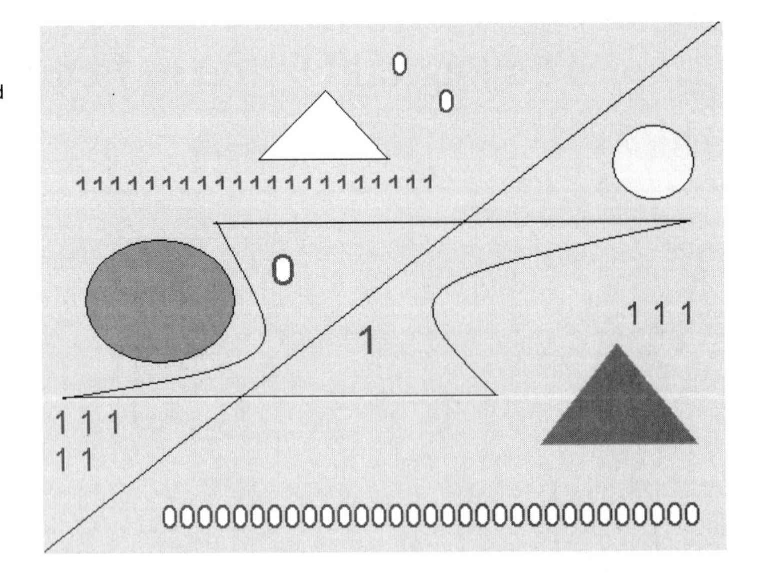

Figure 1
The digital divide.
Source: An abstract painting by Ahmed S. Khan.

part of the school's innovative curriculum on indigenous plants and foods. These small-scale technological developments are being utilized to augment education, one of the cornerstones to improving human development in rural communities around the globe.

In a small town north of Quito there is a very innovative mayor who is harnessing technology with education and political participation to improve socio-economic conditions in his town and surrounding region. In 2002, the city of Cotacachi received one of 20 UNESCO Cities for Peace prizes for its urban, social, and administrative development. One of the programs that has been instituted is a literacy campaign focused on training teachers in a technology-based curriculum which was developed in Cuba and has been recognized by UNICEF for its effectiveness. With Cuban consultants and donated equipment, the town of Cotacachi announced in April of 2005 that 100% literacy had been attained. The program trains teachers in a methodology that utilizes television, video, and software that can be used to teach large numbers of people. The literacy campaign in Cotacachi has been so successful that, in February of 2005, the Eduadorian government signed an agreement with Cuba to expand and promote this program throughout the country.

Bringing literacy to Cotacachi is potentially significant to women, who formed a large part of the illiterate residents before the program was implemented. This, together with a municipal program geared toward increasing the participation of women in the political and decision-making process, could help to incorporate women into the development process as well. As in many areas of the world without access to computers, the municipal government has also built community computer centers with public access. Thus, the development process in this region has focused on technology as a tool for local education and civic participation rather than on foreign direct investment and large-scale projects that would bring capital and jobs into the area. Ecuador, in fact, has one of the lowest rates of foreign direct investment in Latin America. Many Ecuadorian political and business leaders, however, believe Ecuador must privatize more of its industries and services, attract more foreign capital, and become more globally competitive. In the meantime, small communities are implementing grass-roots efforts to improve the lives of their citizens, with technology as a catalyst.

REFERENCES

Cuba around the World; http://www.cubaminrex.cu; retrieved 6/22/05.

Interview by author with Auki Tituana, Cotacachi, Ecuador, January 2006.

The Millennium Development Goals: A Latin American and Caribbean Perspective, bibliography.pdf; retrieved 6/22/05.

UNESCO, Laureate Cities; http://portal.unesco.org/culture/en/ev.php-URL_ID=2550&URL_DO=DO_TOPIC&URL_SECTION=201.html; retrieved 6/22/05.

United Nations Development Program, 2004 Development Indicators, www.undp.org; retrieved 6/22/05.

QUESTIONS

1. What are some of the reasons for the digital divide in Ecuador?
2. Is the small-scale energy solution implemented in Salasaca adequate for the needs of the school?
3. What do you think of the development plan in Cotacachi and its use of technology?
4. Why is Ecuador receiving technology assistance from Cuba? (Go to the UNDP Web site and compare development data among countries in Latin America, including Cuba and Ecuador.)
5. What might be some results of the empowerment of women and other minorities through literacy, technology access, and political participation?
6. What should the role of the Ecuadorian government be regarding the digital divide and development in the country? (Go to the Ecuadorian government Web page and research policy toward technology).
7. Discuss the benefits of both foreign direct investment in job creation and technology transfer versus small-scale projects utilizing technology as an educational tool.

Primary school in Salasaca, Ecuador.
Photo courtesy of Mary Jane C. Parmentier

Applications of Telecommunications Technologies in Distance Learning

AHMED S. KHAN

ABSTRACT

Recent advances in telecommunications technologies (computer networks, satellite communications, fiber-optic systems, the Internet, and so forth) have transformed the modes of learning and teaching. The dissemination of knowledge is no longer confined to constraints of physical premises, with an instructor and a textbook, and it is no longer the only educational resource. The use of telecommunications technologies in distance learning overcomes the barriers of distance and time and allows students in developed as well as in developing countries to learn in a synchronous or asynchronous manner. This paper presents an overview of various telecommunications technologies that are used in distance education. The discussion includes the evolution of distance education, synchronous and asynchronous learning, and characteristics of broadcast television, instructional television fixed service (ITFS), microwave systems, satellite systems, direct broadcast satellite (DBS), cable systems, private fiber, and the Internet.

INTRODUCTION

Distance education or learning is a discipline that links people with information through a variety of technologies (www.fsu.edu). Distance learning has been practiced for over 100 years, primarily in the form of correspondence courses. These print-based courses solved the problem of the geographic dispersal of students for specialized courses of instruction. In the late 1960s and early 1970s, emerging technologies afforded the development of a new generation of distance learning, aided by open universities that combined television, radio, and telephone with print.

Today, advances in telecommunications technologies are again changing the face of distance learning. With the explosion of Internet tools over the last few years, distance learning has become more accessible to people whether they live in urban or rural areas. In this time of increasing technological advances, it is no surprise that college enrollment is on the rise. Increasing enrollment, coupled with yearly budget cut requests, opens the door for distance learning.

With the advent of distance-learning technologies, the dissemination of knowledge is no longer confined to the constraints of physical premises. Distance-learning technologies overcome the barrier of distance to allow face-to-face communication between students and teachers from different locations. With rapid technological change in developed countries, educational institutions are challenged with providing increased opportunities without increased budgets. In developing countries, policymakers face a dilemma: how to provide educational opportunities to an increased population while considering lack of resources. Distance education offers a solution to the problems of educators in developed as well as developing countries. In developed countries, there are three major

Reprinted with permission from The Annual Review of Communications. Copyright by the International Engineering Consortium (IEC).

motivations for an educational institution to incorporate distance learning into its program offerings:

- to compensate for the lack of specialist courses at its own institutions
- to supplement an existing curriculum to make it more enriched
- to compete with other institutions which offer distance-learning courses via the Internet

For developing countries, distance education offers a unique approach to promote literacy and enhance higher education in a cost-effective way.

MODES OF DISTANCE LEARNING

Distance learning can be employed in a synchronous as well as asynchronous mode. In a synchronous mode, the student and teacher interact in real time, whereas the student learns at a convenient time and place in an asynchronous mode.

Advances in computer and communications technologies have greatly contributed to asynchronous access. Asynchronous learning is time and place independent; learning takes place at the convenience and pace of the learner. A student can contact a college or teacher via e-mail or engage in discussion with a group through a conferencing system or bulletin board. A learner can participate interactively with other students in a team project that requires problem analysis, discussion, spreadsheet analysis, or report preparation through modern commercial groupware packages. Similarly, lectures can be transmitted through the Internet, computer networks, videotape, or CD-ROM/DVD. Learning becomes a distributed activity, and participants in these distributed classes access resources and interact asynchronously, more or less at their own convenience (Mayadas and Alfred 1996). Asynchronous learning has been categorized into three different levels: on campus, near campus, and far from campus. The on campus level meets in the traditional classroom, but students participate through computer labs in asynchronous communication through listservs, bulletin boards, and the Internet. The near-campus level (50–60 miles from campus) requires students to meet on campus occasionally for tests, teacher consultation, or labs, but uses the Internet or another electronic medium for lectures, class information, and communication. The far-from-campus level is the "true" asynchronous delivery, where the student never sets foot on the campus and the entire course is delivered through tools such as the Internet, videos, or CD-ROMs/DVDs. All three levels meet the needs of different learners. One of the benefits of asynchronous learning is that it meets the needs of many "nontraditional" students who would normally not have access to an education due to distance or time constraint (Office of Technology Assessment 1989).

From Britain to Thailand and Japan to South Africa, distance learning is an important part of national strategies to educate large numbers of people rapidly and efficiently. In the United States, educators are finding distance learning effective not only for outreach to new populations, but also as an important medium for new instructional models. Distance learning, used as a term associated with new technologies offering a full-fledged alternative to classroom education, got its biggest boost internationally with the founding of the British Open University (BOU) in 1969. BOU gained rapid visibility and recognition by broadcasting its video course components weekly throughout the United Kingdom on the BBC network (Granger).

Table 1 lists the enrollment for distance learners seeking university education in various developed and developing countries. The addition of distance learners, who receive programming in elementary, secondary, training, and noncredit areas, would almost quadruple these numbers (Brown and Brown 1994).

Many agree that distance learning is the fastest-growing instructional pattern in the world. Technologies of delivery, particularly those related to telecommunications, have had a "greening" effect on the distance-learning enterprise. The potential to solve access, cost, time, place, and interactivity considerations that have plagued education since the beginning of time has never been greater (Brown and Brown 1994).

TELECOMMUNICATIONS TECHNOLOGIES

Telecommunications technologies provide opportunities for basic as well as advanced education for those disadvantaged by time, distance, physical disability, and resources. Table 2 lists the evolution of telecommunications technologies. Table 3 lists the breakdown of voice, data, and video technologies used in distance education (Willis 1994).

Distance learning blends technological infrastructure and learning experiences. The experiences and applications that can be provided depend on the appropriate infrastructure. Current technologies have favored applications in which a central source controls information flow. Emerging networked technologies allow for both centralization and

Table 1
Major Open Universities in Developed and Developing Countries

Institution	*Country*	*Year Established*	*Enrollment*
University of South Africa	South Africa	1951	250,000
Open University	United Kingdom	1969	180,000
Universidad Nacional	Spain	1972	200,000
Fernuniversitat	Germany	1974	37,000+
Open University of Israel	Israel	1974	12,000+
Allama Iqbal Open University	Pakistan	1974	448,512
Athabasca University	Canada	1975	10,000+
Universidad Nacional Abierta	Venezuela	1977	29,000+
Universidad Estatal a Distancia	Costa Rica	1977	11,000+
Sukhothai Thammathirat OU	Thailand	1978	200,000
Central Radio and TV University	China	1978	1,000,000
Open University of Sri Lanka	Sri Lanka	1981	18,000+
Open Universiteit	Netherlands	1981	33,000+
Andrha Pradesh Open University	India	1981	41,000+
Korean Air and Correspondence University	South Korea	1982	300,000
University of the Air of Japan	Japan	1983	22,000+
Universitas Terbuka	Indonesia	1984	350,000
Indira Ghandi Open University	India	1986	1,400,000
National Open University of Taiwan	Taiwan	1986	48,000
Al-Quds Open University	Jordan	1986	not available
Universidade Aberta	Portugal	1988	3,800+
Open University of Bangladesh	Bangladesh	1992	600,000

decentralization. While centralized and decentralized approaches each have advantages and disadvantages, networked computer technologies and the distributed-learning applications they enable have not only practical benefits for information management, but they also have attributes that are more congruent with newer educational paradigms based on cognitive approaches to learning (Locatis and Weisberg 1997).

Broadcast Television

Broadcast TV involves the transmission of video and audio over standard VHF and UHF channels to reach a large number of sites in a limited geographic area (e.g., campus, metropolitan, county, multi-county, and so forth). Most educational programming broadcast over television does not allow real-time interaction with the television instructor, but interactively can be designed into live or recorded telecourses using the telephone, computer networks, and Internet tools. This mode of transmission lacks security and confidentiality.

Instructional Television Fixed Service (ITFS)

ITFS refers to a band of microwave frequencies originally set aside by the Federal Communications Commission (FCC) in 1963 exclusively for the transmission of educational and cultural programming. ITFS is similar to broadcast television but is used in a more limited geographic area for simplex point-to-multipoint transmission. ITFS uses omni-directional microwave signals in the 2.5-GHz band to transmit standard 6-MHz video signals to remote locations. An ITFS network can serve as a standalone distance education-delivery system, transmitting locally originated programming directly to local schools or cable

Table 2
Evolution of Telecommunication Technologies

Year	*Technological Development*
1832	Telegraph
1875	Telephone
1895	Radio
1945	Audiotape
1953	Broadcast television
1960	Videotape
1960	Audio teleconferencing
1965	Cable television
1975	Computer-assisted instruction
1980	Audiographic teleconferencing
1980	Satellite delivery
1980	Facsimile
1980	Videoconferencing
1984	Videodisc
1985	CD-ROM compact disc
1988	Compressed video
1989	Multimedia
1990s	Lightwave systems, LANs, HDTV, Internet
2000s	Web-based/online course delivery software platforms/systems (e.g Blackboard, Desire2Learn, etc.). Broadband wired and wireless links, Fiber-to-the-home (FTTH), and Fiber-to-the-Curb (FTTC)

Table 3
Telecommunication Technologies in Distance Learning

	Technology
Voice	Telephone
	Radio
	Short-wave
	AM
	FM
	Audiotapes
Data	LANs
	WANs
	Internet
	Computer-assisted instructions (CAI)
	Computer-mediated education (CME)
	Computer-managed instruction (CMI)
Video	Pre-produced videotapes
	Digital video disc (DVD)
	Videoconferencing
	Broadcast television
	ITFS
	Satellite system
	DBS
	Internet

DBS = direct broadcast satellite, ITFS = instructional television fixed service, LANs = Local area networks, WANs = wide area networks

companies for redistribution through their network. It offers limited or no interaction and a moderate level of security and confidentiality.

Microwave Systems

Terrestrial microwave systems require a line-of-sight transmission between the transmitter and the receiver site. Microwave systems allow simplex or half duplex/duplex point-to-point audio, data, and video transmission. There are two types of point-to-point microwave systems: short haul and long haul. Short-haul systems typically have a range of 5–15 miles, suitable for local communications between two schools or campuses. Long-haul systems typically have a range of up to 30 miles between repeaters, depending on transmitter power, terrain, dish size, and receiver sensitivity. Microwave systems offer a high level of interaction, security, and confidentiality.

Satellite Communications

Satellite communication involves the transmission of a broadcast signal from an earth station via an uplink signal to a geosynchronous satellite, where the signal is processed by the transponder and sent, via a downlink signal, to a receiving dish antenna. Satellite communication is used to cover a large number of sites over a wide geographic area (e.g., statewide, nationwide, continent-wide) for simplex point-to-multipoint audio and video transmission. Analog satellite systems offer low levels of security and confidentiality, and digital systems offer high levels of security and confidentiality. Satellite systems offer limited or no interaction.

Direct Broadcast Satellite (DBS)

These high-powered satellites transmit programming directly to the general public. The received dish antennas used in DBS systems are very small (<1 meter). DBS systems allow programmers to beam educational programming directly to homebound students, providing an alternative to over-the-air broadcast or cable television. DBS technology employs data-compression technologies that enhance the efficiency of video channels.

Televised education offers the following advantages (Richard 1997):

- Televised courses can actually enhance instruction by allowing instructors to present material in novel ways and thus to increase interest in it.
- Televised courses provide access to education for those who might not otherwise have it.
- Televised courses can help the instructor become a better teacher.
- Televised courses are often better organized and more highly developed than traditional courses.
- Televised courses offer an opportunity for faculty members to learn from their students as well as teach them.

Satellite communications is one of the most cost-effective modes to promote literacy and enhance higher education in developing countries. Table 4 lists the typical cost for leasing a video channel for satellites that provides large footprints in Asia and Africa.

Cable Systems

Cable-television systems use coaxial and fiber-optic cable to distribute video channels to local subscribers. Programming is received from local broadcast channels and national programming services at the cable "head end" and is sent out over the cable in a tree configuration. A cable head end can receive many types of signals, such as satellite or microwave transmission, which can then be retransmitted to schools over the normal cable system. Cable-television systems are primarily one-way (simplex) broadcast (point-to-multipoint)–type transmission systems. Many systems also have a limited number of reverse channels, providing some measure of two-way interactivity. There is high-level inter-system security and confidentiality.

Private Fiber

Optical fiber is used for full-duplex point-to-point audio and video transmission. It allows a high degree of interaction

Table 4
Typical Cost for Leasing a Video Channel for Satellites Providing Large Footprints in Asia and Africa

Satellite	*Footprint Coverage Area*	*Typical Hourly Cost for Leasing a Video Channel*
Panamsat	South Asia	$1,850
TDRS-5	South Asia	$1,680
Chinasat 1	Southeast Asia	$ 990
Asiasat 2	Southeast Asia	$ 990
Measat	Southeast Asia	$ 990
Palapa-B2	South Asia	$ 885
Intelsat-k	Western Africa	$2,000
GESTAR-4	Western Africa	$2,000
Inmarsat-3	Western Africa	$2,000
Intelsat	South Africa	$1,000
Orion 2	South Africa	$1,000
Panamsat PAS3	South Africa	$1,000

and offers a high level of security and confidentiality. It also has the highest information-carrying capacity. The bandwidth for mono-mode step index fiber is 10–100 Gbps; multimode step index fiber is around 200 Mbps. By using wavelength division multiplexing (WDM) technique higher data speeds (100+ Gbps) can be achieved.

Public Telephone Service

Public telephone service is used over a limited to wide geographic area (local, regional, national, and international) for full-duplex point-to-point or point-to-multipoint audio transmission. It is used in conjunction with other technologies, such as television and satellite, to provide a feedback channel for students at remote sites to interact with the instructor and other sites. Table 5 compares the characteristics of distance-learning technologies.

The Internet

The Internet offers full-duplex point-to-point and point-to-multipoint audio, video, and data transmission over a wide geographic area (regional, national, and international). It allows a high degree of interaction. Generally, it has a low level of security and confidentiality but, with the use of data encryption, it has a high level of security and confidentiality.

With a computer, minimal software, and an Internet connection, people anywhere in the world can access online sites and programs. In addition, these programs can be accessed almost instantaneously through a user-friendly "point-and-click" interface, using almost any type of computer without producing and distributing printed materials, disks, or CD-ROMs/DVDs and with no training delivery cost. Furthermore, training can be delivered to a potentially unlimited audience, updates can be easily made online, and programs can link to a vast collection of other online resources (Collis 1996).

Although the Internet can be accessed worldwide, current access for many people in developing countries is still limited. Even when access is available, bandwidth for accessing the Internet is frequently low. Low bandwidth severely limits capabilities, such as online interactive multimedia training. Other disadvantages are that students need a moderate degree of computer literacy, sensitive/classified training requires additional security measures, and the high level of current Internet "hype" makes it difficult to determine what is practical for the near future (Collis 1996).

In addition to these technologies, a number of high-speed nonswitched and switched lines could be incorporated in wide-area networking strategies in distance learning. Table 6 lists various switched and nonswitched services provided by U.S. carriers.

A report titled "Distance Education in Higher Education Institutions" issued by the National Center for Education Statistics highlights the following facts (*ASEE Prism* 1997):

- By fall 1998, 90 percent of all institutions with 10,000 or more students and 85 percent of those with enrollments of 3,000 to 10,000 expect to offer at least some distance education courses. Among the smallest institutions (those with fewer than 3,000 students), only 44 percent plan to offer distance education courses.
- More than 750,000 students were enrolled in distance education courses in 1994–95. That year, some 3,430 students received degrees exclusively through distance education.
- More students enroll in distance education courses through public two-year institutions than through any other type of institutions.

In terms of the type of technology used to deliver courses, 57 percent of the institutions offering distance education courses in 1995 used two-way interactive video, and 52 percent used one-way prerecorded video (institutions frequently used more than one type of distance-learning technology). Twenty-four percent used two-way audio and one-way video, 14 percent used two-way on-line interactions, 11 percent used two-way audio, 10 percent used one-way audio, and 9 percent used one-way live way.

A key point to note, however, is that because of the many logistical factors involved, distance education courses typically require considerable investments of money and time. Distance education courses can be cost effective when a large number of students take the courses over the long run. It is important to realize that such courses are costly to produce and deliver. The costs rise even more when the courses are presented with advanced one-way or two-way real-time video (Martin et al. 1997).

The incorporation of new telecommunications technologies in distance education has abolished the barriers of location and time. However, the success of distance education in higher education depends on (*ASEE Prism* 1997):

- the availability of high-quality, full-motion video via the Internet
- the electronic accessibility at a reasonable cost of content currently available only in print format

Table 5
Comparison of Distance-Learning Technologies

Technology	*Characteristics*
Broadcast television	Simple mode of transmission; one site to multi-site video and audio transmission (point-to-multipoint transmission); limited or no interaction; public system; lack of security or confidentiality
Instructional television fixed site (ITFS)	Similar to broadcast TV; semi-public system; moderate level of security and confidentiality
Broadband cable	Similar to broadcast TV; public and private systems; high degree of interaction possible; high level of inter-system security and confidentiality
Microwave	Point-to-point audio and video transmission over limited area (one site to another); FD mode of transmission; high level of interaction; private system; high level of security and confidentiality
Satellite	Simple mode of transmission; point-to-multipoint audio and video transmission over wide area; limited or no interaction; private or public system; low level of security and confidentiality for older analog systems; higher level of security and confidentiality for digital systems
Private fiber	Point-to-multipoint transmission over limited area; FD audio/video transmission; high degree of interaction possible; private system; high level of security and confidentiality
Public switched digital service	Point-to-point and point-to-multipoint audio and video transmission over short and long distances; moderate to high degree of interaction possible; public system; high level of security and confidentiality
Public telephone service	Point-to-point and point-to-multipoint audio transmission over short and long distances
Public packet network services	Point-to-point SP transmission; used in conjunction with other technologies to provide a way for students to interact with the instructor

FD = Full Duplex SP Simplex
Source: http://www.oit.itd.umich.edu/reports/DistanceLearn/sect3.html.

Table 6
High-Speed Wide-Area Networks

Nonswitched (leased)	
Analog	4.8–19.2 kbps
Digital data service (DDS)	2.4–56 kbps
T-1	1.54 Mbps
T-3	44.736 Mbps
Frame relay	1.54 Mbps–44.736 Mbps
Synchronous optical network (SONET)	51.84 Mbps–2.488 Gbps
Switched	
Dial-up/modem	1.2–28 kbps
X.25 packet switching	2.4–56 kbps
Integrated services digital network (ISDN)	64 kbps–1.544 Mbps
Frame relay	1.54 Mbps–44.736 Mbps
Switched multimegabit data service (SMDS)	1.54 Mbps, 44.736 Mbps
Asynchronous transfer mode (ATM)	25 Mbps–155 Mbps
B-ISDN (broadband ISDN)	155 Mbps, 600 Mbps

- development and documentation of effective teaching/learning processes that rely on advanced technology
- the ability of distance learning to comply with an accreditation process for engineering education that is outcome based

CONCLUSION

Distance education has come far since its humble beginnings as correspondence courses conducted by mail. To excel in the twenty-first century, higher education must undergo a paradigm shift. It must be transformed from an environment and culture that defines learning as a classroom process shaped by brick-and-mortar facilities and faculty-centered activities to an environment defined by "learner-centered" processes and shaped by telecommunications networks with universal access to subject content material, learner support services, and technology-literate resource personnel (Dubols 1996). The use of telecommunications technologies in distance learning will promote literacy and enhance higher education. Thus, distance learning will narrow the techno-economic gap between the developed and developing world in the twenty-first century.

Education sows not seeds in you,
but makes your seeds grow.
KHALIL GIBRAN

A technical man must not be lost in his own technology. He must be able to appreciate life; and life is art, drama, music, and most importantly, people.

FAZLUR REHMAN KHAN, world-renowned structural engineer who designed the 100-story John Hancock Center and 110-story Sears Tower

Source: Mir M. Ali, *Art of the Sky Skyscraper: The Genius of Fazlur Khan*, New York: Rizzoli International Publications,. 2001. p. 181.

Technology should enrich the experience of learning. E-learning technologies may save some costs and add a measure of convenience, but if they do not deepen the learning experiences of students, they are not worth much.

VAN B. WEIGEL

Source: Weigel, Van. (2002). *Deep Learning For a Digital Age: Technology's Untapped Potential to Enrich Higher Education*, Jossey-Bass, San Francisco, CA.

Satellite dish antennas are popping up on rooftops everywhere in developing countries. However, satellite broadcasts are used purely for entertainment purposes. The educational use of satellite technology is very limited. The use of satellite communications coupled with computer networks and the Internet is not only a very cost-effective delivery mode to educate millions, but also has the potential to revolutionize the educational structure in developing countries to narrow the technoeconomic gap that exists between the First and Third Worlds. The appropriate use of telecommunications technology is the key to solving problems of illiteracy, poverty, and disease in the developing world.
Photo courtesy of Ahmed S. Khan.

BOX 1 Deep Learning Versus Surface Learning

Attributes of Deep Learning	*Attributes of Surface Learning*
Learners relate ideas to previous knowledge and experience.	Learners treat the course as unrelated bits of knowledge.
Learners look for patterns and underlying principles.	Learners memorize facts and carry out procedures routinely.
Learners check evidence and relate it to conclusions	Learners find difficulty in making sense of new ideas presented
Learners examine logic and argument cautiously and critically.	Learners see little value or meaning in either courses or tasks.
Learners are aware of the understanding that develops while learning.	Learners study without reflecting on either purpose or strategy.
Learners become actively interested in the course content.	Learners feel undue pressure and worry about work.

Source: Weigel, Van. (2002). *Deep Learning For a Digital Age: Technology's Untapped Potential to Enrich Higher Education*, Jossey-Bass, San Francisco, CA, p.6.

REFERENCES

ASEE Prism, February 1997, p. 4.

ASEE Prism, February 1998.

Brown, B., and Brown Y. (1994). Distance Education Around the World. *Distance Education Strategies and Tools*, Barry Willis (ed). Englewood Cliffs, NJ: Educational Technology Publications.

Collis, B. (1996). *Tele-Learning in a Digital World: The Future of the Distance Learning*. London: International Thompson.

Dubols, Jacques R. (1996). Going the Distance, A National Distance Learning Initiative. *Adult Learning* (September/October).

Granger, D. Open Universities: Closing the Distance to Learning, Change. *The Magazine of Higher Learning* 22 (4):42–50.

Linking for Learning: A New Course for Education. (1989). Washington, D.C.: Office of Technology Assessment (OTA), U.S. Congress.

Locatis, C., and Weisberg, M. (1997). Distributed Learning and the Internet. *Contemporary Education* (Winter) 68(2).

Martin, B., Moskal, P., Foshee, N., and Morse, L. (1997). So You Want to Develop a Distance Education Course? ASEE Prism (February), p. 18.

Mayadas, F., and Alfred, P. (1996). Alfred P. Sloan Foundation Home Page. Available at www.sloan.org/Education/ ALN.new.htm.

Richard, Larry G. (1997). Lights, Camera, Teach. *ASEE Prism* (February), p. 26.

Willis, B. (1994). *Distance Education Strategies and Tools*. Englewood Cliffs, NJ: Educational Technology Publications.

www.fsu.edu/<lis/distance/brochure·html#program.

QUESTIONS

1. Define the following terms:
   a. Distance learning
   b. Asynchronous learning
   c. Open university
   d. DBS
   e. Synchronous learning
   f. Academic integrity
2. What telecommunications technologies could be used to educate millions of illiterate children and adults in the developing countries?
3. Compare the pros (+) and cons (–) of distance education.
4. Discuss the academic integrity of online/web-based educational programs/degrees.

(+) Pros	*(–) Cons*

Use the Internet Resource to answer the following question:

5. E-learning is swiftly transforming the education landscape. A number of schools are offering E-learning programs in the areas of business, engineering, library science, nursing, public health, and technology. More than 3 million students (*U.S. News & World Report, Oct 16, 2006)* are pursuing degrees online from institutions of higher learning in the United State. Would E-learning lead to the re-emergence of Diploma mills? Discuss the pros and cons of E-learning in contrast to

onsite learning (traditional learning) considering the following factors:

a. Rigor of curricula
b. Academic integrity of programs
c. Tuition cost savings
d. Convenience (Anytime/Anywhere availability)
e. Accreditation by national agencies
f. Acceptance by employers
g. Pre-requisite technological/computer skills
h. Social interaction (isolation)
i. Teaching and learning styles

BOX 2 Matters of Scale

The Plight of the Displaced	
Number of people in the past decade (1990-2000) displaced by infrastructure projects, such as road and dam construction	80 to 90 million
Number of people in the past decade left homeless by natural disasters including floods, earthquakes, hurricanes, and landslides (based on an annual average over 25 years)	50 million
Number of people in 1981 who were landless or near-landless	938 million
Number of people who were landless or near-landless in 2000	1.24 billion
Number of people per square mile in the five boroughs of New York City	23,700
Number of people per square mile in Jakarta, Indonesia	130,000
Number of people per square mile in Lagos, Nigeria	143,000
Number of international refugees in the early 1960s	1 million
in the mid-1970s	3 million
in 1995	27 million
Number of refugees displaced within the borders of their own countries in 1985	9.5 million
in 1995	20 million
Number of people currently living in coastal areas vulnerable to flooding from storm surges	46 million
Number of people living in vulnerable areas if global warming produces a 50-centimeter rise in sea level	92 million
Number of people living in vulnerable areas if global warming produces a 1-meter rise in sea level	118 million

Reprinted with permission from Worldwatch Institute, Washington, D.C. (*World Watch* Jan/Feb 1996).

BOX 3 Matters of Scale

Human Health and the Future	
Number of people now living within the reach of malaria-transmitting mosquitoes (potential transmission zone), worldwide	2.5 billion
Number of people who will live in the transmission zone by the latter half of the 21st century, after projected expansion of the transmission zone by global warming	4.8 billion
Average number of heat-related deaths in Atlanta each summer	78
Average number projected by 2050, assuming no change in the city's population size or age profile, but with projected global warming	293
Predicted decline in production of cereal grains (rice, corn, wheat as a result of climate change (climate effects only) by 2060	
in the developed countries	23.9 percent
in the developing countries	16.3 percent
Additional decline in grain production from physiological effects of CO_2, in developed countries	3.6 percent
in the developing countries	10.9 percent
Number of people at risk of hunger in 2060, without global warming	640 million
Number at risk with projected warming	680 to 940 million
Life expectancy in the most developed countries in 2000	79
Life expectancy in the least developed countries in 2000	42
Share of worldwide AIDS/HIV cases in developing countries	90 percent
Share of AIDS stories with non-U.S. settings in the U.S. media	4 percent

Reprinted with permission from the Worldwatch Institute, Washington, D.C. (*World Watch* September/October 1996).

BOX 4 Who Owns Indigenous Peoples' DNA?

Aboriginal leaders have long struggled to control native lands. Now some have begun to worry that they may have to fight for control of native genes.

In August 1993, Pat Mooney, the President of Rural Advancement Foundation International (RAFI), a non-profit concerned with third-world agriculture, discovered that the U.S. government was trying to patent a cell line derived from a 26-year-old Guaymi woman. The Guaymi people are native to western Panama. The cell line, a type of culture that can be maintained indefinitely, came from a blood sample obtained by a researcher from the U.S. National Institutes of Health in 1990. The application claimed that the cell line might prove useful for the treatment of the Human T-lymphotropic virus, or HTLV, which is associated with a form of leukemia and a degenerative nerve disease.

RAFI notified Isidro Acosta, President of the Guaymi General Congress, who demanded that the United States withdraw its claim and repatriate the cell line. Acosta also appealed to the General Agreement on Tariffs and Trade and to an intergovernmental meeting on the Rio Biodiversity Convention. But GATT does not forbid the patenting of human material, and Acosta's case before the Biodiversity Convention fared no better. The convention does provide for sovereign rights over genetic resources, but the meeting did not rule on whether the Guaymi cell line came within its jurisdiction. As a growing number of nongovernmental organizations voiced their disapproval, however, the United States dropped its patent claim last November (1993).

The story might have ended there had not a European researcher uncovered two similar claims in January of this year (1994). Miges Baumann, an official at Swissaid, a Swiss NGO that supports rural initiatives in developing countries, discovered that the U.S. government had filed applications on a cell line derived from the Hagahai people of Papua New Guinea and another from the Solomon Islanders. These lines might also prove useful for treating HTLV. Baumann's discovery came as a shock to the governments concerned but, despite their protests, the United States has refused to withdraw the applications. In a letter dated March 3, 1994, Ron Brown, the U.S. Secretary of Commerce, explained the U.S. position to a Solomon Island official. "Under our laws, as well as those of many other countries," Brown wrote, "subject matter relating to human cells is patentable and there is no provision for considerations relating to the source of the cells that may be the subject of a patent application."

Patenting indigenous peoples' genes invites an obvious comparison with the patenting of the developing world's other biological resources, and native leaders have tended to take a dim view of the entire trend. "I never imagined people would patent plants and animals. It's fundamentally immoral, contrary to the Guaymi view of nature," said Acosta, who considers the patenting of human material a violation of "our deepest sense of morality."

But the rapid growth of biotechnology is driving a boom in human patents that may prove difficult to resist. The patenters are looking for genes that could be used to produce substances with commercial potential, usually for treating a disease. To patent a "product of nature," patent laws generally require some degree of human alteration. But in the United States, the simple act of isolating a DNA sequence removes it from nature, as far as the law is concerned.

The accessibility of the patent has fueled a growing commercial interest in the field. Companies that prospect in the human genome use a highly automated process called sequencing to decode bits of DNA from large numbers of samples. One company, Human Genome Sciences of Rockville, Maryland, is reported to have sequenced over 200,000 chunks of DNA thus far. Patent claims may follow if the sequences obtained look novel—and in some cases, even if they don't. In one of the more spectacular instances of "driftnet patenting," as critics call the practice, Incyte Pharmaceuticals of Palo Alto, California, filed claims on 40,000 sequences.

Observers say it's a good bet that other applications on indigenous DNA have already been filed. "I'm not aware of any others," says Hope Shand, RAFI's Research Director, "but it would surprise me if there weren't any more of them."

Reprinted with permission from Worldwatch Institute, Washington, D.C. (*World Watch* November/December 1994).

Remember Rwanda?

JAMES GASANA

The genocide of 1994 seemed inexplicable. But a study of links between extreme environmental degradation and the enormous violence that occurred between Hutus and Tutsis could have important implications for stressed populations in other regions.

EDITOR'S INTRODUCTION

On April 6, 1994, a plane carrying the presidents of two African countries was struck by a missile and crashed. Both presidents—Juvenal Habyarimana of Rwanda and Cyprian Ntaryamira of Burundi—were killed. Both were members of the Hutu ethnic group. Counting the murder of Burundi's president Melchior Ndadaye the previous October, a total of three Hutu presidents had been assassinated in six months.

The crash of the plane was described by a Rwandan official as being "like pouring fuel on a burning house." The country exploded into genocidal conflict between the Hutu and the rival Tutsi, who had been out of power in Rwanda but who had established a base in neighboring Uganda from which they had been launching attacks against the regime that had ousted them. Hutu bands killed large numbers of Tutsi in an effort to forestall the invasion. But within weeks, the Tutsi regained control and waged retaliatory attacks on the Hutu, hundreds of thousands of whom were by then fleeing the country.

The exchanges of massacres were so horrific that people in other parts of the world, who had paid little attention to Rwanda until news of the genocide broke, were bewildered as to what could have caused such fury. The conflict was portrayed in the media as one of deep ethnic hatred. But to those who were on the scene during the years preceding, the story is far more complicated than that. The real causes of the blowup are rooted in a half-century history of rapid population growth, land degradation, inequitable access to resources, political power struggles, famine, and betrayal.

James Gasana, who was Rwanda's Minister of Agriculture and Environment in 1990-92, and Minister of Defense in 1992-93, at one point tried to warn his government of the coming conflagration but to no avail. In the following article, adapted from a paper he wrote for the IUCN's Task force on Environment and Security, he analyzes what happened as environmental and economic decline set the stage for a social collapse. It's a story that has important implications not only for Rwanda, but for every region where population pressure threatens to exceed what the resource base can maintain.

Before the end of the 1950s, it was the Tutsis who dominated Rwanda, both sociologically and politically. Tutsis constituted only 10 to 15 percent of the population, but they owned most of the arable land and accounted for more than 95 percent of the chiefs and 88 percent of the bureaucracy. In 1959, however, a revolution by the Hutu peasants of southern Rwanda brought the Hutu to power and resulted in a redistribution of land to previously landless people. Many of the Tutsi aristocracy fled to neighboring countries, particularly to Uganda, from which they launched counterattacks against the Rwandan regime in the 1960s.

The Hutu, enforcing a one-party regime in which the Tutsi had no voice, lived from then on with the specter of counter-revolution. The hostilities between the two groups were exacerbated by the Cold War, as the Communist countries helped arm the counterattacks of the Tutsi refugees, while the Western countries provided support to the the Hutu regime.

In 1973, under pressure from both internal dissent and external attack, the regime was toppled by a coup d'état.

Major General J. Habyarimana, supported by a northern faction of the army, took control from the southern-based group that had progressively assumed power after independence in 1962. Habyarimana was to hold power for the next 20 years, but under increasingly difficult conditions. It's the story of those two decades that explains the otherwise incomprehensible events of 1994.

The story begins with a country undergoing a population explosion that was to increase it from 1,887,000 people in 1948 to 7,500,000 in 1992—making it the most densely populated country in Africa. Most of the people were poor farmers, and in the 1980s, many of the poor got even poorer, as a result of what I call "*pembenization*"—from the Swahili word "*pembeni*," or "aside," as used in the Rwandan expression "*gushyira i pembeni*"—"to push aside."

One of the root causes of pembenization was, ironically, the land tenure program established by the 1959 revolution as a means of giving the peasants a more equitable share in the country's assets. The revolutionaries did not foresee what would happen as children inherited their parents' land and divided it up equally. With the population expanding, the inherited pieces—many of them very small to begin with—got smaller.

At the same time, the land holdings of the elite who were in power got larger, as wealthy northern Hutus and their allies spent much of the 1970s and 1980s accumulating land for their own estates. Of course, this further reduced the amount of land available for peasant farmers. Many of the peasants moved to marginal land—to steep slopes and acidic soil, where crops barely grew.

By 1989, an estimated 50 percent of Rwanda's cultivated land was on slopes of 10 degrees or higher. Slopes this steep eroded severely when tilled, and the cycle of poverty worsened (see Table 1).

By 1990, the erosion was washing away the equivalent of 8,000 hectares per year, or enough to feed about 40,000 people for a year. Moreover, because demand for land outstripped supply, virtually all the cultivatable land (other than that being hoarded by the elite) was being used, and there was little opportunity to let fields lie fallow and regenerate. As a result, soil fertility declined faster yet.

Of course, as population grew, the demand for energy increased as well. Rwanda has been heavily dependent on biomass for energy—either wood or crop waste. Most of the energy in those years was provided by firewood. But with more people trying to get more firewood from smaller pieces of land, the country's trees were disappearing at an increasing rate. Deforestation on the steep-sloped lands made the ground more exposed to running water, and increased erosion still more.

In 1991, we estimated that annual tree growth would allow for about 1.9 million cubic meters of wood to be cut. Yet, actual wood consumption by then had reached nearly 4.5 million cubic meters. This heavy overharvesting had yet another impact on farm output: with the firewood supply diminishing, people were forced to increase their reliance on straw and other crop residues for fuel. That meant the residues were no longer going back into the soil. The loss amounted to approximately 1.7 tons of organic matter per hectare each year.

The compounding of all these factors led to a disastrous shortfall in food production. Two-thirds of the population of Rwanda was unable to meet even the minimum food energy requirement of 2,100 calories per person per

Table 1
The Erosion: Classification of Cultivated Land by Slope Category

	Category of Slope	*Area of Cultivated Land*	
Description	*Slope (in degrees)*	*Percentage*	*Hectares*
Level to undulating	0-5	34	382,500
Undulating	5-10	16	180,000
Sloping	10-25	32	360,000
Steep	25-30	7	90,000
Very steep	30-35	6	67,000
Stiff slope	>35	4	45,000
Total		100	1,125,000

day. The average person was getting just 1,900 calories—becoming gradually weaker and at the same time more desperate. Nor were there any readily available alternatives to subsistence farming. By the end of the 1980s, the unemployment rate for rural adults had reached 30 percent.

Throughout the 1980s, the worsening of the rural situation, especially in the south where most of the poor farmers lived, had generated increasing resentment against the Hutu government, which was accumulating wealth for its mostly northern elite. It's important to keep in mind that the peasants and the people in power were both mainly Hutu, so this resentment was an economic, not ethnic, concern. At the end of the decade, however, with internal strife splitting the Hutus, the Tutsi-led rebels in Uganda judged that this would be a good time to declare full-scale war against the regime. By 1990, then, the Rwandan peasants were being stricken by both starvation and war. In an interview with Radio Rwanda, representatives of a peasant association named Twibumbe Bahinzi declared:

> "There is a generalized famine in the country, that is difficult to eradicate because it is only the cultivators-pastoralists [peasants] who are bearing its impacts while the 'educated' [the elite] are enjoying its side effects. Those who should assist us in combating that famine are of no use to us.... It will require no less than a revolution similar to that of 1959.... On top of this there is war. Even if the cultivators-pastoralists can still till the land, it is very difficult for them to work in good conditions when they have spent the night guarding the roadblocks, and are not sure that they are going to harvest...."

In retrospect, this statement confirms that even under the added stress of war, the peasants did not at this point consider ethnicity to be the issue. It was still an issue of rich and poor, or north and south.

In 1991, with divisions among the Hutus getting worse, president Habyarimana was forced to abandon the one-party rule and allow a multi-party government. But he continued to hold on to the presidency. Some of the splinter groups tried to weaken him by recruiting bands of disaffected youths based in the south, who mounted a sporadic uprising and perpetrated acts of vandalism aimed at destabilizing the regime. The groups were called *Inkuba*, or "thunder," and *Abakombozi*, or "liberators."

The splinter group leaders spurred on these youths by linking their deprivation to the accumulation of land by the northern elite and its allies. It wasn't that simple, of course. There were other factors, including a collapse in the world market for coffee in the 1980s, which dropped the value of Rwandan coffee exports from $60 per capita in the late 1970s to $13 by 1991. But the political targeting evidently succeeded. A study of the patterns of *Inkuba* and *Abakombozi* acts of violence shows that these acts occurred most frequently in the areas with lowest income, most often in places where daily food energy intake had fallen below 1,500 calories per person. In fact, a table showing the average food energy production in each of Rwanda's 10 prefectures shows that incidents of socio-political violence occurred in 18 communes (communities) where food production was under 1,600 calories per day, but in none where it was above that level (see Table 2).

The Hutus in power, fearful of losing their government positions and properties, also recruited young men for their protection. A youth wing of the governing party, the *Interahamwe*, was organized to protect the politicians and their lands from the opposition youths and from the large numbers of squatters who had fled their impoverished

BOX 1 The Warning

"It can be concluded that if the country does not operate profound transformations in its agriculture, it will not be capable of feeding adequately its population under the present growth rate. Contrary to the tradition of our demographers who show that the population growth rate will remain positive over several years in the future, one can not see how the Rwandan population will reach 10 million inhabitants unless important progress in agriculture as well as other sectors of the economy were achieved. Consequently it is time to fear the Malthusian effects that could derive from the gap between food supply and the demand of the population, and social disorders which could result from there."

—Report of the National Agriculture Commission (1990-1991), chaired by James Gasana

Table 2
The Hunger . . . and the Violence: Food-Energy Deprivation and Incidence of Ethnic Strife, 1991-92

Food Energy per Person per Day	*Communes Where Violence Occurred, 1991-92*	*Communes Where Violence Did Not Occur, 1991-92**
657	Kivu, Nshili Nyamagabe, Rwamiko	Mudasomwa, Mubuga, Muko, Musange, Musebeya,
846	Gatare, Gishoma, Kagano, Karengera	Kamembe, Kirambo
1,056	Runyinya	Huye, Kigembe, Maraba, Ngoma, Nyabisindu,
1,097	Gishyita, Rwamatamu	Bwakira, Gitesi, Kivumu, Mabanza, Mwendo, Rutsiro
1,187	Nyarugenge	Bicumbi, Butamwa, Gikomero, Gikoro, Kanombe, Kanzenze, Mbogo, Mugambazi, Musasa, Rubungo, Rushashi, Rutongo, Shyorongi, Tare
1,219	Masango, Nyabikenke, Nyakabanda, Taba	
1,230	Giciye, Kayove, Kibilira, Ramba, Satinsyi	Nyamyumba
1,595	No violence occurred in communes where average food energy was over 1,500 calories per person, per day	Nyakinama
1,763	No violence occurred in communes where average food energy was over 1,500 calories per person, per day	Muhuru, Murambi, Tumba, Cyumba, Kibali, Kivuye
2,086	No violence occurred in communes where average food energy was over 1,500 calories per person, per day	Birenga, Kabarondo, Kayonza, Kigarama, Muhazi, Mugesera, Rukira, Rukara, Rusumo, Rutonde, Sake

*Data were reported for 72 communes throughout the country.

hillsides. In some cases, the Interahamwe "re-liberated" land that the youth groups of the opposition parties based in the south had seized or occupied.

Habyarimana worked hard to deflect the peasant opposition, personally lobbying farmer representatives to rally the peasant movement to his side and to abandon their rhetoric about rural poverty. He accomplished this by promising them that their concerns would be addressed and by letting his supporters help them to deflect their anger from the elite Hutus to the attacking Tutsis. By 1991, the Uganda-based Tutsi army was making that strategy easy for Habyarimana, as it was targeting Hutus in its guerrilla attacks. By now, thousands of Hutus were fleeing the war and the famine and had become "internally displaced persons" (IDPs) gathering in refugee camps. The Tutsi rebels were more than happy to treat the camps as military targets. By the time a cease-fire took place in 1992, the IDP population had reached 500,000. But the cease-fire was short-lived, as the plane crash that killed Habyarimana immediately reignited the war. By 1993, the number of refugees had reached 1 million, and by the end of the war about 100,000 had died. It was during this post- assassination period that the worst of the genocidal acts occurred.

The internally displaced persons, rather than finding themselves taken in and protected by fellow Hutus whose districts they had fled to, found themselves resented. There were too many of them, and they put impossible strains on traditional hospitality. Where population pressure had become increasingly unbearable on the farms, it became worse around the refugee camps, with IDPs adding heavily to host populations. As the war escalated, food energy dropped to 1,100 calories per person. And while the IDPs were increasingly resented by their fellow Hutus (the host populations, too, were now hungry), they were increasingly attacked and killed by invading Tutsis.

In the two years before his plane was shot down, the embattled Habyarimana and his political enemies both took political advantage of the Hutu refugees' desperate circumstances. IDP children and teenagers, with no schools to occupy them and often no parents to guide them, became the principal recruiting base for the *Interahamwe* militias—the ones bent on sabotaging and destabilizing the regime. At the same time, as the Tutsi invaders drove more Hutus from their homes and killed more of them as they fled to the camps, the IDPs also provided a base for Habyarimana's retaliation against the Tutsi and enabled him to reclaim some of the support he'd lost in the rich-poor conflict. For many of the Hutu IDPs, the harsh reality was that they were forced to choose between two warring camps: the camp of those who wanted them to die before voting, and the camp of those who wanted their votes before they died.

When the presidential plane crashed, it in a sense prefigured the crash of Rwandan society. Extremist Hutu politicians seized on the shock and fear of the moment, using the presidential guard and the *Interahamwe*, comprising mostly the Hutu youths from IDP camps near Kigali, to perpetrate the murder of rival Hutu politicians and the mass slaughter of the Tutsi. Their efforts to turn back the Tutsi failed, and by mid July 1994, the Tutsi-led RPF had taken over. Following this takeover, more than 2 million Hutu refugees fled to neighboring countries, including 1.2 million to the Democratic Republic of Congo (DRC). The mass exodus, predictably, had a devastating environmental, social, and political impact on the DRC.

In the report I wrote for the IUCN's Task Force on Environment and Security, I suggested that four lessons be learned from this tragic chapter in Africa's history: First, rapid population growth is the major driving force behind the vicious circle of environmental scarcities and rural poverty. In Rwanda it induced the use of marginal lands on steep hillsides, shortening of fallow, deforestation, and soil degradation—and resulted in severe shortages of food.

Second, conserving the environment is essential for long-term poverty reduction. Consequently, it is essential for the long-term elimination of links between environment scarcity and conflict. In the long term, this is possible only if Rwanda adopts a bold population policy with aggressive family planning programs aimed at reducing the country's fertility rate. The pressures that produce conflict can also be reduced by adopting more sustainable forms of agriculture, based on techniques that improve soil fertility and increase fuel wood production.

Third, to break the links between environmental scarcities and conflict, win-win solutions—providing all sociological groups with access to natural resources—are essential. The winner-take-all model results in a society gripped by fear, which too easily is exploited by unscrupulous politicians, leading to ethnic enmity and violence.

And fourth, to prevent a bipolar ethnic conflict of the kind that ravaged Rwanda will require a rethinking of what national security really means. Certainly, it means placing human and environmental security ahead of the security of ethno-political regimes.

QUESTIONS

1. What is the root cause of Hutu-Tutsi conflict?
2. Discuss the contribution of the following factors to 1994 genocide.
   a. Population growth
   b. Soil erosion
   c. Deforestation
   d. Hunger
   e. Presidential plane crash
   f. Lack of concern and preventive actions by the international community and the United Nations.
3. What does the international community need to do to prevent future Rwandas?

BOX 2 Nelson Mandela on Africa . . .

"Teach the children that Africans are not one iota inferior to Europeans."

"Take your guns, your knives, and your pangas, and throw them into the sea."

"Africa, more than any other continent, has had to contend with the consequences of conquest in a denial of its own role in history, including the denial that its people had the capacity to bring about change and progress."

"For centuries, an ancient continent has bled from many gaping sword wounds."

"All of us, descendants of Africa, know only too well that racism demeans the victims and dehumanizes its perpetrators."

"It would be a cruel irony of history if Africa's actions to regenerate the continent were to unleash a new scramble for Africa which, like that of the nineteenth century, plundered the continent's wealth and left it once more the poorer."

Nam Theun Dam: The World Bank's Watershed Decision

DAVID F. HALES

For the last thousand years, as kingdoms and countries have fought for sovereignty over Laos' Nakai Plateau, the people there have learned the lessons of the grasses—to bend before the wind. Life has been relatively predictable, marked by continuity from one generation to the next. But the winds of change are blowing again, and this time the strategy of the grasses may not work. By April 2005, the countries on the governing board of the World Bank will consider a proposed high dam on the Nam Theun River. Their decision will not only affect those who live here, but will also set a pattern for decisions regarding hydroelectric dams around the world for years to come.

The Nam Theun 2 dam (NT2) is a project of the Lao People's Democratic Republic (PDR), in partnership with Électricité de France, the Electricity Generating Company of Thailand, and Ital-Thai Development Public Company (jointly, the Nam Theun 2 Power Company Limited, or NTPC). The US$1.3 billion, 1,070-megawatt project would divert 93 percent of the Nam Theun's flow into the adjacent Xe Bang Fai River basin, generating power for Thailand's electrical grid. It would also submerge nearly 40 percent of the Nakai Plateau beneath a 450-square-kilometer reservoir, drastically alter the character of two rivers, displace thousands of desperately poor residents, and disrupt the livelihoods of tens of thousands more, among the other transformations typical of such hydropower projects.

The initial concept was promising. Laos is painfully poor, with few development options. Early estimates understated the number of displaced people and ignored the impacts on those who live downstream, especially on the Xe Bang Fai, so compensation and resettlement seemed manageable. Thailand seemed to need the power, the engineering is not difficult, and the plan included an attractive conservation component.

What has emerged, however, is much less promising. As documents have been made available and proponents have made their case in public, the inadequacy of the preparation and the lack of response to fundamental questions is surprising, especially after the heavy investments of staff at the World Bank to assist the Lao PDR and its partners to develop a coherent proposal. Proponents admit that no comprehensive analysis of alternatives to the project has been done. They have refused to disclose NT2's financial framework (the Concession and Power Purchase Agreements), despite the urging of Bank officials. They are unprepared to document the overall cost of the project, or even the cost of construction. After 15 years of preparation, NTPC calls the proposal "not a fixed package" and insists that it will be responsible "no matter what the cost is."

All parties agree on the need for development in Laos. Yet Laos and NTPC provide few specifics on how NT2 will contribute to this goal. The "strategy" for managing project revenues to alleviate poverty and protect the environment is said to be still "emerging." No revenue management plan has been finalized, nor are there specific measures to ensure good project management and oversight, transparency of finances, or accountability.

At best, the World Bank estimates, the revenue to Laos from this project will not exceed $250 million (net present value) over the 25-year life of the Concession. These revenues are largely deferred and will not exceed 5 percent of

David F. Hales is Worldwatch's Counsel for Sustainability Policy. He chaired the World Bank's U.S. public workshop on Nam Theun 2 in September 2004.

Lao governmental revenue until at least 2020. If revenue sags, if expenses increase, if implementation is as poorly managed as development, if funds are diverted—then benefits to the poor become diminishingly small and risks intolerably large.

The future with the dam has been described to those who have been contacted as one of new homes with electricity, new opportunities for irrigated and mechanized agriculture, a diverse economy with reservoir fishing replacing reliance on river fishing, aquaculture replacing capture fishing along the Xe Bang Fai, and intensive livestock raising replacing the traditional extensive grazing pattern. This vision may have been plausible—if the developers had conducted good scientific studies, developed accurate information about resource baselines, and objectively assessed investments necessary to compensate the villagers. But the hard work and serious investment necessary to make the project a success has mostly not been done.

Government professionals from more than a few countries who are trying to assess the project are nonplussed by the lack of project documentation in critical areas and the poor quality of what is available. There are literally no hydrological data available from NTPC, for example. Claims that there are adequate flows to generate enough power for the project to make money are only slightly more believable than the assurances regarding the risks of downstream flooding. The promises about aquaculture ignore inconvenient facts: only one species of fish to be found naturally in either river basin has been cultivated successfully, and there is a serious risk that introduction of non-native species could have devastating impacts on native ecosystems. Even if the invasive-species problems can be solved, there is no local source of nutrition for farmed fish.

The NTPC posters show tractors and irrigated land—but the company's own Social Development Plan states that the use of tractors generally should be avoided throughout the resettlement area due to "heavily leached and infertile" soils, which are explicitly described as "poorly suited to crop production." Without significant inputs of organic and inorganic fertilizer, financially viable agriculture will be impossible. NTPC plans to help pay for fertilizer for only three years.

New fishing opportunities are promised to replace lost ones, even though the Nam Theun basin is so shallow that dam operations will shrink the reservoir by as much as 80 percent during the dry seasons, and many experts predict that there will be no substantial reservoir fishery. The claim that the loss of fisheries due to increased water flows in the Xe Bang Fai will be replaced by increased fisheries in the river's tributaries is contradicted by project documents that foresee a collapse in the aquatic food chain.

Although detail can be piled upon detail, one conclusion is inescapable. This is exactly the kind of project proposal that the safeguard procedures of the World Bank and the Asian Development Bank are designed to prevent. The risk of further impoverishment of the people, of corruption and mismanagement, and of financial failure of the project is just too high. If Nam Theun is approved, with its inadequate and incomplete documentation, it is hard to imagine a project that would not be. The World Bank will have written a brand new definition of "blank check."

The Bank seems to grasp what is at stake. It has announced a decision framework for Nam Theun stating that the project must be embedded in a development framework aimed at poverty reduction and environmental conservation; must be technically, financially, managerially, and economically sound and adhere to the Bank's environmental and social safeguard policies; and must have understanding and support within the international donor community and civil society. In a background paper for an informal Bank governing board meeting in November 2004, senior Bank staff identified in detail the serious flaws in the proposal and concluded, "The project's ability to have a positive effect...cannot be ensured without major reforms in governance, sufficient progress on human development and natural resources management, and compliance with the Bank's safeguards policies." The necessary changes should, of course, be agreed upon before any decision is made. But even if they are, the Bank understands that the risks are so significant that it is pursuing an unprecedented agreement with Laos and key donor governments stipulating that any failure by Laos to abide by its obligations would jeopardize all external support.

NTPC needs the financial backing of the World Bank and the Asian Development Bank. But the dam is a critical test of the World Bank's own policies (as well as a direct challenge to the recommendations of the World Commission on Dams). Bank officials should release a detailed analysis of how the proposal meets its fundamental standards, and the governments represented on the board, which will make the final decision, should also be fully transparent with their own assessments of compliance. Finally, legislatures should insist on hearing the positions of their governments in public session, in advance of consideration at the Bank. It is time that the Bank board makes clear its position on massive hydroelectric projects. NT2 will be the example others follow.

QUESTIONS

1. Discuss the social and environmental implication of the construction of Nam Theun dam.
2. If you were a consultant to the World Bank, what recommendations would you make so that the World Bank official would do a detailed analysis of required compliance of standards before making the final decision regarding construction of NT2?

Use Internet resources to answer the following question:

3. What is the present status of Nam Theun dam project? What actions have been taken by the World Bank to mitigate the major social and environmental concerns entailed by the construction of the dam?

Case Study 1
Sri Lanka: Technology and Agriculture, Tractor Versus Buffalo

AHMED S. KHAN

The replacement of traditional animal power by modern tractors could pose serious ecological and environmental problems in Third World countries. The use of modern technology in agriculture offers the benefits of efficient farming, but also poses the danger of eliminating jobs and a threat to the environment. A good example of the ramifying environmental consequences of technology is the case presented by Senanayake. He studied the effects of replacing buffalo power by tractor power for agriculture (see Figure P8.1). At first sight, the substitution of tractor for buffalo power in the villages of Sri Lanka seems to appear a straightforward efficient trade-off between timely planting and labor saving at the expense of milk and manure. But it is not only milk and manure at stake. The substitution of buffalo with tractors could lead to a chain of events that could have an adverse impact on the environment. The buffalo and buffalo's wallows offer a number of benefits that the tractors could not offer. Figure P8.1 illustrates these benefits.

In the dry season, the wallows are refuge for fish, who then move back to the rice fields in the rainy season. Some fish are caught and eaten by the farmers; others eat the larvae of mosquitoes that carry malaria. The thickets harbor snakes that eat rats that eat rice, and lizards that eat the crabs that make destructive holes in the ricebunds. The wallows are also used by the villagers to prepare coconut fronds for thatching. If the wallows go, so do these benefits.

If pesticides are used to kill rats, crabs, or mosquito larvae, then pollution or pesticide resistance can become a potential problem. Similarly, if tiles are substituted for the thatch, forest destruction may be hastened since firewood is required to fire the tiles.

REFERENCES

Senanayake, R. (1984). The Ecological, Energetic and Agronomic Systems of Ancient and Modern Sri Lanka, Douglas, G. K. (ed.). *Agricultural Sustainability in a Changing World Order*. Boulder, Colorado: Westview Press.

Gordon, C. (1990). Agriculture and the Environment: Concepts and Issues, Huq, S., et al. (eds.). *Environmental Aspects of Agricultural Development in Bangladesh,* Dhaka: University Press Limited.

QUESTIONS

1. What strategy would you suggest for employing tractors for farming in Sri Lankan villages?

Use Internet resources to answer the following question:

2. In Sri Lanka and its neighboring countries, Bangladesh, India and Pakistan, the excessive use of pesticide and fertilizer is creating environmental problems and polluting the underground water table. As a result the degradation of land is becoming a major problem. It could decrease the production of staple crops like wheat and rice, and thus could adversely impact the food supply in

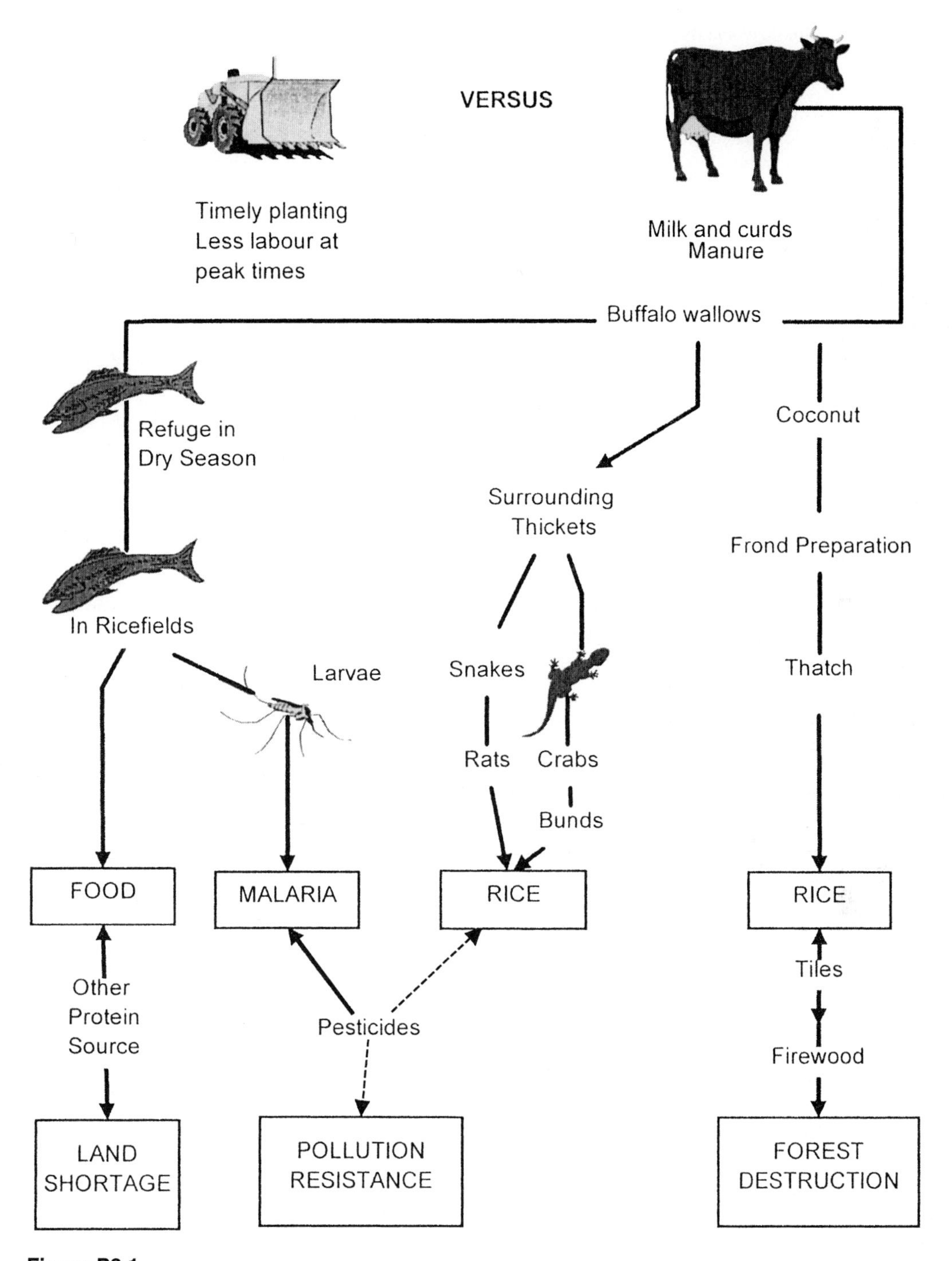

Figure P8.1
Technology and Agriculture: tractor vs. buffalo.

Source: Agriculture and the Environment: Concepts and Issues, *Environmental Aspects of Agricultural Development in Bangladesh*, Dhaka, Bangladesh: University Press Limited.

these countries. Discuss the effectiveness of policies that have been formulated and implemented by the governments in Sri Lanka, Bangladesh, India, and Pakistan to control the degradation of agricultural lands.

3. Complete and discuss the following flowchart.

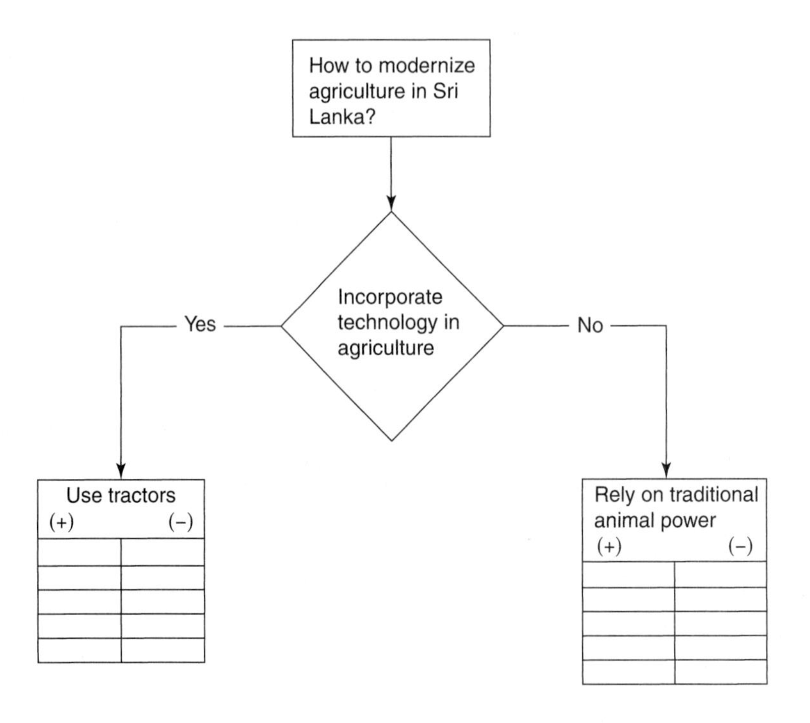

Case Study 2

The Asian Tsunami Tragedy and Beyond

AHMED S. KHAN

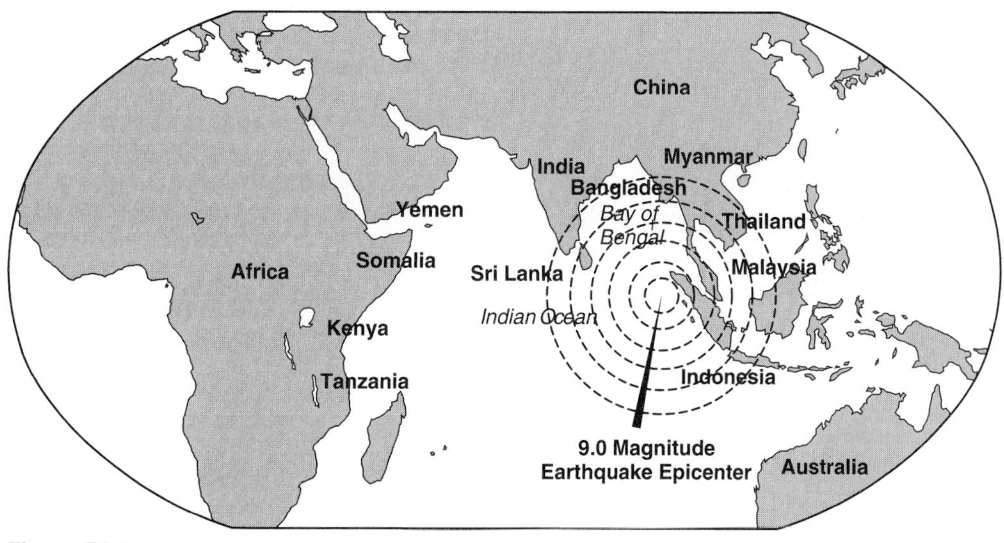

Figure P8.2
The epicenter and path of the Asian tsunami.
Dateline: December 26, 2004

The magnitude of devastation from the tsunami has exposed the limits of scientific knowledge and lack of cooperation among nations in using sophisticated tools to prevent human suffering. In today's era of breathtaking technological innovations, where computers and the Internet have revolutionized the way people communicate globally, it is shocking and hard to believe that so many people would perish instantaneously. Had there been a global tsunami-warning network in place (with the help of today's state-of-the-art telecommunications technologies, in just a few milliseconds, a warning signal could have traversed throughout the world); and thus the death of more than 250,000 people and the displacement and suffering of millions of people could have been prevented.

The occurrence of natural disasters (earthquakes, floods, typhoons, cyclones, tornadoes) cannot be predicted with certitude, but with the help of appropriate technological tools, their impact can be minimized by providing advance warning to the people. To minimize the impact of future natural disasters, especially in the developing world, it is incumbent on the international community to develop a global early-warning protocol and a system to inform people of any looming threat. The academic community needs to collaborate with the scientific community to develop mathematical/computer models to predict the probability of the tsunami generation in the event of an earthquake in an ocean. Wave-monitoring sensor stations also need to be established in all oceans to detect the generation of tsunamis. International aid agencies should collaborate with the United Nations to establish permanent regional disaster-relief centers (with adequate inventory of relief supplies) in the developing countries to minimize the transportation time of relief goods and provide effective and just-in-time relief to the disaster victims. The application of appropriate technologies, coupled with international collaboration between nations, is the key to effectively manage future natural disasters and alleviate human suffering.

Table P8.1
World's Worst Disasters

Year	*Type*	*Location*	*Fatalities*
1887	Flood	China	1 million
1556	Quake	China	830,000
1737	Quake	Kolkata, India	300,000
1970	Cyclone	Bangladesh	300,000
1976	Quake	China	255,000
1138	Quake	Syria	230,000
1920	Quake	China	200,000
1893	Quake	Iran	150,000
1923	Quake	Japan	143,000*
1991	Cyclone	Bangladesh	138,000
1948	Quake	Turkmenistan	110,000
1908	Quake/Floods	Italy	100,000
1815	Volcanic eruption	Indonesia	92,000
2005	Quake	Kashmir/Pakistan	79,000+
1902	Eruption	Martinique	40,000
1935	Quake	Quetta, Pakistan	35,000
1883	Eruption/Tsunami	Indonesia	36,000
2003	Quake	Bam, Iran	31,000

* Many thousands were killed in the Great Tokyo fire caused by the earthquake.

BOX 1 Tsunami

Tsunami: Japanese word for harbor wave. It is generated due to vertical disturbance of the ocean due to an earthquake's landslide or volcanic eruption.

Chronology of Events

1. December 26, 2004: A massive earthquake of magnitude 9 on the Richter scale occurs near Banda Ache, Indonesia.
2. The earthquake causes a massive displacement of water in the Indian Ocean, which generates tsunami waves.
3. The tsunami wave spreads in all directions, moving with a speed of 500 miles per hour (in the deep ocean, the waves have minimal height and may be imperceptible, but as they approach the shallow waters of shores, their height increases).
4. A tsunami wave struck the Island of Sumatra, Indonesia, Thailand, Malaysia, and Burma (Myanmar), causing the death of more than 250,000 people.
5. After one hour, the tsunami wave hit the Islands of Nicobar and Bangladesh.
6. After two hours, the tsunami wave struck the coast of Sri Lanka and Tamil Nadu, India.
7. After three hours, the tsunami wave reached Maldives Island.
8. After seven to ten hours, the tsunami wave hit the African coastal countries of Somalia, Kenya, and Tanzania.

REFERENCES

Elliot, Michael. (2005). Sea of Sorrow. *Time* (January 10), Vol. 165, No. 2.

Khan, Ahmed S. (2005). Letter to the Editor. *Time* (January 31).

QUESTIONS

1. Complete the following flowchart.

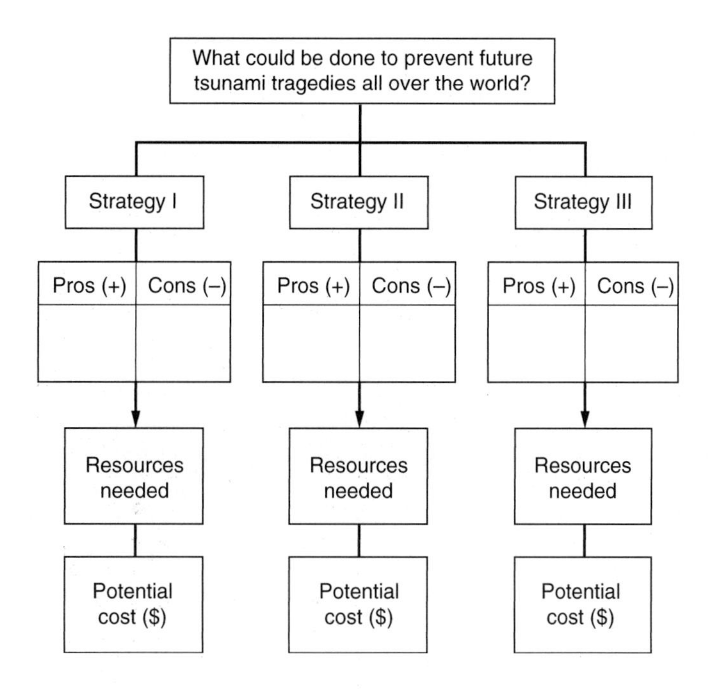

2. Discuss the present state of technological knowledge in predicting natural disasters.

3. Use Internet resources to compare the adequacy of relief efforts for the natural disasters listed in the following table:

Natural Disaster	*Comment on the Adequacy of Relief Efforts*
Asian tsunami (2004)	
Iran (Bam) earthquake (2003)	
Pakistan earthquake (2005)	
Katrina hurricane (2005)	

Epilogue

When, to leave earth, I gathered what was mine,
To have known me through and through was each man's claim;
But of this traveler none knew truly what he
Spoke, or to whom he spoke, or whence he came.

IQBAL

Source: V.G. Kiernan (translator), *Poems from Iqbal*, Karachi: Oxford University Press, Karachi, 2004, p. 280.

SCENARIO I: Turkey

Istanbul (Golden Horn, European side)
Photo courtesy of Mary Jane C. Parmentier

Turkey's geographic and strategic location makes it a unique cultural and commercial gateway. Turkey serves as a bridge between the East and the West. The recent increased demand in oil consumption has also amplified the importance of Turkey as an "energy bridge," connecting the oil fields of the East to the consumers of the West.

Turkey is a developing country. It is trying to recover from the economic contraction it suffered due to the financial crisis of 2001. Turkey, once a part of the great Ottoman Empire of the past, is presently a neighbor to the European Union, a great economic powerhouse in the making.

Turkey aspires to become a member of the European Union. Turkey has campaigned to become a part of Europe ever since the creation of the European Economic Community in 1958. Turkey and the European Union's economic picture complement each other (see Table P8.2). Turkey has a large, well-educated, and skilled workforce that could help the European Union stay competitive in the global market place. Turkey joined the European Customs Union in 1996. It became an official candidate for membership three years later and was promised in 2002 that membership talks would begin if it met the European Union's political criteria by the end of 2002. However, many Europeans believe that Turkey would need at least 10 years to revise its regulations and improve its economy enough to actually join the union. It is interesting to observe that while the Europeans continue to delay the membership talks with Turkey—a NATO member—they have granted full membership status to a number of eastern European and former Soviet Union states.

Many Europeans, especially the French, feel threatened by the possibility of Turkey becoming a full member of the European Union. They feel that Turkey, with its relatively high birthrate and low income levels, could end up as the most populous country in the union and could claim as much as $55 billion in European Union subsidies.

French leaders and the public have repeatedly expressed their concerns in this regard. A poll published in the French daily newspaper *Le Parisien*

Table P8.2
Turkey vs. European Union: Economic Development Indicators

Series	European Monitory Union 2002	Turkey 2002
Agriculture, value added (% of GDP)	2	13
CO_2 emissions (metric tons per capita)	3.3*	8.5*
Electric power consumption (kWh per capita)	5,912	1,458
Energy use (kg of oil equivalent per capita)	3,895	1,083
Exports of goods and services (% of GDP)	36	29
Fertility rate, total (births per woman)	1	2
Fixed line and mobile phone subscribers (per 1,000 people)	1,335	629
GDP (current US$)	6,662,332,088,320	183,888,297,984
GDP growth (annual %)	1	8
GNI per capita, Atlas method (current US$)	20,320	2,510
High-technology exports (% of manufactured exports)	17	2
Internet users (per 1,000 people)	336	62
Personal computers (per 1,000 people)	317	45
Population growth (annual %)	0	2
Population, total	306,088,928	69,626,000
Present value of debt (current US$)	..	134,433,800,192
Trade in goods (% of GDP)	..	48

GDP = gross national product; GNI = gross national income
* 2000 data from http://www.wri.org
Source: *World Development Indicators* database, http://devdata.worldbank.org/data-query/.

(November 2002) stated that 55 percent of the French oppose the arrival of Turkey as a new member of the expanded European Union. Former French President Valery Giscard d'Estaing, 76, head of the Convention on the Future of Europe, has also expressed his concerns. He told *Le Monde* newspaper that those who backed Ankara's candidacy were "the adversaries of the European Union." Alluding to Turkey's Muslim population and high birthrate, Giscard said the country had "a different culture, a different approach, a different way of life" and its demographic dynamism would potentially make it the biggest EU member state. He further said that "its capital is not in Europe, 95 percent of its population lives outside Europe, it is not a European country." The French President Jacques Chirac has also expressed his concerns and has cautioned that Turkey might never reach the standards required for European Union membership and that the bloc might have to find an alternative way to tie it to Europe. The French are against offering a full member status for Turkey; they prefer a "special" status.

The Turkish Prime Minister, Mr. Erdogan, is against the granting of "special status." He observed that Turkey had met the requirements set by the European Union by rewriting the country's laws and amending its Constitution to preserve human rights and promote democracy. He believes that a rejection of Turkey would contradict the European Union's own standards and values and that European leaders would also forfeit a chance to show that they are not at war with Islam.

There are others in Europe who support Turkey's entry as a member to the EU. The German Chancellor Gerhard Schroeder believes that Turkey should be allowed into the European Union because it is a bridge between East and West. He observed that "Turkey can, if its people want, be an important bridge, perhaps the most important, between continental Europe and the eastern Mediterranean."

Dateline: October 29, 20XX

After prolonged talks regarding Turkey's European Union membership, the EU has officially declined to grant full membership status to Turkey.

Response

1. Going back to the 1970s, draw a timeline. Label it by proposing appropriate economic policies implemented in each decade that could have promoted the economic and social uplifting of the masses in Turkey.
2. Propose policies that would promote economic growth in Turkey.
3. Develop a blueprint for economic development of Turkey through the use of new and emerging technologies.

REFERENCES

French Opposed to Turkey in EU: Poll. (2002). Karachi *Dawn*, Nov. 22. Available at http://dawn.com.

"Turkey: Country analysis briefs." http://www.eia.doe.gov.

Turkey's Entry Will End EU: d'Estaing: Ex-French President Sparks Furor. (2002). Karachi *Dawn*, Nov. 9. Available at http://dawn.com.

Turkey May Not Make It to EU: Chirac. (2002). Karachi *Dawn*, Nov. 22. Available at http://dawn.com

Sachs, Susan. (2004). Turkey Insists on Equal Terms in European Union. October 3. Available at http://www.nytimes.com.

Schroeder Pleads Turkey's case, again. Dawn, Karachi, Nov. 22, 2002, http://dawn.com

Smith, Craig. (2003). Turks Say to Europe: Can't We Just Come as We Are? November 24. Available at http://www.nytimes.com.

SCENARIO 2: Brazil

Brazil is the largest and most populous country in South America. It is also South America's leading economic power. However, despite its economic progress, chronic inflation, massive foreign debt, and poverty pose serious challenges to the country's overall development and social welfare. Like many other developing countries, the income disparity between the rich and the poor continues to grow, resulting in extreme poverty. According to the World Bank, "one-fifth of Brazil's 173 million people account for only a 2.2 percent share of the national income. Brazil is second only to South Africa in a world ranking of income inequality."

Poor families do not have means to take care of children, and thus millions of children have no choice but to live on the streets. It has been estimated that about 30 million children live in conditions that are subhuman and inadequate for their development. These conditions lead children to lives of violence and crime. The street children also experience violence at the hands of authorities. A Brazilian government report states that, between 1988 and 1990, 4,661 children under 17 were murdered—an average of four a day. Of the victims, 52 percent were killed by the police or private security guards; 82 percent were black; and 67 percent were males between the ages of 15 and 17.

The majority of the poor and the street children live in the slums, known as *Favelas*, around Rio de Janeiro and Sao Paulo. These Favelas have become such terrible places that the police are afraid to enter, and authorities are thinking of isolating these areas with walls.

> "Street children are a reminder, literally on the doorsteps of rich Brazilians and just outside the five-star hotels where the development consultants stay, of the contradictions of contemporary social life: the opulence of the few amid the poverty of the majority, the plethora of resources amid the squandering of opportunities. They embody the failure of an unacknowledged social apartheid to keep the poor out of view... They are painful reminders of the dangerous and endangered world in which we live."

Kumar Rupesinghe and Rubio C. Marcial, eds., The Culture of Violence, New York: United Nations University Press, 1994, pg. 214

Dateline: March 1, 20XX

Brazilian authorities have finalized their plans to build walls around the slums (Favelas) of Rio and Sao Paulo. The construction of the wall will segregate the poor and the rich and will create a "social apartheid" in Brazil.

Response:

Discuss the implications of the construction of a wall surrounding the slums of Rio and Sao Paulo in promoting social and economic equality in Brazilian society. Is isolating the poor a just and viable solution to combating the problem of poverty? Develop strategies that could be implemented to promote literacy, reduce poverty and social inequity, curb crime, and to establish rule of law for all in Brazil.

A homeless family sleeping on the footpath at the Copacabana Beach area of Rio de Janeiro.
Photo courtesy of Ahmad Rashidianfar.

REFERENCES

Hunger, Poverty Create Breeding Ground for Social Ills. Inter Press Service News Agency. Available at http://ipsnews.net.

Poverty and Violence in the Slums of Brazil. (2004). *Emilia R. Pfannl*, May.

Rupesinghe, Kumar, and Marcial, Rubio C. (eds.). (1994). *The Culture of Violence*. New York: United Nations University Press, pg. 214. Soca, Ricardo (2005).

Conclusion

No great improvement in the lot of mankind are possible until a great change takes place in fundamental constitution of their modes of thought.

John Stuart Mill, philosopher

The world may have become a global village as a result of technological advancements, but billions of people in the developing countries will continue to live with poverty, disease, and illiteracy. Changes in the Third World will occur only when both the intrinsic and extrinsic factors inhibiting its development are addressed. The effects of extrinsic factors (e.g., political and economic instability caused by international financial institutions, economic exploitation of resources by multinational companies) can be rectified only when a great change takes place in the modes of thought of the policymakers of the developed world. These leaders must begin to believe that all people are created equal, no one is superior to another, and treating people with equality and justice is the key to solving the man-made economic, political, and environmental dilemmas that the world faces today.

To address the intrinsic factors (e.g., illiteracy, poverty, social injustice, corruption, lack of resources, population pressures, and power-hungry political and military elite, the key is education. Education is a great equalizer for changing socio-economic conditions. It could provide solutions to many of the problems of developing countries and set them free from economic bondage, autocratic rule, poverty, and disease. The developing nations should give a higher priority to educating their people. By incorporating innovative telecommunications technologies, the developing countries could increase their literacy rates and therefore produce a humanistic techno-elite that could then promote social justice, economic growth, and a sustainable environment and thus narrow the techno-economic gap with the developed countries.

Throughout this book, a spectrum of issues related to a wide array of topics (e.g., history of technology, ethics and technology, energy, ecology, population, war and technology, social responsibility, health and technology, and technology and the Third World) has been presented. The next and final part of this book, Part IX, discusses the state and impact of emerging technologies of the future.

BOX 1 Solitude by Iqbal

Solitude

I stood beside the ocean
And asked the restless wave—
To what eternal troubling,
To what quest are you slave?
With orient pearls by thousands
Your mantle's edge shine,
But is there in your bosom
One gem, one heart, like mine?
—It shuddered from the shore and fled,
It fled, and did not speak.
I stood before the mountain,
And said—Unpitying thing!
Could sorrow's lamentation
Your hearing never wring?
It hidden in your granite
One ruby, blood-drop lie,
Do not to my affliction
One answering word deny!
—Within its cold unbreathing self
It shrank, and did not speak.

IQBAL-THE POET OF EAST

Source: V. G. Kiernan (translator), *Poems from Iqbal*, Karachi: Oxford University Press, 2004, p. 272.

BOX 2 The Sealed Book

The human's progress is that of one who has been given a sealed book, written before he was born. He carries it inside himself until he 'dies.' While man is subject to the movement of Time, he does not know the contents of that sealed book.

HAKIM JAMI

Source: Idrees Shah, The Way of Sufi, New York: E.P. Dutton, 1970, p. 100.

BOX 3 The World Is Not Flat

Not only is the world not flat, but also there is growing inequity around the world, and there is a growing gap between the rich and the poor. The world is becoming less flat as the inequality grows.

Jospeph Stiglitz, 2001 Economics Nobel Laureate

Source: The World is Not Flat, Q & A: Jospeh Stiglitz, by Alan Kingsbury, U.S. News & World Report, September 18, 2006.

INTERNET EXERCISES

1. Visit NASA's Multimedia Gallery at http://www.nasa.gov. (The site contains a large collection of photos of the Earth, taken from space by numerous space shuttle missions, depicting the impact of various human interactions such as deforestation, desertification, pollution, oil well fires, soil erosion, mining, urbanization, irrigation, and oil drilling.) Complete the following table by listing the impact of human interaction in the developed and developing countries listed in the left-hand column.

Country	*Impact of Human Interaction*
Argentina	
Australia	
Bangladesh	
Brazil	
Bolivia	
Cameroon	
Canada	
China	
Ethiopia	
Egypt	
India	
Indonesia	
Japan	
Kuwait	
Libya	
Madagascar	
Malaysia	
Mexico	
Nepal	
Panama	
Philippines	
Sudan	
Tajikistan	

(Continued)

Tanzania	
Turkey	
USA	
Venezuela	

2. The Chinese government is building an enormous dam—The Three Gorges Dam—on the Yangtze River to control flooding and transform its hydropower into electric power. The project is expected to be completed in 2009 and to have a power-generating capacity of 18,200 megawatts. This enormous capacity will enable China to move into the twenty-first century with a hydropower bang.

Use any Internet search engine (e.g., Alta Vista, Yahoo, Google) to answer the following questions:

   a. How many people will be displaced due to The Three Gorges Dam project?
   b. How many towns and villages will be resettled?
   c. What are the environmental costs of this project?
   d. What are the benefits and drawbacks of this project?

Use any Internet search engine (e.g., Alta Vista, Yahoo, Google) to answer the following questions:

3. Define the following terms:
   a. GNP
   b. GDP
4. Compare the per-capita GNP for the following groups of developed and developing countries:

Developed Country	*Per-Capita GNP*	*Developing Country*	*Per-Capita GNP*
United States		Afghanistan	
France		Bangladesh	
United Kingdom		China	
Japan		Ethiopia	
Australia		Ghana	
Norway		India	
Sweden		Indonesia	

(Continued)

Developed Country	*Per-Capita GNP*	*Developing Country*	*Per-Capita GNP*
Canada		Kenya	
Switzerland		Malaysia	
Italy		Pakistan	
Singapore		Venezuela	

5. Describe the factors that are responsible for the wide gap between the per-capita GNP of developed and developing nations.
6. List five benefits of recycling.
7. Compare the following groups of developed and developing countries in terms of their recycling efforts:

Developed Country	*Garbage Produced per Year*	*% of Garbage Recycled*	*Developing Country*	*Garbage Produced per Year*	*% of Garbage Recycled*
United States			Afghanistan		
France			Bangladesh		
United Kingdom			China		
Japan			Ethiopia		
Australia			Ghana		
Norway			India		
Sweden			Indonesia		
Canada			Kenya		
Switzerland			Malaysia		
Italy			Pakistan		
Singapore			Venezuela		

8. What strategies are being used at personal, national, and international levels to promote recycling of garbage?

Personal Level	*National Level*	*International Level*

Issues for the 21st Century and Beyond

1. The history of the twentieth century reveals that the First World has always imposed its will via its technological might on the Third World in dealing with economic and political problems. (a) Do you believe that the level of confrontation between the First and Third World has increased during the last decade? Explain your answer. (b) How could the conflicts between the Third World and the First World be resolved to achieve a win–win situation in the twenty-first century?

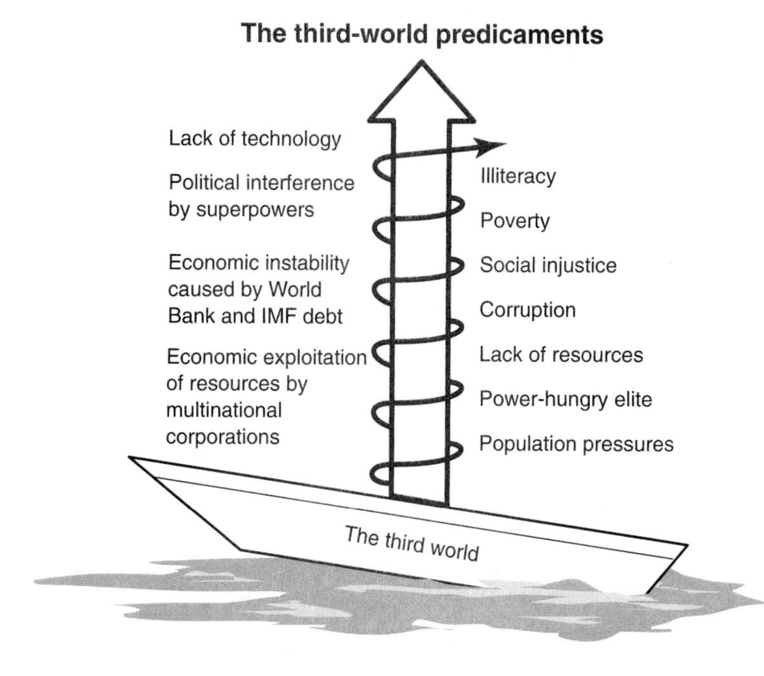

2. What actions must be taken by the developed countries as well as the developing countries to alleviate human suffering in the Third World?
3. What kind of relationship (cooperation vs. confrontation) do you envision will emerge between the First and the Third World in the twenty-first century? Explain your answer.
4. What types of energy technologies are most appropriate for improving the general well-being of Third World countries?
5. Compare the energy consumption (kW/hour per capita) for the following countries:
   a. USA vs. Bangladesh
   b. UK vs. China
   c. France vs. Nepal
   d. Germany vs. Yemen
6. List the advantages (+) and limitations (–) of various energy technologies available to the developing countries on the table below.

Energy Technology	(+)	(–)
Solar		
Wind		
Hydroelectric		
Nuclear		
Biomass		

7. A number of developing countries are planning to construct nuclear power plants to meet growing energy needs. Due to scarce resources in the Third World, nuclear energy seems to be the best viable option. The developed countries, however, are trying to prevent the transfer of nuclear technology to the Third World. The developing countries consider this attitude to be one of technological imperialism.
   a. Do you agree that Third World nations should be denied access to nuclear technology?
   b. What alternative sources of energy would you recommend for Third World countries instead of nuclear technology?
8. What is the future of solar power in developing countries? What obstacles to its implementation and use exist?

9. Discuss the status of solar power projects in the following nations:
   a. Indonesia
   b. Chad
   c. Gambia
   d. Mali
   e. Niger
   f. Senegal
   g. Guinea-Bissau
   h. Mauritania

USEFUL WEB SITES

http://www.worldbank.org	World Bank
http://www.un.org	United Nations
http://www.worldwatch.org	Worldwatch Institute
http://www.envirolink.org	Environment-related links
http://www.igc.apc.org/worldviews	Issues of peace and justice in world affairs
http://www.sipri.se	Military expenditure statistics
http://www.citinet.com	Country information databases
http://www.sunsite.unc.edu/lunarbin/worldpop	World population estimates
http://www.undp.org	United Nations Development Program
http://www.nrel.gov/business/international	Alternative energy sources
http://www.lib.umich.edu/libhome/Documents.center	Statistical resources on the Web
http://www.worldvillage.org	What if the world were a village?
http://www.asiantsunamivideos.com	Amateur Asian tsunami video footage

Technology of the Future

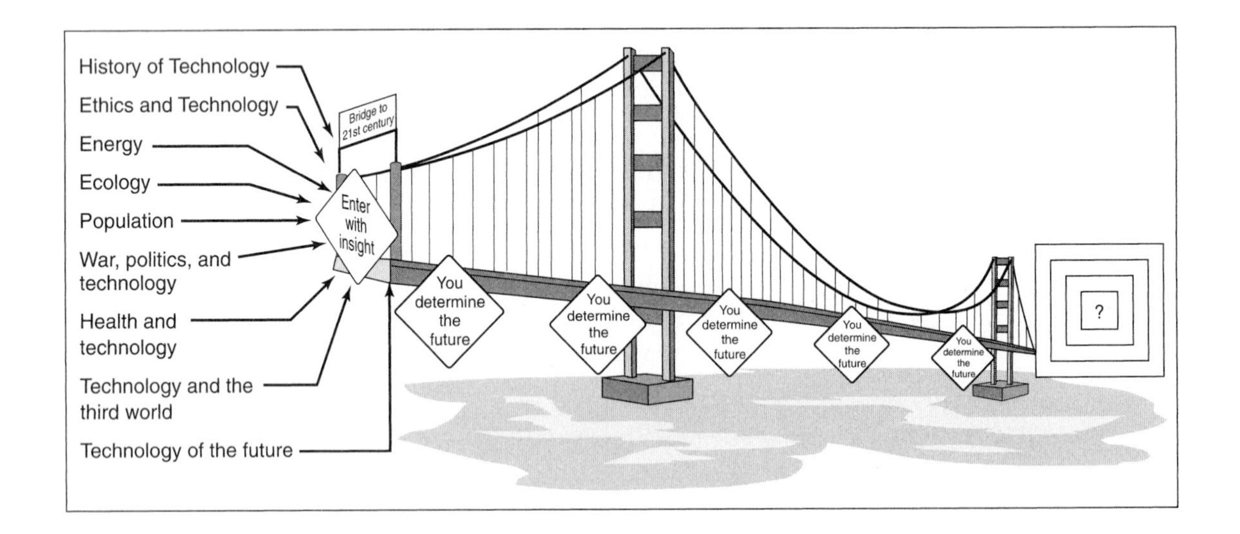

OBJECTIVES

After reading Part IX, Technology of the Future, *you will be able to*

1. Formulate future impacts in society from the application and growth of today's emerging technologies.
2. Review specific new and emerging technologies and understand their growth implications.
3. Apply ethical considerations in the use and expansion of present and future-oriented technologies.
4. Appraise the changing trends in technology and their implications to the workplace, education, and world markets at personal, local, national, and global levels.
5. Assess the strength and potential of planned future technological and social strategies.
6. Appreciate the importance and potential contributions of individual awareness and ethical action to the planning of the future.

INTRODUCTION

As humankind proceeds into the future with unprecedented technological powers, it is time to pause and ponder the technological and ethical issues that bridge the past to the present and the present to the future. The decisions made today will determine the directions of decision-making as we proceed further into the twenty-first century. These decisions will carry implications of historical, scientific, ecological, and sociological significance as humankind moves rapidly through space and time. Part IX presents a time-and-space approach to our examination of technology, enabling the reader to reflect on the issues of the previous eight parts and project their future development and applications to the future. The purpose of this approach is to refocus and converge these thoughts across the "bridge of now" so that they guide us well into the "bridge of the twenty-first century."

The changing landscape of the twenty-first century: a problem of balance.
Photo by Ahmed S. Khan, B. Eichler.

Part IX begins by crossing over the bridge into the twenty-first century and envisioning and creating a preferred future. Specific case studies examine new technological applications that will affect the medical, military, and ecological aspects of our lives. The readings also focus on predictions of future developments. In "Creating the Future," Norman Myers discusses ways to evaluate technology and its "multifarious impacts" in the new century. This selection is followed by "Predictions: Technology of the New Century and the New Millennium," which presents a broad overview of the major growth technologies and their associated industries of the twenty-first century. These first two selections thereby focus and paint a future horizon and perspective as to the future's values, needs, predictions, and technologies.

Part IX turns to exploring specific technologies in information technology and medicine as well as the military and concludes by examining the progress of global future world development efforts and re-examining population growth estimates. "New Horizons of Information Technology" delivers up-to-date information on fiber optics and its applications. "The Amazing Brain" uncovers new information as the brain functions and patterns itself to develop into a "mind." Understanding the capabilities and perspectives of a soldier in 2025 is described in "In the Year 2025." Ethics is applied to the new science of the miniature with ethical considerations brought to the forefront in "Why the Future Doesn't Need Us." J. Kenneth Smail discusses the idea that human global population must be considerably rethought for future stability. Two case studies end this section with the application for understanding the demands of new technologies for use by students and faculty and an overview of the development and progress of the important global social project of the UN 2000 Millennium Project.

These issues and their analysis relate to such technological issues of the future such as microprocessor miniaturization, optical networks, gene alteration, nanotechnology, and artificial intelligence, along with pressing ethical and social issues of the twenty-first century. Part IX comes to a close with discussion of the UN 2000 Millennium Project and other world recovery efforts and perspectives. This book concludes with commentaries from each of the authors, describing our own views on how to interpret our steps into the future as we walk across our most significant bridge, the bridge of the twenty-first century . . . the bridge that determines the future of our modern civilization.

The river of change: directing the future.
Photo courtesy of Barbara A. Eichler.

Creating the Future*

NORMAN MYERS

The future can still be *our* future. We can possess the future, provided we do not let it possess us. All we have to do is to choose. Or rather, to choose to choose.

To do that we shall need to keep an eye better fixed on the long-term future. This is hard, since we have grown adapted to thinking about the short-term future. When the marketplace decrees an interest rate of 10 percent, investors must recoup their investment within seven years, which is akin to saying there is no future to bother about thereafter. Our political systems proclaim the future extends only till the next election. To enable us to cater for the future beyond this foreshortened perspective, we shall have to engage in much restructuring of society. Otherwise we run the risk of an unguided future descending on us, a future with upheaval abounding. Make no mistake: for the first time in human history, the next generation could find itself worse off than the previous one to an extent without parallel. A worst-case outcome—a prospect far from impossible—would entail the end of civilization as we know it. The only way to avoid that is to reconstruct our civilization from top to bottom.

Many of our options are already foreclosed through our penchant for muddling through. As a result, we face an immediate bottleneck of restricted choices. Much of Part IX presents ways for us to squeeze through this bottleneck. We look at measures to contain the greenhouse effect, to move to new energy paths, to reverse biomass shifts, to halt deforestation, to defuse the population bomb, to refashion our economies, to promote "soft" technologies, to build an organic agriculture, to establish green governance, to devise planetary medicine, to foster dynamic peace, and to undertake the many other prerequisites of a sustainable world.

Most of all, these changes in our outer world depend on still more revolutionary changes in our inner world. So the concluding pages look at the greatest transitions of all: from knowledge to wisdom and from constant chatter to a readiness to listen—to listen to each other, to ourselves, and to Gaia.

We shall gain the future we deserve. Part IX points the way, a way with portents, pitfalls, and hopes. Whatever tough times lie immediately ahead, this book, like the expansive promise of the longer-term future, is for optimists.

TOWARD 2000

As we head toward the new millennium, a number of organizations have been making ambitious plans. Well might they be ambitious: only by thinking a great deal bigger than we have in the past will we be ready to tackle the challenges ahead.

From THE GAIA ATLAS OF FUTURE WORLDS by Norman Myers. Copyright © 1990 by Gaia Books Ltd, London. Used by permission of Doubleday, a division of Random House, Inc.

*Editor's note: "Creating the Future" was written in 1990, and the goals were constructed for the year 2000. However, these goals and the philosophy of the Gaia still remain relevant and urgent. They provide current vital priorities and goals to construct our future for the decade *from* the year 2000!

Health for All

This World Health Organization plan aims to: counter the malnutrition suffered by at least 1 billion people; supply primary health care to everyone; ensure life expectancy reaches 60 years; cut infant mortality to 50 per 1,000 live births; ensure all infants and children enjoy proper weight (half of developing-world children are poorly nourished); and supply safe water within 15 minutes' walk. All this is to be accomplished by the year 2000. Just to provide primary health care for all would cost $50 billion a year, and safe water plus sanitation would cost $30 billion a year. The plan also proposes much greater attention to the greying communities in the developed world.

Education for All

This UNESCO plan aims to supply basic education for every person on Earth by the year 2000. Today, 105 million developing-world children enjoy no schooling at all, and more than 900 million adults, or one out of every four, are illiterate. Cost: around $5 billion a year.

Population 2000

According to the International Forum on Population in the 21st Century, convened in late 1989 by the UN Fund for Population Activities, the aim is to "hold the line" on the medium population projection of 6.25 billion people by the year 2000. To achieve this, we need to: reduce developing-world fertility to 3.2 children per family; increase developing-world use of contraceptives to 56%; upgrade the status of women, especially their literacy; and achieve universal enrollment of girls in primary schools. In order to accomplish just the first two items, we shall need to increase the budget for family-planning services in developing countries to $9 billion annually, or twice as much as today.

Anti-Desertification Action Plan

The United Nations' plan seeks to halt the spread of human-made deserts, now threatening one-third of the Earth's land surface. There has been next to no progress on this issue since it was first formulated in 1978, due to government indifference. Yet it would be an unusually sound investment at $6 billion a year for 20 years, against the continuing costs of inaction worth $32 billion a year in agricultural losses alone. In 1986, the world spent more than $100 billion in drought-relief operations that saved lives but did nothing to halt further desertification.

Tropical Forestry Action Plan

The plan, already in action, seeks to slow deforestation by supplying: tree farms for commercial timber; village woodlots and agroforestry plots for fuelwood; more protected areas for threatened species; more research and training, plus public education; and better management all round. Cost: $1.6 billion per year. But the plan does next to nothing to address the main source of deforestation, the slash-and-burn cultivator, who reflects an array of non-forestry problems such as population growth, pervasive poverty, maldistribution of existing farmlands, lack of agrarian reform, and hopelessly inadequate development for rural communities generally.

Climate Change Programme

Focused on global warming and undertaken by the Intergovernmental Panel on Climate Change leading up to a proposed global-warming treaty, the Programme proposes to stabilize the concentration of greenhouse gases in the atmosphere by cutting emissions of carbon dioxide by 80%, of methane by 15–20%, of nitrous oxide by 70–80%, and of CFCs and related HCFCs by 70–85%. In the case of carbon dioxide, which accounts for roughly half the greenhouse effect, developed nations would have to reduce their fossil-fuel burning by 20% by 2005, by 50% by 2015, and by 75% by 2030. Developing nations would have to stabilize their emissions by 2010, allowing them to rise to no more than double today's levels as they continue to industrialize (developing nations would also have to halt deforestation by 2010 and engage in grand-scale tree planting). All this would mean that global emissions of greenhouse gases would fall to one-quarter of their present levels by 2050, limiting the rate of warming to 0.1° C per decade, and eventually holding it at 2° C above today's average temperatures —still a highly disruptive outcome.

Global Change Programme

Formerly known as the International Geosphere-Biosphere Programme, this research effort aims to provide the information we need to assess the future of the planet for the next 100 years, with emphasis on such areas as biogeochemical cycles, the upper layer of the oceans, Earth's soil stocks, and the sun's radiation. With an operational phase beginning in the early 1990s, the plan will last for ten years.

Beyond 2000?

As worthy as all these organizations' plans are, they all suffer from a serious shortcoming. They are targeted at the year 2000: what about longer-term goals, which must be established *now* if we are to avoid much costlier efforts later on, with much less prospect of success? Regrettably, the plans reflect the timid and short-term attitudes of the governments that approve the strategies. It is not in the nature of governments to see beyond the end of their noses, let alone beyond the end of the decade.

GREENHOUSE TACTICS: OPTIONS FOR RESPONSE

The greatest environmental upheaval of all, the greenhouse effect, offers much scope for positive response. While it is not possible to tame the problem entirely—there is already too much change in the pipeline—there is still time to slow it, perhaps even to stabilize it, and buy ourselves precious decades in which to devise longer-term answers. But this will require immediate and vigorous action on a broad front, plus international collaboration and individual initiatives on an unprecedented scale.

First and foremost, we need to consume less energy through energy conservation and more efficient use of energy. In fact, these two strategies represent our best energy source: from 1973 to 1978, some 95 percent of all new energy supplies in Europe came from more efficient use of available supplies—an amount 20 times more than from all other new sources of energy combined.

Since the oil crises of 1973 and 1979, most of the fuel-guzzling industrialized countries have expanded their economies by a full 30 percent while actually cutting back on total energy consumption. In 1985 came a third oil shock, this one of plunging oil prices, whereupon we quickly reverted to our energy-wasting ways.

But the energy-saving technologies are still there, waiting to be mobilized (see below). And at the same time as we cut back on fossil-fuel burning, so we reduce acid rain, among many other forms of environmental pollution.

Burning fossil fuel for energy, however, is only about one-third of the greenhouse problem. Another source is agriculture: nitrous oxide from nitrogen-based fertilizers; methane from rice paddies and ruminant livestock. With more mouths to feed around the world, we cannot allow food production to plateau—though we could do it differently and more productively with genetically engineered breeds. A halt to tropical deforestation would prevent at least 2.4 billion tons of carbon from entering the atmosphere each year, while reforestation would soak up carbon as well as restore vital watershed functions among many other purposes.

Through multiple linkages, then, the greenhouse problem reflects the myriad ways we all live. It will only be through myriad shifts in economic sectors and lifestyles alike that we shall get on top of the problem.

Public Perception

Look out of a window and you view a world in the thrall of climatic upheaval. Although nothing can be seen, the world is undergoing an environmental shift of a type and scale to rival a geological cataclysm—and one of the most rapid ever to overtake the Earth. To confront it, we need a parallel change in our inner world, our world of perception and understanding.

Changing Energy Policies

To mitigate the effects of the greenhouse, we must reduce fossil-fuel burning. In the short term, the greatest savings will come from buildings and products that use energy more efficiently. Many U.S. electricity utilities offer their customers energy surveys, rebates on energy-efficient appliances, and loans to finance energy-saving improvements. In the long term, we need to replace fossil fuels with environmentally benign renewables, such as wind, wave, and solar. As for nuclear power, and leaving aside the unacceptable risks in its use plus its uncompetitive price in the marketplace, to replace the world's coal-burning stations would entail building one nuclear plant every three days for the next 36 years at a cost of $150 billion annually.

A Climate Convention

A worldwide convention will need to decide how much greenhouse effect we are prepared to live with and the degree of remedial action to be taken, and by whom, in order to reduce the problem. This will mean a cap on greenhouse-gas emissions by all nations. But how hefty a cap for such disparate cases as Britain and Bangladesh, the U.S. and Brazil, Germany and China, the Soviet Union, and India? Should the industrialized nations indemnify the industrializing nations in order to safeguard the climate of all nations? Should there be special dispensations on the part of the entire community of nations to help those worst affected? The questions multiply and ramify, and we have no precedents to guide us.

Rethinking Agricultural Practices

Population growth and the intensification of food production are likely to set limits to reductions in certain greenhouse gases. Actively reducing the extent of intensive cattle-rearing operations would curtail some methane production; reducing fertilizer use, or implementing more efficient means of fertilizer application, would do the same for nitrous oxide. An even more important answer lies with new varieties of crops that require less fertilizer.

DYNAMIC PEACE

Just as health is more than the absence of disease, so peace is much more than the absence of war. It is a state of dynamic stability that has to be actively maintained. It is an end to violence among human beings and, by extension, to the planet. When this relationship is understood and accepted and when basic needs are met, there exists the opportunity for positive peace. While many people talk about peace building, they rarely specify what it means. Not that they have had much chance to practice it, since peace-building activities have not received even 1 percent of 1 percent of what has been spent on war activities.

First and foremost, peace on Earth means peace with the Earth. We are engaged in World War III, a war against the planet, a war that is no contest. We must negotiate terms of surrender to the new environmental dictates, recognizing that victory over the Earth would be a no-win outcome. In turn, this postulates a new form of security, environmental security. As past president Gorbachev was one of the first to point out, the threat from the sky is no longer missiles but climate change. Stealth bombers cannot be launched against this uniquely threatening adversary—just as tanks cannot be mobilized to counter rising sea levels, nor troops dispatched to block the advance of the encroaching desert.

So peace with the Earth implies peace with each other, a peace comprising three essential components: relations of harmony among people in society; co-operation for the common good; and justice based on the concept of equity. Much environmental ruin derives from inequity. Economic and political imbalance drives the marginal Third Worlder into marginal environments—ones that are too dry, too wet, too steep for sustainable livelihoods, and where vast environmental damage is caused through desertification, deforestation, and soil erosion.

To achieve dynamic peace at the global level, there must be a submerging of the narrow interests of individual nations in an effort to conduct international affairs in a manner that befits a global community. But unless it is based on general consensus and it is environmentally and economically sound, there is scant hope of success.

Green Security

With the thawing of the Cold War, some of the erstwhile military outlays could be directed toward building a more secure world. According to Robert McNamara and other experts, NATO countries could soon release $175 billion a year from military budgets. Soviet leaders indicate they could release perhaps $100 billion a year, funds needed to rebuild the Soviet economy. According to Lester Brown, President of the Worldwatch Institute, to restore environmental security would cost: protecting topsoil around $9 billion in 1991; restoring forests $3 billion; halting the spread of deserts $4 billion; supplying clean water $30 billion; raising energy efficiency $10 billion; developing renewable energy $5 billion; slowing population growth $18 billion; and retiring developing world debt $30 billion. This total of $109 billion would need to rise to $170 billion a year by 2000—still no more than eight weeks' military spending at late 1980's levels.

The Role of Nonviolence

Nuclear power has forced us to contemplate both extinction and new forms of conflict resolution. The latter could soon become much more common. Nonviolence, however, goes far beyond ensuring peace by reducing military budgets or halting environmental destruction. As well as the weapons, we must also eliminate the "mind" that brought them into being and contemplated their use.

"Vulnerable" to Peace

The arms race has tied the economies of many nations to the military. Now there is a fear that an outbreak of peace will take jobs and shareholders' dividends too. To counter this, communities are developing "economic conversion projects." The objective is to retrain workers and adapt factories to civilian production. To this end, citizens, workers, and managers plan their future with local politicians, technocrats, and economists. In the US,

70 cities have passed "Jobs with Peace" initiatives and are beginning the conversion process.

PLANETARY CITIZENSHIP

We started out on our human enterprise with loyalty to a hunting band of a few dozen people. From there, we successively expanded our allegiance to the village, town, city, region, and eventually the nation. At each stage, our sense of community grew, until today we feel a part of societies of millions. Yet the greatest loyalty leap awaits us. Can we now raise our vision to embrace the whole of humanity? So great is this challenge that it will rank as the second true step away from the cave's threshold.

First, we need to identify with a global community of individuals whose names we do not know, whose faces we shall never see, whose traditions we may not share, and whose hopes we shall not know, but who are all, whether they are aware of it or not, *de facto* members of a single society. Second, we must foster a super-allegiance to Gaia, and frame our actions accordingly. Can we learn in time to identify with these two ultimate communities?

This need not present any conflict of interest. When we salute our countries' flags, we are not thereby denying bonds with families or neighborhoods. We are simply acknowledging a greater context of kindred spirits, a loftier level of allegiance. So our new planetary loyalty will not diminish established links; rather, it will enhance them by adding perspective to local attachments.

There could, however, be some exceptions. What happens when a country asks us to do something against the global good? To hold back for the sake of the national economy on antiglobal-warming measures, for example, or to consume for the sake of the balance of payments, when that means supporting unduly polluting activities? As we head farther into an interdependent future, there may well arise numerous occasions when we shall feel torn by planetary obligations that should outweigh "state dues." On such occasions, it will not be easy to rise above past practices or to keep our attention fixed firmly on the universal need. We cannot live in isolation, whether from one another or from the planetary home. Either we shall become involved through joint effort as global citizens, or we shall become involved through jointly suffering global catastrophe.

Transcending Politics

Traditional politics concerns itself with the managing of social systems. This view will need to be considerably broadened to take into account the concept of planetary citizenship. The best politics will enable people and communities to create their own solutions to their own problems within a larger context. But a prerequisite will be the molding of a new environmental/political consensus from the present anti-environmental world. Among initiatives that could soon become commonplace is an Earth Corps, an organization providing a framework for people, young and old, to make a personal contribution to the planet. The potential is vast, not only in terms of work to be done, but the reservoir of people and energy waiting to be mobilized on behalf of the Earth.

Rising Above the Nation-State

We are suspended at a hinge of history, between two ages—that of competitive nationalism and that of cooperative internationalism. Nationhood is becoming a pernicious anachronism, a primitive phenomenon like feudalism or slavery. Future generations will surely consider that nationhood was a transient phase in society's development, a holding measure until the emergence of planetary citizenship.

Our Evolutionary Conditioning

When we attempt to identify with the global community, we may find that our evolutionary conditioning stands in the way. Our individual nurturing has derived primarily from the 99% of human history spent as hunter-gatherers, dependent for survival on sinking individual interests with those of a few dozen others. So we have inherited a set of sensibilities that, however capable they were for our formative years, may have left us deficient for our future worlds. Can we develop the extra faculties we need to cooperate as a band of more than 5 billion people?

A LAW UNTO OURSELVES

In a new-age world, we shall learn to be our own legal experts in that we shall have to devise rules for living in a crowded global community without treading on anybody else's prerogatives. We shall need to recognize that planetary citizenship entails responsibilities as well as rights, and we shall have to learn that it is in our own best interests to be our own private law enforcer. Not that this concept is anything new in itself. The rule of law has always depended on a strong supporting consensus of the citizenry, without which the best-intentioned laws fail.

Consider, for example, the debacle of the Prohibition years in America. But this time around, there will be such an abundance and complexity of laws, regulations, and rules, whether formal or unwritten, that there cannot be enough police to keep everyone on the straight and narrow. We must devise our own path ahead and follow it because it is in our own best interests.

This mode of behavior will be in stark contrast to the free-wheeling years of a simpler and less vulnerable world, one where there was no threat of terminal breakdown in society through outright environmental collapse. With multiplying numbers of people, multiplying demands, and multiplying linkages through our increasing interdependency, both environmental and economic, there will be a multiplying scope for disruption and dislocation. The answer will lie with the dictum of "mutual coercion, mutually agreed upon." But this need not be a fraction as "Big Brotherish" as it may sound. It will replace the unlicensed liberty of yesterday with new forms of freedom for tomorrow—an expansive and disciplined freedom.

Above all, the new "world of laws" will be all the more acceptable in that it will not be a top-down affair, by contrast with the situation in the past. It will be a grass-roots process, a home-grown homage to largely local imperatives. Communal laws will be more like social codes, finely tuned to local needs—a world away from the rigidity of conventional laws.

Local Control in Sweden

Flexible, local systems of sharing and control are becoming a feature of codes of conduct of certain smaller nations. In Sweden, for example, most of the unionized workforce is solidly behind its government's new program of economic decentralization. In each one of Sweden's two dozen counties, a proportion of the workforce's earnings is automatically paid into a public fund administered by an elected board. The accumulated funds are used to purchase shares in local industries, which are then publicly controlled. Local economic control is thus back in the hands of the community, where decisions can be made in light of local needs.

Roundabouts or Traffic Lights?

Law-abiding societies of the future must operate by consensus rather than by the threat of punishment. Communally agreed codes will be the order of the day, not dictates passed down from some distant national assembly. When we all agree to drive on the same side of the road, we make no sacrifice of personal freedom—rather our freedom is enhanced by communal consensus. As we progress into the future, we shall find that we will be driving ever more sophisticated models (of the figurative type) at higher speeds and sharing the road with ever more drivers. This will necessitate more "traffic control" to facilitate everybody's journey. These controls can take the form of roundabouts (communal codes with which people have licenses to assess, evaluate, and make decisions) or traffic lights (specific laws that require simple acquiescence).

THE GENIE OUT OF THE BOTTLE

Our technological capacities are such that it is now possible to reshape our world from top to bottom—not only our planetary living space, but our social relations, our individual inner worlds, all that we are and do. The record to date does not presage a future as constructive and bountiful as we might wish. Unbridled technology already threatens our very life-support systems. Yet technology could be one of the greatest boons for the human condition, provided we ensure that it serves our overall interests. We need to take a long look at the role of technology in our future world and to ask how we can take systematic control of its multifarious impacts. We have released the genie from its bottle, and even if we perversely wished, we cannot return it. The present challenge is to control the genie before it controls us in unwitting ways that we cannot remotely discern.

What then is to be the future role of technology, as of its scientific underpinnings? The new physics shows us that what scientists observe in nature is intimately conditioned by their minds. Hence, scientific and technological applications are also conditioned by the mind and thus by human values. Scientists are not only intellectually responsible for their research, they are also morally responsible. In turn, should not scientists and technologists now be required to take methodical cognizance of the impacts of their endeavors, whether environmental, social, or even political?

True, this would mark a profound change in our attitudes to science, as to its role in society. Many scientists would be aghast at the notion that their research should somehow be trammeled. But this is not to assert that all science should be subject to detailed constraints. There must always be abundant place for the pursuit of knowledge and understanding, whatever it reveals. Yet the overall context of science should surely be examined to see whether we can determine some limits to its unfortunate technological by-products. It is this new dimension, placing a check on undesirable fall-out, that is the key to controlling the genie.

Human Hubris

Our overweening attitude toward the natural world is a recurring theme in cultures right from the Ancient World. For stealing fire from the gods, Prometheus was chained and tortured. It seems always to have been accepted that there are some areas of enquiry that are simply off limits in view of the potential costs they entrain. The question now is to determine which precisely these areas are. While we understand so little of the world about us, especially the expanding world of the future, we must move from hubris to humility and practice a cautious rather than a Promethean approach.

Self-Imposed Audits

Already some ecology professors are proposing that students' dissertations should include a chapter on the social implications of their findings. In some cases, these implications will be virtually nil, in many others they will be of marginal consequence, in certain others they will be significant. Whatever the outcome, the exercise will induce an explicit awareness that science and technology can no longer be practiced in a vacuum.

Genetic Engineering

To date, genetic engineering has caused no regrettable spillovers onto the environment. The potential, however, is certainly there as new organisms are introduced into natural communities at an increasing rate. After all, an earlier effort at improved breeding of livestock in the form of a goat variety adapted to harsh environments—hailed as an undoubted success at the time—led to semi-arid lands becoming arid lands. The new breeders, for example, gene splicers using recombinant DNA technology, need to exercise far more scrupulous care as they release multitudes of entirely new organisms. While it is unlikely that the newcomers could ever become dominant on a broad scale, they could well cause local ecological disasters that could not readily be controlled. Unfortunately, there is a tendency for expectations and benefits to be overestimated, while costs and problems receive short shrift—as is often the case during the early stages of the development of any new technology. The prospect of hosts of newly minted organisms warrants exceptional caution rather than a "rush-in" approach. Yet we have scarcely started to assess the legal, let alone ethical, issues at stake.

RESPONSIBLE TECHNOLOGY

Modern society is hooked on technology, one that gives us life-saving hospital units and liberating communications, also nuclear weapons and soulless production lines. We are led to believe that there is a technofix for every social ill and global problem—and if technology goes wrong, then technology will put it right again. Yet we continue to feed off ecodestructive agriculture, to use products from energy-wasteful industries, and to live lives in technology-ravaged environments.

Our notions of wealth and welfare, competition and efficiency, are grounded in technology. The progress of technology is supposed to be unstoppable: it represents the crowning achievement of human enterprise, and other notions of progress come second. But our resource-intensive, overcentralized technology is making itself obsolete. Petrochemical agribusiness, nuclear power, and fuel-guzzling cars are environmental disasters. Yet we still have to devise guidelines for our present technology, let alone the fast-expanding technology of the future. What technology is "right?"

The answer lies with technology that supports humankind in our need to live in accord *with* the planet rather than in dominion *over* it: technology with a planetary face. Many technologies for sustainability already exist in the form of so-called "soft technologies," for instance those that utilize renewable resources and recycled materials. A good number of these technologies are already familiar to us: energy-saving devices; ultra-efficient motors; solar energy conversion; wind, wave, and tidal power utilization; organic farming; semiconductors; and superchips. The challenge is to create a flexible, benign, and humane technology—a process that will allow us to exercise our full creativity. Indeed, a crucial factor in this new technology will be its readiness to draw much more on a resource we already possess in abundance—human ingenuity.

In this regard, the soft technologies are often ideal since they tend to be small scale and decentralized. And, being generally labor intensive, they help to establish a local economy, one that is flexible and sensitive to local conditions.

A New Orientation

To solve the multiple crises we face, we don't need more energy-intensive technologies. We need to shift our emphasis from nonrenewable to renewable resources, from hard to soft technologies. But this alone will not be sufficient without a thorough-going cultural change—a move away from the mechanistic "we can fix it" mentality to a more careful and caring approach. This in turn will require the most basic retooling of all: wholescale shifts in our attitudes, lifestyles, and values.

The Solar Future

The planet's principal energy source is the sun, with its potential for limitless, nonpolluting energy. Life has evolved to make optimum use of this form of energy. Plants photosynthesize food with the aid of the sun's radiation and in turn provide the conditions, directly or indirectly, for practically all other forms of life. Solar energy can either be "passive," as when the fabric of a building heats up and then releases its energy, or "active," as in the example of solar collectors. As well, the sun's energy raises air masses that drive wind turbines, and it also energizes the water cycle which we harness as hydropower. Bear in mind that solar energy is not only available everywhere, it is a source of diverse energy types, hence adapted to decentralized technologies in local communities.

People's Technology

Grass-roots technology is not always accorded the recognition it deserves. In the 1960s, traditional fishermen of the Arabian Sea protested that the introduction of mechanized trawling would destroy local fishing stocks. Their detailed ecological insights were dismissed by fisheries experts. Today, and through their own technological expertise, the fishing communities are pioneering the use of artificial reefs and species-specific baits, exploiting them in a manner that is both sustainable and finely tuned to local circumstances. More, their efforts are starting to be supported by the government, strengthening their organization.

FROM KNOWLEDGE TO WISDOM

As T. S. Eliot once asked: "Where is the knowledge we have lost in information? Where is the wisdom we have lost in knowledge?" Obviously, knowledge is an essential component of wisdom, but there is more to it than that. While we possess knowledge in abundance, the world about us proclaims that we are falling woefully short of the step from knowledge to wisdom.

Our long tradition of reductionist analysis has brought us to a state where we often find that we are learning more and more about less and less. Result: our attention is diverted from the whole that is more than the sum of its parts. Confronted with the disjunct pieces of a watch, would we guess that when reassembled they would make up a mechanism for telling the time? Yet we meddle with the intricate components of our planetary ecosystem, no less, and suppose we can do so with impunity, even concluding that certain pieces are forever dispensible. We are unwittingly engaged in global-scale experiments, not only environmental but technological and social experiments, without a thought for their global consequences. Yet these grandiose experiments demand the most scrupulous care and would tax the faculties of the wisest.

Moreover, there is much that we, whether scientist or not, can discern in terms of values without understanding the details. We do not need to understand the physical or chemical make-up of a weapon to realize it is designed to destroy. We do not need endless reports on the influence of the media to see their power to misinform.

Thus far, we have tended to head blithely toward a "universal horizon" and have limited ourselves to doing a better job in getting from here to there—"there" being something that reflects the common understanding of most people. We make sure that we travel along smoothly in a car that is in good order and with enough fuel. Beyond our present crossroads, however, that will not suffice. Often enough, we shall need to consider the proper road to follow. The landscape ahead is no longer preordained by "where we have always been going before." Increasingly, we shall find that we need to choose a different route, change the car, even ask whether we want to travel at all. All the fast-growing knowledge in the world will not help us if we do not have the wisdom to look out all over the world and to decide where we feel most at home.

Dinosaur of the Mind

"But we learn no lesson, give no moral to our children. We go on not having understood what is meant by our technological act of knowing. Some day, in the far, far future, the evolutionary processes will perhaps have gotten rid of this failing attempt at gaining knowledge; either we will be the ancestral progenitors (of a wiser species), or else we will be discarded, an unsuccessful offshoot, a dinosaur of the mind, one of nature's failures." Professor Phillip Siekevitz, Rockefeller University.

Analysis and Vision

The knowledge business—not only science but telecommunications, media, advertising—means we can generate more information than ever before and mobilize it more efficiently. This is far from enough, even though it is the credo of government, business, and the general ethos of "advancement." For we see only what we are disposed to see. When we analyze information to produce new ideas, we do no more than find new slots for old ones. To generate truly new ideas—to strive toward "wise" insights—we must bring into play our creativity, our powers of inspiration, above all our sense of vision. All these are a world beyond the nose-to-the-microscope approach of conventional understanding.

Knowledge as Private Property

The trend toward privatized knowledge is regrettable. While the patenting of ideas is a valid safeguard, the wholesale locking away of knowledge as private property, walled by legal defenses, can only redound to our collective detriment. Instance the North-South gap: whether wittingly or not, the technological underpinnings of the knowledge explosion in the developed nations are acting to exclude developing nations from one of the most productive phenomena of our age. How ironic if the knowledge resources in the North serve to impoverish further the South. Surely, we will recognize that knowledge should be generally available in the public domain, serving the needs of everyone. Only as the walls around knowledge crumble away will the shared experience of all contribute to dawnings of wisdom.

THE FUTURE OF THE FUTURE

For however much is uncertain about the future, one thing is definite: new worlds will constantly unfold, with their new problems *and* their new possibilities. In some senses, in fact, there will be as many possible futures as there are people. We are all deeply involved, whether we appreciate it or not; we shall all play our individual parts. Indeed, there has never been a time like the present when the individual can *count.* So a prime aim has been to show the reader how to envision the future and to decide how to contribute—if only at the local level, which will often be the best level.

The shadows over our future remind us that the optimist proclaims this is the best of all possible worlds, while the pessimist responds that is regrettably true. We must stay hopeful, otherwise we might as well go to the beach until the sky falls in. But let us not be seduced by an airy hope that somehow all will work out in the wash: the laundry water likely contains too many pollutants. There are prophets who assert we should not worry about the prospect of feeding 10 billion people when it is theoretically possible to feed four times as many. As Paul and Anne Ehrlich point out, it is theoretically possible too for your favorite football team to win every game for the next ten years. In any case, the Ehrlichs continue, what is the sense in converting the Earth into a gigantic human feedlot? How about more quality of life for fewer people? Moreover, a future of "the same as usual, only more so" would be a future that for many people would simply not arrive.

If further persuasion is necessary, recall that of the 31 major civilizations in history, only one remains a dominant force—so-called Western civilization. As the historian Arnold Toynbee demonstrates, the rest disappeared because they tried to become dominant on every side—over their neighbors, their environments, whatever else they cast their eyes on. Western civilization, materialist to beat any other, shows plenty of signs of dominating the entire world into an ultimate crunch.

No doubt about it, we stand at a hiatus in the human enterprise. The present is so different from the past, and the present so different from the future, that it is as if we are at a hiatus in the course of human affairs. It is a unique time: a time of breakdown or breakthrough.

To break through into a future of undreamed potential, we must enable a new sort of society to be born. Indeed, the stresses of the present are like the stresses of being born, a time of utmost threat yet with new life ahead. Or, to shift the analogy, our society is like a human being growing up. From the start, it shows a boundless appetite for resources of every type. This appetite continues throughout the first two decades of physical growth, expanding all the while. Then it suddenly levels off. This does not mean the person's growth is at an end. On the contrary, the richest stage of growth begins—mental, intellectual, emotional, and spiritual growth, growth that extends many times longer than the early phase.

Our global society is still adolescent—lusty, vigorous, and assertive. Can we move on to maturity—assured, stable and sensitive, displaying all the attributes of adulthood?

Are we ready to grow up? To shift from egosystem to ecosystem, to social compact and whatever else is needed for us to become citizens of the globe? Are we ready, in fact, to engage in the most salient experience of adulthood, the mutuality of love? Nothing less will do. Finally, we can recognize that there is no longer any "we" and "they." For the first time, and for all time, there is only "us"—all of us humans, together with all our fellow species.

This will be the greatest of our global experiments. As we measure up to the challenge, we shall need to become giants of the human condition. As we approach our climacteric —never attempted before, surely never to be repeated with such instantaneous speed—let us count ourselves fortunate to be living at this hour. No generation of the past has been presented with such a chance to rise above the tide of human history. No generation of the future will have our chance because if we do not do the job, they will have little left but to pick up the pieces. What a privileged generation we are.

WHAT'S POSSIBLE; WHAT'S PROBABLE

What lies ahead? Part IX looks at an array of possibilities that may prompt us to speculate on what else is in store. The following list is highly selective and no more than illustrative. The items derive from analyses of experts who use techniques such as scenario planning to think methodically about the future. Futurists based at universities have set up entire departments for futures studies. There are commissions on the future, established by individual governments, United Nations agencies, the City of Tokyo, and the like. The World Future Society, based in Washington DC, organizes regular conferences on all aspects of the future. But the most remarkable feature of this futurist community is not that there are so many people thinking about what the future holds, but that there are so few. Among some prospective developments are the following.

Global Environment

As the greenhouse effect takes hold, we could witness persistent droughts over much of North America, sub-Saharan Africa, and China. Together with repeated failures of the Indian monsoon and other climatic quirks, plus the most expansive phase of the population explosion, there could ensue a greater outbreak of starvation in a single year early next century than throughout a decade of the twentieth century, even culminating in the deaths of one billion people in just one decade. This worse-case scenario, by no means implausible, would amount to a human catastrophe of unique proportions. In the longer-term future, a one-meter rise in sea level by the middle of next century could, when combined with storm surges reaching far inland plus saltwater intrusions up rivers, threaten a total area of 5 million sq km—or one-third of today's croplands and home to one billion people already. Also on the cards is the prospect of mass migrations in the wake of the greenhouse effect. On the Chinese side of the Sino-Soviet border, there are already acute pressures from 1,300 persons per square kilometer, by contrast with only one person to every 2.5 square kilometers on the Siberian side. What happens when China starts to suffer the full rigors of global warming?

Geopolitics

Consider the following scenario. In the year 2000, the world has 350 billionaires, at least 4 million millionaires, and 250 million homeless. The average income of the top 1 billion people has reached 50 times more than that of the bottom 1 billion. More than 50 nations no longer qualify as developing nations; they are disintegrating nations. Americans spend $10 billion per year on slimming diets, while 1 billion people are so undernourished that they are semi-starving. Water from a single spring in France is still shipped to the affluent around the world, while a full 2.5 billion people lack access to clean water for basic needs. It has become plain that poverty is a luxury we can no longer afford.

Furthermore, there has been a series of crises on top of widespread starvation. Chernobyl-type accidents have occurred in four nations. The North Sea has been declared beyond foreseeable recovery. The most bountiful marine ecosystem on Earth, the Southern Ocean, has collapsed after UV-B radiation knocks out the phytoplankton. A nuclear terrorist has destroyed Cairo. Drug barons have declared jurisdiction over much of South America. The Pope has been added to Willy Brandt, Stevie Wonder, and Steffi Graf as hostages held by extremists. The latest Live World concert has been watched by two and a half billion people and has led to mass protests throughout the world.

This all pushes governments into finally acknowledging that there is only one track ahead: global collaboration with a vengeance, and for the first time ever. As European President Joan Ruddock puts it at the World Conference for the World: "The biggest problem is no longer others, it is ourselves. We are suspended at a hinge of history, between two ages—that of competitive nationalism and that of cooperative internationalism. At long glorious last, we recognize that traditional nationhood is an anachronism, a pernicious phenomenon like feudalism or slavery."

What emerges is a system of government based on concentric circles: local councils, regional assemblies, national governments, groupings such as the European Community, and global bodies such as the United Nations (supplied with teeth). This vertical structure is paralleled by a horizontal structure of NGOs with real power, made up of professions, trade unions, Friends of the Earth, major charities, academics, service clubs, Oxfam, and the like. These NGOs receive collective representation through the long proposed Second Assembly of the Untied Nations. Under these twin structures of government, citizens can cheer equally for Edinburgh, Scotland, Britain, Europe, the Commonwealth, the world, tropical forests, Antarctica, and Action Aid.

Science and Technology

As a result of scientific advances, techno-jumps could include a breakthrough with photovoltaics that transforms the energy prospect from top to bottom, especially for tropical countries; backpack nuclear weapons, with all that means for terrorism; sex selection on the cheap, leading to a massive majority of male babies; and genetically engineered trees that sprout like mushrooms ("plant the seedling and jump aside"), allowing reforestation to do a better job of soaking up excess carbon dioxide from the atmosphere. But note that genetic engineering could lead to some unfortunate consequences of economic and social sort. It will soon be possible to "grow" cocoa in the laboratory, which could devastate those developing-world countries that now earn $2.6 billion a year from the field-grown crop.

A related technology is nanotechnology, enabling us to redesign cellular structures. This will lead to, for example, exceptionally strong and lightweight alloys, leading in turn to organically manufactured aircraft that fly much more speedily and cheaply than today's dinosaur-style devices. Among more "way out" applications of nanotechnology could be steaks from hay, without the help of cows. These, like many other techno-jumps, will derive in part from the fast-growing capacity of supercomputers. Already the latest Cray model, standing no taller than its human operator, can solve problems at a sustained rate better than one "gigaflop," or one billion calculations every second (the term gigaflop comes from "giga" meaning one billion, and FLOP for "floating point operation," a common form of computerized arithmetic). Soon to become available is a computer capable of 22 gigaflops per second, while we should soon see a machine speeding along at 128 gigaflops per second.

As for the car of the future, that is already with us. A Volvo prototype, with lightweight synthetic materials, weighs only half as much as a conventional model. Its lean-and-clean engine, backed up by a continuously variable transmission and a flywheel energy-storage unit, achieves almost 150 km per gallon in average traffic conditions. A further prospect for the petrol-driven car is that there will simply be far fewer of them in urban areas, their place having been taken by vehicles powered by electricity or hydrogen fuel. In any case, there will be far less need for them in the face of competition from efficient and cheap mass-transit systems (in Tokyo today, only 15% of commuters drive cars to the office). Moreover, there will be an increasing trend for offices to be connected by electronic lines rather than crowded highways. Note too that if China were to devote as much land to asphalt as the United States does per head, it would lose over 40% of its croplands.

Health

What price an end to drug addiction? There is prospect of a final solution in the form of "opiate antagonists" that block the effects of, for example, cocaine for a month or so, whereafter the euphoric impact dissipates and the addict loses interest.

More broadly, we can anticipate a growing disaffection with established ideas about health. As more people recognize that health is intrinsically a holistic affair, so they will be inclined to accept personal responsibility for cures of "disease." There will thus be an increase in wholesome diets, exercise, and sports, stress-reducing activities, self-rehabilitation, and recreation in the sense of recreation. Sooner than we might suppose, self-help health will become mainstream.

In developing countries, the technique of oral rehydration therapy (ORT) could soon become as familiar as cola—otherwise we shall witness 25 million children die of dehydration during the 1990s. Ironically, as we bring about an end to the human hemorrhage represented by child mortality in developing countries—one of the great success stories in human history—we may well witness another scourge overtaking hundreds of millions—the ravages of tobacco, as cigarette corporations of the developed world, losing their clientele at home, peddle their wares to the last corners of the developing world.

Lifestyles

Already many households feature a personal computer. A child has hardly learned to read and write before he or she starts to work with a device that, a couple of decades earlier, would have had to be as big as a house to contain the computing power of the tabletop model. It is this new skill, backed by global-scope telecommunications networks, that is opening up worlds for youngsters way beyond the dreams of their parents.

Within just a few years, a majority of women will be engaged in paid work, whether in a formal workplace or in homes linked to offices by computer networks. At the same time, many men will accept a greater role in the family and home. In response to the cocktail party question, "What do you do?" the answer will increasingly be, "I'm a househusband."

QUESTIONS

1. What are the problems and challenges with our current thinking for the future?
2. What are some of the immediate needs that we must address for the future?
3. For population growth to be held at around 6.25 billion people of the year 2000, what was the average number of children per family worldwide? What is the current average?
4. What is the problem with current plans for the year 2000? How relevant are they for the years of 2010–2015? These plans were written in 1990; do you feel that they are realistic? Explain from our current historic view.
5. Compare the Climate Change Program with the outlined plans of the Kyoto Climate Treaty of December 1997. What are the differences between the two?
6. When discussing the greenhouse effect or global warming, what is the main issue that has to be addressed? Name three suggested approaches to deal with the global warming issue.
7. Identify three approaches for developing world peace strategies.
8. Explain two views of global politics that expand planetary political unity.
9. What are some of the considerations and concepts behind planet-wide laws?
10. What is being suggested of engineers, scientists, and technologists when they design new systems and technologies?
11. What is meant by the term "responsible technology?" Provide two examples.
12. How can we attain wisdom for the future?
13. What is the biggest challenge to sustaining our world and ourselves in the future?
14. Comment on your impressions of the predictions of the future with regard to the global environment, geopolitics, science and technology, health, and lifestyles. Do you agree with these future predictions? Give some of your own predictions as a contrast.

Predictions: Technology of the New Century and the New Millennium

BARBARA A. EICHLER

Simply said but profound in impact, progress is proceeding exponentially in everything technological. . . . "The pace of technological change 'advances (at least) exponentially' . . . So we won't experience 100 years of progress in the 21st Century—it will be more like 20,000 years of progress (at today's rate)."
RAY KURZWEIL, Inventor, Film maker, Member 2002 National Inventor's Hall of Fame

The history of technological change is exponential, unlike a linear, numerical view. Since our technological advancements have become increasingly complex and sophisticated, this century of advanced systems and technology will quickly interface and merge our advanced knowledge into multi-systems and create metasystems of capability into every field of technology and discipline. The way we think and do things, a change in the "paradigm," according to Ray Kurzweil, is doubling every decade. Computers have increasingly more processing power and are increasingly more compact. The Internet easily brings information and connections everywhere. Telecommunications of satellites and cell phones transfer the information and computations to all sorts of communication devices. This is not even mentioning the new sciences and applications of nanotechnology, genetic engineering, neuro-technologies, and advanced communication approaches. Knowledge is increasing at an astronomical rate. It is estimated that the world's knowledge base doubles every 18 months (Reinhold, 2004), but specifically in science the information is doubling every five years (Information Today and Tomorrow, 2001). It is increasingly difficult to predict the future, because of this increasing, exponential momentum, but the following represent some of the exciting predictions. We are very fortunate to live in these times of such changing and imaginative new technological approaches and systems.

THE CONVERGING TECHNOLOGY MULTI-REVOLUTION

"Converging Technologies (CT) result from the merging of (a) Nanoscience and nanotechnology; (b) Biotechnology and biomedicine, including genetic engineering; (c) Information technology, including advanced computing and communication; (d) Cognitive science, including cognitive neuroscience—(all referred to as NBIC)" (University of Montreal, 2004). These Converging Technologies seem surreal, but CT is becoming recognized as a bioengineering discipline. CT is expected to reach world policy levels by 2010 and their condition at present is not much different than the state of recombinant rDNA research in the 1970s. There will be many innovations in preventive medicine such as implants, miniaturized diagnostics, nanosensors, etc. Disease treatment will be at new levels of applications such as artificial skins, and organ replacements. Disability alleviation will allow therapies and enhancements to more closely replace natural function such as in cochlear and retinal implants. Military and Security Applications could be used in all sorts of selective chemical or biological agents affecting only people with certain characteristics. This is only a beginning of the range and depth of the developments from the combinations of these four advanced technologies (University of Montreal, 2004).

THE DIGITAL REVOLUTION

Digital World

The supercomputer is here! This has profound implications in the area of super computing, gaming, and virtual reality. The next generation processor has the capability of three trillion mathematical operations per second. This new processor will have 681 milllion transistors, more than twice as many as the current fast processors on the market

(Houle, 2006). Microchips are also extremely small, which means complicated systems can be contained on a chips which are about the size of Lincoln's nose on a penny. Therefore, many applications can be used and implanted easily in most environments and have implications for information processing ubiquitously for systems in training, entertainment, health, and all fields using information systems.

Also by 2010, it is predicted that voice recognition will be available in computers and all sorts of digital devices. We should be able to carry on direct conversations with computers, Internet browsers, automobiles, thermostats, TVs, VCRs, microwaves, and other pieces of equipment.

Personalized World Through Computers and Hyperlinks

In the future, bandwidth, the capacity of fiberoptics and other "pipelines" to carry digital communications, will become freer, which means that all types of content—including data, all forms of shows, books, and home videos—will be online and available to anyone in the world. The practical applications of this ability would allow marketing and subscriptions to be tailored to the specific needs and tastes of individuals. This ability to customize information then would allow the products to be specialized for each user and will change the world from a mass-market world to a more personalized one (Isaacson 1998).

Further Miniaturization of Microprocessor Design

The following items represent future advances in the design of microprocessors:

- The use of quantum dots and other single-electron devices, for which electrons can be used as binary registers to represent the 0 or 1 of a data bit.
- The development of data storage systems using biological molecules, resulting in molecular computing. Optical computers for which streams of photons could replace electrons would be possible if a biological light-sensitive molecule were used. With biological molecules, self-synthesis would be possible through the life systems of the microorganisms. For some estimates, some photonically activated biomolecules could be linked to a three-dimensional memory system that would have a capacity 300 times greater than that of today's CD-ROMs.
- Nanomechanical logic gates where beams one atom wide would be physically moved to carry out logical operations (Patterson 1995).

All-Optical Networks

Already, fiberoptic networks transmit data 10 to 100 times faster than standard electronic wiring, but in the future, light waves will be used for most of the transmission of voice, video, and other data along the networks' pathways. The speed and capacity of fiberoptics in the future will be used such that—in theory—a single fiber has the capacity to transmit 25 terabits (trillions of bits) each second, which is comparable to transmitting all U.S. Mother's Day telephone calls simultaneously in a second. The only time the signal would become electronic would be when it moves into the circuits of a computer or into a lower-speed electronic network (Chan 1995). At present, the state of technology allows data rates of the order of tens of billions of bits per second. Once the potential capacity is developed, the transmission rates will allow hundreds of trillions of bits per second. The future potential of optical networks will allow the transmission of 25 billion telephone calls simultaneously. In other words, consider that with the population of the earth at 6 billion people, every person on earth could talk at the same time using one fiber whose dimensions are ten times smaller than a human hair (Khan 2006).

THE BIOTECHNOLOGICAL REVOLUTION

In the future, we will have the technological capacity to engineer and change human DNA and build on the emerging capabilities that we have already demonstrated with plants and animals. The challenges will not only be scientific, but also moral.

The Human Genome Project

With the completion of the mapping of all of our genetic material, scientists are beginning a new age for biology in which we will understand biological processes and the causes of disease and behavior from the level of genes. Central control genes for the formation of organs, behavior characteristics, and other genetic human traits can be studied and understood. Genetic flaws connected to cancers, leukemia, dyslexia, Alzheimer's disease, and even alcoholism and obesity will be traceable and their discovery contribute to the development of more effective treatments for these disorders. The moral issues of knowing genetic codes and their outcome, predicting susceptibility to disease, the relation of genetic testing to employment and insurance issues, manipulating genes, and knowing the genetic makeup of potential mates are just some of the complex ethical topics that accompany genetic knowledge and its effects (Olesen 1995).

Plant and Animal Gene Alteration

Some scientists feel that the alteration of genes to modify and create new plants and animals has the potential to alter the twenty-first century even more dramatically than computers altered the twentieth century. The alteration of plant and animal lifeforms has the potential to feed the hungry, alleviate human suffering, provide new medicines and transplant organs, and open possible new sources of energy. But the moral decisions and consequences of such genetic procedures are complex and include the issues of genetic determination, discrimination, and unforeseen long-range problems due to the release of new genetically engineered life forms. As an example of such a public policy and moral issue, cell biologist Stuart A. Newman and Jeremy Rifkin, president of the Washington-based Foundation on Economic Trends, have applied for a patent on genetically engineered lifeforms that fuse human and animal embryonic cells. The purpose behind their patent application is to prevent or delay the use of such technology to better control and regulate such use and to give the public time to develop its moral policy on these issues (Travis 1998).

THE INFORMATION REVOLUTION

Artificial Intelligence

In the future, computers will be able to make thinking decisions and create new thoughts from stored information. Software will also become intelligent and will act independently to ease burdens on computers. Within just a few years, computers will build information and conclusions and automatically coordinate applications and output such as spreadsheets, databases, document preparation systems, and online search engines. Content will be automatically checked by word processors to determine completeness, accuracy, and the existence of adequate explanations. Searches for specific content will be possible not just from key word searches, but also from actual content searches (Lenat 1995).

The Internet

Information will continue to be increasingly available on the Internet and rapidly expand into all phases of personal, information, economic, and business aspects. Cyberspace will continue to transform with new "non-PC devices" such as Internet TVs and Internet telephones. Commerce, all types of information, games and entertainment, and personal buying-pattern databases will all continue to become even more available and part of our everyday lives as cyberspace continues to develop into its global potential (The Web: Infotopia or Marketplace? 1997). Additionally, we will have the mix of various types of realities of virtual

BOX 1 Telecommunications

In this decade, cellular services and wireless networks will increasingly provide personalized service all over the globe, regardless of the remoteness of the location or the previous lack of telephone services. Cellular networks and other personal communications services will use digital air-interface standards and will shrink the size of the cell station, providing more compactness, quality, and efficiency. Every year and a half, the digital chips required for wireless operations shrink by 50 percent so that applications can also continue to shrink. Smaller and smaller cellular devices, such as high-quality wireless wristphones, tiny telephone devices in computers, and radio modems are part of this outcome. In the next few years, wireless faxes and video mail will be commonplace globally. Telephone service and cell phones will be available everywhere; today, half of the people in the world have never made a telephone call, but in the future, anyone in any location will be able to do so. Developing countries will eliminate the traditional network foundation and begin with a wireless infrastructure (Zysman 1995).

reality, physical realities, and imaginary realities which can be used for training, entertainment, and research and development (University of Montreal, 2006).

Additional Leaps

As time goes by, software will become increasingly autonomous with the use of software "agents" that help complete users' goals. These agents will be like many digital proxies, simultaneously guiding users through technical complexities, teaching them, searching for information, performing transactions, or even representing people in their absence. They could even perform as personal secretaries—helping to remind, coordinate, and extend capabilities. Rather than manipulating a keyboard or a mouse, the user would use his or her voice or gestures to communicate with the software, and the agents would appear on the screen as people with simplified facial expressions to communicate their current state (Maes 1995).

Virtual reality (VR) will also be used to a greater extent in a variety of applications since it permits people to respond in a re-created environment. Presently, VR provides perceptions of sight, hearing, touch, and response to movement. This range will be extended to perceptions of force, resistance, texture, and smell as well. Applications of VR will include the performance of complex and delicate tasks in hazardous or alien environments, extensive and detailed training operations, the design of prototypes and models of all sorts of systems, and the creation of new artistic and social humanistic expressions and dimensions (Laurel 1995).

THE WORLD OF WORK

In the future, the equivalent of today's upper crust of white-collar workers will need a graduate degree, and they will be classified as some type of "symbolic analyst," possessing some of the most intensive knowledge in the future economy. At the bottom of the ladder will be low-skill and low-paid personal service workers whose jobs will not have been too affected by the computer or information explosion. Most of the changes will occur in the middle class, made up of "technicians"—the mechanics of the computer age; their positions will require education beyond the high school diploma. This middle class will require more education and training than today's middle class. The highly educated, high-earning professionals to be known as "knowledge workers" will be exposed to downsizing with the arrival of more consultant-type, "paid by the job" employment arrangements. These knowledge workers will telecommute to their various offices and contract sites. Many people feel that workers might move to a 25-hour workweek, with higher pay and benefits allowing more time for cultural activities and nonprofit work (McGinn and Raymond 1997–98).

NANOTECHNOLOGY

Many futurists feel that nanotechnology will provide man with the ability to participate in the creation of materials on the atomic level and that some of this capability will be realized in the next 10 to 20 years. Nanotechnology is technology that occurs on the microscopic scale of nanometers. (One billion nanometers are in a meter.) Atoms at this microscopic size level join together to form molecules that, in turn, form materials. In the nanotechnological universe, atoms can be positioned in various configurations to design different specified materials according to the material's specific atomic formula. The technology of MEMS—microelectromechanical systems—already exists. These systems are tiny micromachines that react to stimuli such as light, sound, and motion. When connected to microchips, MEMS can be programmed to take actions, such as those of sensing devices for human health needs, and can perform the unlimited applications of control devices. "Microprocessors defined the 1980s, and cheap lasers allowed the telecommunications revolution of the 90s. MEMS and sensors generally will shape the first decades of the next century," states Paul Saffo, an analyst with the Institute for the Future in Menlo Park, California. The process of MEMS becoming less expensive and smaller will result in such applications as smart materials that react to stimuli, houses programmed for energy needs, unlimited technological sensing devices, and tiny robots that detect disease in the bloodstream—and, of course, this is only the beginning (Rogers and Kaplan 1997–98).

THE MEDICAL FUTURE

New Surgical Techniques

In the twenty-first century, surgeons will develop more precise and less invasive techniques. Radio waves and lasers will replace scalpels, while light-activated drugs will pinpoint cancer treatments to be more specific and less harmful. Surgery involving small incisions and keyhole surgery are becoming developed procedures, using thin probes, cameras, miniaturized lights, and tiny and remotely controlled surgical tools. Even organ extraction and transplants can be accomplished with less invasive surgery requiring smaller incisions. Doctors are also discovering that directing radio waves by the use of electrodes and radio-wave generators can diminish injuries without the pain and difficulties of surgery. The possibilities of the use of this medicine include the tightening of ligaments, the destruction of tumors and other unwanted tissues, the cessation of abnormal uterine bleeding, the treatment of benign prostate enlargement, and the treatment of heart arrhythmias. Prescription medicines for the twenty-first century will also be more narrowly targeted and will have few side effects (Cowley and Underwood 1997–98).

Gene Therapy

Gene therapy will also be part of the new technological medicine bag of the twenty-first century, and it will cause another medical revolution since the use of selected corrective genes in a patient's cells can potentially cure or ease many of today's diseases. This new approach can be very effective, as so many diseases derive from the malfunctioning of one or more genes. For example, more than 4,000 conditions, such as severe combined immunodeficiency and cystic fibrosis, are caused from the developmental damage of a single gene. With gene therapy, new genes can be delivered through modified viruses to provide for new genetic coding to prevent or treat diseases. Other methods of delivery involve directly introducing the corrective gene into the tissue where it is needed. While not possible now, in the future it is hoped that physicians will simply be able to inject the new gene carriers into the bloodstream, where the new genetic carriers will locate their targets cells and simply transfer the new genetic information. Within 20 years, it is predicted that gene therapy will be part of the regular approaches to treat and cure diseases. Today, many diseases are already being treated in clinical trials of gene therapy (Anderson 1995).

OTHER NEW SCIENTIFIC TECHNOLOGIES

Microchip Sensor

A microchip sensor is an optical version of a microchip with tiny lenses and filters that use laser light. The chip is an inexpensive sensor that can be made to detect or sense just about anything. Food-processing plants are interested in using this sensor to detect potentially deadly bacteria. It also could be used to detect chemicals in medical patients, replacing their need to undergo frequent blood tests. Other applications of microchip sensors would be to detect pollutants in smokestacks, streams, and automobile exhaust systems and to detect pesticides and fertilizers in farming areas (Scientists Say Tiny Sensor Has Giant Potential 1996).

Molecular Design of Materials

Material design and manufacturing in the future will involve manipulating molecules to change their characteristics. A material could be made stronger or lighter or could be altered to make a new material by changing its molecular structure. The strengths of this approach to design and construction are its access to unlimited resources and its versatility since the number of molecules in the world is virtually unlimited (Olesen 1995).

FORECASTS FOR THE TWENTY-FIRST CENTURY

Climate change, the quantity and quality of water resources, and deforestation are the United Nation's four foremost future global environmental issues for the twenty-first century, which had polled 200 scientists in 50 countries (Schuelke 1999). Steward Brand, founder of the Whole Earth Catalog, predicts that in ten years there will be obvious changes in population growth, urbanization, genetically engineered organisms, and nuclear power. Population growth will continue to decline because of the world's increasing trend to move to cities since urbanizaton generally results in smaller-sized families. Cities will become larger, but also allow wildlife and trees to return in rural areas. Genetically modified crops will increase because they provide specific solutions for invasive species and also can be protected from disease and produce high yields. Climate change must be addressed, so we need a good alternative to fossil fuels. Nuclear redesigned plants can provide that transition while we develop another abundant type of future clean energy (Brand 2005). Some of the "Top 10 forecasts" from *The Futurist Magazine* for 2000 and beyond are the following: by 2010, biomonitoring devices that look like wristwatches will provide wearers with up-to-the-minute data about their health status; widespread infertility, increase of the aging population, and falling birthrates; farmers will become genetic engineers, growing vaccines as well as food and also sending wind power to utilities; 90 percent of the world's 6,000 languages could go extinct by 2100; water scarcity could threaten 1 billion people by 2025; and the human population will level off by 2035, while pet population will rapidly increase (*The Futurist* as quoted in IACT 2000).

Schuelke mentions that a special report of *The Futurist Magazine* gave further provoking forecasts such as nanotechnology that will soon provide such new breakthroughs as teeth that are as hard as diamonds, a baldness cure, and wrinkled skin made smooth. Other forecasts from the special report include more attacks on information systems and information infrastructures, schools being transformed by interactive multimedia, and intelligent tutoring systems (Schuelke 1999). The twenty-first century will undoubtedly very rapidly advance with its increasingly faster rates of new information, science, and combinations of these systems for even newer capabilities. The prospects of this century are among the most exciting in the history of humankind, but as far as human values, resources, ecology, and the very directions of humanity, this century remains the most challenging and determining.

REFERENCES

Anderson, W. F. (1995). Gene Therapy. *Scientific American: Key Technologies for the 21st Century* (September):pp. 124–128.

Brand, S. (2005). Environmental Heresies. *Technology Review* (May):60–63.

Chan, V. W. S. (1995). All-Optical Networks. *Scientific American: Key Technologies for the 21st Century* (September):72–76.

Cowley, G., and Underwood, A. (1997–98). Surgeon, Drop That Scalpel. *Newsweek Extra 2000: The Power of Invention* (Winter):77–78.

Houle, D. (2006). Sometimes It Is Easy to See the Future. Accessed November 15, 2006. Available http://www.evolutionshift.com/blog/category/predictions/

IACT. (2000). Inventors Association of Connecticut. Accessed August 11, 2005. Available at http://www.inventus.org/JAN.htm.

Information Today and Tomorrow. (2001). *The Looking Glass.* 19 (1), 1-2. Available http://vkv.tripod.com/itt20001/edt12ko1.htm.

Isaacson, W. (1998). Our Century and the Next One. *Time*, April 13, pp. 70–75.

Khan, A. (2006). New Horizons of Information Technology. *Issues of the 21st Century and Beyond.* Upper Saddle River, NJ; Prentice Hall.

Laurel, B. (1995). Virtual Reality. *Scientific American: Key Technologies for the 21st Century* (September):90.

Lenat, D. B. (1995). Artificial Intelligence. *Scientific American: Key Technologies for the 21st Century* (September):80–82.

Maes, P. (1995). Intelligent Software. *Scientific American: Key Technologies for the 21st Century* (September):84–86.

McGinn, D., and Raymond, J. (1997–98). Workers of the World, Get Online. *Newsweek Extra 2000: The Power of Invention* (Winter):32–33.

Olesen, D. E. (1995). The Top Technologies for the Next Ten Years. *The Futurist* (September/October):9–13.

Patterson, D.A. (1995). Microprocessors in 2020. *Scientific American: Key Technologies for the 21st Century* (September):62–67.

Rogers, A., and Kaplan, D. (1997–98). Get Ready for Nanotechnology. *Newsweek Extra 2000: The Power of Invention* (Winter):52–53.

Schuelke, R. (1999). The 21st Century: Global Forecasts. *World Economic News.* Accessed August 11, 2005. Available at http://www.sonic.net/~schuelke/GlobalForescasts (21stC).htm.

Scientists Say Tiny Sensor Has Giant Potential. (1996). *CNN Interactive* (November 20). Available at http://www.CNN.com.

Texas Industry Cluster Initiative. (2005). *TSTC Emerging Technology and the 5th World.* Accessed November 15, 2006. Available at http://www.twc.state.tx.us/new/ti_5thworld.pdf.

Reinhold, B. (2004) Who needs a CLO? Available at http://management.bet.monster.com/articles/cio.

Travis, J. (1998). Patenting the Minotaur? *Science News* (May 9):299.

University of Montreal (2004). Converging Technologies: The Next Challenge. Accessed November 15, 2006. Available at http://agora.qc.ca/colloque/gga.nsf/Conferences/Converging_technologies.

The Web: Infotopia or Marketplace? (1997). *Newsweek: Beyond 2000* (January 27):84–88.

Zysman, G. I. (1995). Wireless Networks. *Scientific American: Key Technologies for the 21st Century* (September):68–71.

QUESTIONS

1. What do you think are the most important technological discoveries mentioned in this reading for the twenty-first century? Rank the five most important ones, explain why they are important, and give the reasons for your particular ranking.
2. Of the subheadings in this reading, only one category is labeled a multi-revolution. Do you think this title is deserved? Explain. Can you think of any other technical multi-revolutions? List some possibilities.
3. Of the subheadings in this reading, only three include the word "revolution": the digital, biotechnical, and information revolutions. Why do think those classifications were made in this way? Do you agree with these classifications and subheadings? Explain why or why not. What would you change about them?
4. Can you name some important developments and technologies for the twenty-first century that were not mentioned in this chapter?

5. Examine the technologies mentioned in this reading, and describe in a brief narrative what would be the cumulative effect of all of these technologies. Give a brief description of what you think life will be like in the future due to the effects of these technologies. What are the pros and cons of this future projection?
6. Think about and describe the specific type of education that will be needed in this future life. How will this education be delivered?
7. Which new technology interests you the most? What applications do you see for it? Describe possible innovative designs or applications for that technology.
8. Research additional specific applications for the new technologies mentioned in this reading in various periodicals and articles.
9. Find and list new projected technologies for the twenty-first century that are not mentioned in this chapter.

New Horizons of Information Technology

AHMED S. KHAN

The exponential growth in telecommunications technologies (satellite communications, cellular communications, fiber optic systems, the Internet, and so forth during the past two decades has impacted people around the globe at personal and national levels. The rapid pace of technological change continues to transform cultures and societies around the globe. Analysts predict that telecommunications technologies will continue to grow at a rapid pace in the coming years (Khan 2006).

Today, fiber is the medium of choice for short- and long-haul broadband applications. Numerous new applications of fiber are proving to be the impetus for the technological developments in the field. With rapid advances in optoelectronic devices, low-loss zero-dispersion fibers, and optical amplifiers—coupled with innovations in optical computing and switching—fiber optic systems are transforming the twentieth-century electronic era of telecommunications toward an optical era of the twenty-first century (Khan 1994).

Oceanic fiber optic systems have revolutionized global telecommunications. Oceanic lightwave systems have been in service since 1988 across the Atlantic Ocean and since 1989 across the Pacific Ocean. These high-capacity digital communications systems have brought about a revolution in available system capacity and service quality compared with prior analog coaxial systems. On the drawing board are systems with capacities of 5,000+ megabits per second (Mbps). Oceanic fiber optic systems are transforming the world into a global village (see Table 1).

Approximately 60 million kilometers of fiber have been installed worldwide. The double-digit growth rate continues unabated, as nations strengthen and extend their infrastructure and as societies enter the information age (Li 1995).

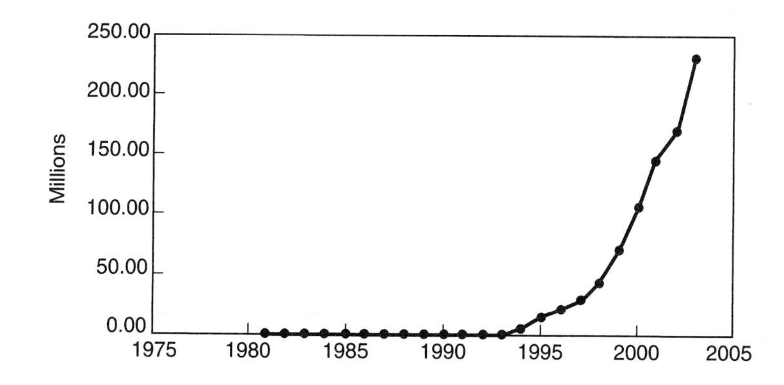

Figure 1
Internet host computers worldwide, 1981–2003.
Source: *Vital Signs 2003*, Worldwatch Institute, Washington, D.C.

Figure 2
Number of transistors/CPU.
Source: Ahmed S. Khan, *The Telecommunication Factbook and Illustrated Dictionary*, 2nd ed., New York: Thomson/Delmar Learning, 2006, p. 353.

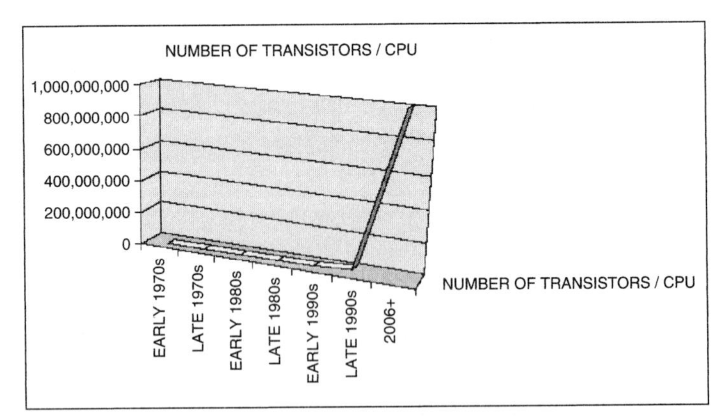

Evolution of the number of transistors on a CPU

The key developments that have revolutionized light-wave communications are the advances in rare-earth optical amplifiers that are used for data transmission. Optical amplifiers have matured into one of the most significant advances in fiber optic technology since the fiber itself. With the advent of erbium-doped fiber amplifiers (EDFA), transmission capacities have increased from hundreds to ten thousands of gigabit-kilometers per second (Desurvire 1992). Optical solitons that exploit both the nonlinearity and dispersion of the fiber medium to maintain a constant pulse shape offer a potential for high-capacity, ultra-long distance systems. Researchers at AT&T have reported achieving error-free soliton data transmission at 10 Gbps over a distance of 1,000,000+ km using a recirculating EDFA loop (Desurvire 1994). There exists a wide technoeconomic gap between the developed and the developing countries, especially in the area of telecommunication networks. In developed countries, telecommunication networks have played a key role in technoeconomic growth. In most developed countries, teledensity is close to 50 percent, and the literacy rate is over 80 percent. In contrast, a vast majority of the developing countries have very low teledensities and low literacy rates (less than 40 percent). More than three billion people living in African, Asian, and South American developing countries have little access to telecommunications services.

With the globalization of business, it has become a universally acknowledged truth that telecommunications will play a critical role in the global economy. Developed nations want a sophisticated telecom infrastructure to provide a competitive edge for their business interests. Developing nations

Figure 3
Top ten nations by number of Internet users.
Source: *Vital Signs 2002*, Worldwatch Institute, Washington, D.C.

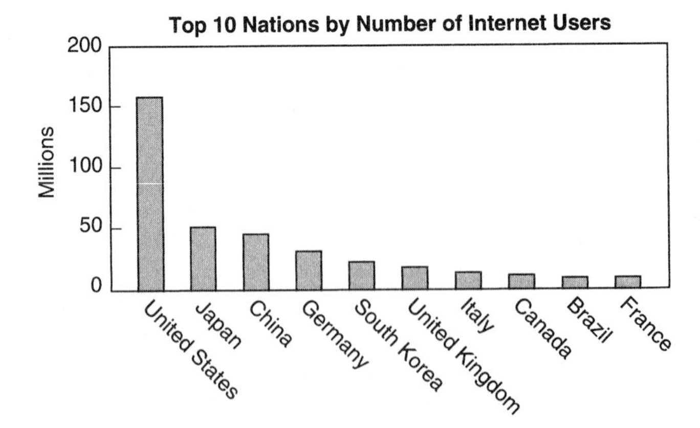

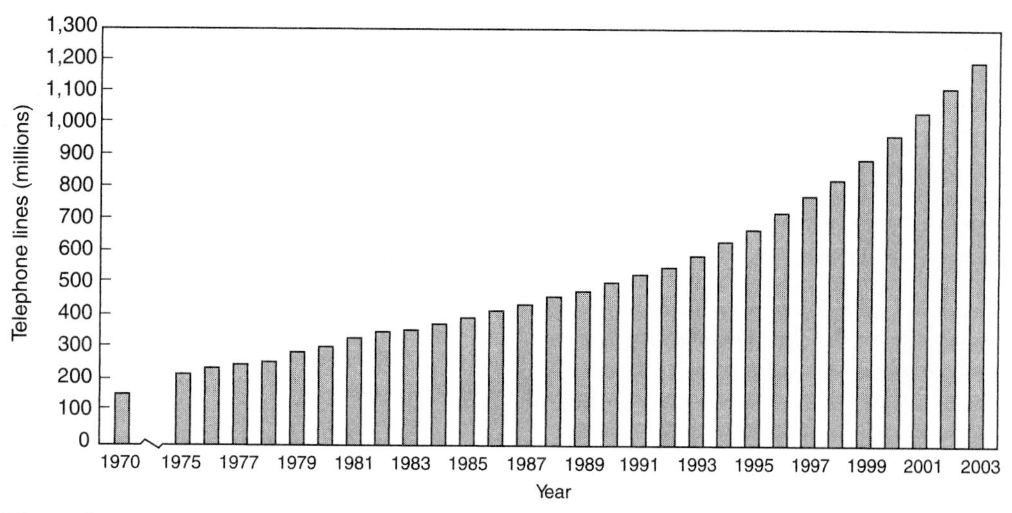

Figure 4
Telephone lines worldwide.
Source: ITU and *Vital Signs 2002*, Worldwatch Institute, Washington, D.C.

Table 1
Major Fiber Oceanic Systems

Cable	*Route*	*Bit Rate/Capacity*	*In-Service Date*
TAT-8	U.S. to U.K. and France	280 Mbps	1988
TAT-9	U.S. to U.K., Spain and France	560 Mbps	1991
PTAT-1	U.S. and to U.K. and Ireland	420 Mbps	1989
TAT-12	U.S. to U.K. and France	5 Gbps (60,000 voice circuits)	1995
TAT-13	U.S. to France	5 Gbps (60,000 voice circuits)	1996
PTAT-2	U.S. to U.K.	420 Mbps	1992
TPC-3	Honolulu to Japan and Guam	280 Mbps	1988
G-P-T	Guam and Philippines to Taiwan	280 Mbps	1989
NPC	Hong Kong to Philippines	420 Mbps	1990
TASMAN-2	New Zealand to Australia	560 Mbps	1991
TPC-4	U.S. and Canada to Japan	560 Mbps	1992
PAC RIM EAST	Honolulu to New Zealand	560 Mbps	1993
PAC RIM WEST	Australia to Guam	560 Mbps	1996
Americas-1	U.S. to the U.S. Virgin Islands, Trinidad, Venezuela, and Brazil	15,000 equivalent voice circuits	1994
Unisur	Brazil to Argentina and Paraguay	15,000 equivalent voice circuits	1994
Sea-Me-We-2	France to Singapore, with landing points in Italy, Algeria, Tunisia, Egypt, Cyprus, Turkey, Saudi Arabia, Djibouti, India, Sri Lanka, and Indonesia	560 Mbps 8000 equivalent voice channels (digital)	1994

Table 1 (*Continued*)

Cable	*Route*	*Bit Rate/Capacity*	*In-Service Date*
RIOJA	Spain with England, Belgium and the Netherlands	2.5 Gbps	1994
FLAG	U.K. to Japan with landing points in Spain, Italy, Egypt, the United Arab Emirates (UAE), India, Thailand, Malaysia, Hong Kong, and Korea	120,000 equivalent voice circuits	1996

Source: Ahmed S. Khan, *The Telecommunications Fact Book & Illustrated Dictionary RE*, 2006.

need a basic telecommunications infrastructure to provide a foundation for economic growth. According to International Telecommunications Union, an investment of $318 billion is required to add 212 million lines in developing countries. The development of a number of global, regional, and national fiber optic systems will play an important role in meeting the demands for developed as well as developing countries and will help in narrowing the technoeconomic gap between the two.

The development of global, regional, and national lightwave systems will not only satisfy the growing telecommunication needs of developing countries, but will also provide the required infrastructure for creating endless opportunities in education (e.g., distance learning), medicine (telemedicine), and business (e-commerce). The lightwave systems will provide an impetus for economic and technological development and enable developing countries to increase their teledensity and literacy rate, improve public health, and increase per-capita income, thus narrowing the technoeconomic gap with the developed countries.

BOX 1 Neurons and Photons: The Limits of Human Knowledge and the Vastness of the Universe

When the potential of fiber optic communication technology is discussed, new perceptions of accumulated human knowledge can be visualized with our new understanding of the laws of physics. Fiber optic technology has a potential data transmission capacity of hundreds of trillions of bits of information per second. We can use the holdings of the Washington, D.C., Library of Congress as an example to provide an understanding of the enormous potential information-carrying capacity of the optical fiber.

In 1997, the Library of Congress held 17,402,100 books. If we assume that each book has 300 pages with 450 words per page, this totals about 135,000 words per book or 2.35 trillion words. If we further assume that each word averages about 7 letters, the information can be digitized using ASCII code, in which each letter represents 7 bits. Therefore, all of the holdings of the Library of Congress would amount to 115.11 trillion bits of information. Using fiber optic technology, an optical fiber that has the capability of transmitting 100 trillion bits per second, the entire Library of Congress book collection could be transmitted anywhere in the world in 1.15 seconds.

If we further assume that the total accumulation of human knowledge over the past 5,000 years was represented by 1,000 Libraries of Congress, all of the accumulation of the recorded knowledge of man could be transmitted through an optical fiber in 19.18 minutes. Therefore, in less than 20 minutes, all known human knowledge could be transmitted anywhere in the world. This analogy helps to build our awareness and appreciation of the finite nature of man's knowledge in contrast to the vast dimensions of the universe and the infinite possibilities of the time-space continuum. It also builds our awareness of how limited our knowledge really is, how finite we are on the infinite time-space axis of the ever-expanding cosmos, and how great our future dimensions of knowledge are.

BOX 2 Fiber Optic Bridge Across the Digital Divide

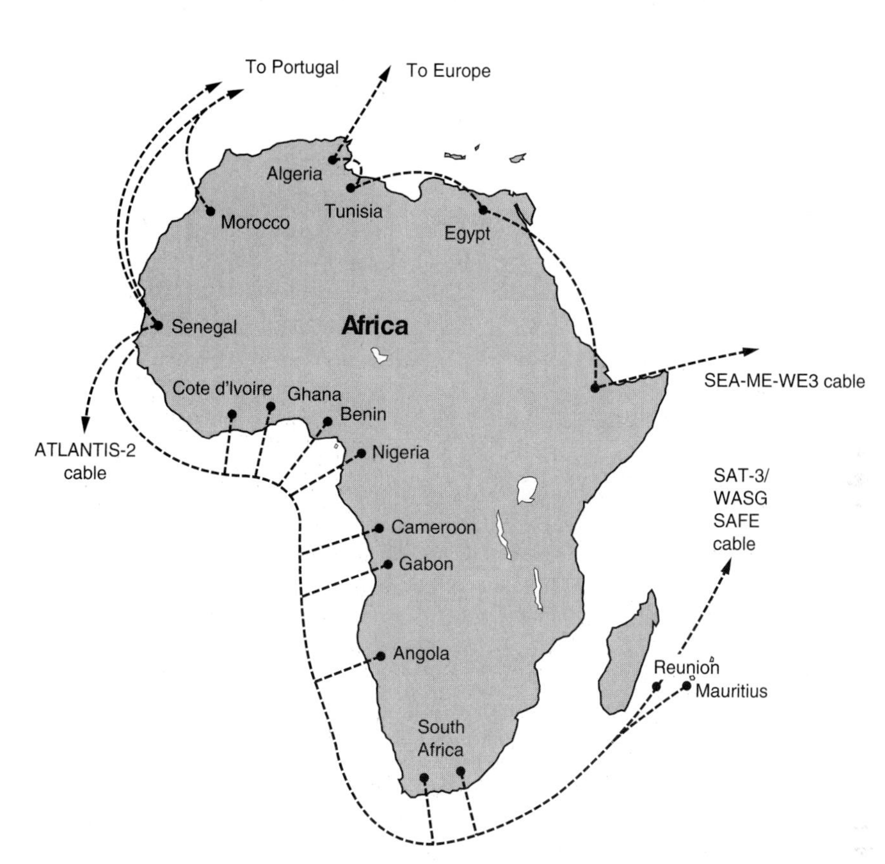

Figure 5
Africa's fiber optic digital bridge to the information age.
Source: Adapted from Harry Goldstein, Surf Africa, *IEEE Spectrum*, February 2004, pp. 48–54.

South Atlantic Telecommunications cable No. 3/West-African Submarine Cable/South Africa-Far East (SAT-3/WASC/SAFE) cable, developed at a cost of $650 million, with the data transmission speed of 120 Gbps, has enabled Africans to enter into the information age. The fiber optic cable offers a great potential for bridging the digital divide in Africa.

The SAT-3 section of fiber optic cable stretches 14,341 km from Sesimbra, Portugal, to Melkbosstrand, South Africa, passing through eight West African countries along the way. At Melkbosstrand, the SAT-3 hooks up with the 13,104-km SAFE segment, which goes through Reunion and Mauritius islands, splitting into two branches which terminate at Cochin, India, and Penang, Malaysia. A consortium of 36 companies owns Cable.

Internet traffic from Africa has increased fivefold during the past four years (2000–2004). With more than 13 percent of the world's population, Africa accounts for only 0.2 percent of the world's total international Internet capacity.

REFERENCES

Desurvire, E. (1992). Lightwave Communications: The Fifth Generation. *Scientific American* (January):114–121.

Desurvire, E. (1994). The Golden Age of Optical Fiber Amplifiers. *Physics Today* (January):20–27.

Khan, A. (2006). *Telecommunication Factbook & Illustrated Dictionary*. New York: Thompson, p. 353.

Khan, A. (1994). Fiber Optics Communications: An Introduction to Technology and Applications. *Annual Review of Communications Vol. 47*, p. 503.

Li, T. (1995). Optical Amplifiers Transform Lightwave Communications. *Photonics Spectra* (January):115.

QUESTIONS

1. Why will telecommunications and fiber optic be the information technology of the future?
2. How soon do you think we will have end-to-end optical communications systems? Do you agree with its progressive advances? How do you think it will impact society?
3. What are some of the social implications of the advancement of telecommunications and fiber optic technologies?

The Amazing Brain

AHMED S. KHAN

"The human brain is the most complex piece of machinery in the universe."
PROF. COLIN BLAKEMORE, *Oxford University*

The recent findings in the field of neuroscience are transforming the process of learning from a speculative exercise toward an exact science. Non-invasive procedures, such as magnetic resonance imaging (MRI) and positron emission tomography (PET), have provided a window on the human brain (see Figure 1). As a result, scientists can actually see a thought occurring, a fear erupting, or a long-buried memory enter into an individual's consciousness. Through these techniques, scientists can distinguish between neuronal groups that are only one millimeter apart (30,000 neurons would fit on a pin-head).

The brain has the ability to grow and change as it learns and experiences its environment. This property is called neuroplasticity. There is growing evidence that both the developing and mature brain are structurally altered when learning takes place. These structural changes are believed to encode learning in the brain. Recent brain research findings also suggest that classroom activities that incorporate motivation, a stimulating environment, and critical thinking promote development of the brain in the form of growth of new dendrites (Rose and Nichol 1997). These findings suggest that the brain is a dynamic organ, shaped to a great extent by experience (see Table 1). To become

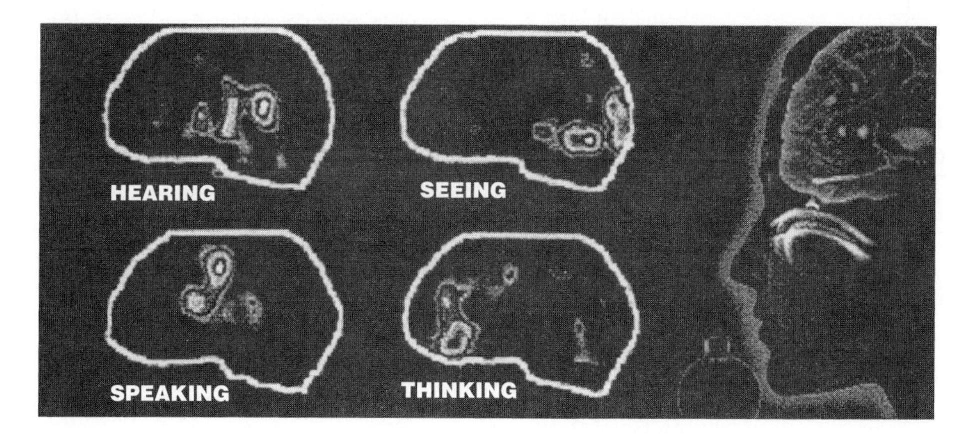

Figure 1
PET scan illustrating different tasks stimulating neural activity in distinct areas of brain.
Source: James Hilton, James and Charles Perdue, MIND MATTERS CDROM 1st Edition © 2000. Reprinted by permission of Pearson Education, Inc., Upper Saddle River, NJ.

Table 1
Brain Facts

Average Brain Weights (in grams)	
Adult human	1,300–1,400 (about 3 pounds)
Newborn human	350–400
Percent brain of total body weight (150-pound human)	2 percent
Average brain width	140 mm
Average brain length	167 mm
Average brain height	93 mm
Electrical activity	• EEG beta wave frequency = 13–30 Hz (active and alert) • EEG alpha wave frequency = 8–13 Hz (relaxation) • EEG theta wave frequency = 4–7 Hz (early stages of sleep) • EEG delta wave frequency = 0.5–4 Hz (occur during sleep)
Blood supply	• Brain utilization of total resting oxygen = 20 percent • Blood flow from heart to brain = 15–20 percent • Blood flow through whole brain (adult) = 750 ml/min • Time until unconsciousness after loss of blood supply to brain = 8–10 sec • Time until reflex loss after loss of blood supply to brain = 40–110 sec
Brain chemicals	• Brain is a giant chemical factory • Neurotransmitters (50 have been identified so far) determine human behavior, personality, and perception
Basic building blocks	**Neurons**: Cells that specialize in communication. They exchange signals with each other and link sense organs, muscles, and glands to the brain. Each neuron is composed of three major parts: **Soma**: The soma or cell body contains the nucleus and structure for keeping the cell alive. It is capable of receiving messages from nearby neurons. **Dendrite**: Dendrites are short, branching fibers that receive messages from nearby neurons. **Axon**: The axon is a single large fiber that conducts electrochemical signals to other neurons, muscles, and organs. Axons are usually only a fraction of an inch long in length, but the longest can be over three feet in humans (Hilton and Perdue 2000). Nerve impulses typically move from the dendrites to the soma and then travel down the axon. Neurons are separated by a tiny gap (a few millionths of an inch across) called a synapse. Neurotransmitters are stored and released at the end of the axon. Proteins, called prions, replicate and strengthen connections with next-door neurons. Prions can flip between two different shapes; researchers believe that they might help in laying down memories (Pearson 2003).

Table 1 (*Continued*)

Types of neurons	Neurons can be classified in three types: **Sensory neurons**: Also called efferent neurons; carry information from the senses to the central nervous systems. **Motor neurons**: Also called efferent neurons; carry signals from the central nervous system to muscles and organs. **Interneurons**: Transmit signals between neurons; most of the neurons in the body are interneurons and are found densely packed together in the brain and the spinal cord (Hilton and Perdue 2000).
Average number of neurons in the brain	100 billion
Potential number of dendrites (neural branches) for a "typical" neuron	1,000–20,000
Potential number of dendrites for 100 billion neurons	100 billion neurons × 20,000 dendrites/neuron = 2,000 trillion dendrites or neural branches
Understanding of the brain (old vs. current)	**1970s–80s**: Brain is a computer (mental model for education: memorization of facts). **1990s–2005**: Brain is a complex network like the Internet (mental model for education: guided experience for meaningful learning). **Brain vs. Computer Network (Internet) Analogy:** The IPv4 (Internet Protocol version 4) addressing scheme (uses a 32-bit unique address to identify a device connected to the Internet), developed in the 1980s, yields around 4 billion possible computer addresses. If the 32-bit address analogy is applied to identify memory entities in the 100 billion neurons with potentially 2,000 trillion dendrites, it yields almost an infinite number of memory locations in the brain.
Key unanswered question in brain (neuroscience) research	How does a human brain transform itself into a mind?

better teachers, educators must (a) understand how the brain learns: the process and the stimuli (environmental, emotional, and physical), and (b) devise teaching strategies that enhance student learning.

Educators must refrain from monotonous and passive activities and should incorporate stimuli that promote learning in students with different learning styles. Motivational activities and stimulating environments promote dendrites to grow, and passive activities apparently do not cause dendrites to grow. Educators must also make sure that they teach students using stress-free teaching strategies and environments. Recent findings suggest that stress inhibits learning because it affects the formation of new memories and hinders higher-order thinking. Over time, stress results in inhibition of neuronal growth in the hippocampus. High levels of stress hormones (cortisol) transiently block hippocampal function and, over time, can produce neuronal death (the number of neurons in the hippocampus is diminished; Rimmele 2003).

Research in neuroscience is advancing at an amazing pace, and literally hundreds of books and journal articles about the human brain have been published in the last few years. With all of this research, there still remains much to be learned. The brain is an extremely complex organ, and it remains an undiscovered universe within the human body. Researchers all over the globe are struggling to answer the key question: *How does a human brain transform itself into a mind?*

BRAIN RESEARCH AND TEACHING/LEARNING

Caine and Caine (1991) state that brain-compatible teaching is based on the following 12 principles:

1. The brain is a parallel processor.
2. Learning engages the entire physiology.
3. The search for meaning is innate.
4. The search for meaning occurs through "patterning."
5. Emotions are critical to patterning.
6. The brain processes parts and wholes.
7. Learning involves both focused attention and peripheral perception.
8. Learning always involves conscious and unconscious processes.
9. The brain has at least two different types of memory: a spatial memory system and a set of systems for rote learning.
10. Humans understand and remember best when facts and skills are embedded in natural, spatial memory.
11. Learning is enhanced by challenge and inhibited by threat.
12. Each brain is unique.

Based on these principles, authors recommend that teachers should incorporate a variety of experiential learning strategies to promote learning. Cardellichio and Field (1997) have suggested seven teaching strategies that encourage neural branching. They are:

1. Hypothetical thinking
2. Reversal
3. Application of different symbols
4. Analogy
5. Analysis of point of view
6. Completion
7. Web analysis

Cardellichio and Field (1991) state that all of these strategies are related to one another; they all provoke divergent thinking and thus extend students' neutral networks and hence enhance student learning.

To incorporate brain-based learning and teaching strategies, educators must understand how the brain learns: the process and the stimuli (environmental, emotional, and physical). The human brain has about 100 billion neurons or nerve cells, and each neuron can generate 20,000 connections or dendrites; therefore, the potential total number of interconnections or development of mind is unlimited. The capacity of the human brain is astonishing (Rose and Nichol 1997; Cardellicho and Field 1997).

Recent brain research findings suggest that classroom activities that incorporate motivation, stimulating environment, and critical thinking promote development of the brain in the form of growth of new dendrites. In this regard, educators can use tools such as multi-media, group exercises, Internet exercises, and group lab projects to enhance and support direct instruction. Educators can enhance student learning by conducting lectures in a friendly manner so that no one feels stress or is afraid to ask a question. Minimal learning can take place in a tense environment.

REFERENCES

Caine, R. N., & Caine, G. (1991). *Making Connections: Teaching and the Human Brain.* Alexandria, VA: Association for Supervision and Curriculum Development, 80–87.

Cardellicho, T. and Field, W. (1997). Seven Strategies that Encourage Neural Branching. *Educational Leadership* (March):33–36.

Hilton, J., and Perdue, C. (2000). *Mind Matters CD-ROM.* Upper Saddle River, NJ: Prentice Hall.

Pearson, H. (2003). Prion Proteins May Store Memories. *Nature* (December 3). Available at http://www.nature.com/news/2003/031229/pf/031229-2_pf.html.

Rimmele,U. (2003) A Primer on Emotions and Learning. Available at http://www.oecd.org/document/12/0,2340,fr_2649_20616271_33813516_1_1_1_1,00.html.

Rose, C., and Nichol., M. (1997). *Accelerated Learning for the 21st Century.* New York: Delacorte Press.

QUESTIONS

1. Explain the function of the following building blocks of the brain:

   a. Neurons
   b. Soma

c. Dendrite
d. Axon

2. Discuss the application of brain-compatible teaching principles for onsite and online education-delivery models.

3. Discuss the role of stress and environment on learning.

4. Compare and contrast the old and current understanding of the brain.

5. Compare the capabilities of a computer with a human brain.

Use Internet resources to answer the following question:

6. Discuss the present state of neuroscience research in the areas of:

   a. Development of nerve cells
   b. Formation of neural circuits and plasticity
   c. Behavioral neuroscience (role of neural circuits in understanding complex behavior, memories, emotions and thought)
   d. Biological nature of perception
   e. Neurogenetics
   f. Functional and molecular imaging

64 In the Year 2025

JOHN MORELLO

Joe Patterson is a sergeant in the U.S. Army's 82nd Airborne Division. The 82nd has seen its share of combat action; World War II, Vietnam, Desert Storm, and most recently Operation Iraqi Freedom, to name just a few. Ask any veteran of the 82nd to tell you what they saw on the battlefield, and the stories will probably make your jaw drop. But it's likely none of them saw anything like Joe Patterson crossing in front of their gun sights. He's a career soldier and proud of the fact that the exploits of the 82nd have made history. But reliving the past isn't what gets Patterson out of bed each day. What sends him out the door is the thought of what awaits U.S. soldiers on the battlefields of tomorrow.

Patterson, like many forward-thinking soldiers, realizes that the wars of the future won't be replays of Desert Storm or even the first phase of Iraqi Freedom. Both of those were largely conventional conflicts in which the objective was to dislodge the enemy in front of advancing American and coalition troops. To do that, the United States and its allies relied on sophisticated technology, massive airpower, armor, and artillery combined with rapid troop deployment. Patterson knows that the wars of tomorrow probably won't be fought behind some desert berm, but in the basements, apartments, and shopping malls of cities around the world. In this nonlinear, asymmetrical environment, the concern won't just be about the enemy in front of you or lurking on your flanks, but also behind you. Buildings offer cover for enemy fire. Mobility and overwhelming firepower are hampered by unfamiliar terrain. Combatants and noncombatants share proximity.

Soldiers in these future wars may not be able to use the traditional types of fire support that their predecessors had. They're going to need new training and new tools, and Joe Patterson wants to make sure it's done right. That's why you won't find him at the 82nd's headquarters in Fort Campbell, Kentucky. Instead, you'll find him at the Army's Soldier Systems Center at Natick, Massachusetts, working with scientists and engineers on a project called Future Soldier 2025. And everything about the future soldier will be different, beginning the moment he suits up for work in the morning. For starters, the helmet he'll be wearing will look nothing like the helmets of old. It will be a lightweight yet very durable motorcycle helmet that will also serve as a personal communication and targeting center. The bones of the skull will help transmit and receive communications from other soldiers as well as commanders in the rear, the information being displayed on a flip-down visor. The data will give the soldier of tomorrow a 180-degree view and 360-degree situational awareness, a necessity when the enemy could be anywhere, and frequently is. And if contact with the enemy is made, the soldier of tomorrow won't have to dangle himself like bait to locate them. The same screen that provides instant communications will double as a targeting display, picking up an enemy soldier's body heat, even if he's hidden. Once location is established, the targeting data is transferred to a five-pound weapons pod mounted on the soldier's forearm. It will carry a clip of 4.6-millimeter bullets, as well as 15-millimeter guided mini-rockets launched by micro electromechanical thrusters. All the soldier has to do is raise his arm and say "fire on target." As the bullets of missiles leave the pod, tiny built-in heat sensors lock onto their target, demonstrating in deadly fashion that in the war of the future, you can run, but you can't hide.

Just like fighting an enemy, fighting the elements will be a different ballgame for the soldier of the future. Too hot? Too cold? No need to shed or add clothing. Just flip a switch and the portable heating and cooling system does the rest. Portable enough to hang at the soldier's waist, the five-pound device will use a micro turbine for power with batteries as a back up.

The helmet, the weapons, and even the cooling system will be complemented by a uniform that will completely integrate the soldier of the future with a vast communications network. It will be a three-layered ensemble that will offer him care, comfort, and communications on a 24/7 basis. The first layer of the uniform, the one closest to his skin, will seem more like a second skin. Laced with sensors, it will monitor heart rate, body temperature, and respiration. The data will be continuously relayed to medics in the rear. That way, if a soldier is wounded, medical personnel will be able to formulate a prompt diagnosis of the extent of the injury and decide whether it can be treated on the spot or if evacuation is in order. If treatment of the wound requires immediate surgery, data relayed to the rear could assist doctors in rehearsing the operation on a hologram rendered from computed tomography images—CAT scans—of the soldier's body. But if the wound is treatable on the battlefield, the soldier's uniform can help. Medics may instruct the soldier to activate a sensor to make the uniform tighten around a wound, stopping bleeding. Or, to treat a broken leg, the uniform could stiffen into a splint, allowing the soldier to crawl off to fight another day instead of lying there, waiting to be found and finished off. Future modifications might provide this first layer of uniform to monitor a soldier's breath for toxins and release the appropriate medicine.

And, if it's true that an army travels on its stomach, the days of K-rations, C-rations, or even MREs (meals ready to eat) will not be making the trip with the soldier of the future. When he's hungry, the future soldier merely activates a food patch woven into his uniform. The patch will never be mistaken for a seven-course meal, but neither for that matter was anything else the Army ever fed its troops. However, the patch will release the necessary nutrients to keep him going while he's in the field.

The uniform's second layer is a virtual power grid, transmitting enough electricity to keep the communication and climate control systems running. The power itself will be generated by micro turbines fuelled by liquid hydrogen cartridges and backed up by battery patches sewn into the lining of the soldier's helmet. This system will supply a steady flow of two watts of power, but can surge to 20 watts on demand, and enable a soldier to remain in the field for nearly a week.

The uniform's third skin, farthest away from the soldier's own but closest to the shooting, will offer him super-hero capabilities. To begin with, the top layer will include a catalyst that will cause the uniform to change color and blend into the surroundings. However, if an enemy gets off a shot before the soldier can activate the chameleon-like features of his uniform, he can rest easy in the fact that the third layer will perform like a skin of steel, designed to withstand small-arms fire at close range. The body armor stitched into the suit surpasses even Kevlar and utilizes ballistic-resistant ceramics backed by a composite material made with carbon nano tubes. And if the uniform makes the soldier of the future feel like a real-life version of the man of steel, the external skeleton, which can fit over his uniform and allow him to be faster and stronger, will make him look like the man of steel. Amazingly, the things he'll be wearing, from helmet to shoes and everything in between, will weigh less than what today's soldiers are lugging around.

So what does all this technological innovation mean to the Joe Pattersons of the world, the ones who'll be going to war? First, each soldier will be more lethal. With more firepower at his fingertips or within the sound of his voice, fewer soldiers may actually be needed. Second, those who do go will be backed up by such a sophisticated medical infrastructure that if someone does get hurt, and someone will, their chances of survival may be better than ever before. Third, the integration of the soldier into the battle plan for the war of tomorrow will give him unimaginable power to shape a battle's outcome instead of just being a participant. In the end, all of this means that, with technology, the soldier of the future will be a walking warning to other nations not to start something they won't be able to finish.

REFERENCES

Military Analysis Network. (2003). "Land Warrior." March 31, 2003. Available at www.fas.org.

Regan, Michael. (2003). "Future Soldier to Have Massive Network." Available at www.channels.netscape.com.

QUESTIONS

1. Discuss how the innovations offered by the Future Soldier project could affect warfare in the future.
2. How will the Future Soldier's uniform protect him or her on the battlefield?
3. If one of the purposes of the Future Soldier project is to limit the size of the armed forces while reducing casualties among those still in service, will that make countries with this technology more or less likely to resort to force to secure their national interests? Discuss.

Why the Future Doesn't Need Us

BILL JOY

Our most powerful 21st-century technologies—robotics, genetic engineering, and nanotech—are threatening to make humans an endangered species.

From the moment I became involved in the creation of new technologies, their ethical dimensions have concerned me, but it was only in the autumn of 1998 that I became anxiously aware of how great are the dangers facing us in the 21st century. I can date the onset of my unease to the day I met Ray Kurzweil, the deservedly famous inventor of the first reading machine for the blind and many other amazing things.

Ray and I were both speakers at George Gilder's Telecosm conference, and I encountered him by chance in the bar of the hotel after both our sessions were over. I was sitting with John Searle, a Berkeley philosopher who studies consciousness. While we were talking, Ray approached and a conversation began, the subject of which haunts me to this day.

I had missed Ray's talk and the subsequent panel that Ray and John had been on, and they now picked right up where they'd left off, with Ray saying that the rate of improvement of technology was going to accelerate and that we were going to become robots or fuse with robots or something like that, and John countering that this couldn't happen, because the robots couldn't be conscious.

While I had heard such talk before, I had always felt sentient robots were in the realm of science fiction. But now, from someone I respected, I was hearing a strong argument that they were a near-term possibility. I was taken aback, especially given Ray's proven ability to imagine and create the future. I already knew that new technologies like genetic engineering and nanotechnology were giving us the power to remake the world, but a realistic and imminent scenario for intelligent robots surprised me.

It's easy to get jaded about such breakthroughs. We hear in the news almost every day of some kind of technological or scientific advance. Yet this was no ordinary prediction. In the hotel bar, Ray gave me a partial preprint of his then-forthcoming book *The Age of Spiritual Machines,* which outlined a utopia he foresaw—one in which humans gained near immortality by becoming one with robotic technology. On reading it, my sense of unease only intensified; I felt sure he had to be understating the dangers, understating the probability of a bad outcome along this path.

I found myself most troubled by a passage detailing a *dys*topian scenario:

THE NEW LUDDITE CHALLENGE

First let us postulate that the computer scientists succeed in developing intelligent machines that can do all things better than human beings can do them. In that case, presumably all work will be done by vast, highly organized systems of machines and no human effort will be necessary. Either of two cases might occur. The machines might be permitted to make all of their own decisions without human oversight, or else human control over the machines might be retained.

If the machines are permitted to make all their own decisions, we can't make any conjectures as to the results because it is impossible to guess how such machines might behave. We only point out that the fate of the human race

"Why the Future Doesn't Need Us" © August 4, 2000, by Bill Joy. This article originally appeared in *Wired* Magazine. Reprinted by permission of the author.

would be at the mercy of the machines. It might be argued that the human race would never be foolish enough to hand over all the power to the machines. But we are suggesting neither that the human race would voluntarily turn power over to the machines nor that the machines would willfully seize power. What we do suggest is that the human race might easily permit itself to drift into a position of such dependence on the machines that it would have no practical choice but to accept all of the machines' decisions. As society and the problems that face it become more and more complex and machines become more and more intelligent, people will let machines make more of their decisions for them simply because machine-made decisions will bring better results than man-made ones. Eventually, a stage may be reached at which the decisions necessary to keep the system running will be so complex that human beings will be incapable of making them intelligently. At that stage, the machines will be in effective control. People won't be able to just turn the machines off because they will be so dependent on them that turning them off would amount to suicide.

On the other hand, it is possible that human control over the machines may be retained. In that case, the average man may have control over certain private machines of his own, such as his car or his personal computer, but control over large systems of machines will be in the hands of a tiny elite—just as it is today, but with two differences. Due to improved techniques, the elite will have greater control over the masses; and because human work will no longer be necessary, the masses will be superfluous, a useless burden on the system. If the elite is ruthless, they may simply decide to exterminate the mass of humanity. If they are humane, they may use propaganda or other psychological or biological techniques to reduce the birth rate until the mass of humanity becomes extinct, leaving the world to the elite. Or, if the elite consists of soft-hearted liberals, they may decide to play the role of good shepherds to the rest of the human race. They will see to it that everyone's physical needs are satisfied, that all children are raised under psychologically hygienic conditions, that everyone has a wholesome hobby to keep him busy, and that anyone who may become dissatisfied undergoes "treatment" to cure his "problem." Of course, life will be so purposeless that people will have to be biologically or psychologically engineered either to remove their need for the power process or make them "sublimate" their drive for power into some harmless hobby. These engineered human beings may be happy in such a society, but they will most certainly not be free. They will have been reduced to the status of domestic animals.[1]

In the book, you don't discover until you turn the page that the author of this passage is Theodore Kaczynski—the Unabomber. I am no apologist for Kaczynski. His bombs killed three people during a 17-year terror campaign and wounded many others. One of his bombs gravely injured my friend David Gelernter, one of the most brilliant and visionary computer scientists of our time. Like many of my colleagues, I felt that I could easily have been the Unabomber's next target.

Kaczynski's actions were murderous and, in my view, criminally insane. He is clearly a Luddite, but simply saying this does not dismiss his argument; as difficult as it is for me to acknowledge, I saw some merit in the reasoning in this single passage. I felt compelled to confront it.

Kaczynski's dystopian vision describes unintended consequences, a well-known problem with the design and use of technology, and one that is clearly related to Murphy's law: "Anything that can go wrong, will." (Actually, this is Finagle's law, which in itself shows that Finagle was right.) Our overuse of antibiotics has led to what may be the biggest such problem so far: the emergence of antibiotic-resistant and much more dangerous bacteria. Similar things happened when attempts to eliminate malarial mosquitoes using DDT caused them to acquire DDT resistance; malarial parasites likewise acquired multi-drug-resistant genes.[2]

The cause of many such surprises seems clear: The systems involved are complex, involving interaction among and feedback between many parts. Any changes to such a system will cascade in ways that are difficult to predict; this is especially true when human actions are involved.

I started showing friends the Kaczynski quote from *The Age of Spiritual Machines*; I would hand them Kurzweil's book, let them read the quote, and then watch their reaction as they discovered who had written it. At around the same time, I found Hans Moravec's book, *Robot: Mere Machine to Transcendent Mind*. Moravec is one of the leaders in robotics research and was a founder of the world's largest robotics research program at Carnegie Mellon University. *Robot* gave me more material to try out on my friends—material surprisingly supportive of Kaczynski's argument. For example:

The Short Run (Early 2000s)

Biological species almost never survive encounters with superior competitors. Ten million years ago, South and North America were separated by a sunken Panama isthmus. South America, like Australia today, was populated by marsupial mammals, including pouched equivalents of rats, deers, and tigers. When the isthmus connecting North and

South America rose, it took only a few thousand years for the northern placental species, with slightly more effective metabolisms and reproductive and nervous systems, to displace and eliminate almost all of the southern marsupials.

In a completely free marketplace, superior robots would surely affect humans as North American placentals affected South American marsupials (and as humans have affected countless species). Robotic industries would compete vigorously among themselves for matter, energy, and space, incidentally driving their price beyond human reach. Unable to afford the necessities of life, biological humans would be squeezed out of existence.

There is probably some breathing room because we do not live in a completely free marketplace. Government coerces nonmarket behavior, especially by collecting taxes. Judiciously applied, governmental coercion could support human populations in high style on the fruits of robot labor, perhaps for a long while.

A textbook dystopia—and Moravec is just getting wound up. He goes on to discuss how our main job in the 21st century will be "ensuring continued cooperation from the robot industries" by passing laws decreeing that they be "nice,"[3] and to describe how seriously dangerous a human can be "once transformed into an unbounded superintelligent robot." Moravec's view is that the robots will eventually succeed us—that humans clearly face extinction.

I decided it was time to talk to my friend Danny Hillis. Danny became famous as the cofounder of Thinking Machines Corporation, which built a very powerful parallel supercomputer. Despite my current job title of Chief Scientist at Sun Microsystems, I am more a computer architect than a scientist, and I respect Danny's knowledge of the information and physical sciences more than that of any other single person I know. Danny is also a highly regarded futurist who thinks long term—four years ago, he started the Long Now Foundation, which is building a clock designed to last 10,000 years, in an attempt to draw attention to the pitifully short attention span of our society. (See "Test of Time," *Wired*, 8/2003, p. 78.)

So I flew to Los Angeles for the express purpose of having dinner with Danny and his wife, Pati. I went through my now-familiar routine, trotting out the ideas and passages that I found so disturbing. Danny's answer—directed specifically at Kurzweil's scenario of humans merging with robots—came swiftly, and quite surprised me. He said, simply, that the changes would come gradually and that we would get used to them.

But I guess I wasn't totally surprised. I had seen a quote from Danny in Kurzweil's book in which he said, "I'm as fond of my body as anyone, but if I can be 200 with a body of silicon, I'll take it." It seemed that he was at peace with this process and its attendant risks, while I was not.

While talking and thinking about Kurzweil, Kaczynski, and Moravec, I suddenly remembered a novel I had read almost 20 years ago—*The White Plague,* by Frank Herbert—in which a molecular biologist is driven insane by the senseless murder of his family. To seek revenge, he constructs and disseminates a new and highly contagious plague that kills widely but selectively. (We're lucky Kaczynski was a mathematician, not a molecular biologist.) I was also reminded of the Borg of *Star Trek,* a hive of partly biological, partly robotic creatures with a strong destructive streak. Borg-like disasters are a staple of science fiction, so why hadn't I been more concerned about such robotic dystopias earlier? Why weren't other people more concerned about these nightmarish scenarios?

Part of the answer certainly lies in our attitude toward the new—in our bias toward instant familiarity and unquestioning acceptance. Accustomed to living with almost routine scientific breakthroughs, we have yet to come to terms with the fact that the most compelling 21st-century technologies—robotics, genetic engineering, and nanotechnology—pose a different threat than the technologies that have come before. Specifically, robots, engineered organisms, and nanobots share a dangerous amplifying factor: they can self-replicate. A bomb is blown up only once—but one bot can become many and quickly get out of control.

Much of my work over the past 25 years has been on computer networking, where the sending and receiving of messages creates the opportunity for out-of-control replication. But while replication in a computer or a computer network can be a nuisance, at worst it disables a machine or takes down a network or network service. Uncontrolled self-replication in these newer technologies runs a much greater risk: a risk of substantial damage in the physical world.

Each of these technologies also offers untold promise: The vision of near immortality that Kurzweil sees in his robot dreams drives us forward; genetic engineering may soon provide treatments, if not outright cures, for most diseases; and nanotechnology and nanomedicine can address yet more ills. Together, they could significantly extend our average life span and improve the quality of our lives. Yet, with each of these technologies, a sequence of small, individually sensible advances leads to an accumulation of great power and, concomitantly, great danger.

What was different in the 20th century? Certainly, the technologies underlying the weapons of mass destruction (WMD)—nuclear, biological, and chemical (NBC)—were powerful and the weapons an enormous threat. But building nuclear weapons required, at least for a time, access to both rare—indeed, effectively unavailable—raw materials and highly protected information; biological and chemical weapons programs also tended to require large-scale activities.

The 21st-century technologies—genetics, nanotechnology, and robotics (GNR)—are so powerful that they can spawn whole new classes of accidents and abuses. Most dangerously, for the first time, these accidents and abuses are widely within the reach of individuals or small groups. They will not require large facilities or rare raw materials. Knowledge alone will enable the use of them.

Thus we have the possibility not just of weapons of mass destruction, but of knowledge-enabled mass destruction (KMD), this destructiveness hugely amplified by the power of self-replication.

I think it is no exaggeration to say we are on the cusp of the further perfection of extreme evil, an evil whose possibility spreads well beyond that which weapons of mass destruction bequeathed to the nation-states, on to a surprising and terrible empowerment of extreme individuals.

Nothing about the way I got involved with computers suggested to me that I was going to be facing these kinds of issues.

My life has been driven by a deep need to ask questions and find answers. When I was 3, I was already reading, so my father took me to the elementary school, where I sat on the principal's lap and read him a story. I started school early, later skipped a grade, and escaped into books—I was incredibly motivated to learn. I asked lots of questions, often driving adults to distraction.

As a teenager, I was very interested in science and technology. I wanted to be a ham radio operator but didn't have the money to buy the equipment. Ham radio was the Internet of its time: very addictive and quite solitary. Money issues aside, my mother put her foot down—I was not to be a ham; I was antisocial enough already.

I may not have had many close friends, but I was awash in ideas. By high school, I had discovered the great science fiction writers. I remember especially Heinlein's *Have Spacesuit Will Travel* and Asimov's *I, Robot,* with its Three Laws of Robotics. I was enchanted by the descriptions of space travel and wanted to have a telescope to look at the stars; since I had no money to buy or make one, I checked books on telescope-making out of the library and read about making them instead. I soared in my imagination.

Thursday nights my parents went bowling, and we kids stayed home alone. It was the night of Gene Roddenberry's original *Star Trek,* and the program made a big impression on me. I came to accept its notion that humans had a future in space, Western-style, with big heroes and adventures. Roddenberry's vision of the centuries to come was one with strong moral values, embodied in codes like the Prime Directive: to not interfere in the development of less technologically advanced civilizations. This had an incredible appeal to me; ethical humans, not robots, dominated this future, and I took Roddenberry's dream as part of my own.

I excelled in mathematics in high school, and when I went to the University of Michigan as an undergraduate engineering student, I took the advanced curriculum of the mathematics majors. Solving math problems was an exciting challenge, but when I discovered computers, I found something much more interesting: a machine into which you could put a program that attempted to solve a problem, after which the machine quickly checked the solution. The computer had a clear notion of correct and incorrect, true and false. Were my ideas correct? The machine could tell me. This was very seductive.

I was lucky enough to get a job programming early supercomputers and discovered the amazing power of large machines to numerically simulate advanced designs. When I went to graduate school at UC Berkeley in the mid-1970s, I started staying up late, often all night, inventing new worlds inside the machines. Solving problems. Writing the code that argued so strongly to be written.

In *The Agony and the Ecstasy,* Irving Stone's biographical novel of Michelangelo, Stone described vividly how Michelangelo released the statues from the stone, "breaking the marble spell," carving from the images in his mind.[4] In my most ecstatic moments, the software in the computer emerged in the same way. Once I had imagined it in my mind, I felt that it was already there in the machine, waiting to be released. Staying up all night seemed a small price to pay to free it—to give the ideas concrete form.

After a few years at Berkeley, I started to send out some of the software I had written—an instructional Pascal system, Unix utilities, and a text editor called vi (which is still, to my surprise, widely used more than 20 years later)—to others who had similar small PDP-11 and VAX minicomputers. These adventures in software eventually turned into the Berkeley version of the Unix operating system, which became a personal "success disaster"—so many people wanted it that I never finished my PhD. Instead, I got a job working for Darpa putting Berkeley Unix on the Internet and fixing it to be reliable and to run large research applications well. This was all great fun and very rewarding. And, frankly, I saw no robots here or anywhere near.

Still, by the early 1980s, I was drowning. The Unix releases were very successful, and my little project of one soon had money and some staff, but the problem at Berkeley was always office space rather than money—there wasn't room for the help the project needed, so when the other founders of Sun Microsystems showed up, I jumped at the chance to join them. At Sun, the long hours continued into the early days of workstations and personal computers, and I have enjoyed participating in the creation of advanced microprocessor technologies and Internet technologies, such as Java and Jini.

From all this, I trust it is clear that I am not a Luddite. I have always, rather, had a strong belief in the value of the scientific search for truth and in the ability of great engineering to bring material progress. The Industrial Revolution has immeasurably improved everyone's life over the last couple hundred years, and I always expected my career to involve the building of worthwhile solutions to real problems, one problem at a time.

I have not been disappointed. My work has had more impact than I had ever hoped for and has been more widely used than I could have reasonably expected. I have spent the last 20 years still trying to figure out how to make computers as reliable as I want them to be (they are not nearly there yet) and how to make them simple to use (a goal that has met with even less relative success). Despite some progress, the problems that remain seem even more daunting.

But while I was aware of the moral dilemmas surrounding technology's consequences in fields like weapons research, I did not expect that I would confront such issues in my own field, or at least not so soon.

Perhaps it is always hard to see the bigger impact while you are in the vortex of a change. Failing to understand the consequences of our inventions while we are in the rapture of discovery and innovation seems to be a common fault of scientists and technologists; we have long been driven by the overarching desire to know that is the nature of science's quest, not stopping to notice that the progress to newer and more powerful technologies can take on a life of its own.

I have long realized that the big advances in information technology come not from the work of computer scientists, computer architects, or electrical engineers, but from that of physical scientists. The physicists Stephen Wolfram and Brosl Hasslacher introduced me, in the early 1980s, to chaos theory and nonlinear systems. In the 1990s, I learned about complex systems from conversations with Danny Hillis, the biologist Stuart Kauffman, the Nobel-laureate physicist Murray Gell-Mann, and others. Most recently, Hasslacher and the electrical engineer and device physicist Mark Reed have been giving me insight into the incredible possibilities of molecular electronics.

In my own work, as codesigner of three microprocessor architectures—SPARC, picoJava, and MAJC—and as the designer of several implementations thereof, I've been afforded a deep and firsthand acquaintance with Moore's law. For decades, Moore's law has correctly predicted the exponential rate of improvement of semiconductor technology. Until last year, I believed that the rate of advances predicted by Moore's law might continue only until roughly 2010, when some physical limits would begin to be reached. It was not obvious to me that a new technology would arrive in time to keep performance advancing smoothly.

But because of the recent rapid and radical progress in molecular electronics—where individual atoms and molecules replace lithographically drawn transistors—and related nanoscale technologies, we should be able to meet or exceed the Moore's law rate of progress for another 30 years. By 2030, we are likely to be able to build machines, in quantity, a million times as powerful as the personal computers of today—sufficient to implement the dreams of Kurzweil and Moravec.

As this enormous computing power is combined with the manipulative advances of the physical sciences and the new, deep understandings in genetics, enormous transformative power is being unleashed. These combinations open up the opportunity to completely redesign the world, for better or worse: the replicating and evolving processes that have been confined to the natural world are about to become realms of human endeavor.

In designing software and microprocessors, I have never had the feeling that I was designing an intelligent machine. The software and hardware is so fragile and the capabilities of the machine to "think" so clearly absent that, even as a possibility, this has always seemed very far in the future.

But now, with the prospect of human-level computing power in about 30 years, a new idea suggests itself: that I may be working to create tools which will enable the construction of the technology that may replace our species. How do I feel about this? Very uncomfortable. Having struggled my entire career to build reliable software systems, it seems to me more than likely that this future will not work out as well as some people may imagine. My personal experience suggests we tend to overestimate our design abilities.

Given the incredible power of these new technologies, shouldn't we be asking how we can best coexist with them? And if our own extinction is a likely, or even possible, outcome of our technological development, shouldn't we proceed with great caution?

The dream of robotics is, first, that intelligent machines can do our work for us, allowing us lives of leisure, restoring us to Eden. Yet in his history of such ideas, *Darwin Among the Machines,* George Dyson warns: "In the game of life and evolution there are three players at the table: human beings, nature, and machines. I am firmly on the side of nature. But nature, I suspect, is on the side of the machines." As we have seen, Moravec agrees, believing we may well not survive the encounter with the superior robot species.

How soon could such an intelligent robot be built? The coming advances in computing power seem to make it possible by 2030. And once an intelligent robot exists, it is only a small step to a robot species—to an intelligent robot that can make evolved copies of itself.

A second dream of robotics is that we will gradually replace ourselves with our robotic technology, achieving near immortality by downloading our consciousnesses; it is this process that Danny Hillis thinks we will gradually get used to and that Ray Kurzweil elegantly details in *The Age of Spiritual Machines.* (We are beginning to see intimations of this in the implantation of computer devices into the human body, as illustrated on the cover of *Wired* 8.02.)

But if we are downloaded into our technology, what are the chances that we will thereafter be ourselves or even human? It seems to me far more likely that a robotic existence would not be like a human one in any sense that we understand, that the robots would in no sense be our children that on this path our humanity may well be lost.

Genetic engineering promises to revolutionize agriculture by increasing crop yields while reducing the use of pesticides; to create tens of thousands of novel species of bacteria, plants, viruses, and animals; to replace reproduction, or supplement it, with cloning; to create cures for many diseases, increasing our life span and our quality of life; and much, much more. We now know with certainty that these profound changes in the biological sciences are imminent and will challenge all our notions of what life is.

Technologies such as human cloning have in particular raised our awareness of the profound ethical and moral issues we face. If, for example, we were to reengineer ourselves into several separate and unequal species using the power of genetic engineering, then we would threaten the notion of equality that is the very cornerstone of our democracy.

Given the incredible power of genetic engineering, it's no surprise that there are significant safety issues in its use. My friend Amory Lovins recently cowrote, along with Hunter Lovins, an editorial that provides an ecological view of some of these dangers. Among their concerns: that "the new botany aligns the development of plants with their economic, not evolutionary, success." (See "A Tale of Two Botanies," page 247.) Amory's long career has been focused on energy and resource efficiency by taking a whole-system view of human-made systems; such a whole-system view often finds simple, smart solutions to otherwise seemingly difficult problems, and is usefully applied here as well.

After reading the Lovins' editorial, I saw an op-ed by Gregg Easterbrook in *The New York Times* (November 19, 1999) about genetically engineered crops, under the headline: "Food for the Future: Someday, rice will have built-in vitamin A. Unless the Luddites win."

Are Amory and Hunter Lovins Luddites? Certainly not. I believe we all would agree that golden rice, with its built-in vitamin A, is probably a good thing, if developed with proper care and respect for the likely dangers in moving genes across species boundaries.

Awareness of the dangers inherent in genetic engineering is beginning to grow, as reflected in the Lovins' editorial. The general public is aware of, and uneasy about, genetically modified foods, and seems to be rejecting the notion that such foods should be permitted to be unlabeled.

But genetic engineering technology is already very far along. As the Lovins note, the USDA has already approved about 50 genetically engineered crops for unlimited release; more than half of the world's soybeans and a third of its corn now contain genes spliced in from other forms of life.

While there are many important issues here, my own major concern with genetic engineering is narrower: that it gives the power—whether militarily, accidentally, or in a deliberate terrorist act—to create a White Plague.

The many wonders of nanotechnology were first imagined by the Nobel-laureate physicist Richard Feynman in a speech he gave in 1959, subsequently published under the title "There's Plenty of Room at the Bottom." The book that made a big impression on me, in the mid-'80s, was Eric Drexler's *Engines of Creation,* in which he described beautifully how manipulation of matter at the atomic level could create a utopian future of abundance, where just about everything could be made cheaply, and almost any imaginable disease or physical problem could be solved using nanotechnology and artificial intelligences.

A subsequent book, *Unbounding the Future: The Nanotechnology Revolution,* which Drexler cowrote, imagines some of the changes that might take place in a world where we had molecular-level "assemblers." Assemblers could make possible incredibly low-cost solar power, cures for cancer and the common cold by augmentation of the human immune system, essentially complete cleanup of the environment, incredibly inexpensive pocket supercomputers—in fact, any product would be manufacturable by assemblers at a cost no greater than that of wood—spaceflight more accessible than transoceanic travel today, and restoration of extinct species.

I remember feeling good about nanotechnology after reading *Engines of Creation.* As a technologist, it gave me a sense of calm—that is, nanotechnology showed us that incredible progress was possible, and indeed perhaps inevitable. If nanotechnology was our future, then I didn't feel pressed to solve so many problems in the present. I would get to Drexler's utopian future in due time; I might as well enjoy life more in the here and now. It didn't make sense, given his vision, to stay up all night, all the time.

Drexler's vision also led to a lot of good fun. I would occasionally get to describe the wonders of nanotechnology to others who had not heard of it. After teasing them with all the things Drexler described I would give a homework assignment of my own: "Use nanotechnology to create a vampire; for extra credit create an antidote."

With these wonders came clear dangers, of which I was acutely aware. As I said at a nanotechnology conference in 1989, "We can't simply do our science and not worry about these ethical issues."[5] But my subsequent conversations with physicists convinced me that nanotechnology might not even work—or, at least, it wouldn't work anytime soon. Shortly thereafter I moved to Colorado, to a skunk works I had set up, and the focus of my work shifted to software for the Internet, specifically on ideas that became Java and Jini.

Then, last summer, Brosl Hasslacher told me that nanoscale molecular electronics was now practical. This was *new* news, at least to me, and I think to many people—and it radically changed my opinion about nanotechnology. It sent me back to *Engines of Creation.* Rereading Drexler's work after more than 10 years, I was dismayed to realize how little I had remembered of its lengthy section called "Dangers and Hopes," including a discussion of how nanotechnologies can become "engines of destruction." Indeed, in my rereading of this cautionary material today, I am struck by how naive some of Drexler's safeguard proposals seem, and how much greater I judge the dangers to be now than even he seemed to then. (Having anticipated and described many technical and political problems with nanotechnology, Drexler started the Foresight Institute in the late 1980s "to help prepare society for anticipated advanced technologies"—most important, nanotechnology.)

The enabling breakthrough to assemblers seems quite likely within the next 20 years. Molecular electronics—the new subfield of nanotechnology where individual molecules are circuit elements—should mature quickly and become enormously lucrative within this decade, causing a large incremental investment in all nanotechnologies.

Unfortunately, as with nuclear technology, it is far easier to create destructive uses for nanotechnology than constructive ones. Nanotechnology has clear military and terrorist uses, and you need not be suicidal to release a massively destructive nanotechnological device—such devices can be built to be selectively destructive, affecting, for example, only a certain geographical area or a group of people who are genetically distinct.

An immediate consequence of the Faustian bargain in obtaining the great power of nanotechnology is that we run a grave risk—the risk that we might destroy the biosphere on which all life depends.

As Drexler explained:

"Plants" with "leaves" no more efficient than today's solar cells could out-compete real plants, crowding the biosphere with an inedible foliage. Tough omnivorous "bacteria" could out-compete real bacteria: They could spread like blowing pollen, replicate swiftly, and reduce the biosphere to dust in a matter of days. Dangerous replicators could easily be too tough, small, and rapidly spreading to stop—at least if we make no preparation. We have trouble enough controlling viruses and fruit flies.

Among the cognoscenti of nanotechnology, this threat has become known as the "gray goo problem." Though masses of uncontrolled replicators need not be gray or gooey, the term "gray goo" emphasizes that replicators able to obliterate life might be less inspiring than a single species of crabgrass. They might be superior in an evolutionary sense, but this need not make them valuable.

The gray goo threat makes one thing perfectly clear: We cannot afford certain kinds of accidents with replicating assemblers.

Gray goo would surely be a depressing ending to our human adventure on Earth, far worse than mere fire or ice, and one that could stem from a simple laboratory accident.[6] Oops.

It is most of all the power of destructive self-replication in genetics, nanotechnology, and robotics (GNR) that should give us pause. Self-replication is the modus operandi of genetic engineering, which uses the machinery of the cell to replicate its designs, and the prime danger underlying gray goo in nanotechnology. Stories of run-amok robots like the Borg, replicating or mutating to escape from the ethical constraints imposed on them by their creators, are well established in our science fiction books and movies. It is even possible that self-replication may be more fundamental than we thought, and hence harder—or even impossible—to control. A recent article by Stuart Kauffman in *Nature* titled "Self-Replication: Even Peptides Do It" discusses the discovery that a 32-amino-acid peptide can "autocatalyse its own synthesis." We don't know how widespread this ability is, but Kauffman notes that it may hint at "a route to self-reproducing molecular systems on a basis far wider than Watson-Crick base-pairing."[7]

In truth, we have had in hand for years clear warnings of the dangers inherent in widespread knowledge of GNR technologies—of the possibility of knowledge alone enabling mass destruction. But these warnings haven't been widely publicized; the public discussions have been clearly inadequate. There is no profit in publicizing the dangers.

The nuclear, biological, and chemical (NBC) technologies used in 20th-century weapons of mass destruction were and are largely military, developed in government laboratories. In sharp contrast, the 21st-century GNR technologies have clear commercial uses and are being developed almost exclusively by corporate enterprises. In this age of triumphant commercialism, technology—with science as its handmaiden—is delivering a series of almost magical inventions that are the most phenomenally lucrative ever seen. We are aggressively pursuing the promises of these new technologies within the now-unchallenged system of global capitalism and its manifold financial incentives and competitive pressures.

This is the first moment in the history of our planet when any species, by its own voluntary actions, has become a danger to itself—as well as to vast numbers of others.

It might be a familiar progression, transpiring on many worlds—a planet, newly formed, placidly revolves around its star; life slowly forms; a kaleidoscopic procession of creatures evolves; intelligence emerges which, at least up to a point, confers enormous survival value; and then technology is invented. It dawns on them that there are such things as laws of Nature, that these laws can be revealed by experiment, and that knowledge of these laws can be made both to save and to take lives, both on unprecedented scales. Science, they recognize, grants immense powers. In a flash, they create world-altering contrivances. Some planetary civilizations see their way through, place limits on what may and what must not be done, and safely pass through the time of perils. Others, not so lucky or so prudent, perish.

That is Carl Sagan, writing in 1994, in *Pale Blue Dot,* a book describing his vision of the human future in space. I am only now realizing how deep his insight was, and how sorely I miss, and will miss, his voice. For all its eloquence, Sagan's contribution was not least that of simple common sense—an attribute that, along with humility, many of the leading advocates of the 21st-century technologies seem to lack.

I remember from my childhood that my grandmother was strongly against the overuse of antibiotics. She had worked since before the first World War as a nurse and had a commonsense attitude that taking antibiotics, unless they were absolutely necessary, was bad for you.

It is not that she was an enemy of progress. She saw much progress in an almost 70-year nursing career; my grandfather, a diabetic, benefited greatly from the improved treatments that became available in his lifetime. But she, like many levelheaded people, would probably think it greatly arrogant for us, now, to be designing a robotic "replacement species," when we obviously have so much trouble making relatively simple things work, and so much trouble managing—or even understanding—ourselves.

I realize now that she had an awareness of the nature of the order of life, and of the necessity of living with and respecting that order. With this respect comes a necessary humility that we, with our early-21st-century chutzpah, lack at our peril. The commonsense view, grounded in this respect, is often right, in advance of the scientific evidence. The clear fragility and inefficiencies of the human-made systems we have built should give us all pause; the fragility of the systems I have worked on certainly humbles me.

We should have learned a lesson from the making of the first atomic bomb and the resulting arms race. We didn't do well then, and the parallels to our current situation are troubling.

The effort to build the first atomic bomb was led by the brilliant physicist J. Robert Oppenheimer. Oppenheimer was not naturally interested in politics but became painfully aware of what he perceived as the grave threat to Western civilization from the Third Reich, a threat surely grave because of the possibility that Hitler might obtain nuclear weapons. Energized by this concern, he brought his strong intellect, passion for physics, and charismatic leadership skills to Los Alamos and led a rapid and successful effort by an incredible collection of great minds to quickly invent the bomb.

What is striking is how this effort continued so naturally after the initial impetus was removed. In a meeting shortly after V-E Day with some physicists who felt that perhaps the effort should stop, Oppenheimer argued to continue. His stated reason seems a bit strange: not because of the fear of large casualties from an invasion of Japan, but because the United Nations, which was soon to be formed, should have foreknowledge of atomic weapons. A more likely reason the project continued is the momentum that had built up—the first atomic test, Trinity, was nearly at hand.

We know that in preparing this first atomic test the physicists proceeded despite a large number of possible dangers. They were initially worried, based on a calculation by Edward Teller, that an atomic explosion might set fire to the atmosphere. A revised calculation reduced the danger of destroying the world to a three-in-a-million chance. (Teller says he was later able to dismiss the prospect of atmospheric ignition entirely.) Oppenheimer, though, was sufficiently concerned about the result of Trinity that he arranged for a possible evacuation of the southwest part of the state of New Mexico. And, of course, there was the clear danger of starting a nuclear arms race.

Within a month of that first, successful test, two atomic bombs destroyed Hiroshima and Nagasaki. Some scientists had suggested that the bomb simply be demonstrated, rather than dropped on Japanese cities—saying that this would greatly improve the chances for arms control after the war—but to no avail. With the tragedy of Pearl Harbor still fresh in Americans' minds, it would have been very difficult for President Truman to order a demonstration of the weapons rather than use them as he did—the desire to quickly end the war and save the lives that would have been lost in any invasion of Japan was very strong. Yet the overriding truth was probably very simple: As the physicist Freeman Dyson later said, "The reason that it was dropped was just that nobody had the courage or the foresight to say no."

It's important to realize how shocked the physicists were in the aftermath of the bombing of Hiroshima, on August 6, 1945. They describe a series of waves of emotion: first, a sense of fulfillment that the bomb worked, then horror at all the people that had been killed, and then a convincing feeling that on no account should another bomb be dropped. Yet of course another bomb was dropped, on Nagasaki, only three days after the bombing of Hiroshima.

In November 1945, three months after the atomic bombings, Oppenheimer stood firmly behind the scientific attitude, saying, "It is not possible to be a scientist unless you believe that the knowledge of the world, and the power which this gives, is a thing which is of intrinsic value to humanity, and that you are using it to help in the spread of knowledge and are willing to take the consequences."

Oppenheimer went on to work, with others, on the Acheson-Lilienthal report, which, as Richard Rhodes says in his recent book *Visions of Technology,* "found a way to prevent a clandestine nuclear arms race without resorting to armed world government"; their suggestion was a form of relinquishment of nuclear weapons work by nation-states to an international agency.

This proposal led to the Baruch Plan, which was submitted to the United Nations in June 1946 but never adopted (perhaps because, as Rhodes suggests, Bernard Baruch had "insisted on burdening the plan with conventional sanctions," thereby inevitably dooming it, even though it would "almost certainly have been rejected by Stalinist Russia anyway"). Other efforts to promote sensible steps toward internationalizing nuclear power to prevent an arms race ran afoul either of US politics and internal distrust, or distrust by the Soviets. The opportunity to avoid the arms race was lost, and very quickly.

Two years later, in 1948, Oppenheimer seemed to have reached another stage in his thinking, saying, "In some sort of crude sense which no vulgarity, no humor, no overstatement can quite extinguish, the physicists have known sin; and this is a knowledge they cannot lose."

In 1949, the Soviets exploded an atom bomb. By 1955, both the US and the Soviet Union had tested hydrogen bombs suitable for delivery by aircraft. And so the nuclear arms race began.

Nearly 20 years ago, in the documentary *The Day After Trinity*, Freeman Dyson summarized the scientific attitudes that brought us to the nuclear precipice:

"I have felt it myself. The glitter of nuclear weapons. It is irresistible if you come to them as a scientist. To feel it's there in your hands, to release this energy that fuels the stars, to let it do your bidding. To perform these miracles, to lift a million tons of rock into the sky. It is something that gives people an illusion of illimitable power, and it is, in some ways, responsible for all our troubles—this, what you might call technical arrogance, that overcomes people when they see what they can do with their minds."[8]

Now, as then, we are creators of new technologies and stars of the imagined future, driven—this time by great financial rewards and global competition—despite the clear dangers, hardly evaluating what it may be like to try to live in a world that is the realistic outcome of what we are creating and imagining.

In 1947, *The Bulletin of the Atomic Scientists* began putting a Doomsday Clock on its cover. For more than 50 years, it has shown an estimate of the relative nuclear danger we have faced, reflecting the changing international conditions. The hands on the clock have moved 15 times and today, standing at nine minutes to midnight, reflect continuing and real danger from nuclear weapons. The recent addition of India and Pakistan to the list of nuclear powers has increased the threat of failure of the nonproliferation goal, and this danger was reflected by moving the hands closer to midnight in 1998.

In our time, how much danger do we face, not just from nuclear weapons, but from all of these technologies? How high are the extinction risks?

The philosopher John Leslie has studied this question and concluded that the risk of human extinction is at least 30 percent,[9] while Ray Kurzweil believes we have "a better than even chance of making it through," with the caveat that he has "always been accused of being an optimist." Not only are these estimates not encouraging, but they do not include the probability of many horrid outcomes that lie short of extinction.

Faced with such assessments, some serious people are already suggesting that we simply move beyond Earth as quickly as possible. We would colonize the galaxy using von Neumann probes, which hop from star system to star system, replicating as they go. This step will almost certainly be necessary 5 billion years from now (or sooner if our solar system is disastrously impacted by the impending collision of our galaxy with the Andromeda galaxy within the next 3 billion years), but if we take Kurzweil and Moravec at their word it might be necessary by the middle of this century.

What are the moral implications here? If we must move beyond Earth this quickly in order for the species to survive, who accepts the responsibility for the fate of those (most of us, after all) who are left behind? And even if we scatter to the stars, isn't it likely that we may take our problems with us or find, later, that they have followed us? The fate of our species on Earth and our fate in the galaxy seem inextricably linked.

Another idea is to erect a series of shields to defend against each of the dangerous technologies. The Strategic Defense Initiative, proposed by the Reagan administration, was an attempt to design such a shield against the threat of a nuclear attack from the Soviet Union. But as Arthur C. Clarke, who was privy to discussions about the project, observed: "Though it might be possible, at vast expense, to construct local defense systems that would 'only' let through a few percent of ballistic missiles, the much touted idea of a national umbrella was nonsense. Luis Alvarez, perhaps the greatest experimental physicist of this century, remarked to me that the advocates of such schemes were 'very bright guys with no common sense.'"

Clarke continued: "Looking into my often cloudy crystal ball, I suspect that a total defense might indeed be possible in a century or so. But the technology involved would produce, as a by-product, weapons so terrible that no one would bother with anything as primitive as ballistic missiles."[10]

In *Engines of Creation,* Eric Drexler proposed that we build an active nanotechnological shield—a form of immune system for the biosphere—to defend against dangerous replicators of all kinds that might escape from laboratories or otherwise be maliciously created. But the shield he proposed would itself be extremely dangerous—nothing could prevent it from developing autoimmune problems and attacking the biosphere itself.[11]

Similar difficulties apply to the construction of shields against robotics and genetic engineering. These technologies are too powerful to be shielded against in the time frame of interest; even if it were possible to implement defensive shields, the side effects of their development would be at least as dangerous as the technologies we are trying to protect against.

These possibilities are all thus either undesirable or unachievable or both. The only realistic alternative I see is relinquishment: to limit development of the technologies that are too dangerous, by limiting our pursuit of certain kinds of knowledge.

Yes, I know, knowledge is good, as is the search for new truths. We have been seeking knowledge since ancient times. Aristotle opened his Metaphysics with the simple statement: "All men by nature desire to know." We have, as a bedrock value in our society, long agreed on the value of open access to information, and recognize the problems that arise with attempts to restrict access to and development of knowledge. In recent times, we have come to revere scientific knowledge.

But despite the strong historical precedents, if open access to and unlimited development of knowledge henceforth puts us all in clear danger of extinction, then common sense demands that we reexamine even these basic, long-held beliefs.

It was Nietzsche who warned us, at the end of the 19th century, not only that God is dead but that "faith in science, which after all exists undeniably, cannot owe its origin to a calculus of utility; it must have originated *in spite of* the fact that the disutility and dangerousness of the 'will to truth,' of 'truth at any price' is proved to it constantly." It is this further danger that we now fully face—the consequences of our truth-seeking. The truth that science seeks can certainly be considered a dangerous substitute for God if it is likely to lead to our extinction.

If we could agree, as a species, what we wanted, where we were headed, and why, then we would make our future much less dangerous—then we might understand what we can and should relinquish. Otherwise, we can easily imagine an arms race developing over GNR technologies, as it did with the NBC technologies in the 20th century. This is perhaps the greatest risk, for once such a race begins, it's very hard to end it. This time—unlike during the Manhattan Project—we aren't in a war, facing an implacable enemy that is threatening our civilization; we are driven, instead, by our habits, our desires, our economic system, and our competitive need to know.

I believe that we all wish our course could be determined by our collective values, ethics, and morals. If we had gained more collective wisdom over the past few thousand years, then a dialogue to this end would be more practical, and the incredible powers we are about to unleash would not be nearly so troubling.

One would think we might be driven to such a dialogue by our instinct for self-preservation. Individuals clearly have this desire, yet as a species our behavior seems to be not in our favor. In dealing with the nuclear threat, we often spoke dishonestly to ourselves and to each other, thereby greatly increasing the risks. Whether this was politically motivated, or because we chose not to think ahead, or because when faced with such grave threats we acted irrationally out of fear, I do not know, but it does not bode well.

The new Pandora's boxes of genetics, nanotechnology, and robotics are almost open, yet we seem hardly to have noticed. Ideas can't be put back in a box; unlike uranium or plutonium, they don't need to be mined and refined, and they can be freely copied. Once they are out, they are out. Churchill remarked, in a famous left-handed compliment, that the American people and their leaders "invariably do the right thing, after they have examined every other alternative." In this case, however, we must act more presciently, as to do the right thing only at last may be to lose the chance to do it at all.

As Thoreau said, "We do not ride on the railroad; it rides upon us"; and this is what we must fight, in our time. The question is, indeed, Which is to be master? Will we survive our technologies?

We are being propelled into this new century with no plan, no control, no brakes. Have we already gone too far down the path to alter course? I don't believe so, but we aren't trying yet, and the last chance to assert control—the fail-safe point—is rapidly approaching. We have our first pet robots, as well as commercially available genetic engineering techniques, and our nanoscale techniques are advancing rapidly. While the development of these technologies proceeds through a number of steps, it isn't necessarily the case—as happened in the Manhattan Project and the Trinity test—that the last step in proving a technology is large and hard. The breakthrough to wild self-replication in robotics, genetic engineering, or nanotechnology could come suddenly, reprising the surprise we felt when we learned of the cloning of a mammal.

And yet I believe we do have a strong and solid basis for hope. Our attempts to deal with weapons of mass destruction in the last century provide a shining example of relinquishment for us to consider: the unilateral US abandonment, without preconditions, of the development of biological weapons. This relinquishment stemmed from the realization that while it would take an enormous effort to create these terrible weapons, they could from then on easily be duplicated and fall into the hands of rogue nations or terrorist groups.

The clear conclusion was that we would create additional threats to ourselves by pursuing these weapons, and that we would be more secure if we did not pursue them. We have embodied our relinquishment of biological and chemical weapons in the 1972 Biological Weapons Convention (BWC) and the 1993 Chemical Weapons Convention (CWC).[12]

As for the continuing sizable threat from nuclear weapons, which we have lived with now for more than 50 years, the US Senate's recent rejection of the Comprehensive Test Ban Treaty makes it clear relinquishing nuclear weapons will not be politically easy. But we have a unique opportunity, with the end of the Cold War, to avert a multipolar arms race. Building on the BWC and CWC relinquishments, successful abolition of nuclear weapons could help us build toward a habit of relinquishing dangerous technologies. (Actually, by getting rid of all but 100 nuclear weapons worldwide—roughly the total destructive power of World War II and a considerably easier task—we could eliminate this extinction threat.[13])

Verifying relinquishment will be a difficult problem, but not an unsolvable one. We are fortunate to have already done a lot of relevant work in the context of the BWC and other treaties. Our major task will be to apply this to technologies that are naturally much more commercial than military. The substantial need here is for transparency, as difficulty of verification is directly proportional to the difficulty of distinguishing relinquished from legitimate activities.

I frankly believe that the situation in 1945 was simpler than the one we now face: The nuclear technologies were reasonably separable into commercial and military uses, and monitoring was aided by the nature of atomic tests and the ease with which radioactivity could be measured. Research on military applications could be performed at national laboratories such as Los Alamos, with the results kept secret as long as possible.

The GNR technologies do not divide clearly into commercial and military uses; given their potential in the market, it's hard to imagine pursuing them only in national laboratories. With their widespread commercial pursuit, enforcing relinquishment will require a verification regime similar to that for biological weapons, but on an unprecedented scale. This, inevitably, will raise tensions between our individual privacy and desire for proprietary information, and the need for verification to protect us all. We will undoubtedly encounter strong resistance to this loss of privacy and freedom of action.

Verifying the relinquishment of certain GNR technologies will have to occur in cyberspace as well as at physical facilities. The critical issue will be to make the necessary transparency acceptable in a world of proprietary information, presumably by providing new forms of protection for intellectual property.

Verifying compliance will also require that scientists and engineers adopt a strong code of ethical conduct, resembling the Hippocratic oath, and that they have the courage to whistleblow as necessary, even at high personal cost. This would answer the call—50 years after Hiroshima—by the Nobel laureate Hans Bethe, one of the most senior of the surviving members of the Manhattan Project that all scientists "cease and desist from work creating, developing, improving, and manufacturing nuclear weapons and other weapons of potential mass destruction."[14] In the 21st century, this requires vigilance and personal responsibility by those who would work on both NBC and GNR technologies to avoid implementing weapons of mass destruction and knowledge-enabled mass destruction.

Thoreau also said that we will be "rich in proportion to the number of things which we can afford to let alone." We each seek to be happy, but it would seem worthwhile to question whether we need to take such a high risk of total destruction to gain yet more knowledge and yet more things; common sense says that there is a limit to our material needs—and that certain knowledge is too dangerous and is best forgone.

Neither should we pursue near immortality without considering the costs, without considering the commensurate increase in the risk of extinction. Immortality, while perhaps the original, is certainly not the only possible utopian dream.

I recently had the good fortune to meet the distinguished author and scholar Jacques Attali, whose book *Lignes d'horizons* (*Millennium,* in the English translation) helped inspire the Java and Jini approach to the coming age of pervasive computing, as previously described in this magazine. In his new book *Fraternités,* Attali describes how our dreams of utopia have changed over time:

"At the dawn of societies, men saw their passage on Earth as nothing more than a labyrinth of pain, at the end of which stood a door leading, via their death, to the company of gods and to *Eternity.* With the Hebrews and then the Greeks, some men dared free themselves from theological demands and dream of an ideal City where *Liberty* would flourish. Others, noting the evolution of the market society, understood that the liberty of some would entail the alienation of others, and they sought *Equality.*"

Jacques helped me understand how these three different utopian goals exist in tension in our society today. He goes on to describe a fourth utopia, *Fraternity,* whose foundation is altruism. Fraternity alone associates individual happiness with the happiness of others, affording the promise of self-sustainment.

This crystallized for me my problem with Kurzweil's dream. A technological approach to Eternity—near immortality through robotics—may not be the most desirable utopia, and its pursuit brings clear dangers. Maybe we should rethink our utopian choices.

Where can we look for a new ethical basis to set our course? I have found the ideas in the book *Ethics for the New Millennium,* by the Dalai Lama, to be very helpful. As is perhaps well known but little heeded, the Dalai Lama argues that the most important thing is for us to conduct our lives with love and compassion for others, and that our societies need to develop a stronger notion of universal responsibility and of our interdependency; he proposes a standard of positive ethical conduct for individuals and societies that seems consonant with Attali's Fraternity utopia.

The Dalai Lama further argues that we must understand what it is that makes people happy, and acknowledge the strong evidence that neither material progress nor the pursuit of the power of knowledge is the key—that there are limits to what science and the scientific pursuit alone can do.

Our Western notion of happiness seems to come from the Greeks, who defined it as "the exercise of vital powers along lines of excellence in a life affording them scope."[15]

Clearly, we need to find meaningful challenges and sufficient scope in our lives if we are to be happy in whatever is to come. But I believe we must find alternative outlets for our creative forces, beyond the culture of perpetual economic growth; this growth has largely been a blessing for several hundred years, but it has not brought us unalloyed happiness, and we must now choose between the pursuit of unrestricted and undirected growth through science and technology and the clear accompanying dangers.

It is now more than a year since my first encounter with Ray Kurzweil and John Searle. I see around me cause for hope in the voices for caution and relinquishment and in those people I have discovered who are as concerned as I am about our current predicament. I feel, too, a deepened sense of personal responsibility—not for the work I have already done, but for the work that I might yet do, at the confluence of the sciences.

But many other people who know about the dangers still seem strangely silent. When pressed, they trot out the "this is nothing new" riposte—as if awareness of what could happen is response enough. They tell me, There are universities filled with bioethicists who study this stuff all day long. They say, All this has been written about before, and by experts. They complain, Your worries and your arguments are already old hat.

I don't know where these people hide their fear. As an architect of complex systems I enter this arena as a generalist. But should this diminish my concerns? I am aware of how much has been written about, talked about, and lectured about so authoritatively. But does this mean it has reached people? Does this mean we can discount the dangers before us?

Knowing is not a rationale for not acting. Can we doubt that knowledge has become a weapon we wield against ourselves?

The experiences of the atomic scientists clearly show the need to take personal responsibility, the danger that things will move too fast, and the way in which a process can take on a life of its own. We can, as they did, create insurmountable problems in almost no time flat. We must do more thinking up front if we are not to be similarly surprised and shocked by the consequences of our inventions.

My continuing professional work is on improving the reliability of software. Software is a tool, and as a toolbuilder I must struggle with the uses to which the tools I make are put. I have always believed that making software more reliable, given its many uses, will make the world a safer and better place; if I were to come to believe the opposite, then I would be morally obligated to stop this work. I can now imagine such a day may come.

This all leaves me not angry but at least a bit melancholic. Henceforth, for me, progress will be somewhat bittersweet.

Do you remember the beautiful penultimate scene in *Manhattan* where Woody Allen is lying on his couch and talking into a tape recorder? He is writing a short story about people who are creating unnecessary, neurotic problems for themselves, because it keeps them from dealing with more unsolvable, terrifying problems about the universe.

He leads himself to the question, "Why is life worth living?" and to consider what makes it worthwhile for him: Groucho Marx, Willie Mays, the second movement of the Jupiter Symphony, Louis Armstrong's recording of "Potato Head Blues," Swedish movies, Flaubert's Sentimental Education, Marlon Brando, Frank Sinatra, the apples and pears by Cézanne, the crabs at Sam Wo's, and, finally, the showstopper: his love Tracy's face.

Each of us has our precious things, and as we care for them we locate the essence of our humanity. In the end, it is because of our great capacity for caring that I remain optimistic we will confront the dangerous issues now before us.

My immediate hope is to participate in a much larger discussion of the issues raised here, with people from many different backgrounds, in settings not predisposed to fear or favor technology for its own sake.

As a start, I have twice raised many of these issues at events sponsored by the Aspen Institute and have separately proposed that the American Academy of Arts and Sciences take them up as an extension of its work with the Pugwash Conferences. (These have been held since 1957 to discuss arms control, especially of nuclear weapons, and to formulate workable policies.)

It's unfortunate that the Pugwash meetings started only well after the nuclear genie was out of the bottle—roughly 15 years too late. We are also getting a belated start on seriously addressing the issues around 21st-century technologies—the prevention of knowledge-enabled mass destruction—and further delay seems unacceptable.

So I'm still searching; there are many more things to learn. Whether we are to succeed or fail, to survive or fall victim to these technologies, is not yet decided. I'm up late again—it's almost 6 am. I'm trying to imagine some better answers, to break the spell and free them from the stone.

QUESTIONS

1. What are some of the implications of the development of machines to be able to make better complex decisions than humans? Name five.
2. What are the implications if humans retain control of ever-increasing complex machines? Name three.
3. What is the argument of parallelism of the free marketplace with the interaction of superior robots and humans? Explain.
4. What are the three most compelling new twenty-first century technologies? What are the more obvious advantages and disadvantages of the three? What are the social implications of each of these technologies?
5. What were the three most compelling twentieth century technologies and how do they differ from the ones of the twenty-first century in scope, power, and availability?
6. Why does the author state that we should be asking how humans can best co-exist with these new technological developments?
7. List some of the exciting promises of genetic engineering. What are the social implications of their use?
8. What are some of the dangers of nanotechnology? Name five.
9. What are some of the new capabilities that nanotechnology will bring to society? Name five.
10. What is the greatest danger of GNR according to the article? Explain why.
11. Explain the reasoning of philosopher John Leslie's prediction of the risk of human extinction.
12. What is the author's solution to the dangers of rapidly accelerating technologies?
13. Give five examples that Mr. Joy suggests to control and plan a positive future.
14. Why is Mr. Joy optimistic about the future? What is the central characteristic that he says is foundational for our positive future? Discuss.

ENDNOTES

1. The passage Kurzweil quotes is from Kaczynski's *Unabomber Manifesto*, which was published jointly, under duress, by The *New York Times* and The *Washington Post* to attempt to bring his campaign of terror to an end. I agree with David Gelernter, who said about their decision:

   "It was a tough call for the newspapers. To say yes would be giving in to terrorism, and for all they knew he was lying anyway. On the other hand, to say yes might stop the killing. There was also a chance that someone would read the tract and get a hunch about the author; and that is exactly what happened. The suspect's brother read it, and it rang a bell.

   "I would have told them not to publish. I'm glad they didn't ask me. I guess."

   (*Drawing Life: Surviving the Unabomber.* Free Press, 1997: 120.)
2. Garrett, Laurie. *The Coming Plague: Newly Emerging Diseases in a World Out of Balance.* Penguin, 1994: 47-52, 414, 419, 452.
3. Isaac Asimov described what became the most famous view of ethical rules for robot behavior in his book, *I, Robot,* in 1950, in his Three Laws of Robotics: 1. A robot may not injure a human being or, through inaction, allow a human being to come to harm. 2. A robot must obey the orders given it by human beings, except where such orders would conflict with the First Law. 3. A robot must protect its own existence, as long as such protection does not conflict with the First or Second Law.
4. Michelangelo wrote a sonnet that begins:

   Non ha l' ottimo artista alcun concetto
   Ch' un marmo solo in sè non circonscriva
   Col suo soverchio; e solo a quello arriva
   La man che ubbidisce all' intelleto.

   Stone translates this as:

   The best of artists hath no thought to show
   which the rough stone in its superfluous shell
   doth not include; to break the marble spell
   is all the hand that serves the brain can do.

   Stone describes the process: "He was not working from his drawings or clay models; they had all been put away. He was carving from the images in his mind. His eyes and hands knew where every line, curve, mass must emerge, and at what depth in the heart of the stone to create the low relief."

(*The Agony and the Ecstasy.* Doubleday, 1961: 6, 144.)

5. First Foresight Conference on Nanotechnology in October 1989, a talk titled "The Future of Computation." Published in Crandall, B. C. and James Lewis, editors. Nanotechnology: Research and Perspectives. MIT Press, 1992: 269. See also www.foresight.org/Conferences/MNT01/Nano1.html.
6. In his 1963 novel *Cat's Cradle,* Kurt Vonnegut imagined a gray-goo-like accident where a form of ice called ice-nine, which becomes solid at a much higher temperature, freezes the oceans.
7. Kauffman, Stuart. "Self-replication: Even Peptides Do It." *Nature*, 382, August 8, 1996: 496. See www.santafe.edu/sfi/People/kauffman/sak-peptides.html.
8. Else, Jon. The Day After Trinity: J. Robert Oppenheimer and The Atomic Bomb (available at www.pyramiddirect.com).
9. This estimate is in Leslie's book *The End of the World: The Science and Ethics of Human Extinction,* where he notes that the probability of extinction is substantially higher if we accept Brandon Carter's Doomsday Argument, which is, briefly, that "we ought to have some reluctance to believe that we are very exceptionally early, for instance in the earliest 0.001 percent, among all humans who will ever have lived. This would be some reason for thinking that humankind will not survive for many more centuries, let alone colonize the galaxy. Carter's doomsday argument doesn't generate any risk estimates just by itself. It is an argument for revising the estimates which we generate when we consider various possible dangers." (Routledge 1996: 1, 3, 145.)
10. Clarke, Arthur C. "Presidents, Experts, and Asteroids." *Science,* June 5, 1998. Reprinted as "Science and Society" in *Greetings, Carbon-Based Bipeds! Collected Essays, 1934-1998.* St. Martin's Press, 1999: 526.
11. And, as David Forrest suggests in his paper "Regulating Nanotechnology Development," available at www.foresight.org/NanoRev/Forrest1989.html, "If we used strict liability as an alternative to regulation it would be impossible for any developer to internalize the cost of the risk (destruction of the biosphere), so theoretically the activity of developing nanotechnology should never be undertaken." Forrest's analysis leaves us with only government regulation to protect us—not a comforting thought.
12. Meselson, Matthew. "The Problem of Biological Weapons." Presentation to the 1,818th Stated Meeting of the American Academy of Arts and Sciences, January 13, 1999. (minerva.amacad.org/archive/bulletin4.htm)
13. Doty, Paul. "The Forgotten Menace: Nuclear Weapons Stockpiles Still Represent the Biggest Threat to Civilization." *Nature,* 402, December 9, 1999: 583.
14. See also Hans Bethe's 1997 letter to President Clinton, at www.fas.org/bethecr.htm.
15. Hamilton, Edith. *The Greek Way.* W. W. Norton & Co., 1942: 35.

Global Population Reduction: Confronting the Inevitable

J. KENNETH SMAIL

Looking past the near-term concerns that have plagued population policy at the political level, it is increasingly apparent that the long-term sustainability of civilization will require not just a leveling-off of human numbers as projected over the coming half-century, but a colossal reduction in both population and consumption.

It has become increasingly apparent over the past half-century that there is a growing tension between two seemingly irreconcilable trends. On one hand, moderate to conservative demographic projections indicate that global human numbers will almost certainly reach 9 billion, perhaps more, by mid-21st century. On the other, prudent and increasingly reliable scientific estimates suggest that the Earth's long-term sustainable human carrying capacity, at what might be defined as an "adequate" to "moderately comfortable" developed-world standard of living, may not be much greater than 2 to 3 billion. It may be considerably less, particularly if the normative lifestyle (level of consumption) aspired to is anywhere close to that of the United States.

As a consequence of this modern-day "Malthusian dilemma," it is past time to think boldly about the midrange future and to consider alternatives that go beyond merely slowing or stopping the growth of global population. The human species must develop and quickly implement a well-conceived, clearly articulated, flexible, equitable, and internationally coordinated program focused on bringing about a *very significant reduction* in human numbers over the next two or more centuries. This effort will likely require a global population shrinkage of at least two-thirds to three-fourths, from a probable mid-to-late 21st century peak in the 9 to 10 billion range to a future (23rd century and beyond) "population optimum" of not more than 2 to 3 billion.

Obviously, a demographic change of this magnitude will require a major reorientation of human thought, values, expectations, and lifestyles. There is no guarantee that such a program will be successful. But if humanity fails in this effort, nature will almost certainly impose an even harsher reality. As a practicing physical anthropologist and human evolutionary biologist, I am concerned that this rapidly metastasizing (yet still partly hidden) demographic and environmental crisis could emerge as the greatest evolutionary/ecological "bottleneck" that our species has yet encountered.

Although the need for population reduction is controversial, it can be tested scientifically. The hypothesis may be falsified if it can clearly be shown that ongoing estimates of global population size over the next few hundred years will not exceed our increasingly accurate projections of both current and future optimal carrying capacities. However, the hypothesis will be confirmed if future global population size continues to exceed those carrying capacity estimates by a significant margin. And even if the 2 to 3 billion optimal carrying capacity estimate turns out to be off by, say, a factor of two, achieving a global population optimum of 4 to 6 billion would still necessitate a very substantial reduction from the 9-plus billion projected for mid-century.

Reprinted with permission from the author from the original article, which appeared in *WorldWatch,* September/October 2004.

BELOW THE RADAR?

It is surprising how little scientific and public attention has been directed toward establishing quantifiable, testable, and socioculturally agreed-upon parameters for what the Earth's long-term human carrying capacity might actually be. Unfortunately, with only a few notable exceptions, many otherwise well-qualified scientific investigators and public policy analysts have been rather hesitant to take a clear and forthright position on this profoundly important matter. One wonders why—inherent caution, concerns about professional reputation, the increasingly specialized structure of both the scientific and political enterprises, or any of several other reasons. Given the issue's global nature and ramifications, perhaps the chief reason is simply "scale paralysis," that enervating sense of individual and collective powerlessness when confronted by problems whose magnitude seems overwhelming.

Certainly the rough-and-ready human carrying capacity estimates of the more distant past show considerable variation, ranging from fewer than 1 billion to over 20 billion. And it is obvious that it will be difficult to engender any sort of effective response to the crisis if the desired future population goals continue to be poorly understood and imperfectly articulated. It is, however, worthy of note that several investigators and organizations have developed reasonably well thought out positions on future global population optima, and those estimates have all clustered in the range of 1 to 3 billion.

I hope my hypothesis is wrong and that various demographic optimists are correct in claiming that human numbers will begin to stabilize and decline somewhat sooner than expected. But this optimism is warranted only by corroborative data, that is, only if the abovementioned "irreconcilable numbers" show unmistakable evidence of coming into much closer congruence.

Clearly, assertions that the Earth might be able to support a population of 10, 15, or even 20 billion people for an indefinite period of time at a standard of living superior to the present are not only cruelly misleading but almost certainly false. Notwithstanding our current addiction to continued and uninterrupted economic growth, humanity must recognize that there are *finite physical, biological, and ecological limits* to the Earth's long-term sustainable carrying capacity. And to judge by the growing concerns about maintaining the quality, stability, and/or sustainability of the Earth's atmosphere, water, forests, croplands, fisheries, and so on, there is little if any doubt that many of these limits will soon be reached, if they haven't already been surpassed. Since at some point the damage stemming from the mutually reinforcing effects of excessive human reproduction and overconsumption of resources could well become irreversible, and because there is only one Earth with which to experiment, it would undoubtedly be better for our species to err on the side of prudence, exercising wherever possible a cautious and careful stewardship.

Perhaps it is time that the burden of proof on these matters, so long shouldered by so-called neo-Malthusian pessimists, be shifted to the "cornucopian optimists." Let them answer: What is the evidence that the Earth can withstand, without irreparable damage, another two or more centuries during which global human numbers and per capita consumption increasingly exceed the Earth's optimal (sustainable) carrying capacity?

In any event, having established a "quantifiable and falsifiable" frame of reference, it is time to make the case that current rhetoric about "slowing the growth of" or even "stabilizing" global human numbers is clearly insufficient. Both the empirical data and inexorable logic suggest that our default position for the next two or three centuries ought to seek a very significant reduction in global human numbers.

ACKNOWLEDGING OUR DILEMMA

Is it naive to hope that, once a critical mass of concerned investigators begins to make a serious case for such a reduction, it would become much easier for scientists, environmentalists, politicians, economists, moralists, and other concerned citizens of the planet to speak forthrightly about humanity's critical need for population stabilization and shrinkage? At the least, they should not feel as though they are committing political, professional, or moral suicide by raising these issues. Time is increasingly precious, and our window of opportunity for effective remedial action may not be open much longer—assuming it has not already closed.

Until demonstrated otherwise, I would therefore argue that insufficiently restrained population growth should be considered the single most important feature in a complex (and synergistic) physical, ecological, bio-cultural, and sociopolitical landscape. Regulating human population size, and confronting the numerous problems that will be engendered by its eventual and inevitable contraction, should thus be accorded a central position within the modern dilemma, and as such should be dealt with much more forthrightly, and promptly, than has heretofore been the case.

More than half a century ago, at the dawn of the nuclear age, Albert Einstein suggested that we would require a new manner of thinking if humankind were to survive. Even though the population explosion is neither as instantaneous nor as spectacular as its nuclear counterpart, the ultimate consequences may be just as real (and potentially just as devastating) as the so-called nuclear winter scenarios promulgated in the early 1980s.

That there will be a large-scale reduction in global human numbers over the next two or three centuries appears to be inevitable. The primary issue seems to be whether this process will be under conscious human control and (hopefully) relatively benign, or whether it will turn out to be unpredictably chaotic and (perhaps) catastrophic. We must begin our new manner of thinking about this critically important global issue now, so that Einstein's prescient and legitimate concerns about human and civilizational survival into the 21st century and beyond may be addressed as rapidly, fully, and humanely as possible.

Don't speak to me of shortage. My world is vast
And has more than enough—for no more than enough.
There is a shortage of nothing, save will and wisdom;
But there is a longage of people.

GARRETT HARDIN (1975)

REFERENCE

Smail, J. K. (Sept.-Oct./2004). Global population reduction: confronting the inevitable. *World Watch* 17 (5), pp.58–59.

QUESTIONS

1. Why is there an urgency to return the "population optimum" of not more than 2 or 3 billion for world population totals? Give several reasons.
2. How could such a number be achieved by the 23rd century?
3. Why can't the earth support a population of 10, 15, or even 20 billion people?
4. Apply and explain the principle of "carrying capacity" from the population chapter to the arguments of this article.
5. What are some of the dire consequences and scenarios if population numbers are not reduced adequately?
6. Present your opinion as to whether such a large-scale reduction in global human population numbers can be a controlled process or some large-scale catastrophe. Discuss.

Case Study 1

Emerging Technologies: New Demands for Students and Faculty

AHMED S. KHAN

The rapid pace of technological growth has placed new demands on the skills, competencies, and knowledgebase of graduates. To be successful in the twenty-first century workplace, graduates are required to acquire digital-age technical literacy. Graduates are not only expected to understand the theory behind state-of-the-art technologies, but also to exhibit hands-on and analytical problem solving, expert thinking, and complex communication skills.

The fast pace of innovation is transforming the global economy into a knowledge-based or innovation-based economy. For nations to acquire or maintain a technological edge, the technical competency of graduates has become of paramount importance. Rapid technological growth coupled with the phenomena of globalization, increased productivity, and outsourcing also puts new demands on engineering and technology curricula and faculty. The global marketplace seeks manpower with an up-to-date technical knowledge base. The pace of technological change also imposes new challenges for faculty development and the technical currency of academic programs.

NEW AND EMERGING TECHNOLOGIES AND AREAS OF POTENTIAL JOB GROWTH IN NORTH AMERICA

Unlike the 1990s, when information technology boomed, no industry has risen to drive employment, but there are plenty of possible candidates for the next technology surge.

Table P9.1 lists emerging and new technologies with related potential employment fields.

IEEE fellows—an elite group of men and women representing the cream of their profession—expect both job prospects and technology investments to pick up over the next decade, according to a technology survey conducted by IEEE. As far as which field and which part of the world will dominate in 10 years' time, the fellows are unequivocal: biomolecular engineering, they say, will have a far greater impact on society than nanotechnology, mega-computing, or robotics, and the United States will remain the world's technology R & D leader.

IMPACT OF COMPUTERS ON JOBS OF THE FUTURE

The use of computers has increased productivity at the workplace. Outsourcing has been described as the reason for unemployment, but the real culprit in today's jobless recovery is productivity, not offshoring. Unlike most previous business cycles, productivity has continued to grow at a fast pace. One point of productivity growth can eliminate 1.3 million jobs a year. Companies are using information technology to cut costs, and that means less labor is needed. Of the 2.7 million jobs lost over the past three years, only 300,000 have been from outsourcing.

Table P9.1
New and Emerging Technologies and Related Potential Employment Fields

Technology Domain	*Potential Employment Fields/Job Growth Potential in North America*
Telecommunications	Broadband and wireless technology manufacturing companies could create new employment opportunities in such fields as animation, video, and other forms of multimedia. Dominance of Internet traffic growth will drive networks toward a more data-centric network architecture over the next 10 years
Biotechnology/biomolecular engineering	Biotech companies will grow into full-blown pharmaceutical companies as their technologies mature. That would create new jobs in research and sales.
Nanotechnology	Advances in building materials and devices on the atomic level could create a new wave of startups, employing experts in computer science, basic materials, and applied physics, to work in the following areas: - Nanotechnology - Carbon nanotubes - Molecular electronics - Nanometrology/nanomanipulation/nanofabrication - Nanomaterials - Nanomagnetics and spintronics - Nanoscale modeling and simulation - Nanophotonics - Nanodevices and systems - Nanoelectronics education - Nano-bio systems - Twelve hundred nano startups have emerged around the world, with half of them in the United States. - Nanotechnology will be used in a wide spectrum of products worth $ 292 billion.
Energy	Due to an increasing gap between the supply and demand for fossil fuel, energy prices will rise. This will change the infrastructure and underlying economy in many ways. There will be an increase in R & D for the alternative forms of energy, and energy shortages will spawn new industries. The R & D effort will aim at increasing the size of fuel cells and improving the efficiency of photovoltaic. Economically viable solar power and hydrogen fuel cells could catalyze a new energy infrastructure, creating growth in electric power, autos, and manufacturing.
Space	The long-awaited commercialization of space could open up new frontiers and jobs in the following areas: - Space tourism - Asteroid mining - Satellite refueling

(Continued)

Table P9.1 *(Continued)*
New and Emerging Technologies and Related Potential Employment Fields

Technology Domain	*Potential Employment Fields/Job Growth Potential in North America*
Semiconductor	- **CPU**: Advances in semiconductor technology will allow fabricating 1 billion transistors on a single CPU chip by 2007. Presently, Intel's Itanium-2 Processor has more than 100 million transistors. - IBM has consolidated its chip manufacturing business to produce a single chip: Power Processor. IBM will supply 50 million Power Processors for all three next-generation video game consoles coming from Sony, Microsoft, and Nintendo by 2007. - **Memory**: During the last three decades, semiconductor memory has increased fourfold in density (from less than 1K-bit in 1972 to more than 1G-bit in 2002). Applications that historically permitted the technology to trade-off speed and power now demand both high speed and low power. While the price has dropped from one dollar per 100 bits to less than one dollar per 100 megabits, the production technology has become so complex and expensive that the average multi-billion-dollar company can no longer afford it. The circuits have reached geometries so small that semiconductor theories are being altered and fundamental limits are foreseen. This has led to new investigation in the following areas: - DNA memories - Light memories - Molecular memories - Carbon nanotube memories - Polymer memories These technologies hold the potential for significant reduction in the manufacturing cost.
Electronic identification/RFID/ biometrics security technologies)	Biometric security technologies: electronic devices that scan physiological or behavior traits, such as faces, voices, fingerprints, handprints, and signatures, will become a $4.64 billion market globally. Governments around the world will move to replace paper-based IDs. New forms of electronic identification used in passports, ID cards, bank cards, and credit cards will include information such as the individuals name, address, nationality, digital photo, and even biometric data. Electronic identification will be principally designed to curb fraud and identify theft, but will also speed up the process of identification and authentication.
IT security	Electronic viruses run rampant. Massive growth in connected technologies, from PCs and mobile phones to PDAs and gaming consoles, will cause a corresponding leap in electronic viruses and other malicious attacks. Nuisances, such as unsolicited e-mail (SPAM) and unsolicited instant

(Continued)

Table P9.1 *(Continued)*
New and Emerging Technologies and Related Potential Employment Fields

Technology Domain	*Potential Employment Fields/Job Growth Potential in North America*
	messages (SPIM), will continue to proliferate. More harmful intrusions, such as viruses, worms, and malware (malicious software) as well as blue-jacking (attacks on Bluetooth-enabled devices) and VoIP SPAM, will become common, and increased use of mobile phones, remote working, and WiFi will give hackers more access to private, corporate, and government networks. This trend will cost businesses worldwide billions of dollars in lost data and downtime. At the same time, it will reveal vast opportunities for companies that sell IT security, and new lines of businesses will spring up from mobile operators, handset makers, service providers and system integrators.
Health care/electronic medical records system	The push toward the use of electronics medical records could trim 20 percent off the nation's $1.6 trillion healthcare bill and could also reduce the alarming number of medical mistakes, as well as improve the quality of the nation's health care. The healthcare industry plans to complete its migration toward electronic medical record systems over the next two years with the help of five innovations: - Computerized physician order entry - Clinical decision support systems - Electronic prescribing systems - Patient tracking and verification systems - National Health Information infrastructure

In the book *In The New Division of Labor: How Computers are Creating the Next Job Market*, Harvard Professor Richard Murnane and MIT economist Frank Levy examined how computers are reshaping the job market and how human skills are rewarded in the marketplace. Core to the argument of their research, Levy and Murnane contend that those jobs growing in numbers share two general skills that the computer cannot replicate: expert thinking and complex communication. The first skill, expert thinking, addresses the ability to solve new problems that cannot be solved by

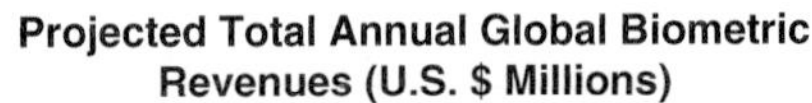

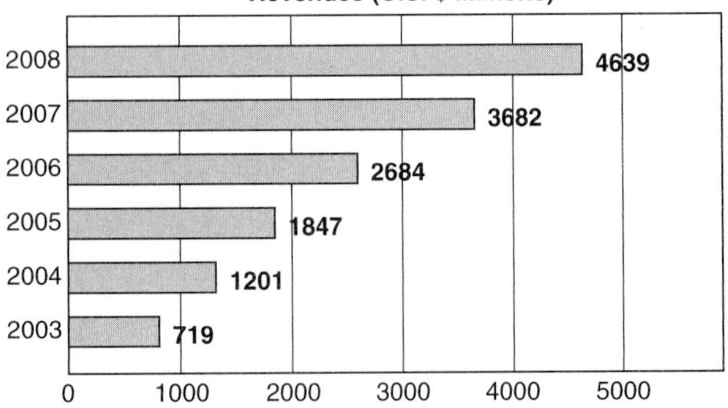

Figure P9.1
Projected total annual global biometric revenues (U.S. $ millions).
Source: American Society of Engineering Education, "Biometric Boom," *Prism*, October 2004, p. 15; A. Applewhite and J. Kurmagai, "Technology Trends 2004," *IEEE Spectrum*, January 2004.

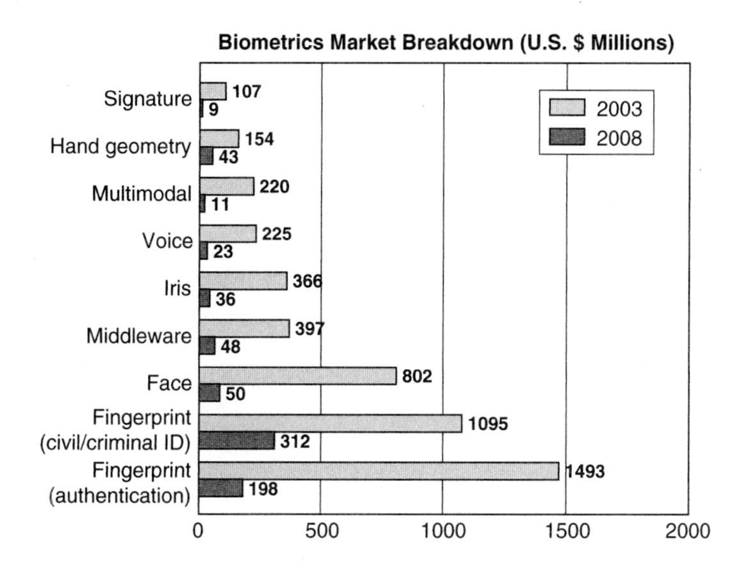

Figure P9.2
Biometrics market breakdown (U.S. $ millions).
Source: American Society of Engineering Education, "Biometric Boom," *Prism*, October 2004, p. 15; A. Applewhite and J. Kurmagai, "Technology Trends 2004," *IEEE Spectrum*, January 2004.

rules. The second general skill, complex communications, addresses the ability not only to transmit information, but to convey a particular interpretation of information to others in professions such as teaching, selling, and negotiation. According to Professor Murnane, "Training all students to engage in complex communication and expert thinking is the challenge American schools have never met, there is an enormous capacity building challenge.... It is the problem solving skills that will trump that digital technology."

IMPACT OF EMERGING TECHNOLOGIES FOR THE TECHNICAL CURRENCY OF FACULTY

To narrow the gap between the state-of-emerging technologies in industry and state-of- technology in academia, faculty are required to revise/enhance curricula frequently and maintain their technical currency. This endeavor is very challenging and requires institutional vision, planning, and allocation of appropriate resources. To narrow this gap, curriculum development/revision activities should be synchronized with faculty development and training activities vis a vis technical currency to optimize teaching/learning using continuous quality improvement. Faculty technical currency will become of utmost importance when the majority of students in a class will belong to "Generation Millenial," who will be highly proficient in Web and wireless technologies (see Tables P9.2 and P9.3).

REFERENCES

American Society of Engineering Education. Biometrics Boom. (2004). *Prism* (October):15.

Table P9.2
Technology Waves and Generations

Teen/Young Adult in	*Generation*	*Technology/Computer Training Received in School*
1970s	Boomers	Mainframe/timeshare (college)
1980s	Gen-X (Atari wave)	Personal computers, video games
1990s	Gen-X/Y (Nintendo wave)	Internet, animation, games
2002–2010s	Millenial	Internet, wireless network devices, laptops/PDAs

Source: P. Markiewiez, Robots and Generations at Robonexus. *Servo*, January 2005, pp. 80–81.

Table P9.3
Future Technology and the Future of Technology Commandments

How Long Will Technology Commandments (technology laws and rules of thumb) Hold?	
Moore's Law	The number of transistors on a chip doubles annually.
Rock's Law	The cost of semiconductor tools doubles every four years.
Machrone's Law	The PC you want to buy will always be $5000.
Metcalfe's Law	A network's value grows proportionately to the number of its users squared.
Wirth's Law	Software is slowing faster than hardware is accelerating.

Source: Phillip E. Ross, 5 Commandments, *IEEE Spectrum*, December 2003, pp. 30–35.

Applewhite, A., and Kumagai, J. (2004). Technology Trends 2004. *IEEE Spectrum* (January).

Baker, Stephen, and Aston, Adam. (2004), The Business of Nanotech. *Business Week* (February 14), pp.164–68.

Beth, P. (2004). *Computers Ate My Job, Economists Educate the Future Work Force at GSE Lecture*. Accessed January 26, 2005. Available athttp://www.news.harvard.edu/gazette/2004/05.06/11-murnane.html.

Cooper, J. (2004). The Price of Efficiency. *Business Week* (March 22), pp. 38–42.

Costlow, T. (2005). EE Job Market Brightens. *IEEE Spectrum* (January).

Computer Software Engineers. (2004). Occupational Outlook Handbook 2004–05 edition. Accessed January 26, 2005. Available at http://bls.gov/oco/ocos267.htm.

Coy, P. (2004). The Future of Work. *Business Week* (March 22), pp. 50–52.

"Deloitte Identifies Top Trends in Technology for 2005." (2005). Accessed February 9, 2005. Available at http://freshnews.com/news/other-tech-areas/article_21266.html.

Economic Policy Institution. (2004). Economic Snapshots (July 21). Accessed January 26, 2005. Available at http://www.epinet.org/content.cfm/webfeatures_snapshots_07212004.

Engineers. (2004). Tomorrow's Jobs. Occupational Outlook Handbook 2004–05 edition. Accessed January 26, 2005. Available at http://bls.gov/oco/ocos027.htm.

Feinberg, C. (2004). Rethinking Education in the Information Age. *HGSE News*. Accessed February 9, 2005. Available at http://www.gse.harvard.edu/news/features/murnane10012004.html.

IBM Discovers the Power of One. (2005). *Business Week* (February 15), pp. 80–81.

Khan, A., Karim, A., Gloeckner, G., and Morgan, G. (2004). *Faculty Technical Currency: Status Report on a National Survey of Engineering Technology Faculty*. Paper presented at 2004 ASEE Annual Conference, June 20–23, Salt Lake City, UT, and published in the conference proceedings.

Khan, A., Gloeckner, G., and Morgan, G. (2005). *Students' Perceptions of the Importance of the Faculty Technical Currency in Their Learning/Success in a Technology-Based Baccalaureate Program.* Paper presented at 2005 ASEE Annual Conference, June 12–15, Portland, OR, and published in the conference proceedings.

Landro, L. (2005). The Informed Patient: Five Innovations Aid to Push to Electronic Medical Records. *The Wall Street Journal* (February 9). Accessed February 18, 2005. Available at http://www.sims.berkeley.edu/~cdclph/doceng/WSJ-ElectronicMedicalRecords.pdf.

Levy, F. and Murnane, R. (2004). *The New Division of Labor: How Computers Are Creating the Next Job Market.* Princeton University Press.

Mandel, M. (2004). Productivity: Who Wins, Who Loses. *Business Week* (March 22), pp. 44–46.

Markiewiez, P. (2005). *Robots and Generations at Robonexus.* Servo (January):80–81.

The New Division of Labor–How to Prepare for America's Changing Job Market. *Harvard Graduate School of Education (HGSE) News.* Accessed February 9, 2005. Available at http://www.gse.harvard.edu/news/features/murnane05132004.html.

Robinson, M. (2005). "Top Jobs for the Future." Accessed January 26, 2005. Available at www.careerplanner.com/Career-Articles/Top_Jobs.htm.

Software Publishers. (2004). Occupational Outlook Handbook 2004–05 edition. Accessed January 26, 2005. Available at http://bls.gov/oco/cg/cgs051.htm.

Technology and Architecture Trends in Optical Networking. (2003). IEEE Distinguished Lecturer Talk, Telecordia Technologies.

Tomorrow's Jobs (2004). Occupational Outlook Handbook 2004–05 edition. Accessed January 26, 2005. Available at http://www.bls.gov/oco/oco2003.htm.

Where Are the Jobs? (2004). *Business Week* (March 22), pp. 36–37.

Who wins, who loses. (2004, March 22). *Business Week*, 44-46.

Wright, S. (2005). "Computers Won't Replace Human Capacity for Problem-Solving in Workplace." *MIT News*. Accessed February 9, 2005. Available at http://web.mit.edu/newsoffice/2005/wrk.html.

QUESTIONS

1. What are the key new and emerging technologies that will shape the future?
2. What is the impact of emerging and new technologies on curricula and the knowledge-base and skills/competencies of graduates?
3. Explain the impact of computers on the jobs of the future.
4. Discuss the impact of emerging technologies for the technical currency of the faculty.
5. Complete the following flowchart:

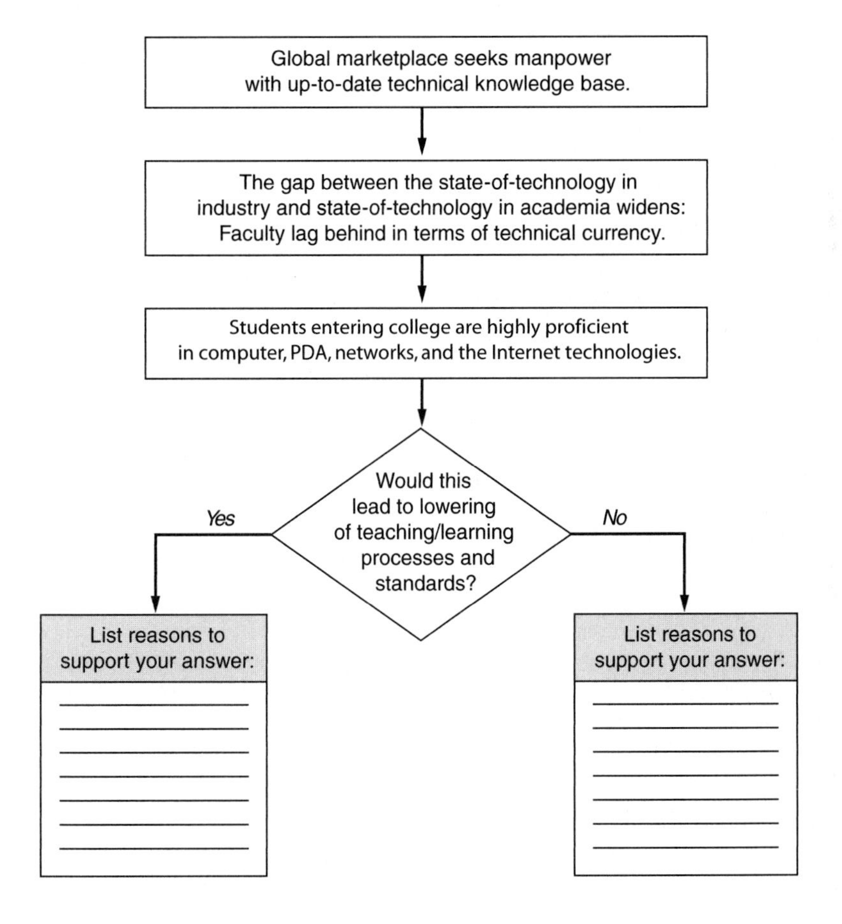

Case Study 2

The Development and Progress of the Ambitious UN 2000 Millennium Project*

EDITED BY BARBARA A. EICHLER

This triumph of the human spirit gives us the hope and confidence that extreme poverty can be cut by half by the year 2015 and indeed ended altogether with the coming years. The world community has at its disposal the proven technologies, policies, financial resources, and most importantly, the human courage and compassion to make it happen.

JEFFREY D. SACHS, UN Millennium Project Director, January 2005

The Millennium Summit in 2000 has given rise to some of the most ambitious realizable goals in the history of the world. These goals are known as the Millennium Development Goals (MDGs) and are the world's time-bound and quantified targets for addressing extreme poverty, hunger, disease, lack of adequate shelter, and exclusion while promoting gender equality, education, and environmental sustainability. They are also basic human rights—the rights of each person on the planet to health, education, shelter, and security as pledged in the Universal Declaration of Human Rights and the UN Millennium Declaration. These goals are to be accomplished by 2015. If successful, more than 300 million will no longer suffer from hunger. There will be dramatic global progress in children's health, education, reduction in maternal mortality , clean drinking water, basic sanitation, and many more basic rights and quality-of-life essentials. These goals are attainable with a framework designed for accomplishment, global cooperation, and bold action. This case study traces development up to the Millennium Project and its current accomplishments and progress.

During the last two decades, there have been a number of significant global summits that have started a momentum for world action in addressing urgent world issues by their definitions and clarification of these specific world issues, by priority and action plans, and by their calls and responses for needed world cooperation. in 1992, the United Nations Conference on Environment and Development (UNCED) in Rio de Janeiro, Brazil, was one of the most ambitious of these summits and was the first global effort devoted to making world progress on issues of environment and poverty. Over 100 heads of state attended, making it one of the largest gatherings of world leaders in history. It raised high hopes for the commitment of the world's communities to save the earth.

The 1992, Earth Summit made history by focusing global attention (new at that time) on the concept that the planet's environmental problems were linked to economic conditions and social justice. It demonstrated that social, environmental, and economic needs must be dealt with in balance with each other by planning for long-term sustainable outcomes. It further demonstrated that when

*UN Millennium Goals data reprinted with permission of the United Nations.

people are poor and national economies weak, the environment suffers; accordingly, if the environment is abused and resources over-consumed, people suffer and economies decline. "The conference also demonstrated that the smallest local actions have potential worldwide ramifications." (United Nations Division for Sustainable Development, 2001) The Rio Summit was a significant milestone that set a new agenda for sustainable development.

The major achievement of the 1992 World Summit was Agenda 21, which was a broad-ranging program of future actions to reach global sustainable development in the twenty-first century. (The full text of AGENDA 21 can be accessed on the World Wide Web through the United Nations Sustainable Development Committee at http://www.un.org/esa/sustdev/agenda21text.htm.) Its recommendations ranged from new ways to educate and care for natural resources to participation in the design of a sustainable economy. "The overall ambition of Agenda 21 was breathtaking, for its goal was nothing less than to make a safe and just world in which all life has dignity and is celebrated." (United Nations Division for Sustainable Development).

Outgrowths of the 1992 Summit included the creation of the Kyoto climate change convention in 1997, a biodiversity convention, a World Summit for Children, a collection of forest principles, and The 2002 Second World Summit on Sustainable Development in Johannesburg, South Africa, which essentially focused on refinement of Agenda 21 and assessment of progress after 10 years toward the goals of Agenda 21. Nitin Desai, Secretary-General of the World Summit on Sustainable Development, stated:

> Of all these conferences, perhaps Rio was the most ambitious, with the most wide-ranging agenda. Its outcomes included Agenda 21, the Rio Declaration, the Statement of Principles on Forests and the launch of a number of major conventions, which, together, have put before the world a truly ambitious agenda, combing the social, economic and environmental dimension of development, and focusing the challenges facing us in three key areas: eradicating poverty; moving us to a pattern of consumption and production that is ecologically more sustainable; and allowing us to handle critical ecosystems such as forests and oceans in a more holistic and integrated way (UN 2002).

From these foundational frameworks, the Millennium Summit in 2000 emerged and created its Millennium Project, which presented to the world a bold and comprehensive set of medium-term realizable goals for the year 2015, focusing on issues of poverty, education, health, and sustainable development. The Millennium Summit in New York was historic, with the largest gathering of world leaders ever assembled agreeing to a real time frame and expressingurgency for ending the world's top global concerns of poverty, hunger, disease, illiteracy, environmental degradation, and discrimination against women.

The UN Millennium Project is an independent advisory body commissioned by the UN Secretary–General, Kofi Annan, to advise the UN on strategies for achieving MDGs. The MDGs are the set of internationally agreed-upon targets for reducing poverty, hunger, disease, illiteracy, environmental degradation, and discrimination against women by 2015. The Millennium Project is headed by Professor Jeffrey Sachs (named by *Time Magazine* as one of the 100 most influential people in the world) and is researched from more than 265 experts from around the world through 10 task forces. Each task force includes independent scientists, development practitioners, parliamentarians, policymakers, and representatives from civil society, UN agencies, the World Bank, the International Monetary Fund, and the private sector. The UN Millennium Project presented its first findings in January 2005 as the first of a series of major global initiatives on the MDGs. In September 2005, there was a high-level summit of the UN General Assembly to make the necessary global policies and plans needed to help the poorest countries achieve the Goals from this first series of data. (UN 2005).

The following are the Millennium Development Goals.

1. **Eradicate extreme hunger and poverty**
   - Reduce by half the proportion of people living on less than a dollar a day.

   Reduce by half the proportion of people who suffer from hunger. 1.2 billion people still live on less than $1 a day. But 43 countries, with more than 60 percent of the world's people, have already met or are on track to meet the goal of cutting hunger in half by 2015.

2. **Achieve universal primary education**
   - Ensure that all boys and girls complete a full course of primary schooling.

   113 million children do not attend school, but this goal is within reach; India, for example, should have 95 percent of its children in school by 2005.

3. **Promote gender equality and empower women**
   - Eliminate gender disparity in primary and secondary education, preferably by 2005, and in all levels of education no later than 2015.

   Two-thirds of the world's illiterates are women, and 80 percent of its refugees are women and children. Since the 1997 Microedit Summit, progress has been made in reaching and empowering poor women, nearly 19 million in 2000 alone.

4. **Reduce child mortality**
   - Reduce by two-thirds the mortality rate among children under five.

   11 million young children die every year, but that number is down from 15 million in 1980.

5. **Improve maternal health**
   - Reduce by three-quarters the maternal mortality ratio.

   In the developing world, the risk of dying in children is one in 48. But virtually all countries now have safe motherhood programs and are poised for progress.

6. **Combat HIV/AIDS, malaria, and other diseases**
   - Halt and begin to reverse the spread of HIV/AIDS.

     Halt and begin to reverse the incidence of malaria and other major diseases.

   Killer diseases have erased a generation of development gains. Countries such as Brazil, Senegal, Thailand, and Uganda have shown that we can stop HIV in its tracks.

7. **Ensure environmental sustainability**
   - Integrate the principles of sustainable development into country policies and programmes; reverse loss of environmental resources.
   - Reduce by half the proportion of people without sustainable access to safe drinking water.

     Achieve significant improvement in the lives of at least 100 million slum dwellers by 2020.

     More than one billion people still lack access to safe drinking water; however, during the 1990s, nearly one billion people gained access to safe water and as many to sanitation.

8. **Develop a global partnership for development**
   - Develop further an open trading and financial system that is rule based, predictable, and non-discriminatory that includes a commitment to good governance, development, and poverty reduction—nationally and internationally.
   - Address the least developed countries' special needs. This includes tariff- and quota-free access for their exports, enhanced debt relief for heavily indebted poor countries, cancellation of official bilateral debt, and more generous official development assistance for countries committed to poverty reduction.
   - Address the special needs of landlocked and small island developing States.
   - Deal comprehensively with developing countries' debt problems through national and international measures to make debt sustainable in the long term.
   - In cooperation with the developing countries, develop decent and productive work for youth.
   - In cooperation with pharmaceutical companies, provide access to affordable essential drugs in developing countries.

     In cooperation with the private sector, make available the benefits of new technologies—especially information and communications technologies.

   Too many developing countries are spending more on debt service than on social services. However, new aid commitments made in the first half of 2002 alone will reach an additional $12 billion per year by 2006.

To realize the necessary basic reforms toward these goals, please refer to BOX 1, "Fast Facts: The Faces of Poverty."

According to UN data, "The world has already made significant progress in achieving many of the Goals. Between 1990 and 2002, average overall incomes increased by approximately 21 percent. The number of people in extreme poverty declined by an estimated 130 million. Child mortality rates fell from 103 deaths per 1,000 live births a year to 88. Life expectancy rose from 63 years to nearly 65 years. An additional 8 percent of the developing world's people received access to water. And an additional 15 percent acquired access to improved sanitation services" (UN 2005).

However, progress is uneven throughout the world and among the goals. There are huge differences across and within countries. Poverty remains greatest for rural areas, although urban poverty is growing and underreported. Sub-Saharan Africa is the center of crisis, with continuing food insecurity, a rise of extreme poverty, high child and maternal mortality, and large numbers of

BOX 1 Fast Facts: The Faces of Poverty

More than one billion people in the world live on less than one dollar a day. In total, 2.7 billion struggle to survive on less than two dollars a day. Poverty in the developing world, however, goes far beyond income poverty. It means having to walk more than one mile every day simply to collect water and firewood; it means suffering diseases that were eradicated from rich countries decades ago. Every year, 11 million children die—most under the age of five—and more than six million die from completely preventable causes such as malaria, diarrhea, and pneumonia.

In some deeply impoverished nations, less than half of the children are in primary school and under 20 percent go to secondary school. Around the world, a total of 114 million children do not even receive a basic education, and 584 million women are illiterate.

Following are basic facts outlining the roots and manifestations of the poverty affecting more than one-third of our world.

Health

- Every year, six million children die from malnutrition before their fifth birthday.
- More than 50 percent of Africans suffer from water-related diseases, such as cholera and infant diarrhea.
- Every day, HIV/AIDS kills 6,000 people, and another 8,200 people are infected with this deadly virus.
- Every 30 seconds, an African child dies of malaria—more than one million child deaths a year.
- Each year, approximately 300–500 million people are infected with malaria. Approximately three million people die as a result.
- TB is the leading AIDS-related killer and, in some parts of Africa, 75 percent of people with HIV also have TB.

Hunger

- More than 800 million people go to bed hungry every day; 300 million are children.
- Of these 300 million children, only eight percent are victims of famine or other emergency situations. More than 90 percent are suffering long-term malnourishment and micronutrient deficiency.
- Every 3.6 seconds, another person dies of starvation, and the large majority are children under the age of 5.

Water

- More than 2.6 billion people—over 40 percent of the world's population—do not have basic sanitation, and more than one billion people still use unsafe sources of drinking water.
- Four of every ten people in the world do not have access to a simple latrine.
- Five million people, mostly children, die each year from water-borne diseases.

Agriculture

- In 1960, Africa was a net exporter of food; today, the continent imports one-third of its grain.
- More than 40 percent of Africans do not even have the ability to obtain sufficient food on a day-to-day basis.
- Declining soil fertility, land degradation, and the AIDS pandemic have led to a 23-percent decrease in food production per capita in the last 25 years, even though population has increased dramatically.
- For the African farmer, conventional fertilizers cost two to six times more than the world market price.

The devastating effect of poverty on women

- More than 80 percent of farmers in Africa are women.
- More than 40 percent of women in Africa do not have access to basic education.
- If a girl is educated for six years or more, as an adult her prenatal care, postnatal care, and childbirth survival rates will dramatically and consistently improve.
- Educated mothers immunize their children 50 percent more often than mothers who are not educated.
- AIDS spreads twice as quickly among uneducated girls than among girls that have some schooling.
- The children of a woman with five years of primary school education have a survival rate 40 percent higher than children of women with no education.

- A woman living in sub-Saharan Africa has a 1 in 16 chance of dying in pregnancy. This compares with a 1 in 3,700 risk for a woman from North America.
- Every minute, a woman somewhere dies in pregnancy or childbirth. This adds up to 1,400 women dying each day—an estimated 529,000 each year—from pregnancy-related causes.

people living in slums without most of the MDGs. Asia is the region with the fastest progress, but there remain hundreds of millions of people in extreme poverty, with even the fastest growing countries not achieving some of the non-income Goals. Latin America, the Middle East, and North Africa have more mixed results, with often slow or no progress on some of the Goals as well as basic inequalities that deter progress on others (UN 2005). Below are some of the graphic results and projections for the goals.

THE MILLENNIUM DEVELOPMENT GOALS: HOW ARE WE DOING?

Under the eight broad Millennium Development Goals, governments have agreed on 18 targets as well as 48 indicators to measure progress. Following are some highlights from the first annual report by the United Nations Secretary-General on the implementation of the Millennium Declaration (A/57/270, www.un.org/millenniumgoals).

Goal 1: Eradicate extreme hunger and poverty

Target: To halve, between 1990 and 2015, the proportion of people whose income is less than $1 a day.

The target has largely been met in East Asia and the Pacific, but Sub-Saharan Africa, Latin America and the Caribbean, and parts of Europe and Central Asia are falling short (see Figure P9.3 on next page).

Target: Between 1990 and 2015, to halve the proportion of people who suffer from hunger.

In the developing world, Eastern and Southeastern Asia, Latin America, and the Caribbean are on track to achieve the target among young children. The prevalence of underweight children is unchanged in Northern Africa. It remains high in South-central Asia and sub-Saharan Africa and is rising in Western Asia (see Figure P9.4 on page 674).

Goal 2: Achieve universal primary education

Target: To ensure by 2015 that children everywhere will complete a full course of primary schooling.

Many regions are on track to achieve the target before 2015, but lower levels of achievement and progress persist in sub-Saharan Africa as well as Western and Southern Asia (see Figure P9.5 on page 674).

Goal 3: Promote gender equality and empower women

Target: Eliminate gender disparity in primary and secondary education by 2005 and in all tertiary education no later than 2015.

In developing countries, gender gaps still exist in enrollment at all levels of education (see Figure P9.6 on page 675).

Goal 4: Reduce child mortality

Target: Between 1990 and 2015, to reduce by two-thirds the under-five mortality rate.

On average in developing countries, more than 100 children die before age five for every 1,000 born (see Figure P9.7 on page 675).

Goal 5: Improve maternal health

Target: Between 1990 and 2015, reduce by three-quarters the maternal mortality ratio.

Based on best estimates for 1995, maternal mortality is far higher in all developing regions compared with developed regions and is fifty times higher in sub-Saharan Africa (see Figure P9.8 on page 676).

Goal 6: Combat HIV/AIDS, malaria, and other diseases

Target: By 2015, to have halted and begun to reverse the spread of HIV/AIDS.

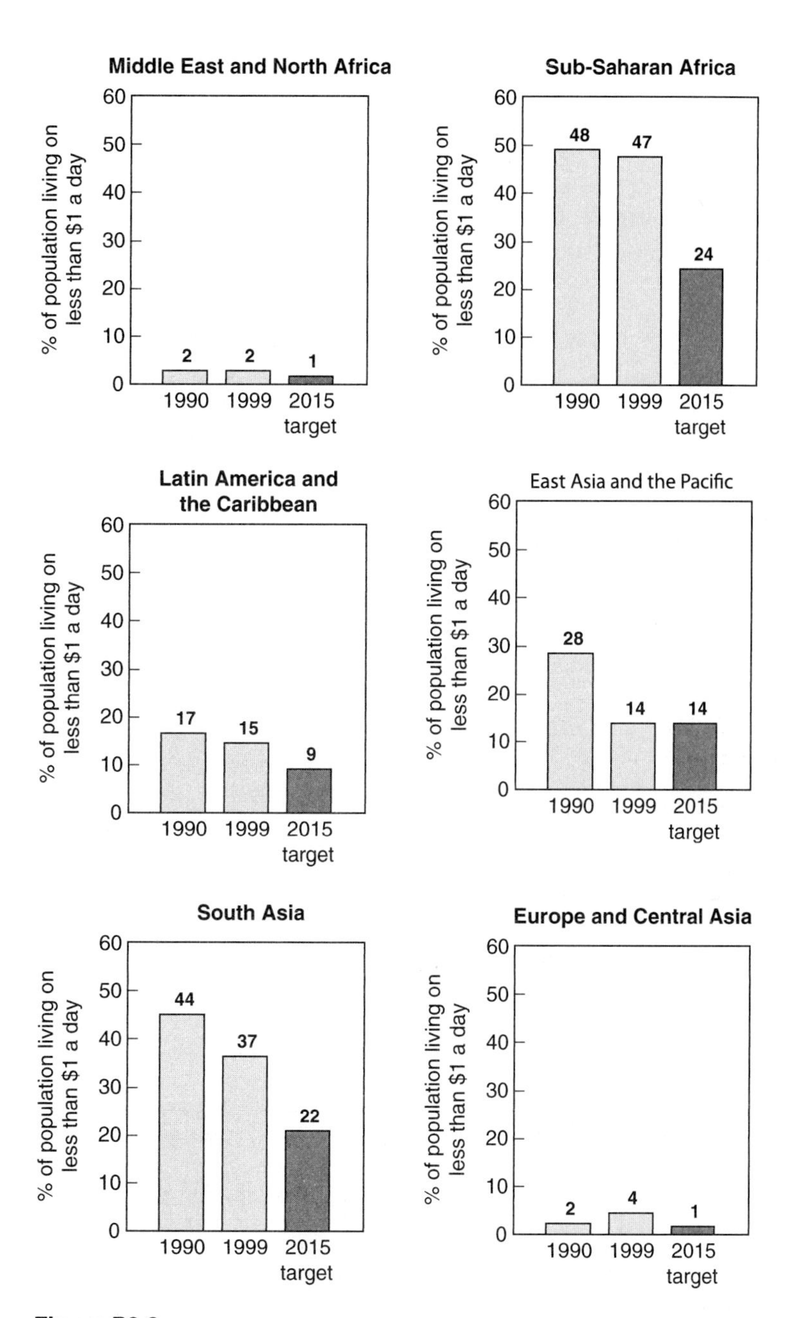

Figure P9.3
Population living in extreme poverty by region.
Source: Report of the Secretary-General on Implementation of the Millennium Declaration. Data based on World Bank estimates.

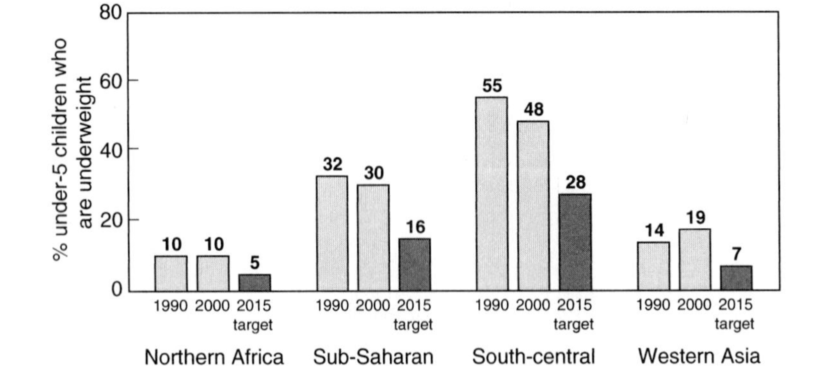

Figure P9.4
Prevalence of underweight children in selected regions.
Source: Report of the Secretary-General on Implementation of the Millennium Declaration. Data based on UNICEF and WHO estimates.

HIV prevalence rates are still increasing for men and women in the developing world. The rate is seven times higher in developing countries than developed countries for women and almost three times higher for men (see Figure P9.9 on page 676).

Goal 7: Ensure environmental sustainability

Target: Reverse the loss of environmental resources.

Carbon dioxide (CO_2) emissions: Worldwide emissions of CO_2—the largest single source of greenhouse gas emissions from human activities—hardly changed on a per capita basis (see Figure P9.10 on page 677).

Chlorofluorocarbons (CFCs): Since the adoption of the Montreal Protocol in 1986, the global consumption of CFCs—the major substances causing depletion of the ozone layer—has decreased significantly (see Figure P9.11 on page 677).

Goal 8: Develop a global partnership for development

Target: Address the special needs of the least developed countries.

Net official development assistance—as a percentage of donor countries' gross national income (GNI)—has

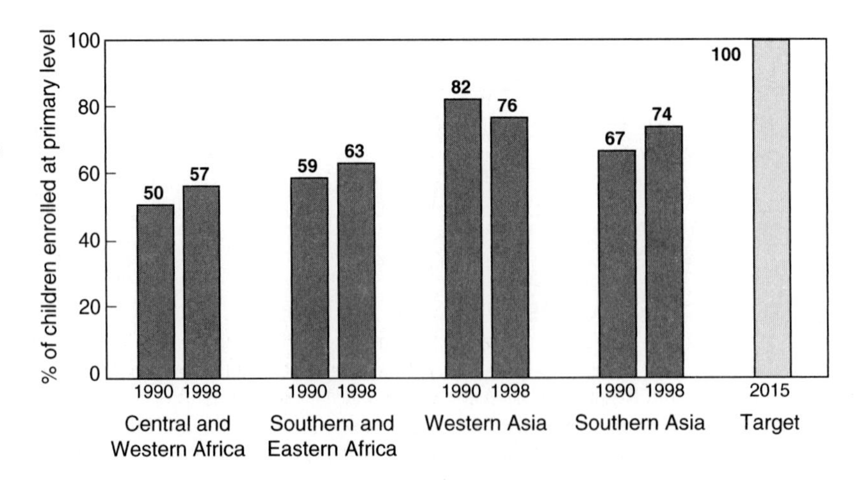

Figure P9.5
Percentage of children enrolled in primary school.
Source: Report of the Secretary-General on Implementation of the Millennium Declaration. Data based on UNESCO estimates.

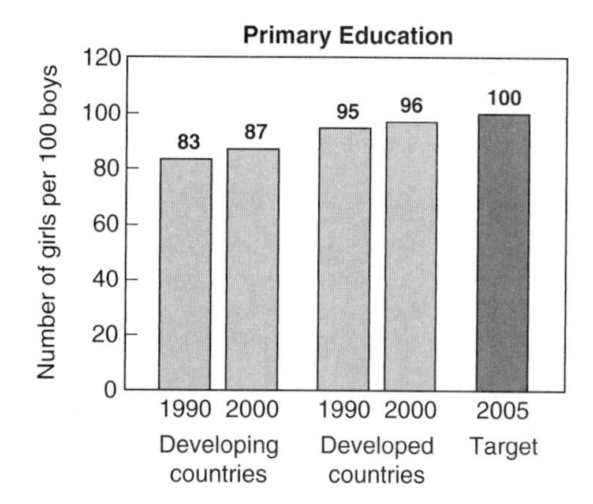

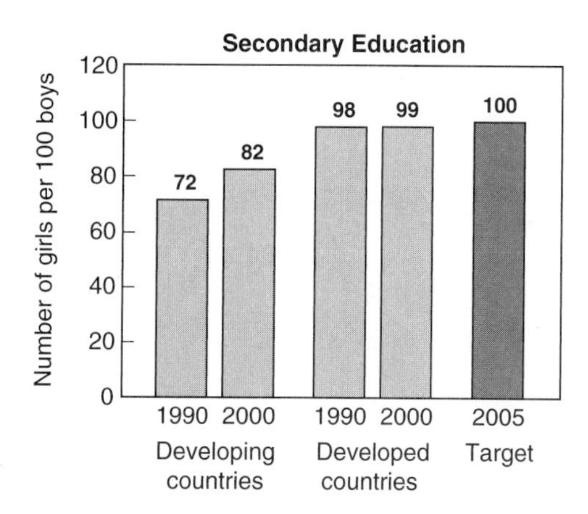

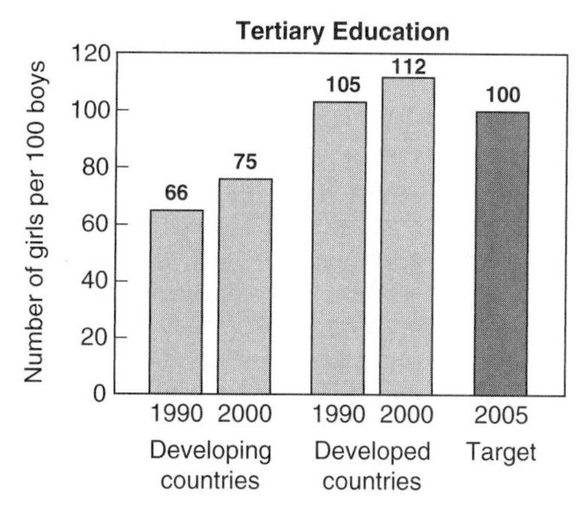

Figure P9.6
Gender disparity in school enrollment.
Source: Report of the Secretary-General on Implementation of the Millennium Declaration. Data based on UNESCO estimates.

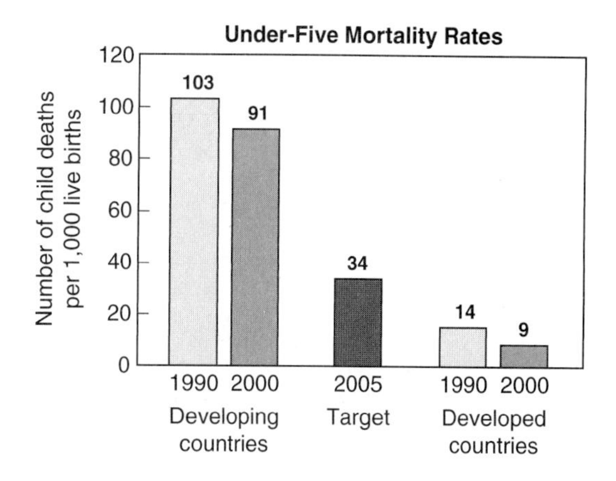

Figure P9.7
Under-five mortality rates.
Source: Report of the Secretary-General on Implementation of the Millennium Declaration. Data based on UNICEF and WHO estimates.

decreased over the last decade and is at an all-time low. The net official development assistance (ODA) to least-developed countries has decreased even more (see Figure P9.12 on page 677).

In recent years, the percentage of imports to developed countries (excluding arms and oil) admitted duty-free from developing countries has increased based on dollar value, yet the percentage of imports from least-developed countries admitted duty-free has decreased (see Figure P9.13 on page 677).

"The UN Millennium Project's analysis indicates that 0.7% of rich world Gross National Income can provide enough resources to meet Millennium Development Goals, but developed countries must follow through on commitment and begin increasing Official Development Assistance (ODA) volumes. If every developed country set and followed through on a timetable to reach o.7% by 2015, the world could make dramatic progress in the fight against poverty and start on a path to achieve the Millennium Development Goals and end extreme poverty by 2015. The UN Millennium Project's costing shows that a comprehensive package to meet the Millennium Development Goals would cost about $75–150 U.S. per person per year over the period, and that less than half of this would need to be financed by ODA." (UN 2005) Please refer to the Web site of the Millennium Project for more details (http://www.unmileniumproject.org).

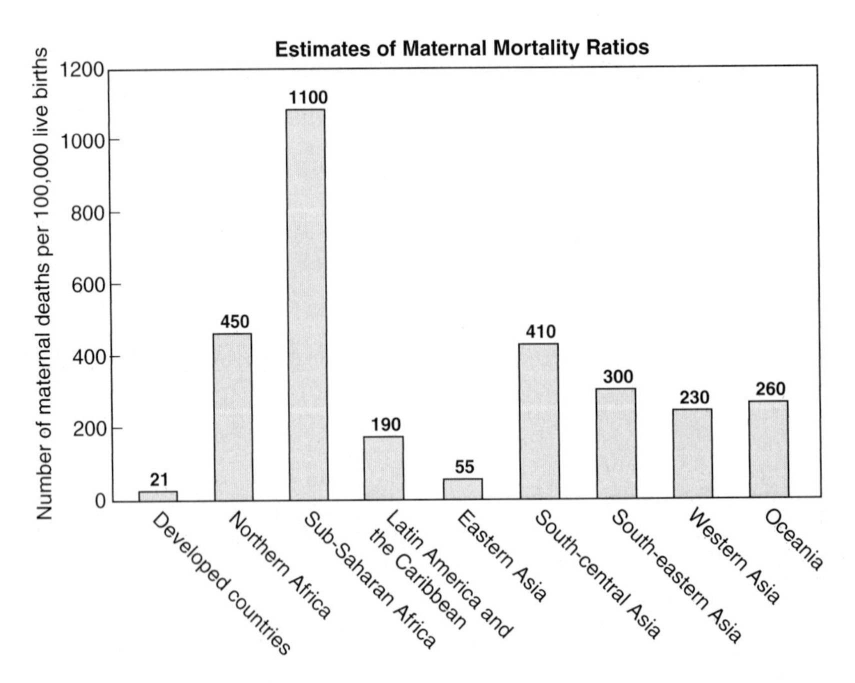

Figure P9.8
Estimates of maternal mortality ratios
Source: Report of the Secretary-General on Implementation of the Millennium Declaration. Data based on UNICEF and WHO estimates.

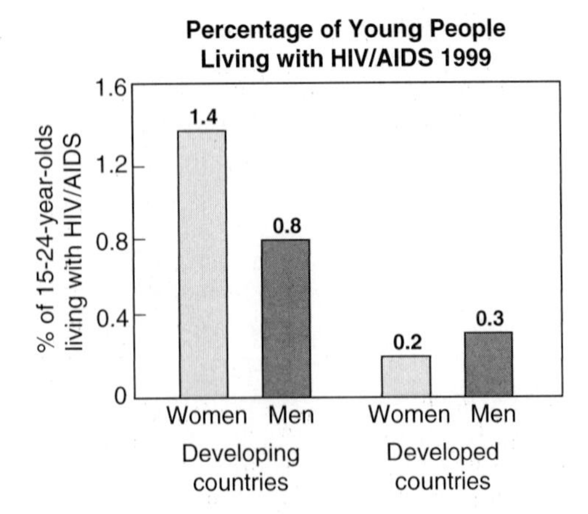

Figure P9.9
Percentage of young people living with HIV/AIDS.
Source: Report of the Secretary-General on Implementation of the Millennium Declaration. Data based on Report on the Global HIV/AIDS Epidemic 2002, UNAIDS.

"If the world achieves the MDGs, more than 500 million people will be lifted out of poverty. A further 250 million will no longer suffer from hunger. Thirty million children and two million mothers who might reasonably have been expected to die will be saved." (UN 2005) As stated in the report, "The Millennium Project has established a political framework and practical solutions to the most broadly supported, comprehensive and specific poverty-reduction targets the world has ever established. And for the first time, the cost is utterly affordable. Whatever one's motivations for attacking the crisis of extreme poverty—human rights, religious values, security, fiscal prudence, ideology—the solutions are the same. All that is needed is action. For the international political system, they are the fulcrum on which development policy is based. For the billion-plus people living in extreme poverty, they represent the means to a productive life. For everyone on Earth, they are a linchpin to the quest for a more secure and peaceful world." (UN 2005, Overview Report)

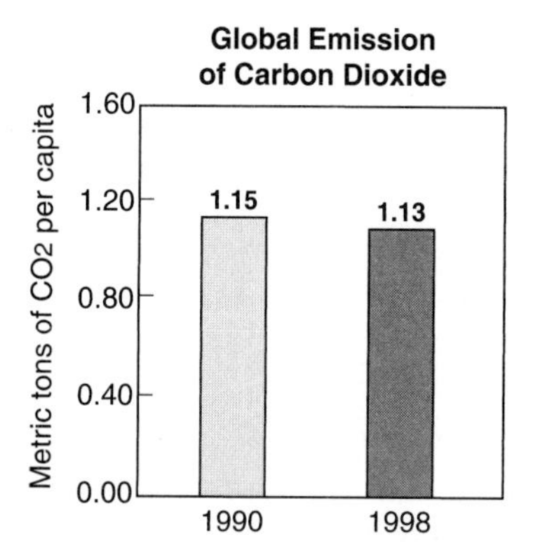

Figure P9.10
Global emission of carbon dioxide.
Source: Report of the Secretary-General on Implementation of the Millennium Declaration. Data based on estimates of the United Nations Framework Convention on Climate Change and the Carbon Dioxide Information Analysis Center (USA).

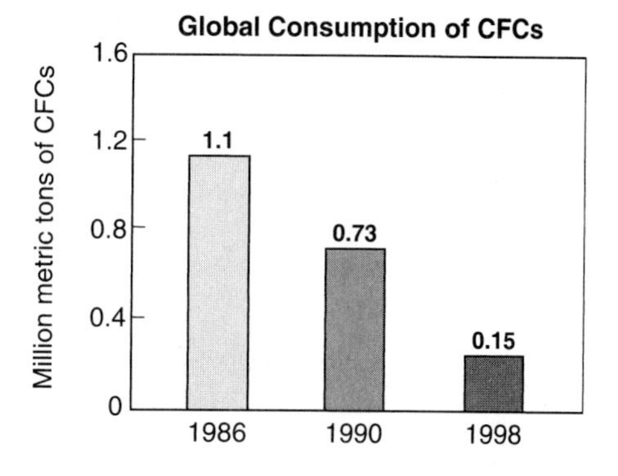

Figure P9.11
Global consumption of CFCs.
Source: Report of the Secretary-General on Implementation of the Millennium Declaration. Data based on estimates of the United Nations Environment Programme, Ozone Secretariat.

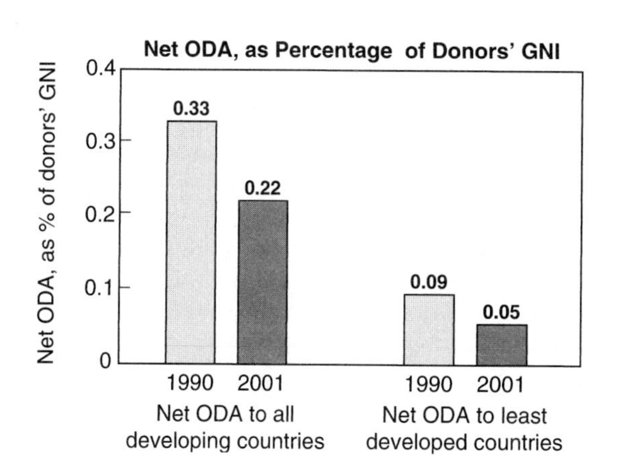

Figure P9.12
Net ODA, as percentage of donors' GNI.
Source: Report of the Secretary-General on Implementation of the Millennium Declaration. Data based on data compiled by the Organisation for Economic Co-operation and Development/ Development Assistance Committee.

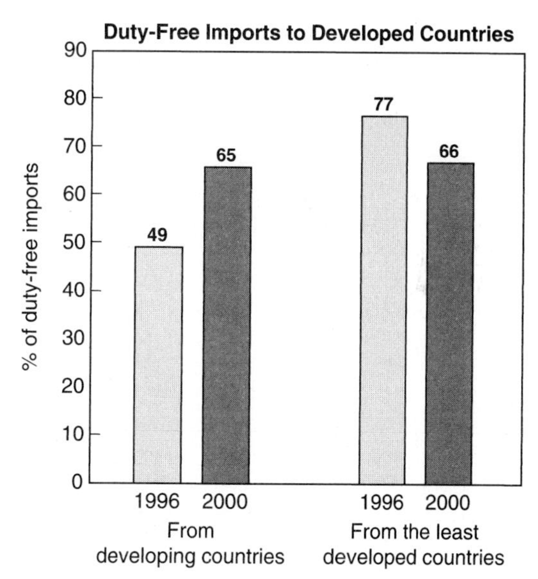

Figure P9.13
Duty-free imports to developed countries.
Source: Report of the Secretary-General on Implementation of the Millennium Declaration. Data based on estimates by UNCTAD in collaboration with WTO and the World Bank.

REFERENCES

United Nations. (2005). Investing in Development: A Practical Plan to Achieve the Millennium Development Goals. *Overview; Millennium Project Report*. New York: United Nations Development Programme.

United Nations. (2005). Millennium Project. Accessed August 13, 2005. Available at http://www.unmillenniumproject.org.

United Nations (2002). *Report of the World Summit on Sustainable Development: Johannesburg, South Africa, 26-August – 4 September.* New York: United Nations. Available at http://www.world-tourism.org/sustainable//wssd/final-report.pdf.

United Nations Division for Sustainable Development. "What Is Johannesburg 2002?" Accessed September 5, 2001. Available at http://www.johannesburgsummit.org/web_pages/rio+10_background.htm.

QUESTIONS

1. Download AGENDA 21 from the UN Web site (http:///www.un.org/esa/sustdev/agenda21text.htm). Review the goals. Comment on their scope and attainability.
2. Review the Millennium Development Goals (MDGs). How do they differ from AGENDA 21? Identify five similarities and five differences.
3. Further inspect the MDGs by visiting the following Web site: http://www.unmillenniumproject.org. Comment on their scope and attainability.
4. Why do you think these MDGs are so historic?
5. Research how much difference these MDGs will make in world patterns if accomplished. Will there still be significant problems on a world scale? What will be the impact of the MDGs?
6. Where are the world and individual countries in their pursuit toward accomplishing these goals, and where is there little or no progress?
7. Predict what will be the global effect if these MDGs are accomplished? How much momentum will this provide toward addressing many of the urgent problems of the twenty-first century?

SCENARIO I: Techno Wars

Dateline: 19:19:01 GMT December 21, 20XX

The Northern Economic Trading Block (NETB) Authority has accused the Southern Economic Trading Block (SETB) of illegal trading practices in winning major trading export contracts worth more than $35 billion from the Eastern Trading Zone (ETZ).

A global videoconference between the leaders of the Northern and Southern trading blocks has resulted in a stalemate and increasing frustration. The leaders of the Northern economic block have accused the Southern block of technoimperialism.

The Southern block leaders think that the Northern block is trying to pressure them to revoke the contracts they signed with the African Economic Block for the development of carefree oil fields, which are estimated to contain 30 percent of the world oil reserves. The carefree oil fields were recently developed by the Southern economic zone after it won the development bid in the international market. The Northern block accused the Southern block of offering billions of dollars to the royal families of the African Trading Zone in exchange for the bid. During the past 10 years, the Northern economic block has lost manufacturing and high-tech jobs to the Southern block. A number of multi-block global corporations have moved operations from North to South. The gross domestic products of the Southern economic block are expected to be equal to those of the Northern economic block. The experts in the North have predicted that if the Southern development is not curbed, it will seriously impact the Northern economy, which stands to lose $100 billion exports globally. The leaders of the North have issued an ultimatum to the South to vacate the carefree oil fields by January 1; otherwise, the North will declare a technowar. The South has refused to succumb to what it calls technoimperialism.

The United Nations has failed to find any solution to this crisis. The last meeting on the GLOBALNET between the five trading blocks resulted in futile attempts to solve the crisis, and all efforts have been in vain. The North has superior high-tech electromagnetic weapons compared with the South, where for the last 50 years there has been little development of technoweaponry. The South has instead emphasized the development of business technology to gain an edge on global consumer businesses.

QUESTIONS

1. Is the Northern Economic Trading Block justified in issuing an ultimatum to the Southern Economic Trading Block? Discuss future economic and political scenarios for this situation.
2. What kind of high-tech electromagnetic weapons might the NETB use against the SETB? Discuss real and prospective weaponry of this type.
3. Do you think that, in the future, business conflicts will be resolved by technowars?
4. What type of actions and resolutions would you recommend to prevent future technowars?

SCENARIO II: Optical Era of Information Society

Date line: 10:01 AM, March 10, 20XX

The MTT Labs has announced that their scientists have successfully developed a micro-optical logic gate. A logic gate is the basic building block of a computer. The state of the current electronics technology allows CPUs to operate at speeds of 5+ GHz. The development of the optical gate will lead to the development of optical computers that will operate at 100+ THz speed ("T" = Tera = 1,000,000,000,000 Hz). During the last decade of the 20th century, high-speed optical fibers have already been developed that allow data transmission rates of Tbps (tera bits per second). The present technology of optical storage permits the storage of 100 terabytes of information per optical compact disc (OCD). The development of a micro-optical logic gate at MTT Labs will enable the development of an end-to-end optical communications system. An end-to-end optical system will use optical computing, optical data storage, and optical data transfer. (See the figure on page 681.) The development of the miniature optical gate has transformed the electronic era of telecommunications into an optical era of information exchange.

QUESTIONS

1. Discuss the pros and cons of the optical computer.
2. The speed of optical computers and optical data transfer will be much faster than human response times and will thereby force users to process information in their brain much faster and to interact more quickly, thus increasing the stress level of the average computer user. Do you think that optical end-to-end information networks will help or harm the health of a computer user in terms of stress levels?
3. Discuss the implications of the development of end-to-end optical information exchange systems on society.
4. Complete the following flowchart.

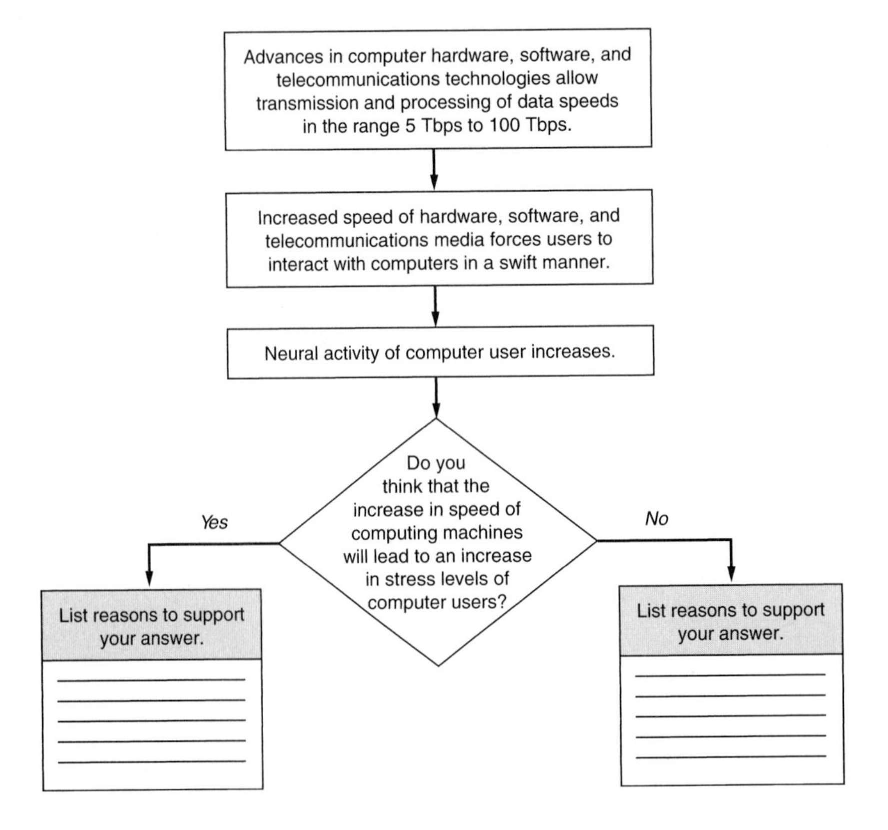

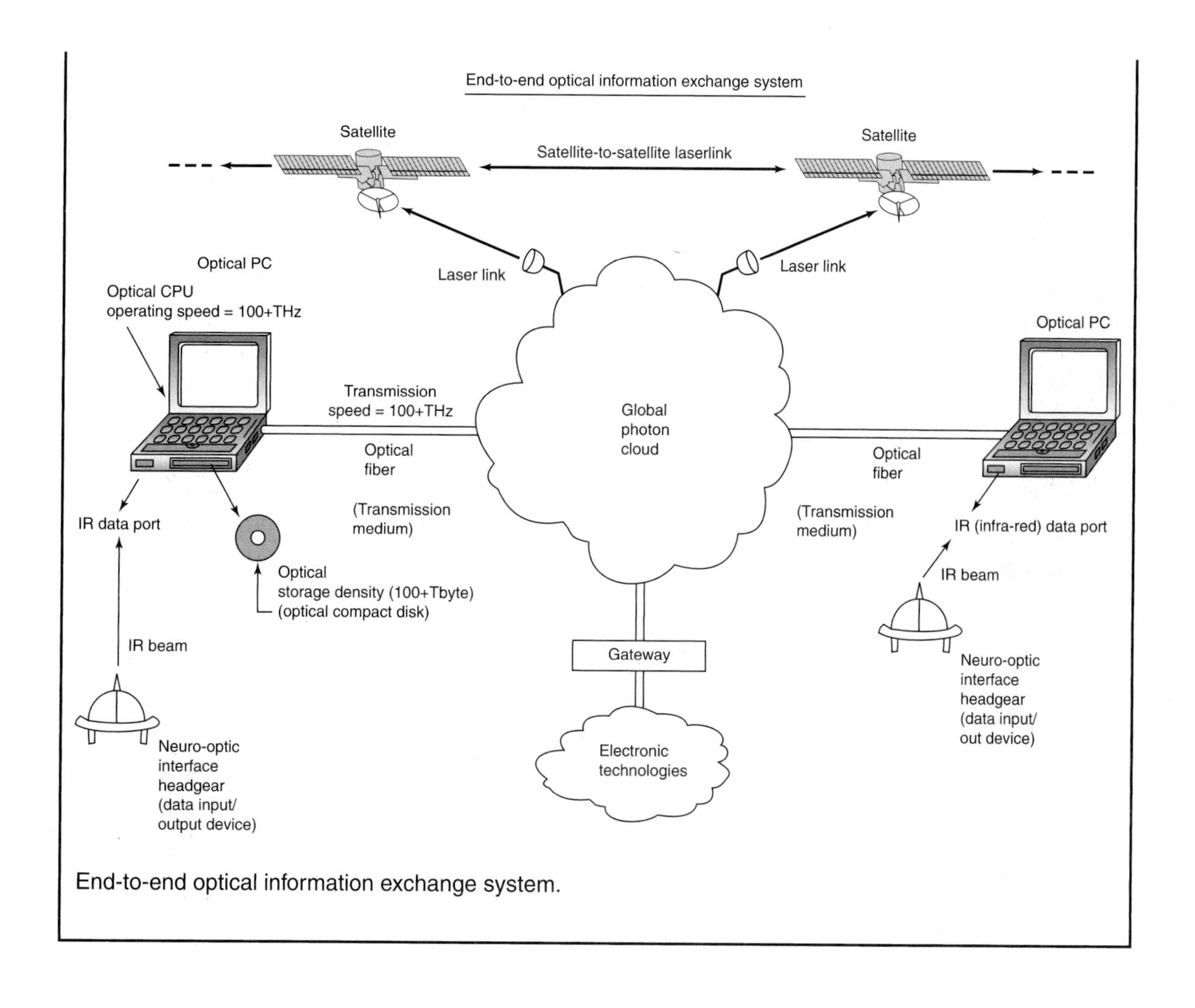

End-to-end optical information exchange system.

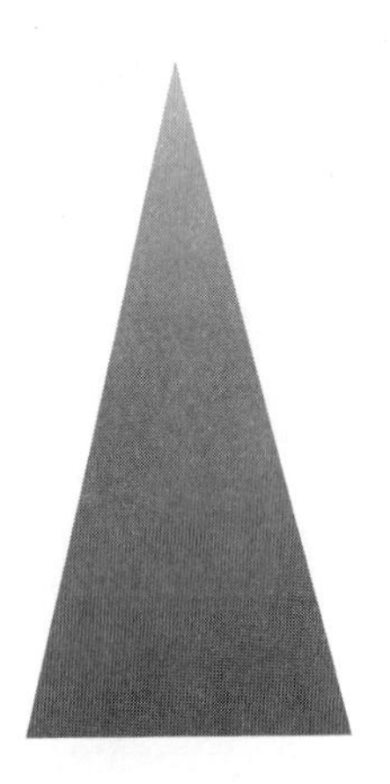

Conclusion

This section is remarkable in that it indicates the excitement of the technological progression of this century as technologies accelerate to develop our capabilities and insight. As Albert Einstein said in his essay, "Science and Society" (1935), "There are two ways in which science affects human affairs. The first is familiar to everyone: Directly, and to an even greater extent indirectly, science produces aids that have completely transformed human existence. The second way is educational in character—it works on the mind" (p. 135). Not only are we progressing with tools and all sorts of new gadgets, but with that comes new capability, understanding, and power.

Part IX began with "Creating the Future" from Norman Myers's *Gaia Atlas of Future Worlds,* which brings an understanding of the planetary urgency for human and ecological sustainability into the twenty-first and twenty-second centuries. It distills our priorities into sustainable, necessary urgencies as we try to understand the new technologies, their implications, and their allure and yet still try to forge ahead with the necessities of sustainable values for this and the next centuries. Part IX then visited some of the twenty-first century's specific concerns and technologies from new information systems, genomics, nanotechnologies, insights into the brain, and dimensions of new learning, the importance of being needed, and population concerns. Part IX ended with the ambitious UN Millennium Project, which tries to make real and accountable the vital humanitarian concerns of the twenty-first century. These UN issues are primary issues for the sustainability of the planet and its inhabitants, and our essential humanitarian values.

Part IX begins with an urgency for humaneness and ends with a program trying to accomplish just that. Bookcased between these readings are all types of technological wonders and concerns. Let's hope that we can keep the path of the future "bookcased" and supported not only by such technological excitement, but also by beginning and ending with the application of help for humanitarian concerns, values, and the good for the people of this planet!

REFERENCE

Einstein, A. (1956). "Science and Society." *Out of My Later Years*. (written in 1935). New York: Citadel Press, pp. 135–137.

Can nature and cities coexist?
Photo courtesy of Ahmed S. Khan

All human beings are in truth akin; All in creation share one origin.
When fate allots a member pangs and pains, No ease for other members then remains.
If, unperturbed, another's grief canst scan, Thou are not worth of the name of man.
—SAADI

Photo courtesy of Ahmed S. Khan

INTERNET EXERCISES

1. Use any of the Internet search engines (e.g., Yahoo, Google, etc.) to research the following topics:
   a. Nanotechnology
   b. Responsible technology
   c. Biotechnology
   d. Neuroscience
   e. Telemedicine
   f. Genetic medicine
   g. Human chromosomes and DNA
   h. Xenotransplants
   i. Virus transfer
   j. Humionics
   k. Bionics
   l. Ergonomics
   m. Tactical Information Assistants (TIAs)
   n. Any words you have listed from your research and reading of future issues

2. Use any of the Internet search engines (e.g., Yahoo, Google, etc.) to research the following issues:
   a. The concept of responsible technology—what does this responsibility entail?
   b. What are the fields of neural engineering and neuroscience?
   c. Research the progress and developments of the Human Genome Project. What are some of the implications of this project?
   d. What is the capacity and capability of the latest, most highly developed computer?
   e. What are the latest developments in human bionics? Are there any animal bionic developments? What are some of the ethical issues involved in bionics?
   f. Research world educational levels, and then identify world educational levels for women. Draw any possible conclusions as to correlations with population levels.
   g. Discover the criteria needed for sustainable world development. What are some of the necessary philosophies involved?
   h. What are some alternative energy systems? How effective can they be?
   i. What is meant by the concepts of population stabilization? When can that happen on a global level?
   j. Explain what is meant by the biotechnological revolution. What are some of the technologies involved, and why are they so revolutionary?
   k. Explain possible ways that ecology and ecological resources can be included in global economic planning.
   l. Research more about military Tactical Information Assistants (TIAs).
   m. Discover the latest pros, cons, and progress in the field of xenotransplants.
   n. What is the U.S. space exploration plan for the next 15 years?
   o. Explain some of the newest developments in telecommunications.
   p. Research the available material on the military power of two or more unindustrialized nations and two or more industrialized nations. Discuss the implications of the differences you find.
   q. Research the ecological priorities for the United States and the world. Explain the differences.

USEFUL WEB SITES

http://www.ufs.org	World Future Society
http://www.ufs.org/futurist.html	The Futurist: official publication of the World Future Society
http://www.discover.com	Discover Magazine
http://www.extropy.org/eo/	Extropy Magazine
http://www.press.umich.edu/jep	Journal of Electronic Publishing
http://www.nanotech.news.com/nanotech.news/nano	Nanotech (nanotechnology) News
http://www.nanozine.com/news.html	Nanotechnology Magazine
http://nanospot.org	Nanotechnology Search Engine
http://www.nasa.gov/today/index.html	Today @ NASA
http://www.nature.com/nature/	Nature (Journal)
http://www.science	Science Magazine
http://www.sciencenow.sciencemag.org	Science Now Newsletter
http://www.science.org	Science Newsletter
http://www.spectrum.ieee.org	IEEE Spectrum –Magazine of the Institute of Electrical and Electronic Engineers
http://radburn.rutgers.edu/andrews/projects/ssit/default.htm	IEEE Society on Social Implications of Technology
http://www.cnn.com/TECH/	CNN News: Science and Technology
http://www.scitechdaily.com	SciTech Daily Review
http://jefferson.village.virginia.edu/pmc	Postmodern culture
http://www.wired.com	Wired Magazine
http://www.worldwatch.org	World Watch
http://www.amsci.org/amsci/	American Scientist

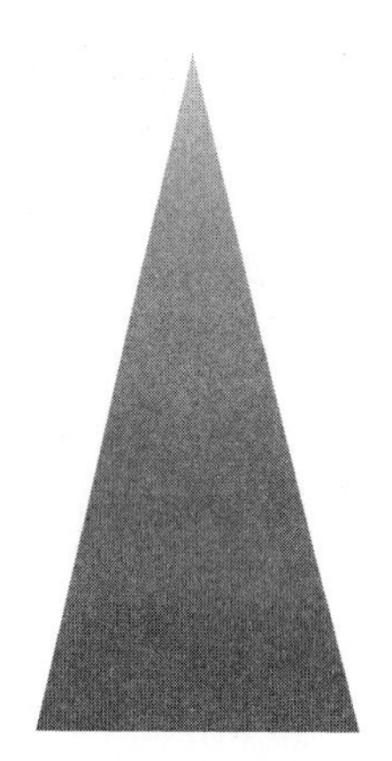

Authors' Commentaries

Authors/Editors: Dr. Ahmed Khan, Dr. Barbara Eichler, Linda Hjorth, Dr. John Morello.
Photographed by Evan Girard

TECHNOLOGY AND THE FUTURE: THE MILITARY PICTURE

When dawn first broke on the typical battlefield several hundred years ago, it must have revealed a surreal picture. Rays of sunlight, cutting through the early morning mist, found their way to the polished steel of the sabres and spears of the armies facing each other. Bright, almost gaudy uniforms stood in stark contrast to the surrounding terrain. At a given signal, usually drums, bugles, or bagpipes, the two forces converged on a designated point, slugging it out until one side disengaged, defeated, but not destroyed. There were always other dawns and other battlefields.

Move to that next battlefield a few generations later. The sun that illuminated that same field of battle might catch not only the glare of sabres and spears, but also the spurs of cavalry and the sight of artillery. The uniforms might still have been garish, given the surrounding terrain. This time, the hostilities began not with the burst from a bagpipe, but most likely with an artillery barrage, as one side tried to use technology to soften up the opposition prior to the assault. And when that assault finally came, it was not led by the screams of men rushing headlong into the breach, but by pounding hooves, as the cavalry, attacking from the flanks, sought to collapse the enemy's perimeter, dash to the rear, and panic the troops headlong into the advancing infantry. There would be no disengagement this time—only total victory, as one side destroyed the other.

The sun that broke over the early battlefields of World War I helped to illuminate not just the familiar past, but also the wave of the future. Men on horseback, sabres and spurs glinting in the sun, prepared to race across an open field to wreak havoc on the enemy's position. However, this time the men on the other side did not quail; instead, they retaliated with machine guns, hand grenades, and poison gas. The technical advantage had shifted, if only temporarily, while the horsemen dismounted and climbed into tanks for a second attack.

And so it has been over the years. In war, technology has proven to be a decisive edge. It has helped to make offensive efforts irresistible, defensive stands impregnable, and for those who found themselves facing superior numbers, it made them equal. However, the edge did not remain so for long. In time, the other side would have the secret, or perhaps go it one better. And so, the search for technological superiority in war has become an escalating factor. Where will it all lead? Here are a few possibilities:

1. *Fewer high-tech wars:* Given the escalating cost and complexity of the new military technologies, only those nations with deep pockets and high-tech capability will be able to use these weapons effectively. Cruise missiles, like the kind used during and since the Gulf War, can cost over $1,000,000 each—not the kind of thing to shoot off willy-nilly. Planes like the F-22 also pack budget-busting price tags. The bottom line is, or should be, that because of the cost of these weapons, nations may want to think twice about the reasons they go to war. Are these reasons sound enough to mortgage their financial present and future?

No one should be naive enough to think that swearing off high-tech weapons will put a stop to warfare. There are plenty of older, more conventional weapons to go around. Millions of antipersonnel devices, also known as land mines, are still buried on battlefields around the world. They are cheap and have a terrific shelf life. Nations that want weapons don't have to go high tech; low tech will do just fine. The U.S. experience in Iraq has proven the lethal effectiveness of home-made bombs known as IEDs. The complexity of high-tech weapons might also be a deterrent to war. Nations wanting sophisticated weapons systems might someday realize that these systems are of no use unless they have the trained personnel to use them or an infrastructure that can support them. And even then, there's no guarantee that the weapons will get the job done. After the glow of praise for the Patriot missile's achievements during Operation Desert Storm had faded, a reevaluation of the system revealed that the missiles were not as effective as everyone had thought they were. Now, if the United States cannot get the expected results out of a weapons system, especially one it built, can we honestly expect anyone else to?

2. *Fewer casualties, less political fallout:* High-tech wars, if they are fought, can produce unexpected benefits. For starters, those personnel committed to the battlefield will be better armed and better prepared than previous fighting forces. Equipment could make one soldier's firepower equal to that of ten in earlier battles. Therefore, fewer troops will have to be put in harm's way. Because of that fact, wars may become more politically sustainable. Taxpayers may be less likely to question their government's intentions if only a few hundred troops are being jeopardized. And, given the high-tech equipment available, those troops will have a better chance of survival and success than ever before. Medical technologies will make it possible for seriously injured troops to receive immediate attention right there on the battlefield, where it really counts. Communications technology will help locate the wounded and get them to a rear area for more care. "Soft Kill" weapons systems will be used not to kill, but to blind, disorient, or otherwise incapacitate the opposition. Having to deal with thousands of sick or wounded soldiers is actually more difficult than tending to the dead ones.

3. *Simulated war as conflict resolution:* North Korea's forces are on alert. All reservists have been called up, and active duty personnel have been mobilized. Nuclear missiles are being repositioned, and have been given new targets. The front page of the *New York Times?* No, the latest data from satellites tracking North Korea's military operations. Those satellites have been watching for some time now, cataloging troop strength and weapons capability since the latest nuclear test and the latest round of failed disarmament talks. The data produced has been sifted through by analysts who conclude what the North Koreans may be up to. Furthermore, those analysts have produced a number of scenarios that the United States might execute to blunt anything North Korean president Kim Il Sung tries. North Korea's intentions and U.S. responses have been loaded on a floppy disk. The U.S. ambassador at the United Nations delivers the disc to the North Korean delegation. "We know what you're doing," he says. "Here's what will happen. We have developed a simulation, and you will lose. Don't believe me? Load this on your computer and make your own decision." Several days later, satellites record that North Korea's military machine has stood down from its alert. Just as high-tech weapons can be the edge in conflict, they can also be an effective deterrent. Just knowing what the other guy has and what he might be contemplating is half the battle. Of course, the reverse of this scenario might be that North Korea vows to upgrade its weapons systems until they are on par with those of the United States before trying anything. But there should always be room for optimism.

JOHN MORELLO, Ph.D.

John Morello, Ph.D.

THOUGHTS ON MEDICAL TECHNOLOGY

I still recall vivid memories from my childhood of Civil War stories as told by my grandfather. His grandfather was a medical doctor in the Civil War. Not only did Gramps talk about the scarcity of medicines and painkillers, gangrene and unbearable pain, his detailed accounts of a time period long gone also revolved around treatments. In his accounts, he described a small black box (approximately 4″ × 2″) with a handle on top and razor blades on the bottom. When a patient had an infection, Dr. O.P. Stevens, my great-great-grandfather, would pull out the box, place it on one of the patient's arteries, and push hard on the handle, forcing the blades into the skin and causing blood to gush; the goal was to extinguish the patient's ills. The theory behind the technologically simple black box was that infection and impurities would be extracted from the body through the relentless blood flow caused by the "bleeder's" gashes on the skin's surface. The treatment for gangrene, often correlated with bullet wounds, was to remove the tainted limb, aided only by primitive forms of anesthesia.

In recalling these stories, I still remain amazed at the dramatic historical changes in medical technology. When thinking of "bleeders" or primitive amputations, I feel revulsion and confusion. Because today's medicine is more humane, complex, and progressive, it remains difficult to imagine the limitations of doctors in past times. If my grandfather had told his grandfather that in the 1990s, antibiotics (penicillin was discovered in 1928), computed tomography (CAT scans were discovered in 1972), test-tube babies (the first one was born in England in 1978), eradication of smallpox (a very serious disease in his time, and one that was wiped out in 1977), organ transplants (including the successful transplant of genetically altered pig hearts into baboons in 1995), and DNA analysis for disease prevention (1999) would be available or had occurred, I am sure that he would have told his grandson that he was either "nuts" or grandiose (Nuland 1996).

As new medical technologies are created, families are grateful because life spans are extended, low-weight babies are saved, and sight can be restored, as can life after death. However, I remain convinced that my great-great-grandfather would still shake his head in dismay at some of the current ethical issues that doctors face today. His stories were gruesome, bold, and primitive, but the stories for the new millennium are amazing, technological, and sometimes ethically disconcerting. If Grandpa Doc were alive today, what would he say about baboon-heart and pig-stomach transplants, tissue engineering, or genetic engineering? I am not sure, but I do know that he would express a concern that all medical technology be used with caution, remembering that its purpose is to increase the quality of life, not necessarily to make life technology-dependent.

Linda Stevens Hjorth

Linda Stevens Hjorth

REFERENCE

Nuland, S.B. (1996, Fall). An Epidemic of Discovery. *Time.* pp. 12–13.

TECHNOLOGY AND THE FUTURE OF THE FUTURE

The future belongs to those who are willing to learn from the experiences of the present and the past. Indeed, no one can predict the future. But with the lessons learned during the gigantic explosion of technology in the twentieth century, we can chart pathways to a future that will enable us to control the fission-like chain reaction of technology's growth for its appropriate, humane, and responsible utilization.

As we continue the journey into the new millennium, the great technological accomplishments of the twentieth century in the areas of telecommunications, computers, energy, agriculture, materials, medicine, genetic engineering, and defense have transformed the world and brought people closer, yet millions of people worldwide still go to bed hungry at night. These technological advances have enabled us to design advanced early warning systems against missile attacks, but we have failed to develop an advanced early warning system to warn against and prevent global famine or the spread of disease. Thanks to state-of-the-art technologies, we are able to design spaceships to explore life on other planets, yet we have failed to preserve life on planet Earth. Millions of children worldwide continue to die due to malnutrition, disease, and poverty.

We have made tremendous strides in science and technology, but at the cost of numerous environmental, social, and moral dilemmas. Technological advances and economic expansion could become meaningless if moral and social implications are not considered in charting the course of the future.

In developed countries, technology has provided an impetus for economic growth at the cost of many social and environmental problems. In developing countries, the lack of advanced technology has resulted in low economic growth. Many developing nations aspire to become technologically advanced, but lack the required infrastructures and skilled manpower and are reluctant to consider the social and environmental costs that the industrialized nations have paid in the course of their industrialization. The key question for developing nations is how to increase the pace of industrialization without paying the same cost that the developed nations have paid for becoming industrialized. The answer to this question lies in the appropriate use of technology. For example, telecommunications technologies can provide endless opportunities in education (e.g., distance learning), medicine (e.g., telemedicine), and business (e.g., e-commerce) in developing countries. These applications could enable developing countries to increase teledensity and literacy, improve per capita income, and narrow the technoeconomic gap with developed countries.

As we continue our journey into the future with our technological might, we ought to be aware not only of the short-term gains and advantages of technology, but also of its long-term impact and associated problems. In the preceding text, we have attempted to present a spectrum of issues related to a wide array of topics (i.e., history of technology, ethics and technology, energy, ecology, population, war and technology, social responsibility, health and technology, technology and the Third World, and technology of the future) that deal with the impact of technology in the developed and the developing world. I hope that our endeavor will serve as a guide to help the reader understand, anticipate, and address the impact of technology on the various facets of society in order to make future decisions for the humane, just, and responsible utilization of technology.

Ahmed S Khan

Ahmed S. Khan, Ph.D.

IT IS NOW THE TIME FOR WISDOM

The *future* is that concept that we who have embraced Western civilization and the concepts of progress, change, improvement, and technology can hardly wait for! The future is ours. We only have to design it, and through technology, it will service us and give us greater pleasures than ever before. The future adventures, the daring, the excitement, and the new luxuries appeal unlimitedly to those of us in the industrialized societies and are dazzling and changing the unindustrialized societies. Anticipating the future gives us a telescopic sight of the accomplishment of our greater hopes, dreams, and visions . . . new reaches for our graceful hand of knowledge.

But this future represents many changes from the futures that were imagined by past generations. This future represents social and economic differences from previously imagined futures, which were more predictable, less multidimensional, less exponential, and less critical. Yes, most futures represent change, growth, and variation from the "comfortable known," but this one is really different. This future is not only a telescopic graceful vision; it is also a multimedia-accelerated production based on technology everywhere, with increasingly faster machines and tools to aid, support, and satisfy mankind. Along with the mach speed of technologies are accompanying critical concerns that have to be addressed before we blindly ride the technological acute paths. We have to ask ourselves some basic questions, for we are at a critical juncture. Critical ecological issues emerge because of our new technological development. Questions and concerns about population emerge because of our progress in the fields of health and medicine and due to our expanded life spans. War can be globally devastating both easily and quickly. Moral and ethical concerns of actions increasingly parallel all of the thoughtful use of technology. The unindustrialized world hopefully tries to adopt some of the technologies to help its economic development, but has a problem with the expense, consequences, and appropriateness of Western technology.

Part IX's objective has been to converge the thoughts of the other eight parts to their implications for the twenty-first century so that issues from those parts can find a responsible bridge to the future, a future that is sustainable and fulfills its potential exciting promise. The path, however, is not automatic. Because of the steeper acceleration of technology and accompanying critical humanitarian and ecological issues, all actions have to include more than economic appetites and a gratification of needs and wants; they must also contain larger understandings of technological responsibility and global sustainability. The concern for the quality of life in the future—for a natural sustainable life—depends on our understanding and accompanying actions. This is the challenge and ultimate flow chart for the twenty-first century: wise decisions and wise and caring actions.

Barbara A. Eichler

Barbara A. Eichler, Ed.D.

Explore Future Options

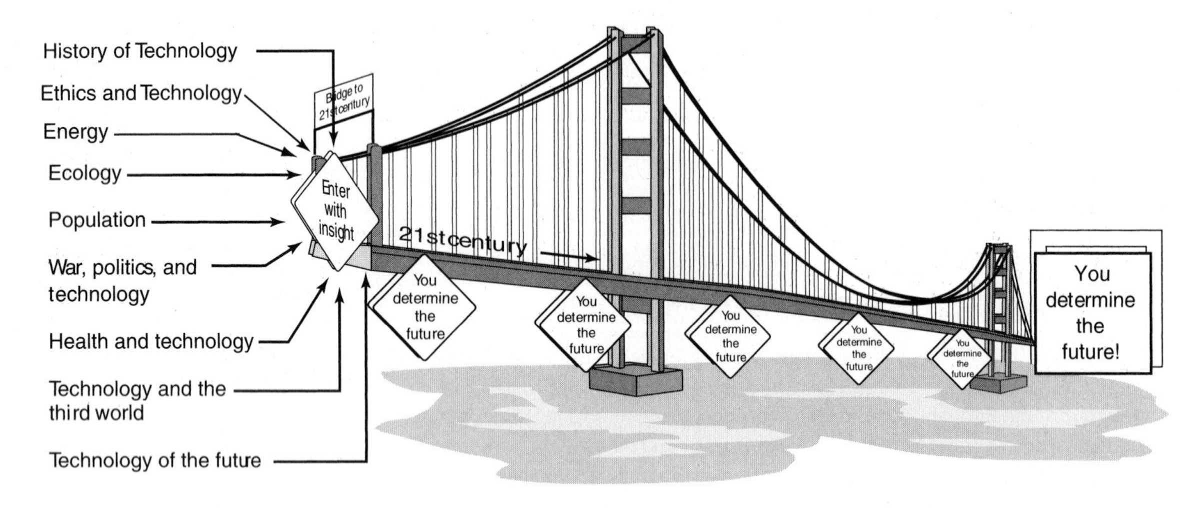

It is a great advantage that man should know his station and not erroneously imagine that the whole universe exists only for him.

—MAIMONIDES

(Dalalat al-ha'irin, Part iii, Chapter xii, c. 1190)

Index

D

E

F

H

I

J

K

L

M

N

T

Q

R

S

U

V

W

X

Y

Z